VOLUME SIX HUNDRED AND SIXTY SEVEN

METHODS IN ENZYMOLOGY

Pseudokinases

METHODS IN ENZYMOLOGY

VOLUME SIX HUNDRED AND SIXTY SEVEN

METHODS IN ENZYMOLOGY

Pseudokinases

Edited by

NATALIA JURA

Cardiovascular Research Institute;
Quantitative Biosciences Institute (QBI);
Department of Cellular and Molecular Pharmacology,
University of California San Francisco,
San Francisco, CA, United States

JAMES M. MURPHY

Walter and Eliza Hall Institute of Medical Research;
Department of Medical Biology, University of Melbourne,
Parkville, VIC, Australia

ELSEVIER

ACADEMIC PRESS
An imprint of Elsevier

Academic Press is an imprint of Elsevier
50 Hampshire Street, 5th Floor, Cambridge, MA 02139, United States
525 B Street, Suite 1650, San Diego, CA 92101, United States
The Boulevard, Langford Lane, Kidlington, Oxford OX5 1GB, United Kingdom
125 London Wall, London, EC2Y 5AS, United Kingdom

First edition 2022

ISBN: 978-0-323-91541-0
ISSN: 0076-6879

For information on all Academic Press publications
visit our website at https://www.elsevier.com/books-and-journals

Publisher: Zoe Kruze
Developmental Editor: Federico Paulo S. Mendoza
Production Project Manager: James Selvam
Cover Designer: Christian Bilbow

Typeset by STRAIVE, India

Contents

Contributors

Christopher M. Barbieri
Leads Discovery and Optimization, Mechanistic Pharmacology, Bristol Myers Squib, Lawrenceville, NJ, United States

Jonah Beenstock
The Lunenfeld-Tanenbaum Research Institute, Mount Sinai Hospital, Toronto, ON, Canada

Stefan Belić
Leads Discovery and Optimization, Mechanistic Pharmacology, Bristol Myers Squib, Lawrenceville, NJ, United States

Richard T. Beresis
Chempartner Co, Ltd., Shanghai, China; Chempartner Co, Ltd., South San Francisco, CA, United States

Miles H. Black
Department of Molecular Biology, University of Texas Southwestern Medical Center, Dallas, TX, United States

Julius Bogomolovas
School of Medicine, University of California, San Diego, La Jolla, CA, United States

Avipsa Bose
Department of Molecular Reproduction, Development and Genetics, Indian Institute of Science, Bangalore, India

Dominic P. Byrne
Department of Biochemistry and Systems Biology, Institute of Systems, Molecular and Integrative Biology, University of Liverpool, Liverpool, United Kingdom

Angus J.M. Cameron
Section of Structural Biology, Department of Medicine, Imperial College London; Kinase Biology Laboratory, Barts Cancer Institute, Queen Mary, University of London, John Vane Science Centre, London, United Kingdom

Annelise M. Cassidy
Department of Medical and Molecular Genetics, Indiana University School of Medicine, Indianapolis, IN, United States

Deep Chatterjee
Structural Genomics Consortium, Buchmann Institute for Molecular Life Sciences; Institute for Pharmaceutical Chemistry, Johann Wolfgang Goethe-University, Frankfurt am Main, Germany

Hanying Chen
Department of Medical and Molecular Genetics, Indiana University School of Medicine, Indianapolis, IN, United States

W. Seth Childers
Department of Chemistry, University of Pittsburgh, Pittsburgh, PA, United States

Arthur Chow
Department of Oncological Sciences, Department of Pharmacological Sciences, The Tisch Cancer Institute, Mount Sinai Centre for Therapeutic Discovery, Icahn School of Medicine at Mount Sinai, New York, NY, United States

Jeroen Claus
Phospho Biomedical Animation, The Greenhouse Studio 6, Unit 17, Millmead Industrial Estate, London, United Kingdom

Audrey Colomba
Alzheimer Research UK- UCL Drug Discovery Institute, University College London, London, United Kingdom

Weiwen Dai
The Walter and Eliza Hall Institute of Medical Research; Department of Medical Biology, University of Melbourne, Parkville, Vic, Australia

Arvin C. Dar
Department of Oncological Sciences, Department of Pharmacological Sciences, The Tisch Cancer Institute, Mount Sinai Centre for Therapeutic Discovery, Icahn School of Medicine at Mount Sinai, New York, NY, United States

Verena Dederer
Structural Genomics Consortium, Buchmann Institute for Molecular Life Sciences; Institute for Pharmaceutical Chemistry, Johann Wolfgang Goethe-University, Frankfurt am Main, Germany

Devan Diwanji
Cardiovascular Research Institute; Medical Scientist Training Program, University of California San Francisco, San Francisco, CA, United States

Sabine Elowe
Biologie Cellulaire et Moléculaire, Faculté de Médicine, Université Laval, Québec; Centre de Recherche du Centre Hospitalier Universitaire (CHU) de Québec-Université Laval, Axe de Reproduction, Santé de la Mère et de l'Enfant, Québec; PROTEO-Regroupement Québécois de Recherche sur la Fonction, l'Ingénierie et les Applications des Protéines, Québec; Département de Pédiatire, Faculté de Médicine, Université Laval et le Centre de Recherche sur le Cancer de l'Université Laval, Québec, QC, Canada

Patrick A. Eyers
Department of Biochemistry and Systems Biology, Institute of Systems, Molecular and Integrative Biology, University of Liverpool, Liverpool, United Kingdom

Emma Fairweather
Department of Biochemistry and Systems Biology, Institute of Systems, Molecular and Integrative Biology, University of Liverpool, Liverpool, United Kingdom

Bradley Falk
Leads Discovery and Optimization, Mechanistic Pharmacology, Bristol Myers Squib, Lawrenceville, NJ, United States

Cheree Fitzgibbon
Walter and Eliza Hall Institute of Medical Research; Department of Medical Biology, University of Melbourne, Parkville, VIC, Australia

Arianna Fornili
Department of Chemistry, School of Physical and Chemical Sciences, Queen Mary University of London, London, United Kingdom

Luciano Gama Braga
Biologie Cellulaire et Moléculaire, Faculté de Médicine, Université Laval, Québec; Centre de Recherche du Centre Hospitalier Universitaire (CHU) de Québec-Université Laval, Axe de Reproduction, Santé de la Mère et de l'Enfant, Québec; PROTEO-Regroupement Québécois de Recherche sur la Fonction, l'Ingénierie et les Applications des Protéines, Québec; Département de Pédiatire, Faculté de Médicine, Université Laval et le Centre de Recherche sur le Cancer de l'Université Laval, Québec, QC, Canada

Forson Gao
Phospho Biomedical Animation, The Greenhouse Studio 6, Unit 17, Millmead Industrial Estate; Section of Structural Biology, Department of Medicine, Imperial College London, London, United Kingdom

Chantal Garand
Centre de Recherche du Centre Hospitalier Universitaire (CHU) de Québec-Université Laval, Axe de Reproduction, Santé de la Mère et de l'Enfant; PROTEO-Regroupement Québécois de Recherche sur la Fonction, l'Ingénierie et les Applications des Protéines, Québec, QC, Canada

Roger George
Structural Biology Science Technology platform, The Francis Crick Institute, London, United Kingdom

John D. Gordan
Division of Hematology Oncology and Quantitative Biosciences Institute, University of California, San Francisco (UCSF), San Francisco, CA, United States

Marcin Gradowski
Department of Biochemistry and Microbiology, Institute of Biology, Warsaw University of Life Sciences, Warsaw, Poland

Peter Gravenhorst
Department of Biology, University of Konstanz, Konstanz, Germany

Joshua M. Hardy
The Walter and Eliza Hall Institute of Medical Research; Department of Medical Biology, University of Melbourne, Parkville, Vic, Australia

John A. Harris
Department of Biochemistry and Systems Biology, Institute of Systems, Molecular and Integrative Biology, University of Liverpool, Liverpool, United Kingdom

Annette V. Jacobsen
Walter and Eliza Hall Institute of Medical Research; Department of Medical Biology, University of Melbourne, Parkville, VIC, Australia

Sam A. Jamieson
Biochemistry Department, School of Biomedical Sciences, University of Otago, Dunedin, New Zealand

Natalia Jura
Cardiovascular Research Institute; Quantitative Biosciences Institute (QBI); Department of Cellular and Molecular Pharmacology, University of California San Francisco, San Francisco, CA, United States

Natarajan Kannan
Department of Biochemistry & Molecular Biology; Institute of Bioinformatics, University of Georgia, Athens, GA, United States

Hanna Karvonen
Cancer Signaling, Faculty of Medicine and Health Technology, Tampere University, Tampere, Finland

Karen Keeshan
Paul O'Gorman Leukaemia Research Centre, Institute of Cancer Sciences, University of Glasgow, Glasgow, SC, United Kingdom

Zaigham M. Khan
Department of Oncological Sciences, Department of Pharmacological Sciences, The Tisch Cancer Institute, Mount Sinai Centre for Therapeutic Discovery, Icahn School of Medicine at Mount Sinai, New York, NY, United States

Stefan Knapp
Structural Genomics Consortium, Buchmann Institute for Molecular Life Sciences; Institute for Pharmaceutical Chemistry, Johann Wolfgang Goethe-University, Frankfurt am Main, Germany

Krzysztof Kochut
Department of Computer Science, University of Georgia, Athens, GA, United States

Kimberly A. Kowallis
Department of Chemistry, University of Pittsburgh, Pittsburgh, PA, United States

Yogesh Kulathu
MRC Protein Phosphorylation and Ubiquitylation Unit, Sir James Black Centre, Dundee, Scotland, United Kingdom

Emin Kuliyev
Department of Immunology, St. Jude Children's Research Hospital, Memphis, TN, United States

Ridhima Lal
Paul O'Gorman Leukaemia Research Centre, Institute of Cancer Sciences, University of Glasgow, Glasgow, SC, United Kingdom

Sven M. Lange
MRC Protein Phosphorylation and Ubiquitylation Unit, Sir James Black Centre, Dundee, Scotland, United Kingdom

Mark A. Lemmon
Department of Pharmacology and Yale Cancer Biology Institute, Yale University School of Medicine, West Haven, CT, United States

Fen Liu
Chempartner Co, Ltd., Shanghai, China; Chempartner Co, Ltd., South San Francisco, CA, United States

Jin Liu
Chempartner Co, Ltd., Shanghai, China; Chempartner Co, Ltd., South San Francisco, CA, United States

Gregory A. Locke
Leads Discovery and Optimization, Mechanistic Pharmacology, Bristol Myers Squib, Lawrenceville, NJ, United States

Isabelle S. Lucet
The Walter and Eliza Hall Institute of Medical Research; Department of Medical Biology, University of Melbourne, Parkville, Vic, Australia

Peter D. Mace
Biochemistry Department, School of Biomedical Sciences, University of Otago, Dunedin, New Zealand

William M. Marsiglia
Department of Oncological Sciences, Department of Pharmacological Sciences, The Tisch Cancer Institute, Mount Sinai Centre for Therapeutic Discovery, Icahn School of Medicine at Mount Sinai, New York, NY, United States

Sebastian Mathea
Structural Genomics Consortium, Buchmann Institute for Molecular Life Sciences; Institute for Pharmaceutical Chemistry, Johann Wolfgang Goethe-University, Frankfurt am Main, Germany

Olga Mayans
Department of Biology, University of Konstanz, Konstanz, Germany

Hamish D. McMillan
Biochemistry Department, School of Biomedical Sciences, University of Otago, Dunedin, New Zealand

Yanxiang Meng
Walter and Eliza Hall Institute of Medical Research; Department of Medical Biology, University of Melbourne, Parkville, VIC, Australia

Dom C. Mitchell
Division of Hematology Oncology and Quantitative Biosciences Institute, University of California, San Francisco (UCSF), San Francisco, CA, United States

Jodi Muckelbauer
Leads Discovery and Optimization, Mechanistic Pharmacology, Bristol Myers Squib, Lawrenceville, NJ, United States

James M. Murphy
Walter and Eliza Hall Institute of Medical Research; Department of Medical Biology, University of Melbourne, Parkville, VIC, Australia

Brady O'Boyle
Department of Biochemistry & Molecular Biology, University of Georgia, Athens, GA, United States

Samara Mishelle Ona
The Lunenfeld-Tanenbaum Research Institute, Mount Sinai Hospital; Department of Molecular Genetics, University of Toronto, Toronto, ON, Canada

Onisha Patel
The Walter and Eliza Hall Institute of Medical Research; Department of Medical Biology, University of Melbourne, Parkville, Vic, Australia

Krzysztof Pawłowski
Department of Molecular Biology, University of Texas Southwestern Medical Center, Dallas, TX, United States; Department of Biochemistry and Microbiology, Institute of Biology, Warsaw University of Life Sciences, Warsaw, Poland

Stephane Pelletier
Department of Medical and Molecular Genetics, Indiana University School of Medicine, Indianapolis, IN, United States

Franziska Preuss
Structural Genomics Consortium, Buchmann Institute for Molecular Life Sciences; Institute for Pharmaceutical Chemistry, Johann Wolfgang Goethe-University, Frankfurt am Main, Germany

Jun Qin
Department of Cardiovascular & Metabolic Sciences, Lerner Research Institute, Cleveland Clinic, Cleveland, OH, United States

Tingting Qing
Chempartner Co, Ltd., Shanghai, China; Chempartner Co, Ltd., South San Francisco, CA, United States

Juuli Raivola
Applied Tumor Genomics, Faculty of Medicine, University of Helsinki, Helsinki, Finland

Laura Richmond
Paul O'Gorman Leukaemia Research Centre, Institute of Cancer Sciences, University of Glasgow, Glasgow, SC, United Kingdom

Jake Ritchie
Paul O'Gorman Leukaemia Research Centre, Institute of Cancer Sciences, University of Glasgow, Glasgow, SC, United Kingdom

Michael J. Roy
The Walter and Eliza Hall Institute of Medical Research; Department of Medical Biology, University of Melbourne, Parkville, Vic, Australia

Joshua B. Sheetz
Department of Pharmacology and Yale Cancer Biology Institute, Yale University School of Medicine, West Haven, CT, United States

Safal Shrestha
Institute of Bioinformatics, University of Georgia, Athens, GA, United States

Frank Sicheri
The Lunenfeld-Tanenbaum Research Institute, Mount Sinai Hospital; Department of Molecular Genetics; Department of Biochemistry, University of Toronto, Toronto, ON, Canada

Minglyanna Surudoi
The Walter and Eliza Hall Institute of Medical Research; Department of Medical Biology, University of Melbourne, Parkville, Vic, Australia

Vincent S. Tagliabracci
Department of Molecular Biology; Harold C. Simmons Comprehensive Cancer Center; Hamon Center for Regenerative Science and Medicine; Howard Hughes Medical Institute, University of Texas Southwestern Medical Center, Dallas, TX, United States

Destinée B. Thomas
Department of Medical and Molecular Genetics, Indiana University School of Medicine, Indianapolis, IN, United States

John S. Tokarski
Leads Discovery and Optimization, Mechanistic Pharmacology, Bristol Myers Squib, Lawrenceville, NJ, United States

Jeffrey Tredup
Leads Discovery and Optimization, Mechanistic Pharmacology, Bristol Myers Squib, Lawrenceville, NJ, United States

Raphael Trenker
Cardiovascular Research Institute, University of California San Francisco, San Francisco, CA, United States

Yuko Tsutsui
Department of Pharmacology and Yale Cancer Biology Institute, Yale University School of Medicine, West Haven, CT, United States

Daniela Ungureanu
Cancer Signaling, Faculty of Medicine and Health Technology, Tampere University, Tampere; Applied Tumor Genomics, Faculty of Medicine, University of Helsinki, Helsinki; Faculty of Biochemistry and Molecular Medicine, University of Oulu, Oulu, Finland

Julia Vaynberg
Department of Cardiovascular & Metabolic Sciences, Lerner Research Institute, Cleveland Clinic, Cleveland, OH, United States

Kliment A. Verba
Quantitative Biosciences Institute (QBI); Department of Pharmaceutical Chemistry, University of California San Francisco, San Francisco, CA, United States

Sandhya S. Visweswariah
Department of Molecular Reproduction, Development and Genetics, Indian Institute of Science, Bangalore, India

Ying-Kai Wang
Leads Discovery and Optimization, Mechanistic Pharmacology, Bristol Myers Squib, Lawrenceville, NJ, United States

Xiaole Yang
Department of Chemistry, University of Pittsburgh, Pittsburgh, PA, United States

Preface

The genome sequencing efforts of the 1990s enabled complete cataloguing of proteomes in humans and the organisms around us, revealing an unexpected prevalence of catalytically dead cousins of conventional enzymes. Subsequently, identification of the predicted catalytically defective protein kinases, termed "pseudokinases" by Manning and colleagues, in their description of the human kinase complement (the kinome) in 2002, has attracted ever-increasing interest. Although David Garbers and colleagues described a pseudokinase domain for the first time—that within the cytoplasmic domain of a sea urchin guanylate kinase receptor—in 1988, most pseudokinases remained relatively understudied until recently. A turning point was the discovery of a pathogenic substitution in the pseudokinase domain of Janus kinase 2 (JAK2) by the Vainchenker, Green, and Lee/Gilliland laboratories as the basis for myeloproliferative neoplasms, which underscored that while pseudokinases might not have catalytic activity, this does not diminish their importance as signaling entities, including as oncogenic drivers. The abundance of pseudokinases in nature underscores their importance to biology, where typically ~10% of mammal kinomes and ~30% of apicomplexan kinomes are pseudokinases. The functions of many of these proteins are yet to emerge, making this a highly topical and rapidly advancing field of study.

Interest and advances in the pseudokinase field have been rapid and global over the past decade. Since the first pseudokinase domain structures of CASK (2008), VRK3, LKB1, and HER3 (2009/2010), our knowledge of the structures and functions of pseudokinase domains has grown astronomically with 17 further structures of human pseudokinases now known, in addition to several others from animals, plants, apicomplexans, and microbes. Although a number of core principles hold regarding their noncatalytic functions—with roles in allosteric regulation; scaffolding the assembly of signaling complexes and; as molecular switches; and substrate sinks or competitors of holoenzyme assembly—their precise modes of action require detailed study on an individual basis. This is because pseudokinases are not a family of proteins per se; what unites them is their divergence from their protein kinase cousins such that one or more of the key motifs required to catalyze phosphoryl transfer have been lost. Furthermore, in some (rare) cases, proteins annotated as pseudokinases, owing to deviations from the canonical catalytic motif sequences, have evolved completely novel enzymatic functions. As a result, studying pseudokinases to deduce their signaling functions, how they are regulated, and whether they might encode a cryptic catalytic function can be incredibly challenging.

Equally challenging are drug discovery efforts aimed at modulation of the aberrant pseudokinase signaling in disease. Several pseudokinases, including JAK2, HER3, MLKL, and STRAD, are of big interest as pharmacological targets due to the documented consequences of their deregulation for disease pathologies. Finding drugs that modulate noncatalytic functions, especially those that are still not fully characterized, is, however, not straightforward. Most of the current pseudokinase-directed drug discovery efforts focus on exploring the nucleotide-binding pockets of pseudokinases as points for allosteric regulation. A successful and exciting example of such approach was the discovery of TYK2 pseudokinase binders that specifically inhibit signaling by the adjacent catalytically active TYK2 kinase domain (Chapter 23). This approach is, however, limited to pseudokinases with conserved nucleotide-binding pockets and whose protein-protein interactions can be modulated via the small molecule engagement in these pockets. When this is not the case, deep knowledge of pseudokinase functions is required to design effective inhibitors. For example, pseudokinases present within the context of membrane receptors, like HER3, can be successfully targeted by the antibodies directed toward their extracellular domains that disrupt signaling-relevant complexes. Here, in Volume 667 of *Methods in Enzymology*, we have collated detailed protocols from leaders in the pseudokinase field to facilitate rigorous study of these proteins, including many with currently unknown functions, thus enabling progress in unveiling the mechanisms by which they regulate signaling while avoiding many of the experimental pitfalls many of us have encountered while studying these fascinating proteins.

Detailed studies of pseudokinases have relied on robust methods for production of recombinant proteins for downstream structural, biophysical, and biochemical studies. The methodologies for protein production, and how they take into consideration the unique properties of individual pseudokinases, are described in Chapter 1 by Lucet et al. for the study of the PEAK pseudokinases, in Chapter 2 by Mace et al. for TRIB1, in Chapter 4 by Eyers et al. for TRIB2, in Chapter 5 by Lange and Kulathu for IRAK pseudokinases, in Chapter 7 by Mayans et al. for Titin, in Chapter 8 by Murphy et al. for MLKL complexes, in Chapter 10 by Childers et al. for bacterial pseudokinases, in Chapter 13 by Dar et al. for

KSR, in Chapter 20 by Verba et al. and in Chapter 21 by Jura et al. for full-length receptor tyrosine pseudokinases, and in Chapter 22 by Mathea et al. for pseudokinase domains from receptor tyrosine pseudokinases.

Utility of recombinant protein preparations in biophysical assays is detailed for ILK by Vaynberg and Qin in Chapter 6 and in HDX-MS for ROR1 by Lemmon et al. in Chapter 11. Structural studies by X-ray crystallography are exemplified by Mace et al. for TRIB1 in Chapter 2, by Lange and Kulathu for IRAK3 in Chapter 5, by Mathea et al. for pseudokinase domains of receptor tyrosine pseudokinases in Chapter 22, and by Locke et al. for TYK2 in Chapter 23. Furthermore, application of cryo-EM to the studies of full-length receptor tyrosine pseudokinases is described for HER3 and its complexes by Verba et al. in Chapter 20 and by Jura et al. in Chapter 21. Considering the lack of catalytic activity exhibited by most pseudokinases, the authors have presented an array of important and broadly adaptable binding assays to evaluate binding to natural ligands (e.g., ATP), such as in Chapter 18 by Bose and Visweswariah for receptor guanylyl cyclases. This widespread interest in drugging pseudokinases and how this might be accomplished has been detailed by Eyers et al. for TRIB2 in Chapter 4, by Lange and Kulathu for IRAK3 in Chapter 5, by Dar et al. in Chapter 13 for KSR, by Gordan et al. in Chapter 15 for STRAD, by Cameron et al. in Chapter 16 for HER3, by Mathea et al. in Chapter 22 for ROR1, and by Locke et al. in Chapter 23 for TYK2. To complement these binding assay approaches, computational biology has emerged as a powerful tool, as exemplified for HER3 in Chapter 16 by Cameron et al. for understanding how small molecule binding might be optimized and more broadly by Kannan et al. who describe methods for identifying pseudokinases from genomic databases in Chapter 14.

Although studies of recombinant protein preparations have proven invaluable for detailed analyses of pseudokinase structure and function, complementary studies of their functions in cellular and organismal contexts remain of enormous significance in the exploration of pseudokinase biological functions in signaling pathways. To this end, Keeshan et al. describe methods for detecting TRIB2 in cells in Chapter 3, Jacobsen and Murphy detail methods for deleting pseudokinases from cultured cells to allow for reconstitution studies in Chapter 9, Ungureanu et al. detail the use of cellular thermal shift assays in establishing the cellular targets of small molecule drugs in Chapter 12, and Elowe et al. present methods for studying the phosphorylation status of BUBR1 in mitotic checkpoint signaling in Chapter 17. These methods are complemented by the capacity to generate mice harboring functional mutations in a pseudokinase of interest using CRISPR-Cas9 editing, as detailed by Pelletier et al. in Chapter 25 using the Scyl1 pseudokinase as an example.

Finally, it remains of immense interest to deduce whether pseudokinases harbor cryptic catalytic functions. In Chapter 19, Tagliabracci and Pawlowski et al. present methods to identify divergent or atypical kinase family members from their sequences and how to characterize their activities biochemically. These are rare, but incredibly interesting, cases of the kinase scaffold evolving to perform unusual functions. Likewise, the yeast protein Bud32, as described by Sicheri et al. in Chapter 24, represents another rare and interesting case, where Bud32 exhibits ATPase activity but does not catalyze the transfer of phosphate from ATP to a substrate protein.

This collection of methods provides an important, contemporary framework of the best practice for further examination of the structure, biochemistry, biophysics, and biological functions of pseudokinases. We thank the authors for their magnificent contributions to this volume, especially considering the disruptions many have experienced over the past 2 years. We also thank the production staff at Elsevier, especially Paulo Mendoza, for support and encouragement to keep on schedule. We expect the methods presented in this volume will serve as an important reference for not just the burgeoning pseudokinase field, but also for others in the broader pseudoenzymes field, and we look forward to the looming expansion in knowledge in these areas as more and more of these understudied proteins become household names.

NATALIA JURA
JAMES M. MURPHY

CHAPTER ONE

Production and purification of the PEAK pseudokinases for structural and functional studies

Onisha Patel[a,b,*], Minglyanna Surudoi[a,b], Weiwen Dai[a,b], Joshua M. Hardy[a,b], Michael J. Roy[a,b], and Isabelle S. Lucet[a,b,*]
[a]The Walter and Eliza Hall Institute of Medical Research, Parkville, Vic, Australia
[b]Department of Medical Biology, University of Melbourne, Parkville, Vic, Australia
*Corresponding authors: e-mail address: patel.o@wehi.edu.au; lucet.i@wehi.edu.au

Contents

Abstract

The PEAK family of pseudokinases, which comprises PEAK1, PEAK2 and PEAK3, are newly identified scaffolds that dynamically assemble oncogenic signaling pathways known to contribute to the development of several aggressive cancers. A striking feature of this unique family of pseudokinase scaffolds is their large multi-domain structure, which allows them to achieve protein complex assemblies through their structural plasticity and functional versatility. Recent structural advances have begun to reveal the critical regulatory elements that control their function. Specifically, the dimer-dependent scaffolding activity of PEAK pseudokinases is emerging as a critical mechanism for their signaling function, in addition to their ability to hetero-associate to form higher-order regulatory networks to diversify and amplify their signaling output. Here, we present a suite of techniques that enable the efficient expression and purification of PEAK proteins for functional characterization.

Methods in Enzymology, Volume 667
ISSN 0076-6879
https://doi.org/10.1016/bs.mie.2022.03.022

1. Introduction

The PEAK family of pseudokinases comprises three members: Sugen kinase 269 (SgK269), also known as Pseudopodium-enriched atypical kinase 1 (PEAK1) (Wang et al., 2010); SgK223 (PEAK2), an orthologue of rat Pragmin (Tanaka, Katoh, & Negishi, 2006) and mouse Notch activation complex kinase (NACK) (Weaver et al., 2014); and the recently identified PEAK3 (C19orf35) (Lopez et al., 2019). PEAK1 and PEAK2 were identified as oncogenic pseudokinases in a number of studies, in which their increased expression altered cell morphology and enhanced migratory and invasive characteristics (Croucher et al., 2013; O'Rourke & Daly, 2018; Patel, Roy, Murphy, & Lucet, 2020; Tactacan et al., 2015; Wang et al., 2010). However, recent studies conducted in a colorectal cancer model suggest a role for PEAK1 as a tumor suppressor, as its overexpression is associated with a suppression of cell growth and metastasis (Ding et al., 2018). Likewise, in gastric cancer models, loss of PEAK1 correlates with tumor growth and invasion (Guo et al., 2014). While less is known about the recently identified PEAK3's role in disease, it has been shown to be upregulated in acute myeloid leukemia (Lopez et al., 2019).

In addition to changes in expression levels, missense mutations of PEAK proteins have been reported in various cancer types, but how these mutations affect their function and signaling is not well understood (Cerami et al., 2012; Gao et al., 2013). Our current understanding of PEAK signaling suggests that each PEAK protein acts as a dynamic scaffold that spatially and temporally recruits a distinct set of signaling molecules and that each member can undergo homo- and hetero-association to form higher molecular assembly to promote diversification and amplification of signaling output, a mechanism likely to be dependent on cellular context (Patel et al., 2020).

PEAK proteins display a unique multi-domain architecture. PEAK1 and PEAK2 are large scaffolds, 193 and 143 kDa respectively, which make them challenging to express and purify as full-length proteins. Both PEAK1 and PEAK2 have the same domain organization consisting of: an N-terminal region of ~200 amino acids with an undefined fold and function; a 50 kDa long intrinsically disordered region (IDR) that harbors a plethora of post-translational modification (PTM) sites and molecular recognition motifs for the recruitment of key signaling proteins; and a C-terminal pseudokinase domain (PsK) that is surrounded by conserved helices (αN1, αJ, αK, αL and αM) (Fig. 1A). PEAK3 also contains a C-terminal pseudokinase domain and surrounding helices, which has a high sequence

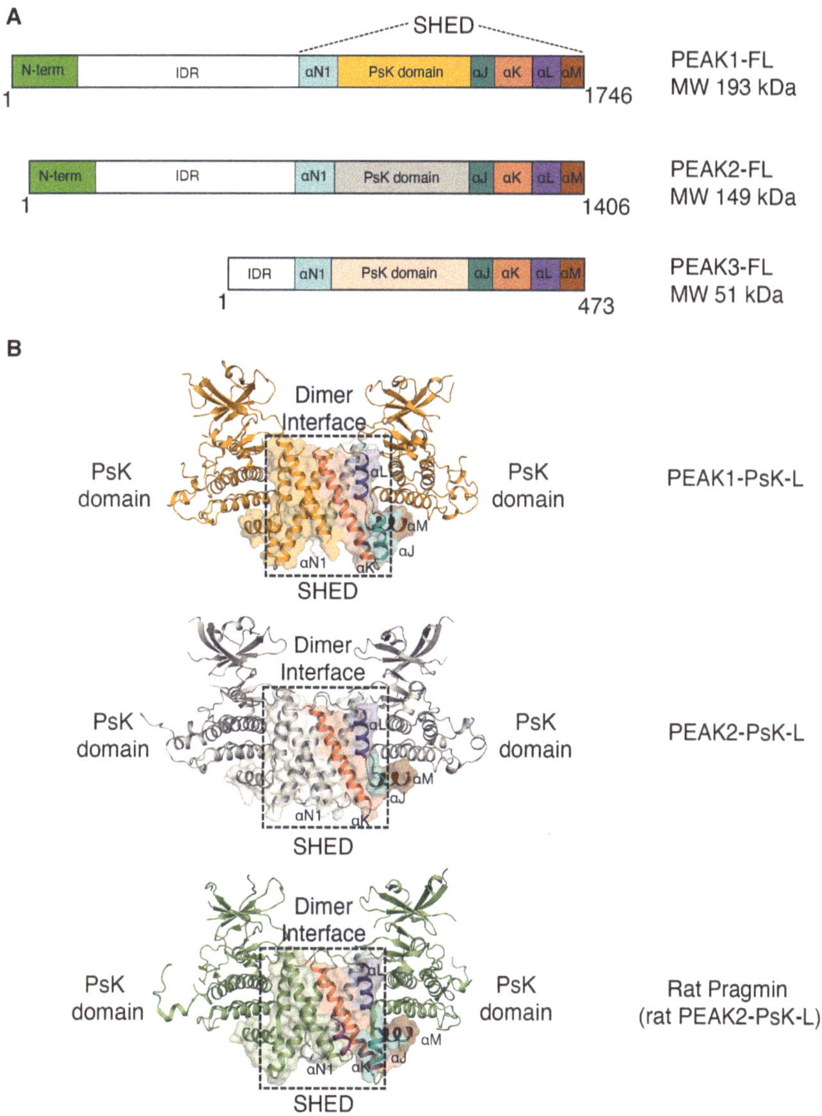

Fig. 1 Structure and schematic representation of PEAK1, PEAK2, PEAK3. (A) Schematic representation of PEAK1, PEAK2 and PEAK3 domain organization. (B) Structure of PEAK proteins pseudokinase and dimerization domains and their theoretical molecular weight (MW). Top, structure of PEAK1 dimer (PDB: 6BHC; Ha & Boggon, 2018), highlighting the PsK domain and the SHED domain comprised of αN1, αJ, αK, αL and αM. Middle, structure of PEAK2 (PDB: 5VE6) displaying the same arrangement as observed in PEAK1 (PDB: 5VE6). Bottom, structure of rat SgK223 dimer (PDB: 6EWX). Structure figures were generated using PyMOL (The PyMOL Molecular Graphics System Version 1.7.4.0, Schrödinger, LLC).

homology and structural conservation with PEAK1/PEAK2, but the intrinsically disordered region at the N-terminus is significantly shorter, indicating PEAK3 likely has distinct signaling properties.

Our first understanding of the scaffolding function of PEAK proteins came with our crystal structure of the PEAK2 pseudokinase domain and its surrounding regulatory helices (αN1, αJ, αK, αL and αM) (Patel et al., 2017). This structure revealed a novel dimerization domain comprised of the regulatory helices, subsequently referred to as the split helical dimerization (SHED) domain (Fig. 1B). This structure, alongside biochemical and cellular data (Liu et al., 2016; Murphy et al., 2014), provided the rationale to classify PEAK proteins as pseudokinases and demonstrated that these scaffolds can homo- and hetero-associate via their unique and conserved SHED domains to regulate signal diversity. The subsequent structures of the human PEAK1 (Ha & Boggon, 2018) and rat Pragmin (PEAK2-rat) (Lecointre et al., 2018) pseudokinase domains and surrounding regulatory helices, confirmed that the SHED domain is a highly conserved dimerization module unique to the PEAK family that can also promote cross-interaction between PEAK family members (Fig. 1B).

All crystal structures solved to date were derived from protein expressed and purified in *E. coli* (Table 1) and encompass only the pseudokinase domain and SHED domain (Fig. 1). While *E. coli* expression provides an efficient platform to express protein in large amounts for structural studies, post-translational modifications (PTMs), such as phosphorylation, are absent. In order to resolve how these PTMs may alter or regulate the conformation and function of PEAK proteins, it is necessary to generate protein or peptides that include these PTMs such as through in vitro phosphorylation, recombinant expression in insect cells, or through semi-synthetic approaches.

Despite these recent advances in unravelling key structural elements that underpin the scaffolding function of PEAK proteins, many challenges remain. The precise mechanisms by which cross-interactions between PEAK proteins are regulated are yet to be elucidated, and the interaction of PEAK proteins with other key signaling components, particularly in oncogenesis, is poorly understood.

In this chapter, we present an array of techniques that have enabled the efficient expression, purification and characterization of the pseudokinase domains and associated regulatory helices of the PEAK proteins in *E. coli*. In addition, and importantly, we describe the design of new PEAK constructs and provide detailed protocols for their successful expression and purification from insect cells. These insect cells-derived PEAK proteins enable the characterization of critical PTM sites and the further identification of PEAK signaling interactors.

Table 1 Summary of PEAK constructs and expression systems.

PEAK constructs	Domain boundaries	Tag	Vector/ expression system	PDB ID	Reference
Bacterial expression					
PEAK1-PsK-L human (SgK269)	Residues 1272–1743 with a deletion of residues 1409–1451	Uncleavable C-ter hexahistidine tag (LEHHHHHH)	pET-28a vector (Novagen) BL21 (DE3) pLysS (Novagen) cells IPTG induction (0.5 mM) at 18 °C	6BHC	Ha and Boggon (2018)
PEAK1-PsK-L human (SgK269)	Residues 1267–1746	N-terminal octahistidine tag with a TEV cleavage site	Modified pCold™ IV vector (Takara) *E. coli* C41(DE3) auto-induction method (Studier, 2005)	N/A	Patel et al. (2017)
PEAK2-PsK-L human (SgK223)	Residues 932–1406	N-terminal octahistidine tag with a TEV cleavage site	Modified pCold™ IV vector (Takara) *E. coli* C41(DE3) auto-induction method (Studier, 2005)	5VE6	Patel et al. (2017)
PEAK2-PsK-L rat (Pragmin)	Residues 906–1368	N-terminal octahistidine tag with a thrombin cleavage site	pET 28b/*E. coli* Rosetta™ (DE3) IPTG induction (0.5 mM) at 37 °C	6EWX	Lecointre et al. (2018)
Insect cell expression					
PEAK1-N_{ext}PsK-L human (SgK269)	Residues 1082–1746	N-terminal octahistidine tag with a TEV cleavage site	pFastBac™ Dual/*Sf*21 insect cells	N/A	To be published
PEAK2-N_{ext}PsK-L human (SgK223)	Residues 802–1406	N-terminal octahistidine tag with a TEV cleavage site	pFastBac™ Dual/*Sf*21 insect cells	N/A	To be published
PEAK3-FL human (C19orf35)	Full-length (1–473)	N-terminal octahistidine tag with a TEV cleavage site	pFastBac™ Dual/*Sf*21 insect cells	N/A	To be published

2. Recombinant protein expression

2.1 Protein expression in bacterial cells

The production of recombinant PEAK1 and PEAK2 proteins containing only the pseudokinase domain and surrounding regulatory helices has been successfully achieved in various *E. coli.* Strains (BL21 (DE3), Rosetta™ (DE3) and C41 (DE3)), cloned either in a standard pET28 vector (Ha & Boggon, 2018) or a pCold™ IV vector (Patel et al., 2017). All PEAK1 and PEAK2 constructs reported to date incorporate a cleavable or non-cleavable His-Tag at the N- or C-terminus (Table 1, Fig. 2).

We detail here what we consider the method of choice: using the pCold™ IV DNA cold shock expression system (Takara Bio) in conjunction with an auto-induction method. The advantage of pCold™ vectors over other standard expression vectors is the inclusion of a cold-shock protein

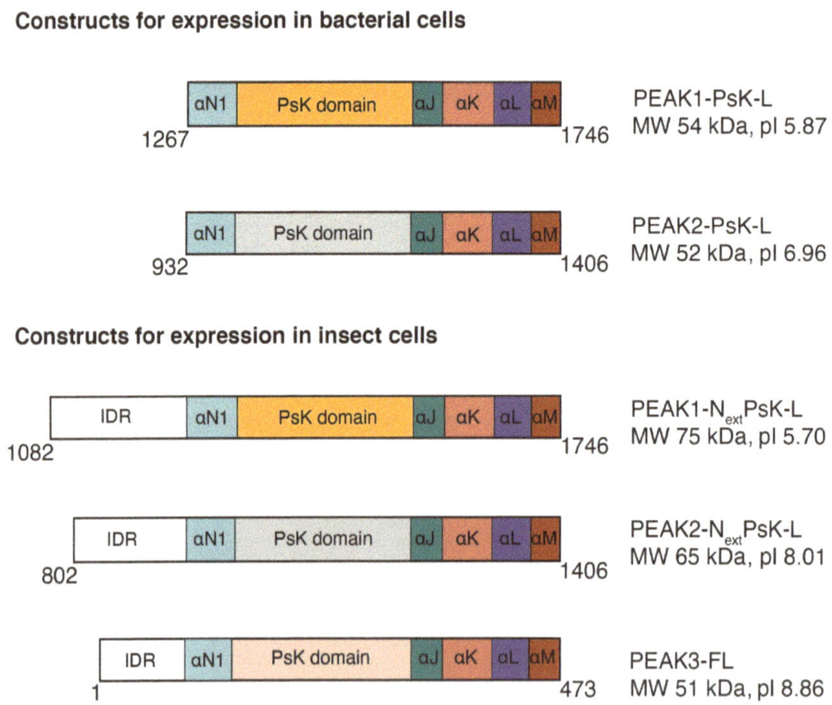

Fig. 2 Schematic representation of the constructs generated for expression and purification of PEAK proteins. Top, constructs generated for expression in *E. coli*. Bottom, constructs generated for expression in insect cells. Theoretical molecular weight (MW) and isoelectric point (pI) are as shown.

A (cspA) promoter that allows the synthesis of target proteins to be selectively induced at a low temperature (15 °C) so that host protein expression is suppressed, and protease activity is decreased. This method has proven valuable for the expression of toxic proteins and has been highly efficient for the production of recombinant PEAK proteins in large amounts, with the exception of PEAK3.

2.1.1 Key resources, reagents, materials and equipment

Key reagents and resources.

Reagent	Supplier	Cat. no.
C41 (DE3) cells	Sigma-Aldrich	CMC0017
pCold™ IV expression vector	Takara Bio	3360
80% (v/v) ethanol (for decontamination of surfaces)	Chem-supply	EL156
Isopropyl β-D-1-thiogalactopyranoside (IPTG) (1 M stock solution in water; 0.22 μm sterile filtered; store aliquots at −20 °C)	Astral Scientific	AST0487
Ampicillin (1000 × stock solution, 100 mg/mL in water; 0.22 μm sterile filtered; store as 1 mL aliquots at −20 °C)	Astral Scientific	BIOAB0028
$MgSO_4$ (1 M stock solution in water; autoclaved)	Sigma Aldrich	M7506
D-Glucose (40% w/v stock solution in water; 0.22 μm sterile filtered)	Sigma Aldrich	G7021

LB agar.

Component	Concentration
Sterile Type II water	–
Tryptone	1% (w/v)
Yeast extract	0.5% (w/v)
NaCl	0.5% (w/v)
D-Glucose	0.2% (w/v)
Tris-HCl pH 7.4	10 mM
$MgCl_2$	1 mM
Agar	1.5% (w/v)

Note: Autoclave to dissolve chemicals at 121 °C for 30 min, store at room temperature.

LB broth.

Component	Concentration
Sterile Type II water	–
Tryptone	1% (w/v)
Yeast Extract	0.5% (w/v)
NaCl	1% (w/v)
NaOH (5 M, for pH adjustment to 7.3–7.4)	–

Note: Autoclave to dissolve chemicals at 121 °C for 30 min, store at room temperature.

NPS (20 × stock).

Component	Concentration (20 ×)	Concentration (1 ×)
Deionized distilled water	–	–
$(NH_4)_2SO_4$	0.5 M	25 mM
KH_2PO_4	1 M	50 mM
Na_2HPO_4	1 M	50 mM

Note: Autoclave to dissolve chemicals at 121 °C for 30 min, store at room temperature. (NPS = 100 mM PO_4, 25 mM SO_4, 50 mM NH_4).

5052 (50 × stock).

Component	Concentration (50 ×)	Concentration (1 ×)
Deionized distilled water	–	–
Glycerol	25% (w/v)	0.5% (w/v)
D-Glucose	2.5% (w/v)	0.05% (w/v)
α-Lactose	10% (w/v)	0.2% (w/v)

Note: Autoclave to dissolve chemicals at 121 °C for 30 min, store at room temperature.

Trace metals mixture (1000 × stock).

Component	Concentration (1000 ×)	Concentration (1 ×)
Deionized distilled water	–	–
$FeCl_3$ (0.1 M)	50 mM	50 μM
$CaCl_2$ (1 M)	20 mM	20 μM
$MnCl_2$ (1 M)	10 mM	10 μM

Trace metals mixture (1000 × stock).—cont'd

Component	Concentration (1000×)	Concentration (1×)
$ZnSO_4$ (1 M)	10 mM	10 μM
$CoCl_2$ (0.2 M)	2 mM	2 μM
$CuCl_2$ (0.1 M)	2 mM	2 μM
$NiCl_2$ (0.2 M)	2 mM	2 μM
Na_2MoO_4 (0.1 M)	2 mM	2 μM
Na_2SeO_3 (0.1 M)	2 mM	2 μM
H_3BO_3 (0.1 M)	2 mM	2 μM

Note: Prepare 0.1 M $FeCl_3$ in 0.1 N HCl, prepare all other metal stock solutions in deionized distilled water. Autoclave the stock solutions of the individual metals at 121 °C for 30 min and cool to room temperature prior to combining. For rich media, it is used at 0.1 ×.

Autoinduction overnight culture media.

Component	Concentration
L Broth	–
$MgSO_4$	1 mM
D-Glucose	0.8% (w/v)
Ampicillin	0.1 mg/mL
NPS (20 × stock)	1 ×

Note: $MgSO_4$ should be added before NPS to avoid precipitation and minimize errors in cell density measurement as well as to maximize the bioavailability of the metal ions. Glucose is included to suppress expression of recombinant protein.

Autoinduction large scale culture media.

Component	Concentration
L Broth	–
1 M $MgSO_4$	1 mM
100 mg/mL Ampicillin	0.1 mg/mL
5052^ (50 × stock)	1 ×
Trace Metals Mixture (1000 × stock)	0.1 ×
NPS (20 × stock)	1 ×

Note: $MgSO_4$ and Trace Metals Mixture should be added before NPS to avoid precipitation and minimize errors in cell density measurement as well as to maximize the bioavailability of the metal ions. ^Glucose is included to suppress expression while growing to high cell density, while lactose is included to induce expression when glucose and glycerol are exhausted.

Materials.
Sterile Petri dishes
Fernbach flasks and plugs
Cuvettes
Sterile tubes of various sizes
Tube racks and holders
Bunsen burner
Pipettes and pipette tips
Sterile serological pipettes and pipette controller
Sterile syringes and 0.22 μm membrane filters
Sterile bottles
Beakers and measuring cylinders
Stir bars and mixers

Equipment.
Static incubator (37 °C) (e.g., Thermoline Scientific—Model: TI-150G)
Temperature controlled orbital shaker (e.g., Ratek Orbital Shaking Incubator—Model: OM15, Infors HT—Multitron Pro Orbital Shaking Incubator)
Refrigerated centrifuge (e.g., Ultracentrifuge: Thermo Scientific™ Sorvall™ LYNX 6000—Rotor F9—6 × 1000 LEX with 1 L centrifuge tubes for harvesting cell cultures, Multifuge: Thermo Fisher Scientific—Multifuge X-Pro Series for 15 or 50 mL tubes, Microcentrifuge: Eppendorf centrifuge 5424R for 1.5 or 2 mL tubes)
Spectrophotometer (e.g., Varian Cary® 50 UV–Visible Spectrophotometer, Thermo Fisher Scientific NanoDrop One/OneC)
Heat Block (e.g., Ratek Dry Block Heaters—Model: DBH10D)
Water bath (e.g., Ratek Waterbath—Model: WB14)
Digital weighing scales (e.g., Pioneer OHAUS—Model: PA4101C, Mettler Toledo—Model: ML204T)
Magnetic stirrer (e.g., Thermoline Scientific CH2081-001)
Autoclave

2.1.2 Cloning into pCold™ IV vector

Expression clones are made by inserting the coding sequence for the target gene into the multiple cloning site of pCold™ IV vector in-frame with the upstream fusion tag. The target gene is under the control of the cspA promoter, which is derived from the cold-shock gene, cspA. The pCold™ IV vector has a lac operator downstream of the cspA promoter for strictly controlled expression. The constructs have an N-terminal 8 × His-tag followed by a TEV recognition site for the removal of the 8 × His-tag after purification.

2.1.3 Transformation of pCold™ IV constructs into E. coli host strain

1. Thaw frozen aliquots of chemically competent cells OverExpress™ C41(DE3) on ice
2. Add 1 μL of plasmid to 30 μL of competent cells in a 1.5 mL Eppendorf tube
3. Incubate mixture on ice for 30 min
4. Heat-shock the cell suspension in a 42 °C water bath for 90 s
5. Transfer tubes back on ice to cool for 2 min
6. Add 300 μL of sterile L Broth to each transformation reaction
7. Recover cells by incubating in a 37 °C shaking incubator for 1 h. While this is optional for transformation using ampicillin, it usually increases yield of transformants
8. Plate 150 μL of the cell suspension onto a LB agar plate with ampicillin (100 μg/mL)
9. Incubate the plated cells in a 37 °C static incubator overnight

Note: It is recommended to use freshly transformed colonies for growing overnight starter cultures.

2.1.4 Setting up overnight starter culture for autoinduction in C41 (DE3) cells

1. Prepare Autoinduction Overnight Culture Media as described in Section 2.1.1. The media should be prepared and autoclaved just prior to performing the expression. All cell culture steps must be done aseptically close to a Bunsen burner flame
2. Inoculate a single colony of transformed cells per 10 mL (or less) of media. The capacity of the sterile flasks or tubes used to set up overnight cultures should be at least double the culture volume to ensure proper aeration and support optimal growth

3. Incubate overnight cultures at 37 °C with constant shaking at 200 rpm for 18–20 h. Leave caps slightly loose to allow for air exchange

Note: Overnight cultures are to be used immediately and any leftover discarded appropriately and not stored for future expressions.

2.1.5 Setting up autoinduction large scale culture in C41(DE3) cells

1. Prepare Autoinduction Large Scale Culture Media as stated in Section 2.1.1
2. Inoculate 300 mL of media with 1% (v/v) overnight culture. The capacity of sterile flasks used to set up overnight cultures should be at least double the culture volume to ensure proper aeration for optimal growth. We used 300 mL in a 2 L sterile flask. The autoinduction method promotes growth to high densities and hence growth cultures can be optimized in 200–1000 mL volume depending on protein expression
3. Incubate the cultures at 16 °C with constant shaking at 220 rpm for approximately 68–70 h. Leave caps slightly loose to allow for air exchange
4. Final cell density can be determined by measuring the Optical Density / Absorbance at 600 nm (OD-A_{600}) on a Varian Cary 50 spectrophotometer using visible cuvettes with an aliquot of suspended cells diluted to 5% (v/v) in 20 mM Tris pH 8.0 buffer. A typical 300 mL culture would have an average OD-A_{600} of about 8
5. Cells can be harvested by ultracentrifugation at 4 °C, 5000 rpm (5422 × *g*) for 20 min. Discard the supernatant and store the pellet in a 50 mL tube at −80 °C. Flash freezing in liquid nitrogen is not necessary. The average pellet size of a typical 300 mL culture is approximately 5 mL or 5 g

2.2 Protein expression in insect cells

While the bacterial expression system offers large scale production of the protein of interest in a relatively short timeframe and requires simple culture conditions at minimum cost, expression of some mammalian proteins can be difficult. In addition, proteins that can be expressed lack critical PTMs that are often required for protein folding and function. The next, most cost-effective and efficient technique to express mammalian proteins in large scale for structural and functional studies is the insect cell expression system as it provides a means to express the protein of interest in its most native form. As the PEAK proteins are relatively large, and their sequence is replete with post-translational modification sites, we sought to develop a methodology to express the PEAK proteins using an insect cell expression system to enable comprehensive structural and functional studies. The methodology is described below.

2.2.1 Key resources, reagents, materials and equipment

Key resource or reagent table.

Cells	Supplier	Cat. no.
*Sf*21 (clonal isolates, derived from *Spodoptera frugiperda*)	ThermoFisher Scientific	11497013
Kit		
Bac-to-Bac Baculovirus Expression System	Invitrogen	10359-16
Vector		
pFastBac™ Dual	Invitrogen	10712-024
Antibiotics		
Ampicillin, sodium salt	Astral Scientific	BIOAB0028
Gentamicin sulfate	Sigma-Aldrich	G1914
Tetracycline	Sigma-Aldrich	87128
Kanamycin sulfate	Astral Scientific	BIOKB0286
Media		
Insect Xpress, protein free medium with L-glutamine	Lonza	BELN12-730Q
Transfection reagent		
Cellfectin II	ThermoFisher Scientific	10362100
Trypan blue solution	Sigma-Aldrich	T8154
Bacmid generation and purification		
5-Bromo-3-indolyl-beta-D-galactopyranoside (Bluo-Gal)	Gold-Bio	B-673
Isopropyl β-D-1-thiogalactopyranoside (IPTG)	Astral Scientific	AST0487
Dimethyl sulfoxide (DMSO)	Sigma-Aldrich	D8418
LG Agar (as described in Section 2.1.1)		
L Broth (as described in Section 2.1.1)		
Ethanol	Chem-supply	EA043
Isopropanol/2-propanol	Sulpeco	1.09634.2511
Plasmid Mini Kit	Qiagen	12123

TE buffer for dissolving bacmid DNA: 10 mM Tris, 1 mM EDTA in Milli-Q water (bring to pH 8.0 with HCl or NaOH and sterile filter using a 0.22 μm filter).

Materials.

1 L culture flasks
T75 EasYFlask, TC Surface, Filter Cap—Thermo Scientific 156499
T25 EasYFlask, TC Surface, Filter Cap—Thermo Scientific 156367
50 mL tubes—Thermo Scientific 339652
2.8 L Fernbach Culture Flasks and Plugs
Schott Bottles
Sterile serological pipettes and pipette controller
Aluminium foil
1.5 mL tubes to store samples for cell counting
Hemocytometer and glass coverslips for cell counting—HIRSCHMANN 8100204
Cryogenic vials for storing P1 viral stock—Corning® 430487Cell Culture Flasks—Thermo Scientific 156367

Equipment.

Class II Biosafety cabinet
Water bath (e.g., Ratek Waterbath—Model: WB14)
Static incubator (27 °C) for 1 L bottles (e.g., SANYO—MIR 253) Orbital shaker for 1 L bottles (e.g., ELMI—analog orbital shaker)
Temperature controlled flat base orbital shaker for 2.8 L flasks (e.g., Ratek, XLarge Cooling Orbital Shaking Incubator)
Benchtop centrifuge for harvesting virus stocks (e.g., Thermofisher—Multifuge X Series)
Refrigerated centrifuge for harvesting cell cultures (e.g., Thermo Scientific™ Sorvall™ LYNX 6000, Rotor F9—6 × 1000: 1 L centrifuge tubes)
Phase-contrast microscope (e.g., Olympus CK2)

2.2.2 Cloning into pFastBac™ dual vector

PEAK1, PEAK2 and PEAK3 (domain boundaries highlighted in Table 1) were cloned in-frame into the multiple cloning sites of pFastBac™ Dual vector under the polyhedrin promoter. The constructs have an

N-terminal 6xHis-tag followed by a Tobacco Etch Virus (TEV) cleavage site for the removal of the 6 × His-tag.

2.2.3 Transformation into DH10Bac™ E. coli cells and generation of bacmid DNA

1. pFastBac™ Dual vector containing various PEAK constructs are transformed in DH10Bac™ cells by mixing 1 μL of plasmid with 30 μL of DH10Bac™ chemically competent cells in a 1.5 mL Eppendorf tube
2. Incubate mixture on ice for 30 min
3. Heat-shock the cell suspension in a 42 °C water bath for 60 s
4. Transfer tubes back on ice to cool for 2 min
5. Add 1000 μL of sterile L Broth to each transformation reaction and transfer the suspension into 15 mL round-bottom polypropylene tube
6. Recover cells by incubating at 37 °C shaking incubator for 4 h
7. Prepare 1: 10 and 1: 100 dilutions of cells and plate 100 μL of each dilution onto an LB agar plate containing kanamycin (50 μg/mL), gentamycin (7 μg/mL), tetracycline (10 μg/mL), Bluo-Gal (200–400 μg/mL) and IPTG (40 μg/mL) to select for DH10Bac™ transformants
8. Incubate plates for 48 h at 37 °C. The presence of Bluo-Gal will permit blue/white selection. Pick white colonies for further analysis
9. Pick a single white colony and inoculate a culture medium of sterile L Broth containing kanamycin (50 μg/mL), gentamycin (7 μg/mL) and tetracycline (10 μg/mL)
10. Isolate recombinant bacmid DNA using the Qiagen Plasmid Mini Kit (Catalog no. 12123) and elute the final bacmid DNA in 1 × TE buffer (10 mM Tris-HCl pH 8.0, 1 mM EDTA). Bacmid DNA can be analyzed by PCR using the M13 primers according to the instructions provided in the Bac-to-Bac® Baculovirus Expression System manual (Invitrogen)

2.2.4 Transfection into Sf21 cells and amplification of recombinant virus

The isolation of baculoviral stocks P1, P2 and P3 in *Sf*21 cells is performed according to the Bac-to-Bac® Baculovirus Expression System manual with the following modifications:

1. The *Sf21* stock culture is maintained by splitting the cells to 0.5×10^6 cells/mL every 2–3 days at 27 °C temperature and shaking at 90 rpm. The cell viability is maintained above ~95% with doubling time every 24 h. To determine the viability of cells, place equal volume

(10 μL) of cells and trypan blue solution on a hemocytometer and count under the microscope the total number of dead cells (stained blue) and total number of cells (dead and live). For the transfection, the stock culture is split the day before to 1×10^6 cells/mL such that on the day of transfection, cells have doubled to 2×10^6 cells/mL

2. Instead of a 6-well tissue culture plates, use sterile T25 filter cap flasks and seed 3×10^6 *Sf*21 cells from the stock culture per flask in 2–3 mL volume
3. Allow cells to attach by incubating at room temperature for 45 min
4. Dilute 30 μL bacmid DNA (3–4 μg) in 300 μL Insect Xpress medium (A)
5. Dilute 24 μL Cellfectin II reagent in 300 μL Insect Xpress medium (B)
6. Combine (A) and (B), mix by gently tapping the tube and incubate at room temperature for 30 min
7. While the DNA:lipid complexes are incubating, gently aspirate the media from the attached cells *Sf*21 cells in the T25 flask and wash with 5 mL Insect Xpress medium
8. Gently aspirate the wash media and add the 600 μL DNA:lipid mix to the T25 flask and incubate the cells at 27 °C for 5 h
9. Gently aspirate the DNA:lipid complexes and add 5 mL of Insect Xpress medium and incubate the cells in a 27 °C humidified static incubator for 72 h
10. Proceed to harvesting the P1 viral stock by transferring the 5 mL medium containing the virus to a sterile 10 mL snap-cap tube. Centrifuge the tube at 3000 × *g* for 10 min to remove cells and large debris. Remove the supernatant: **this is the P1 viral stock**. The P1 viral stock can be aliquoted into 1 mL cryotubes for storage at −80 °C. Determination of the titer of the viral stock can be performed according to the instructions in the Bac-to-Bac® Baculovirus Expression System manual
11. To amplify the baculoviral stock, inoculate 100 mL of *Sf*21 cells at a density of 2×10^6 cells/mL in a 1 L flask with 1 mL P1 viral stock. Incubate the cells in a 27 °C shaker incubator and monitor for signs of infection. Infected cells can be visualized under the microscope and are characterized by an increased cell diameter and a granular appearance. Optimal harvest times can vary and should be determined for each construct (normally between 4 and 7 days). At the time of harvest, transfer supernatant into sterile 50 mL Falcon tubes and centrifuge the tubes at 3000 × *g* for 10 min to remove cells and large debris. **This is the P2 viral stock**

12. To further amplify the baculoviral stock, inoculate 500 mL of *Sf*21 cells at a density of 2×10^6 cells/mL in a 2.8 L flask with 20–30 mL of P2 viral stock. Incubate the cells in a 27 °C shaker incubator and monitor for infection as described previously. Optimal harvest times can vary and should be determined for each construct. Harvest the cells as described before. **This is the P3 viral stock**
13. P2 and P3 baculoviral stocks are stored in sterile T75 filter cap flasks in the dark (wrapped in foil) at 4 °C for up to 2–6 months (longer storage will decrease viral titer efficiency)

2.2.5 Protein expression: Infection of Sf21 cells

1. The day prior to infection, split *Sf*21 cells to a density of $\sim 1 \times 10^6$ cells/mL in 0.5 L Insect Xpress medium per 2.8 L flask
2. Incubate overnight at 27 °C with gentle shaking (90–100 rpm)
3. The following day check cell density. Cells should be at a density of $\sim 2\text{–}2.5 \times 10^6$ cells/mL, with ~98–100% viability as described in Section 2.2.4
4. Infect cells with 20–50 mL P3 baculoviral stock per 0.5 L of cells and return cells to the 27 °C shaker for 48–72 h. The optimal volume of the P3 baculoviral stock for expression should be determined for each construct. A series of test expressions in 100 mL media with varying amount of P3 stock can be performed to determine an optimal infection volume between 48 and 72 h. Protein expression can be evaluated using a Western blot. Once the optimal P3 stock for infection is determined, protein expression can be scaled up by preparing multiple 0.5 L culture volume to get 1–3 L final expression volume
5. Check cell density and viability upon harvest. Cell density should be at $\sim 2\text{–}3 \times 10^6$ cells/mL, with around 80–90% viability
6. Centrifuge the cells at $1100 \times g$, 4 °C for 20 min. Cell pellets can be stored at −80 °C. The typical pellet size from a standard 1 L culture is approximately 10 mL or 10 g

3. Recombinant protein purification

This section describes in detail the methodology used to purify **PEAK1** (PEAK1-PsK-L) or **PEAK2** (PEAK2-PsK-L) expressed in bacterial cells, and **PEAK1** (PEAK1-N_{ext}PsK-L), **PEAK2** (PEAK2-N_{ext}PsK-L) or **PEAK3** (PEAK3-FL) expressed in insect cells (Fig. 2). Samples are to be kept at 4 °C whenever possible, unless otherwise stated. All PEAK purifications start with an immobilized metal affinity chromatography (IMAC) step followed by a typical size-exclusion chromatography (SEC) step using a HiLoad 16/600 Superdex 200 pg column. It is recommended to pass the samples eluted after IMAC through SEC as soon as possible rather than keeping the eluted fractions in the His-tag purification elution buffer for extended periods. Depending on the purity of the samples obtained after SEC using a HiLoad 16/600 Superdex 200 pg column, subsequent optional purification steps, such as anion exchange chromatography using a MonoQ™ 5/50 GL column and/or passing through another SEC step using a Superdex 200 Increase 10/300 GL column might be needed to achieve the desired purity.

3.1 Key resources, reagents, materials and equipment

Reagent or resource	Supplier	Cat. no.
Buffer reagents		
HEPES	Sigma-Aldrich	H4034
Tris-hydrochloride	Invitrogen	15504020
Sodium chloride	Ajax-Finechem (Univar)	AJA465
Glycerol	Ajax-Finechem (Univar)	AJA242
Imidazole	Merck	1.04716.1000
Nonaethylene glycol monododecyl ether (Thesit)	Sigma-Aldrich	P9641
Dithiothreitol (DTT)	Astral Scientific	C-1029
Tris(2-carboxyethyl)phosphine hydrochloride (TCEP)	AK Scientific	X4741
EDTA disodium salt dihydrate	Amresco	0105
Sodium hydroxide	Merck	1.06498.0500
Hydrochloric acid	Ajax-Finechem (Univar)	AJA256
Potassium chloride	Sigma-Aldrich	P9541
Deoxyribonuclease I (DNaseI)	Sigma-Aldrich	DN25

—cont'd

Reagent or resource	Supplier	Cat. no.
Lysozyme	Sigma-Aldrich	L6876
cOmplete™, EDTA-free Protease Inhibitor Cocktail	Roche	5056489001
cOmplete™ His-Tag Purification Resin	Roche	5893801001
Quick Start Bradford 1 × Dye Reagent	Biorad	5000205
SDS-PAGE reagents		
Tris-hydrochloride	Invitrogen	15504020
Glycine	Sigma-Aldrich	G7126
Sodium dodecyl sulfate (SDS)	VWR Life sciences	0227
Bromophenol blue	Progen Industries Limited	2000050
Glycerol	Ajax-Finechem (Univar)	AJA242
DTT	Astral Scientific	C-1029
Precision Plus Unstained Protein Standards	Biorad	1610363
Gel Filtration Calibration Kit LMW	Cytiva	28403841
Gel Filtration Calibration Kit HMW	Cytiva	28403842
12% Mini-PROTEAN® TGX™ Precast Protein Gels	Biorad	4561046

SDS-PAGE 4 × sample loading dye.

Component	Amount for 5 mL
Milli-Q water	Top up to 5 mL
0.5 M Tris-HCl (pH 6.8)	2.5 mL
SDS	0.4 g
100% glycerol	2.0 mL
Bromophenol blue	2 mg
DTT	0.31 g

Note: Mix all components together and store at −20 °C as single use aliquots.

SDS-PAGE running buffer (10× stock).

Component	Amount per 1 L	Concentration
Milli-Q water	Top to 1 L	–
Tris	30.28 g	250 mM
Glycine	144.13 g	1.92 M
SDS	10 g	1% (w/v)

Note: Mix all components together and store at room temperature.

SDS-PAGE staining solution.

Component	Amount per 1 L	Concentration
Milli-Q water	–	–
Coomassie Brilliant Blue R	0.25 g	0.025% (w/v)
Methanol	400 mL	40% (v/v)
Acetic acid	70 mL	7% (v/v)

SDS-PAGE destaining solution.

Component	Amount per 1 L	Concentration
Milli-Q water	Top to 1 L	–
Methanol	400 mL	40% (v/v)
Acetic acid	70 mL	7% (v/v)

Materials.

Sterile serological pipettes and pipette controller
Micropipettes and pipette tips
Tubes and tube racks—various sizes
50 mL Nalgene™ oak ridge high-speed PPCO centrifuge tubes (ThermoFisher Scientific 3119-0050)
Lab bottles—various sizes
Beakers and measuring cylinders
Chromatography column (e.g., Biorad Glass Econo-Column® Chromatography Columns 7374151)

Materials.—cont'd

0.22 μm filters (e.g., Merck Millipore Millex-GV filter unit SLGV013SL; Sartorius Ministart Syringe filter 16532-K; Pall Nanosep MF Centrifugal Devices ODM02C34; Merck Durapore PVDF membrane filter GVWP04700) 0.45 μm filters (e.g., Sartorius Ministart Syringe filter 16533-K; Merck Durapore PVDF membrane filter HVLP04700)
Protein concentrators (e.g., Amicon® Ultra 15 mL Centrifugal Filter Units 10 kDa, Merck UFC901024; Amicon® Ultra 4 mL Centrifugal Filters Units 10 kDa, Merck UFC801024; Amicon® Ultra 0.5 mL Centrifugal Filters Units 10 kDa, Merck UFC5010BK)
Disposable syringes—various sizes
Injection needles (e.g., 0.70 × 50 mm Unimed 13.205)
SGE manual fixed needle glass syringes (e.g., Trajan (500 μL—#007000, 250 μL—#006000, 100 μL—#005000)
SEC column (e.g., HiLoad 16/600 Superdex 200 pg Cytiva, product code 28989335; Superdex 200 Increase 10/300 GL (Cytiva, product code: 28990944) MonoQ™ anion exchange chromatography column: MonoQ™ 5/50 GL (Cytiva, product code: 17516601)

Equipment.

Digital weighing scale (e.g., Pioneer OHAUS—PA4101C, Mettler Toledo—ML204T)
pH meter (e.g., Mettler Toledo SevenCompact™ S220) and standards for calibration (e.g., Mettler Toledo, pH 9.21—51302070, pH 7.00—51302047)
Magnetic stirrers (e.g., Thermoline Scientific CH2081-001)
Sonicator (e.g., Bandelin Sonopuls, Probe: UW3200)
Peristaltic Pump (e.g., Gilson minipuls3)
Refrigerated centrifuges: Ultracentrifuge (e.g., Thermo Scientific™ Sorvall™ LYNX 6000; F21 Rotor-8 × 50y for 50 mL Nalgene™ centrifuge tubes) Centrifuge (e.g., Thermo Fisher Scientific—Multifuge X-Pro Series for 15 or 50 mL tubes) Microcentrifuge (e.g., Eppendorf centrifuge 5424R for 1.5 or 2 mL tubes)
Fast protein liquid chromatography (FPLC) (e.g., ÄKTA pure protein purification system with Unicorn software; ÄKTA pure accessory unit Fraction Collector, F9-R, Cytiva, USA)

Continued

Equipment.—cont'd
Heat block (e.g., Ratek Dry Block Heaters—Model: DBH10D)
Mixers and tube rollers (e.g., Ratek Tube Roller BTR5-12V, Ratek Orbital Mixer, Ratek Rocking Platform Mixer)
Spectrophotometer (e.g., Thermo Fisher Scientific NanoDrop One/OneC)
Biorad protein electrophoresis equipment
Gel scanner (e.g., Biorad Gel Doc XR + Imaging System with Image Lab software)
AKTA pure protein purification system (Cytiva)

3.2 Buffers for protein purification

Refer to Table 2 for the list of purification buffers to use for purifying PEAK proteins. The same buffer composition is used for purifying PEAK1 expressed in *E. coli* and insect cells, likewise for PEAK2. All PEAK proteins expressed from the constructs detailed are stable in both Tris and HEPES buffers and can be buffer exchanged into either if needed. While all PEAK proteins are relatively stable between pH 7.5–8.5, we recommend using buffers at pH 8.0 to purify PEAK1, at pH 8.2–8.5 to purify PEAK2 and at pH 7.5 to purify PEAK3. Glycerol can help to stabilize the proteins, during initial purification steps in particular, but is not critical and can be omitted if necessary (e.g., to assist in concentrating the samples). Varying the concentration of imidazole during IMAC binding and washing steps can aid in minimizing impurities that non-specifically bind to the His-tag purification resin.

3.3 Purification of PEAK1/PEAK2 from *E. coli*

Duration: 2 to 3 days, samples should be kept at 4 °C where possible unless specified.

3.3.1 Immobilized metal affinity chromatography: cOmplete™ his-tag purification resin

The methodology described below is applicable for PEAK1 (PEAK1-PsK-L) and PEAK2 (PEAK2-PsK-L) constructs (refer to Fig. 2).

1. Thaw the tube(s) containing the pellets in a beaker of cold water and transfer to ice when fully thawed

Note: cell lysates should be processed straight away.

Table 2 Compositions of buffers used for PEAK1/PEAK2/PEAK3 protein purification.

	Lysis/resuspension buffer	IMAC	SEC	Ion exchange
PEAK1	20 mM Tris pH 8, 500 mM NaCl, 2 mM EDTA, 5 mM DTT, 10 mM Imidazole, 10% Glycerol, 0.1% Thesit	*Wash buffer:* 20 mM Tris pH 8, 500 mM NaCl, 2 mM EDTA, 5 mM DTT, 10 mM Imidazole, 10% Glycerol *Elution buffer:* 20 mM Tris pH 8, 500 mM NaCl, 2 mM EDTA, 5 mM DTT, 250 mM Imidazole, 10% Glycerol	HiLoad 16/600 Superdex 200 pg *Buffer:* 20 mM Tris pH 8.0, 200 mM NaCl, 5% Glycerol, 0.5 mM TCEP	MonoQ™ 5/50 GL *Buffer A:* 20 mM Tris pH 8.0, 5% Glycerol, 1 mM TCEP *Buffer B:* 20 mM Tris pH 8.0, 5% Glycerol, 1 M NaCl, 1 mM TCEP
PEAK2	20 mM Tris pH 8.5, 500 mM NaCl, 2 mM EDTA, 10 mM DTT, 5 mM Imidazole, 10% Glycerol, 0.1% Thesit	*Wash buffer:* 20 mM Tris pH 8.5, 500 mM NaCl, 2 mM EDTA, 10 mM DTT, 5 mM Imidazole, 10% Glycerol *Elution buffer:* 20 mM Tris pH 8.5, 500 mM NaCl, 2 mM EDTA, 10 mM DTT, 250 mM Imidazole, 10% Glycerol	HiLoad 16/600 Superdex 200 pg *Buffer:* 20 mM Tris pH 8.5, 200 mM NaCl, 5% Glycerol, 1 mM TCEP Superdex 200 increase 10/300 GL *Buffer:* 20 mM HEPES pH 8.2, 200 mM NaCl, 1 mM TCEP	N/A
PEAK3	20 mM Tris pH 7.5, 500 mM NaCl, 10 mM DTT, 20 mM Imidazole, 10% Glycerol, 0.1% Thesit	*Wash buffer:* 20 mM Tris pH 7.5, 500 mM NaCl, 10 mM DTT, 20 mM Imidazole, 10% Glycerol *Elution buffer:* 20 mM Tris pH 7.5, 500 mM NaCl, 10 mM DTT, 250 mM Imidazole, 10% Glycerol	HiLoad 16/600 Superdex 200 pg *Buffer:* 20 mM Tris pH 7.5, 250 mM NaCl, 1 mM TCEP	MonoQ™ 5/50 GL *Buffer A:* 20 mM Tris pH 7.5, 1 mM TCEP *Buffer B:* 20 mM Tris pH 7.5, 1 M NaCl, 1 mM TCEP

2. Resuspend the cell pellet(s) with resuspension buffer to a total of 40 mL per L of culture (a pellet size of approximately 5 g) per tube

Note: The volume of lysis buffer can play a critical role for successful protein purification. Although the ideal volume of lysis buffer should always be determined experimentally, typically 5–8 mL of lysis buffer per gram of cells is sufficient.

3. Add 0.5 mg of Dnase-I, 10 mg of Lysozyme and 1 tablet of cOmplete™ EDTA-free Protease Inhibitor Cocktail pre-dissolved in 1 mL of Milli-Q water to each tube of resuspended cells
4. Incubate with constant agitation (e.g., on a roller) at room temperature for 10 min
5. Keeping sample tubes on ice, sonicate to lyse the resuspended cells (e.g., 40% amplitude for 25 s followed by resting on ice for 25 s). Repeat sonication process until cells are lysed (usually two to five times per tube, inverting tubes a few times between cycles to mix and avoid heating of the sample) while keeping the sample on ice and making sure the temperature of the sample stays close to 4 °C
6. Centrifuge the sample in Nalgene™ high-speed centrifuge tubes at 18,000 rpm (38,759 × *g*), 4 °C for 30 min to pellet cell debris
7. Decant the cell lysate supernatant and filter through 0.45 μm syringe filter using gentle pressure
8. Pre-equilibrate the cOmplete™ His-Tag Purification Resin pre-washed with Milli-Q water with the resuspension buffer and pass the filtered supernatant twice through with a peristatic pump and a slow flow rate (~1 mL/min). Keep samples at 4 °C throughout. It is recommended to use 0.5 mL resin per L culture of PEAK1 and PEAK2
9. Using the peristaltic pump, wash the resin with 50 mL of resuspension buffer followed by 25 mL of wash buffer
10. Elute bound protein in elution buffer containing 250 mM imidazole either by gravity flow or using a peristaltic pump (flow rate ~0.5 mL/min) collecting 1 mL fractions. Keep eluted fractions on ice
11. To quickly identify fractions containing protein, a simple Bradford assay can be useful. To a clear 96-well plate, add 40 μL of 1 × Bradford reagent per well, followed by 2 μL of each elution fraction. Gently mix the plate on the bench top and fractions containing protein will turn blue

3.3.2 Size-exclusion chromatography: HiLoad 16/600 Superdex 200 pg column

The method below is applicable for PEAK1 (PEAK1-PsK-L) and PEAK2 (PEAK2-PsK-L) constructs expressed in *E. coli* (Fig. 2). Refer to Table 2 for the list of purification buffers to use for each PEAK protein and to Fig. 3 for elution profiles and SDS-PAGE.

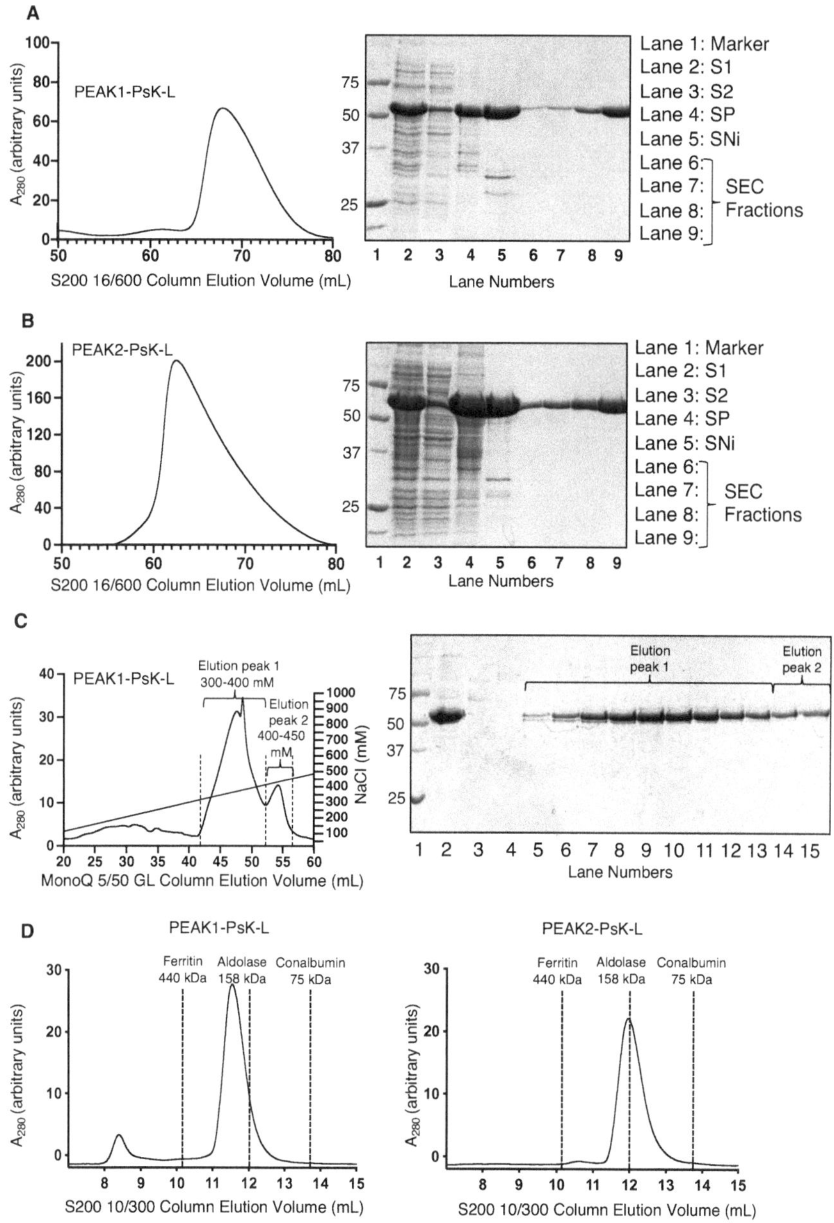

Fig. 3 See figure legend on next page.

1. Pool all fractions containing protein from the IMAC purification and concentrate if required (see *Note* below). Filter the sample (0.22 μm syringe filter) and load onto a HiLoad 16/600 Superdex 200 pg column pre-equilibrated in SEC running buffer. Perform SEC at a flow rate of 1 mL/min and maximum pressure of 0.5 MPa. Store eluted fractions on ice

   Note: If the total sample volume from IMAC purification exceeds the standard 5 mL loop volume, samples can be concentrated to below 5 mL in a centrifugal concentrator unit prior to filtering (e.g., Amicon® Ultracentrifugal Filter Unit, 10 kDa molecular weight cut-off; 4000 g at 4 °C for 5–10 min per spin) with gentle mixing between spins.
2. Perform SDS-PAGE to analyze sample purity of the eluted fractions from IMAC and SEC. Mix 4 μL of 4 × SDS-PAGE sample loading dye with a maximum of 10.5 μL of samples in separate tubes. Heat sample tubes in a heating block at 95 °C for 3 min. Centrifuge heated samples briefly and load all samples and protein standards into separate wells of 12% (w/v) gel(s) and perform SDS-PAGE with 1 × SDS-PAGE running buffer at 200 V for 35–45 min
3. Upon completion of the run, stain the gels with SDS-PAGE staining solution for 45 min followed by repeated washing with SDS-PAGE destaining solution until the background is clear
4. Transfer the gel to Milli-Q water

Fig. 3 Example purification results for PEAK proteins expressed in C41(DE3) *E. coli*. (A) PEAK1 (PEAK1-PsK-L) purification on SEC using a HiLoad 16/600 Superdex 200 pg column and the corresponding SDS-PAGE analysis. Lane 1, protein marker; lane 2, supernatant after cell lysis (S1); lane 3, filtered supernatant prior to the IMAC step (S2); lane 4, cell pellet after ultracentrifugation step (SP); lane 5, pooled IMAC elution fractions prior to loading on SEC (SNi); lanes 6–9, SEC elution fractions. (B) PEAK2 (PEAK2-PsK-L) purification on SEC using a HiLoad 16/600 Superdex 200 pg column and corresponding SDS-PAGE analysis. Samples are arranged in lanes 1–9 in the same order as (A). (C) PEAK1 (PEAK1-PsK-L) purification on MonoQ™ 5/50 anion exchange column and the corresponding SDS-PAGE analysis. PEAK1 was eluted by applying a linear gradient of 50–600 mM NaCl (indicated by the diagonal line in the Mono Q profile) which resolved in two separate elution peaks (elution peak 1 and elution peak 2). Elution peak 1 eluted between 300 and 400 m NaCl and elution peak 2 eluted between 400 and 450 mM NaCl. SDS-PAGE analysis of PEAK1 is as shown. Lane 1, protein marker; lane 2, sample before loading on MonoQ; Lanes 3–13, elution peak 1; Lanes 14–15, elution peak 2. (D) PEAK1 (PEAK1-PsK-L) and PEAK2 (PEAK2-PsK-L) elution profiles on an analytical SEC Superdex 200 Increase 10/300 GL column when loaded at a concentration of 51 μM. This elution profile represents a mixture of dimers (~108 kDa) and oligomers when compared to molecular weight standards (dotted lines) of ferritin (440 kDa), aldolase (158 kDa) and conalbumin (75 kDa).

5. Image the gel using the Bio-Rad Gel Doc XR+ Imaging System with Image Lab software
6. Pool fractions containing pure protein and concentrate by centrifugation *(e.g., Amicon® Ultracentrifugal Filter Unit, 10kDa molecular weight cut-off;* sequential spins at 4000 × *g*, 4 °C for 5–10 min each, with gentle mixing between spins) to the final desired working concentration. Purified PEAK1 (PEAK1-PsK-L) and PEAK2 (PEAK2-PsK-L) from bacterial expression can be stably concentrated to ~4 mg/mL
7. Final protein concentrations of PEAK1/PEAK2 samples can be measured using a NanoDrop spectrophotometer at OD A280 nm and applying the mg/mL conversions, using the theoretical extinction coefficients. The extinction coefficients are calculated from the ProtParam tool at Expasy (https://www.expasy.org) on the Swiss Bioinformatic Resource Portal based on cleaved or uncleaved protein sequences
8. The purified protein is relatively stable at 4 °C for a week but flash freezing in liquid nitrogen and storing at −80 °C is recommended for long term storage

3.3.3 Sample results for PEAK1/PEAK2 expressed in bacterial cells

PEAK1/PEAK2 protein samples expressed in bacterial cells and purified through IMAC and SEC are typically sufficiently pure for use in subsequent assays. SEC elution profiles (HiLoad 16/600 Superdex 200 pg column) and SDS-PAGE analysis of PEAK1-PsK-L and PEAK2-PsK-L expressed in bacterial cells are shown in Fig. 3A and B, respectively.

Removal of the His-tag is not necessarily required but can be performed as described for the PEAK proteins expressed in insect cells (see Section 3.4.3), followed by anion exchange chromatography (Fig. 3C).

Analytical SEC can also be performed to further assess the oligomerization status of PEAK proteins as a function of concentration. For example, PEAK1 and PEAK2 expressed from *E. coli* will mainly elute between 11.5 and 13 mL using a S200 Increase 10/300 GL gel filtration column, when loaded at a concentration of around 2.7 mg/mL (50 μM) (Fig. 3D); this corresponds to a mixture of dimer and tetramers. Higher-order oligomers can also be present at this concentration and typically elute at ~8 mL post-injection (corresponding to the column void volume). Our previous analysis of PEAK2 at concentrations between 0.03 and 3.9 mg/mL (0.6 and 74 μM) on analytical ultracentrifugation showed that concentrations above 0.2 mg/mL (4.6 μM) results in the formation of tetramers as well as higher-order oligomers (Patel et al., 2017).

PEAK1 (PEAK1-PsK-L construct) purified from C41(DE3) *E. coli* expressed through the autoinduction method can yield approximately 0.7 mg of protein per g of cell pellet while PEAK2 (PEAK2-PsK-L construct) typically yields approximately 2.3 mg of protein per g of cell pellet.

3.4 Purification of PEAK1/PEAK2/PEAK3 from insect (*Sf*21) cells

3.4.1 *Immobilized metal affinity chromatography: cOmplete™ His-Tag purification resin*

The methodology described is applicable for PEAK1 (PEAK1-N_{ext}PsK-L), PEAK2 (PEAK2-N_{ext}PsK-L) and PEAK3 (PEAK3-FL)

1. Thaw the tube(s) containing the pellets in a beaker of cold water and transfer to ice when fully thawed.
2. *Note: cell lysates should be processed straight away.*
3. Resuspend the cell pellet(s) with resuspension buffer to a total of 80 mL per L of culture (a pellet size of approximately 10 g).

   Note: The volume of lysis buffer can play a critical role in the successful completion of protein purification. Although the ideal volume of lysis buffer should always be determined experimentally, typically 8 mL of lysis buffer per gram of cells is sufficient.
4. Add 0.5 mg of DNase-I and a tablet of cOmplete™, EDTA-free Protease Inhibitor Cocktail pre-dissolved in Milli-Q water to each tube of resuspended cells.
5. Keeping sample tubes on ice, sonicate to lyse the resuspended cells (e.g., 40% amplitude for 25 s followed by resting on ice for 25 s). Repeat sonication process until cells are lysed (usually two to five times per tube), inverting tubes a few times between cycles to mix and avoid heating of the sample by keeping the samples on ice and making sure the temperature stays close to 4 °C. Sample should become less viscous after sonication.
6. Centrifuge sample in the Nalgene™ high-speed centrifuge tubes at 19,500 rpm (45,487 × *g*), 4 °C for 60 min to pellet cell debris. Recover the cell lysate (cell supernatant) and filter (0.45 μm filter).
7. For IMAC purification via batch method, incubate the clarified cell lysate in 50 mL tubes with Roche cOmplete™ His-Tag purification resin (0.6 mL resin per 2 L cells pre-equilibrated in resuspension buffer) and rotate at 4 °C for 1.5 h
8. Centrifuge the resin/lysate mixture (50 mL tubes) at 1500 rpm (1537 × *g*), 4 °C for 5 min to pellet the resin. Remove the supernatant.

9. Wash the protein-bound resin by resuspending in resuspension buffer (20 mL per tube) and centrifuge the tubes again at 1500 rpm (1537 × g), 4 °C for 5 min. Remove the supernatant.
10. Repeat the wash twice more, then combine the resin into a single tube and centrifuge the tube again at 1500 rpm (1537 × *g*), 4 °C for 5 min.
11. Resuspend the resin in 10–20 mL of resuspension buffer and transfer to a Bio-Rad Glass Econo-Column Chromatography Column.
12. Allow the resin to settle in the column and wash with an additional 25 mL of resuspension buffer either by gravity flow or with the help of a peristaltic pump using a flowrate of 1–2 mL/min.
13. Wash the resin with 40 mL of wash buffer, either by gravity or with the help of a peristaltic pump using a flowrate of 1–2 mL/min.
14. Elute bound protein with elution buffer containing 250 mM imidazole either by gravity or using a peristaltic pump with a flow rate of approximately 0.5 mL/min. Collect 1 mL per fraction, and store eluted samples on ice immediately.
15. Perform a quick Bradford assay as described in Section 3.3.1 (step 11).

3.4.2 Size-exclusion chromatography: HiLoad 16/600 Superdex 200 pg column

Following IMAC, apply the same SEC methodology described in Section 3.3.2 for bacterial samples to further purify insect cell derived PEAK1-N_{ext}PsK-L, PEAK2-N_{ext}PsK-L or PEAK3-FL. Refer to Table 2 for the list of buffers to use for each PEAK protein and to Fig. 4 for example chromatograms and SDS-PAGE gels.

3.4.3 Ion exchange column chromatography: MonoQ™ 5/50 GL column

PEAK1/PEAK3 proteins samples expressed in insect cells typically contain some impurities after purification by SEC with HiLoad 16/600 Superdex 200 pg column. These can be removed by performing an additional anion exchange column chromatography purification step using a MonoQ™ 5/50 GL column. Refer to Table 2 for the buffers to use for PEAK1 or PEAK3 samples. Due to the typically lower yield of protein obtained for PEAK3, it is recommended to combine at least two batches of PEAK3 expression (a total of 4–5 L of culture) which is purified by IMAC and SEC before performing anion exchange chromatography.

Removal of the His-tag can be performed prior to the anion exchange chromatography step by adding purified recombinant TEV protease to the

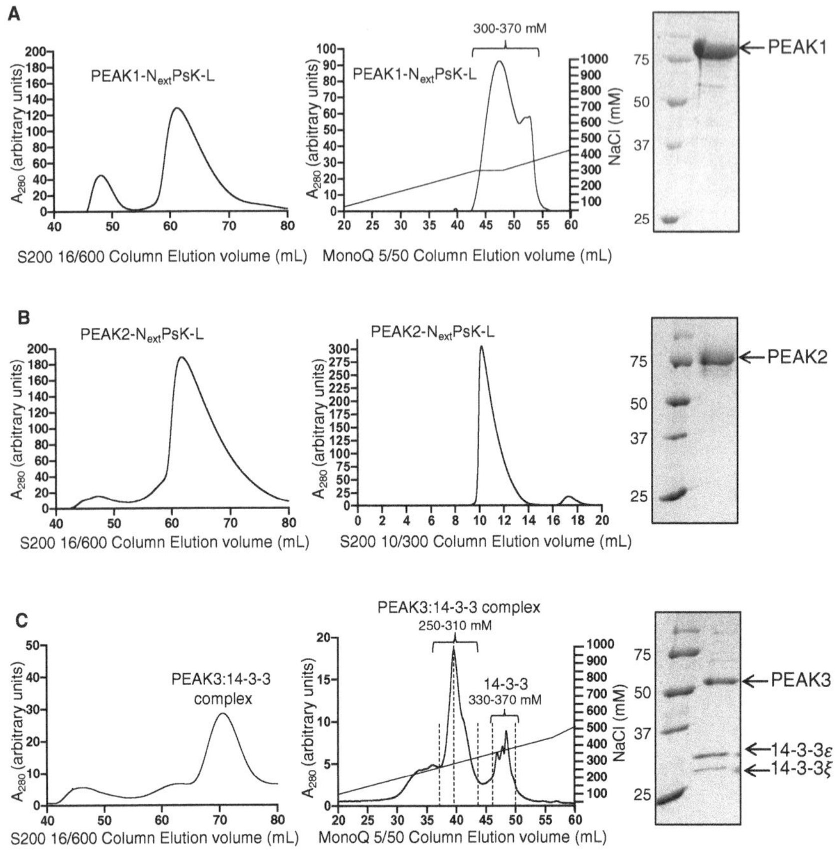

Fig. 4 Example purification results for PEAK proteins expressed in insect cells (*Sf*21). (A) PEAK1 (PEAK1-N_{ext}PsK-L) purification on SEC using a HiLoad 16/600 Superdex 200 pg column (left) followed by MonoQ™ 5/50 anion exchange (middle) and the SDS-PAGE analysis of the final purified sample (right). PEAK1 was eluted by applying a linear gradient of 50–600 mM NaCl. Insect cell PEAK1 eluted between 300 and 370 mM NaCl. SDS-PAGE analysis of purified PEAK1 following the MonoQ step is shown in the figure. (B) PEAK2 (PEAK2-N_{ext}PsK-L) purification on SEC using a HiLoad 16/600 Superdex 200 pg column (left) followed by a Superdex 200 Increase 10/300 GL column (middle) and the SDS-PAGE analysis of the final purified sample (right). (C) PEAK3 (PEAK3-FL) purification on SEC using a HiLoad 16/600 Superdex 200 pg column (left) followed by MonoQ™ 5/50 anion exchange (middle) and the SDS-PAGE analysis of the final purified sample (right). PEAK3 eluted between 250 and 310 mM NaCl and SDS-PAGE analysis of purified PEAK3 shows co-elution with insect cell derived 14-3-3 heterodimer (14-3-3ε ~28 kDa and 14-3-3ξ ~27 kDa). Dissociated 14-3-3 heterodimer eluted between 330 and 370 mM NaCl. SDS-PAGE analysis of the final purified sample is as shown (right).

concentrated SEC sample in a ratio of 1 μg of TEV protease per 25 μg of sample. To ensure complete cleavage, samples are typically co-incubated with TEV protease for 2 h at room temperature, then overnight at 4 °C, followed by a further of 0.2 μg of TEV per 25 μg of sample and incubation for 1 h at room temperature. All of the incubation steps with TEV were performed without shaking. Sample incubation at room temperature for more extended periods is generally avoided as this may promote protein degradation or precipitation.

The following describes the methodology to perform anion exchange chromatography for PEAK1 (PEAK1-N_{ext}PsK-L) and PEAK3 (PEAK3-FL). However, it can also be applied to the purification of other PEAK constructs, using corresponding buffers at a suitable pH.

1. Dilute cleaved or uncleaved PEAK1/PEAK3 protein samples with Anion Exchange Buffer A to lower the NaCl concentration to a final concentration of 50 mM. Perform this immediately prior to purification by anion exchange chromatography, as an extended period at low salt concentration may promote aggregation or precipitation of PEAK1/PEAK3 proteins
2. Filter the diluted sample through a 0.22 μm filter and load the filtered sample on a MonoQ™ 5/50 GL column pre-equilibrated with 5% Anion Exchange Buffer B attached to the AKTA pure protein purification system
3. Perform anion exchange chromatography with gradient elution from 50 to 600 mM NaCl over 60 column volumes at flowrate of 1 mL/min. To minimize the coelution of the protein of interest with other impurities and reduce the final sample's salt concentration, hold the gradient when the protein starts to elute (approximately 280–300 mM NaCl for PEAK1-N_{ext}PsK-L and PEAK3-FL) until the peak plateaus
4. Analyze and concentrate eluted fractions after passing through Anion exchange chromatography as described for SEC samples in Section 3.3.2
5. Measure the final protein concentration at OD A280 nm using a NanoDrop spectrophotometer. The concentration is calculated by applying the respective mg/mL conversion, using theoretical extinction coefficients which is calculated from the ProtParam tool at Expasy (https://www.expasy.org), based on the corresponding protein sequences
6. Store purified PEAK1/PEAK3 samples at 4 °C if used within 7 days or flash freeze with liquid nitrogen and store at −80 °C in 10–50 μL aliquots for long term storage. Purified PEAK1 (PEAK1-N_{ext}PsK-L), and PEAK3 (PEAK3-FL) expressed in insect cells can be stably concentrated up to 5 and 3 mg/mL respectively

3.4.4 Analytical size-exclusion chromatography: Superdex 200 increase 10/300 GL column (PEAK2)

Purified PEAK2 protein (PEAK2-N_{ext}PsK-L) sample from insect cells after SEC using a HiLoad 16/600 Superdex 200 pg column may still contain low levels of impurities that can be removed by passing through an additional SEC polishing step (e.g., using an analytical scale Superdex 200 Increase 10/300 GL column). TEV Cleavage of the His-tag can also be performed as described in Section 3.4.3 prior to this polishing step, however, PEAK2 will elute immediately before the TEV protease, and there may be some degree of coelution where the two elution peaks overlap. An additional anion exchange column chromatography purification step using a MonoQ™ 5/50 GL column can also be used to remove the TEV but may lower the final yield of PEAK2.

The following describes the analytical SEC methodology for PEAK2 (PEAK2-N_{ext}PsK-L). This methodology can be applied to all PEAK constructs as an additional purification step and/or to analyze the oligomerization status of the expressed proteins. For the correct buffer composition and pH for each PEAK protein, refer to Table 2.

1. Further concentrate the cleaved or uncleaved protein sample (typically about 2–6 mg) from the first round of SEC to between 100 and 500 μL
2. Filter the concentrated sample through a 0.22 μm filter and immediately inject on a Superdex 200 Increase 10/300 GL column pre-equilibrated with SEC buffer using sample loops and syringes of appropriate sizes. Use a flowrate of 0.5 mL/min, collect 0.5 mL elution fractions
3. Analyze eluted fractions by SDS-PAGE (as described in Section 3.3.2) and pool fractions with the desired purity
4. Measure final protein concentration at OD A280 nm using a NanoDrop spectrophotometer, and applying the respective mg/mL conversion which is calculated from the ProtParam tool at Expasy (https://www.expasy.org) based on the corresponding protein sequences
5. Store purified PEAK2 samples at 4 °C or snap freeze in liquid nitrogen and store at −80 °C in 10–50 μL aliquots for long term storage. Purified PEAK2 (PEAK2-N_{ext}PsK-L) expressed in insect cells can be stably concentrated up to 5 mg/mL

3.4.5 Sample results for PEAK1/PEAK2/PEAK3 expressed in insect (Sf21 cells)

PEAK1 (PEAK1-N_{ext}PsK-L) and PEAK2 (PEAK2-N_{ext}PsK-L) purified from insect cells expression (*Sf*21) using IMAC and SEC can yield approximately 300 and 400 μg of purified protein per gram of cell pellet, respectively. However, the yield from PEAK3 (PEAK3-FL) purified from insect cells expression (*Sf*21) is typically much lower, resulting in approximately 2 μg purified protein per gram of cell pellet. SDS-PAGE analysis of purified PEAK3-FL revealed that PEAK3 co-purifies with insect cell derived 14-3-3 isoforms, 14-3-3ε (~28 kDa) and 14-3-3ξ (~27 kDa). The identities of these co-purified proteins were confirmed by mass spectrometry, as well as the positive phosphorylation status of residue Ser69, the predicted 14-3-3 binding site in PEAK3. Fig. 4 shows the UV elution profiles for PEAK1/PEAK2/PEAK3 purified by SEC (HiLoad 16/600 Superdex 200 pg column), for PEAK1/PEAK3 purified by anion exchange chromatography (MonoQ™ 5/50 GL column) and for PEAK2 purified by analytical SEC (Superdex 200 Increase 10/300 GL column) as well as the SDS-PAGE analysis of PEAK1/PEAK2/PEAK3 purified from insect cells expression.

4. Overview and conclusion

PEAK proteins expressed in *E. coli* or insect cells tend to form concentration-dependent higher-order oligomers. For this reason, it is highly recommended that the oligomerization status of the purified samples be confirmed prior to subsequent assays. The simplest method to conduct this is routine analytical SEC as described in Section 3.4.4 (Patel et al., 2017). Other methods can also be applied to characterize the oligomerization status of purified PEAK proteins, such as analytical ultracentrifugation, as described in Patel et al. (2017), SEC coupled with multi-angle laser light scattering (SEC-MALLS) or small-angle X-ray scattering (SAXS).

The methods presented here provide a reliable strategy for the expression, purification, and characterization of PEAK pseudokinase proteins. We have provided detailed protocols for the expression and purification, using a *E. coli* expression system, of recombinant PEAK1 and PEAK2 (pseudokinase and dimerization domains). Importantly, for the first-time, we describe a strategy to express and purify longer constructs of PEAK proteins, including full-length PEAK3, that contain additional regulatory/PTM sites using an insect cells expression system. The methodology that we have described will provide tools to quantitatively characterize each of the PEAK proteins, their PTMs and interactions with signaling adaptors to subsequently interrogate the PEAK interactome.

Acknowledgments

This work was supported by the National Health and Medical Research Council, Australia (Project Grant APP1162058) with additional support from the Victorian State Government Operational Infrastructure Support Program, from the Australian Cancer Research Foundation and the Walter and Eliza Hall Institute. J.M.H was supported by an NHMRC Investigator Grant (2008096). The contents of this published material are solely the responsibility of the individual authors and do not reflect the views of the NHMRC or funding partners.

References

Cerami, E., Gao, J., Dogrusoz, U., Gross, B. E., Sumer, S. O., Aksoy, B. A., et al. (2012). The cBio cancer genomics portal: An open platform for exploring multidimensional cancer genomics data. *Cancer Discovery*, *2*(5), 401–404. https://doi.org/10.1158/2159-8290.CD-12-0095.

Croucher, D. R., Hochgrafe, F., Zhang, L., Liu, L., Lyons, R. J., Rickwood, D., et al. (2013). Involvement of Lyn and the atypical kinase SgK269/PEAK1 in a basal breast cancer signaling pathway. *Cancer Research*, *73*(6), 1969–1980. https://doi.org/10.1158/0008-5472.CAN-12-1472.

Ding, C., Tang, W., Fan, X., Wang, X., Wu, H., Xu, H., et al. (2018). Overexpression of PEAK1 contributes to epithelial-mesenchymal transition and tumor metastasis in lung cancer through modulating ERK1/2 and JAK2 signaling. *Cell Death & Disease*, *9*(8), 802. https://doi.org/10.1038/s41419-018-0817-1.

Gao, J., Aksoy, B. A., Dogrusoz, U., Dresdner, G., Gross, B., Sumer, S. O., et al. (2013). Integrative analysis of complex cancer genomics and clinical profiles using the cBioPortal. *Science Signaling*, *6*(269), pl1. https://doi.org/10.1126/scisignal.2004088.

Guo, Q., Qin, W., Li, B., Yang, H., Guan, J., Liu, Z., et al. (2014). Analysis of a cytoskeleton-associated kinase PEAK1 and E-cadherin in gastric cancer. *Pathology, Research and Practice*, *210*(12), 793–798. https://doi.org/10.1016/j.prp.2014.09.013.

Ha, B. H., & Boggon, T. J. (2018). The crystal structure of pseudokinase PEAK1 (Sugen kinase 269) reveals an unusual catalytic cleft and a novel mode of kinase fold dimerization. *The Journal of Biological Chemistry*, *293*(5), 1642–1650. https://doi.org/10.1074/jbc.RA117.000751.

Lecointre, C., Simon, V., Kerneur, C., Allemand, F., Fournet, A., Montarras, I., et al. (2018). Dimerization of the Pragmin Pseudo-kinase regulates protein tyrosine phosphorylation. *Structure*, *26*(11), 1563. https://doi.org/10.1016/j.str.2018.10.020.

Liu, L., Phua, Y. W., Lee, R. S., Ma, X., Jenkins, Y., Novy, K., et al. (2016). Homo- and heterotypic association regulates signaling by the SgK269/PEAK1 and SgK223 Pseudokinases. *The Journal of Biological Chemistry*, *291*(41), 21571–21583. https://doi.org/10.1074/jbc.M116.748897.

Lopez, M. L., Lo, M., Kung, J. E., Dudkiewicz, M., Jang, G. M., Von Dollen, J., et al. (2019). PEAK3/C19orf35 pseudokinase, a new NFK3 kinase family member, inhibits CrkII through dimerization. *Proceedings of the National Academy of Sciences of the United States of America*, *116*(31), 15495–15504. https://doi.org/10.1073/pnas.1906360116.

Murphy, J. M., Zhang, Q., Young, S. N., Reese, M. L., Bailey, F. P., Eyers, P. A., et al. (2014). A robust methodology to subclassify pseudokinases based on their nucleotide-binding properties. *The Biochemical Journal*, *457*(2), 323–334. https://doi.org/10.1042/BJ20131174.

O'Rourke, R. L., & Daly, R. J. (2018). The pseudokinases SgK269 and SgK223: A novel oncogenic alliance in human cancer. *Cell Adhesion & Migration*, *12*(6), 524–528. https://doi.org/10.1080/19336918.2017.1394570.

Patel, O., Griffin, M. D. W., Panjikar, S., Dai, W., Ma, X., Chan, H., et al. (2017). Structure of SgK223 pseudokinase reveals novel mechanisms of homotypic and heterotypic association. *Nature Communications*, *8*(1), 1157. https://doi.org/10.1038/s41467-017-01279-9.

Patel, O., Roy, M., Murphy, J. M., & Lucet, I. S. (2020). The PEAK family of Pseudokinases, their role in cell signalling and cancer. *The FEBS Journal*, *287*, 4183–4197. https://doi.org/10.1111/febs.15087.

Studier, F. W. (2005). Protein production by auto-induction in high density shaking cultures. *Protein Expression and Purification*, *41*(1), 207–234. Retrieved from https://www.ncbi.nlm.nih.gov/pubmed/15915565.

Tactacan, C. M., Phua, Y. W., Liu, L., Zhang, L., Humphrey, E. S., Cowley, M., et al. (2015). The pseudokinase SgK223 promotes invasion of pancreatic ductal epithelial cells through JAK1/Stat3 signaling. *Molecular Cancer*, *14*, 139. https://doi.org/10.1186/s12943-015-0412-3.

Tanaka, H., Katoh, H., & Negishi, M. (2006). Pragmin, a novel effector of Rnd2 GTPase, stimulates RhoA activity. *The Journal of Biological Chemistry*, *281*(15), 10355–10364. https://doi.org/10.1074/jbc.M511314200.

Wang, Y., Kelber, J. A., Tran Cao, H. S., Cantin, G. T., Lin, R., Wang, W., et al. (2010). Pseudopodium-enriched atypical kinase 1 regulates the cytoskeleton and cancer progression [corrected]. *Proceedings of the National Academy of Sciences of the United States of America*, *107*(24), 10920–10925. https://doi.org/10.1073/pnas.0914776107.

Weaver, K. L., Alves-Guerra, M. C., Jin, K., Wang, Z., Han, X., Ranganathan, P., et al. (2014). NACK is an integral component of the notch transcriptional activation complex and is critical for development and tumorigenesis. *Cancer Research*, *74*(17), 4741–4751. https://doi.org/10.1158/0008-5472.CAN-14-1547.

CHAPTER TWO

Structural and biophysical characterization of Tribbles homolog 1

Sam A. Jamieson, Hamish D. McMillan, and Peter D. Mace*
Biochemistry Department, School of Biomedical Sciences, University of Otago, Dunedin, New Zealand
*Corresponding author: e-mail address: peter.mace@otago.ac.nz

Contents

Abstract

Tribbles proteins are pervasive pseudokinases in cellular signaling. They play a major role in the differentiation of myeloid cells, hepatocytes and adipocytes, and more widely in immune function, metabolism and cancer. Like many other pseudokinases, an inherent lack of catalytic activity has meant that a specialized cadre of techniques has been required to investigate Tribbles function. A prerequisite to most *in vitro* biochemistry has been robust methods for purifying useful quantities of Tribbles protein, which can sometimes exhibit non-optimal behavior upon recombinant expression. For instance, structural studies of the Tribbles family have largely focused on TRIB1, in part because of more readily available protein. Here we describe methods we have

Methods in Enzymology, Volume 667
ISSN 0076-6879
https://doi.org/10.1016/bs.mie.2022.03.023

developed to routinely produce milligram quantities of TRIB1, and specific considerations when employing TRIB1 protein for various downstream analyses. Namely, we describe preparation and crystallization of TRIB1 for structural studies, and using fluorescence polarization and isothermal titration calorimetry to analyze interactions with TRIB1. We hope that applying these considerations can facilitate further understanding of TRIB1 function, specifically, and can be selectively applied to improve studies of other Tribbles proteins and pseudokinases more generally.

1. Introduction

The Tribbles homolog family is a pseudokinase-only family, containing four proteins, TRIB1, TRIB2, TRIB3, and the divergent STK40 (Eyers, Keeshan, & Kannan, 2017; Mace & Murphy, 2021). While each of the Tribbles genes encodes a distinct protein, they are all characterized by the loss of the DFG catalytic motif and the truncation of a glycine rich loop, characteristic of most active kinases (Kwon et al., 2019). Like many pseudokinases, Tribbles proteins play key roles in signal transduction and regulation of cell biology, by acting as signaling scaffolds that nucleate protein complexes (Mace & Murphy, 2021; Murphy, Mace, & Eyers, 2017). Specifically, Tribbles proteins play distinct and overlapping roles in modulating kinase signaling, cell proliferation and gene transcription.

Both TRIB1 and TRIB2 have been implicated as oncogenes in acute myeloid leukemia (Dedhia et al., 2010; Keeshan et al., 2006; Richmond & Keeshan, 2019). In mice, TRIB1 or TRIB2 overexpression alone is sufficient to drive leukemogenesis, by driving the excessive degradation of C/EBPα and inhibiting myeloid cell regulation. The mechanism behind this oncogenic capability is the most extensively studied TRIB1 interaction, and underpins TRIB1 function in a variety of contexts. To facilitate the degradation of C/EBPα, TRIB1 scaffolds an interaction between C/EBPα and the E3 ubiquitin ligase, COP1, leading to the ubiquitination of C/EBPα (Richmond & Keeshan, 2019). Conservation of the key C/EBPα and COP1 binding residues in TRIB1 and TRIB2, suggests both proteins mediate C/EBPα degradation through the same mechanism, which is evolutionarily conserved from the *Drosophila* protein (Dobens, Nauman, Fischer, & Yao, 2021). The ubiquitination of C/EBPα by COP1 results in proteasomal degradation of C/EBPα, but binding between TRIB1 and COP1 also controls subcellular location of the COP1 protein (Kung & Jura, 2019a).

Over recent years several snapshots of TRIB1 structure have been obtained in different states—two different crystal structures of the pseudokinase domain alone (Murphy et al., 2015), a structure of the TRIB1-C/EBPα complex (Jamieson et al., 2018), and structures of the TRIB1 C-terminal tail bound to the substrate-recruiting WD40 domain of COP1 (Uljon et al., 2016). Together, these structures suggest a mechanism of TRIB1 allosteric activation. In the inactive state the C-terminus of TRIB1 binds onto the back of the αC-helix of the pseudokinase domain, and prevents an interaction with COP1. Upon C/EBPα binding, the C-terminal tail is released from the pseudokinase domain, removing the auto-inhibition, and promoting binding to COP1. Of particular mechanistic significance is the movement of the SLE-motif, which occupies the sequence position of the DFG-motif in conventional kinases (Müller, Chaikuad, Gray, & Knapp, 2015; Taylor & Kornev, 2011; Taylor, Shaw, Hu, Meharena, & Kornev, 2013). In the inactive and active states the SLE-motif at the N-terminus of the activation loop switches between structures that resemble DFG-out and DFG-in states respectively (Jamieson et al., 2018). This raises the possibility that the "SLE-in" and "SLE-out" conformations of Tribbles proteins may be targeted selectively by small molecules (Kung & Jura, 2019b). There is already some evidence that this may be possible for the TRIB2 pseudokinase (Foulkes et al., 2018). However, selective targeting of different states of TRIB1, and perhaps Tribbles proteins in general, is likely to be crucial to successful pharmacological treatments.

Like many pseudokinases, the lack of catalytic activity and a routine enzymatic assay for TRIB1 means that: (a) highly pure protein is required for biochemical characterization, and (b) structural studies and specialist biophysical assays must often be developed to characterize specific aspects of protein function. To complicate matters slightly, our studies of TRIB1 have uncovered some quirks of protein behavior, where it differs relative to the majority of proteins that we purify. Nonetheless, we convey the detailed methods that have enabled functional investigation of various aspects of TRIB1 function. A combination of these related methods, combined with cellular studies, are likely to be key to future efforts to derive small molecules that can antagonize TRIB1 function, and may be applied to other Tribbles pseudokinases. We present detailed methods describing the purification of recombinant TRIB1 as has been applied for determining TRIB1 crystal structures, and that have been used to characterize binding of C/EBPα peptides and small molecules to TRIB1.

2. Key resources table

Reagent or resource	Source	Identifier
Bacterial and virus strains		
BL21(DE3) competent cells	(In house, chemically competent)	
Biological samples		
Miller Luria broth	Formedium	LMM 103
Kanamycin	Various	
Isopropyl β-D-1-thiogalactopyranoside, IPTG	A&A Biotechnology	
Chemicals, peptides, and recombinant proteins		
Peptides		
C/EBPα FITC-GGICEHETSIDISAYIDPAAFND-OH (>90%-purity)	Mimotopes	
C/EBPβ	Mimotopes	
H-GSIGDHERAIDFSPYLEPLGA-OH (>90%-purity)		
Tris	BioFroxx	1125KG001
NaCl	Sigma-Aldrich	1.06404.1000
3C protease	In-house preparation	
Lysozyme (25 mg/mL stock)	Hampton Research	HR7-110
Sucrose	PanReac AppliChem	A3935,1000
Glycerol	LabServ	BSPGL885.2.5
Imidazole	PanReac AppliChem	A1073,1000
HEPES	PanReac AppliChem	A1069,0500

—cont'd

Reagent or resource	Source	Identifier
Deposited data		
PDB structures		
Recombinant DNA		
His tagged TRIB1 constructs	In-house preparation	
pET28::TRIB1(84–343)		
pET28::TRIB1(84–372)		
Software and algorithms		
PRISM	GraphPad	

3. Materials and equipment

Beckman Coulter Avanti J-26 XP Centrifuge
Eppendorf 5424R centrifuge
Eppendorf 5804R centrifuge
Sonifier (Bronson Sonic Power Company)
Rotary tube mixer (Ratek, RSM7DC)
Poly-Prep gravity column (Biorad, #7311550)
Fluorometer Spectrophotometer (Denovix DS-11)
HIS-Select® Nickel Affinity Gel (Sigma, P6611)
AKTA Pure 25 (GE Life Science)
Tricorn 10/20 Column (GE Healthcare, 28406413)
Source 15Q (GE healthcare, 17094720)
S200 Increase 10/300 column (GE Healthcare)
5 mL HiTrap Desalting (GE Life Healthcare, 17-1408-01)
SWISSCI 3 Lens Crystallization Plate (Hampton Research, HR3-206)
Crystal Clear Sealing Film (Hampton Research, HR3-609)
Mosquito crystallization robot (sptlabtech)
RockImager 1000 (Formulatrix)
MicroCal VP-ITC (Malvern)
Black 384-well microplates (Greiner Bio-One, 781076)

Alternatives: Most of these reagents can be substituted with generic equivalents. Attention should be paid to commercial syntheses of C/EBPα peptides, as we have experienced poor results with batches from some manufacturers.

4. Expression and purification of TRIB1

For recombinant work with the TRIB1 pseudokinase domain, several different constructs can be successfully purified at high yield (Fig. 1). We recommend using one of two construct lengths with a 3C protease-cleavable His_6-tag at the N-terminus—residues 88–342 (the core pseudokinase domain) or residues 88–372 (the pseudokinase domain with the C-terminal tail). In general the longer construct is slightly more stable, can be more readily concentrated, exhibits less precipitation upon storage, and may be more relevant if interactions of the C-terminal tail are relevant to downstream experiments. However, in some cases the core pseudokinase domain has been more amenable to crystallization. For crystallization of the TRIB1–C/EBPα complex, an approach of fusing the pseudokinase domain to the degron sequences from C/EBPα using a glycine-serine linker can be used. The purification procedures for each construct are largely equivalent (as described below).

4.1 Storage and protein stability

Constructs of TRIB1 pseudokinase domain in buffers without stabilizing additives (either glycerol or sucrose) often do not tolerate freeze-thawing when at high concentration (>5 mg/mL), with precipitation being observed upon thawing. For crystallization studies we snap freeze aliquots in liquid nitrogen, and thaw as rapidly as possible to room temperature by holding the tube in your fingers. After this procedure we recommend centrifuging rapidly at 4 degrees to remove any precipitated material before setting up

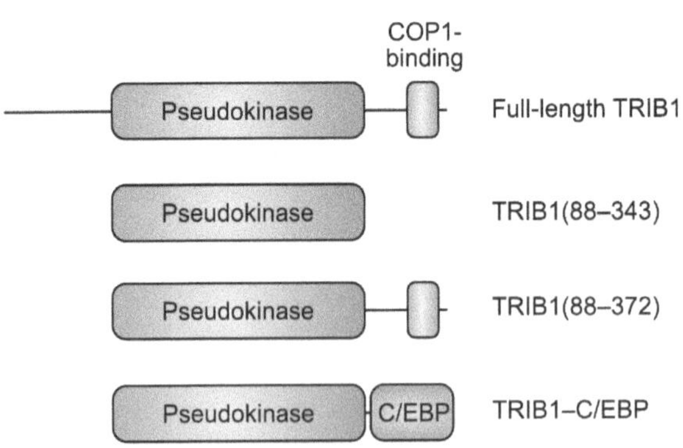

Fig. 1 Schematic illustration of the TRIB1 protein, and constructs that have been successful for structural and biophysical studies.

crystal trials. The fusion approach used for solving the TRIB1-C/EBPα structure was much more stable, and amenable to freeze-thawing.

4.2 Time frames

The time frame for preparation of recombinant TRIB1 largely depends on the intended application. For isothermal titration calorimetry and fluorescence polarization we recommend purification by size-exclusion chromatography (as below), followed by freezing aliquots at a concentration comfortably higher than the desired final concentration (approximately 2 mg/mL). The number of TRIB1 aliquots required for one day's experiments can be recovered as required.

Although not described here, fusion proteins intended for pulldown assays (for example GST-fused TRIB1 pseudokinase domain) can be purified using similar conditions but are best used soon after purification. For optimal results we do not store immobilized fusion proteins more than ~24 h at 4 °C.

4.2.1 Expected yield

Expression levels of soluble TRIB1 pseudokinase are generally quite robust, and we routinely expect between 10 and 25 mg of purified TRIB1 (using either ion-exchange or size-exclusion chromatography) from each liter of *E. coli* cells grown following this procedure.

4.3 Culture growth

Timing: Allow ~48 h, including two overnight steps, for expression of TRIB1 to a stage where cell pellets can be stored for later purification.

1. Add 0.5 μL TRIB1 plasmid DNA (~80 ng/μL) to 25 μL of competent BL21(DE3) *E. coli* in a 1.5 mL Eppendorf tube and incubate on ice for 10 min.
2. Heat shock cells by incubating at 42 °C for 45 s in a water bath.
3. Incubate the tube on ice for a further 2 min, then add 350 μL Miller Luria Broth (Formedium, LMM 103).
4. Outgrow cells for 45 min in a shaking incubator at 37 °C (see note below).
5. Spread cells on 1.5% agar plate supplemented with kanamycin (50 μg/mL), and grow overnight at 37 °C in a static incubator. After overnight growth there should be robust growth of colonies on the plate,

but the exact number will highly depend on quality and contraction of plasmid DNA and competent cell quality.

Note: We generally perform a fresh transformation for each expression of TRIB1, but expect that growth from glycerol stocks and/or overnight cultures would also work robustly.

If expressing from an Ampicillin resistant plasmid (e.g. GST-TRIB1 from a pGEX plasmid), the outgrowth step (4) can be omitted.

6. Add 1 mL of Kanamycin (50 mg/mL) to each liter of sterile Miller Luria Broth in a baffled Erlenmeyer flask, giving a final concentration of 50 μg/mL.
7. Re-suspend colonies from the agar plate using 3 mL sterile LB medium, and inoculate each 1 L of Miller Luria Broth with 1 mL of resuspended cells. Inoculation volume can be adjusted based on density of colonies on the plate and desired volume of cultures to be grown—using less innoculant for very successful transformations, or more for less successful transformations.
8. Incubate cultures in the shaking incubator (200 rpm) at 37 °C and check OD600 after 2 h. When cultures reach OD600 ~ 0.6 move the culture to 18 °C shaking incubator.
9. After around 45 min growth at 18 °C, add 0.2 mL of IPTG (1 M stock) to each 1 L of culture. Leave culture growing overnight, shaking for between 16 and 20 h.
10. The following morning, harvest cells by centrifugation in 1 L Beckman Coulter bottles (A98813) at 3500 x*g* for 30 min 4 °C (JLA8.1 rotor, Beckman Coulter)
11. Cell pellets were then re-suspended using Freeze buffer (10 mL of 20 mM Tris (pH 8.0) 300 mM NaCl) per liter, transferred to a 50 mL Falcon tube to be frozen at −25 °C. Adding 10 mL of buffer generally results in a final storage volume of approximately 15 mL per liter of cell culture.

Pause Point: Frozen pellets of *E. coli* that have expressed TRIB1 can be stored for months at −20 °C until purification.

4.4 Affinity capture of His-tagged TRIB1

Early purifications of TRIB1 showed that once it was isolated from the cell content, protein would precipitate when kept on ice or at 4 °C for prolonged periods. Thus, we recommend performing initial lysis by sonication on ice and clarification of lysate in a centrifuge at 4 °C, but subsequent steps on the bench at room temperature.

1. Defrost frozen pellets by placing 50 mL tubes into a beaker of distilled water at room temperature. Once defrosted adjust the volume to approximately 30 mL using purification buffer consisting of 50 mM Tris pH 8.0, 300 mM NaCl, 10% sucrose, 10% glycerol, 10 mM imidazole.
2. Add 200 μL 25 mg/mL lysozyme, mix and leave to incubate on ice for 30 min
3. Lyse cells by sonication. Instruments may vary, however our standard practice uses a Sonifier (Bronson Sonic Power Company) instrument set at 70% amplitude, 7 output. Pulsed sonication in an ice bath, for two rounds of 1 min with at least 1 min cooling on ice in between treatment normally results in excellent lysis.
4. Lysed cells were then transferred into centrifuge tubes and 40 mL centrifuge tubes (Nalgene, T4793) and centrifuged at 21,500 x*g* at 4 °C (JA17 rotor, Beckman Coulter) for 30 min. While the lysate is centrifuging, pre-equilibrate 2 mL of nickel resin slurry (HIS-Select nickel affinity gel, Sigma), per liter of culture to be purified.
5. After centrifugation, carefully pipette the clarified lysate onto the nickel resin and bind in batch on a rotator, at 4 °C, for 45 min
6. Recover the resin by centrifugation in a 50 mL Falcon tube (2500 rpm, 4 °C for 5 min; Eppendorf 5804 bench top), and remove the supernatant. Retain a sample of the supernatant for later gel analysis.
7. Wash the resin three times in batch by adding 20 mL purification buffer, gently resuspending, then centrifuging. After the third batch wash, transfer into a Poly-Prep gravity column (Biorad) and wash with three 2 mL applications of purification buffer.
8. Elute TRIB1 protein by adding purification buffer supplemented with 300 mM imidazole in 1 mL fractions, and collecting corresponding 1 mL fractions as they elute. Quantify fractions using measuring Absorbance at 280 nm (A_{280}) by spectrophotometry (Denovix DS-11). Keep fractions for analysis by SDS-PAGE.
9. Pool elution fractions with a concentration of more than 0.5 mg/mL (usually elution fractions 2 and 3). With the pooled sample, the His tag can be cleaved overnight at 15 °C using 3C protease (ratio of 1:100 by weight), in purification buffer supplemented with 2 mM DTT

4.5 Option 1-anion-exchange chromatography for crystallization

TRIB1 has a tendency to precipitate upon concentration in a centrifugal concentrator. To achieve the higher concentrations required for crystallization trials, the protein is best purified using ion-exchange chromatography to take advantage of its concentrating effect. We use a home-poured Source15Q 2mL Tricorn column (GE Life Science), and have previously used prepacked SourceQ columns without issue.

1. Equilibrate the SourceQ column using an AKTA Pure 25 (GE Life Science). We alternate between three column volumes of low- (50mM Tris pH 8.0, 10mM NaCl, 2mM DTT) and three column volumes of high-NaCl buffer (50mM Tris pH 8.0, 1M NaCl, 2mM DTT), repeated three times, finally equilibrating the column with five column volumes of low-NaCl buffer at 5mL/min.
2. Rapidly dilute the TRIB1 sample that cleaved overnight at a ratio of 1 in 10, using 50mM Tris pH 8.0, 10mM NaCl, 5% glycerol, 5% sucrose, 2mM DTT (dilution buffer). This dilution step ensures effective capture on Source Q resin.
3. Load the diluted protein sample onto the column using a GE Healthcare sample pump (S9H) attached the AKTA Pure system, and wash with 10 column volumes of low-NaCl buffer.
4. Eluted using an AKTA Pure program consisting of a linear gradient from low NaCl buffer (50mM Tris pH 8.0, 10mM NaCl, 2mM DTT) to 50% high-NaCl buffer (50mM Tris pH 8.0, 1M NaCl, 2mM DTT) over 25 column volumes, then a gradient from 50% to 100% high-NaCl buffer over 5 column volumes. The protein should elute in the first half of the gradient. (Fig. 2B).
5. Analyse peak protein fractions using A_{280} nano-spectrophotometry (Denovix DS-11) and on a 10–20% gradient reducing SDS-PAGE gel using a Tris-glycine buffer system, Precision Plus molecular weight marker (BioRad, P4X99QE367), and visualization with Coomassie Brilliant blue staining (Fig. 2B). Pool fractions of suitable concentration/purity (generally >2mg/mL) and snap-freeze 35μL aliquots in liquid N_2 for later use in crystallization.

4.6 Option 2-purification for biophysical studies

For biochemical and biophysical measurements, which do not require as high a protein concentration as crystallization, size-exclusion chromatography following affinity chromatography is preferred to gain maximum control of buffer conditions. Starting from affinity-purified protein, with the His-tag cleaved overnight, we recommend purification using a S200 Increase 10/300 column (GE Healthcare) and standard AKTA FPLC system.

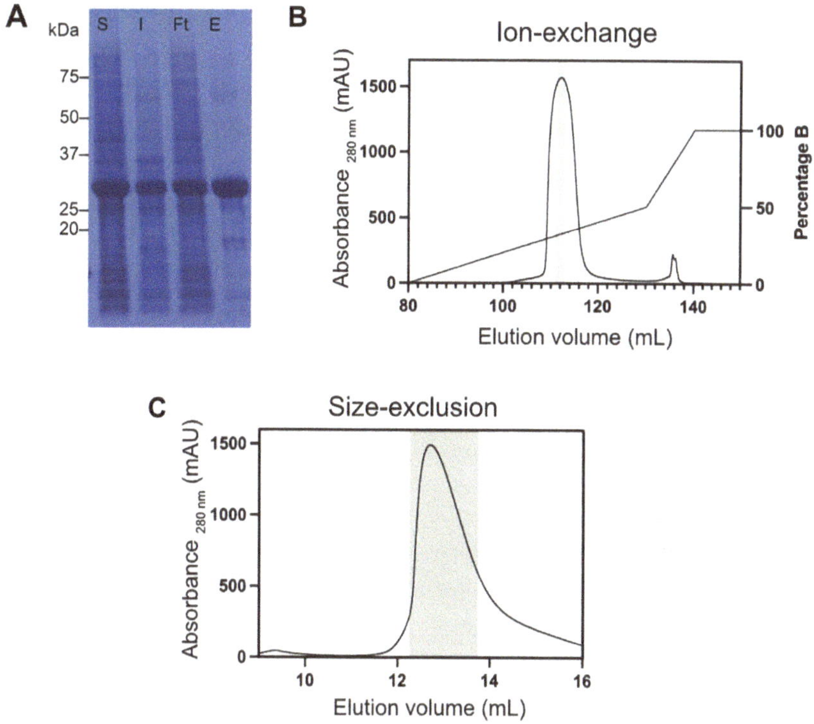

Fig. 2 Purification of TRIB1(88–372). (A) Example of TRIB1 purification by affinity chromatography. Lanes correspond to (S) soluble, (I) insoluble, (Ft) flowthrough of nickel resin binding, and (E) protein eluted from nickel resin. (B/C) Example chromatography traces for TRIB1 purification by (B) ion-exchange chromatography, and (C) size-exclusion chromatography.

Concentrate cleaved TRIB1 using a 10 kDa cutoff centrifugal concentrator (GE Life Science, 28932294) at 3220 x*g* at 15 °C (Eppendorf 5810R). In the permissive purification buffer noted for affinity purification (which contains 10% glycerol and 10% sucrose), concentrations of 20 mg/mL can be reached.

1. While the protein is concentrating, pre-equilibrate the S200 Increase column with two column volumes of filtered and degassed water followed by two column volumes of filtered and degassed SEC buffer 10 mM Hepes pH 7.6, 0.3 M NaCl, 0.5 mM TCEP.
2. Centrifuged concentrated TRIB1 at 21,000 x*g* for 5 min at 15 °C (Eppendorf 5424R) to pellet any precipitated protein.
3. Carefully withdraw TRIB1 sample into a 500 μL Hamilton syringe and load into a 1 mL loop attached to the AKTA Pure system.

4. Run an isocratic elution program, at 0.5 mL/min, equilibrating first with 1 mL, emptying the loop with 3 mL, followed by a 35 mL isocratic elution.
5. Assess peak protein fractions using A_{280} nano-spectrophotometry (Denovix DS-11) and 10–20% gradient SDS-PAGE gel, using a Tris-glycine buffer system, Precision Plus molecular weight marker (BioRad), and visualization with Coomassie Brilliant blue staining. Snap-freeze desired fractions in liquid nitrogen for storage at −80 °C. Pool fractions of suitable concentration/purity (generally 1–2 mg/mL) and snap-freeze 100 μL aliquots in liquid N_2 for later use in crystallization.

The peak that we observe for preparative runs of TRIB1 on size-exclusion chromatography is not conventionally ideal (Fig. 2C), although we have not observed notable differences when using only the front of the peak compared to the entire pool. As the peak is often not ideally shaped and elutes over a relatively large number of fractions, pooling a broader range of fractions means that the protein is diluted sufficiently (to a final ~2 mg/mL) to minimize protein precipitation upon thawing. We conventionally also snap freeze ~200 mL of SEC buffer and store at −80 °C for buffer matching assays, and dissolving synthetic peptides in equivalent buffer.

5. Crystallization of TRIB1 pseudokinase domain

We have optimized three different conditions for crystallization of TRIB1 from three main constructs (88–343, 88–372 and the C/EBPα fusion construct; Fig. 1). The general theory of crystallization and crystal improvement is well covered elsewhere (McPherson & Cudney, 2014; McPherson & Gavira, 2014). However, depending on the exact protein concentration and behavior of the specific preparation, it is often still necessary to screen coarsely around these conditions to optimize crystallization of that particular preparation. Here we describe an example coarse screen that can be set up for rapidly checking each of the respective constructs shown in Fig. 1.

Instead of preparing a grid of precipitant concentrations manually, we use a rapid method of preparing one stock solution ('100 %' precipitant) which is diluted down the vertical-axis of the plate, and then recommend screening different protein to precipitant drop ratios across the horizontal direction of the crystallization plate (Fig. 3A). Using this system means labor creating mother liquor solutions is minimized, and a crystallization robot

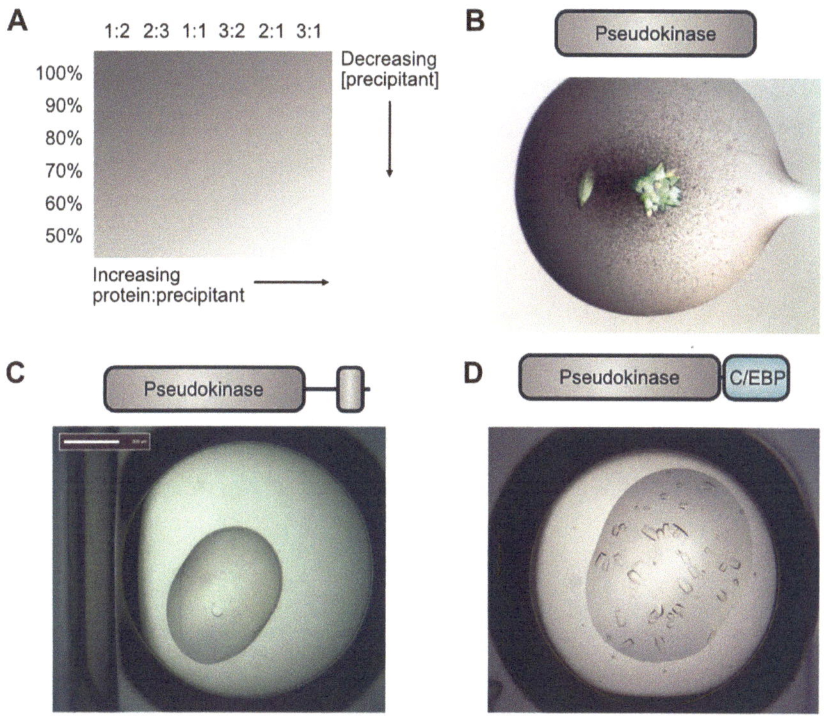

Fig. 3 Crystallization of TRIB1. (A) Schematic illustrating setup of a coarse grid to optimize crystallization of a particular batch of purified TRIB1. The top row is filled 100% concentration of optimal precipitant. Rows below are diluted with water to indicated final concentrations. Different drop ratios are easily set up across the plate. using a crystallization robot. (B–D) Example crystals of each TRIB1 construct, as indicated.

(in our case a Mosquito crystallization robot) can set up different drop ratios to sample different crystallization kinetics. Initial drop sizes are typically ~500 nL total for known TRIB1 crystallization conditions, but can be reduced when screening for new conditions. Even when setting up a single drop per crystallization well we routinely use three-drop capable plates (Swisssci 3 well plate), as it allows for easier crystal harvesting, cryoprotection and handling in adjacent wells.

5.1.1 Solutions to prepare

- For TRIB1(88–343): 0.1 M Bis-Tris (pH 5.5), 2–3 M NaCl, 0.1 M sodium citrate, 0.01 M reduced L-glutathione, and 0.01 M oxidized L-glutathione. [Target protein concentration: 8 mg/mL]

- For TRIB1(88–372): 0.1 M Bis-Tris pH 5.5, 1.0 M Ammonium sulfate, 1% w/v Polyethylene glycol 3,350 [Target protein concentration: 5 mg/mL]
- For TRIB1–C/EBPα fusion: 2.8 M sodium acetate and 0.1 M bis-tris propane (pH 7) with 5% sodium citrate [Target protein concentration: 10 mg/mL]

1. Prepare appropriate solution above for crystallization of specific TRIB1 construct. Make a series of dilutions from 100 down to 50% using 10% steps, conveniently performed in a PCR strip tube. Use a multichannel pipette to distribute 30 μL into the reagent wells of SWISSCI 3 Lens Crystallization Plate in rows. Volume in reagent well may differ depending on format of the plate
2. Remove TRIB1 protein from −80 °C storage, and thaw by holding.
3. Once defrosted pool aliquots in a 0.6 mL microfuge tube, and pellet any precipitated TRIB1 by centrifugation at 21,000 x*g* for 5 min at 15 °C (Eppendorf 5424R).
4. Using a Mosquito robot (sptLabTech) or similar dispense protein and precipitants in accordance with drop ratios indicated in Fig. 3A. Seal and incubate at 16 °C. Note the values in Fig. 3A relate to ratios, so actual volumes dispensed can be tailored to match crystallization plate and available protein. For known TRIB1 crystallization conditions we use drops with a total volume of 500 nL as a starting point.
5. Monitor crystal growth, daily, using a Formulatrix RockImager system (or similar) if available, or manually using a light microscope. Protein crystals should generally appear within 3 days and have finished growing by approximately 7 days (Fig. 3B–D).

All three TRIB1 crystal forms were cryoprotected with a mixture of mother liquor with 25% glycerol before snap-freezing in liquid nitrogen for data collection. Practically, we recommend adding a small amount of this cryoprotectant onto the drop containing the crystal, and letting the glycerol containing solution diffuse into crystals, before looping the crystal and very briefly immersing in pure cryoprotectant solution. This method was particularly useful for crystals grown in high salt, which were otherwise very fragile.

Using these procedures all three TRIB1 constructs reliably grew crystals, however very few of the crystals of TRIB1(88–343) or the TRIB1–C/EBPα fusion diffracted to or beyond 3 Å. Dozens of each crystal form had to screened to solve the structure to 2.7/2.8 Å resolution. In contrast, the TRIB1(88–372) crystals routinely approached 2 Å resolution. If crystals that grow are not of sufficient size for harvesting and data collection, we have had success in using streak seeding to improve crystal size. In general we use what we understand to be 'conventional' streak-seeding methods. For us, this consists of setting up crystallization drops using conditions from the grid in Fig. 3A which have slightly less precipitant than drops that do give crystals, but fail to crystallize themselves. We then seal the crystallization drops within the plate and leave for a few hours. Streak-seeding is then performed by reopening the plate and brushing a cat's whisker (preferably from a silver tabby) across non-ideal crystals, then streaking through drops and re-sealing the plate.

In some scenarios, if intending to crystallize a new complex, or ligand with TRIB1, we return to screening sparse-matrices using commercial screens, favoring Index HT, SaltRx HT, Peg/Ion HT (Hampton Research) and JCSG-plus HT-96 (Molecular Dimensions). The TRIB1 pseudokinase domain especially seems to favor high salt conditions, and both constructs saw multiple hits in the SaltRx HT screen. To cover a broad range of conditions we routinely establish four high throughput screens with ratios of 2:1, 1:1 and 1:2 precipitant to protein, as protein allows.

6. Biophysical measurement of TRIB1-binding

We have performed various downstream biochemical studies using proteins purified using the above method. The three major assays we routinely use are differential scanning fluorimetry (DSF), isothermal titration calorimetry (ITC), and fluorescence polarization. Because it has proved highly informative and is widely applied in the pseudokinase field (Foulkes et al., 2018; Mace & Murphy, 2021; Murphy et al., 2014), we have not provided in depth methodology for DSF application to TRIB1 here. Nonetheless, applying the same principle DSF methods as published in other studies to TRIB1 has been extremely useful for analyzing both nucleotide and small molecule binding (Jamieson et al., 2018; Murphy et al., 2015). Here we describe in detail the use of fluorescence polarization and ITC for analyzing TRIB1 protein-protein interactions.

6.1 Fluorescence polarization

Fluorescence polarization for studying TRIB1 protein binding has proved useful since identification of a 23 amino acid peptide derived from C/EBPα, which serves as a TRIB1-binding degron. We have used a fluorescein-labelled version of this degron peptide as a probe to monitor

C/EBPα binding in two different fluorescence polarization formats. For direct measurement of binding affinity we recommend keeping the concentration of FITC-C/EBPα constant, and incubating with increasing concentrations of TRIB1 protein. Measuring polarization of circularly-polarized light then allows quantitation of how much light is depolarized by the rapidly-tumbling free peptide, compared to the TRIB1-bound peptide which is effectively part of a larger complex and so tumbles more slowly and does not depolarize light as effectively. This format of fluorescence polarization experiment is useful to ascertain the binding affinity of a particular peptide for TRIB1, especially when it is already known to bind. To execute direct measurement of affinity for a moderately binding peptide, a stock of TRIB1 at atleast 5.3 mg/mL (160 μM) should be used.

For ease of setup, it is convenient to use PCR strip tubes with a multichannel pipette to transfer and mix samples. Be gentle to avoid the bubbles that are easy to introduce. Using one in two dilutions makes an easy set up starting point for assays. Final assay volumes are 30 and 60 μL for 384 and 96 well plates respectively. All samples are measured using POLARstar Optima (BMG LabTech) at room temperature. The reader should be prepared with a 485p excitation filter and two 520p emission filters for detecting fluorescence polarization by FITC, according to the manufacturer's instructions for fluorescence polarization.

6.1.1 Direct measurement of peptide affinity by fluorescence polarization

1. Defrost 200 μL of TRIB1 (at at least 5.3 mg/mL), an aliquot of FITC-cEBPα peptide (previously snap frozen as 50 μL aliquots at 37 μM concentration) and a 5 mL aliquot of buffer (10 mM Hepes pH 7.6, 0.3 M NaCl, 0.5 mM TCEP)
2. Add 10 μL of 10% Tween20 stock to buffer to give a 0.02% v/v final concentration
3. Aliquot the buffer to each tube in the 8 tube PCR strip. Follow with TRIB1 protein, adding to the first well and transferring to the next tube. By mixing each tube in turn with protein from the previous tube, establish a dilution series from 160 μM down to 2.4 nM, leaving the 8th tube as a buffer blank.
4. Use the multichannel pipette to transfer either 15 or 30 μL to each well of a black flat bottomed 384 or 96 well plate (Greiner Bio-One, 781209 or 655209)

5. Dilute FITC-cEBPα peptide to 50 nM (final concentration 25 nM) and dispense into a PCR strip to serve as a reservoir for pipetting to the 384 or 96 well plates.
6. Use the multichannel to transfer and mix either 15 or 30 μL to each well
7. Cover the plate in foil and leave at room temperature for 30 min for it to equilibrate
8. Measure the samples using the POLARstar Optima (BMG LabTech) making sure to set gain on the well with the highest free peptide (without TRIB1)
9. Read the output file in Microsoft Excel. Calculate the change in fluorescence polarization value for each well by subtracting the value of the free peptide (no TRIB1), to account for residual fluorescence polarization
10. Perform final analysis and output a graph using GraphPad Prism, with non-linear regression fitting specifying a single-site binding model.

6.1.2 Competition fluorescence polarization assay

A competition-based experiment can be performed even if it is not clear that a protein of interest is a TRIB1 binder. As implied, increasing concentrations of unlabeled protein or peptide are titrated into a fixed concentration of TRIB1-FITC-C/EBPα mixture. If the protein is a suitable competitor, it will displace the labelled peptide from TRIB1, allowing it to diffuse into solution and depolarize light to a greater extent. In unpublished studies we have tested the ability of other folded proteins to displace FITC-C/EBPα, to infer whether they may block C/EBPα recruitment. Here we describe in detail the application of this method to test whether phosphorylated and non-phosphorylated C/EBPβ degron peptides to displace FITC-C/EBPα.

We have found that it is useful to perform a preliminary direct fluorescence polarization measurement of affinity with the exact protein and FITC-C/EBPα stocks that you intend to use for a competition experiment. Batch-to-batch variation in both TRIB1 and peptide stocks can affect the optimal protein-peptide ratio to use. From the direct binding measurement, we ideally aim for a TRIB1 concentration that is at approximately the same concentration in μM as the directly measured dissociation constant. From this starting point at the mid-point of the curve, fluorescence polarization values have the potential to go either up or down, depending on if the titrated protein is hypothesized to be a competitor, or may potentially

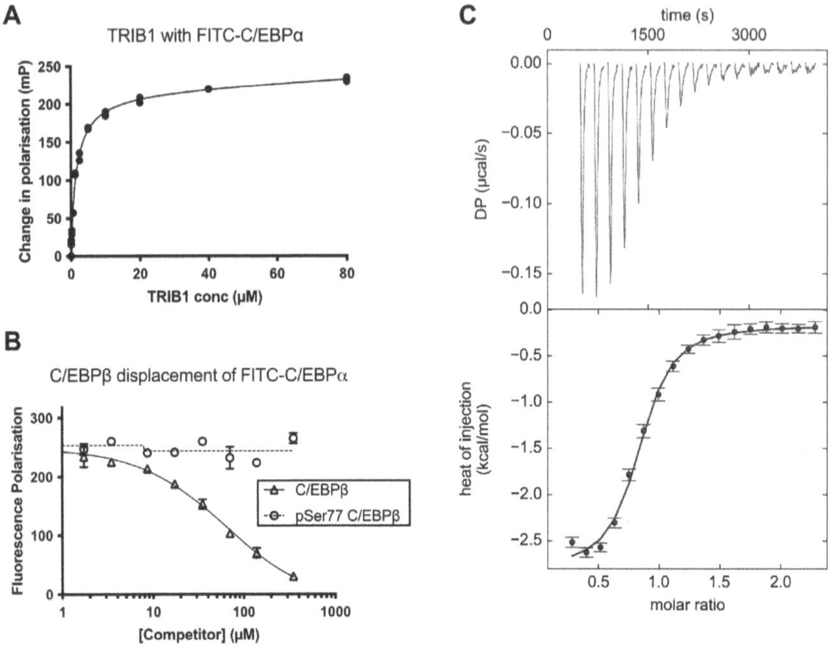

Fig. 4 Biophysical measurement of TRIB1 binding. (A) Example direct measurement of FITC-C/EBPα peptide binding affinity to TRIB1 measured by fluorescence polarization, following fitting of a one-site binding model. (B) Example fluorescence polarization competition assay, showing that the unmodified C/EBPβ peptide (triangles) competes with FITC-C/EBPα (causing a decrease in fluorescence polarization), whereas the Ser77 phosphorylated peptide (circles) does not. (C) Example ITC trace of an exemplar TRIB1-specific nanobody binding to TRIB1, following data processing in NITPIC, fitting in SEDPHAT and display in the GUSSI software.

enhance FITC-C/EBPα binding. Note this design means that competition experiments can be performed from lower concentration stocks of TRIB1 (~2 mg/mL).

This protocol is assumed to continue from the measurement shown in Fig. 4A, where the dissociation constant was calculated as 1.5 μM. Based on this value, a final concentration of 1.5 μM TRIB1 was chosen for our competition assay, with 25 nM FITC-C/EBPα peptide.

1. Defrost 100 μL TRIB1 (~2 mg/mL; 60 μM), one 50 μL aliquot of FITC-C/EBPα peptide (37 μM), peptides for competition and a 5 mL aliquot of buffer (10 mM Hepes pH 7.6, 0.3 M NaCl, 0.5 mM TCEP)
2. Add 10 μL of 10% Tween20 stock to 5 mL of buffer to give 0.02% v/v final concentration

3. For one experiment in triplicate, mix 600 μL of TRIB1 (0.2 mg/mL; 6 μM) with 600 μL of the FITC-cEBPα peptide (0.28 μg/mL; 100 nM) and incubate together at room temperature for 30 min.
4. Prepare a two-fold serial dilution of competing protein across 11 PCR strip tubes (spanning 300–0.3 μM), reserving a twelfth tube for a buffer only control. Prepare at least 120 μL of each dilution, in order to add 30 μL per well to each reaction, in triplicate.
5. Dispense 30 μL of the stock from step 3, into each well across three rows of a black 96 well plate (Greiner Bio-One, 655209)
6. Next dispense 30 μL of each of the peptide serial dilutions and mix with TRIB1-FITC-C/EBPα mixture in the plate, in three triplicate rows.
7. Cover the plate with foil and leave to incubate for 30 min at room temperature
8. Measure the samples using the POLARstar Optima (BMG LabTech) making sure to set gain on the well with the highest free peptide (without TRIB1)
9. Read the output file in Microsoft Excel. In this case, raw fluorescence polarization can be compared to the no-competitor control.
10. Perform final analysis and output a graph using Prism 9, again fitting a one-site binding model.

Expected results can be seen in Fig. 4B, showing displacement, or lack of displacement, of FITC-C/EBPα peptide by non-phosphorylated or phosphorylated C/EBPβ degron peptides, respectively (Jamieson et al., 2018).

6.2 Isothermal titration calorimetry

Isothermal titration calorimetry can be used to measure binding of peptides, proteins, or small molecules to TRIB1. A general rule-of-thumb for ITC experiments is to have a 10-fold higher ligand concentration, than the concentration of protein "receptor" in the cell. Because of the propensity of TRIB1 to oligomerize and/or aggregate, we have only ever successfully performed ITC with TRIB1 held at a steady concentration in the cell, while injecting ligands at a higher concentration. Generally starting with 20 μM TRIB1 (~0.60 mg/mL) in the cell and 200 μM ligand (variable depending on molecular weight) has allowed us to determine accurate dissociation constants, but higher or lower concentrations may be possible depending on the affinity and nature of binding of the intended protein:ligand pair. Preparation for experiments involves purifying the ligand using the same size exclusion buffer as the TRIB1 described above, ideally at the same time to ensure exact buffer matching. If TRIB1 and ligand were not prepared at the same time, either dialysis overnight or desalting using 5 mL HiTrap Desalting column (GE Life Science) should be considered to minimize buffer dilution effects that could confound ITC measurement.

Here we describe an experiment measuring the binding affinity of a nanobody (to be published) selected to bind TRIB1 (Fig. 4C). The tight affinity of this TRIB1-nanobody pair means that ample signal is observed using 16 μM TRIB1 (0.53 mg/mL), and 174 μM nanobody (2.2 mg/mL) as titration concentrations. These experiments were carried out at 30 °C using a MicroCal VP-ITC (Malvern), but other models of calorimeter can be substituted as required.

1. Ensure the ITC is well equilibrated before starting experiments. Turn on the machine well in advance, to let it warm up while buffer and proteins defrost.
2. Once samples are defrosted, quantify all proteins using nano-spectrophotometry (Denovix DS-11) and adjust protein concentrations as required using matched buffer (for instance if the ligand is stored at high concentration, and needs to be brought down to 200 μM)
3. Degas the TRIB1, ligand, buffer and milliQ water, using degassing apparatus with ITC machine. While degassing is in progress, flush the ITC sample chamber three times with milliQ water followed by three washes with buffer
4. Once the protein/buffer have finished degassing, carefully load the reference chamber with degassed milliQ and ensure bubbles are not introduced to the chamber. Place the loading syringe onto the lip of the chamber and remove excess water
5. Wash the syringe with buffer
6. Fill the syringe with minimum 2 mL TRIB1 solution, and load it into the experimental chamber ensuring bubbles are not introduced. Place the loading syringe onto the lip of the chamber and remove excess sample.
7. Load the injectant-syringe with ligand (method may vary based on the calorimeter).
8. Purge and refill the stirring syringe to safeguard from bubbles
9. Place the syringe into the main chamber

In a larger volume instrument such as the MicroCal VP-ITC the titration program we execute consists of an equilibration period, followed by (i) an initial injection of 7.5 μL over 15 s duration, (ii) 28 x 15 μL injections of 30 s duration with a delay of 210 s between injections. The stirring syringe was set at 307 rpm. Rather than using the default software, we frequently use NITPIC for unbiased baseline calibration (Scheuermann & Brautigam, 2015), and fit isotherms to a single site-binding model using SEDPHAT (Zhao, Piszczek, & Schuck, 2015). In the case of a tight binder, the binding isotherm should proceed toward, and remain relatively constant following full titration (Fig. 4C), whereas weaker binders such as C/EBPα may struggle to reach a flat post-titration baseline due to rapid association and dissociation (Murphy et al., 2015).

7. Conclusion

While TRIB1 is not an ideally behaved protein in all cases, with some adaptation it has proven to be relatively amenable to structural and biochemical characterization. Closely related proteins can frequently exhibit quite different behavior—adapting similar techniques to those described here has also allowed work soon to be published regarding TRIB2, whereas TRIB3 in our hands has required more major adaptation (insect cell culture expression for instance). Nonetheless we hope that the downstream applications described here may prove useful either directly with Tribbles proteins, or with some adaptation for other pseudokinases.

Acknowledgments

We thank various postgraduate students in the lab who have helped us optimize methods for working with Tribbles proteins over several years, particularly Michael Pudjihartono, Jasper Struwig, and Jack Curry. Funding support for TRIB1 research has been provided by a Project Grant from the Health Research Council of New Zealand, and Hamish McMillan was supported by a University of Otago Postgraduate Scholarship.

References

Dedhia, P. H., Keeshan, K., Uljon, S., Xu, L., Vega, M. E., Shestova, O., et al. (2010). Differential ability of tribbles family members to promote degradation of C/EBPalpha and induce acute myelogenous leukemia. *Blood, 116*, 1321–1328.

Dobens, L. L., Nauman, C., Fischer, Z., & Yao, X. (2021). Control of cell growth and proliferation by the tribbles pseudokinase: Lessons from Drosophila. *Cancers, 13*.

Eyers, P. A., Keeshan, K., & Kannan, N. (2017). Tribbles in the 21st century: The evolving roles of tribbles pseudokinases in biology and disease. *Trends in Cell Biology, 27*, 284–298.

Foulkes, D. M., Byrne, D. P., Yeung, W., Shrestha, S., Bailey, F. P., Ferries, S., et al. (2018). Covalent inhibitors of EGFR family protein kinases induce degradation of human Tribbles 2 (TRIB2) pseudokinase in cancer cells. *Science Signal, 11*.

Jamieson, S. A., Ruan, Z., Burgess, A. E., Curry, J. R., McMillan, H. D., Brewster, J. L., et al. (2018). Substrate binding allosterically relieves autoinhibition of the pseudokinase TRIB1. *Science Signal, 11*.

Keeshan, K., He, Y., Wouters, B. J., Shestova, O., Xu, L., Sai, H., et al. (2006). Tribbles homolog 2 inactivates C/EBPalpha and causes acute myelogenous leukemia. *Cancer Cell, 10*, 401–411.
Kung, J. E., & Jura, N. (2019a). The pseudokinase TRIB1 toggles an intramolecular switch to regulate COP1 nuclear export. *The EMBO Journal, 38*.
Kung, J. E., & Jura, N. (2019b). Prospects for pharmacological targeting of pseudokinases. *Nature Reviews. Drug Discovery*.
Kwon, A., Scott, S., Taujale, R., Yeung, W., Kochut, K. J., Eyers, P. A., et al. (2019). Tracing the origin and evolution of pseudokinases across the tree of life. *Science Signal, 12*.
Mace, P. D., & Murphy, J. M. (2021). There's more to death than life: Non-catalytic functions in kinase and pseudokinase signaling. *The Journal of Biological Chemistry, 100705*.
McPherson, A., & Cudney, B. (2014). Optimization of crystallization conditions for biological macromolecules. *Acta Crystallographica. Section F, Structural Biology Communications, 70*, 1445–1467.
McPherson, A., & Gavira, J. A. (2014). Introduction to protein crystallization. *Acta Crystallographica Section F Structural Biology Communications, 70*, 2–20.
Müller, S., Chaikuad, A., Gray, N. S., & Knapp, S. (2015). The ins and outs of selective kinase inhibitor development. *Nature Chemical Biology, 11*, 818–821.
Murphy, J. M., Mace, P. D., & Eyers, P. A. (2017). Live and let die: Insights into pseudoenzyme mechanisms from structure. *Current Opinion in Structural Biology, 47*, 95–104.
Murphy, J. M., Nakatani, Y., Jamieson, S. A., Dai, W., Lucet, I. S., & Mace, P. D. (2015). Molecular mechanism of CCAAT-enhancer binding protein recruitment by the TRIB1 pseudokinase. *Structure, 23*, 2111–2121.
Murphy, J. M., Zhang, Q., Young, S. N., Reese, M. L., Bailey, F. P., Eyers, P. A., et al. (2014). A robust methodology to subclassify pseudokinases based on their nucleotide-binding properties. *The Biochemical Journal, 457*, 323–334.
Richmond, L., & Keeshan, K. (2019). Pseudokinases: A tribble-edged sword. *The FEBS Journal*.
Scheuermann, T. H., & Brautigam, C. A. (2015). High-precision, automated integration of multiple isothermal titration calorimetric thermograms: New features of NITPIC. *Methods, 76*, 87–98.
Taylor, S. S., & Kornev, A. P. (2011). Protein kinases: Evolution of dynamic regulatory proteins. *Trends in Biochemical Sciences, 36*, 65–77.
Taylor, S. S., Shaw, A., Hu, J., Meharena, H. S., & Kornev, A. (2013). Pseudokinases from a structural perspective. *Biochemical Society Transactions, 41*, 981–986.
Uljon, S., Xu, X., Durzynska, I., Stein, S., Adelmant, G., Marto, J. A., et al. (2016). Structural basis for substrate selectivity of the E3 ligase COP1. *Structure, 24*, 687–696.
Zhao, H., Piszczek, G., & Schuck, P. (2015). SEDPHAT—A platform for global ITC analysis and global multi-method analysis of molecular interactions. *Methods, 76*, 137–148.

CHAPTER THREE

Detecting endogenous TRIB2 protein expression by flow cytometry and Western blotting

Ridhima Lal, Jake Ritchie, Laura Richmond, and Karen Keeshan*
Paul O'Gorman Leukaemia Research Centre, Institute of Cancer Sciences, University of Glasgow, Glasgow, SC, United Kingdom
*Corresponding author: e-mail address: karen.keeshan@glasgow.ac.uk

Contents

Abstract

Protein kinases catalyze the transfer of a phosphate group thereby activating proteins and initiating signaling cascades. Their cousins, the pseudokinases, are enzymatically nonactive counterparts of protein kinases that can be considered zombie enzymes. Interestingly, pseudokinases, which constitute about 10% of the human kinome, have been implicated in many cancers, despite their sequences predicting a lack of catalytic activity. Owing to recent research, it has been demonstrated that dysregulation of many pseudokinases triggers changes in cell signaling, proliferation, and drug resistance. This review is aimed at describing methods that can be used for detection of Tribbles family of pseudokinases, specifically TRIB2. We describe intracellular staining by flow cytometry and Western blotting techniques for the detection of endogenous TRIB2 protein.

Methods in Enzymology, Volume 667
ISSN 0076-6879
https://doi.org/10.1016/bs.mie.2022.03.024

1. Introduction

Signal transduction is a tightly regulated process that helps to maintain tissue homeostasis. Cancer cells with various genetic and epigenetic changes escape signal transduction regulation, leading to abnormal cell proliferation, migration and function. Dysregulation of a signaling cascade could occur by hyperactivation of a cascade or by the silencing of critical tumor suppressor genes. Hyperactivation of a signaling cascade can occur due to mutations that convert proto-oncogenes to oncogenes, such as within receptor tyrosine kinases (RTKs), phosphoinositide 3-kinases (PI3Ks), Wingless integrated (Wnt), Notch and Hedgehog pathway (Komiya & Habas, 2008; Mittal et al., 2014; Sever & Brugge, 2015). Reduced expression or inhibition of negative regulators of signaling can also occur due to inactivation of tumor suppressor genes such as p53 that regulates cell cycle and apoptosis (Sever & Brugge, 2015).

Protein kinases are enzymes that play a vital role in cell signaling by phosphorylating members of signaling cascades, thereby activating/inhibiting them. While kinases transfer a phosphate group from ATP to a target protein to regulate their function, pseudokinases are considered catalytically defective or "zombie" enzymes due to their inability or significantly impaired ability to catalyze phosphorylation, which corresponds with their distinct, non-enzymatic signaling functions (Kung & Jura, 2016). The human kinome encodes about 540 kinases with $\sim$10% categorized as pseudokinases (Zhang et al., 2012).

The Tribbles protein family are pseudokinases that were first discovered in *Drosophila* to play significant roles in morphogenesis and proliferation (Seher & Leptin, 2000). In humans, Tribbles comprise three members, namely TRIB1, TRIB2 and TRIB3 (Seher & Leptin, 2000). TRIB2 is recognized as the ancestor from which TRIB1 and TRIB3 evolved. TRIB1 and TRIB2 share 71%, TRIB2 and TRIB3 share 53%, and TRIB1 and TRIB3 share 53.7% amino acid homology (Yokoyama & Nakamura, 2011). Gene expression analysis revealed diverse roles exhibited by the Tribbles family in hemopoiesis (Salomé, Hopcroft, & Keeshan, 2018). Myeloid cells displayed highest TRIB1 expression while the lowest was observed in lymphoid cells and natural killer cells. On the contrary, TRIB2 levels were highest in lymphoid cells and lowest in myeloid cells, while TRIB3 expression was mainly found in the quiescent hematopoietic stem cells (Salomé, Hopcroft, & Keeshan, 2018).

Tribbles consist of three domains, the N-terminal domain, a central bilobal pseudokinase core, and the C terminal domain. The ATP binding status of Tribbles proteins is currently debated: structural studies have confirmed that TRIB1 does not possess an intact ATP binding pocket. However, TRIB2 has been previously reported to have low or no ATP binding capacity, which necessitates further investigation (Bailey et al., 2015; Durzynska et al., 2017; McMillan et al., 2021), while ATP binding of TRIB3 remains unstudied. The C-terminal domain comprises the constitutive photomorphogenesis protein 1 (COP1) binding site where the COP1 E3 ubiquitin ligase docks on Tribbles and ubiquitinates other protein substrates recruited by Tribbles, resulting in their proteasomal degradation (Eyers, 2015; Kung & Jura, 2019a; Richmond & Keeshan, 2020). The C-terminal domain also consists of the MEK1-binding site which regulates RAS/mitogen-activated protein kinase (MAPK/ERK) pathway (Richmond & Keeshan, 2020).

Tribbles proteins are often found to be dysregulated (up- or down-regulated) in cancer and can be stimulated by range of factors, such as different mitogens and stress (Eyers, Keeshan, & Kannan, 2017; Salomé et al., 2018). As mentioned above, Tribbles proteins play important roles in scaffolding or in the initiation of substrate degradation, which contributes to their roles in disease progression (Sakai et al., 2016). The best studied pathway targeted by TRIB1 and TRIB2 involves the degradation of CEBPα, an important tumor suppressor in acute myeloid leukemia (AML), whose abundance correlates with poor prognosis (Hasemann et al., 2014). This degradation occurs via the recruitment of COP1 to the COP1 binding domain in the C-terminus, with COP1-mediated ubiquitination of CEBPα inducing its proteasomal degradation (Dedhia et al., 2010; Keeshan et al., 2006, 2010; Kung & Jura, 2019a; Murphy et al., 2015; O'Connor et al., 2016; Uljon et al., 2016). Depending on context, elevated or suppressed levels of Tribbles protein expression have been associated with drug resistance through the activation of antiapoptotic and pro-survival pathways (Salomé et al., 2018). Lack of TRIB2 contributes to higher resistance to chemotherapies such as daunorubicin in AML cells due to a defective cell cycle response (Salomé et al., 2018). Daunorubicin is a routinely used chemotherapy drug that, together with growth factor deprivation, significantly increase TRIB2 levels (Salomé et al., 2018). High TRIB2 levels are known to activate protein kinase B (AKT) and consequently downregulate p53 and FOXO, thereby contributing to drug resistance and disease progression (Hill et al., 2017). TRIB2, as well as TRIB1 and TRIB3 control MAPK signaling by phosphorylating ERK

(Foulkes et al., 2018; Kiss-Toth et al., 2004; Ma et al., 2018). TRIB3 is involved in leukemogenesis via stabilization of the PML-RARα oncogene in acute promyelocytic leukemia (Keeshan et al., 2016).

So far the pseudokinase field has achieved only a tiny proportion of attention compared to research efforts on the catalytically active protein kinases. The awareness of the possible druggability of protein kinases resulted in an upswing of kinase inhibitors with more than 65 kinase inhibitors approved for clinical use as of 2020 (Cohen, Cross, & Janne, 2021). However, although pseudokinases were first comprehensively described by Manning and colleagues two decades ago (Boudeau et al., 2006; Manning et al., 2002), they remain challenging drug targets (Byrne, Foulkes, & Eyers, 2017; Kung & Jura, 2019b; Mace & Murphy, 2021). No drug inhibitors of Tribbles proteins have been identified to date, although a recent study curiously revealed that TRIB2 is destabilized by binding of afatinib, a multi-ErbB receptor tyrosine kinase inhibitor (Foulkes et al., 2018).

The investigation of Tribbles pseudokinases is impeded by the sparsity of tools for their detection (Shrestha et al., 2020). Studies so far have predominantly used Tribbles overexpression or detection of mRNA levels due to the difficulty in detecting endogenous protein levels, especially for TRIB2 (Corish & Tyler-Smith, 1999). The main reason for this can be attributed to the fact that TRIB1 and TRIB2 proteins are highly unstable due to the presence of the PEST region in the N-terminal domain (Eyers et al., 2017). The PEST domain is Proline-, Glutamine-, Serine- and Threonine-rich, and is a hallmark of proteins with short half-lives, such that removal of this domain from proteins can increase their stability (Corish & Tyler-Smith, 1999). As a result of lack of suitable tools for Tribbles protein levels' detection, their half-life and post translational modifications are poorly characterized, and only the crystal structure of TRIB1 has been solved (Jamieson et al., 2018; McMillan et al., 2021; Murphy et al., 2015). Thus, access to robust TRIB protein detection methods will pave the way for their better functional characterization and for the development of targeted therapeutic approaches, or drug repurposing.

In this review we describe two methods, namely Western blotting and flow cytometry, for detection of endogenous TRIB2 protein levels in mammalian AML cells. Flow cytometry is a commonly used method for the detection of both intracellular and cell surface molecules and Western blotting is a gold standard technique based on the use of specific antibodies to identify target proteins from a concoction of other proteins.

2. Flow cytometry-based method for staining of the intracellular TRIB2

Flow cytometry is a versatile tool, employed in various fields of research such including immunology, cancer studies and virology. A flow cytometer consists of multiple lasers and cells are passed through the path of these lasers leading to perturbations in emitted light. This emission depends on the size, shape, and granularity of the cells. The light scattering is measured in forward and side scatter (at 90 degrees) which predominantly depends on refraction due to size and complexity of the cell respectively.

Different fluorescence lasers can be used to excite tagged or conjugated fluorochromes (e.g., green fluorescent protein or other fluorescent protein tags fused to a protein of interest, or fluorescent tags conjugated to secondary antibodies) at different wavelengths to identify specific cell populations (McKinnon, 2018; Schmit, Klomp, & Khan, 2021). The signals are recorded in Flow Cytometry Standard (FCS) file format which can be easily analyzed using specialized software such as FlowJo. Owing to the extensive use of flow cytometry in the immunology field to identify different cell types/states, we employed this method to identify and quantify intracellular TRIB2 levels across different AML cell lines. We used M07E, MONOMAC1 and AP1060 cells to assay the TRIB2 levels using a commercially available antibody.

2.1 Equipment required

1. BD FACSCanto II
2. FlowJo software (FlowJo™ v10.8.1)
3. Centrifuge: Sigma 3–16 K 10280 with 11,180 swinging-bucket rotor

2.2 Reagents needed

1. 1 × PBS (Phosphate Buffered Saline): 137 mM NaCl, 2.7 mM KCl, 10 mM Na_2HPO_4, 1.8 mM KH_2PO_4. Adjust pH to 7.4 with hydrochloric acid. Note, always use PBS for making any solution for flow cytometry as it maintains the pH of the cell suspension. Avoid using water as it can cause cell lysis and therefore cell death.
2. 10% Triton X100 (Sigma, Cat# 9036-19-5).
3. 37% Formaldehyde (Sigma, Cat# 47608). Dilute 1/10 with PBS to obtain a 3.7% formaldehyde solution

4. Primary antibody: TRIB2 monoclonal antibody (ProteinTech, Cat# 67171-1-Ig).
5. Secondary antibody: IRDye 800CW donkey anti-mouse secondary antibody (Licor, Cat#926-32212, Licor).
6. Isotype control antibody: Mouse IgG (Abcam, Cat# ab37355).
7. Cell suspension: We used AML cells (AP1060, M07E, MONOMAC1) obtained from DSMZ (Germany) which are suspension cells, but adherent cells can also be used by trypsinising the cells and spinning down as described for suspension cells. All cells were grown at 37 °C and 5% CO_2, using aseptic techniques in a class II biosafety cabinet. Cells were counted using hemocytometer.
8. 0.4% Trypan blue (ThermoFisher, Cat# 15250061).
9. 12 × 75 mm Polystyrene, non sterile FACS tubes (Elkay Labs, Cat# 2052-004).

2.3 Controls needed

It is important to include necessary controls to correct for background staining and nonspecific binding, aiding in accurate interpretation of the results. For detection of endogenous TRIB2, we choose a validated primary mouse monoclonal antibody supplied by ProteinTech, used in our lab to detect exogenously expressed TRIB2 and used to confirm loss of TRIB2 expression using a multi-ErbB receptor tyrosine kinase inhibitor (not shown), an inhibitor previously shown to specifically bind and destabilize endogenous TRIB2 with other TRIB2 antibodies (Foulkes et al., 2018). The ProteinTech clone is our chosen antibody as it is commercially available and in good supply.

1. *Unstained control*—This control corrects for any background or auto fluorescence emerging from the cells of interest. It is also essential to use unstained cells for setting the voltage and gating strategy. Unstained cells are fixed and permeabilized like the stained cells but in the absence of the anti-TRIB2 antibody and the secondary antibody.
2. *IgG/Isotype control*—The signal obtained from the experimental samples stained with the primary antibodies can be normalized to the isotype control. To perform staining with the IgG/Isotype control, cells of interest are fixed, permeabilized and stained with an IgG control antibody solution diluted to be at the same concentration as the anti-TRIB2 antibody solution. All cells are incubated with the secondary antibody solution.

3. *Secondary antibody only control*—Including a secondary antibody only control will allow exclusion of background binding contributed by that antibody. In this control, cells of interest that are fixed and permeabilized are incubated with PBS instead of the anti-TRIB2 antibody solution, followed by incubation with the secondary antibody solution. Note: It is important that controls are performed for all different cell lines to enable comparisons to be made.

2.4 Protocol

1. AP1060, M07E, MONOMAC1 cells were seeded at 2×10^5 cell/mL density in a T75 flask a day prior to the flow cytometry. The cells were counted using hemocytometer and live cells measured by trypan blue exclusion (1:1 mix of cell suspension and trypan blue solution).
2. Harvest cells into a 15 mL polypropylene Falcon tube—at least 1×10^5 cells per condition; more is preferred as many cells will be lost during subsequent fixation and washing steps. Begin with higher starting cell numbers (1×10^6 cells per condition) if intending to look at a rare cell subpopulation.
3. Centrifuge cells at $300 \times g$ for 5 min at room temperature and discard media.
4. Wash cells with 5 mL PBS, centrifuge cells at $300 \times g$ for 5 min and discard PBS by decanting (can aspirate supernatant if pellet is visible and avoid disruption) leaving cells in minimum residual PBS.
5. Fix cells for 10–15 min with 500 μL of final 3.7% formaldehyde (37% formaldehyde diluted in 1 × PBS) to the cell pellet and incubate at room temperature. Vortex cells briefly after adding formaldehyde, using a standard lab bench vortex mixer (at medium speed for about 2 s) to obtain a homogenous solution (can be assessed visually) and to prevent cell clumps. After fixing, add 2 mL 1 × PBS to wash off the formaldehyde, centrifuge cells at $300 \times g$ for 5 min and discard PBS by decanting.
6. Wash cells with 2 mL 1 × PBS again, centrifuge cells at $300 \times g$ for 5 min at room temperature and discard PBS by decanting.
7. Optional: Cells can be stored at 4 °C at this point for a week (if storing at 4 °C, do not fully remove PBS to prevent drying out of the sample). To proceed, aliquot the cells from the 15 mL Falcon tube into four FACS tubes—one to be stained with anti-TRIB2 antibody, one for the IgG control, one for unstained cells and one for secondary control.

These controls are sufficient for one experimental timepoint and may need to be included if different timepoints are being used due to shifting of background signals.

8. Dilute anti-TRIB2 antibody (1 mg/mL) 1/50 dilution in PBS and IgG control (5 mg/mL) 1/250 dilution in PBS and prepare 0.5% Triton X100 (diluted in 1X PBS) to permeabilize the cells. Add 50 μL of 0.5% Triton to each sample and to the control, then add 50 μL of diluted antibody to each sample and 50 μL of diluted IgG control to each IgG control sample. Leave cells to permeabilize for 1 h at room temperature. Vortex cells briefly after addition of diluted antibody/control and the 0.5%Triton X100 solution.
9. Wash cells with 2 mL PBS, centrifuge cells at 300° × *g* for 5 min, and discard the supernatant.
10. Dilute secondary antibody 1/1000 dilution in PBS and add 100 μL of diluted secondary antibody to each sample, IgG control sample and to the secondary only control. Vortex cells at medium speed for 2 s after addition of the antibody solution. Incubate for 30 min at room temperature in the dark.
11. Wash cells by adding 1 mL PBS, centrifuge cells at 300° × *g* for 5 min at room temperature, and discard the supernatant by decanting.
12. Run samples and controls on a flow cytometer and analyze the results using appropriate software (we have used FlowJo™ v10.8.1 for our analysis). Guidance on settings: The voltages are set using unstained cells (control), first based on FSC (forward scatter) and SSC (side scatter), and then from the secondary antibody fluorophore controls. The unstained cells should appear in the first quartile of the axis, i.e., should be at about 10^2 on the scale. Cells stained with isotype control or with only secondary antibody will appear in the first quadrant like the unstained cells. Cells with both primary and secondary antibody will appear in the other quadrants depending on the abundance of the protein. Once the voltages are set they should not be changed during the experiment. Voltages will need to be set individually for every cell line to be use. At least 10,000 events per sample were acquired/recorded.

2.5 Data analysis using FlowJo software

1. Load all the FCS files into the FlowJo software, gate unstained cells for SSC-A/FSC-A to select live cell population (see Fig. 1A).

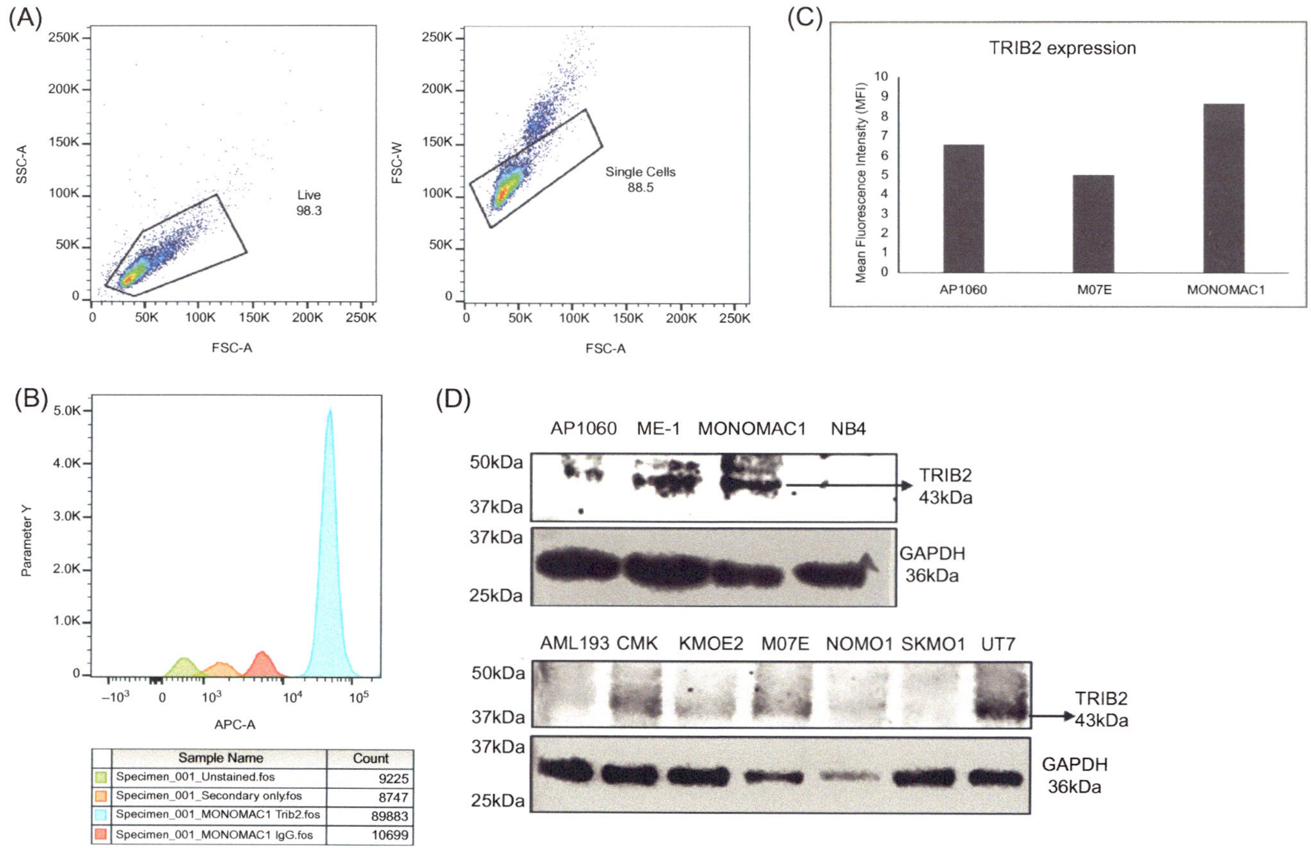

Fig. 1 Representative data for a flow cytometry-based method of staining for intracellular endogenous TRIB2 and western blot detection of endogenous human TRIB2 expression (A) Representative gating strategy for the live cell population and single cell population. (B) Histogram showing TRIB2 expression on the x-axis (APC fluorescence), each cell line represented with different colors. (C) Graph depicting the Mean Fluorescence Intensity (MFI) of intracellular TRIB2 expression levels measured by flow cytometry in the indicated cell lines, normalized to control stains. (D) Western blot showing TRIB2 (~43 kDa, always runs below the 50 kDa protein marker) and GAPDH expression in the indicated human AML cell lines.

2. Select the population and then gate for single cells to exclude doublets by separating the cell population using FSC-W/FSC-A gates. Doublets are predominately two cells passing very close together during a flow cytometry experiment and the machine cannot exclude them as two different events and hence result in distorted size therefore leading to inaccurate analysis. It is imperative in any flow cytometry analysis to exclude this population by separating them using the FSC-W/FSC-A gating strategy which is based on the size and shape of the cell. The same settings are applied to all samples (see Fig. 1A).
3. Use the FlowJo statistics tab to calculate geometric mean (see Fig. 1B) between the samples to report the mean fluorescence intensity (MFI). This calculation should be performed for the fluorescence channel corresponding to the antibody of interest, in our case APC-A channel. We used the APC-A channel since our secondary antibody is conjugated to the APC fluorochrome. The MFI can be used when your entire population of cells shifts on the x-axis rather than a separate histogram peak (population of positive cells). If a positive and negative population appears on the histogram, an MFI would be taken for the positive and negative populations separately.
4. MFI can be calculated by dividing the geometric mean of IgG control with the geometric mean of the sample (see Fig. 1C).

3. TRIB2 identification by Western blotting method

Identification of proteins using Western blotting is a standard method in molecular biology. Proteins are resolved on a denaturing gel according to molecular weight, and are then transferred onto a nitrocellulose membrane, which is incubated with the specific primary antibody targeted toward protein of interest. The protein antibody complex is washed several times to remove any unbound primary antibody followed by incubation with the secondary antibody (Mahmood & Yang, 2012). The entire complex is washed again to remove unbound secondary antibody and is visualized by one of many detection methods. In the method described here we used LICOR, although other methods such as chemiluminescent detection using X-ray film or exposure to a chemiluminescent camera can be used. However, quantitation using the latter methods are unreliable because signals can often be overexposed and not in the linear range for detection; accordingly we recommend the LICOR Odyssey for quantitative analyses.

We performed Western blotting to examine endogenous TRIB2 expression across different AML cell lines. The calculated molecular weight of TRIB2 is ~39 kDa however it runs slightly higher ~43 kDa on polyacrylamide gels. This variation in endogenous TRIB2 molecular weight may relate to charge properties or post-translational modifications in these myeloid cells. The molecular weight of the protein can be calculated using AAT Bioquest (https://www.aatbio.com/tools/calculate-peptide-and-protein-molecular-weight-mw) by providing the UniProtKB ID (Q92519 for TRIB2) (Bioquest, 2022).

3.1 Equipment required

1. Mini-PROTEAN® Tetra Handcast Systems (Biorad, Cat# 1658001FC)
2. PowerPac™ Basic Power Supply (Biorad, Cat# 1645050)
3. Mini Gel Tank and Blot Module Set (Thermo Fisher Scientific, Cat# NW2000)
4. BD Microlance 3 Needles Orange 25 g × 5/8″ (Microlance, Cat# ND600)
5. LICOR Fc ODYSSEY V 1.0
6. SpectraMax®M5 microplate reader to measure visible absorbance for calculating protein concentration
7. Centrifuge: Sigma 1–16 K refrigerated centrifuge 25 × 1.5/2 mL Rotor

3.2 Materials needed

1. Modified radioimmunoprecipitation assay buffer (RIPA buffer): 50 mM Trizma® hydrochloride (pH 7.4) (Sigma Aldrich, Cat# T3253), 1% NP40 (Thermo Fisher Scientific, Cat# FNN0021), 0.25% sodium deoxycholate (Sigma-Aldrich, Cat# D6750), 150 mM NaCl (Sigma-Aldrich, Cat# S9625888-500G), 1 mM Ethylenediaminetetraacetic acid (EDTA) (Sigma-Aldrich, Cat# 60-00-4). Modified RIPA can be pre-made and stored in aliquots at −20 °C. It is supplemented with 1 tablet each of cOmplete™, EDTA-free Protease Inhibitor Cocktail (Roche, Cat# 04693132001) and PhosSTOP™ (Roche, Cat# 4906837001), added fresh.
2. NuPAGE™ LDS Sample Buffer (4 ×) (Thermo Fisher Scientific, Cat# NP0007).
3. Western blot buffer (10 ×): 30 g Trizma® base (Sigma-Aldrich, Cat# T1503,), 144 g glycine (Sigma -Aldrich, Cat# G8898). Prepare up to 1000 mL with milli Q water.

4. Running buffer: 10 mL 10% Sodium dodecyl sulfate (SDS) (Sigma-Aldrich, Cat# 151-21-3), 100 mL Western blot buffer. Prepare up to 1000 mL with water.
5. Transfer buffer: 10% methanol (Honeywell, Cat# 24229,), 5% NuPAGE™ MOPS SDS Running Buffer (20 ×) (Thermo Fisher Scientific, Cat # NP000102).
6. TBST: 100 mL of 10 × TBS, 1 mL Tween 20 (Sigma-Aldrich, Cat # P1379. Prepare to 1000 mL with water.
7. 5% milk TBST: 2.5 g milk powder (Marvel dried skimmed milk powder, any supermarket) and TBST to 50 mL final volume.
8. Anti-TRIB2 monoclonal antibody, 1000 μg/mL (ProteinTech, Cat# 67171-1-Ig).
9. IRDye 800CW donkey anti-mouse secondary antibody (LICOR, Cat# 926-32212).
10. Loading control, e.g., anti-GAPDH antibody (Cell Signalling Technologies, Cat# 2118 L).
11. 10% Sodium Dodecyl Sulfate—Poly Acrylamide Gel Electrophoresis (SDS-PAGE) gel. Note: we chose 10% SDS -PAGE gel, which is most optimal for separation for the molecular weight of the proteins we studied, resolving our protein of interest approximately half-way down the gel relative to the dye front.
12. Nitrocellulose membrane 0.45 μm (Bio-rad, Cat# 1620115).
13. SuperSignal™ West Pico PLUS Chemiluminescent Substrate (Thermo scientific, Cat# 34577).
14. Precision Plus Protein Dual Color Standards (Bio-rad, Cat# 161-0374).
15. Bio-Rad Protein Assay Dye Reagent Concentrate (Bio-rad, Cat# 5000006).
16. 30% Acrylamide/Bis Solution (Bio-rad, Cat# 1610158).

3.3 Controls required

1. *Positive control*—It is important to include a positive control for TRIB2 detection while performing Western blot analysis to ensure that the protocol, buffers and antibodies are in working order. We have consistently measured a high level of TRIB2 expression in the MONOMAC-1 cell line using both flow cytometry and Western blotting detection methods described here. Thus, to confirm that TRIB2 detection is occurring, we suggest using this cell line as a positive control.

2. *Loading control*—Once TRIB2 visualization by Western blotting is complete, the membrane should be probed for a housekeeping protein such as actin or GAPDH as a loading control. We used a Bradford assay to measure protein concentration to ensure equal loading of our samples. As housekeeping genes, GAPDH and actin are expected to express at the same levels across all samples, irrespective of the treatment conditions. This step also allows us to examine whether samples are sufficiently and uniformly transferred from the gel to the membrane across all samples.

3.4 Protocol

1. Prepare the SDS-PAGE gel as per the instructions in Fig. 2A.
2. To prepare the protein lysate: Typically 1×10^6 AML suspension cells is sufficient to yield 50 μg protein. The cells are counted using a hemocytometer slide with Trypan Blue exclusion, and harvested, as per the intracellular staining protocol. The cell pellet is then resuspended in 1 mL of ice cold PBS, and vortexed using a standard lab bench vortex until a homogenous solution is obtained and transferred to a 2 mL clear microcentrifuge tube. Cells are centrifuged at 300° ×*g* for 5 min at 4 °C and media discarded by aspiration or decanting, being careful not to disrupt the cell pellet. Wash cells again with ice cold 1 mL PBS, centrifuge cells at 300° ×*g* for 5 min and discard PBS. To lyse the cells, add 10–30 μL per 10^6 cells modified RIPA buffer, vortex vigorously and leave on ice for 30 min, with intermittent vortexing (for 5 s) to mix the samples and aid lysis. The volume of RIPA buffer is variable depending on how much protein is obtained from 1×10^6 cells to ensure you can load at least 50 μg per lane. Centrifuge the lysate for 15 min at 4 °C at 21,000° ×*g*. Discard the pellet (debris) and transfer the supernatant containing the protein to a cold, labeled, 0.5 or 2 mL tube and proceed with protein quantification. It is important that the protein lysate is kept cold (on ice) from this point onwards.
3. Quantify the protein concentration in each lysate sample using a commercial kit, for example, the Bio-Rad Protein Assay Dye Reagent Concentrate (Bio-Rad protein assay reagent kit II, Cat# 5000002). Prepare protein standards of concentrations of 0.25, 0.5, 0.75, 1 and 1.5 mg/mL Bovine serum albumin (BSA) (Sigma-Aldrich, Cat# A2153). Dilute the protein assay dye reagent 1:5 to make a working solution. 10 μL each of the protein standards and 2 μL of the sample

A

Resolving gel composition	10ml	Stacking gel composition	4ml
Water	4 ml	Water	2.7 ml
30% Acrylamide mix	3.3 ml	30% Acrylamide mix	0.67 ml
1.5M Tris (pH 8.8)	2.5	1.5M Tris (pH 8.8)	0.5
10% SDS	0.1	10% SDS	0.04
10% ammonium per sulphate	0.1	10% ammonium per sulphate	0.04
TEMED	0.004	TEMED	0.004

B

Fig. 2 Western blot method. (A) 10% SDS-PAGE gel composition (B) Representative figure of a transfer sandwich.

lysate are added to a 96 well plate. 200 μL of the protein assay dye is then added to each well and incubated at room temperature for 20 min in the dark. The absorbance is measured at 595 nm using SpectraMax®M5.

4. A standard curve was generated by plotting the absorbance obtained for each standard concentration against protein concentration using the Bradford assay absorbance assay in an automated microplate reader (e.g., SpectraMax SoftMax Pro software). Thermo Fischer Scientific tools brochures can be used as a reference when applying the standard curve to quantify the protein concentration in the experimental samples (Scientific, 2012).
5. Aliquot 50 μg of protein, 7.5 μL of 4 × sample buffer and water, to prepare 30 μL samples for electrophoresis. Surplus protein lysate can be stored long-term at −20 °C.
6. Heat samples for 10 min at 90 °C, centrifuge briefly (10–20 s in a benchtop centrifuge). At this point, samples can be stored at −20 °C. It is helpful to reheat the samples at 90 °C for 5–10 min before running the gel, if they were stored at −20 °C.

7. Insert the 10% SDS-PAGE gel into the chamber of the electrophoresis box and fill with running buffer. Note, we are using 10% SDS-PAGE gel because the molecular weight at which TRIB2 protein is resolved on a gel is ~43 kDa, about half way down a 10% gel.
8. Load samples and the protein ladder onto a 10% SDS-PAGE gel using gel loading tips and run the gel at 160 V for 1 h until the dye front has run off the gel.
9. Transfer proteins from resolved gel to membrane at 20 V for 1 h. Make sure to avoid air bubbles in the transfer sandwich. Transfer sandwich is created by placing the membrane on the gel and sandwiching them between filter papers and sponges with no air bubbles (Fig. 2B).
10. Block membrane for 1 h on a rocker at room temperature with 5% milk TBST in an appropriately sized dish/container, e.g., a plastic empty pipette tip box or lid, or sealed hybridisation bags. The membrane can also be inserted into a sealed 50 mL falcon tube for the blocking and antibody incubations in a tube rotator. The volume of primary antibody solution used can be reduced to 1 mL if using the sealed bag method.
11. Incubate membrane overnight on a rocker at 4 °C with anti-TRIB2 antibody diluted 1/300 (3 μg) in 5% milk TBST.
12. Next day, incubate the membrane with the primary antibody for additional 1 h on a rocker at room temperature. We do not routinely recover the primary antibody for further use due to the dilution in milk, but it can be stored at −20 °C by adding 0.05% sodium azide to prevent bacterial growth.
13. Wash the membrane with TBST (5 mL, or enough to cover the membrane) three times for 5 min each on a rocker.
14. Incubate membrane for 1 h on a rocker in the dark at room temperature with the secondary antibody diluted 1/5000 in 5% milk TBST.
15. Wash membrane with TBST (5 mL, or enough to cover the membrane) three times for 5 min, each time on a rocker.
16. Use a LICOR Odyssey instrument to visualize protein for 10 min using the 800 channel and the ladder for 30 s using the 700 channel. SuperSignal™ West Pico PLUS Chemiluminescent Substrate was used to detect GAPDH.

    Note: since the anti-TRIB2 antibody was incubated with IRDye 800CW secondary antibody the signal can be read at wavelength of 800 in the LICOR machine.
17. Wash membrane with TBST (5 mL, or enough to cover the membrane) three times for 5 min, each time on the rocker.

18. Incubate the membrane with the loading control antibody of choice (anti-GAPDH in our case) and proceed as recommended for that antibody (see Fig. 1D).

4. Discussion

Success of flow cytometry analysis is dependent on factors such as cell viability, the optimum fixing buffer composition for the cells, the optimal detergent concentration for permeabilization of the cells, and antibody specificity. There are two main steps in an intracellular cell staining protocol. The first step is cell fixation for which we use formaldehyde. The second step is cell permeabilization which is very important for the antibody binding and identification, and for which we use Triton X100. Other fixing reagents such as paraformaldehyde (Lanier & Warner, 1981), and permeabilization solutions such as NP40 can be used (Amidzadeh et al., 2014). The size of the cells and the location of your protein are determining factors in what solutions work best for fixing and permeabilizing cells for intracellular staining by flow cytometry and through a process of elimination/optimization we settled on the presented methods of choice for TRIB2. Ready to use kits, such as BD Cytofix/Cytoperm™ Fixation/Permeabilization Solution Kit, can also be used, however, we found success with homemade buffers. We have also determined that storage of fixed cells at 4 °C, prior to staining for a week, has no adverse effect on the fluorescence levels. To exclude any background nonspecific staining in flow cytometry experiments, it is important to use the relevant IgG controls. The mean fluorescence intensity of the IgG control staining is used to determine the background fluorescence contributed by staining that is not specific to the TRIB2 signal (Radcliff & Jaroszeski, 1998; Schmit et al., 2021). Overall, flow cytometry using the intracellular staining method proved to be a versatile and reproducible method to look at the endogenous TRIB2 levels in AML cell lines.

Western blotting is a standard technique routinely used in molecular biology studies to measure protein expression levels in lysed cells. We used Western blotting to measure levels of endogenous TRIB2 in lysates from acute myeloid leukemia suspension cells. This method can also be applied to tissues, if needed, which would need to be sonicated prior to the analysis. The cell lysates need to be heated to denature the protein prior to loading on a gel. For a successful Western blotting analysis, it is important to ensure loading of positive and negative controls that supports clear identification of protein of interest. We found that the optimum percentage resolving gel for TRIB2 was 10% due to its molecular weight, but a 12% gel could also be used if smaller size proteins were being detected on the same gel. We optimized the blocking of non-specific binding with milk in the blocking buffer and for the antibody solutions but alternatively a solution with BSA or other blocking agents could be trialed. An appropriate molecular weight housekeeping protein (i.e., one that is resolved distanced from the protein of interest by SDS-PAGE) should also always be quantified on the same membrane to determine protein loading across each of the wells and in our case, on the 10% gel, we had good resolution for detection with GAPDH housekeeping protein (36 kDa). We would like to highlight that Western blot quantification is a semi quantitative method. While Western blotting-based detection of endogenous TRIB2 in cells works generally well, intracellular staining by flow cytometry would be preferable for cells expressing lower levels of TRIB2 (e.g., MO7e cells) as the bands on the Western blot were clearer in cell lines with higher TRIB2 expression than the ones with lower TRIB2 expression, at least when using 50 μg of protein. An estimation of how much protein to use for Western blotting could be made from a combination of both detection methods, as TRIB2 expression levels across the cell lines are recapitulated by both Western blotting and intracellular staining, as shown in Fig. 1. In conclusion, the anti-TRIB2 antibody we used in our experiments worked well for both described methods allowing detection of endogenous TRIB2. The preferred method for cell lines with lower levels of TRIB2 is the flow cytometry method. Increasing the amount of protein loaded in the Western blotting protocol (e.g., can troubleshoot up to about 150 μg) may improve band detection with lower expressing cell lines although this would need to be optimized to avoid over-saturation and background signals (protocols for this available from, e.g., Bio-Rad).

Acknowledgments

We are thankful to Proteintech who provided us with different clones of TRIB2 antibody to trial in our experimental methods. This work was supported by Blood Cancer UK (project grant to K.K.) and Medical Research Council (Confidence in Concept grant to K.K.)

References

AAT Bioquest. (2022). Quest Calculate™ peptide and protein molecular weight calculator. AAT Bioquest, Inc.

Amidzadeh, Z., et al. (2014). Assessment of different permeabilization methods of minimizing damage to the adherent cells for detection of intracellular RNA by flow cytometry. *Avicenna Journal of Medical Biotechnology*, *6*(1), 38–46.

Bailey, F. P., et al. (2015). The tribbles 2 (TRB2) pseudokinase binds to ATP and autophosphorylates in a metal-independent manner. *The Biochemical Journal*, *467*(1), 47–62.

Boudeau, J., et al. (2006). Emerging roles of pseudokinases. *Trends in Cell Biology*, *16*(9), 443–452.

Byrne, D. P., Foulkes, D. M., & Eyers, P. A. (2017). Pseudokinases: Update on their functions and evaluation as new drug targets. *Future Medicinal Chemistry*, *9*(2), 245–265.

Cohen, P., Cross, D., & Janne, P. A. (2021). Kinase drug discovery 20 years after imatinib: Progress and future directions. *Nature Reviews. Drug Discovery*, *20*(7), 551–569.

Corish, P., & Tyler-Smith, C. (1999). Attenuation of green fluorescent protein half-life in mammalian cells. *Protein Engineering*, *12*(12), 1035–1040.

Dedhia, P. H., et al. (2010). Differential ability of tribbles family members to promote degradation of C/EBPalpha and induce acute myelogenous leukemia. *Blood*, *116*(8), 1321–1328.

Durzynska, I., et al. (2017). STK40 is a Pseudokinase that binds the E3 ubiquitin ligase COP1. *Structure*, *25*(2), 287–294.

Eyers, P. A. (2015). TRIBBLES: A twist in the pseudokinase tail. *Structure*, *23*(11), 1974–1976.

Eyers, P. A., Keeshan, K., & Kannan, N. (2017). Tribbles in the 21st century: The evolving roles of tribbles Pseudokinases in biology and disease. *Trends in Cell Biology*, *27*(4), 284–298.

Foulkes, D. M., et al. (2018). Covalent inhibitors of EGFR family protein kinases induce degradation of human tribbles 2 (TRIB2) pseudokinase in cancer cells. *Science Signaling*, *11*(549), 7951.

Hasemann, M. S., et al. (2014). C/EBPα is required for long-term self-renewal and lineage priming of hematopoietic stem cells and for the maintenance of epigenetic configurations in multipotent progenitors. *PLoS Genetics*, *10*(1), e1004079.

Hill, R., et al. (2017). TRIB2 confers resistance to anti-cancer therapy by activating the serine/threonine protein kinase AKT. *Nature Communications*, *8*, 14687.

Jamieson, S. A., et al. (2018). Substrate binding allosterically relieves autoinhibition of the pseudokinase TRIB1. *Science Signaling*, *11*(549).

Keeshan, K., et al. (2006). Tribbles homolog 2 inactivates C/EBPalpha and causes acute myelogenous leukemia. *Cancer Cell*, *10*(5), 401–411.

Keeshan, K., et al. (2010). Transformation by tribbles homolog 2 (Trib2) requires both the Trib2 kinase domain and COP1 binding. *Blood*, *116*(23), 4948–4957.

Keeshan, K., et al. (2016). Co-operative leukemogenesis in acute myeloid leukemia and acute promyelocytic leukemia reveals C/EBPα as a common target of TRIB1 and PML/RARA. *Haematologica*, *101*(10), 1228–1236.

Kiss-Toth, E., et al. (2004). Human tribbles, a protein family controlling mitogen-activated protein kinase cascades. *The Journal of Biological Chemistry*, *279*(41), 42703–42708.

Komiya, Y., & Habas, R. (2008). Wnt signal transduction pathways. *Organogenesis*, *4*(2), 68–75.

Kung, J. E., & Jura, N. (2016). Structural basis for the non-catalytic functions of protein kinases. *Structure*, *24*(1), 7–24.

Kung, J. E., & Jura, N. (2019a). The pseudokinase TRIB1 toggles an intramolecular switch to regulate COP1 nuclear export. *The EMBO Journal*, *38*(4), e99708.

Kung, J. E., & Jura, N. (2019b). Prospects for pharmacological targeting of pseudokinases. *Nature Reviews. Drug Discovery*, *18*(7), 501–526.

Lanier, L. L., & Warner, N. L. (1981). Paraformaldehyde fixation of hematopoietic cells for quantitative flow cytometry (FACS) analysis. *Journal of Immunological Methods*, *47*(1), 25–30.

Ma, X., et al. (2018). TRIB2 knockdown as a regulator of chemotherapy resistance and proliferation via the ERK/STAT3 signaling pathway in human chronic myelogenous leukemia K562/ADM cells. *Oncology Reports, 39*(4), 1910–1918.

Mace, P. D., & Murphy, J. M. (2021). There's more to death than life: Noncatalytic functions in kinase and pseudokinase signaling. *The Journal of Biological Chemistry, 296*, 100705.

Mahmood, T., & Yang, P. C. (2012). Western blot: Technique, theory, and trouble shooting. *North American Journal of Medical Sciences, 4*(9), 429–434.

Manning, G., et al. (2002). The protein kinase complement of the human genome. *Science, 298*(5600), 1912–1934.

McKinnon, K. M. (2018). Flow cytometry: An overview. *Current Protocols in Immunology, 120*, 5.1.1–5.1.11.

McMillan, H. D., et al. (2021). Structure vs. function of TRIB1-myeloid neoplasms and beyond. *Cancers (Basel), 13*(12), 3060.

Mittal, S., et al. (2014). Coordinate hyperactivation of Notch1 and Ras/MAPK pathways correlates with poor patient survival: Novel therapeutic strategy for aggressive breast cancers. *Molecular Cancer Therapeutics, 13*(12), 3198–3209.

Murphy, J. M., et al. (2015). Molecular mechanism of CCAAT-enhancer binding protein recruitment by the TRIB1 pseudokinase. *Structure, 23*(11), 2111–2121.

O'Connor, C., et al. (2016). The presence of C/EBPα and its degradation are both required for TRIB2-mediated leukaemia. *Oncogene, 35*(40), 5272–5281.

Radcliff, G., & Jaroszeski, M. J. (1998). Basics of flow cytometry. *Methods in Molecular Biology, 91*, 1–24.

Richmond, L., & Keeshan, K. (2020). Pseudokinases: A tribble-edged sword. *The FEBS Journal, 287*(19), 4170–4182.

Sakai, S., et al. (2016). Tribbles-related protein family members as regulators or substrates of the ubiquitin-proteasome system in cancer development. *Current Cancer Drug Targets, 16*(2), 147–156.

Salomé, M., et al. (2018). A Trib2-p38 axis controls myeloid leukaemia cell cycle and stress response signalling. *Cell Death & Disease, 9*(5), 443.

Salomé, M., Hopcroft, L., & Keeshan, K. (2018). Inverse and correlative relationships between TRIBBLES genes indicate non-redundant functions during normal and malignant hemopoiesis. *Experimental Hematology, 66*. 63–78.e13.

Schmit, T., Klomp, M., & Khan, M. N. (2021). An overview of flow cytometry: Its principles and applications in allergic disease research. *Methods in Molecular Biology, 2223*, 169–182.

Seher, T. C., & Leptin, M. (2000). Tribbles, a cell-cycle brake that coordinates proliferation and morphogenesis during Drosophila gastrulation. *Current Biology, 10*(11), 623–629.

Sever, R., & Brugge, J. S. (2015). Signal transduction in cancer. *Cold Spring Harbor Perspectives in Medicine, 5*(4), a006098.

Shrestha, S., et al. (2020). Cataloguing the dead: Breathing new life into pseudokinase research. *The FEBS Journal, 287*(19), 4150–4169.

Thermo Fisher Scientific. (2012). *How to use a protein assay standard curve*. https://tools.thermofisher.com/content/sfs/brochures/TR0057-Read-std-curves.pdf.

Uljon, S., et al. (2016). Structural basis for substrate selectivity of the E3 ligase COP1. *Structure, 24*(5), 687–696.

Yokoyama, T., & Nakamura, T. (2011). Tribbles in disease: Signaling pathways important for cellular function and neoplastic transformation. *Cancer Science, 102*(6), 1115–1122.

Zhang, H., et al. (2012). The role of pseudokinases in cancer. *Cellular Signalling, 24*(6), 1173–1184.

CHAPTER FOUR

Analysis of human Tribbles 2 (TRIB2) pseudokinase

John A. Harris, Emma Fairweather, Dominic P. Byrne, and Patrick A. Eyers*

Department of Biochemistry and Systems Biology, Institute of Systems, Molecular and Integrative Biology, University of Liverpool, Liverpool, United Kingdom

*Corresponding author: e-mail address: patrick.eyers@liverpool.ac.uk

Contents

Abstract

Human Tribbles 2 (TRIB2) is a cancer-associated pseudokinase with a broad human protein interactome, including the well-studied AKT, C/EBPα and MAPK modules. Several lines of evidence indicate that human TRIB2 promotes cell survival and drug-resistance in solid tumors and blood cancers and is therefore of interest as a potential therapeutic target, although its physiological functions remain relatively poorly understood. The unique TRIB2 pseudokinase domain lacks the canonical 'DFG' motif, and subsequently possesses very low affinity for ATP in both the presence and absence of metal ions. However, TRIB2 also contains a unique cysteine-rich αC-helix, which interacts with a conserved peptide motif in its own carboxyl-terminal tail. This regulatory flanking region drives regulated interactions with distinct E3 ubiquitin ligases that serve to control the stability and turnover of TRIB2 client proteins. TRIB2 is also a low-affinity target of several known small-molecule protein kinase inhibitors, which were originally identified using

Methods in Enzymology, Volume 667
ISSN 0076-6879
https://doi.org/10.1016/bs.mie.2022.03.025

purified recombinant TRIB2 proteins and a thermal shift assay. In this chapter, we discuss laboratory-based procedures for purification, stabilization and analysis of human TRIB2, including screening procedures that can be used for the identification of both reversible and covalent small molecule ligands.

1. Introduction

1.1 Pseudokinases

Pseudokinases are a group of understudied protein kinases that lack the ATP-dependent catalytic signaling outputs associated with their canonical counterparts (Boudeau, Miranda-Saavedra, Barton, & Alessi, 2006; Jacobsen & Murphy, 2017; Kwon et al., 2019). However, the degraded ATP site can still be targeted with appropriate small molecules, including repurposed drugs (Byrne, Foulkes, & Eyers, 2017; Kung & Jura, 2019; Pushpakom et al., 2019), and drug-like allosteric pseudokinase modulators are of intense interest for studying TRIB2 modulation. Of note, deucravacitinib, an orally available drug that was designed to specifically target the pseudokinase domain of the tyrosine kinase TYK2, has shown significant clinical promise as an immune system modulator (Burke et al., 2019). This breakthrough is likely to hasten the uptake of new approaches to analyze and modulate cell signaling driven through pseudokinase-based mechanisms (Murphy, Mace, & Eyers, 2017). Pseudokinases are part of a much broader pseudoenzyme group of atypical enzymes whose molecular analysis requires special attention due to a lack of catalytic function (Ribeiro et al., 2019). They typically function as scaffolds or allosteric activators, and can be rate-limiting for the fine-tuning of canonical signaling pathways. Pseudokinases are found throughout life and in the absence of tools and small molecules with which to probe their function, they form a significant subset of the understudied 'dark' kinome (Berginski et al., 2020; Shrestha, Byrne, Harris, Kannan, & Eyers, 2020).

Some 60 pseudokinases are encoded in the human kinome, where they were classified based on the absence of one or more of the conserved amino acid residues found in other kinases (Eyers & Murphy, 2013; Manning, Whyte, Martinez, Hunter, & Sudarsanam, 2002; Wilson et al., 2018). These residues are known to be crucial for catalytic activity in canonical kinases, and their absence means that the majority of pseudokinases have lost the ability to phosphorylate substrates in an ATP-dependent manner, as is considered the norm for canonical protein kinases (Boudeau et al., 2006;

Scheeff, Eswaran, Bunkoczi, Knapp, & Manning, 2009). Thermal analysis employing purified protein kinase or pseudokinase domains is a mainstay for the analysis of kinome members, and in the case of pseudokinases, has permitted their sub-classification based on the effects of various nucleotides and small molecules (Foulkes et al., 2018; Lucet, Babon, & Murphy, 2013; Murphy et al., 2014). The discovery of stabilizing or destabilizing TRIB2 ligands, which can aid biophysical and structural workflows, is also providing pertinent information on their conformation and druggability, with potential clinical impact (Byrne et al., 2017). Small molecule screening can also be combined with cellular thermostability assays (CETSA) to reveal target-engagement in complex mixtures (Molina et al., 2013), including several TRIB2 ligands (Foulkes et al., 2018).

1.2 Tribbles pseudokinases

The Tribbles gene was originally identified in flies as a protein kinase regulator of differentiation, where it controls signaling through the AKT module (Seher & Leptin, 2000). The human kinome contains four related Tribbles pseudokinase family members, termed TRIB1, TRIB2, TRIB3, and STK40, and evolutionary sequence analysis has established that human TRIB2 is the ancestral pseudokinase within the vertebrate lineages (Eyers, Keeshan, & Kannan, 2017). Despite there being no experimental structure for TRIB2, architectural inferences can be made from related proteins such as TRIB1 and STK40, which possess a truncated and distorted αC helix which is predicted to be incompatible with ATP binding (Durzynska et al., 2017; Jamieson et al., 2018). In addition, instead of containing the magnesium positioning 'DFG' motif, TRIB2 contains a unique SLE motif that is one of the 'hallmarks' of Tribbles proteins (Fig. 1).

While Tribbles retains several of the canonical residues found in model Ser/Thr kinases such as PKA, which together are required for high-affinity Mg^{2+}-ATP binding and substrate phosphorylation (Fig. 1), TRIB2 is best classified as a pseudokinase owing to the divergent activation segment motif (DFG > SLE) (Eyers & Murphy, 2013). Vertebrate TRIB2 polypeptides are composed of three domains: an N-terminal PEST domain, which likely controls the rate of proteasome turnover within the cell; the central pseudokinase domain, and a C-terminal tail, which contains the binding motifs: ILLHPW and DQLVPE that interface with MEK (mitogen-activated protein kinase kinase) and COP1 (constitutive photomorphogenesis protein 1), respectively. Interest in evaluating the signaling functions

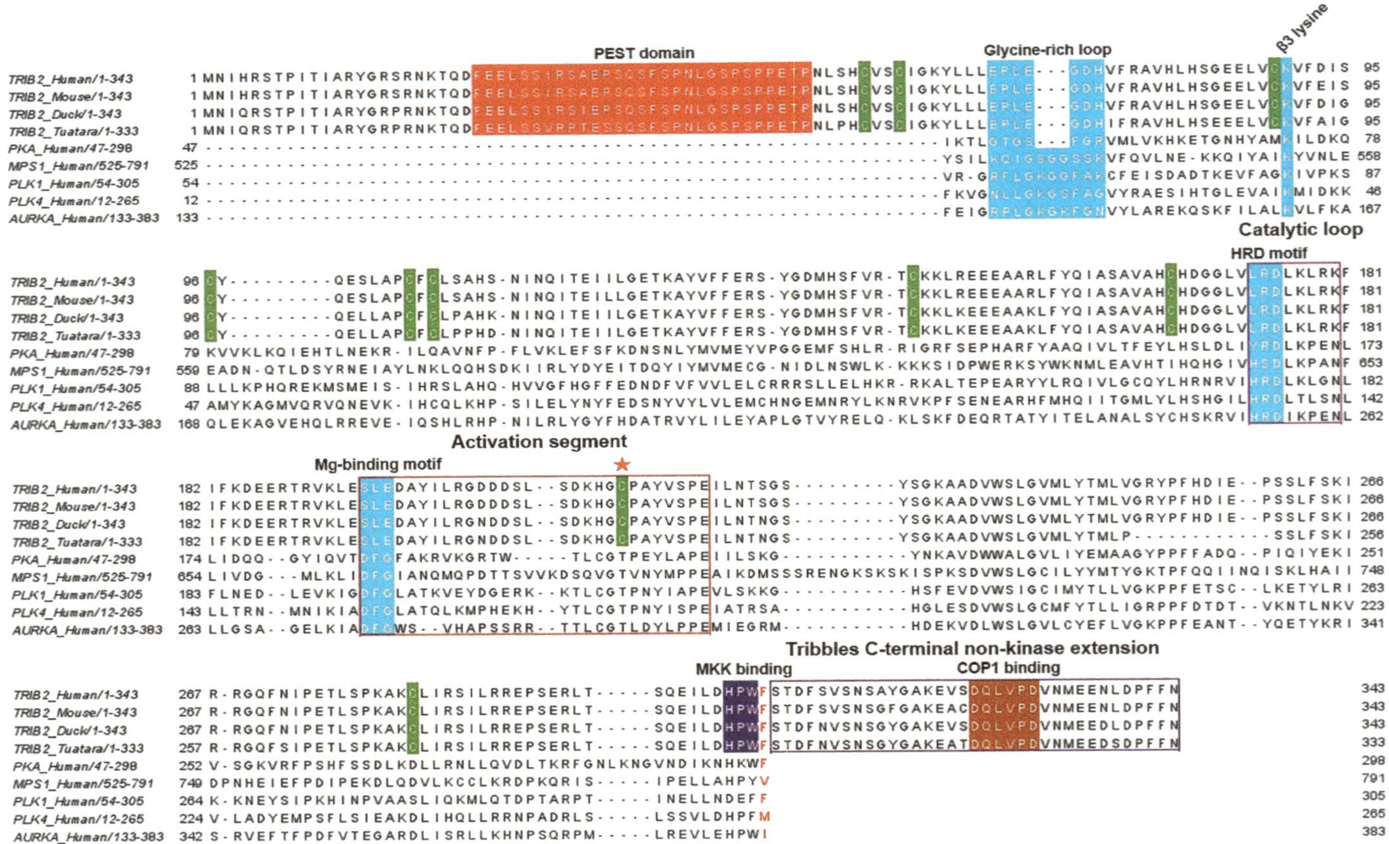

Fig. 1 See figure legend on opposite page.

of Tribbles pseudokinases has been mounting over the previous decade, since it became clear that their expression and turnover are linked to a variety of malignancies. Cellular experiments clearly support a central role for TRIB2 in cancer cell proliferation, survival and drug-resistance, consistent with its role as a homeostatic regulator of multiple signaling pathways (Grandinetti et al., 2011; Hill et al., 2017; Hou et al., 2018; Zanella et al., 2010). TRIB2 is involved in 'scaffolding' and activating or inhibiting other protein kinases in cells, which can be reconstituted in vitro with purified components for biochemical analysis (Foulkes et al., 2018). In terms of downstream impact, TRIB2 interfaces with ubiquitination-based signaling nodes, which act to destabilize target proteins that can be considered as TRIB2 'clients.' TRIB2 complexes form a signaling 'network' in cells whereby the TRIB2 pseudokinase can mediate signaling outputs through a number of protein:protein interactions. Indeed, in the presence of repurposed covalent kinase inhibitors, TRIB2 becomes rapidly destabilized in human cells, suggesting a new way of degrading intracellular TRIB2 protein with drugs such as kinase inhibitors (Guo et al., 2021).

In this chapter, we describe procedures for optimizing the expression, purification and stabilization of full-length human TRIB2 produced in an *E. coli* hosts, and provide a protocol for a thermal stability screening method that has revealed TRIB2 as a target for known kinase inhibitors. These are now being exploited to further evaluate the signaling roles of TRIB2 in human cells, and are a useful starting point for the development of new TRIB2 chemical tools and pharmaceutical agents.

Fig. 1 Comparative analysis of vertebrate TRIB2. (A) Alignment of TRIB2 protein sequences with canonical kinase catalytic subunits; the end of the kinase domain is colored in red. TRIB2 proteins are defined by three functional domains: an N-terminal 'PEST' domain (24–52, shaded red), a central pseudokinase domain (61–308), and a flexible C-terminal extension (blue box). TRIB2 residues that align with canonical features of protein kinase domains are shaded blue and correspond to the GxGxxG motif (in TRIB2 this loop is shorter and has the sequence EPLEGD); the β3 lysine residue (K90); the catalytic loop HRDXXXXN motif (in TRIB2 this is LRDXXXXK); and the magnesium-positioning DFG motif (in TRIB2 this is SLE). Of note, 10 cysteine residues are conserved in vertebrate TRIB2 homologues (green), including a Cys residue (red star) in the middle of the 'activation segment' (red box), in the vicinity of phosphorylatable Thr and Tyr amino acids residing in the 'P+1' loop of canonical kinases such as MPS1, PLK1, PLK4 and Aurora A (Byrne, Clarke, et al., 2020; Byrne, Shrestha, et al., 2020; Caron et al., 2016; Chu et al., 2010; Tyler et al., 2009). MEK and COP1-binding motifs are also highlighted. The alignment was generated using MUSCLE in the Jalview software suite (Troshin, Procter, & Barton, 2011).

1.3 Production of recombinant pseudokinases

Meaningful analysis of pseudokinase domain-containing proteins requires purification of the complete polypeptide, which includes additional domains or intrinsically-disordered regions that may actively regulate pseudokinase conformations as well as the formation of intra- and intermolecular interactions with partner proteins. Many of the experimental approaches that drive this process begin with the decision to express the pseudokinase in a prokaryotic or eukaryotic cell-type, or both. This chapter is focused on prokaryotic expression of full-length human Tribbles 2 (TRIB2), although the vast majority of human pseudokinase domains have been expressed by exploiting a variety of bacterial or insect-cell sources (Murphy et al., 2014), success in generating large quantities of full-length pseudokinases using HEK-293T transfection systems has also been successful in our laboratory.

For prokaryotic-based expression, several initial considerations are required. These include gene sequence optimization and plasmid selection (particularly in regards to selection of affinity and/or solubility tags, which may include cleavage motifs for proteases) and need to be partnered with an appropriate host-strain and expression conditions, which may also need to be supplemented with quality-control elements such as co-factors or co-expression with molecular chaperones to facilitate proper folding of the target protein. Assuming that a correctly folded and stable target protein can be purified to homogeneity, this workflow can then be linked to biochemical analysis, including assessment of protein: protein interactions, enzymatic assays, and small-molecule inhibitor screens.

The selection of appropriate buffering conditions which maintain long-term protein stability is also critical for any protein purification strategy, and is the primary focus of this chapter. The basic approaches described here are guided by well-established screening platforms designed to identify optimal buffer conditions and aid downstream structural analysis (Vedadi et al., 2006) and have been crucial for the mechanistic characterization of TRIB1 and TRIB2 (Foulkes et al., 2018; Jamieson et al., 2018). As described below, monitoring the relative change of TRIB2 stability in the presence of varying pH and salt conditions, and common additives (including osmolytes and small molecules) provides useful information which can be exploited to improve yields of the purified pseudokinase. These methods remain highly cost-effective, and can be readily implemented in modestly equipped research laboratories (Fedorov, Niesen, & Knapp, 2012; Vedadi et al., 2006).

Our investigations build on a broad study in which the purification of a stable TRIB2 pseudokinase domain was made possible through comparative buffer screening (Vedadi et al., 2006). Subsequently, we generated full-length human TRIB2 with either an N-terminal 6His- or GST-tag (Bailey et al., 2015; Foulkes et al., 2018), and demonstrated that purified human TRIB2 was relatively stable in a buffer containing 20 mM Bicine, pH 9.0, 100 mM NaCl,1 mM DTT and 10% (*v*/v) glycine (Bailey et al., 2015). We also observed that the TRIB2 pseudokinase domain was stabilized by the presence of an intact C-terminal tail (Foulkes et al., 2018). The benefit of the approach described below is that it is highly reproducible between TRIB2 batches, and both rapid and simple, even for a relatively unstable protein like TRIB2.

As exemplified in Fig. 2A, we have established high level full-length TRIB2 expression in several commercial strains of *E. coli*. Notably, we regularly employ either BL21(DE3)pLysS, which minimizes 'leaky' expression of TRIB2, or co-express TRIB2 in the presence of the chaperone cluster dnaK-dnaJ-grpE groES-groEL encoded by BL21(DE3) stably transformed with pG-KJE8 (TaKaRa). Following a two step-purification procedure (affinity and size-exclusion chromatography (SEC)), both expression systems produce high yields of full-length human TRIB2 (~5 mg/L of culture medium). Furthermore, the elution profile of TRIB2 on an S200 SEC column is consistent with the predicted molecular weight of soluble monomeric protein (Fig. 2B, 'Peak 2'), and is >95% pure as judged by SDS-PAGE and Coomassie blue staining (Fig. 2C). The purified protein is stable over long-term storage at −80 °C, and is suitable for thermal denaturation analysis (often referred to as Differential Scanning Fluorimetry, DSF) as represented by the osmolyte-additive screen depicted in Fig. 3. A description of the purification (and mass spectrometric analysis) of the PKA catalytic subunit, which serves as a useful comparative control when studying TRIB2 due to the similar size of its kinase domain and flanking regions, is reported elsewhere (Byrne et al., 2016).

1.4 Ligand analysis of recombinant human Tribbles 2

In addition to the generation and purification of full-length TRIB2, two procedures for small-molecule screens using DSF are also described in this chapter. In the first, full-length purified TRIB2 is screened against a panel of different biochemical solutions (Fig. 3).

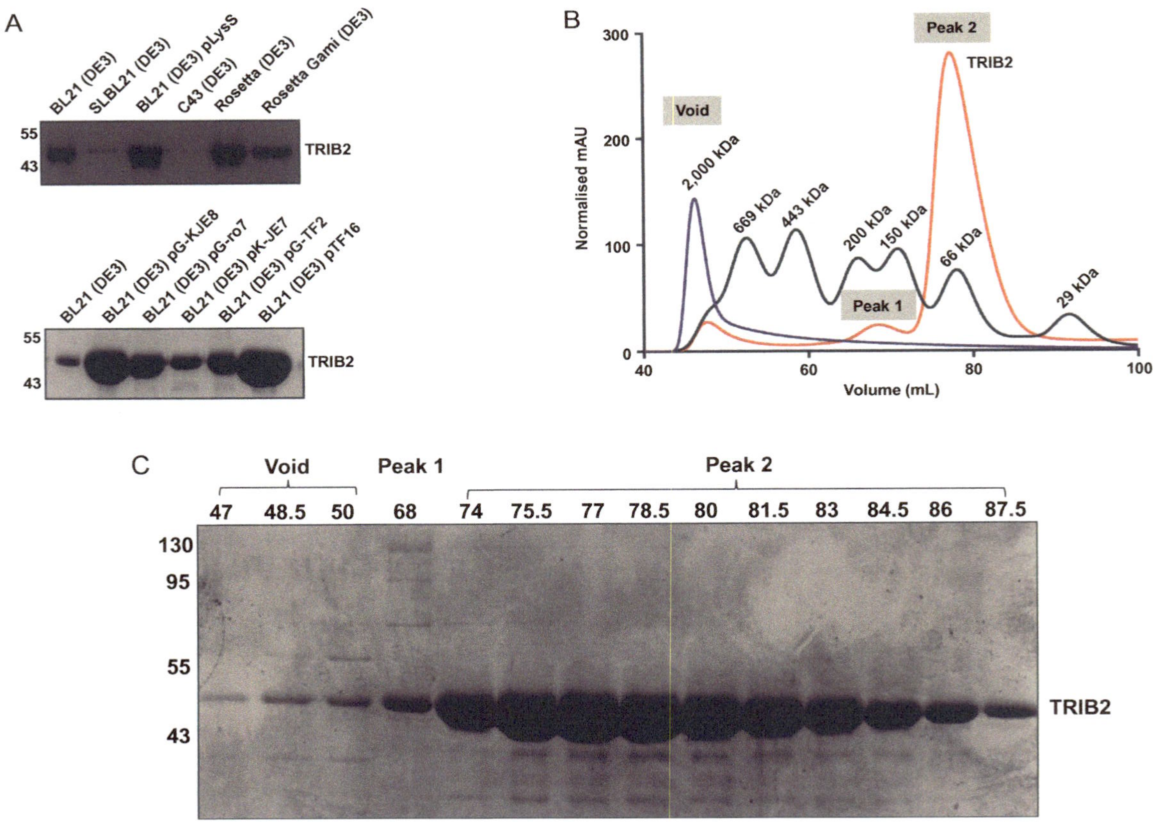

Fig. 2 See figure legend on opposite page.

The second protocol is designed to indirectly analyze binding to two types of ligands: physiologically-relevant nucleotides and metal cofactors, and a selection of previously known kinase inhibitors (Fig. 4). Several applications of these procedures have been reported by various groups for both kinases and pseudokinases, and detailed information can be obtained in these reports (Byrne, Clarke, et al., 2020; Byrne, Shrestha, et al., 2020;

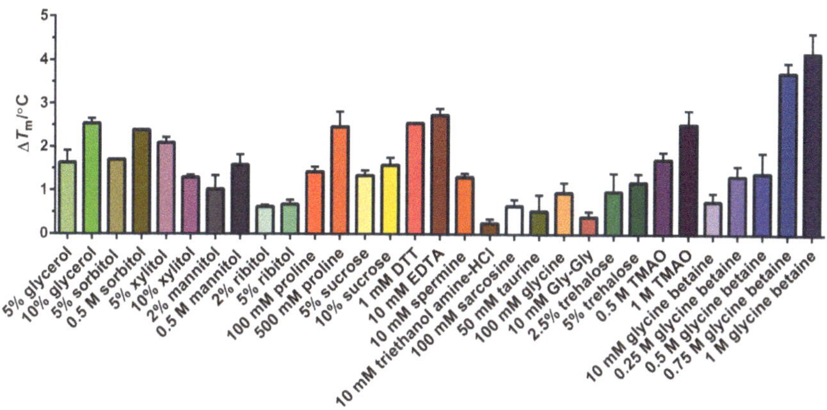

Fig. 3 TRIB2 thermal stability analysis using an osmolyte panel. Thermal shift analysis of 5 μM TRIB2 (purified as in Fig. 2) was assayed the presence of 20 mM Bicine pH 8.0, 0.1 M NaCl, and the indicated chemical additives. The relative change in melting temperature (T_m (°C)) was calculated by comparison with TRIB2 in buffer only, and is displayed as ΔT_m/°C.

Fig. 2 Recombinant human TRIB2 expression and purification. (A) Analysis of soluble TRIB2 protein expressed and purified from a variety of commercial *E. coli* strains. Expression of N-terminally tagged 6xHis-TRIB2 was induced using IPTG, and bacterial cells were collected by centrifugation. TRIB2 was initially affinity purified on a nickel column, and samples collected following elution with imidazole. Samples were separated by reducing SDS-PAGE followed by Coomassie Blue staining. (B) The elution profile of TRIB2 from BL21 (DE3)-pG-KJE8 generated by a HiLoad 16/600 Superdex 200 pg gel-filtration column (GE Healthcare) equilibrated with 20 mM Bicine pH 9.0, 0.1 M NaCl, 1 mM DTT and 10% (*v*/v) glycerol. TRIB2 elutes from the column as two major UV-absorbing peaks: a minor peak (Ve = 67–73 mL, labelled **1**) and a major peak (Ve = 79–90 mL, **2**). An aggregated TRIB2 species sometimes appears in the void (**V**) volume between 47 and 50 mL, and this can be minimized by performing all experimental procedures at 4 °C. Chromatograms for molecular weight markers are also represented; blue dextran (2000 kDa), thyroglobulin (669 kDa), apoferritin (443 kDa), β-amylase (200 kDa), alcohol dehydrogenase (150 kDa), serum albumin (66 kDa), and carbonic anhydrase (29 kDa). (C) TRIB2 containing fractions were collected following SEC and were separated by reducing SDS-PAGE and stained with Coomassie Blue. Monomeric TRIB2 with an N-terminal 6His tag is indicated, with a predicted molecular weight of 50 kDa.

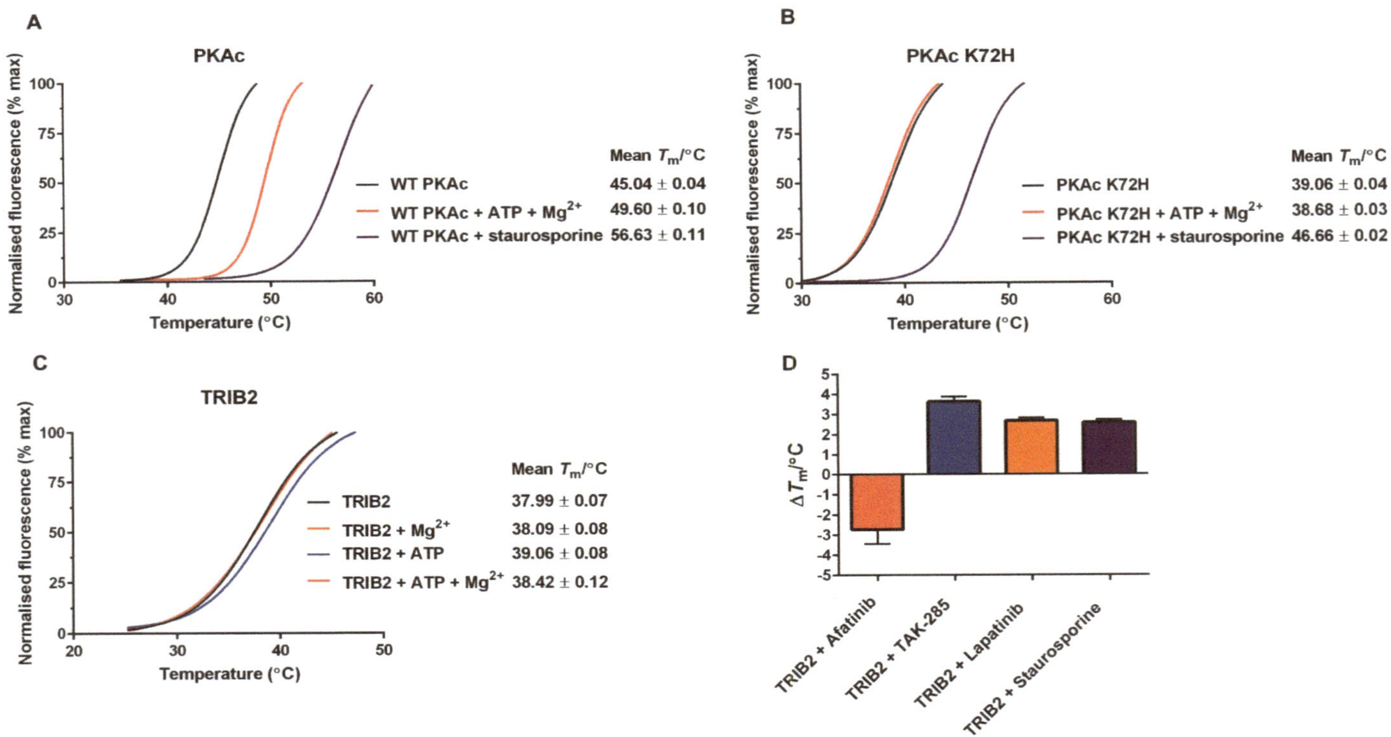

Fig. 4 Thermal stability analysis of TRIB2 nucleotide and inhibitor binding. (A) Catalytically-active full-length PKA catalytic domain was purified to homogeneity and subjected to DSF in the presence or absence ATP-Mg^{2+} or 1 mM staurosporine. (B) Catalytically-inactive full-length PKA catalytic domain (K72H) was purified to homogeneity and then subject to fluorimetry in the presence or absence ATP-Mg^{2+} or 1 mM staurosporine. (C) Full-length TRIB2 purified from BL21 (DE3) pG-KJE8 *E. coli* cells was analyzed in the presence and absence of the indicated nucleotide complexes. (D) Thermal shift analysis (presented as change in T_m value, ΔT_m) for TRIB2 in the presence of the destabilizing ligand afatinib or the stabilizing ligands TAK-285, lapatinib, or staurosporine (all 20 μM). All kinases and pseudokinases were assayed at 5 μM.

Elkins et al., 2016; Luis et al., 2021; Rudolf, Skovgaard, Knapp, Jensen, & Berthelsen, 2014). DSF measures the fluorescent emission from a commercial dye (Sypro Orange) upon interaction with hydrophobic surfaces in the protein that become exposed as it unfolds. Originally established in 1997 (Poklar, Lah, Salobir, Macek, & Vesnaver, 1997) and subsequently miniaturized into a higher-throughput microplate format (Pantoliano et al., 2001), these assays are both cheap and adaptable, and most purified kinases and pseudokinases can be analyzed using a range of common buffers (Murphy et al., 2014). The plot of fluorescence intensity vs temperature has a hyperbolic shape for a two-state unfolding mechanism (Fig. 4), As adopted in our lab, this method requires relatively small amounts (low μg) of purified polypeptide, and when employed in 96 or 384 well format, the assay can be set up, and data analyzed, in a matter of hours.

The *Tm* value (defined as the temperature at which there is an equilibrium between unfolded and folded protein and recognized as the midpoint of the protein melt transition) provides a reproducible marker of protein quality and intrinsic stability and most critically, can be adopted for screening to identify ligands that modulate *Tm* values. A *Tm* can be derived for most protein kinases, and their polypeptide complexes, the exception being highly thermostable proteins such as PKI, whose biophysical properties provide an additional opportunity for analysing its targets among conventional PKA holoenzyme complexes (Byrne et al., 2016; Smith et al., 2017). This simple approach is very attractive for pseudokinase research, partly because each kinase and pseudokinase possess a broad spectrum of nucleotide-affinities (for those that have been analyzed), which cannot necessarily predict the potential for interaction with non-physiological ligands, such as kinase inhibitors.

Perhaps most importantly, due to a lack of meaningful catalytic turnover, pseudokinases and catalytically-inactive versions of canonical kinases (including those that have been inactivated by mutation) are frequently not amenable to enzyme-based analysis, which typically employs direct phosphorylation of a proxy substrate, or quantification of ADP produced during hydrolysis of the ATP co-factor during catalysis. In this regard, there is increasing interest in the non-catalytic functions of kinases and pseudokinases, which are also drug targets due to the unique conformations that they adopt as part of physiological mechanisms (Mace & Murphy, 2021). In this context, differential phosphorylation can also exert effects on the conformational stability of protein kinases, as established for recombinant PKA, Aurora A, PLK4 and through mass-screening of cellular kinases in vitro (Byrne, Clarke, et al., 2020; Byrne, Shrestha, et al., 2020; Byrne et al., 2016; Tsuchiya et al., 2020) and in cells (Huang et al., 2019; Potel et al., 2020). The prevention of ATP binding through single amino acid substitution (e.g., K72H PKAc, Fig. 4B) is readily confirmed by thermal analysis, demonstrating that 'inactive' (and in this case, non-phosphorylated) kinases can possess *Tm* values that are similar to pseudokinases (indicating relatively low changes in thermal stability), while retaining binding to small molecule ligands such as staurosporine (Fig. 4B).

Of note, DSF analysis of TRIB2 isolated from a variety of bacterial strains reveals weak, but detectable, interaction with an excess of ATP, but only in the absence of metal ions (Bailey et al., 2015). Furthermore, two distinct binding modes for small molecule ligands were discovered for TRIB2 using the same assay. The first (destabilization) was evident for TRIB2 in the presence of covalent inhibitors such as Afatinib, whose acrylamide moiety modifies several of the Cys residues conserved in TRIB2 (Fig. 1A). This is predicted to physically uncouple the N-lobe from the C-tail docking peptide (Foulkes et al., 2018), which leads to overall TRIB2 destabilization (Fig. 4D). In contrast, classical ATP-competitive kinase inhibitors, such as Lapatinib and Staurosporine, induce reproducible stabilization, which is evidence of an interaction in the degraded nucleotide binding site (Fig. 4D). Importantly, translation of this screening information was subsequently linked to the ability of compounds such as Afatinib and Lapatinib to modulate TRIB2 turnover and signaling in cells. In the case of Afatinib, mutation of the Cys residues in the TRIB2 αC-helix abolishes the effects of the repurposed EGFR inhibitor, consistent with 'on-target' off-target binding (Foulkes et al., 2018). Similar screens have also established that TRIB1 is a target of small molecule kinase inhibitors (Jamieson et al., 2018), and together, these approaches have driven forward the biophysical analysis of these important pseudokinases.

1.5 Equipment

The following laboratory equipment is needed for the experimental procedures described.

- Table-top centrifuge capable of holding 1.5 mL plastic microcentrifuge tubes (e.g., Microcentrifuge 24, Starlab)
- Centrifuge that can be chilled to 4 °C and is capable of holding 1.5 mL Eppendorf tubes (e.g., Centrifuge 5427R, Eppendorf)
- Centrifuge capable of reaching 43,000 × *g* with temperature control and rotors capable of holding 1 L and 50 mL centrifuge tubes (e.g., Sorvall Lynx 6000 centrifuge with the rotors F10-4 × 1000 Lex and F21-8x50g, Thermoscientific).

- Centrifuge capable of holding microplate (e.g., Centrifuge 5810 with plate rotor S-4-104, Eppendorf)
- 1.5 mL Eppendorf or equivalent plastic microcentrifuge tubes
- Shaking incubator with temperature control—needs to hold 8 × 2 L flasks
- 8 × 2 L Erlenmeyer Flasks (standard or baffled)
- 250 mL conical flask (standard or baffled)
- Static bacterial incubator at 37 °C
- Spectrophotometer (e.g., Biowave 3 +, Biochrom WPA)
- Temperature adjustable water bath
- Plastic chromatography column with stand
- Automatic liquid chromatography system (e.g., ÄKTA pure, Cytiva)
- Superdex 200 (16/60) column (GE Healthcare Life Sciences, UK)
- Gel plates, stands, combs, electrophoresis system (Bio-Rad laboratories, UK)
- StepOnePlus Real-Time PCR System (Thermo Fisher, UK)
- 96-well clear microplate with adhesive cover

1.6 Chemicals

The following analytical grade chemicals can be found in most research laboratories. Our standard stock solutions are described. Antibiotics are stored at −20 °C.

1. 3 M Sodium chloride
2. 500 mM Bicine pH 9.0
3. 1 M Dithiothreitol (DTT)
4. 50% Glycerol
5. 100 mM Ethylenediaminetetraacetic acid (EDTA)
6. 100 mM *Ethylene glycol tetraacetic acid (EGTA)*
7. *50% Triton X-100*
8. Complete™ EDTA Protease Inhibitor tablets (Sigma-Aldrich, UK)
9. Chemically competent BL21 (DE3) pLysS *E. coli*
10. Super Optimal broth with Catabolite Repression (SOC) media
11. Lysogeny broth (LB) agar plates, containing 50 μg/mL kanamycin and 35 μg/mL chloramphenicol
12. 1 × 50 mL LB (broth), containing 50 μg/mL kanamycin and 35 μg/mL chloramphenicol
13. 8 × 800 mL LB (broth) held in 2 L Erlenmeyer flasks, containing 50 μg/mL kanamycin, 35 μg/mL chloramphenicol, 0.5 mg/mL L-arabinose, and 5 ng/mL tetracycline

14. 100 mM Isopropyl β-D-1-thiogalactopyranoside (IPTG)
15. Nickel-NTA resin
16. 1.5 M Tris·HCl pH 8.8 (20 °C)
17. 0.5 M Tris·HCl pH 6.8 (20 °C)
18. 10% (*w*/*v*) Ammonium persulfate
19. 30% Bis-acrylamide
20. 20% (w/v) sodium dodecyl sulfate (SDS) solution
21. 10 × SDS polyacrylamide gel electrophoresis (SDS-PAGE) running buffer: 25 mM Tris, 192 mM glycine, 3.5 mM SDS
22. *N′N′N′N′*-tetramethyl-ethylenediamine (TEMED)
23. 5 × Laemmli Sample Buffer: 250 mM Tris·HCl pH 6.8, 10% (w/v) SDS, 50% (*v*/v) glycerol, 0.5% (v/v) bromophenol blue, 500 mM DTT
24. Coomassie Blue staining solution: 20% (v/v) methanol, 10% (v/v) acetic acid, 0.1% Coomassie Blue R350
25. Destaining solution: 50% (v/v) methanol, 10% (v/v) acetic acid
26. Prestained molecular weight standard
27. 10 mM ATP solution
28. 50 mM Magnesium chloride
29. 1:1000 dilution of Sypro-Orange (Invitrogen, UK)
30. 200 μM stock Staurosporine (dissolved in DMSO) (Selleckchem)
31. 200 μM stock Afatinib (dissolved in DMSO) (LC labs)
32. 200 μM stock Lapatinib (dissolved in DMSO) (LC labs)
33. 200 μM TAK-285 (dissolved in DMSO) (Selleckchem)

2. Protocols

2.1 Expression of human full-length 6His-tagged TRIB2

1.1. Working on ice and using a sterilized 1.5 mL microfuge tube, add 1 μL of 10 ng/μL purified plasmid DNA to 50 μL BL21 (DE3) pLysS *E. coli* cells. Incubate on ice for 30 min. In this example, we are using TRIB2 cloned in to pET30.

1.2. Heat shock cells in microfuge tube in a 42 °C water bath for 30 s. Return to ice for 2 min.

1.3. Add 200 μL SOC media to the tube, and incubate at 37 °C in shaking incubator (240 rpm) for 30 min.

1.4. Evenly spread the transformation onto 10 cm^3 LB agar plate containing X antibiotics to select for the transformed plasmid. Incubate overnight at 37 °C in a static incubator.

1.5. Pick a colony from the agar plate using a sterile inoculation loop and add to a 250 mL conical flask containing 50 mL LB broth with appropriate antibiotics. Incubate overnight in a shaking incubator (~240 rpm) at 37 °C.

1.6. Add 5 mL of the overnight culture to 8 × 2 L Erlenmeyer flasks containing 800 mL LB broth supplemented with the appropriate antibiotics and incubate at 37 °C in a shaking incubator (~240 rpm).

1.7. Once OD_{600} reaches 0.6–0.8, remove flasks from incubator and chill in cold room (4 °C). Cool shaking incubator to 18 °C.

1.8. Once the shaking incubator reaches 18 °C, add 3 mL of 100 mM IPTG (final concentration 0.4 mM IPTG) to the bacterial culture and transfer to shaking incubator. Incubate at 18 °C for 16–18 h (~240 rpm).

1.9. Collect cells by centrifugation in a 1 L centrifuge tubes using an appropriate rotor (e.g., F10-4 × 1000 Lex rotor) at 7000 × *g*, 5 min at 4 °C. Discard the supernatant and decontaminate with a Virkon. Transfer the pellet to a 100 mL glass beaker and seal with parafilm. The pellet can immediately be snap-frozen on dry-ice or liquid N_2, and placed in freezer for short term storage at −20 °C (for up to a week).

2.2 Purification of human full-length 6His-tagged TRIB2

All steps should be performed on ice or at 4 °C, since full-length TRIB2 to improve protein yields and stability.

2.1. Prior to starting purification, equilibrate a Superdex 200 (16/60) column with 200 mL de-gassed, and filtered gel filtration buffer (50 mM Bicine pH 9.0, 10% (*v*/v) glycerol, 100 mM NaCl, and 1 mM DTT) at a flow-rate of 1 mL/min (0.5 mPa).

2.2. Resuspend cells in 10 mL ice-cold lysis buffer (50 mM Bicine pH 9.0, 150 mM NaCl, 1 mM imidazole, 100 μM EDTA, 100 μM EGTA, 10% glycerol, 1% Triton X-100, 1 mM DTT, and protease inhibitor). We typically employ 10 mL buffer per 1 g of wet cell pellet.

2.3. Lyse the cell suspension by pulsed sonication at an amplitude of 16 (30 s on, 1 min off) until homogenous lysate solution is obtained (5–10 cycles). To prevent excessive heating of the lysate, continuous sonication is not recommended and all sonication should be performed in an ice bath.

2.4. Transfer the lysates in to a 50 mL centrifuge tube and balance to within 0.5 g. Centrifuge lysates at 43000 × *g* for 60 min at 4 °C.

2.5. During the centrifugation step, collect and equilibrate a 1 mL packed volume of Nickel-NTA resin in plastic chromatography gravity column, by washing (×3) with 5 mL wash buffer (50 mM bicine pH 9.0, 500 mM NaCl, 2 mM imidazole).

2.6. After centrifugation, pass the supernatant through a 0.45 μm filter syringe and collect in a 100 mL glass beaker. This step is important to remove any residual aggregated material that could block the Ni-NTA column.

2.7. Resuspend the equilibrated Ni-NTA beads in the clarified cell lysate and incubate for 1 h with gentle agitation using a magnetic stirrer (4 °C).

2.8. Sequentially load the suspended bead-lysate mixture onto a reusable plastic chromatography column to collect the TRIB2-containing Ni-NTA beads. Collect the mobile (non-binding) phase as it passes through the column.

2.9. Sequentially wash the beads with ~50 mL wash buffer (50 mM bicine pH 9.0, 500 mM NaCl, 2 mM imidazole) in 5 mL increments to remove none specifically bound material.

2.10. Add elution buffer (50 mM Bicine pH 9.0, 150 mM NaCl, 10% glycerol, 0.5 M imidazole, and 1 mM DTT) to the column in 1 mL increments (5 ×), collecting 500 μl eluent fractions in labelled (1–8) 1.5 mL Eppendorf tubes. Analyze the fractions by SDS-PAGE to identify TRIB2 containing fractions.

2.11. In order to remove any aggregated material, centrifuge TRIB2-containing fractions for 20 min at 16000 × *g* (4 °C) using a chilled Eppendorf 5427R centrifuge. Pool the resulting supernatants together.

2.12. Load up to 3 mL TRIB2 eluate onto Superdex 200 (16/60) column using a 5 mL loading loop. Multiple loading or a concentration step may be necessary prior to this step if a larger volume of eluted TRIB2 is obtained in step 2.11.

2.13. Further purify the protein on an automatic liquid chromatography system at a flow-rate of 1 mL/min (0.5 mPa), collecting 1.5 mL fractions at 4 °C. If this step cannot be performed at 4 °C, promptly transfer eluted fractions to ice upon completion of the run to maintain protein stability. Analyze 10–20 μL of each fraction by running on a 12% SDS-PAGE gel (200 V, ~1 h), staining with Coomassie blue and compare to protein standard (e.g., Pageruler™ Prestained Protein Ladder, Thermofisher). The His-TRIB2 band should migrate at an apparent molecular mass of 45 kDa, above the 40 kDa standard.

Tip: TRIB2 can also be purified with an N-terminal GST-tag, which was used to confirm TRIB2 interaction with AKT in vitro (Foulkes et al., 2018).

Purification involves the same steps and buffers as described above, except glutathione resin is employed instead of Ni-NTA resin and the incubation step (2.7) is increased from 1–3 h.

2.3 Differential scanning fluorimetry (DSF) analysis

3.1. Thermal shift analysis is carried out on the StepOnePlus Real-Time PCR System using a validated thermal ramping procedure (Vedadi et al., 2006).

3.2. The assay is carried out in triplicate, in a final volume of 25 μL per well of 96-well clear PCR-compatible microplate.

3.3. For each reaction TRIB2 is assayed at a final concentration of 5 μM in 50 mM Bicine, 100 mM NaCl alongside the desired additive, and Sypro-Orange (final concentration: 1:10,000). Control reaction with no additive are also included. An example is shown below, where five conditions are being tested, each in triplicate—TRIB2 + 10 μM afatinib (in DMSO), 10 μM neratinib (in DMSO), 10 μM staurosporine (in DMSO), 10 μM Tak-285 (in DMSO), and DMSO control (final DMSO concentration in all assays is 4%):

Make up the following master mix (enough for 17 reactions).

- 2.5 μL of 500 mM Bicine pH 9.0 per well (42.5 μL for 17)
- 2.5 μL of 1 M NaCl (42.5 μL for 17)
- 5 μL of 25 μM TRIB2 (85 μL for 17)
- 2.5 μL of 50 × Sypro-Orange (42.5 μL for 17)
- Make up to 24 μL with nuclease-free water (195.5 μL for 17)

Prepare the assay plate by adding 1 μL of 25 × inhibitor to the appropriate well (for example, add 1 μL of 250 μM afatinib to first three wells, 1 μL of 250 μM neratinib to next three wells, and so on till all conditions complete).

Complete the assay set up by adding 24 μL of the TRIB2 containing master mix to each well.

3.4. Once the plates are prepared, seal with adhesive cover, mix briefly and centrifuge at 1000 × *g* for 1 min at room temperature (this can be carried out using the Centrifuge 5810 with plate rotor S-4-104, Eppendorf).

3.5. Set the StepOnePlus Real-Time PCR System to measure fluorescence at 530 nm, with the temperature ramping at 0.3 °C/min, between 25 and 95 °C.

3.6. Raw data is exported onto a portable USB stick, and imported as a Microsoft Excel file, then data normalized using GraphPad Prism software.

3.7. The normalized data are fitted to the Boltzmann equation to generate a sigmoidal denaturation curve. The melting temperature (T_m/°C) is the temperature at which 50% of the protein has unfolded. The change in melting temperature (ΔT_m) can be calculated by subtracting the control conditions from the matched experimental condition T_m. The mean ΔT_m can be calculated by averaging the ΔT_m across all the biological replicates (typically $n = 3$).

3. Summary

In this chapter, we provide an overview of some of the experimental workflows and assay protocols that have supported biochemical analysis of human TRIB2 over the last decade. TRIB2 is a relatively-unstable pseudokinase, especially when compared to the hundreds of kinase paralogues that are familiar to researchers and pharmaceutical companies. This is not unusual among pseudokinase domains, which have evolved to act as flexible molecular scaffolds and allosteric regulators rather than as phosphotransferases. The presence of N- and C-terminal regions outside the TRIB2 pseudokinase domain are well-tolerated when these proteins are expressed in bacteria, and further information about the use of various deletion constructs is available (Foulkes et al., 2018). A folded and ligand-responsive TRIB2 pseudokinase that binds to both its canonical kinase target AKT and a variety of small molecules can be prepared in a day or so, as long as all procedures take place using chilled buffers and a cold cabinet, with all samples stored on ice. This purification strategy is invaluable for the variety of downstream interaction, screening and assay procedures that have been developed for TRIB2 (Bailey et al., 2015; Foulkes et al., 2018), and is also suited to biophysical analysis of the complete TRIB2 polypeptide in complexes with peptides, allosteric partners or small molecules.

Acknowledgments

We thank Daniel Foulkes for TRIB2 reagents, the MRC DiMEN programme for DTP studentships (project 1961582, to E.F. and project 2274741, to J.A.H.), the Royal Society, BBSRC (BB/S018514/1) and the NIH (1U01CA239106-01) *for funding to supports our kinase and pseudokinase research programmes. We are happy to supply TRIB2 and PKA plasmids, proteins and relevant antibodies upon reasonable request.*

References

Bailey, F. P., Byrne, D. P., Oruganty, K., Eyers, C. E., Novotny, C. J., Shokat, K. M., et al. (2015). The tribbles 2 (TRB2) pseudokinase binds to ATP and autophosphorylates in a metal-independent manner. *Biochemical Journal, 467*(1), 47–62.

Berginski, M. E., Moret, N., Liu, C., Goldfarb, D., Sorger, P. K., & Gomez, S. M. (2020). The dark kinase knowledgebase: An online compendium of knowledge and experimental results of understudied kinases. *Nucleic Acids Research, 49*(D1), D529–D535.

Boudeau, J., Miranda-Saavedra, D., Barton, G. J., & Alessi, D. R. (2006). Emerging roles of pseudokinases. *Trends in Cell Biology, 16*(9), 443–452.

Burke, J. R., Cheng, L., Gillooly, K. M., Strnad, J., Zupa-Fernandez, A., Catlett, I. M., et al. (2019). Autoimmune pathways in mice and humans are blocked by pharmacological stabilization of the TYK2 pseudokinase domain. *Science Translational Medicine, 11*(502), eaaw1736.

Byrne, D. P., Clarke, C. J., Brownridge, P. J., Kalyuzhnyy, A., Perkins, S., Campbell, A., et al. (2020). Use of the polo-like kinase 4 (PLK4) inhibitor centrinone to investigate intracellular signalling networks using SILAC-based phosphoproteomics. *Biochemical Journal, 477*(13), 2451–2475.

Byrne, D. P., Foulkes, D. M., & Eyers, P. A. (2017). Pseudokinases: Update on their functions and evaluation as new drug targets. *Future Medicinal Chemistry, 9*(2), 245–265.

Byrne, D. P., Shrestha, S., Galler, M., Cao, M., Daly, L. A., Campbell, A. E., et al. (2020). Aurora a regulation by reversible cysteine oxidation reveals evolutionarily conserved redox control of Ser/Thr protein kinase activity. *Science Signaling, 13*(639), eaax2713.

Byrne, D. P., Vonderach, M., Ferries, S., Brownridge, P. J., Eyers, C. E., & Eyers, P. A. (2016). cAMP-dependent protein kinase (PKA) complexes probed by complementary differential scanning fluorimetry and ion mobility-mass spectrometry. *The Biochemical Journal, 473*(19), 3159–3175.

Caron, D., Byrne, D. P., Thebault, P., Soulet, D., Landry, C. R., Eyers, P. A., et al. (2016). Mitotic phosphotyrosine network analysis reveals that tyrosine phosphorylation regulates polo-like kinase 1 (PLK1). *Science Signaling, 9*(458), rs14.

Chu, M. L., Lang, Z., Chavas, L. M., Neres, J., Fedorova, O. S., Tabernero, L., et al. (2010). Biophysical and X-ray crystallographic analysis of Mps1 kinase inhibitor complexes. *Biochemistry, 49*(8), 1689–1701.

Durzynska, I., Xu, X., Adelmant, G., Ficarro, S. B., Marto, J. A., Sliz, P., et al. (2017). STK40 is a pseudokinase that binds the E3 ubiquitin ligase COP1. *Structure, 25*(2), 287–294.

Elkins, J. M., Fedele, V., Szklarz, M., Abdul Azeez, K. R., Salah, E., Mikolajczyk, J., et al. (2016). Comprehensive characterization of the published kinase inhibitor set. *Nature Biotechnology, 34*(1), 95–103.

Eyers, P. A., Keeshan, K., & Kannan, N. (2017). Tribbles in the 21st century: The evolving roles of tribbles pseudokinases in biology and disease. *Trends in Cell Biology, 27*(4), 284–298.

Eyers, P. A., & Murphy, J. M. (2013). Dawn of the dead: Protein pseudokinases signal new adventures in cell biology. *Biochemical Society Transactions, 41*(4), 969–974.

Fedorov, O., Niesen, F. H., & Knapp, S. (2012). Kinase inhibitor selectivity profiling using differential scanning fluorimetry. *Methods in Molecular Biology, 795*, 109–118.

Foulkes, D. M., Byrne, D. P., Yeung, W., Shrestha, S., Bailey, F. P., Ferries, S., et al. (2018). Covalent inhibitors of EGFR family protein kinases induce degradation of human tribbles 2 (TRIB2) pseudokinase in cancer cells. *Science Signaling, 11*(549), eaat7951.

Grandinetti, K. B., Stevens, T. A., Ha, S., Salamone, R. J., Walker, J. R., Zhang, J., et al. (2011). Overexpression of TRIB2 in human lung cancers contributes to tumorigenesis through downregulation of C/EBPα. *Oncogene, 30*(30), 3328–3335.

Guo, S., Chen, Y., Yang, Y., Zhang, X., Ma, L., Xue, X., et al. (2021). TRIB2 modulates proteasome function to reduce ubiquitin stability and protect liver cancer cells against oxidative stress. *Cell Death & Disease*, *12*(1), 42.

Hill, R., Madureira, P. A., Ferreira, B., Baptista, I., Machado, S., Colaço, L., et al. (2017). TRIB2 confers resistance to anti-cancer therapy by activating the serine/threonine protein kinase AKT. *Nature Communications*, *8*(1), 14687.

Hou, Z., Guo, K., Sun, X., Hu, F., Chen, Q., Luo, X., et al. (2018). TRIB2 functions as novel oncogene in colorectal cancer by blocking cellular senescence through AP4/p21 signaling. *Molecular Cancer*, *17*(1), 172.

Huang, J. X., Lee, G., Cavanaugh, K. E., Chang, J. W., Gardel, M. L., & Moellering, R. E. (2019). High throughput discovery of functional protein modifications by hotspot thermal profiling. *Nature Methods*, *16*(9), 894–901.

Jacobsen, A. V., & Murphy, J. M. (2017). The secret life of kinases: Insights into non-catalytic signalling functions from pseudokinases. *Biochemical Society Transactions*, *45*(3), 665–681.

Jamieson, S. A., Ruan, Z., Burgess, A. E., Curry, J. R., McMillan, H. D., Brewster, J. L., et al. (2018). Substrate binding allosterically relieves autoinhibition of the pseudokinase TRIB1. *Science Signaling*, *11*(549), eaau0597.

Kung, J. E., & Jura, N. (2019). Prospects for pharmacological targeting of pseudokinases. *Nature Reviews. Drug Discovery*, *18*(7), 501–526.

Kwon, A., Scott, S., Taujale, R., Yeung, W., Kochut, K. J., Eyers, P. A., et al. (2019). Tracing the origin and evolution of pseudokinases across the tree of life. *Science Signaling*, *12*(578), eaav3810.

Lucet, I. S., Babon, J. J., & Murphy, J. M. (2013). Techniques to examine nucleotide binding by pseudokinases. *Biochemical Society Transactions*, *41*(4), 975–980.

Luis, A. S., Jin, C., Pereira, G. V., Glowacki, R. W. P., Gugel, S. R., Singh, S., et al. (2021). A single sulfatase is required to access colonic mucin by a gut bacterium. *Nature*, *598*(7880), 332–337.

Mace, P. D., & Murphy, J. M. (2021). There's more to death than life: Noncatalytic functions in kinase and pseudokinase signaling. *The Journal of Biological Chemistry*, *296*, 100705.

Manning, G., Whyte, D. B., Martinez, R., Hunter, T., & Sudarsanam, S. (2002). The protein kinase complement of the human genome. *Science*, *298*(5600), 1912–1934.

Molina, D., Jafari, R., Ignatushchenko, M., Seki, T., Larsson, A., Dan, C., et al. (2013). Monitoring drug target engagement in cells and tissues using the cellular thermal shift assay. *Science (New York, N.Y.)*, *Vol. 341*, 84–87.

Murphy, J. M., Mace, P. D., & Eyers, P. A. (2017). Live and let die: Insights into pseudoenzyme mechanisms from structure. *Current Opinion in Structural Biology*, *47*, 95–104.

Murphy, J. M., Zhang, Q., Young, S. N., Reese, M. L., Bailey, F. P., Eyers, P. A., et al. (2014). A robust methodology to subclassify pseudokinases based on their nucleotide-binding properties. *The Biochemical Journal*, *457*(2), 323–334.

Pantoliano, M. W., Petrella, E. C., Kwasnoski, J. D., Lobanov, V. S., Myslik, J., Graf, E., et al. (2001). High-density miniaturized thermal shift assays as a general strategy for drug discovery. *Journal of Biomolecular Screening*, *6*(6), 429–440.

Poklar, N., Lah, J., Salobir, M., Macek, P., & Vesnaver, G. (1997). pH and temperature-induced molten globule-like denatured states of equinatoxin II: A study by UV-melting, DSC, far- and near-UV CD spectroscopy, and ANS fluorescence. *Biochemistry*, *36*(47), 14345–14352.

Potel, C. M., Kurzawa, N., Becher, I., Typas, A., Mateus, A., & Savitski, M. M. (2020). Impact of phosphorylation on thermal stability of proteins. *Nature Methods*, *18*(7), 757–759.

Pushpakom, S., Iorio, F., Eyers, P. A., Escott, K. J., Hopper, S., Wells, A., et al. (2019). Drug repurposing: Progress, challenges and recommendations. *Nature Reviews. Drug Discovery*, *18*(1), 41–58.

Ribeiro, A. J. M., Das, S., Dawson, N., Zaru, R., Orchard, S., Thornton, J. M., et al. (2019). Emerging concepts in pseudoenzyme classification, evolution, and signaling. *Science Signaling*, *12*(594), eaat9797.

Rudolf, A. F., Skovgaard, T., Knapp, S., Jensen, L. J., & Berthelsen, J. (2014). A comparison of protein kinases inhibitor screening methods using both enzymatic activity and binding affinity determination. *PLoS One*, *9*(6), e98800.

Scheeff, E. D., Eswaran, J., Bunkoczi, G., Knapp, S., & Manning, G. (2009). Structure of the pseudokinase VRK3 reveals a degraded catalytic site, a highly conserved kinase fold, and a putative regulatory binding site. *Structure*, *17*(1), 128–138.

Seher, T. C., & Leptin, M. (2000). Tribbles, a cell-cycle brake that coordinates proliferation and morphogenesis during Drosophila gastrulation. *Current Biology*, *10*(11), 623–629.

Shrestha, S., Byrne, D. P., Harris, J. A., Kannan, N., & Eyers, P. A. (2020). Cataloguing the dead: Breathing new life into pseudokinase research. *The FEBS Journal*, *287*(19), 4150–4169.

Smith, F. D., Esseltine, J. L., Nygren, P. J., Veesler, D., Byrne, D. P., Vonderach, M., et al. (2017). Local protein kinase A action proceeds through intact holoenzymes. *Science*, *356*(6344), 1288–1293.

Troshin, P. V., Procter, J. B., & Barton, G. J. (2011). Java bioinformatics analysis web services for multiple sequence alignment—JABAWS:MSA. *Bioinformatics*, *27*(14), 2001–2002.

Tsuchiya, Y., Byrne, D. P., Burgess, S. G., Bormann, J., Baković, J., Huang, Y., et al. (2020). Covalent Aurora a regulation by the metabolic integrator coenzyme a. *Redox Biology*, *28*, 101318.

Tyler, R. K., Chu, M. L., Johnson, H., McKenzie, E. A., Gaskell, S. J., & Eyers, P. A. (2009). Phosphoregulation of human Mps1 kinase. *The Biochemical Journal*, *417*(1), 173–181.

Vedadi, M., Niesen, F. H., Allali-Hassani, A., Fedorov, O. Y., Finerty, P. J., Jr., Wasney, G. A., et al. (2006). Chemical screening methods to identify ligands that promote protein stability, protein crystallization, and structure determination. *Proceedings of the National Academy of Sciences of the United States of America*, *103*(43), 15835–15840.

Wilson, L. J., Linley, A., Hammond, D. E., Hood, F. E., Coulson, J. M., MacEwan, D. J., et al. (2018). New perspectives, opportunities, and challenges in exploring the human protein kinome. *Cancer Research*, *78*(1), 15–29.

Zanella, F., Renner, O., García, B., Callejas, S., Dopazo, A., Peregrina, S., et al. (2010). Human TRIB2 is a repressor of FOXO that contributes to the malignant phenotype of melanoma cells. *Oncogene*, *29*, 2973–2982.

CHAPTER FIVE

Purification, crystallization and drug screening of the IRAK pseudokinases

Sven M. Lange* and Yogesh Kulathu

MRC Protein Phosphorylation and Ubiquitylation Unit, Sir James Black Centre, Dundee, Scotland, United Kingdom

*Corresponding author: e-mail address: slange001@dundee.ac.uk

Contents

Abstract

Pseudokinases are emerging as critical components of cell signaling pathways. Consequently, the ability to obtain large quantities of pure protein for structural characterization and drug discovery efforts has become essential for the study of these proteins. Small molecules binding to pseudokinases may induce allosteric changes and serve as valuable tools to study the physiological roles of these "dead" enzymes. The IRAK family of kinases are key components of the innate immune response and the active IRAK family members, IRAK-1 and -4, have been extensively studied. However, the other two IRAKs, IRAK-2 and IRAK-3, are classified as pseudokinases and their detailed functions and roles remain to be described. In this chapter, we present comprehensive protocols for the purification of IRAKs, the crystallization of the pseudokinase domain of IRAK3, and a high-throughput drug screening pipeline using thermal shift and biolayer-interferometry assays to identify small molecule binders.

Methods in Enzymology, Volume 667

ISSN 0076-6879

https://doi.org/10.1016/bs.mie.2022.03.026

1. Introduction

The Toll-like receptors (TLRs) and interleukin-1 receptors (IL-1R) are critical components of the innate immune system that recruit the adapter protein MyD88 (myeloid differentiation primary response gene 88), which serves as an assembly platform for members of the interleukin-1 receptor associated kinase (IRAK) family to form the Myddosome, an oligomeric signaling complex (Lin, Lo, & Wu, 2010; Motshwene et al., 2009). Subsequently, IRAKs recruit and activate several E3 ubiquitin ligases that trigger downstream signaling events to induce the production of pro- and anti-inflammatory mediators (Cohen, 2014; Cohen & Strickson, 2017). The emergence of crystal structures and development of specific inhibitors of the catalytically active IRAK family members, IRAK1 and IRAK4, have advanced our understanding of their roles in the TLR/IL-1R signaling pathway (Ferrao et al., 2014; Kuglstatter et al., 2007; Wang et al., 2017). The other two IRAK family members, IRAK2 and IRAK3, are classified as pseudokinases, as they both lack residues in the kinase active site that are essential for phosphorylation in canonical kinases. IRAK3 lacks the Mg^{2+}-binding Asp in the DFG, and IRAK2 lacks both this Asp and the catalytic Asp of the HRD motif. However, both pseudokinases retain the ability to bind ATP, although IRAK3 is a weak binder with ATP binding affinity in the high micromolar range (Lange, Nelen, Cohen, & Kulathu, 2021).

Extensive research has revealed that pseudokinases exert their effects on signal transduction through unusual means. For instance, some pseudokinases retain catalytic activity through non-canonical phosphoryl-transfer reaction mechanisms (Bailey et al., 2015; Min, Lee, Cobb, & Goldsmith, 2004; Mukherjee et al., 2008; Shi, Telesco, Liu, Radhakrishnan, & Lemmon, 2010; Villa et al., 2009; Xu et al., 2000), or have even developed entirely different enzymatic activities (Black et al., 2019; Sreelatha et al., 2018). In contrast, other pseudokinases have also emerged with new non-enzymatic functions, for example, as scaffolding proteins or as allosteric regulators of active kinases (Lavoie et al., 2018; Scheeff, Eswaran, Bunkoczi, Knapp, & Manning, 2009; Zeqiraj, Filippi, Deak, Alessi, & van Aalten, 2009). Despite these advances of our understanding of pseudokinase mechanisms, the roles of the two IRAK pseudokinases in innate immune signaling still remain largely unknown. IRAK2 has been shown to operate redundantly with IRAK1 during early TLR signaling and to only become rate-limiting in the late phase of signaling (2–8 h) in mouse bone marrow-derived macrophages (Pauls et al., 2013). IRAK3, previously known as IRAK-M, has been proposed as a negative regulator of innate immune signaling based on enhanced inflammatory responses and increased susceptibility to LPS-induced septic shock in IRAK3−/− mice (Kobayashi et al., 2002). The recent emergence of a crystal structure of the pseudokinase domain of IRAK3 has revealed that mutations in IRAK3 associated with Early-onset asthma cluster at an evolutionarily conserved surface on IRAK3 (Balaci et al., 2007; Lange, Nelen, Cohen, & Kulathu, 2021). This conserved surface on IRAK3 superimposes well with the surface utilized by IRAK4 for homodimerization during the IRAK4 trans-autophosphorylation reaction and a model for the negative regulation of IRAK4 through binding by IRAK3 has been proposed (Lange, Nelen, Cohen, & Kulathu, 2021).

In this chapter, we describe detailed protocols to produce milligram amounts of recombinant IRAK pseudokinase proteins, for the crystallization of IRAK3 and for the high-throughput small molecule binding assays with IRAK2 and IRAK3. We outline two approaches to assay small molecule binding in a high-throughput manner, thermal shift assay and biolayer interferometry (BLI). In thermal shift assays, the protein of interest is slowly heated in the presence of a hydrophobic dye and upon denaturation of the protein, the dye binds to the exposed hydrophobic residues, which leads to an increase in the fluorescence signal. Small molecules that bind and thereby stabilize the protein of interest can lead to a shift in the thermal denaturation and serve as a qualitative measure in high-throughput binding screens (Cummings, Farnum, & Nelen, 2006). Of note, this method has also been used to assess the nucleotide and kinase inhibitor binding properties of a wide range of pseudokinases (Murphy et al., 2014). In addition to thermal shift assays, we present BLI as a quantitative secondary screen to further characterize small molecule hits identified in the primary thermal shift assay. The BLI signal is generated by the interference pattern of white light reflected from an internal reference layer and the sample surface of a sensor tip (Abdiche, Malashock, Pinkerton, & Pons, 2008). Binding of small molecules to immobilized proteins on the biosensor are detected in real time in 96-well or 384-well microplate formats and this high-throughput capability of BLI is a significant advantage over similar techniques such as surface plasmon resonance (SPR).

2. Before you begin

Timing: ∼3 weeks.

The pseudokinase domains of IRAK2 and IRAK3 express well in insect cell cultures. Generally, the expression of proteins in insect cells requires the integration of the expression cassettes into a suitable baculoviral genome, followed by initial generation and subsequent amplification of baculovirus stocks. We routinely use the DH10EMBacY *E. coli* strain that contains a YFP co-expression reporter on the baculoviral genome to monitor the infection and expression efficiencies with a standard GFP-fluorescence cell counter. Detailed background information and procedures for the generation of baculoviruses and protein expression in insect cells can be found in the MultiBac Expression System User Manual (Berger & Craig, 2010) and elsewhere in this volume (see chapters "Co-expression of recombinant RIPK3:MLKL complexes using the baculovirus-insect cell system" by Murphy et al. and "Production and Purification of the PEAK pseudokinases for structural and functional studies" by Lucet et al.). The basic steps and timings for the protein expression from insect cells are:

1. IRAK2 and/or IRAK3 expression plasmids can be obtained from MRC PPU Reagents (https://mrcppureagents.dundee.ac.uk):
   a) 6His-TEV-GST-3C-IRAK2 192–507 (DU51963),
   b) 6His-TEV-GST-3C-IRAK3 145–454 (DU56013).

   Note: The here described expression and purification protocols of IRAK pseudokinase domains can be directly applied to the kinase domains of IRAK1 and IRAK4, or for IRAK-1, -2 and -3 constructs that include the C-terminal TRAF6-binding region. Relevant expression plasmids are available from the MRC PPU Reagents plasmid catalog.
2. Transform DH10EMBacY *E. coli* with expression plasmids and blue/white selection to identify clones with successfully integrated expression cassettes (2–3 days).
3. Extract bacmid DNA and transfect insect cells (2 days).
4. Generate initial baculovirus generation (P0 stock) (1 week).
5. Amplify baculovirus (P1 stock) (3–4 days).
6. Set up large scale ($1-3\,L$) insect cell expression cultures (3–4 days).

3. Key resources table

Reagent or Resource	Source	Identifier
Bacterial and Virus Strains		
DH10EMBacY	Geneva Biotech	N/A
Chemicals, Peptides, and Recombinant Proteins		
4-(2-aminoethyl) benzene sulphonyl fluoride hydrochloride (AEBSF)	Apollo Scientific	Cat#BIMB2003

—cont'd

Reagent or Resource	Source	Identifier
Benzamidine	Sigma Aldrich	Cat#B6506
Dithiothreitol	Melford	Cat#D11000
Sf-900 II SFM insect cell medium	Thermo Fisher Scientific	Cat#10902096
Antibiotic/Antimycotic solution (100 ×)	Thermo Fisher Scientific	Cat#15240062
GSH 4B Sepharose	GE Healthcare	Cat#17-0756-01
Ni-NTA agarose	Qiagen	Cat#30210
Glycerol	VWR	Cat#24388.320
SYPRO orange (5000 ×)	Sigma-Aldrich	Cat#S5692
4-(2-Hydroxyethyl) piperazine-1-ethanesulphonic acid (HEPES)	Sigma-Aldrich	Cat#H3375
β-Mercaptoethanol (BME)	Sigma-Aldrich	Cat#M3148
Magnesium chloride	Sigma-Aldrich	Cat#M1028
Seed Bead	Hampton Research	Cat#HR2–320
Seeding Tool	Hampton Research	Cat#HR8–133
BirA	MRC PPU reagents	Cat#DU43486
Imidazole	Sigma-Aldrich	Cat#56749
NuPAGE SDS sample loading buffer 4 ×	Thermo Fisher	Cat#NP0007
DMSO	Sigma-Aldrich	Cat#D8418
Ethyl mercury phosphate	Hampton Research	Cat#HR2-446-02
AmSO4 Suite crystallization screen	Qiagen	Cat#130905
Sodium citrate tribasic dihydrate	Sigma-Aldrich	Cat#C8532
Ammonium sulfate	Sigma-Aldrich	Cat#A4418
Biocytin	Sigma-Aldrich	Cat#B4261
Streptavidin	IBA	Cat#2–0203-010
Pierce Nuclease	Thermo Fisher	Cat#88701
SYPRO orange, 5000 ×	Sigma-Aldrich	Cat#S5692

Continued

—cont'd

Reagent or Resource	Source	Identifier
Staurosporine	Apexbio	Cat#A8192
Silicone oil	Sigma-Aldrich	Cat#378348
ATP	Apollo Scientific	Cat#BIB3003
D-Biotin	Apollo Scientific	Cat#BIB0603
Experimental Models: Cell Lines		
Sf21 insect cells	Invitrogen	Cat#11497013
Recombinant DNA		
6His-TEV-GST-3C-IRAK2 192–507	MRC PPU Reagents	Cat#DU51963
6His-TEV-GST-3C-IRAK3 145–454	MRC PPU Reagents	Cat#DU56013
Software and Algorithms		
Octet Data Analysis Software	ForteBio	NA
DSF analysis v3.0′ excel sheet	SGC Oxford	(Niesen, Berglund, & Vedadi, 2007)
Prism	GraphPad	v8.0

4. Materials and equipment

- Buffer A (ice-cold): 50 mM HEPES pH 7.5, 500 mM NaCl, 20% Glycerol, 5 mM β-Mercaptoethanol (BME)
- Lysis buffer (ice-cold): Buffer A supplemented with 1 mM AEBSF, 1 mM Benzamidine, 20 ppm Pierce Nuclease
- SEC buffer: 20 mM HEPES pH 7.5, 500 mM NaCl, 10% glycerol, 5 mM DTT
- IRAK3 crystallization cocktail: 0.1 M sodium citrate pH 5.0, 1.62 M ammonium sulfate
- BLI assay buffer: 50 mM HEPES pH 7.5, 300 mM NaCl, 0.001% Tween-20, 2 mM BME, 1 mg/mL BSA
- Glass tissue homogenizer (e.g., KIMBLE, Dounce tissue grinder set Cat#D0189)
- Amicon Ultra-15 10 kDa MWCO (Merck)

- Cell counter with GFP fluorescence channel (e.g., Denovix, CellDrop FL)
- Hanging drop crystallization plates (e.g., Hampton Research, 48-Well VDX Plate Greased)
- Sitting drop crystallization plates (e.g., Swissci MRC 2 Lens plate Cat#MRC96T-UVP)
- White 384-well low-volume PCR microplates (e.g., BioRad, Cat#HSP3805)
- Echo 550 acoustic liquid handler (Labcyte)
- Super-streptavidin biosensor tips (ForteBio)
- Octet Red384 system (ForteBio)
- Thermomixer (e.g., Eppendorf, Thermomixer comfort)
- PD-10 desalting columns (Cytiva, Cat#17-0851-01)
- FPLC system (e.g., Cytiva, AKTA pure 25, Cat#29-0211-97)
- HiLoad 16/600 Superdex 75 pg (Cytiva, Cat#28-9893-33)

5. Step-by-step method details

5.1 Purification of IRAK pseudokinase domains

Timing: ~2 days.

Tandem-affinity purification of IRAK pseudokinase domain containing proteins followed by size-exclusion chromatography (SEC).

1. Check optimal timing for harvesting expression culture by determining the fraction of YFP reporter positive cells with a fluorescent cell counter using standard GFP detection channel (i.e., >70% YFP positive).
2. Spin down large-scale insect cell expression cultures at 2500 × g and 4 °C for 30 min
3. Decant supernatant and resuspend cells in 40 mL lysis buffer per 1 L expression culture, flash freeze cell suspension in liquid nitrogen and store at −80 °C till further use

   Critical: Use ice-cold buffers and work on ice or at 4 °C.
4. Thaw cell pellets in a room temperature water bath and lyse cells on ice by 30 pushes in a glass tissue homogenizer
5. Clarify lysate by centrifugation at 40,000 × g and 4 °C for 40 min
   a) Successful lysis can be detected through bright yellow YFP in the clarified lysate and absence of yellow color in the debris pellet
6. Meanwhile, equilibrate 1 ml 50% Ni-NTA agarose slurry with buffer A in a gravity flow column

7. Incubate equilibrated Ni-NTA resin with clarified lysate for 30 min on ice, gently resuspending the resin every 5–10 min

   Note: In our experience, constant agitation of the resin by rolling or end-over-end rotation leads to decreases in protein yield and quality.
8. Pour the suspension into a gravity flow column and wash with 5 × 50 mL buffer A
9. Elute with 5 × 1 mL buffer A supplemented with 300 mM Imidazole (pH 8).
10. Equilibrate 2 mL 50% slurry GSH-4B Sepharose with buffer A
11. Incubate equilibrated GSH resin with pooled Ni-NTA elution fraction for 1 h on ice, gently resuspending the resin every 10 min
12. Pour the suspension into a gravity flow column and wash with 10 × 50 mL buffer A
13. Close gravity flow column outflow and add 5 mL buffer A supplemented with ~100 μg of GST-fusion PreScission/3C-protease
14. Carefully agitate resin to mix and incubate overnight (~16 h) at 4 °C
15. Elute cleaved protein and rinse resin with 3 × 1 mL to collect residual protein
16. Concentrate protein in 10 kDa MWCO centrifugal filter units to ~1.5 mL

   Note: The following size-exclusion chromatography step is recommended prior to crystallization.
17. Meanwhile, equilibrate Superdex 75 16/600 pg column with SEC buffer
18. Transfer concentrated protein to a 1.5 mL Eppendorf tube and centrifuge at 17,000 × *g* and 4 °C for 10 min
19. Inject protein on equilibrated Superdex 75 16/600 pg column with a flow rate of 1 mL/min
20. Collect fractions of 1.8 ml and analyze by SDS-PAGE

   Note: IRAK3 exists in a concentration-dependent equilibrium of monomeric and dimeric species, therefore a doublet of peaks is observed at higher protein concentrations and fractions from the two peaks may be pooled (**Figure 1**).
21. Concentrate peak fractions in a 10 MWCO centrifugal filter unit to ~8 mg/mL
22. Prepare aliquots and flash-freeze in liquid nitrogen, store at −80 °C for future use

Pause Point: Frozen protein aliquots can be stored at −80 °C for at least 1 year without sample deterioration.

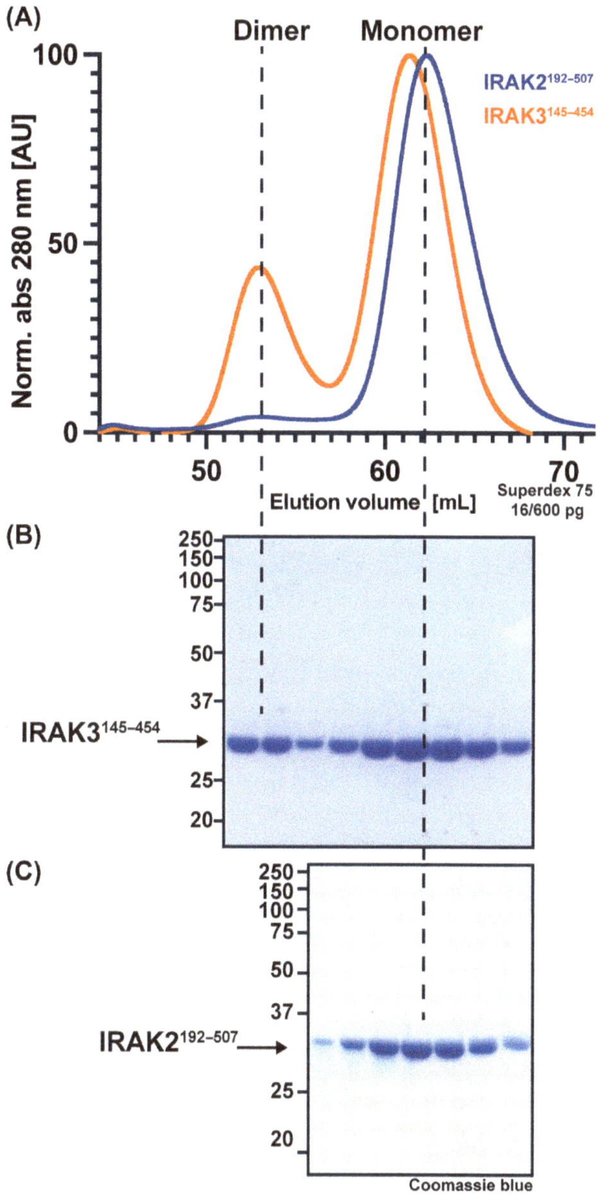

Fig. 1 Purification of IRAK2 and IRAK3 pseudokinase domains. (A) Elution profiles from size-exclusion chromatography (SEC) with IRAK2$^{192-507}$ (*blue/dark*) and IRAK3$^{145-454}$ (*orange/light*). (B and C) SDS-PAGE of SEC elution fractions of IRAK-3 and -2, respectively. *Dashed lines* indicate positions of dimeric and monomeric elution peaks. Molecular mass markers are in kDa.

5.2 Crystallization of the IRAK3 pseudokinase domain

Timing: 1–2 weeks.

Protocol to obtain initial crystals of IRAK3 pseudokinase domain (res. 145–454) and optimization crystal growth by macroseeding. The crystal structure of the IRAK3 pseudokinase homodimer in space group P21212 obtained following this protocol is deposited as PDB ID 6RUU (Lange et al., 2020).

23. Prepare crystallization cocktail of 0.1 M sodium citrate pH 5.0 and 1.62 M ammonium sulfate for vapor diffusion crystal trays

Note: Optimal crystallization results were obtained with the above precipitant mix in our hands, however, we recommend starting with a 2-dimensional grid screen around these conditions, as small variations in the pH and concentration of reagents can greatly affect crystallization. For example, start by varying the pH of 0.1 M sodium citrate between 4.8 and 5.5 in one dimension and ammonium sulfate concentration between 1.5 and 1.7 M in the second dimension.

24. Use fresh IRAK3 protein or quickly thaw protein aliquots in a water bath (~2 min), keep on ice until further use

25. Centrifuge protein at 17,000 × *g* and 4 °C for 10 min immediately before setting up crystallization trays

26. Set up sitting drop crystal trays with a 2:1 (protein:precipitant) drop ratio and incubate at 20 °C

27. Initial IRAK3 crystals should appear within 2–3 days and can be further optimized by macroseeding:

- **a)** Prepare crystal seed stock by collecting ~5 drops with small or clustered crystals from the initial crystal screen in 50 μL crystallization cocktail
- **b)** Vortex for 30 s in a "Seed bead" tube to crush crystals
- **c)** Prepare 1:100, 1:1000 and 1:3000 seed stock dilution in crystallization cocktail
- **d)** Set up fresh crystallization drops, dip the seeding tool (Hampton Research) into the seed stock dilution and introduce crystal seeds into drops with a smooth motion. Store crystal trays at 20 °C
- **e)** Seed stocks can be aliquoted, flash-frozen in liquid nitrogen and stored at −80 °C for future use
- **f)** Well-shaped crystals should be visible after 2–3 days and reach maximum size in 5–7 days (Fig. 2).

28. Transfer single crystals to new drops of mother liquor supplemented with 5 mM ethyl mercury phosphate and 30% (*v*/v) glycerol for 16 h. In addition, small molecule binders of interest can be added at this stage for soaking

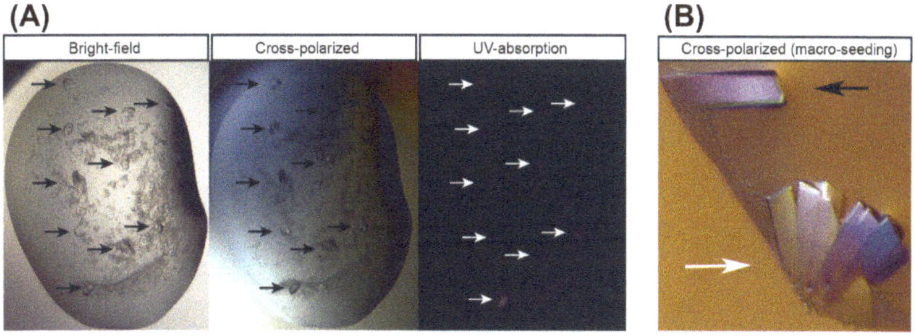

Fig. 2 Initial IRAK3 crystallization and macroseeding outcomes. (A) Initial IRAK3 crystallization hit in a 200 nL drop recorded in bright-field, cross-polarized light, and UV-absorption imaging. *Arrows* indicate positions of small crystal clusters. (B) Macroseeding resulted in the formation of well-formed crystals of approximately $20 \times 20 \times 50$ μM in size (*black arrow*), alongside crystal clusters (*white arrow*). IRAK3 crystals typically formed after 2–3 days in 0.1 M sodium citrate pH 5.0 and 1.62 M ammonium sulfate and reached maximum size in 5–7 days.

Note: In our experience, IRAK3 crystals showed relatively weak diffraction properties with strong anisotropy and we empirically determined that soaking with ethyl mercury phosphate can improve diffraction up to 2.9 Å. The formation of vitrified ice was achieved by addition of 30% glycerol to the mother liquor as a cryoprotectant.

29. Mount crystals on a nylon-loop and vitrify in liquid nitrogen for X-ray crystallography analysis

Note: An alternative crystal form of IRAK3 in space group I4132 has been reported, which was obtained by sitting drop vapor diffusion in 0.1 M citric acid pH 5.5, 0.1 M lithium sulfate, 0.5 M ammonium sulfate at 4 °C (PDB ID: 6ZIW; Mathea, Chatterjee, Preuss, Kraemer, & Knapp, 2020). These crystallization conditions may be explored in parallel to increase chances of success, however, we have not tested this alternative crystallization condition.

5.3 High-throughput thermal shift assay to identify small molecule binders

Timing: variable.

Protocol for high-throughput screening of small molecule libraries by thermal shift assay to identify binders of IRAK pseudokinases. Multiple 384-well plates can be prepared and screened per day with standard lab equipment, however, the overall timing strongly depends on the small molecule library size, availability of automized or semi-automized pipetting robots and number of detection systems.

30. Spot small molecule libraries, positive control (e.g., Staurosporine) and equal volumes of vehicle control (e.g., DMSO) onto 384-well low-volume PCR microplates using an Echo 550 acoustic liquid handler
   a) Recommended starting concentration of small molecules is 10 μM in a final reaction volume of 4 μL
   b) Include 24 wells of vehicle control and 8 wells of positive control per plate
   c) Prepare plates in duplicate or triplicate and seal until further use

Pause Point: Heat-sealed plates can be stored for several days.

31. Add 4 μL of 2.5 μM protein, 2.5 × SYPRO orange in 50 mM HEPES pH 7.5, 200 mM NaCl, 0.001% Tween-20 to each well and spin microplate at 200 × *g* for 1 min

Note: We recommend to test combinations of protein and dye concentrations to validate optimal thermal shift results, as optimal conditions may be dependent on dye quality and equipment.

Critical: Perform dye in buffer (no protein) controls to test the quality of a new batch of microplates. We have observed false-positive fluorescence peaks around 60 °C in the absence of protein in a batch-dependent manner. If this is observed, pre-heating empty microplates to 95 °C for 30 min prior to use can eliminate false-positive fluorescence peaks.

32. Add 1 μL silicone oil to each well and spin at 200 × *g* for 1 min

33. Transfer plate to CFX384 Real-Time PCR detection system and record fluorescence every minute (FRET channel) while heating from 25 °C to 95 °C at a rate of 1 °C/min

Alternative: Other real-time 384-well PCR plate readers capable of detecting SYPRO orange (ex. 491 nm, em. 586 nm) can be used. We also successfully used 60 μM 8-Anilinonaphthalene-1-sulfonic acid (ANS; ex. 360 nm, em. 400 nm) as an alternative dye to SYPRO orange.

34. Determine melting temperatures T_m and thermal shifts $T_{m\Delta}$ with the "DSF analysis v3.0" excel sheet and Boltzmann sigmoidal fitting function in GraphPad Prism (v8.0) as described in detail by the Structural Genomics Consortium, Oxford (Niesen et al., 2007).

35. Determine the standard deviation σ of T_m in vehicle control wells per plate and use vehicle $T_m + 3\sigma$ as the threshold for primary hits with significantly increased thermal shifts

$$\sigma = \sqrt{\frac{\sum (T_{mi} - \mu)^2}{n.}}$$

Where T_{mi} is the vehicle control melting temperature, μ is the mean T_m of all vehicle controls in the plate and n is the number of vehicle control wells (i.e., 24).

Alternative: Newer algorithms are available to analyze thermal shift data, such as the DSFworld web application (Wu et al., 2020), which may improve data analysis speed for large datasets.

5.4 Biolayer-interferometry (BLI) assay to characterize small molecule binders

Timing: variable

A secondary screen using BLI to validate and characterize binding kinetics of primary hits identified in thermal shift assays. This protocol describes the in vitro biotinylation of Avi-tagged IRAK proteins for immobilization on BLI sensor tips and subsequent BLI binding assays with small molecules.

36. To immobilize IRAK pseudokinases on super-streptavidin (SSA) BLI sensor surfaces, IRAK constructs with N-terminal AVI-tags (GLNDIFEAQKIEWHE) are used to enable site-specific biotinylation of the target proteins:

- **a)** Prepare 40 μM of avi-tagged protein with 1 μM BirA enzyme in 50 mM HEPES pH 7.5, 500 mM NaCl, 20% (*v*/v) glycerol, 5 mM BME, 5 mM MgC12, 2 mM ATP, 150 μM D-Biotin
- **b)** Incubate for 1 h at 25 °C and 300 rpm in a thermomix
- **c)** Add additional 150 μM D-Biotin and 1 μM BirA and incubate for 1 h as above
- **d)** Buffer exchange biotinylated protein using PD-10 desalting columns to remove free biotin
- **e)** Aliquot biotinylated protein, flash-freeze in liquid nitrogen and store at −80 °C

Pause point: Biotinylated IRAKs are stable at −80 °C for at least 1 year.

Alternative: Protein immobilization can also be achieved through other affinity tags or covalent coupling to BLI sensor surfaces (e.g., NiNTA or amine reactive surfaces). However, SSA BLI sensors have a high density of Streptavidin that offers increased BLI signals, which is crucial to detect small molecule binding efficiently.

37. Confirm biotinylation of target proteins with a Streptavidin shift assay:

- **a)** Mix 1 μg of biotinylated target protein with 4 μg Streptavidin and incubate for 5 min

  b) Add SDS sample loading buffer (no boil) and run SDS-PAGE of biotinylated target protein with and without Streptavidin added
  c) Successful biotinylation can be visualized through the formation of SDS-resistant Streptavidin complexes with increased molecular masses of ~60 kDa (Fig. 1A).

38. Prepare fresh BLI assay buffer: 50 mM HEPES pH 7.5, 300 mM NaCl, 0.001% Tween-20, 2 mM BME, 1 mg/mL BSA
39. Equip OctetRed384 BLI system with 96-well microplate with 200 μL assay buffer per sensor well to hydrate and store sensor tips

Critical: Hydrate sensor tips for at least 5 min before loading.

40. For sensor loading, prepare a 384-well microplate wells with 80 μL of the following buffers per sensor tip:
  a) BLI buffer (baseline/first wash),
  b) 1 μM biotinylated protein in BLI buffer (loading),
  c) 10 μg/mL biocytin in BLI buffer (blocking),
  d) BLI buffer (second wash).
41. Establish a baseline signal with SSA sensors in the first well for 60 s
42. Transfer sensors to second well to load biotinylated protein for 15 min or until a BLI response of maximum 5 nm is reached

Critical: The final response signal should be in the linear range of the loading curve to avoid adverse crowding effects caused by overloaded sensors.

43. Following loading, wash tips for 5 min in first buffer well
44. Transfer SSA sensors to biocytin solution to block unbound sensor surface for 2 min
45. Perform a final wash in the second buffer well for 5 min and return to storage position
46. We strongly recommend using the double referencing method to subtract non-specific binding of analytes to the biosensor and to correct for signal fluctuation during measurements
  a) Prepare one reference sensor per loaded sensor by repeating the above protocol, but without biotinylated protein in the loading step
  b) Include a parallel measurement using a protein-loaded sensor without analyte to subtract signal fluctuation

Note: Dose responses of up to 14 compounds with a 10-point dilution series can be measured on a single plate as outlined in Fig. 3B.

Critical: The Octet Data Analysis software (ForteBio) requires reference sensors and wells to be in specific positions to enable double referencing. It is therefore important to familiarize yourself with the manufacturer's manual and the assay layout needed to process the data.

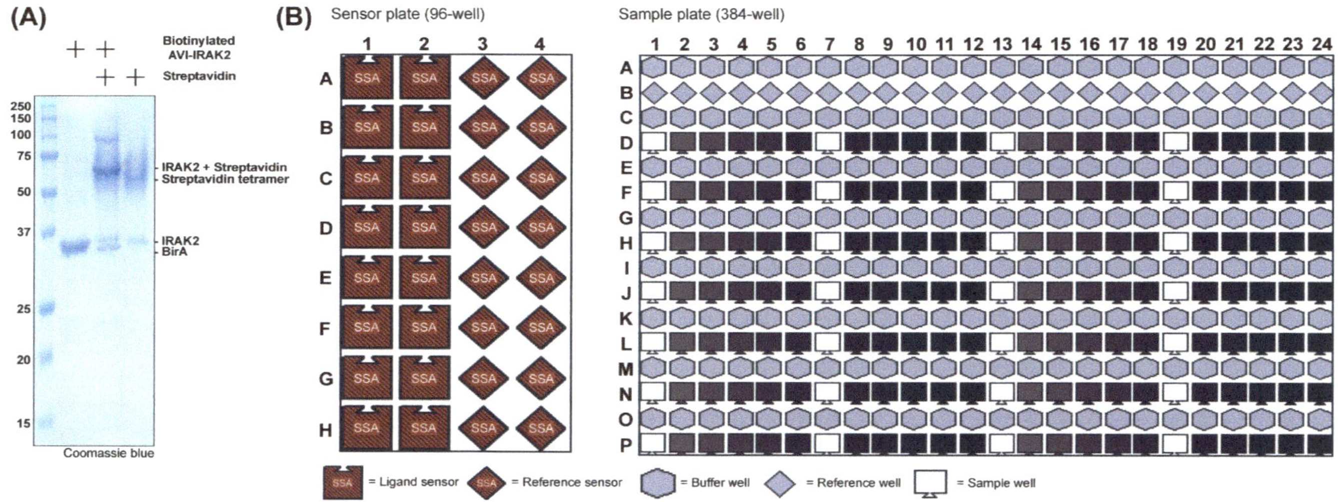

Fig. 3 Example of Streptavidin shift assay and BLI assay plate layouts for dose response measurement. (A) Biotinylated AVI-tagged IRAK2$^{192-507}$ is shifted to a higher molecular mass through formation of an SDS-resistant biotin-streptavidin complex. (B) *Left*: Sensor storage plate with 16 SSA ligand sensors (loaded with biotinylated protein) and 16 SSA reference sensors (empty). *Right*: Sample plate layout for 10-fold serial dilution dose response with up to 14 compounds and reference wells. *White* sample wells represent no compound controls and *gray to black shading* indicates increasing compound concentrations.

47. Add 80 μL of BLI buffer to all wells in every other row for wash steps (A, C, E, etc.) and in the reference row without analyte (B).
48. Prepare twofold dilution series in the remaining rows (i.e., D, F, H, etc.):
    a) Pipette 160 μL of the highest compound concentration in BLI buffer into wells at position 12 and 24
    b) Add 80 μL of BLI buffer to the remaining empty wells
    c) Take up 80 μL of the highest concentration and mix sequentially with neighboring wells to create a twofold dilution series, skipping one internal well (e.g., 7 and 19) and the first wells (1 and 13) as control wells. Discard the excess 80 μL of the lowest dilution
49. Create BLI measurement by cycling through the following steps for each analyte well:
    a) Baseline: 60 s in new buffer well
    b) Association: 60 s in analyte well
    c) Dissociation: Return for 300 s to buffer well

Critical: Always move from lower to higher concentrations to minimize the effect of bleed over.

Note: Sensor tips can be re-used for multiple measurements if full dissociation of compounds is observed (non-covalent binders).

50. Save and import data into Octet Data Analysis software
51. Assign well and sensor positions, select double reference method with automatic baseline adjustment and inter-step correction for analysis
52. Use global fit with 1:1 binding model to calculate affinity constants (Fig. 4).

Note: The presented BLI protocol is not limited to small molecule interactions and can also be adapted to measure protein-protein interactions.

6. Expected outcomes

We present detailed protocols for the expression and purification of IRAK-2 and -3, crystallization of IRAK3, and high-throughput screening methods to identify small molecules that bind to the IRAK pseudokinases.

Insect cell expression has proven to be an ideal system to obtain high-quality recombinant protein preparations of IRAK family members in milligram amounts. The presented tandem-affinity and SEC purification method yields pure material (Fig. 1) suitable for delicate downstream processes, such as crystallization (Fig. 2). High salt and glycerol concentrations in the purification buffers are required to prevent unspecific interaction with resin and plastic materials causing significant loss of protein. It should be

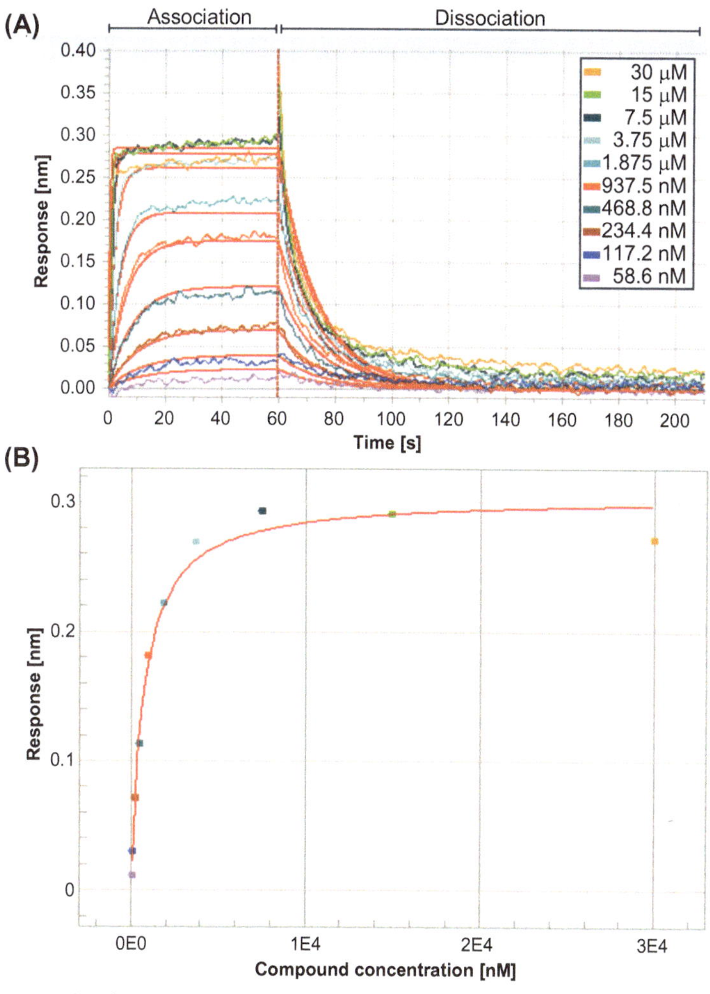

Fig. 4 Example of 10-point dose response BLI assay and binding analysis. (A) BLI traces acquired with immobilized $IRAK3_{145–454}$ on SSA biosensors and 10-point dose response with a small molecule binder in a concentration range from 30 μM to 58.6 nM. The raw BLI data was processed in the Octet Data Analysis software using the double reference method and fitted with a global 1:1 binding model (*red lines*). (B) Steady-state equilibrium plot of the mean association response at 45–55 s against the compound concentration. A 1:1 binding model was fitted to obtain binding kinetics (*red lines*).

noted that IRAK3 exists in a concentration-dependent equilibrium of monomeric and dimeric species (Lange et al., 2020), therefore a doublet of peaks is observed during SEC at higher protein concentrations and fractions of the two peaks may be pooled.

Initial IRAK3 crystals are typically small multi-crystal clusters and macro-seeding is required to obtain single, well-shaped crystals suitable for X-ray crystallography (Fig. 3). Seed-stocks prepared from initial crystals need to be tested empirically in a dilution series to determine optimal seeding density that yields crystals of sufficient number and quality. Prepared seed-stocks and dilutions can be aliquoted, flash-frozen and stored at −80 °C for future application. In our experience, IRAK3 crystals showed relatively weak diffraction properties with strong anisotropy and required soaking with ethyl mercury phosphate for diffraction up to 2.9 Å. An alternative crystal form of IRAK3 with improved diffraction properties is mentioned in the above protocol. In certain cases, soaking of IRAK3 crystals with small molecule binders might improve diffraction quality while providing molecular details of the mode of small-molecule interaction.

The presented high-throughput thermal shift assay provides the means for a rapid, low-cost qualitative screen of large small molecule libraries to identify primary hits. These hits can then be cross-validated in the outlined secondary biolayer-interferometry (BLI) assay in a quantitative manner to obtain detailed binding kinetics (Fig. 4). BLI binding curves reveal important binding properties such as the on and off rate of compounds, and binding affinities.

7. Discussion

The pharmacological inhibition of immunological pathways can cause drug-induced secondary immunodeficiencies, for example, the treatment with the immunosuppressant rapamycin to prevent organ transplant rejection can cause severe and even fatal side effects in patients due to an increased risk of infection (Mahé et al., 2005). It would therefore be a valuable strategy to explore important but less essential components of immune pathways as potential drug targets. While current data suggests that the two IRAK pseudokinases are catalytically inactive, they might play important roles, for example, as allosteric regulators of IRAK1 or IRAK4 activity (Lange et al., 2020). The identification of small molecules that allosterically modulate the conformation of the pseudokinase domains by stabilization of the open or closed state, might therefore fine tune the IRAK-1/-4-mediated immune response. The crystal structure of IRAK3 revealed a closed, pseudo-active conformation with the αC-helix in a central position of a homo-dimer interface. The αC-helix is near the ATP binding pocket and bulky ATP-competitive inhibitors can potentially induce repositioning of the helix. Compounds that stabilize the open conformation of IRAK3 might therefore prevent the formation of the IRAK3 homodimer. In contrast, compounds that stabilize the closed conformation might exhibit properties that favor the homo-dimeric state. While the physiological role of the IRAK3 homodimer is unclear to date, small molecules that bind to IRAK2 and IRAK3 have the potential to modulate MyD88-dependent signaling and may be used as tools to study the roles and signaling mechanisms of the pseudokinases in the innate immune system.

A similar mechanism was previously observed for the pseudokinase STRADα, which requires a pseudo-active, ATP-bound conformation of the pseudokinase domain to engage with LKB1 and stimulate its kinase activity (Zeqiraj et al., 2009). Point-mutations in the G-loop of STRADα that hindered ATP-binding and thereby promoted the open conformation of the pseudokinase, prevented the activation of LKB1. Therefore, small molecules that stabilize either the open or closed conformation of the IRAK pseudokinases would help to dissect their roles within the complex MyD88-signaling network.

Acknowledgments

We thank Prof. Philip Cohen for his support and advice during this research and for comments on this manuscript. This work has been supported by Medical Research Scotland and Janssen Pharmaceutica in the form of a PhD studentship awarded to SML (PhD-770-2014). YK is supported by MRC grant MC_UU_00018/3, the EMBO Young Investigator Program and the Lister Institute of Preventive Medicine.

References

Abdiche, Y., Malashock, D., Pinkerton, A., & Pons, J. (2008). Determining kinetics and affinities of protein interactions using a parallel real-time label-free biosensor, the octet. *Analytical Biochemistry*, *377*(2), 209–217. https://doi.org/10.1016/j.ab.2008.03.035.

Bailey, F. P., Byrne, D. P., Oruganty, K., Eyers, C. E., Novotny, C. J., Shokat, K. M., et al. (2015). The tribbles 2 (TRB2) pseudokinase binds to ATP and autophosphorylates in a metal-independent manner. *The Biochemical Journal*, *467*(1), 47–62. https://doi.org/10.1042/BJ20141441.

Balaci, L., Spada, M. C., Olla, N., Sole, G., Loddo, L., Anedda, F., et al. (2007). IRAK-M is involved in the pathogenesis of early-onset persistent asthma. *The American Journal of Human Genetics*, *80*(6), 1103–1114. https://doi.org/10.1086/518259.

Berger, I., & Craig, A. (2010). *MultiBac expression system user manual*. https://doi.org/10.13140/2.1.2563.6645.

Black, M. H., Osinski, A., Gradowski, M., Servage, K. A., Pawłowski, K., Tomchick, D. R., et al. (2019). Bacterial pseudokinase catalyzes protein polyglutamylation to inhibit the SidE-family ubiquitin ligases. *Science (New York, N.Y.)*, *364*(6442), 787–792. https://doi.org/10.1126/science.aaw7446.

Cohen, P. (2014). The TLR and IL-1 signalling network at a glance. *Journal of Cell Science, 127*(Pt 11), 2383–2390. https://doi.org/10.1242/jcs.149831.

Cohen, P., & Strickson, S. (2017). The role of hybrid ubiquitin chains in the MyD88 and other innate immune signalling pathways. *Cell Death and Differentiation, 24*(7), 1153–1159. https://doi.org/10.1038/cdd.2017.17.

Cummings, M. D., Farnum, M. A., & Nelen, M. I. (2006). Universal screening methods and applications of ThermoFluor®. *Journal of Biomolecular Screening, 11*(7), 854–863. https://doi.org/10.1177/1087057106292746.

Ferrao, R., Zhou, H., Shan, Y., Liu, Q., Li, Q., Shaw, D. E. E., et al. (2014). IRAK4 dimerization and trans-autophosphorylation are induced by Myddosome assembly. *Molecular Cell, 55*(6), 891–903. https://doi.org/10.1016/j.molcel.2014.08.006.

Kobayashi, K., Hernandez, L. D., Galán, J. E., Janeway, C. A., Medzhitov, R., & Flavell, R. A. (2002). IRAK-M is a negative regulator of toll-like receptor signaling. *Cell, 110*(2), 191–202. https://doi.org/10.1016/S0092-8674(02)00827-9.

Kuglstatter, A., Villaseñor, A. G., Shaw, D., Lee, S. W., Tsing, S., Niu, L., et al. (2007). Cutting edge: IL-1 receptor-associated kinase 4 structures reveal novel features and multiple conformations. *Journal of Immunology, 178*(5), 2641–2645. https://doi.org/10.1016/j.febslet.2011.08.036.

Lange, S. M., Nelen, M. I., Cohen, P., & Kulathu, Y. (2020). Dimeric structure of the Pseudokinase IRAK3 suggests an allosteric mechanism for negative regulation. *Structure, 29*(3), 238–251. e4. https://doi.org/10.1016/j.str.2020.11.004.

Lavoie, H., Sahmi, M., Maisonneuve, P., Marullo, S. A., Thevakumaran, N., Jin, T., et al. (2018). MEK drives BRAF activation through allosteric control of KSR proteins. *Nature, 554*(7693), 549–553. https://doi.org/10.1038/nature25478.

Lin, S.-C., Lo, Y.-C., & Wu, H. (2010). Helical assembly in the MyD88-IRAK4-IRAK2 complex in TLR/IL-1R signalling. *Nature, 465*(7300), 885–890. https://doi.org/10.1038/nature09121.

Mahé, E., Morelon, E., Lechaton, S., Sang, K.-H. L. Q., Mansouri, R., Ducasse, M.-F., … Bodemer, C. (2005). Cutaneous adverse events in renal transplant recipients receiving sirolimus-based therapy. *Transplantation, 79*(4), 476–482. https://doi.org/10.1097/01.tp.0000151630.25127.3a.

Mathea, S., Chatterjee, D., Preuss, F., Kraemer, A., & Knapp, S. (2020). *PDB ID: 6ZIW. The IRAK3 Pseudokinase domain bound to ATPgammaS*. https://doi.org/10.2210/pdb6ziw/pdb.

Min, X., Lee, B.-H., Cobb, M. H., & Goldsmith, E. J. (2004). Crystal structure of the kinase domain of WNK1, a kinase that causes a hereditary form of hypertension. *Structure, 12*(7) (http://www.ncbi.nlm.nih.gov/pubmed/15242606).

Motshwene, P. G., Moncrieffe, M. C., Grossmann, J. G., Kao, C., Ayaluru, M., Sandercock, A. M., et al. (2009). An oligomeric signaling platform formed by the toll-like receptor signal transducers MyD88 and IRAK-4. *The Journal of Biological Chemistry, 284*(37), 25404–25411. https://doi.org/10.1074/jbc.M109.022392.

Mukherjee, K., Sharma, M., Urlaub, H., Bourenkov, G. P., Jahn, R., Südhof, T. C., et al. (2008). CASK functions as a Mg2+-independent neurexin kinase. *Cell, 133*(2), 328–339. https://doi.org/10.1016/J.CELL.2008.02.036.

Murphy, J. M., Zhang, Q., Young, S. N., Reese, M. L., Bailey, F. P., Eyers, P. A., et al. (2014). A robust methodology to subclassify pseudokinases based on their nucleotide-binding properties. *The Biochemical Journal, 457*(2), 323–334. https://doi.org/10.1042/BJ20131174.

Niesen, F. H., Berglund, H., & Vedadi, M. (2007). The use of differential scanning fluorimetry to detect ligand interactions that promote protein stability. *Nature Protocols, 2*(9), 2212–2221. https://doi.org/10.1038/nprot.2007.321.

Pauls, E., Nanda, S. K., Smith, H., Toth, R., Arthur, J. S. C., & Cohen, P. (2013). Two phases of inflammatory mediator production defined by the study of IRAK2 and IRAK1 knock-in mice. *Journal of Immunology (Baltimore, Md.: 1950)*, *191*(5), 2717–2730. https://doi.org/10.4049/jimmunol.1203268.

Scheeff, E. D., Eswaran, J., Bunkoczi, G., Knapp, S., & Manning, G. (2009). Structure of the pseudokinase VRK3 reveals a degraded catalytic site, a highly conserved kinase fold, and a putative regulatory binding site. *Structure (London, England: 1993)*, *17*(1), 128–138. https://doi.org/10.1016/j.str.2008.10.018.

Shi, F., Telesco, S. E., Liu, Y., Radhakrishnan, R., & Lemmon, M. A. (2010). ErbB3/HER3 intracellular domain is competent to bind ATP and catalyze autophosphorylation. *Proceedings of the National Academy of Sciences of the United States of America*, *107*(17). https://doi.org/10.1073/pnas.1002753107.

Sreelatha, A., Yee, S. S., Lopez, V. A., Park, B. C., Kinch, L. N., Pilch, S., et al. (2018). Protein AMPylation by an evolutionarily conserved Pseudokinase. *Cell*. https://doi.org/10.1016/J.CELL.2018.08.046.

Villa, F., Capasso, P., Tortorici, M., Forneris, F., de Marco, A., Mattevi, A., et al. (2009). Crystal structure of the catalytic domain of Haspin, an atypical kinase implicated in chromatin organization. *Proceedings of the National Academy of Sciences of the United States of America*, *106*(48), 20204–20209. https://doi.org/10.1073/pnas.0908485106.

Wang, L., Qiao, Q., Ferrao, R., Shen, C., Hatcher, J. M., Buhrlage, S. J., et al. (2017). Crystal structure of human IRAK1. *Proceedings of the National Academy of Sciences*, *114*(51), 13507–13512. https://doi.org/10.1073/pnas.1714386114.

Wu, T., Yu, J., Gale-Day, Z., Woo, A., Suresh, A., Hornsby, M., et al. (2020). Three essential resources to improve differential scanning fluorimetry (DSF) experiments. *BioRxiv*. https://doi.org/10.1101/2020.03.22.002543.

Xu, B., English, J. M., Wilsbacher, J. L., Stippec, S., Goldsmith, E. J., & Cobb, M. H. (2000). WNK1, a novel mammalian serine/threonine protein kinase lacking the catalytic lysine in subdomain II. *The Journal of Biological Chemistry*, *275*(22), 16795–16801. https://doi.org/10.1074/jbc.275.22.16795.

Zeqiraj, E., Filippi, B. M., Deak, M., Alessi, D. R., & van Aalten, D. M. F. (2009). Structure of the LKB1-STRAD-MO25 complex reveals an allosteric mechanism of kinase activation. *Science (New York, N.Y.)*, *326*(5960), 1707–1711. https://doi.org/10.1126/science.1178377.

CHAPTER SIX

Characterization of pseudokinase ILK-mediated actin assembly

Julia Vaynberg and Jun Qin*
Department of Cardiovascular & Metabolic Sciences, Lerner Research Institute, Cleveland Clinic, Cleveland, OH, United States
*Corresponding author: e-mail address: qinj@ccf.org

Contents

Abstract

Pseudokinase ILK is a key protein involved in regulating focal adhesion assembly and cell-extracellular matrix adhesion. By forming a tight trimeric complex (IPP) with adaptor proteins PINCH and Parvin that both contain an actin binding WH2 motif, IPP can promote the formation of unique actin bundles thereby linking the focal adhesions and actin cytoskeleton and triggering cytoskeleton reassembly and dynamic cell adhesion processes such as cell spreading and migration. This chapter describes detailed characterization of the IPP-mediated actin binding and bundle formation.

1. Introduction

Actin cytoskeleton is a highly dynamic system comprising a network of actin polymers and actin binding proteins. It plays a central role in

Methods in Enzymology, Volume 667
ISSN 0076-6879
https://doi.org/10.1016/bs.mie.2022.03.027

regulating diverse biological functions such as maintenance of cell structure, cell shape change, cell migration, organelle transport, and many others (Cooper, 2000). In cells, actin exists in two forms: globular/monomeric form (G-actin) and filamentous/polymeric form (F-actin). Both forms are in static equilibrium, which can be regulated by certain actin-binding proteins. Under physiological conditions, G-actin readily polymerizes into F-actin, which can crosslink into higher order structures—bundles and bundle networks with different morphologies. Dozens of actin binding proteins are known to regulate the kinetics and morphology of actin assembly by engaging with either G-actin or F-actin (Winder & Ayscough, 2005). Studying the mechanisms of these actin regulators and their engagement with actin is of critical importance for understanding how the cytoskeleton is dynamically controlled in cells and how it is able to respond quickly to various stimuli. Extensive studies have shown that many actin regulators communicate between the cytoskeleton and extracellular matrix (ECM), thereby regulating cell adhesion and diverse cell adhesion-dependent processes such as cell shape change and migration (Parsons, Horwitz, & Schwartz, 2010).

Integrin-Linked Kinase (ILK) is one of the key components of focal adhesions (FAs) known to mediate communication between intracellular actin cytoskeleton and the ECM (Legate, Montanez, Kudlacek, & Fässler, 2006; Qin & Wu, 2012). Originally thought to act as a kinase, ILK was later shown by rigorous biochemical and structural studies to be a pseudokinase incapable of performing catalytic function (Fukuda, Gupta, Chen, Wu, & Qin, 2009; Fukuda, Knight, Piszczek, Kothary, & Qin, 2011). Rather than acting as a kinase, ILK actually works as a hub that binds numerous proteins in cells and relay biochemical information (Qin & Wu, 2012). In particular, it serves as a key platform of a very tight obligate ternary complex with focal adhesion (FA) adaptor proteins PINCH and Parvin (IPP), which forms before complex localization to FAs (Zhang et al., 2002). Using a combination of GST-based pull-down experiments, NMR-based structural analysis, and sequence alignment, we recently identified two actin binding WH2 motifs in IPP: one in the C-terminal region of PINCH1 and the other in the N-terminal region of parvin (Vaynberg et al., 2018). By assembling these two actin binding motifs containing proteins into one trimeric complex, ILK triggers formation of actin bundles and the extent of bundle formation and the morphology of the bundles depend on Mg^{2+}-ATP binding to ILK (Vaynberg et al., 2018). Here we describe detailed procedures and protocols we used to pursue these actin assembly studies.

2. Preparation of ILK-PINCH1-α-parvin (IPP) complex

In order to understand how IPP utilizes the two WH2 domains to engage with actin filaments, we used a bacterial tricistronic coexpression system to express the IPP complex. Briefly, genes encoding the full-length human ILK (residues 1–452), human PINCH-1 (residues 1–325), and human α-Parvin (residues 1–372) were PCR-amplified and subcloned into a polycistronic coexpression vector pST39 (Tan, 2001) according to the inventor's protocol. Two surface cysteine residues of ILK were substituted by serine residues (C346S; C422S) to improve the protein solubility of ILK, as previously demonstrated (Fukuda et al., 2009). The tricistronic coexpression construct contains a hexahistidine-tag sequence followed by a thrombin cleavage site prior to the sequence of PINCH-1 for conventional affinity purification. The hexahistidine-tagged IPP complex was coexpressed in *E. coli* BL21(DE3) pLysS and purified from the bacterial cell lysate by Ni-affinity chromatography column followed by HiLoad 16/60 Superdex 200 gel filtration and HiTrap SP cation-exchange chromatography columns (all from GE Healthcare). It is important to note that IPP needs to be extremely pure since tiny amount of impurity proteins, such as chaperones, that might weakly associate with IPP could interfere with the detection of the actin bundles. Therefore, we only selected the purest fractions of the eluted peak during gel filtration for actin binding/bundling experiments.

3. F-actin co-sedimentation assay to detect IPP binding to F-actin

Since IPP contains two WH2 actin binding domains, we first performed this assay with each individual domain and then with an entire IPP complex. The assay was based on the fact that F-actin can be precipitated by high speed centrifugation for either 20 min at 400,000 × *g* or for 1 h at 150,000 × *g* at 20 °C while G-actin stays in the supernatant. In this assay, actin is polymerized and then incubated with test protein and centrifuged at high speed. Proteins that bind F-actin are expected to be in the pellet with F-actin. Special care should be taken that test protein is well behaved in solution and is not aggregated. The sample needs to be pre-spun and pre-cleared of possible large aggregates. Since many actin-binding proteins have moderate (micromolar range) affinities, the highest possible concentration of test protein should be used to increase chances for the binding to be detected. Test protein should be prepared in a buffer compatible with actin buffers. In absence of G-actin sequestering proteins, G-actin polymerizes with addition of ions, therefore all G-actin buffers contain no salt and many proteins are not stable in this condition. F-actin tolerates salt well and so binding to F-actin is easier to test. A control sample of test protein buffer with actin (without the test protein) should be prepared to test whether components of the buffer affect actin protein integrity and polymerization. Rabbit Skeletal Muscle Actin (lyophilized) for all the tests we performed (unless stated otherwise) was purchased from Cytoskeleton, Inc., and we followed the manufacturer's protocol to prepare actin stock solutions. For detailed description of such assay, please see Hertzog and Carlier (2005). Binding to G-actin may be tested by such techniques as NMR or ITC but then it should either be done in presence of G-actin sequestering protein such as DNAse I (Kabsch, Mannherz, Suck, Pai, & Holmes, 1990), gelsolin (McLaughlin, Gooch, Mannherz, & Weeds, 1993), or profilin (Schutt, Myslik, Rozycki, Goonesekere, & Lindberg, 1993) that will prevent polymerization (but may also obscure the binding site). Alternatively, non-polymerizable mutant AP-actin (Joel, Fagnant, & Trybus, 2004) should be used instead of wild-type G-actin.

When preparing buffers for working with actin, special attention should be paid to accuracy of buffer pH (actin is stable in range of 6.5–8.0) and concentrations of ATP, DTT and MgCl2 as they may affect actin stability. If a test protein binds metal, DTT may be replaced by the gentler reductant, TCEP (trisphosphine), with the same concentration. The test protein needs to be tested to make sure that it is still bound to the metal, ensuring integrity of the test protein's native conformation. In our case, PINCH1 LIM5-T is zinc finger protein so we used NMR to confirm that the protein conformation was not altered in a TCEP-containing buffer. Addition of the test protein should not dilute any of these additives, so either they should be added into test protein buffer, or more should be added into the assay samples to compensate for the dilution by the test protein solution. General actin buffer may be prepared in advance and stored at 4 °C, and ATP and DTT (or TCEP) must be added fresh before each experiment.

It may be convenient to first try commercial kits with all the prepared reagents and then prepare your own reagents. We found Cytoskeleton, Inc. "Actin Binding Protein Spin-Down Assay Biochem Kit: Rabbit Skeletal Muscle Actin (Cat # BK001)" quite helpful, which provides actin and well-behaved control proteins such as α-Actinin (a positive control) and bovine serum albumin (BSA, negative control).

ATP and DTT stocks should be aliquoted and kept at −20 °C (better freshly prepared) and should not undergo multiple freeze-thaw cycles to avoid hydrolysis. Because the ratios of F-actin and test protein in the samples may vary, we describe the starting conditions in the following protocol. However, one can reduce the amount of F-actin and increase volumes of test protein as long as F-actin and control proteins are detected well by SDS-PAGE. Coomassie dye staining is the easiest and is probably the most used in-gel protein detection method, but it is not the most sensitive. Silver staining, fluorescent staining, and Western blot detection may be also suitable if a protein band is too weak with Coomassie stain and protein amounts are limited. More details on protein visualization methods in gels can be found in Steinberg (2009).

Protein/actin concentration ratio and pH (within 6.5–8.0 range) may be varied too in order to find optimal binding conditions. Actin concentration is measured at 290 nm not 280 nm (due to bound ATP interference at 280 nm) with an extinction coefficient of 26,600 M^{-1} cm^{-1}.

Fig. 1A shows resulting SDS-PAGE of our experiment testing binding of LIM5-T (PINCH1 WH2 containing construct) to F-actin. Only pellets are shown on this gel, and all samples were run in duplicates.

The following are detailed information for materials and procedures.

Materials

- 1 mg aliquot of skeletal muscle actin (Cytoskeleton, Inc., Catalog# AKL99)
- α-Actinin as positive control (Cytoskeleton, Inc. Catalog# AT01)
- BSA as negative control (Cytoskeleton, Catalog# AACX1)
- General Actin Buffer (5 mM Tris-Cl, pH 8.0, 0.2 mM CaCl2, 0.2 mM ATP, 0.5 mM DTT or TCEP)
- Actin Polymerization Buffer, 10× (50 mM KCl, 2 mM MgCl2, 1 mM ATP)
- F-actin Buffer: 90% G-buffer +10% of 10× Actin Polymerization Buffer
- ATP (Adenosine 5′ triphosphate disodium salt), 100 mM stock (Cytoskeleton, Inc., Catalog#BSA04). Reconstitute according to manufacturer's instructions. Aliquot and store at −20 °C. Thaw on ice and do not reuse the aliquots
- $MgCl_2$, 100 mM stock (VWR Catalog# 97064-150). Dissolve 9.5 mg in 1 mL of purified water. Aliquot and store at −20 °C
- DTT (Dithiothreitol), 100 mM stock. (VWR, Catalog# EM-3860). Weigh 15.4 mg and dissolve in 1 mL of purified water. Aliquot and store at −20 °C. Thaw on ice. Do not reuse the aliquots
- Laemmli Sample Buffer (BioRad, Catalog# 161-0737)

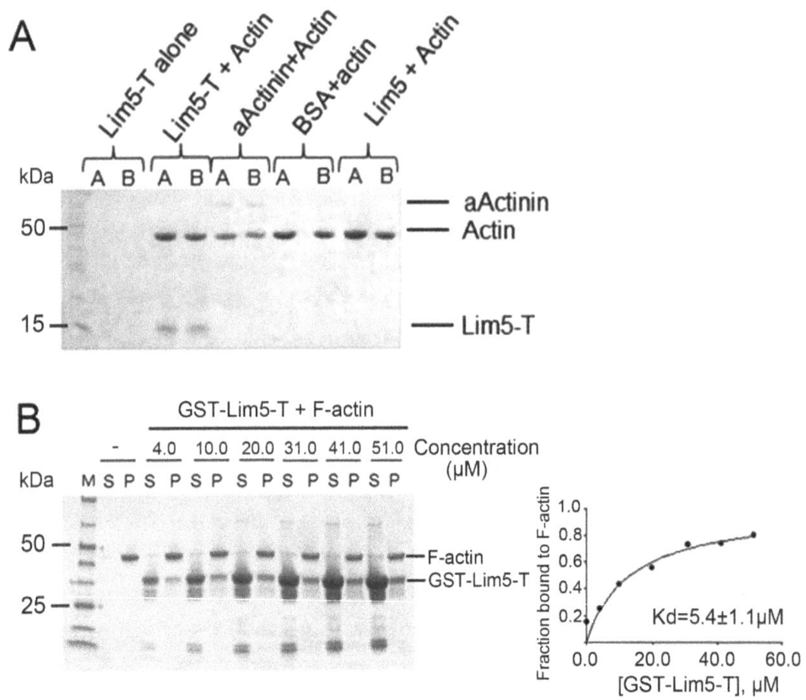

Fig. 1 F-actin co-sedimentation assays. (A) SDS-PAGE gel stained with Coomassie stain of LIM5-T. LIM5-T demonstrates that LIM5-T spins down with F-actin by high-speed centrifugation, along with positive control (α-actinin) but not with BSA. All samples were tested in duplicates. SDS-PAGE gel of pellets containing F-actin and bound to it LIM5-T are shown. (B) SDS-PAGE gel stained with Coomassie stain illustrates dose-dependent binding of GST-LIM5-T to F-actin allowing determination of protein–actin binding K_d. S—supernatants, P—pellets. All samples were run on the same gel for quantification by the ImageJ software. Only test protein bands of Pellets were included in K_d determination. Fraction of bound protein was derived from [Intensity of pellet band (protein + F-actin) − Intensity of pellet band (protein)] × $Bmax^{-1}$. Bmax was obtained from the projected saturation point in Graphpad Prism that estimated K_d by one-site total binding fit model. *Adapted from Vaynberg, J., Fukuda, K., Lu, F., Bialkowska, K., Chen, Y., Plow, E. F., et al. (2018). Non-catalytic signaling by pseudokinase ILK for regulating cell adhesion.* Nature Communications *9(1), 4465.*

- PageRuler Broad Range Unstained Protein Ladder (Thermo Fischer Scientific, Catalog# 26630)

Equipment:

- Centrifuge capable of 150,000 × *g* at 4 °C and 24 °C, 50–200 μL volume capacity. We used Optima TLX Ultracentrifuge (Beckman), rotor TLA-100

- Polycarbonate thick-walled centrifuge tubes (Beckman, Catalog# 343775)
- SDS-PAGE system with proper test protein detection technique (Coomassie, western blot or silver stain; see Steinberg, 2009, for overview of possible methods). We used precast BioRad Mini-PROTEAN TGX gradient 4–15% gels either 15-well comb/15 μL per well (BioRad, Catalog# 4561086) or 10 well comb/30ul per well (BioRad, Catalog# 456-1083). Running buffer: 25 mM Tris, 192 mM glycine, 0.1% (*w/v*) SDS, pH 8.3. 10 × stock of the buffer is available at BioRad (Catalog# 1610772). We stained the proteins with Coomassie stain
- Fisherbrand™ gel loading tips 1–200 μL (Fisher Scientific, Catalog# 02-707-81)
- Spectrophotometer suitable for measuring absorbance in UV/Vis range with a cuvette for protein concentration determination
- Gel or good quality photo scanner (such as Epson perfection V370 Photo) to scan gels, ImageJ software to quantify band intensities (optional for Protocol 1, necessary for Protocol 2).

3.1 Protocol 1

1. Prepare stock solution of test protein in a buffer that is compatible with Actin and with highest possible concentration. It is good to start with 2–4 times molar access relative to final F-actin concentration. Following this protocol, good starting point for final F-actin concentration is 18 μM (40 μL of 23 μM stock diluted to 50 μL of final sample volume, see steps 3–6 of this protocol for preparation of F-actin stock and step 7 for suggested mixing conditions), so final test protein concentration should be 40–100 μM. As the test protein stock gets diluted 5 times in the suggested protocol, the initial test protein stock solution should be 200–500 μM.
   Centrifuge at either 150,000 × *g* for 1 h or 400,000 × *g* for 20 min
2. Carefully remove supernatant with a long pipette tip, taking care not to touch the bottom of the centrifuge tube where a pellet of aggregated protein may be located. Determine the concentration of the test protein in the supernatant by either measuring UV–vis absorbance at 280 nM or other methods (for review of various methods of protein concentration determination, please see Goldring, 2012). Keep this protein stock on ice

3. Thaw 1 mg aliquot of Actin skeletal muscle lyophilized actin (Cytoskeleton, Inc) by inverting the tube on the lab bench and placing some room temperature water on the top of the bottom of the inverted tube, leave for 2 min then place on ice
4. Add 100 μL of ddH_2O and resuspend lyophilized powder well. Briefly vortex and then spin down for 30 s at 10,000–13,000 rpm. This will be 10 mg/mL stock of Actin. It can be aliquoted and frozen in liquid nitrogen at −80 °C. Actin is more stable at 10 mg/mL than at more diluted concentrations
5. Dilute 10 mg/mL actin solution to 1 mg/mL (~23 μM) by diluting 10-fold with G-actin buffer (make sure ATP and DTT were added). If using all of the above made stock, add 900 μL of G-buffer to 100 μL of 10 mg/mL stock from Step 4. Place on ice for 30 min
6. Add 10 × Actin Polymerization Buffer, 1/10th of the sample volume (so the resulting concentration is 1 ×) to the actin solution from step 5 and incubate for 1 h at Room Temperature. If total volume from Step 5 is 1 mL, add 100 μL of Actin polymerization Buffer. Discard left-over Actin Polymerization Buffer as it cannot be refrozen. This is F-actin solution at roughly 23 μM
7. Mix the samples in duplicates (two of each) in labeled ultracentrifuge tubes. Total Volume 50–100 μL. For 50 μL sample volume, mix 40 μL of either F-actin solution or F-buffer (no actin control) with 10 μL of test protein solution (or test protein buffer for a buffer control) or 10 μL of control protein solution (BSA, actinin, and fusion protein if used). Incubate all samples at Room Temperature for 30 min
8. Centrifuge all tubes at either 150,000 × *g* for 1.5 h at 24 °C or 400,000 × *g* for 20 min by ultracentrifugation
9. Using long pipette tip carefully remove supernatants taking special care to not disturb the pellets. It is very difficult to see the pellets so it is helpful to label the position of the top of tube as it is placed in the rotor. The pellet will be located on that side. Remove supernatant from the other side of the tube to a fresh Eppendorf tube
10. Add 10 μL of 5 × SDS-PAGE gel loading buffer (Laemmli Sample Buffer) to each supernatant tube. Heat for 10 min at greater than 95 °C, spin down for 10 min at 13,000 rpm
11. Resuspend the pellets in 60 μL of 1 × SDS-PAGE gel loading buffer (Laemmli Sample Buffer). Heat for 10 min at greater than 95 °C, spin down for 10 min at 13,000 rpm

12. Analyze the samples by running an SDS-PAGE gel of supernatants and pellets, staining with Coomassie blue or other method, and scanning the gel to perform densitometry if needed by using ImageJ (or analogous) software. Protein Ladder should be used to identify test protein and Actin band (Actin MW is 43 kDa)

4. Co-sedimentation assay to measure binding affinity of a protein to F-actin

Materials and equipment are same as previous section. The example of the results of this experiment is shown on Fig. 1B.

4.1 Protocol 2

Repeat steps 1–6 of protocol 1

7. Prepare a series of 8–10 test protein dilutions with concentrations ranging between approximately 0–50 μM in test protein buffer. Include buffer only control
8. Mix in duplicates F-actin with the same total volume of test protein + test protein buffer with ranging concentrations to a total volume of 50 – 100 μL. Include Actin Polymerization Buffer to a total of 1/10th of volume. Test protein buffer and test protein together should amount to the same volume in all samples. F-actin stock already contains Actin Polymerization buffer in it, 1/10th of its volume (see Step 6), this needs to be accounted for when adding Actin Polymerization Buffer to the sample. Please see Table 1 for example of assay samples composition Incubate at Room Temperature for 30 min
9. Repeat steps 8–12 of protocol 1
10. Use ImageJ software to analyze the densitometry data. Scan the gel and load image file into ImageJ software. Convert to grayscale mode, adjust brightness/contrast if needed (on the entire image, never on separate parts of the gel), to ensure that all the bands are visible but not saturated. Using rectangular tool, draw a box around test protein band. Make sure you only include the band that is in that lane. If bands are overlapping between the lanes, rerun the gel with loading less sample. Record intensities of each band and corresponding concentrations. Normalize binding data by subtracting the blank. Plot the normalized intensities as a function of the protein concentration. Use GraphPad Prism (GraphPad Software, Inc) or analogous software to plot and fit

Table 1 Sample composition example for Protocol 2.

C [Lim5-T] final, μM	Lim5-T buffer, μL	Actin polym buffer, μL	Lim5-T stock, μL	F-actin stock, μL
2	36	4	1	9
4	35	4	2	9
7	33.5	4	3.6	9
10	32	4	5	9
15	29.5	4	7.5	9
20	27	4	10	9
15	24.5	4	12.5	9
30	22	4	15	9
40	17	4	20	9

Total sample volume: 50 μL; Lim5-T stock: 100 μM; actin polymerization buffer: 10 ×; total volume of test protein buffer and Lim5-T should be the same in all samples.

the data and determine binding affinity. Fig. 1 is an example showing how PINCH1 LIM5-T binds to F-actin in a concentration-dependent manner yielding a $K_D \sim 5.4$ μM (Vaynberg et al., 2018).

5. Pyrene-actin polymerization: Direct observation of initial polymerization rates

The above protocols examine the F-actin binding to a protein. The same protein may also bind to G-actin as we found for PINCH-1 and α-Parvin that act through their WH2 motifs (Vaynberg et al., 2018). We used NMR spectroscopy to examine the PINCH/parvin domains for binding to G-actin using non-polymerizable G-actin mutant, which has been described in detail in Vaynberg et al. (2018). Here we provide detailed description how we examine whether a G-actin binding protein, like PINCH or parvin, may affect F-actin polymerization. Under physiological conditions, actin is at a steady state equilibrium between its G and F forms and factors, like ionic strength or presence of certain actin regulating proteins, can influence the kinetics and ratio of F- to G-actin. This experiment is very simple, quick and offers wealth of information. It is based on the fact that when G-actin is conjugated to pyrene its fluorescence is significantly enhanced upon polymerization, therefore the process of polymerization in the absence or presence of a test protein can be observed in real time.

In order to allow room for signal enhancement, we used ~3–8 μM 15% pyrenyl-labeled actin and observed its polymerization upon addition of Actin Polymerization Buffer in the absence and presence of ~3 μM IPP complex. IPP was always premixed with Actin Polymerization Buffer before addition to G-actin solution. Useful controls are test protein buffer without test protein at the same volume or a known actin nucleator Arp2/3 and VCA (WASP VCA domain) that cause sharp increase of polymerization rate. These proteins may also be purchased commercially. In our case, the final sample concentrations of Arp2/3 and VCA were 20–30 nM and 70 nM, respectively. Polymerization rate is monitored *in situ* by measuring fluorescence intensity, over a period of time, with excitation and emission wavelengths 365 and 407 nm, respectively, using a Fluorescent Plate Reader.

The concentration of the test protein depends on the above determined/estimated binding affinity. A 1:1 ratio of protein:actin may be a suitable point to start. We used 4 μM of IPP since IPP tends to precipitate at higher concentrations. If a protein potently affects polymerization, like Arp2/3, nM concentrations are enough to see the effect. Some newer instruments are sensitive enough that less than 5% of pyrene labeled actin may be used in G-actin solution preparation. It is good to perform a pilot experiment starting with 5% pyrenyl labeled actin and then adjust the percentage if needed. Good rule of thumb is that during polymerization, the signal should increase 3–4-fold before reaching polymerization plateau. Lyophilized actin can be purchased from Cytoskeleton, Inc. Pyrene-labeled actin must be stored in the dark. Fig. 2 shows polymerization curves of G-actin. Arp2/3 and VCA act as powerful enhancers of polymerization rate due to their nucleating and branching functions. IPP has a mild G-actin sequestering effect due to binding of the WH2 domains to G-actin as well as F-actin, but the total effect on the polymerization effect is modest.

Materials

- 1 mg aliquot of skeletal muscle actin (Cytoskeleton, Inc., Catalog# AKL99).
- 1 mg aliquot of pyrene-conjugated skeletal muscle actin (Cytoskeleton, Catalog# AP05).
- Arp2/3 (Cytoskeleton, Inc., Catalog# RP01P) and WASP protein VCA domain (Cytoskeleton, Inc., Catalog# VCG03) proteins as control
- General Actin Buffer (5 mM Tris-Cl, pH 8.0, 0.2 mM CaCl2, 0.2 mM ATP, 0.5 mM DTT)

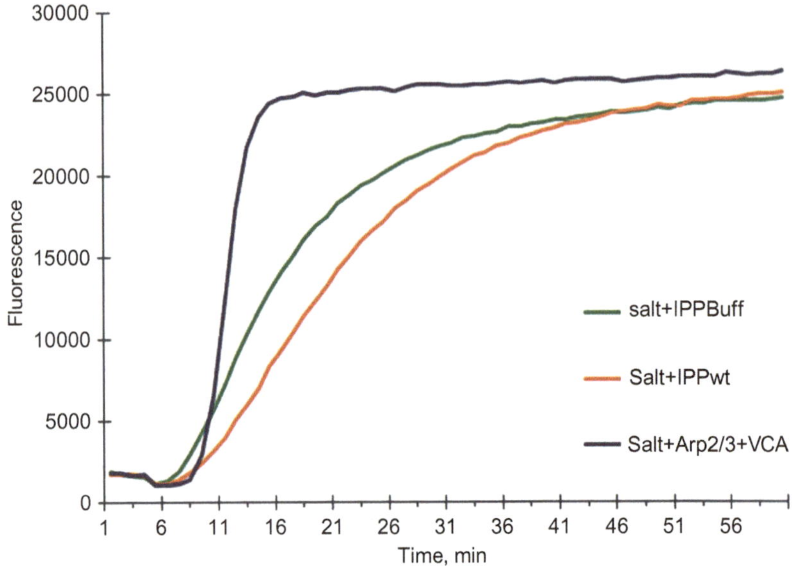

Fig. 2 Polymerization of pyrene-labeled actin with addition of Actin Polymerization Buffer ("salt") in real time demonstrating possible effects of proteins on its kinetics. "Salt + IPPBuff" curve is a control condition where G-actin was polymerized by salt in presence of IPP buffer. "Salt + IPP" curve is polymerization of G-actin by salt in presence of IPP protein complex (IPP:G-actin = 4.3 μM:4.3 μM = 1:1). There is a slight delay in polymerization rate due to IPP containing two WH2 domains that bind G-actin and partially sequester it. "Salt + Arp2/3 + VCA" shows rate of G-actin polymerization by salt in presence of Arp2/3 protein and VCA domain or WASP protein (Arp2/3:G-actin = 30 nM:4.3 μM = 1:143). Due to strong nucleating activity of Arp2/3-VCA mix, G-actin polymerizes much faster. *Adapted from Vaynberg, J., Fukuda, K., Lu, F., Bialkowska, K., Chen, Y., Plow, E. F., et al. (2018). Non-catalytic signaling by pseudokinase ILK for regulating cell adhesion.* Nature Communications *9(1), 4465.*

- Actin Polymerization Buffer, 10 × (50 mM KCl, 2 mM MgCl2, 1 mM ATP)
- F-actin Buffer: 90% G-buffer + 10% of 10 × Actin Polymerization Buffer
- ATP, 100 mM stock (Cytoskeleton, Inc., Catalog#BSA04). Reconstitute according to manufacturer's instructions. Aliquot and store at −20 °C. Thaw on ice and do not reuse the aliquots
- $MgCl_2$, 100 mM stock. (VWR, Catalog# 97064–150). Dissolve 9.5 mg in 1 mL of purified water. Aliquot and store at −20 °C
- DTT (Dithiothreitol), 100 mM stock (VWR, Catalog# EM-3860). Weigh 15.4 mg and dissolve in 1 mL of purified water. Aliquot and store at −20 °C. Thaw on ice. Do not reuse the aliquots

Equipment

- Fluorescence spectrophotometer plate reader able to measure at 365 nm excitation wavelength and 407 emission wavelength with 4–10 nm bandwidth such as PerkinElmer EnSpire2300 multimode plate reader
- 96 well black walled plate with flat bottom suitable for fluorescence such as Corning polystyrene microplate (Catalog# 3915).
- Centrifuge capable of 150,000 × *g* at 4 °C and 24 °C, 50–200 μL volume capacity such as Optima TLX Ultracentrifuge (Beckman), rotor TLA-100
- Polycarbonate thick walled centrifuge tubes (Beckman, Catalog# 343775)
- Fisherbrand™ gel loading tips 1–200 μL (Fisher Scientific, Catalog# 02-707-81)
- Spectrophotometer suitable for measuring absorbance in UV/Vis range with a cuvette for protein concentration determination
- Eppendorf tubes 500, or 1000, or 1500 μL volume (such as ThermoFisher Scientific, Catalog# 90410)

5.1 Protocol 3

1. Perform Steps 1–2 of Protocol 1
2. Perform Steps 3 and 4 of Protocol 1 for pyrene-labeled and non-labeled lyophilized actin aliquots to prepare 10 mg/mL stock solutions
3. To prepare 15% pyrene labeled actin solution, mix 10 μL of 10 mg/mL labeled stock with 54 μL of 10 mg/mL unlabeled stock. Add 1600 μL of G-actin buffer (resulting in ~4 mg/mL; ~9 μM actin solution).
4. Incubate for 40 min at room temperature
5. Spin at 150,000 g at 4 °C for 1 h (if the centrifuge tubes are smaller volume—use several). This step is optional but we found that eliminating actin aggregates greatly improves reproducibility of the assay
6. Using a long pipette tip remove 90% of supernatant from all the tubes and combine them in one Eppendorf tube. Determine concentration of G-actin by measuring OD290 (OD280 has significant interference from bound ATP) and determine actin concentration. Actin has an extinction coefficient at 290 nm of 26,600 M^{-1} cm^{-1}, and 1 mg/mL of Actin has OD290 of 0.63. In our example on Fig. 2, OD290 was 0.13 which equaled to 0.22 mg/mL (5 μM). We used 150 μL of this solution per sample in a final sample volume of 200 μL which resulted in final concentration of Actin 3.8 μM

Table 2 Sample composition for actin polymerization kinetic assay.

Sample	Salt, 10× stock, μL	15% Pyrenyl G-actin, 5μM stock, μL	IPP wt, 4.7μM stock, μL	IPPwt buffer, μL	Arp2/3, μL+VCA, μL
Salt + IPP Buffer	20	150	-	17	-
Salt +IPP wt	20	150	17	-	-
Salt + Arp2/3 + VCA	20	150	-	8.5	2.6 + 6.0

Total sample volume 200 μL. Final concentration of actin was 3.8 μM. Final concentration of IPPwt was 4 μM; Arp2/3 and VCA were reconstituted according to manufacturer's instructions and mixed to the final concentrations of 30 and 70 nM.

7. Mix samples and controls of 150–200 μL volumes, in triplicates. Mix 1/10th volume of Actin Polymerization buffer with test protein first. Last step is to add G-actin to final concentration of 3–10 μM. Please see Table 2 for the example of mixing the samples we used for the experiment presented in Fig. 2

Immediately start recording fluorescence in kinetic mode for 1 h or until saturation is reached in all wells, with intervals of 30 or 60 s. Excitation wavelength is 365 nm and emission is 407 nm.

6. Detection and visualization of IPP formed bundles

Once we found that IPP complex contains two WH2 domains that bind G- and F-actin, we needed to test whether IPP is able to crosslink actin filaments and form bundles. The initial assay for bundling is identical to the Actin Binding assay in Section 3 except that samples should be spun down at low speed of 10,000 × *g* for 15 min. Almost no F-actin will precipitate at this speed but bundles do. A control containing test protein buffer should be run alongside test protein. Analyzing the pellet by western blot with an anti-actin antibody would help to confirm the presence of actin in the pellet (Fig. 3).

One of the limitations of the described assay is that only large bundles spin down at low speed. Bundle morphology also cannot be judged solely by spin down. Are bundles long and thin? Short and thick? Straight or branched out? Are there many smaller bundles or a few large ones? To answer these questions, one needs to visualize the bundle using microscopy. One of the best methods is *in vitro* total internal reflection microscopy

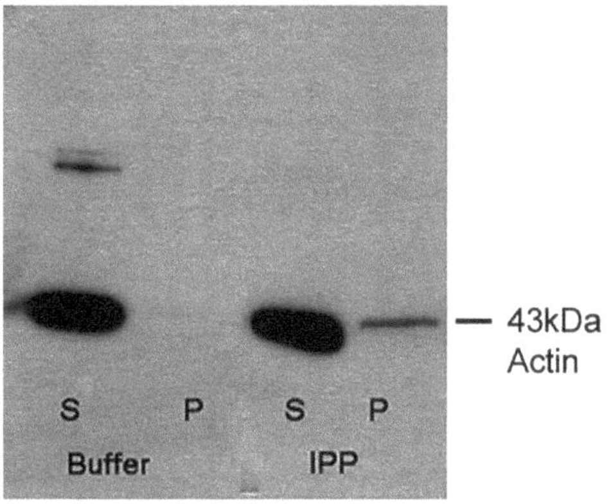

Fig. 3 Low speed co-sedimentation assay showing that IPP induces F-actin bundle formation. Here, G-actin was polymerized by addition of Actin Polymerization Buffer ("salt") in presence of either IPP Buffer ("Buffer" on the figure) or IPP ("IPP"). Samples were centrifuged at low speed (10,000 × *g* for 15 min), supernatants (S) separated from pellets (P) and all were analyzed by Western Blot with anti-actin antibody (we used Anti-Actin mAb from Calbiochem, Catalog# CP01). Actin bundles usually spin down at 10,000 × *g* but not F-actin fibers. In presence of salt and Buffer, no actin band is apparent in the pellet but there is a clear actin band in pellet in presence of IPP indicating that IPP induces formation of actin bundles (There is a nonspecific band or a gel artifact in Supernatant of Buffer control sample). *Adapted from Vaynberg, J., Fukuda, K., Lu, F., Bialkowska, K., Chen, Y., Plow, E. F., et al. (2018). Non-catalytic signaling by pseudokinase ILK for regulating cell adhesion.* Nature Communications *9(1), 4465.*

(TIRF). It allows one to visualize actin polymerization and bundling in real time. However, it requires TIRF microscope, which at the time was not available to us. A great detailed description of the method and its protocol is in Breitsprecher, Kiesewetter, Linkner, and Faix (2009). Confocal or widefield microscopy are simpler techniques that do not require special microscopes or intensive sample preparation and still allow visualization of the actin bundles albeit without kinetic information. Confocal microscopy resolves the individual actin bundles better but it focuses on a very narrow space and thus does not provide representative bundle morphology (shape, size distribution) of the solution as even if looking at tens of areas, one can only closely focus on several specific bundles at a time. Widefield microscopy is suitable for larger scale actin bundle observation and gives a better representation of the overall bundle composition in solution. Using this type of microscope, we can scan a slide with a large enough area to evaluate overall bundle morphology (Fig. 4), which was successfully done for IPP (Vaynberg et al., 2018). Visualizing a larger area of a slide, rather than

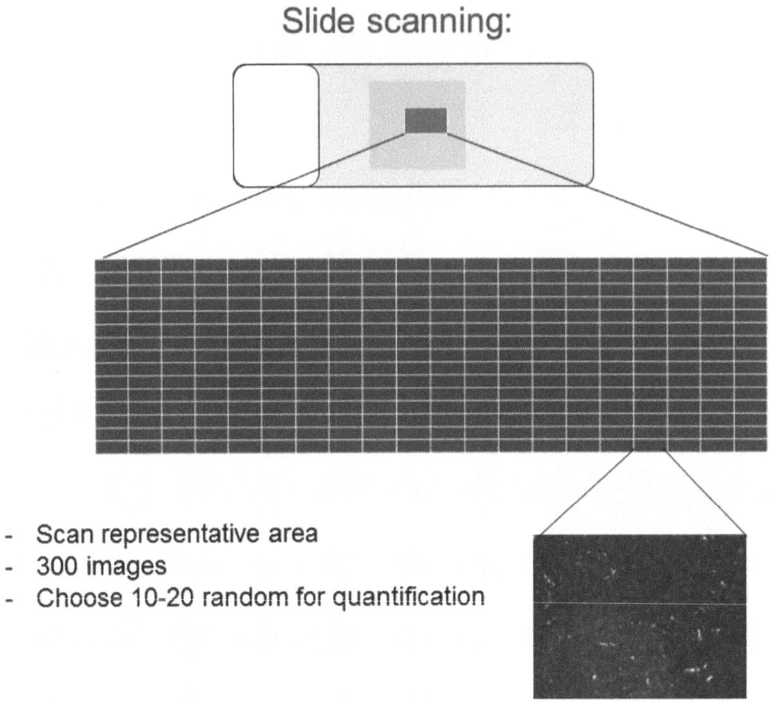

Fig. 4 Diagram illustrating the large-scale imaging of F-actin bundles using widefield fluorescence microscopy. Here we show a diagram of a slide with blue shaded area representing cover glass and gray rectangle—a region that is being scanned. The scanned area contains 300 snapshots, each was taken as an individual image/picture. The lower insert shows an example of one such snapshot image. Later, we randomly choose 10–20 of such snapshots that are spread out over the scanned area for quantification (please see more detailed description in the text of Protocol 4). *Adapted from Vaynberg, J., Fukuda, K., Lu, F., Bialkowska, K., Chen, Y., Plow, E. F., et al. (2018). Non-catalytic signaling by pseudokinase ILK for regulating cell adhesion.* Nature Communications *9(1), 4465.*

focusing on individual bundles, allowed us to clearly demonstrate that a mutation in the WH2 domain of PINCH1 that weakens its binding affinity to actin affects bundling ability of IPP. Additionally, we were able to quantify the difference in bundle sizes and the amount of formed bundles under different conditions (Fig. 5).

To visualize actin fibers and bundles, actin needs to be labeled with a fluorescent probe conjugated to phalloidin. It is critical for this experiment to remove even the smallest actin aggregates; monomeric G-actin should be purified by size exclusion chromatography. Since G-actin will polymerize in solution even in the absence of ions at concentrations over 7–8 μM, the

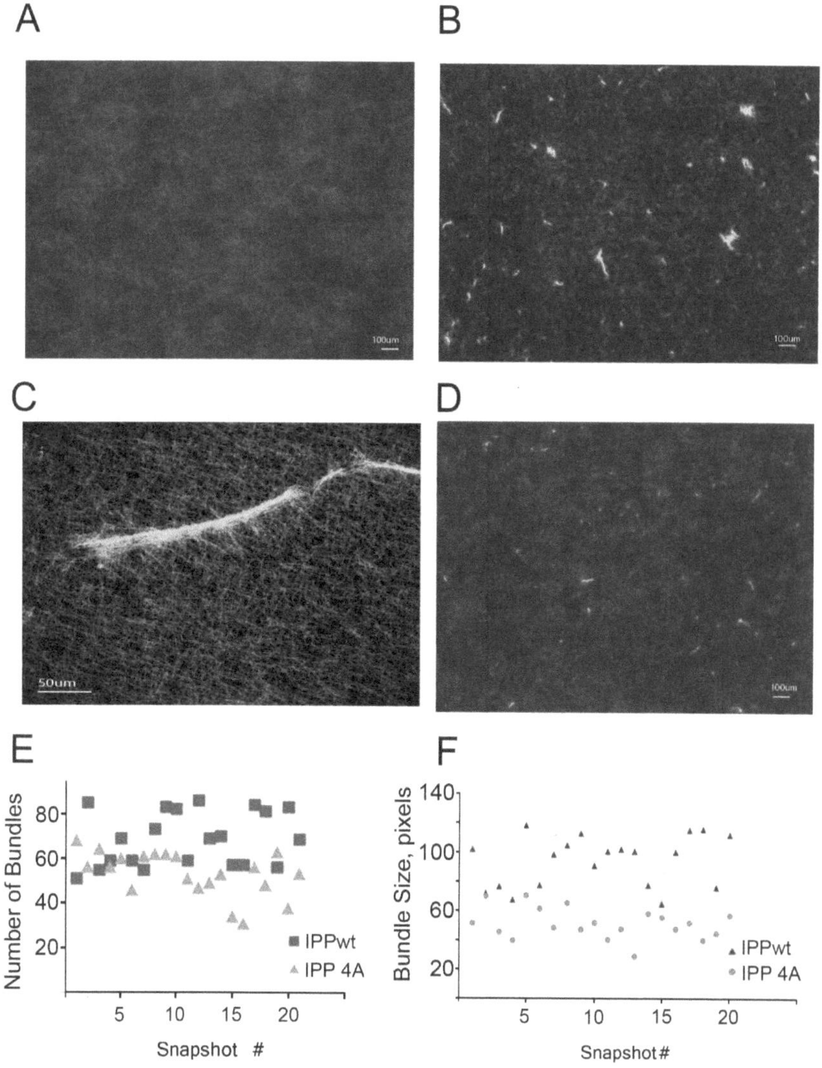

Fig. 5 Microscopy imaging and quantification of IPP induced actin bundles. This figure demonstrates comparison of confocal and widefield imaging. Here, G-actin was polymerized with addition of actin polymerizing buffer in presence of either IPP buffer, or IPPwt or IPP-4A mutant protein where actin binding WH2 domain of PINCH1 was disabled by mutating 4 of the actin binding residues into Alanines (Vaynberg et al., 2018). F-actin was labeled with Alexa488-phalloidin. (A) Selected microscopic image (a snapshot) showing the uniform F-actin filament network in the presence of the buffer alone taken with a widefield microscope. (B) Selected microscopic image (a snapshot) showing F-actin bundles in the presence of IPP taken with a widefield microscope. (C) A high resolution image of IPP-induced F-actin bundle

(Continued)

solutions should be at a concentrations below 7–8 μM, such as 5 μM. This concentration is fortunately enough for microscope visualization. Actin purchased from Cytoskeleton Inc. should be further purified using a smaller volume (24 mL) size exclusion chromatography column. Again, test protein/F-actin ratio and dilution factors have to be adjusted and incubation time may also matter. Sometimes actin may quickly polymerize but it may take couple hours for the bundles to come to equilibrium with the solution. Coverslips need to be coated with Poly-L-Lysine—positively charged molecules will fix negatively charged actin fibers much better and prevent floating that makes it difficult to focus. The sealing of a coverslip on a slide should be done with a nonvolatile chemical, similar to wax, that will prevent evaporation over hours (scanning of a slide may take time) yet not allow any organic solvent vapors to leak inside and disturb the bundles. We used VALAP sealant which is an equal part mixture of Vaseline, Lanolin and Paraffin wax.

Reagents

- 1 mg aliquot of skeletal muscle actin (Cytoskeleton, Inc., Catalog# AKL99)
- General Actin Buffer (5 mM Tris-Cl, pH 8.0, 0.2 mM $CaCl_2$, 0.2 mM ATP, 0.5 mM DTT).
- Actin Polymerization Buffer, 10× (50 mM KCl, 2 mM MgCl2, 1 mM ATP).
- F-actin Buffer: 90% G-buffer +10% of 10 × Actin Polymerization Buffer

Fig. 5—Cont'd detected by confocal microscopy. (D) Selected image (a snapshot) taken with a widefield microscope showing that IPP-4A mutations led to drastically reduced F-actin bundle formation as compared to (B). Both IPPwt and IPP-4A induce actin bundles with a range of sizes since mutated WH2 domain of PINCH1 still retains some binding to F-actin. If we only use confocal microscope, we are only able to closely image very small area (although at a higher resolution like on panel (C) and since panels (B) and (D) both contain bundles of similar sizes, we may or may not be only randomly focusing on similar sized bundles and not detect that size distributions and the number of bundles in the solution are very different between the two samples. (E) and (F) Quantitative comparison on the number (E) and size (F) of the F-actin bundles induced by IPP and IPP-4A, respectively. Bar = 100 μm except in (C) where bar = 50 μm. *Adapted from Vaynberg, J., Fukuda, K., Lu, F., Bialkowska, K., Chen, Y., Plow, E. F., et al. (2018). Non-catalytic signaling by pseudokinase ILK for regulating cell adhesion.* Nature Communications *9(1), 4465.*

- ATP, 100 mM stock (Cytoskeleton, Inc., Catalog#BSA04). Reconstitute according to manufacturer's instructions. Aliquot and store at −20 °C. Thaw on ice and do not reuse the aliquots
- $MgCl_2$, 100 mM stock. (VWR Catalog# 97064-150). Dissolve 9.5 mg in 1 mL of purified water. Aliquot and store at −20 °C
- DTT (Dithiothreitol), 100 mM stock. (VWR, Catalog# EM-3860). Weigh 15.4 mg and dissolve in 1 mL of purified water. Aliquot and store at −20 °C. Thaw on ice. Do not reuse the aliquots
- Poly-L-Lysine MW 70,000-150,000; 0.01% solution, (Sigma, Catalog# P4707)
- Alexa Fluor 488 Phalloidin (Thermo Fischer Scientific, Catalog# A12379), reconstituted stock according to manufacturer's instructions. Aliquot and store at −20 °C in a foil wrapped box
- VALAP sealant (prepare by mixing Vaseline, Lanolin and Paraffin wax flakes- combine equal parts by weight and gently heat until melts) and q-tips for application. VALAP is solid at room temperature. Prewarm on an electric plate heater on a low to medium heat setting until liquid before using
- Laemmli Sample Buffer (BioRad, Catalog# 161-0737).
- PageRuler Broad Range Unstained Protein Ladder (Thermo Fischer Scientific, Catalog# 26630).

Equipment

- FPLC system with UV detection and automatic fraction collector such as AKTA Purifier FPLC UPC10 system with Frac-920 collector (Cytiva, Catalog# 28406268)
- Superdex 75 10/300 GL column (Cytiva, Catalog# 29148721). It is better to use smaller volume column since we are dealing with a rather diluted concentration of actin—10/300GL has 25 mL volume. In a large 120 mL volume column, the small peak may not be as apparent
- Spectrophotometer suitable for measuring absorbance in UV/Vis range with a cuvette for protein concentration determination
- Superfrost Microscope Slides 75x25mm (Cardinal Health, Catalog# M6156).
- Microscope Cover Glass 22 × 22-2 (FisherBrand, Catalog# 12-540-B).
- Widefield upright fluorescent microscope with capability of slide scanning. We used (Leica Microsystems, GmbH, Wetzlar, Germany), equipped with a Retiga SRV camera and QCapture Pro software (QImaging, Surrey, BC Canada).

- Optional: Confocal microscope. We used Leica TCS-SP5 II upright confocal microscope (Leica Microsystems, GmbH, Wetzlar, Germany).
- Falcon tubes, 15 mL Volume (Millipore Sigma, Catalog# CLS430791)
- Eppendorf tubes 500, or 1000, or 1500 μL volume (such as ThermoFisher Scientific, Catalog# 90410)
- Nunc Petri Dishes (ThermoFisher Scientific, Catalog# 263991).

6.1 Protocol 4

Preparation of monomeric G-actin. (Steps 1–8) may be done several days in advance

1. Thaw 1 mg aliquot of Actin skeletal muscle lyophilized actin (Cytoskeleton, Inc.) by inverting the tube on the lab bench and placing some room temperature water on top of the inverted tube, leave for 2 min then place on ice
2. Add 100 μL of ddH2O and resuspend lyophilized powder well by pipetting several times up and down. The solution is quite viscous so we cut off several millimeters off the tip to create a wider opening for easier pipetting. Briefly vortex and then spin down for 30 s at 10,000–13,000 rpm. This will be 10 mg/mL stock of Actin. It can be aliquoted and frozen in liquid nitrogen at −80 °C. Actin is more stable at 10 mg/mL then at more diluted concentrations
3. Add 1 mL of General Actin Buffer to the above tube (final actin concentration now is 1 mg/mL; ~23 μM), mix well by pipetting up and down several times and keep on ice for 30 min. Transfer the solution to a 15 mL falcon tube
4. Add further 2.5 mL of General Actin buffer to final concentration of 0.4 mg/mL (~10 μM).
5. Meanwhile, at 4 °C equilibrate Superdex 75 10/300 GL column with 3 column volumes of General Actin Buffer (make sure pH is 7–8 after adding fresh ATP and DTT) with 0.5 mL/min flow using an FPLC instrument. It is acceptable to run the FPLC system at room temperature since the runs are short (~50 min long) but it is preferable that fraction collector (and collected fractions) is located at 4 °C. If this is impossible, remove collected fractions right after each run and transfer them immediately to either ice or 4 °C until analyzed
6. Load Actin solution, about 800–1000 μL at per Size Exclusion Chromatography run. Collect 0.5 mL fractions. Repeat the runs two more times and collect all the fractions in the same tubes. Actin elutes

between 7 and 10 mL. If actin aggregates are present, there may be a shoulder at or before 7 mL, but only major peak—monomeric actin-is useful

7. Mix 20 μL of each fraction with 5 μL of Laemmli Sample Buffer. Boil the samples for 10 min at 95 °C and then spin for 10 min at 13,000 rpm. Load the entire sample onto SDS-PAGE gel (anything between 10 and 15% or gradient 4–15%) of fractions. Use commercial protein ladder for identifying Actin band (Actin MW is 43 kDa).
   Stain the gel with Coomassie stain. Please refer to equipment section of protocol 1 for detailed description of the gel system we used and stain alternatives
8. Measure OD290 of all the fractions of actin peak using General Actin Buffer as a blank and determine actin concentration (Actin has an extinction coefficient at 290 nm of 26,600 M^{-1} cm^{-1}. 1 mg/mL of actin has OD290 0.63).

Aliquot the fractions with the highest concentrations of actin (usually 3–5 μM) into new Eppendorf tubes, 100 μL/tube, do not mix them, label with the fraction number snap freeze in Liquid Nitrogen, and store at −80 °C until ready to use

6.2 Microscope visualization of bundles

9. At least a day in advance, soak several coverslips in poly-L-lysine solution for 1 h at room temperature in a Petri dish. Rinse twice with distilled water by filling the dish enough to cover the glasses and then removing the water with any pipette. Use a tweezer to pick up the glasses from the dish. Spread out on clean paper towel to dry
10. Prepare samples of 10–20 μL total volume, in duplicates: generally, 1/10th volume of 10 × Actin Polymerization Buffer + test protein; then add G-actin to the final concentration of ~1 μM. Let it polymerize at Room Temperature for about an hour, then add 2 μL of Alexa488-phalloidin, wrap the samples in foil and let it stand for another hour. It is useful to test various concentrations of test protein. Also, it is crucial to prepare a test protein buffer control. We found in our case that 1 μM final actin concentration is a good starting point. Then, in case of 5 μM G-actin stock and 20 μM test protein stock, we would mix in 1:1 actin: IPP ratio:2 μL of Actin Polymerization Buffer, 4 μL of G-actin, 1 μL of IPP and 13 μL of IPP buffer. For test protein buffer control we would only add 14 μL of the IPP buffer without IPP

It is important to note that these are the starting conditions.
The exact actin:test protein ratio will have to be optimized. Factors like strength of interaction with actin, nature of its effect on actin polymerization will affect the optimal amount of test protein. At the beginning, a range of concentrations such as 0.1 μM, 1 μM and 3 μM may be mixed with 1 μM G-actin. It will then become clear whether one needs to try lower or higher concentrations next

11. Mix a sample gently by pipetting, take 1 μL and add to 9 μL of F-actin buffer.
Mix gently again and place all 10 μL on the cover glass. Carefully turn it over on a slide and seal with prewarmed VALAP around the perimeter of the cover glass by applying the VALAP mixture with a q-tip (it will almost immediately solidify). Keep the slides in the dark always

12. Scan a representative portion of a slide to get an idea of bundle distribution and to be able to quantify formed bundles. We scanned an area of 24 mm × 16 mm which resulted in 300 snapshots per sample. Neither IPP buffer control (Fig. 5A), nor PINCH1 or Parvin alone were able to induce bundles (Vaynberg et al., 2018), only actin fibers are seen in those samples yet IPP complex induced formation of bundles (Fig. 5B).

6.3 Quantification

Bundles may be analyzed by randomly choosing 15–20 snapshots and counting the formed bundles. With 300 snapshots, we chose every 15th snapshot (#1, #15, #30, etc.). In our case of snapshot order taking, this would ensure that no chosen to quantify snapshots were adjacent to each. Generally, it is best to choose snapshots that would be well spread out over the analyzed area rather than focused on one corner of the scanned area to maximize the slide area quantified.

Software like ImagePro Plus allows one to count either all bundles or set limits by length and width. Quantification allows to evaluate effects of different variables either in solution environment or test protein (mutations) on bundling ability and morphology of obtained bundles. For our purposes we collected information on the number of bundles (number of objects over a chosen threshold of intensity) and their length (ImagePro Plus allows to choose longest dimension of an object).

7. Summary

In this article, we provide a detailed description of how we used various assays to assess how IPP engages with actin and induce bundle formation. We hope this information will be useful for researchers in the field to continue the investigation of ILK function. The information may be also useful for other researchers who study the actin cytoskeleton regulation.

Acknowledgment

This work was supported by NIH R01 HL58758.

References

Breitsprecher, D., Kiesewetter, A. K., Linkner, J., & Faix, J. (2009). Analysis of actin assembly by *in vitro* TIRF microscopy. In T. Jin, & D. Hereld (Eds.), *Vol. 571. Chemotaxis, methods in molecular biology* Springer.

Cooper, G. M. (2000). *The cell: A molecular approach* (2nd ed.). Sunderland, MA: Sinauer Associates. ISBN 10:0-87893-106-6.

Fukuda, K., Gupta, S., Chen, K., Wu, C., & Qin, J. (2009). The pseudoactive site of ILK is essential for its binding to alpha-Parvin and localization to focal adhesions. *Molecular Cell, 36*, 819–830.

Fukuda, K., Knight, J. D., Piszczek, G., Kothary, R., & Qin, J. (2011). Biochemical, proteomic, structural, and thermodynamic characterizations of integrin-linked kinase (ILK): Cross-validation of the pseudokinase. *The Journal of Biological Chemistry, 286*, 21886–21895.

Goldring, D. J. P. (2012). Protein quantification methods to determine protein concentration prior to electrophoresis. *Methods in Molecular Biology, 869*, 29–35.

Hertzog, M., & Carlier, M. F. (2005). Functional characterization of proteins regulating actin assembly. *Current Protocols in Cell Biology*. Chapter 13: Unit 13.6.

Joel, P. B., Fagnant, P. M., & Trybus, K. M. (2004). Expression of a nonpolymerizable actin mutant in Sf9 cells. *Biochemistry, 43*, 11554–11559.

Kabsch, W., Mannherz, H. G., Suck, D., Pai, E. F., & Holmes, K. C. (1990). Atomic structure of the actin:DNase I complex. *Nature, 347*, 37.

Legate, K. R., Montanez, E., Kudlacek, O., & Fässler, R. (2006). ILK, PINCH and parvin: The tIPP of integrin signalling. *Nature Reviews Molecular Cell Biology*, 7, 20–31.

McLaughlin, P. J., Gooch, J. T., Mannherz, H. G., & Weeds, A. G. (1993). Structure of gelsolin segment 1-actin complex and the mechanism of filament severing. *Nature, 364*, 685.

Parsons, J. T., Horwitz, A. R., & Schwartz, M. A. (2010). Cell adhesion: Integrating cytoskeletal dynamics and cellular tension. *Nature Reviews. Molecular Cell Biology, 11*(9), 633–643.

Qin, J., & Wu, C. (2012). ILK: A pseudokinase in the center stage of cell-matrix adhesion and signaling. *Current Opinion in Cell Biology, 24*(5), 607–613.

Schutt, C. E., Myslik, J. C., Rozycki, M. D., Goonesekere, N. C., & Lindberg, U. (1993). The structure of crystalline profilin-beta-actin. *Nature, 365*, 810.

Steinberg, T. (2009). Protein gel staining methods: An introduction and overview. *Methods in Enzymology*, *463*, 541–563.
Tan, S. (2001). A modular polycistronic expression system for overexpressing protein complexes in *Escherichia coli*. *Protein Expression and Purification*, *21*, 224–234.
Vaynberg, J., Fukuda, K., Lu, F., Bialkowska, K., Chen, Y., Plow, E. F., et al. (2018). Non-catalytic signaling by pseudokinase ILK for regulating cell adhesion. *Nature Communications*, *9*(1), 4465.
Winder, S. J., & Ayscough, K. R. (2005). Actin binding proteins. *Journal of Cell Science*, *118*(Pt 4), 651–654.
Zhang, Y., Chen, K., Tu, Y., Velyvis, A., Yang, Y., Qin, J., et al. (2002). Assembly of the PINCH-ILK-CH-ILKBP complex precedes and is essential for localization of each component to cell-matrix adhesion sites. *Journal of Cell Science*, *115*(Pt 24), 4777–4786.

CHAPTER SEVEN

Production and analysis of titin kinase: Exploiting active/inactive kinase homologs in pseudokinase validation

Julius Bogomolovas[a], Peter Gravenhorst[b], and Olga Mayans[b,*]
[a]School of Medicine, University of California, San Diego, La Jolla, CA, United States
[b]Department of Biology, University of Konstanz, Konstanz, Germany
*Corresponding author: e-mail address: olga.mayans@uni-konstanz.de

Contents

Abstract

Protein pseudokinases are key regulators of the eukaryotic cell. Understanding their unconventional molecular mechanisms relies on deciphering their putative potential to perform phosphotransfer, their scaffolding properties and the nature of their regulation. Titin pseudokinase (TK) is the defining member of a family of poorly characterized muscle-specific kinases thought to act as sensors and transducers of mechanical signals in the sarcomere. The functional mechanisms of TK remain obscure due to the challenges posed by its production and analysis. Here, we provide guidelines and tailored research approaches for the study of TK, including profiting from its close structure-function relationship to the catalytically active homolog twitchin kinase (TwcK) from *C. elegans*. We describe a methodological pipeline to produce recombinant

Methods in Enzymology, Volume 667
ISSN 0076-6879
https://doi.org/10.1016/bs.mie.2022.03.028

TK and TwcK samples; design, prioritize and validate mutated and truncated variants; assess sample stability and perform activity assays. The strategy is exportable to other pseudokinase members of the TK-like kinase family.

1. Introduction

The contractile cytoskeleton of the muscle cell—the sarcomere—contains various gigantic filamentous proteins of the titin family, which act as passive elastomers that sense and translate mechanical signals derived from exercise. Members of this family are titin and obscurin in vertebrates; twitchin/UNC-22, the obscurin homolog UNC-89 and the small titin TTN-1 in nematodes; twitchin in mollusks; projectin, UNC-89 and stretchin in insects (Bullard, Linke, & Leonard, 2002; Kontrogianni-Konstantopoulos, Ackermann, Bowman, Yap, & Bloch, 2009). Distinctive of these proteins is a conserved multi-domain region located near their C-terminus that includes one or two kinase domains. Despite the invariable presence of such sarcomeric kinases across the muscle biodiversity, their physiological function is poorly understood and most members of the family remain uncharacterized. Notably, several kinases of this family lack canonical active site motifs, which has led to their classification as pseudokinases predicted to be inactive (Bogomolovas et al., 2014; Manning, Whyte, Martinez, Hunter, & Sudarsanam, 2002; Mayans, Benian, Simkovic, & Rigden, 2013). The main representative of the family, human titin kinase (TK), is regarded as a pseudokinase whose catalytic motifs are moderately degenerated when compared to other family members (Bogomolovas et al., 2014; Manning et al., 2002). TK contains YMAK and EFG motifs (instead of canonical VAIK and DFG sequences) in the ATP-binding pocket and the Mg^{2+}-binding site, respectively. The VAIK motif, located in strand-β3, contains the essential catalytic lysine that ion-pairs and optimally positions the non-transferable α- and β-phosphates of ATP. Among active kinases, small hydrophobic residues dominate position 2 of this motif, as this position forms the bottom of the cavity that accommodates the purine moiety of ATP (Endicott, Noble, & Johnson, 2012). The methionine residue in TK is speculated to sterically hinder ATP from binding in a productive fashion and, thereby, to hamper phosphotransfer activity (Bogomolovas et al., 2014). In the DFG motif, aspartate chelates the magnesium ion commonly required for catalysis and its substitution for glutamate is known to inactivate phosphotransfer (*e.g.*, Bogomolovas et al., 2014), which is also expected to occur in TK. These deviating methionine and glutamate residues are conserved in TK sequences across vertebrate (sequence alignments in Supplementary Materials to Bogomolovas et al., 2014). However, these substitutions do not necessarily abolish phosphotransfer in pseudokinases. ATR kinase, which contains a methionine positioned as that of TK, is active (Kim, Lim, Canman, & Kastan, 1999) and the CASK CaM-pseudokinase, which lacks the DFG motif, has been proposed to perform phosphotransfer in a Mg^{2+}-independent manner (Mukherjee et al., 2008). Thus, the question of whether in TK the YMAK/EFG motifs result in inactivity or, in contrast, constitute a compensatory residue exchange that retains phosphotransfer capability requires careful investigation.

Importantly, the titin kinase family also comprises active members with canonical active sites. The best characterized of these is twitchin kinase from *C. elegans* (*ce*TwcK), which displays robust levels of phosphotransfer activity on peptide model substrates *in vitro* (Heierhorst et al., 1996; von Castelmur et al., 2012). *ce*TwcK phosphotransfer has been linked to the regulation of muscle contraction by assaying a K to A inactive mutant in nematodes (Matsunaga et al., 2017). Despite differences in their catalytic motifs, TK and *ce*TwcK share a remarkable structural similarity. Both kinases are part of a *Fn-linker-kinase-tail-Ig* segment (where Fn and Ig refer to fibronectin and immunoglobulin domains, respectively) that share domain composition and overall 3D-architecture (Fig. 1A and B). Crystal structures of this multi-domain segment from human titin (Bogomolovas et al., 2021) and *C. elegans* twitchin (von Castelmur et al., 2012) have shown that it adopts a compact arrangement, where the kinase domain is enwrapped by N- and C-terminal tail extensions with the flanking Fn and Ig domains protruding from the assembly. The N-terminal linker (NL) joining the Fn domain to the kinase wraps around the N-terminal kinase lobe, packing against the inter-lobular hinge region and hindering lobe motions. The C-terminal linker (CRD; C-terminal Regulatory Domain), joining the kinase to the subsequent Ig domain, wraps around the C-terminal kinase lobe, blocking both the ATP binding pocket and the peptide substrate binding site. In *ec*TwcK, both NL and CRD extensions have been shown to be inhibitory and to act in synergy to silence phosphotransfer activity (von Castelmur et al., 2012). In addition to overall multi-domain architecture, TK and *ec*TwcK share a close sequence conservation of their kinase domains (40% sequence identity; 65% similarity) and very high structural agreement ($RMSD_{C\alpha}=1.29\,Å$), which spans ATP-binding pockets, β3 strands containing the essential lysine, and the $^{D}/_{E}$FG motifs (Fig. 1C and D). This suggests that *ce*TwcK could act as a suitable surrogate to investigate the functional output of TK's pseudokinase motifs on phosphotransfer catalysis.

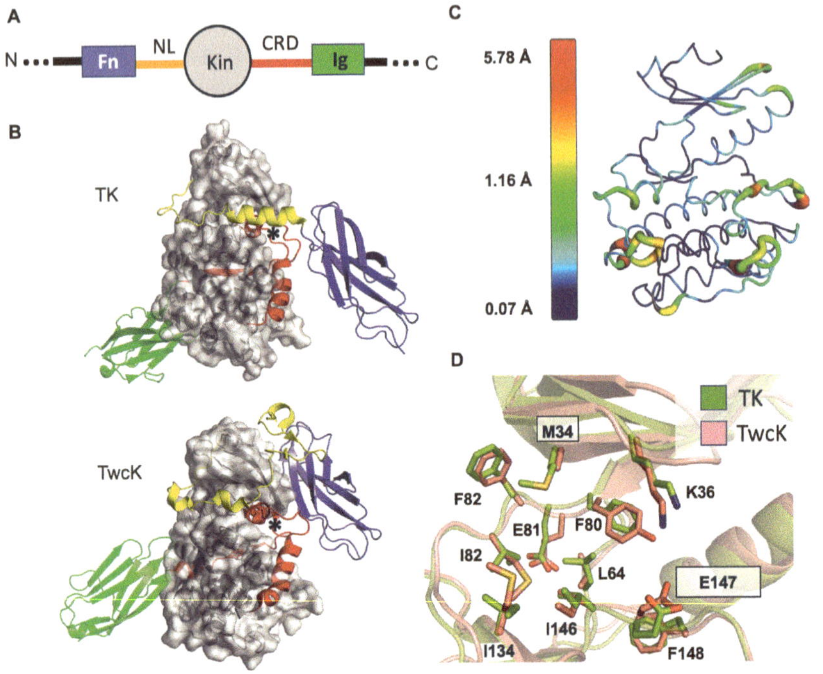

Fig. 1 Structural features of human TK and *C. elegans* TwcK. (A) Domain composition of the conserved kinase region of human titin and nematode twitchin (Fn: fibronectin domain; NL: N-terminal linker; kin: kinase domain; CRD: C-terminal regulatory domain; Ig: immunoglobulin domain); (B) Crystal structure of human TK (PDB: 6YGN; Bogomolovas et al., 2021) and *ce*TwcK (PDB: 3UTO; von Castelmur et al., 2012) color-coded as in (A). The occupation of the ATP-binding pocket by a helical component of the CRD tail is indicated with an asterisk; (C) Structural superposition of TK and *ce*TwcK. Ribbon thickness and coloring indicate the RMSD values of the superposition as given in the accompanying scale (minimum, maximum and average values are stated on the scale). The structural agreement is excellent overall. Only divergences displayed in orange/red are moderately significant. These are confined to mostly non-conserved residues located in outer, flexible loop positions and occasionally involved in crystallographic packing. An examination of divergences in the activation loop shows them to fall in this category and indicate them not to be functionally relevant. (D) Comparison of the ATP-binding pockets and $^{D}/_{E}$FG motifs of TK (green) and *ce*TwcK (pink). Methionine and glutamate residues responsible for the pseudokinase character of the TK are boxed (Figure created with PyMOL using the "cartoon putty" command and the ColorbyRMSD plug-in; https://pymol.org). *Panels (C and D) are reproduced from Bogomolovas, J., Gasch, A., Simkovic, F., Rigden, D.J., Labeit, S., & Mayans, O. (2014). Titin kinase is an inactive pseudokinase scaffold that supports MuRF1 recruitment to the sarcomeric M-line.* Open Biology, 4, *140041.*

The activation of TK and TwcK requires the removal of NL and CRD tails from the kinase domain. In the sarcomere *in vivo*, such removal appears to be induced by the pulling forces that emerge during muscle function. The mechanical unwinding of the tails is supported by *in vitro* data acquired by atomic force microscopy (Greene et al., 2008; Puchner et al., 2008) and, most recently, by *in vivo* data that monitored conformational changes in FRET-labeled *ec*TwcK in live, freely swimming *C. elegans* (Porto et al., 2021). The mechanical unwinding of the tails is expected to leave the kinase domain transiently exposed and, thereby, enable phosphotransfer activity (Greene et al., 2008; Porto et al., 2021; Puchner et al., 2008). However, a kinase catalytic output induced by stretch-unfolding of the regulatory tails is yet to be demonstrated experimentally. To this effect, a single *in vitro* study has been reported that aimed to test the mechanoactivation of molluscan TwcK by using smooth muscles from *Mytilus* that had been permeabilized and soaked in a model peptide substrate (with sequence KKRAARATSNVFA). Here, a 10% stretch resulted in a two-fold increase in phosphorylation of the model substrate (Butler & Siegman, 2011).

Aside from phosphotransfer considerations, TK and TK-like kinases are known to act as scaffolds for a variety of signaling proteins (Kontrogianni-Konstantopoulos et al., 2009; Mayans et al., 2013). Human TK recruits several turn-over factors to the sarcomere M-line; namely, the E3 ubiquitin ligases MuRF1 and MuRF2, and the autophagosomal receptors Nbr1/p62 (Bogomolovas et al., 2021; Centner et al., 2001; Lange et al., 2005). These interactions have been linked to cellular pathways that regulate gene expression in the myofibril as well as being speculated to participate in the onset of myofibrillar atrophy. In parallel, *ce*TwcK has been shown to bind and be phosphorylated by MAK-1/p38 MAPK kinase that is linked to cellular stress, possibly linking the latter to the regulation of sarcomere contraction (Matsunaga, Qadota, Furukawa, Choe, & Benian, 2015). In both TK and *ce*TwcK scaffolding appears to be regulated by the mechanical unraveling of the NL and/or CRD tail extensions. Hence, TK-like kinases are thought to act as mechanosensory nodes in the sarcomere that mediate cellular responses to mechanical stress.

Despite the physiological and clinical significance of muscle adaptation to mechanical load, TK-like kinases remain understudied. Mechanistic studies of their phosphotransfer activity and mechanosensory scaffolding are hampered by the difficulty of their production and analysis in activated states. Building on the latest structural and biochemical knowledge of these kinases, we describe here a practical strategy for the exploration of human

TK aided by *ce*TwcK as a kinase surrogate. This strategy is applicable to other complementary pseudokinase/kinase pairs members of the TK-like kinase family.

2. Challenges in the mechanistic analysis of the titin pseudokinase

The mechanistic characterization of a kinase minimally requires its production in an active state and the availability of a test substrate. At present, these experimental conditions are poorly met by TK. An early crystal structure of TK (Mayans et al., 1998) suggested that its active site is dually inhibited by (1) the CRD extension and (2) a tyrosine residue (Y170; residue 32350 in UniProtKB entry Q8WZ42) that hydrogen bonds the catalytic aspartate base (HFD) (Fig. 2). TK activation is thought to require both the mechanical removal of the CRD tail and the release by phosphorylation

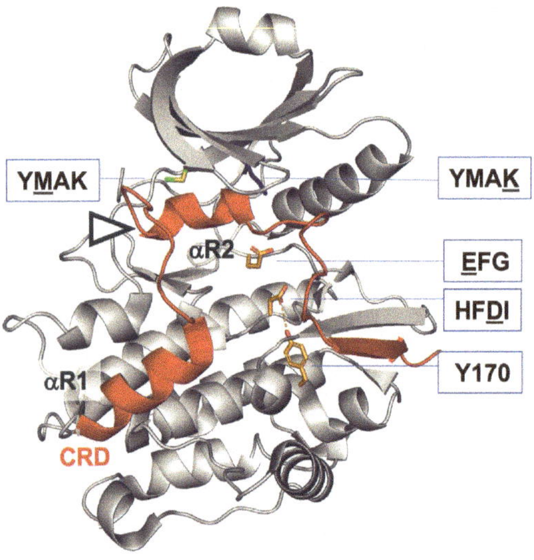

Fig. 2 Dual inhibition of titin pseudokinase. Crystal structure of TK (PDB code 1TKI; Mayans et al., 1998), where essential and divergent functional residues are displayed. Those include the essential lysine and atypical methionine residues in the YMAK motif, the atypical glutamate in the EFG (rather than the conventional DFG) motif, the catalytic aspartate in the HFDI sequence and residue Y170 in the P+1 loop. The blockage of the ATP binding site by α-helix αR2 from the CRD tail is marked with a triangle. The C-terminal end of the CRD tail forms an antiparallel β-strand with the activation loop of TK, which is in open conformation, and thereby allegedly blocks the binding of the peptide substrate. In the CRD, helix αR1 that is essential for TK stability and helix αR2 that blocks the ATP-binding site are labeled.

of residue Y170. However, the production of recombinant TK variants mimicking these processes is poorly attainable. The CRD inhibitory tail, by means of its tight packing against the kinase domain (Figs. 1 and 2), also plays a structural role and is crucial for kinase stability. Thus, its truncation as to mimic the postulated mechanically-open state leads to an unstable, insoluble sample. In addition, the requirement for phosphorylation of one or both tyrosine residues in the APEY$\underline{Y}^{170}$APE motif in the P + 1 loop (which supports the binding of the peptide substrate) has been proposed but it is currently insufficiently analyzed. Initially, Y170 was suggested to be the target of an unidentified kinase in developing C2C12 cells (Mayans et al., 1998), while later studies proposed it as a site of autophosphorylation (Puchner et al., 2008). In all instances, Y170 phosphorylation has been assumed to activate TK's phosphotransfer activity.

Interestingly, *ce*TwcK contains an equivalent threonine-rich motif in its P + 1 loop, KVTTG$\underline{T}^{316}$AE, which undergoes autophosphorylation at residue T316 (residue 6430 in UniProtKB entry Q23551) (Williams et al., 2018). Multiphosphorylated *ce*TwcK as well as variants TwcK^{T316A} and TwcK^{T316D} were largely inactive, while dephosphorylated *ce*TwcK that had been treated with λ-phosphatase showed the highest level of activity. Thus, it was concluded that T316 phosphorylation in *ce*TwcK is inhibitory. Phosphorylation sites in the P + 1 loop are also known in other kinases, where they can have either inhibitory or activating effects on phosphotransfer, as exemplified by ERK1 (Lai & Pelech, 2016; further examples are in the Supplementary Table S1 of that publication). Thus, at the current time, the activating or inhibitory nature of the proposed Y170 phosphorylation site in TK remains unestablished and, thereby, the constitution of TK's activated state unclear.

An additional difficulty in characterizing the phosphotransfer activity of TK is the lack of an established substrate. To this date, efforts to identify TK candidate substrates have yielded the sarcomeric protein Tcap/telethonin (Mayans et al., 1998) and the autophagy receptor Nbr1 (Lange et al., 2005). Tcap is an intrinsically disordered protein that is largely insoluble in isolation. It becomes structured upon binding to the N-termini of two titin molecules, which it cross-links by forming an intermolecular β-sheet (Zou et al., 2006). This interaction occurs at the sarcomeric Z-disk located *ca.* 1 μm away from the TK locus in the M-line. As both TK and Tcap are non-diffusible proteins in mature muscle, the physiological relevance of this substrate (originally tested in C2C12 myocytes) was thought to be confined to developmental muscle stages (Mayans et al., 1998). A later study

(Bogomolovas et al., 2014), however, crucially showed that the insect cell preparations commonly used to produce recombinant TK contained a kinase contaminant that phosphorylated Tcap on a Ca^{2+}/calmodulin-enhanced manner, as originally attributed to TK. Concurrently, the phosphorylation site in Tcap, S157, assigned to TK (Mayans et al., 1998; Zou et al., 2003) was found to be a target of Ca^{2+}/calmodulin-dependent kinase II and protein kinase D (Candasamy et al., 2014). Adding to this, TK variants produced in bacterial systems revealed no detectable phosphotransfer activity on Tcap, even though crystallographic validation and mass spectrometry data showed TK samples of bacterial and eukaryotic origin to be in exact agreement (PDB: 1TKI, 4JNW; $RMSD_{C\alpha}=0.29$ Å) (Bogomolovas et al., 2014). Taken together, the data led to the view that Tcap is a questionable, if not artifactually assigned, substrate of TK.

As for Nbr1, its relationship to TK remains poorly explored. Nbr1 has been proposed to bind TK through its PB1 domain and to become thereby phosphorylated in a disordered neighboring sequence at sites S115 or S116 (Lange et al., 2005). However, attempts to characterize the TK/Nbr1 (PB1) interaction were unsuccessful (Müller, Kursula, Zou, & Wilmanns, 2006). Furthermore, recent work has shown that Nbr1 binds TK when the latter is in an ubiquitinated state, with Nbr1 interacting with the ubiquitin modifier through its C-terminal ubiquitin-binding domain (UBA) (Bogomolovas et al., 2021). To date, Nbr1-based peptides are yet to be tested in TK assays and, thus, Nbr1's potential utility for *in vitro* mechanistic investigations of TK remains unclear. To address the lack of a validated substrate for experimentation, studies were also directed to test TK phosphotransfer on the general kinase substrates myelin basic protein and casein, but no activity was revealed in either case (Bogomolovas et al., 2014). TK activity could also not be detected when extracts of either gastrocnemius muscle from adult mice or differentiating murine C2C12 myocytes were incubated with recombinant TK samples (Bogomolovas et al., 2014). Taken together, these results supported the view that TK is a pseudokinase with little or no phosphotransfer activity.

In stark contrast, *ce*TwcK benefits from an established *in vitro* model substrate (peptide KKRARAATSNVFS derived from chicken myosin light chain; Heierhorst et al., 1996; von Castelmur et al., 2012; Williams et al., 2018) that supports robust levels of phosphotransfer activity and its production in active states is possible using bacterial cultures. *ce*TwcK is also amenable to alteration by mutagenesis and re-engineering, being a robust acceptor scaffold for grafted motifs (Bogomolovas et al., 2014, 2021) and even exogenous protein moieties (Porto et al., 2021). Given its close similarity to TK (Fig. 1), *ce*TwcK offers a viable avenue to explore the functional output of TK pseudokinase features.

3. Impact of domain composition on sample production

In order to remove inhibition by flanking components, mechanistic studies of TK and its *ce*TwcK homolog require the generation of variants with truncated domain borders. However, the presence or absence of flanking elements has proven to be the most critical determinant of solubility and stability in recombinant samples of these kinases (production yields are shown in Table 1 and melting temperatures for *ec*TwcK variants of diverse composition are given in Fig. 3). Multi-domain variants of both TK and *ec*TwcK comprising the full-length *Fn-NL-kinase-CRD-Ig* segment are most stable and amenable to high yield expression, in agreement with the stabilizing contacts held by flanking regions revealed by crystal structures (Fig. 1B). Trimmed variants lacking the flanking Fn and Ig domains (*i.e.*, the fragment *NL-kinase-CRD*) or the NL tail (fragment *kinase-CRD*) can

Table 1 Recombinant production yields of constructs from TK and *ec*TwcK.

	TK		*ec*TwcK	
Construct	**Host**	**Yield**	**Host**	**Yield**
A168-M2[a]	Sf9	6 mg/L[b]	–	–
Fn-NL-Kin-CRD-Ig	*E. coli* soluBL21	4 mg/L	*E. coli* Rosetta (DE3)	60 mg/L[c]
NL-Kin-CRD	–	–	*E. coli* Rosetta (DE3)	10 mg/L
NL-Kin	–	–	*E. coli* Rosetta (DE3)	6–9 mg/L
Kin-CRD	*E. coli* soluBL21	4 mg/L	*E. coli* Rosetta (DE3)	6–9 mg/L
Kin-αR1	Sf9	0.02 mg/L[b]	*E. coli* Rosetta (DE3)	6–9 mg/L
Kin	Sf9/ *E. coli*	~	*E. coli* Rosetta (DE3)	6 mg/L

[a]Corresponding to Ig-Ig-Fn-NL-TK-CRD-Ig-Ig in titin.
[b]Puchner et al. (2008).
[c]von Castelmur et al. (2012).
(–): not tested; (~): trace amounts.

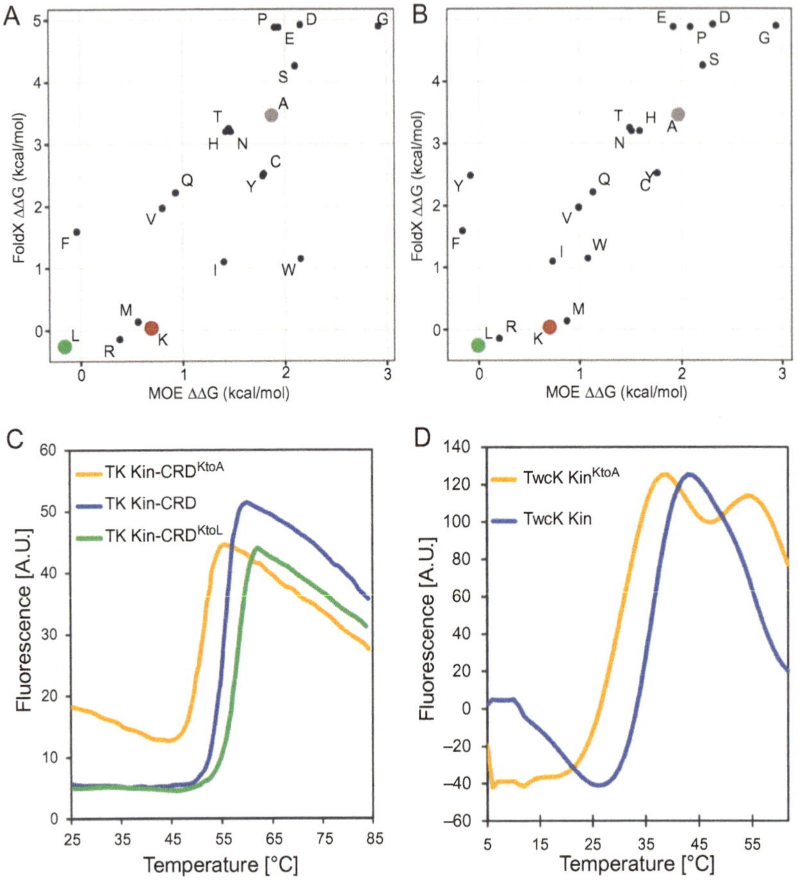

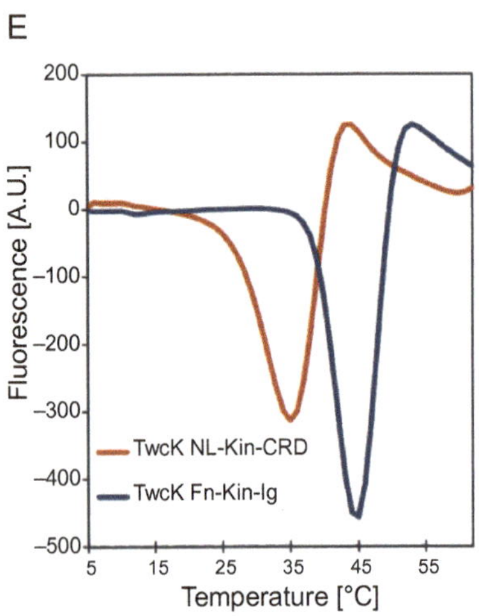

F

TwcK Construct	***Tm [°C]***
Fn-NL-Kin-CRD-Ig	48.39 ± 0.037
NL-Kin-CRD	39.18 ± 0.024
Kin	35.76 ± 0.025
Kin^{KtoA}	29.95 ± 0.021

Fig. 3 See figure legend on opposite page.

be produced with losses in yield. However, a particular challenge is the removal of the CRD tail, which drastically compromises stability to the point that the recombinant production of the single, exposed kinase domain of TK is unviable (Table 1).

A current strategy to overcome the above problems is to produce TK variants that carry a partially truncated CRD retaining some of its stabilizing interactions. The retention of the CRD's N-terminal helix, helix αR1 (Fig. 2), is minimally required, although the yields of production of such variant continue to be unpracticable for analytical characterization. An alternative is to generate TK variants where sections of the CRD tail are substituted for flexible dummy sequences. For example, to unblock the ATP-binding pocket, a version of TK has been reported where helix αR2 (Fig. 2) was substituted for a flexible GGSGGSA sequence in addition to carrying the phosphomimetic Y170E exchange ($TK^{\Delta R2/Y170E}$ in Bogomolovas et al., 2014). The production yield of such sequence substituted variants is limited, but viable (approx. 0.5–1 mg/L *E. coli*). In contrast to TK, the isolated kinase domain of *ec*TwcK (lacking regulatory tails) is obtainable in *E. coli* cultures and suitable for biochemical and biophysical analysis, although it displays modest half-life times (see Section 5). The minimal *ec*TwcK construct with best preserved stability and expression yield corresponds to the kinase domain carrying the first 9 amino acids of the CRD tail (residues 6108–6685 in UniProtKB entry Q23551), which contain a hydrophobic motif that packs against and stabilizes the kinase (von Castelmur et al., 2012; Williams et al., 2018). In summary, to date, the production of a fully exposed, kinase domain in sufficient yields to permit mechanistic studies is only achievable for *ec*TwcK.

Fig. 3 *In silico* and experimental analysis of TK and *ec*TwcK variants. (A and B) Predicted Gibbs-energy changes (ΔΔG) upon mutation of the essential lysine (TK: residue K53 in PDB 1TKI; *ec*TwcK: residue K185 in PDB 3UTO) to every other amino acid using MOE and FoldX for TK (A) and *ec*TwcK (B). The exchange to alanine (gray) is predicted as destabilizing. Replacement of this residue by leucine is predicted to retain structural stability; (C and D) DSF thermal denaturation curves recorded at λem = 590 nm for TK carrying the CRD extension (TK-CRD) (C) and the isolated kinase domain of *ec*TwcK lacking inhibitory tails (D). In agreement with *in silico* predictions, the substitution to alanine has a destabilizing effect on TK-CRD (ΔTm = −6.6 °C) and *ec*TwcK (ΔTm = −5.8 °C). In TK, substitution to leucine yields a sample of stability comparable to the wild-type (ΔTm = 2.2 °C); (E) DSF thermal denaturation curves for multi-domain forms of *ec*TwcK. An initial decrease in fluorescence is likely to reflect aggregation of the samples as first response event to increasing temperature, which is then followed by unfolding; (F) Overview of melting temperatures for diverse TwcK variants showing the stabilizing effect of regions flanking the kinase domain (domain nomenclature is as given in Fig. 1A).

4. Prediction of mutants for pseudokinase validation

In addition to truncated flanking elements, variants of TK (as well as the *ec*TwcK reference) carrying selected residue exchanges are required to assess the impact of individual motifs on functional output. In pseudokinases, residues counterpart to catalytic residues in conventional active kinases exhibit great variability and can carry an associated co-variance of other neighboring residues with compensatory functional effects. As an example, WNK kinase lacks the essential lysine residue, but a functional catalytic role is performed by an alternative nearby lysine (Min, Lee, Cobb, & Goldsmith, 2004). Testing the role of such unconventional features *via* mutagenesis can cause undesirable effects on sample stability, not always leading to an integral kinase scaffold that faithfully reports on the desired functional event. For example, in the integrin-linked pseudokinase domain (ILK) the instability of a "kinase-dead" K to M mutation, where the essential lysine in the ATP-binding pocket had been exchanged, was misinterpreted as proof of the biological relevance of ILK phosphotransfer activity (Fukuda, Knight, Piszczek, Kothary, & Qin, 2011). In TK, mutagenesis has been employed to test the role of individual residues in phosphotransfer activity both by its own mutation and by transferring test residues to the active *ec*TwcK homolog (Bogomolovas et al., 2014, 2021). Given the reduced stability of TK and *ec*TwcK variants carrying minimal or sequence-substituted CRD extensions, mutagenesis in this case has proven to benefit from the *virtual* prediction of the impact of residue exchanges on sample stability.

In mutagenesis prediction, the change in molecular stability that results from an amino acid substitution is estimated by calculating the difference in Gibbs energy of unfolding between the wild-type and the mutated structures, ΔΔG (kcal/mol), where $\Delta\Delta G = \Delta G_{mutant} - \Delta G_{wt}$ (Sanavia et al., 2020). Software commonly used to this effect are, *e.g.*, FoldX (Schymkowitz et al., 2005), the Rosetta routine ddg_monomer (Kellogg, Leaver-Fay, & Baker, 2011) and the drug discovery software platform Molecular Operating Environment (MOE; https://www.chemcomp.com). These programs use a protein structure as template to calculate positional effects in virtual mutational screens, where individual residues are mutated *in silico* into every other amino acid type. The protocols applied vary across software platforms, primarily relating to (i) the force fields and approximations applied; (ii) the conformational sampling of side-chain rotamers and (iii) the degree to which backbone adjustments of the protein chain are modeled to accommodate the new side chains. An increase in ΔΔG indicates in all cases a decrease in fold stability and, thereby, a destabilizing residue exchange.

It is important to note that ΔΔG values are not directly comparable across programs and that there is also no established threshold to classify a residue exchange as destabilizing enough so that it can compromise a given protein sample. The critical ΔΔG destabilization threshold is different for each individual protein as it is relative to the unfolding energy of the wild type molecule, with more stable proteins being able to tolerate a larger insult (Bromberg & Rost, 2009). Thus, ΔΔG values must be merely regarded as a guidance and interpreted in relative terms. To assist the interpretation of calculated ΔΔG values, a consensus of tendencies across programs is often used as evaluation criterion. Here, we illustrate the application of stability calculations to explore the tolerance of TK and *ec*TwcK to residue exchanges by identifying a suitable substitute for their essential lysine using FoldX and MOE.

The popular K to A mutation (TK: residue K53 in PDB 1TKI; *ec*TwcK: residue K185 in PDB 3UTO) is destabilizing for both TK and *ec*TwcK as experimentally demonstrated (see Section 6 for experimental validation of sample stability) and effectively predicted as such *in silico* (Fig. 3A–D). However, the exchange is nevertheless tolerated by the more robust *ec*TwcK sample that can be experimentally produced as a *ec*TwcK^{KtoA} variant without major yield losses. For both TK and *ec*TwcK, virtual screening suggested the exchange K to L as best tolerated by the protein structure (Fig. 3A). Effectively, the exchange K to L benefited TK production as to obtain a sample that experimentally exhibited improved stability over the K to A exchange (Bogomolovas et al., 2014) (Section 6). In the calculations applied in this study, FoldX calculates the energy of the mutated structure after only a small adjustment of the residue rotamers around the mutation point. To investigate if the residue exchange could be accommodated better upon local structural rearrangements, we employed MOE that energy minimizes the mutated region of the model. We queried the agreement of the FoldX and MOE predictions by means of a 2D representation of the respective ΔΔG values (Fig. 3A and B). Values located along the diagonal of the graph indicate a consensus of the two programs. Differences (off-diagonal data points) are more commonly observed for substitutions to bulky residues (*e.g.*, aromatics), which receive better scores in MOE as they can be better accommodated by model refinement. This approach can be extended to the generation of other functional variants of TK directed to test phosphotransfer or scaffolding interactions.

4.1 Protocol

We provide a protocol for the prediction of mutant stability using FoldX and MOE. We have selected these two packages among available alternatives as they offer simplicity of use to non-specialists, require only modest computing power, and integrate model preparatory steps and stability calculations into graphical interfaces.

4.1.1 Selection of structural template

For mutant stability prediction, an accurate 3D-structure of the protein under study is required. Experimentally determined structures of TK and TwcK can be obtained from the Protein Data Bank (https://www.rcsb.org/) under the codes listed in Table 2. For the analysis of TK and TwcK from other organisms and in the absence of experimental structures, computational models such as those available in the AlphaFold Protein Structure Database (https://alphafold.ebi.ac.uk/) could be employed or generated through homology modeling profiting from the high conservation of these kinases. It must be considered, however, that the correctness of side chain rotamers (and not just protein backbone) is critical for the calculation of such stability

Table 2 Crystal structures of TK and TwcK.

PDB code	Release	Resolution (Å)	Description	Reference
Human TK				
6YGN	2021	2.40	Fn-NL-TKin-CRD-Ig	Bogomolovas et al. (2021)
4JNW	2014	2.06	TKin-CRD (*E. coli*)	Bogomolovas et al. (2014)
1TKI	1999	2.00	TKin-CRD (sf9)	Mayans et al. (1998)
C. elegans TwcK				
3UTO	2012	2.40	Fn-NL-TwcKin-CRD-Ig	von Castelmur et al. (2012)
1KOA	1996	3.30	TwcKin-CRD-Ig	Kobe et al. (1996)
Aplysia TwcK				
1KOB	1996	2.30	TwcKin-CRD	Kobe et al. (1996)

predictions. Side chain correctness is often unmet by predicted protein models and, thus, these must be employed with great caution.

4.1.2 Preparation of PDB file

PDB files of experimental crystal structures often contain features that need to be removed before proceeding to *in silico* stability calculations. These are, for example, alternative side chain conformations and non-protein components such as bound cofactors, small molecules or metal ions. It is also recommendable to use a single protein copy in the PDB. The preparation of structures in this way can be performed automatically in YASARA (Krieger & Vriend, 2014) by executing the command "Clean" or, alternatively, in MOE executing the command "QuickPrep". If needed, the protein model can also be computationally trimmed at this stage to fit the length of the construct under study.

4.1.3 In silico *mutagenesis and stability calculation*

- **FoldX** (http://foldxsuite.crg.eu) can be executed through a dedicated graphical interface to YASARA as a plugin (YasaraPG). Before starting modeling, execute the "Analyze > FoldX > Repair Object" command, which will adjust the structure to the FoldX force field through a brief minimization step. Next, select "Analyze > FoldX > Position Scan" and select the residue you wish to mutate in the subsequent window. Finally, choose what amino acid (or set of) the selected residue should be mutated into (amino acid sets are described in http://foldxsuite.crg.eu/command/PositionScan). Multi-site mutagenesis can be performed by selecting "Analyze > FoldX > Mutate multiple residues". A table with the resulting predicted changes in the free energy of unfolding (denoted "ddg") for each mutant will appear in the YASARA console window. Data from the console can be directly copied for further processing or display
- **MOE** is a commercial drug discovery software platform that contains a module to predict protein stability upon mutagenesis. To perform the calculation, select "Protein > Design > Residue Scan". In the new window, select residue(s) to be mutated and type the set of residues to mutate into. When performing multi-site mutagenesis the number of concurrently calculated point mutations can be controlled using the Site Limit option. Calculation results will appear in the Database Viewer window from where they can be exported for further processing. Column "dStability" provides predicted changes in free energy of unfolding ($\Delta\Delta G$).

4.1.4 Assessing the consensus on stability predictions

$\Delta\Delta G$ values obtained from FoldX and MOE suites can be graphically displayed in a 2D-plot using R (https://www.r-project.org) as shown in Fig. 3A and B. Residue exchanges will lie approximately in the diagonal of the plot if the stability trend is equivalent in both software, with off-diagonal positions reflecting a discrepancy in values. When plotted as in Fig. 3, positions above the diagonal indicate that MOE is more optimistic than FoldX and *vice versa*.

5. Recombinant protein production

Upon variant design, TK and *ec*TwcK must be recombinantly produced. Expression in eukaryotic hosts is often favored for kinase production as these offer the protein maturation machinery often required to attain a functional kinase state, *e.g.* upstream kinases. Yet, the plethora of protein kinases produced by eukaryotic expression systems can contaminate recombinant preparations and trouble kinase characterization (*e.g.* Fukuda et al., 2011). In effect, human TK can be successfully produced in Sf21/Sf9 insect cells *via* baculovirus transfection (Bogomolovas et al., 2014; Lange et al., 2005; Mayans et al., 1998; Puchner et al., 2008), but the contaminating Tcap-phosphorylating activity in the insect cell lysates has conflicted TK characterization (Bogomolovas et al., 2014). In addition to assessing kinase-dead mutants, a valuable - but little employed - procedure to identify contaminant kinase activities in preparations is the measurement of phosphotransfer activity on a selected substrate throughout the purification process, and not just for the final sample. If phosphotransfer on the substrate under study is due to a contaminating kinase, the activity will decrease as the purity of the target kinase increases during purification, as it was observed for TK (Bogomolovas et al., 2014). Similarly, testing lysates from non-transfected cultures is a necessary control often overlooked. This was decisive in revealing contaminating activities in the case of TK (Bogomolovas et al., 2014). This conflict led to pursuing TK production in prokaryotic hosts. *Escherichia coli* is a commonly used host for recombinant protein expression due to its simplicity, high yields, and low cost. It can be a suitable expression host for both human TK (Bogomolovas et al., 2014, 2021) and *ce*TwcK (Heierhorst et al., 1996; von Castelmur et al., 2012; Williams et al., 2018) when employing customized protocols.

5.1 Protocol

Here we describe key optimization steps for the production of the most demanding truncated constructs, mutagenized or sequence-substituted variants of human TK (*i.e.*, kinase carrying CRD components) and *ce*TwcK (kinase domain lacking the NL and/or CRD extensions) in *E. coli* bacterial strains that result in yields in the milligram scale per L of bacterial culture (Table 1):

5.1.1 Bacterial production of TK

- Use of synthetic DNA to increase yield: The cDNA-derived coding sequence for human TK results in low overall levels of expression, even though a rare codon bias is not overtly apparent in the sequence. The use of a synthetic, codon-optimized DNA version of the TK gene results in a significant increase in expression in *E. coli*. This increase, however, only leads to the formation of insoluble inclusion bodies, with the soluble TK fraction remaining limited
- Use of thioredoxin as a fusion tag to increase solubility: both sample solubility and yield are notably improved when expressing the synthetic gene of TK fused to thioredoxin. Thioredoxin has a compact, highly soluble, thermally stable fold that, when fused to a target protein, is speculated to prevent its aggregation as a chimeric nascent protein, thereby supporting its adoption of a correct tertiary fold (LaVallie, Lu, Diblasio-Smith, Collins-Racie, & McCoy, 2000). In the study of TK, we observe it to produce results superior to the use of other established solubility tags, including GST and MBP. Thioredoxin tagging of TK is uncomplicated when using the vector pET-Trx1a (J. Bogomolovas). This is a non-commercial derivative of the vector pETM-11 (EMBL collection) and fuses a His_6-tag, thioredoxin and a tobacco etch virus (TEV) protease cleavage site N-terminally to the target construct
- Choice of bacterial strain and growth medium: both solubility and yields improve further when expressing the synthetic gene pET-Trx1a construction in *E. coli* strain SoluBL21 (Genlantis) grown in Terrific Broth (TB) medium, compared to the use of the native TK cDNA, maternal *E. coli* BL21 DE3 or Rosetta/2 (DE3) strains, and LB medium

For protein expression, the pET-Trx1a plasmid carrying the synthetic, codon optimized gene of TK is to be transformed into commercially available, chemically competent SoluBL21 *E. coli* cells (Genlantis). A single vial of competent cells is thawed on ice, 1 μL of pET-Trx1a expression plasmid

(50–100 ng) added, and the vial incubated on ice for 30 min. This is then transferred to a warm water bath at 42 °C for 45 s and returned to ice for another 1 min. 500 μL of SOC medium (ThermoFisher Scientific) is added to the vial and this shaken for 1 h in a table top shaker (37 °C, 900 rpm). Subsequently, the cells are pelleted on a tabletop centrifuge for 15 s at 13,000 × *g*, supernatant loosely decanted, cells resuspended in remaining medium and plated on TB-agar plate supplemented with 100 μg/mL kanamycin and incubated overnight at 37 °C.

Next morning, numerous colonies (scraping 1–2 full bacteriological loops) are added to 1 mL of prepared liquid TB medium (TB medium +100 μg/mL kanamycin +0.01% Antifoam 204), resuspended and then released to 1 L of prepared TB medium in 2.5 L TUNAIR™ shake flask. Bacteria are grown at 37 °C, with constant shaking (200–250 rpm). TB medium contains a high content of nutrients, uses glycerol as carbon source and offers buffering, which allows to achieve significantly higher cell densities than LB medium. Upon bacterial growth, when the cell density reaches OD_{600} = 3–4, the temperature is reduced to 18 °C and protein expression induced by the addition of IPTG to 0.2 mM final concentration. Expression is allowed to proceed overnight. Next morning, the *E. coli* biomass is pelleted for 20 min at 3000 × *g* in a swinging rotor S-4-104 at 4 °C. The supernatant is discarded and cells resuspended in 100 mL prechilled PBS buffer on ice. The suspension is then transferred into 50 mL conical tubes and repelleted on FA-45-6-30 rotor for 10 min at 18,000 × *g* at 4 °C. The suprenatant is then discarded and the remaining pellet shock frozen in liquid N_2 and stored at −80 °C until further processing. Reagents and equipment employed in this work are provided in Table 3. Cell lysis and purification can then follow standard procedures as described in (Bogomolovas et al., 2014).

The combined steps described above have been applied to the production of TK-CRD and a range of mutated variants as described in Bogomolovas et al. (2014).

5.1.2 Bacterial production of ecTwcK

*ec*TwcK variants can be expressed in *E. coli* with ease. However, forms of TwcK lacking either of the inhibitory domains (NL or CRD) result in highly autophosphorylated samples that are largely inactive (Williams et al., 2018) (Fig. 4A). As kinase phosphorylation is an incomplete reaction that results in phosphorylation sites at partial occupancy, autophosphorylated *ec*TwcK samples commonly obtained through production in *E. coli* suffer from notable batch-to-batch variability. Treatment with λ-phosphatase yields a

Table 3 Reagents and equipment for the bacterial production of TK variants.

Reagent	Recipe or manufacturer	Comments
SoluBL21™ Competent *E. coli* Cells	Amsbio Inc., C700200	Significantly improved BL21 host strain for soluble TK expression.
Terrific Broth (TB)	23.6 g/L Yeast Extract 11.8 g/L Tryptone 9.4 g/L K_2HPO_4 2.2 g/L KH_2PO_4 4 mL/L Glycerol	Nutrient rich, buffered, glycerol-as-carbon-source based liquid media for *E. coli* growth and protein expression.
Innova 24 Incubator Shaker	New Brunswick	Cooling incubator shaker for the culturing of *E. coli*.
TUNAIR™ Shake Flasks, 2.5 L	IBI Scientific, SS-5013	Baffled reusable flasks providing intense aeration and mixing for efficient growth and recombinant protein expression.
Antifoam 204	Sigma-Aldrich, A8311	Antifoam agent added to TB medium (0.01% *v*/v) during *E. coli* growth to prevent excessive foaming and thus improve aeration.
IPTG	Fisher Bioreagents, BP1755	Inducer of recombinant protein expression using pET expression system.
PBS	37 mM NaCl 2.7 mM KCl 8 mM Na2HPO4 2 mM KH2PO4 pH 7.4	Phosphate Buffered Saline for the wash of *E. coli* biomass.
Eppendorf™ Model 5810R Centrifuge	VWR, 05–413-325	Tabletop refrigerated centrifuge for the harvest of *E. coli* biomass.
Rotor S-4-104 incl. 4 × 750 mL round buckets	VWR, 89,305–186	Swinging rotor for the initial harvest of *E. coli* biomass.
Rotor FA-45-6-30	VWR, 97,058–952	Fixed angle for secondary pelleting of washed *E. coli* cells.

dephosphorylated *ec*TwcK sample (Fig. 4B) that presents maximal activity; this should be employed in phosphotransfer activity assays instead (Williams et al., 2018). The production of active, dephosphorylated *ec*TwcK is described below and materials are listed in Table 4.

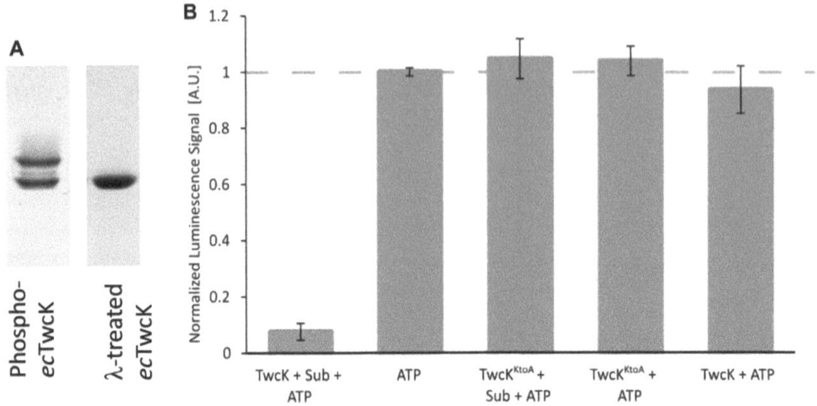

Fig. 4 (A) Luciferase-based phosphotransfer activity assay on *ec*TwcK SDS-PAGE (12%) of the recombinant catalytic kinase domain of *ec*TwcK expressed in *E. coli* pre- and post- treatment with λ-phosphatase, stained with Coomassie Brilliant blue G-250; (B) Representative results of a phosphotransfer activity assay using Kinase-Glo® (Promega) where the kinase domain of *ec*TwcK lacking regulatory tails is assayed on the model peptide substrate kMLC11–23. The luminiscence signal has been normalized to the complete luciferase reaction in the absence of reaction mixture (marked "ATP"). The signal reports on the amount of ATP remaining in solution after manually applied reaction mixtures (in triplicate) were incubated for 8 min at 25 °C (*i.e.*, signal is inverse to kinase activity). Error bars indicate standard deviation. Negative controls demonstrate the absence of contaminating ATPase activities.

*ec*TwcK can be expressed in *E. coli* strain Rosetta (DE3) (Merck Millipore) at good yields (Table 1) using His_6-tagged TwcK constructs and standard *E. coli* culturing protocols (Williams et al., 2018). Transformation is performed in the same way as for TK, followed by culture growth at 37 °C and 180 rpm in LB media supplemented with respective antibiotics to OD_{600} of 0.6–0.8. Protein expression is induced by adding 0.5 mM IPTG (final concentration) to bacterial cultures and grown at 18 °C for further 18 h. Cells are then harvested at 5000 × *g* for 20 min at 4 °C. The bacteria pellet is resuspended in lysis buffer (50 mM Tris-HCl, 500 mM NaCl, 1 mM DTT, pH 7.9) supplemented with one complete EDTA-free protease inhibitor tablet (Roche) and sonicated (three cycles, each 30 s duration time, alternating pulses and pauses of 1 s at an amplitude of 60% of the Sonifier's maximal power) on ice and, subsequently, centrifuged at 39,000 × *g* for 45 min at 4 °C. The supernatant of the bacterial lysate is syringe filtered (0.22 μm) and applied to a HisTrap HP column (Cytiva) equilibrated in lysis buffer (50 mM Tris-HCl, 500 mM NaCl, pH 7.9) supplemented with 20 mM imidazole. Elution of the His_6-tagged TwcK is best done using continuous imidazole gradient, by increasing the ratio of elution buffer

Table 4 Reagents and equipment for the production of dephosphorylated ecTwcK.

Reagent	Recipe or manufacturer	Comments
Rosetta™(DE3) Competent E.coli Cells	Merck Millipore	Suitable host strain for soluble TwcK expression in high yields.
Lysogeny Broth (LB)	5 g/L Yeast Extract 10 g/L Tryptone 10 g/L NaCl	
IPTG	Fisher Bioreagents, ref. no. BP1755	Inducer of recombinant protein expression using pET expression system.
Thermo Scientific™ Sorvall Lynx 4000 Superspeed-Centrifuge	Thermo Scientific™ 75006580	Superspeed centrifuge for the harvest of *E. coli* biomass.
Fiberlite™ F10–4 × 1000 LEX Rotor	Thermo Scientific™ 096-041075	Superspeed rotor for the initial harvest of *E. coli* biomass.
Fiberlite™ A27–6 × 50 Rotor	Thermo Scientific™ 75003007	Superspeed rotor for cell fractionation after lysis.
Eppendorf™ Centrifuge 5804 R	EP022628146	Tabletop refrigerated centrifuge for secondary pelleting of *E. coli* cells.
Eppendorf™ A-4-44 Rotor	RA44450-EP	Swinging rotor for secondary pelleting of washed *E. coli* cells.
Branson 450 Digital Sonifier	Branson Ultrasonics	Sonifier with intensity and time control as well as a probe.
Eppendorf™ Centrifuge 5430 R	EP5427000011	Tabletop refrigerated centrifuge for pelleting aggregates before SEC.
Eppendorf™ FA-45-30-11 Rotor	RFA4530113	Fixed angle rotor for pelleting aggregates before SEC.
Amicon® Ultra-15 Centrifugal Filter	Merck Millipore Ltd., R5NA55226	Centrifugal filter for concentrating proteins.

(50 mM Tris-HCl, 500 mM NaCl, 500 mM Immidazole) towards lysis buffer, in an FPLC instrument (*e.g.*, Äkta Start protein purification system; Cytiva) as this improves the segregation of contaminants.

For dephosphorylation, the eluted sample must be supplemented with 1 mM $MnCl_2$ (to account for the manganese dependence of the phosphatase) and incubated with no less than 400 units (per kinase product obtained from 1 L *E. coli* culture) of lambda protein phosphatase (NEB) for 1 h at room temperature. After 1 h incubation, His_6-tagged TEV protease (iXCells Biotechnologies) (1 mg per L *E. coli* equivalent kinase product) is to be added to the mixture for His_6-tag removal from *ec*TwcK and incubated in the presence of λ-phosphatase overnight at 4 °C. Subtractive Ni^{2+}-affinity purification (procedure as described above) can then be performed on the mixture to remove uncleaved kinase, remaining phosphorylated *ec*TwcK (as phosphate groups have certain affinity for Ni^{2+}-IMAC resin), phosphatase and TEV protease (both His_6-tagged). Dephosphorylated kinase so obtained is to be buffer exchanged to 50 mM Tris pH 7.9, 50 mM NaCl and 0.5 mM TCEP for size exclusion chromatography (SEC) that enhances conformational and oligomeric purity. For SEC, the dephosphorylated kinase sample is to be injected (*e.g. via* a loading loop) into a FPLC instrument like an Äkta Pure protein purification system (Cytiva) with an attached HiLoad® 16/600 Superdex® 75 PG column (Cytiva) equilibrated in SEC-buffer. At a flow-rate of 1 mL per minute, *ec*TwcK is expected to elute at 60–80 mL from the mentioned column. The dephosphorylated state of the resulting *ec*TwcK sample can be confirmed using preferentially, but not necessarily, PhosTag™ (Wako Pure Chemical Industries) 12% SDS-PAGE containing 100 μM $ZnCl_2$ (running buffer: 500 mM Tris, 500 mM MOPS, 0.5% SDS, 0.5 M $NaHSO_3$). During the run, a prestained molecular weight marker (P7719S, BioLabs) act as a running time control while Coomassie blue is used for staining. The kinase sample resulting from this production can be stored at 4 °C and will display high levels of activity for max. 7–9 days, a period after which phosphotransfer decays notably. For longer term use, *ec*TwcK samples can be flash frozen in LN_2 (no glycerol is required) after concentrating using centrifugal filters (*e.g.* Amicon® Ultra-15 centrifugal filter with a molecular mass cut-off of 10 kDa) to a concentration of 2–4 mg/mL and stored at −80 °C.

6. Experimental assessment of sample stability

Molecular uniformity, stability, and a native fold are the key features of a protein preparation of analytical quality. Such features should be assessed for recombinant kinase samples and can serve as guiding criteria for further construct optimization. For the analysis of TK and *ec*TwcK variants, we use biomolecular NMR and differential scanning fluorimetry to evaluate kinase preparations. 1D proton NMR (^{1}H NMR) is performed on unlabeled samples at low concentrations and provides qualitative information on the fold state of a protein. The loss of peaks or changes in the dispersion of the ^{1}H NMR spectrum from a variant sample with respect to the wild-type strongly indicates that the amino acid exchange has a detrimental effect on kinase fold and/or that it induces the aggregation of the sample.

Differential scanning fluorimetry (DSF) reports on fold stability and can be successfully applied to the analysis of protein kinases, including TK and *ec*TwcK (Bogomolovas et al., 2014; Fukuda et al., 2011). DSF measures the changes in fluorescence of dyes sensitive to the molecular environment, such as the popular SYPRO-Orange (Invitrogen). During thermal unfolding, the dye interacts with the progressively exposed hydrophobic core of the kinase fold leading to a sharp increase in fluorescence, which is sometimes followed by signal decay due to macromolecular aggregation (Fig. 3). Such fluorescence profiles permit calculating the kinase melting temperature (T_m). By comparing the T_m value of a variant to that of a wild-type reference, the impact of the introduced variation (*e.g.* domain trimming, residue exchange or phosphorylation event) on sample stability can be assessed. A decrease in T_m indicates a destabilizing effect.

Applied to TK and *ec*TwcK, DSF confirms the destabilizing character of the essential K to A exchange predicted by *in silico* $\Delta\Delta G$ calculations as well as the tolerability of the K to L exchange by TK (Fig. 3C and D). DSF also confirms the stabilizing character of regions flanking the kinase domain exemplified by constructs of *ec*TwcK (Fig. 3E and F). Similarly, the method can be employed to establish the effect of buffer conditions on sample stability (Ristic, Rosa, Seabrook, & Newman, 2015). It is our experience that DSF screening, including the testing of media conditions and small molecule ligands, of truncated or mutated constructs of TK and *ec*TwcK provides strong guidance towards optimal construct design and optimization (where a few terminal residues or residue exchanges can influence sample stability notably). It also supports the identification of suitable media conditions for storage and biochemical testing of these kinase samples, and crucially guides structural characterization efforts. DSF can be performed conveniently in regular real-time PCR machines, the protein does not require labeling, uses μg amounts of material, and is easily parallelized utilizing 96-well plates. Below we provide a protocol of DSF as applied to experimentally identifying destabilizing changes in TK and *ec*TwcK as exemplified by stability measurements on wild-type TK and the TK^{KtoL} and TK^{KtoA} variants (Fig. 3C).

6.1 Protocol

Reagents and materials to implement the protocol described are listed in Table 5.

1. Before the experiment, prepare the working buffer by diluting commercially available 5000 × SYPRO-Orange concentrate (Invitrogen) to a final 5 × concentration in assay buffer (10 mM HEPES pH 7.5, 150 mM NaCl). The working buffer can be stored at room temperature protected from intense light
2. Dispense the purified kinase sample directly onto a RT-PCR machine compatible 96-well plate (*e.g.*, Bio-rad). Optimal protein concentration for this assay should be determined empirically, for TK samples it was determined to be 5 μM in the final assay solution. It is advisable to use concentrated protein samples in order to minimize dilution of the working buffer. Here, from a 50 μM TK stock solution, 2.5 μL were dispensed per well in triplicates and 22.5 μL of working buffer were added. Then, the plate is sealed with optically clear seal such as Microseal 'B' Film (Bio-rad), briefly vortexed, spun down for 1 min at 600 × *g* at room temperature and transferred to a RT-PCR machine
3. Fluorescence is typically monitored ($\lambda_{ex} = 465$ nm, $\lambda_{em} = 590$ nm) from 25 °C to 85 °C at 1 °C/min increments using *e.g.* a Mx3005p RT-PCR machine (Stratagene). Fluorescence values from individual wells at each cycle are exported into a comma-separated value (csv) Excel file

Table 5 Reagents and materials to implement DSF measurements.

Component	Manufacturer/ recipe	Comments
Assay Buffer	10 mM HEPES 150 mM NaCl pH 7.5	
SYPRO™ Orange Protein Gel Stain	Invitrogen™ S6650	5000 × Concentrate in DMSO.
Microseal "B"	Bio-rad MSB1001	Adhesive PCR Plate Sealing Film compatible with optical readout.
Multiplate™ 96-Well PCR Plate	Bio-rad MLP9601	96-well PCR plate compatible with an optical readout.
Mx3005p RT-PCR Machine	Agilent	Real time PCR machine capable of monitoring SYPRO-Orange fluorescence ($\lambda_{ex} = 465$ nm, $\lambda_{em} = 590$ nm).

4. Curves recorded in triplicate that show good agreement are averaged and T_m values determined. Curve fitting or derivative based methods for T_m estimation can be performed using general standalone data analysis software packages such as Prism using a modified Boltzmann equation with linear correction of baseline drifts (Matulis, Kranz, Salemme, & Todd, 2005) or employing specialist applications for melting temperature analysis in DSF experiments, such as *dsf-fitting* (Bai, Roder, Dickson, & Karanicolas, 2019; https://sourceforge.net/projects/dsf-fitting/files), SimpleDSFviewer (Sun, Li, Yates, & Fernig, 2020; https://github.com/hscsun/SimpleDSFviewer-5.0), Meltdown (Rosa et al., 2015; https://github.com/C3-CSIRO/Meltdown) or MTSA (Schulz, Landström, & Hubbard, 2013; https://www.ysbl.york.ac.uk/fragments/MTSA). For simple cases, the T_m value can be rapidly extracted from fluorescence profiles by identifying the maximum value of the first derivative using the "Melt curve" option built-in within the RT-PCR machine software

7. Evaluating phosphotransfer activity

Understanding the biological relevance of TK relies on clarifying if it holds the capability to perform phosphotransfer activity and, if so, identifying candidate substrates. This, in addition to revealing its scaffolding binding partners. Various methodologies are available to explore (pseudo)kinase activity *in vitro* including, among others, the use of peptide chip microarrays derived from known phosphorylation sites and the screening of systematically generated libraries of artificial sequences (Shah & Kim, 2019). However, more commonly, in order to reveal substrate candidates of potential physiological relevance, lysates of pertinent biological tissues or cells are incubated with the recombinantly produced, purified kinase of interest. The presence of intrinsic protein kinases in such tissue or cell lysates can obscure experimental readouts. This obstacle can be solved by inactivating endogenous kinases in the lysate *via* pretreatment with fluorosulfonylbenzoyl 5′-adenosine (FSBA). FSBA is an ATP analog that irreversibly inhibits kinases by covalently modifying the essential lysine in the ATP-binding site. Unreacted FSBA can be removed from lysates by dialysis or centrifugal ultrafiltration, leaving a protein mixture of biological protein substrate candidates that can be used in *in vitro* kinase assays (Menzel, Chari, Fischer, Linder, & Raabe, 2007). Among other pseudokinases, this method has been applied to the search of candidate substrates for TK using adult muscle as well as C2C12 myocytes (Bogomolovas et al., 2014). Alternatively, kinase-binder complexes can be captured from such lysates using bead-immobilized kinase and directly applied to phosphorylation assays (Amano et al., 2015).

In cell lysate-based assays, kinase activity can be monitored by detecting phosphogroups on protein substrates. A sensitive detection method is commonly required as protein substrates are found at endogenous cellular levels, which are often low. The most sensitive detection method relies on the use of radioactive ATP, where the ATP molecule carries the isotope ^{32}P or ^{33}P at the γ-phosphate position. This is transferred to the substrate by the kinase. The reacted lysate mixture can be separated on SDS-PAGE and the gel imaged using a phosphorimager or X-ray film (Carter, 2001). Because of the hazardous nature of radio-labeled ATP, γS-ATP can be used as an alternative. γS-ATP can be utilized by most kinases, resulting in thiophosphorylation of the substrate. Thiophosphorylated substrates can be efficiently alkylated with *p*-nitrobenzylmesylate and detected with an antibody specific for the resulting thiophosphate esters (Allen et al., 2007).

7.1 Protocol for assaying TK activity on FSBA-inactivated protein lysates

A biologically relevant tissue or cell population should be lysed (*e.g.* using mechanical procedures such as applying ultrasound and/or grinding of frozen tissue) in mild lysis buffer such as 50 mM Tris-HCl pH 7.4, 150 mM NaCl and 1% NP-40 supplemented with protease inhibitor cocktail (Halt™ Protease Inhibitor Cocktail; Thermo Scientific™). A good starting point for cell material would be a single confluent 10 cm dish, with $6\text{-}9 \times 10^6$ cells yielding approximately (depending on cell type) 1 mg of total protein when lysed. Place the cell dish on ice and wash once with pre-chilled PBS, aspirate PBS and add 250–500 μL of cold lysis buffer and scrape cells using a cell scraper. Then, transfer the cell suspension to a microcentrifuge tube. To facilitate lysis, sonify suspension on ice using a pre-chilled probe; 5 J delivered in 1–3 five-second-long pulses are sufficient to efficiently lyse mammalian cells. When using tissue as starting material, 500 μL of lysis buffer per 10 mg of tissue is a good starting point. Tissues are most efficiently lysed using mechanical homogenizers such as ULTRA-TURRAX® or POLYTRON®. The lysate is to be clarified by centrifugation and can be additionally dialyzed or desalted on a PD-10 (lysate volume 1.0–2.5 mL) or G-25 (0.1–0.5 mL) columns (Cytiva) to remove endogenous nucleotides that could compete with FSBA inactivation. Briefly, the desalting column is equilibrated in lysis buffer and the clarified lysate exchanged into fresh lysis buffer using gravity flow or centrifugal force (according to manufacturer's protocols per column type). Dialysis should be performed overnight at 4 °C into lysis buffer without detergent (NP-40 is not dialyzable detergent), ideally using low molecular cutoff membrane (for example 3.5 kDa) in order to avoid loss of potential kinase substrates. Protein concentration should be measured in prepared lysates using Bradford assay, a Lowry assay or a bicinchoninic acid (BCA) assay.

Next, the desalted lysate is adjusted to 2 mg/mL concentration and treated with an excess of FSBA (Sigma-Aldrich). A good starting point for treatment is the dilution of the initial FSBA stock (*i.e.*, FSBA dissolved in dimethyl sulfoxide (DMSO)) directly into the prepared lysate to a final concentration of 20 mM and incubation at 30 °C for 1 h under mild agitation in a heated table top microcentrifuge tube shaker with constant shaking at 700–900 rpm. Unreacted FSBA can be removed by buffer exchange into phosphorylation assay buffer (20 mM Tris-HCl pH 7.4, 10 mM magnesium acetate, 0.05% NP40, 0.1 mM DTT) either *via* dialysis or using a desalting column in similar fashion as described for nucleotide removal. Lysate so prepared can be supplemented with 0.4 mM ATP (0.2 μCi/reaction of [γ-^{33}P] ATP) and purified TK. *In vitro* phosphorylation reaction is performed in 30 μL. Typically, the reaction mixture consists of: 0.2 μL of 10 mCi/mL [γ-^{33}P]ATP (for example: Perkin Elmer NEG002A100UC), 1.2 μL of 10 mM ATP stock, 1 μL of 10 mg/mL TK stock solution and 27.6 μL of FSBA treated lysate. As a negative control, heat-inactivated TK can be used. The type of positive control depends on the scientific question being addressed, but a closely related kinase with documented catalytic activity, such as *ec*Twck, can serve as a positive control. The reaction should be allowed to proceed at *e.g.* 30 °C for 30 min (although various temperatures and times should be tested). The reaction can be terminated by adding Laemmli buffer and heating the sample for 5 min at 95 °C. Proteins in the mixture can then be resolved by SDS-PAGE. For detection, the gel can be dried, exposed to a phosphoscreen, imaged with *e.g.* a Fujifilm BAS 2500 phosphoimager, and images processed using AIDA (Raytest). Reagents and materials to implement the protocol described are given in Table 6.

Identification of candidate substrates in this way is often followed by mass spectrometry analyses and *in vitro* phosphorylation assays aimed to identify and validate substrates and their sites of phosphorylation. This then permits the synthesis of peptides based on the modified sequence, which can be used in mechanistic *in vitro* phosphorylation assays geared to characterize phosphotransfer activity assisted by engineered kinase variants. Such assays

Table 6 Reagents and materials for the testing of kinase activity in FSBA-inactivated cell lysates.

Component	Manufacturer/recipe	Comments
Lysis Buffer	50 mM Tris-HCl 150 mM NaCl 1% NP-40 Protease inhibitors pH 7.4	Strong detergents should be avoided, due to possible kinase activity inhibition.
FSBA 5′-O-(4-Fluorosulfonylbenzoyl)-adenosine	Jena Biosciences N-1086	
PD-10 desalting column	Cytiva life sciences 17085101	Desalting column prepacked with Sephadex G-25 resin for volumes up to 2.5 mL. For smaller volumes (0.1–0.5 mL) PD MiniTrap G-25 can be used.
Slide-A-Lyzer™ Dialysis Cassettes, 3.5 K MWCO	Thermo Scientific™ 66333 for volumes up to 0.5 mL 66330 for volumes up to 3 mL	Dialysis can be used as alternative to desalting columns for nucleotide and FSBA removal.
Phosphorylation assay buffer	20 mM Tris-HCl 10 mM magnesium acetate, 0.05% NP40, 0.1 mM DTT pH 7.4	
[γ-^{33}P]ATP	Perkin Elmer NEG002A100UC	
ATP stock	10 mM ATP pH 7.5	When preparing ATP stock, use deionized water and adjust pH with NaOH after dissolving ATP salt (for example Adenosine 5′-triphosphate disodium salt hydrate A7699 Sigma-Aldrich).
Fujifilm BAS 2500 phosphoimager		
Ultrasonic processor	Sonics & Materials Inc. VCX 130 PB	Ultrasonic processor equipped with a probe suitable for small volumes.
Mechanical homogenizer	IKA T 25 digital ULTRA-TURRAX®	Mechanical homogenizer capable of processing small volumes of 1 mL.

can employ, as discussed before, radiolabeled ATP to monitor the degree of modification of the peptide substrate under the experimental conditions. Peptide substrates so tested in a reaction mixture can be transferred to chromatographic P81 paper or LSA-50 paper (Appelmans et al., 2021), the remaining reaction components washed out with a phosphoric acid solution, and the retained radioactive phosphopeptides quantified with a scintillation counter, phosphorimager, or X-ray film (Asensio & Garcia, 2003). This method is well established for the study of *ec*TwcK activity (*e.g.* Williams et al., 2018). A less hazardous alternative is to monitor ATP consumption as an indicator of kinase activity using luminescence. ATP consumption serves as an accurate proxy readout of kinase phosphotransfer activity subject to appropriate controls. The most common method to assess ATP consumption makes use of luciferase-based luminescence as provided in the commercial Kinase-Glo® Kinase Assay (Promega). This method successfully reports on the phosphotransfer activity of *ec*TwcK using model peptides.

In the absence of established substrates for TK, *ec*TwcK variants carrying TK-specific motifs can be assayed for activity with the goal of evaluating the impact of selected features on phosphotransfer output. In this context, *ec*TwcK has been mutated to carry the TK-specific exchanges A34M and/or D147E (variants *ce*TwcK^{A34M}, *ce*TwcK^{D147E} and *ce*TwcK$^{A34M/D147E}$) in the ATP-binding pocket and DFG motif and phosphorylation assays performed (Bogomolovas et al., 2014). The double mutant *ce*TwcK$^{A34M/D147E}$ was generated to account for a possible compensatory coevolution of these substitutions in TK. Both single and double *ce*TwcK mutants lacked measurable phosphotransfer. A positive control variant, *ce*TwcK^{A34V}, had the alanine in the VAIK motif replaced by a valine residue, which is found in the catalytically active TwcKs from mollusks as well as 5% of the active members of the human kinome (Scheeff, Eswaran, Bunkoczi, Knapp, & Manning, 2009). Meeting expectations, the *ce*TwcK^{A34V} retained notable levels of catalysis. These results supported the view that the two atypical active site residues M34 and E147 in TK are sufficient to inactivate it. The methodological pipeline described in this manuscript permits now a more extensive grafting of the active site environment of TK onto *ce*TwcK, where secondary residues lining the ATP-pocket and other mechanistic interfaces can be grafted onto *ce*TwcK as a robust carrying scaffold with a phosphotransfer output readout.

To quantitate ATP consumption by recombinant *ec*TwcK samples, luciferase-based kinase activity assays can be performed using the Kinase-Glo® Max system or, alternatively, the ADP-Glo™ Kinase Assay (both from Promega). Our protocols in use employ the former. ATP serves as a substrate for the Kinase-Glo® reagent resulting in a luminescence signal that is proportional to the amount of ATP present in a solution. When this is applied to a reaction mixture where a kinase reaction is taking place, phosphotransfer activity will be terminated at that point and the luminescence signal will report on the amount of ATP not consumed by the kinase, *i.e.*, the signal will inversely report on ATP depletion by kinase phosphotransfer activity. In such assay, redundant negative controls are of the upmost importance as to confirm that kinase preparations do not carry contaminants with ATPase activity, which could conflict the interpretation of the measurements, or the presence of others chemical sources of luminescence in the medium. In this regard, we provide below a protocol for quantitating the catalysis of *ce*TwcK variants applying this method.

7.2 Protocol for luciferase-based kinase activity assay

A synthetic peptide derived from chicken MLC, kMLC11–23, with sequence KKRARAATSNVFS (Heierhorst et al., 1996; von Castelmur et al., 2012) in powder form is to be dissolved in 50 mM Tris-HCl pH 7.9, 50 mM NaCl, 0.5 mM tris(2-carboxyethyl)phosphine (TCEP) to a concentration of 4 mg/mL. Assays are then set up in triplicate in white 96-well plates (Greiner Bio-One) with 50 μL reaction volume in each well.

1. A 5 μM ATP solution, a 2 mg/mL *ec*TwcK solution, a 4 mg/mL peptide substrate solution and a mastermix containing 5 mg/mL BSA and 2.4 mM $MgCl_2$, all in the buffer given above, are to be prepared
2. 20 μL of mastermix is to be added to each well, followed by 10 μL of each respective reaction partner (*ec*TwcK, peptide substrate, ATP). The final reaction mixture in the buffer above consists of 0.4 mg/mL *ec*TwcK variant under analysis, 1 μM ATP, 0.96 mM $MgCl_2$, 2 mg/mL bovine serum albumin (BSA), and 0.8 mg/mL peptide substrate. Control experiments in which one component is absent are made up to 50 μL by addition of buffer. ATP must be added last, as it initiates the reaction
3. Reaction mixtures are typically incubated for 8 to 10 min at 25 °C. The assay can be adapted to perform time courses as desired

4. Following incubation, 50 μL of Kinase-Glo® reagent (Promega) is added to all wells, which stops reaction, and luminescence is measured by using *e.g.* a Varioskan™ Flash (Thermo Scientific™) plate reader and SkanItTM 2.4.3 RE software

Negative control reactions can be only buffer or only peptide, only ATP (to confirm performance of kit reagents, particularly after storage periods), both peptide substrate and ATP, only kinase, kinase + ATP, kinase + peptide and the inactive kinase variant $\text{TwcK}^{\text{KtoA}}$. A positive control reaction is required and should consist of the dephosphorylated, wild-type isolated kinase domain of *ec*TwcK (lacking both regulatory tails), ATP, and peptide substrate; this mixture will yield the maximal catalytic performance under the given experimental conditions.

Reactions are initiated by the addition of *ec*TwcK to the wells and typically incubated for 8 to 10 min at 25 °C. The assay can be adapted to perform time courses as desired. Following incubation, 50 μL of Kinase-Glo® reagent (Promega) is added to all wells, which stops reaction, and luminescence is measured by using *e.g.* a VarioskanTM Flash (Thermo ScientificTM) plate reader and SkanItTM 2.4.3 RE software. We have confirmed that under these experimental conditions the luminescence signal is stable over time and that, once the reaction mixture has been supplemented with the Kinase-Glo® reagent, luminescence can be measured up to 30 min after application (maximal time tested) without losing any performance. For optimal results, it is recommended to perform the assay using a dispensing robot (*e.g.*, the Freedom EVO® dispenser; Tecan) to minimize variations derived from manual pipetting errors and time delays.

8. Summary

Protein pseudokinases contain variegated active sites of unknown phosphotransfer capability and their modes of signaling and regulation often involve unconventional mechanisms (Mace & Murphy, 2021). Unveiling such mechanisms is complex, particularly when the occurrence of phosphotransfer appears unlikely or is uncertain. These considerations apply to sarcomeric TK and related (pseudo)kinases of the titin family. These kinases are thought to act as elastic scaffolds, that sense the mechanical state of the sarcomere by means of the mechanically-induced unfolding of its flanking elements, which regulate their signaling. However, the functional output of tail unfolding in terms of phosphotransfer (if any occurs in pseudokinases of this family) and scaffolding remains poorly understood and further data are needed on the contribution of these events to their signaling function. In the case of TK, its structural and mechanistic similarity to the closely-related *ec*TwcK, which is well characterized and for which a plethora of mechanistic tools are available, can be exploited to aid deciphering its functional principles. To answer outstanding questions particularly in terms of TK phosphotransfer and regulation, we provide here guidelines on how to produce recombinant sample variants and examine kinase catalytic (in)activity for human TK and *C. elegans* TwcK, offering suggestions on how to overcome technical limitations and prevent potential sources of false outcomes. While TK and *ec*TwcK are the main representatives of this family, the workflow presented here can be easily adapted to other non-canonical (pseudo)kinase members. In particular, a strategy based on kinase/pseudokinase homolog pairs is applicable to the canonical obscurin kinases and their invertebrate pseudokinase counterparts, UNC-89 kinases. Similar to TK, the production and analysis of obscurin- and UNC89- kinases is problematic and, thus, their function and mechanisms remain poorly understood at this time. Methodological pipelines to advance the characterization of TK will, thereby, progress the prospects of unveiling the unified functional principles of this (pseudo)kinase family.

Acknowledgments

We acknowledge the financial support of DFG SFB969. We thank Gert Matthias van der Berg (University of Konstanz) for his contribution to phosphotransfer activity assays of *ec*TwcK.

References

Allen, J. J., Li, M., Brinkworth, C. S., Paulson, J. L., Wang, D., Hübner, A., et al. (2007). A semisynthetic epitope for kinase substrates. *Nature Methods*, *4*, 511–516.

Amano, M., Hamaguchi, T., Shohag, M. H., Kozawa, K., Kato, K., Zhang, X., et al. (2015). Kinase-interacting substrate screening is a novel method to identify kinase substrates. *Journal of Cell Biology*, *209*, 895–912.

Appelmans, O., Kashyap, R. S., Gilles, P., De Borggraeve, W. M., Voet, A., & Van Lint, J. (2021). LSA-50 paper: An alternative to P81 phosphocellulose paper for radiometric protein kinase assays. *Analytical Biochemistry*, *630*, 114313.

Asensio, C. J., & Garcia, R. C. (2003). Determination of a large number of kinase activities using peptide substrates, P81 phosphocellulose paper arrays and phosphor imaging. *Analytical Biochemistry*, *319*, 21–33.

Bai, N., Roder, H., Dickson, A., & Karanicolas, J. (2019). Isothermal analysis of ThermoFluor data can readily provide quantitative binding affinities. *Scientific Reports*, *9*, 2650.

Bogomolovas, J., Fleming, J. R., Franke, B., Manso, B., Simon, B., Gasch, A., et al. (2021). Titin kinase ubiquitination aligns autophagy receptors with mechanical signals in the sarcomere. *EMBO Reports*, *22*, e48018.

Bogomolovas, J., Gasch, A., Simkovic, F., Rigden, D. J., Labeit, S., & Mayans, O. (2014). Titin kinase is an inactive pseudokinase scaffold that supports MuRF1 recruitment to the sarcomeric M-line. *Open Biology*, *4*, 140041.

Bromberg, Y., & Rost, B. (2009). Correlating protein function and stability through the analysis of single amino acid substitutions. *BMC Bioinformatics*, *10*, S8.

Bullard, B., Linke, W. A., & Leonard, K. (2002). Varieties of elastic protein in invertebrate muscles. *Journal of Muscle Research and Cell Motility*, *23*, 435–447.

Butler, T. M., & Siegman, M. J. (2011). A force-activated kinase in a catch smooth muscle. *Journal of Muscle Research and Cell Motility*, *31*, 349–358.

Candasamy, A. J., Haworth, R. S., Cuello, F., Ibrahim, M., Aravamudhan, S., Krüger, M., et al. (2014). Phosphoregulation of the titin-cap protein telethonin in cardiac myocytes. *Journal of Biological Chemistry*, *289*, 1282–1293.

Carter, N. (2001). Assays of protein kinases using exogenous substrates. *Current Protocols in Protein Science*, 7. Chapter 13: Unit 13.

Centner, T., Yano, J., Kimura, E., McElhinny, A. S., Pelin, K., Witt, C. C., et al. (2001). Identification of muscle specific ring finger proteins as potential regulators of the titin kinase domain. *Journal of Molecular Biology*, *306*, 717–726.

Endicott, J. A., Noble, M. E., & Johnson, L. N. (2012). The structural basis for control of eukaryotic protein kinases. *Annual Review of Biochemistry*, *81*, 587–613.

Fukuda, K., Knight, J. D., Piszczek, G., Kothary, R., & Qin, J. (2011). Biochemical, proteomic, structural, and thermodynamic characterizations of integrin-linked kinase (ILK): Cross-validation of the pseudokinase. *Journal of Biological Chemistry*, *286*, 21886–21895.

Greene, D. N., Garcia, T., Sutton, R. B., Gernert, K. M., Benian, G. M., & Oberhauser, A. F. (2008). Single-molecule force spectroscopy reveals a stepwise unfolding of Caenorhabditis elegans giant protein kinase domains. *Biophysical Journal*, *95*, 1360–1370.

Heierhorst, J., Tang, X., Lei, J., Probst, W. C., Weiss, K. R., Kemp, B. E., et al. (1996). Substrate specificity and inhibitor sensitivity of Ca2+/S100-dependent twitchin kinases. *European Journal of Biochemistry*, *242*, 454–459.

Kellogg, E. H., Leaver-Fay, A., & Baker, D. (2011). Role of conformational sampling in computing mutation-induced changes in protein structure and stability. *Proteins: Structure, Function, and Bioinformatics*, *79*, 830–838.

Kim, S. T., Lim, D. S., Canman, C. E., & Kastan, M. B. (1999). Substrate specificities and identification of putative substrates of ATM kinase family members. *Journal of Biological Chemistry*, *274*, 37538–37543.

Kobe, B., Heierhorst, J., Feil, S. C., Parker, M. W., Benian, G. M., Weiss, K. R., et al. (1996). Giant protein kinases: Domain interactions and structural basis of autoregulation. *EMBO Journal*, *15*, 6810–6821.

Kontrogianni-Konstantopoulos, A., Ackermann, M. A., Bowman, A. L., Yap, S. V., & Bloch, R. J. (2009). Muscle giants: Molecular scaffolds in sarcomerogenesis. *Physiological Reviews*, *89*, 1217–1267.

Krieger, E., & Vriend, G. (2014). YASARA view - molecular graphics for all devices - from smartphones to workstations. *Bioinformatics*, *30*, 2981–2982.

Lai, S., & Pelech, S. (2016). Regulatory roles of conserved phosphorylation sites in the activation T-loop of the MAP kinase ERK1. *Molecular Biology of the Cell*, *27*, 1040–1050.

Lange, S., Xiang, F., Yakovenko, A., Vihola, A., Hackman, P., Rostkova, E., et al. (2005). The kinase domain of titin controls muscle gene expression and protein turnover. *Science*, *308*, 1599–1603.

LaVallie, E. R., Lu, Z., Diblasio-Smith, E. A., Collins-Racie, L. A., & McCoy, J. M. (2000). Thioredoxin as a fusion partner for production of soluble recombinant proteins in Escherichia coli. *Methods in Enzymology*, *326*, 322–340.

Mace, P. D., & Murphy, J. M. (2021). There's more to death than life: Noncatalytic functions in kinase and pseudokinase signaling. *Journal of Biological Chemistry*, *296*, 100705.

Manning, G., Whyte, D. B., Martinez, R., Hunter, T., & Sudarsanam, S. (2002). The protein kinase complement of the human genome. *Science*, *298*, 1912–1934.

Matsunaga, Y., Hwang, H., Franke, B., Williams, R., Penley, M., Qadota, H., et al. (2017). Twitchin kinase inhibits muscle activity. *Molecular Biology of the Cell*, *28*, 1591–1600.

Matsunaga, Y., Qadota, H., Furukawa, M., Choe, H. H., & Benian, G. M. (2015). Twitchin kinase interacts with MAPKAP kinase 2 in Caenorhabditis elegans striated muscle. *Molecular Biology of the Cell*, *26*, 2096–2111.

Matulis, D., Kranz, J. K., Salemme, F. R., & Todd, M. J. (2005). Thermodynamic stability of carbonic anhydrase: Measurements of binding affinity and stoichiometry using ThermoFluor. *Biochemistry*, *44*, 5258–5266.

Mayans, O., Benian, G. M., Simkovic, F., & Rigden, D. J. (2013). Mechanistic and functional diversity in the mechanosensory kinases of the titin-like family. *Biochemical Society Transactions*, *41*, 1066–1071.

Mayans, O., van der Ven, P. F., Wilm, M., Mues, A., Young, P., Fürst, D. O., et al. (1998). Structural basis for activation of the titin kinase domain during myofibrillogenesis. *Nature*, *395*, 863–869.

Menzel, N., Chari, A., Fischer, U., Linder, M., & Raabe, T. (2007). A 5′-fluorosulfonylbenzoyladenosine-based method to identify physiological substrates of a Drosophila p21-activated kinase. *Analytical Biochemistry*, *368*, 178–184.

Min, X., Lee, B. H., Cobb, M. H., & Goldsmith, E. J. (2004). Crystal structure of the kinase domain of WNK1, a kinase that causes a hereditary form of hypertension. *Structure*, *12*, 1303–1311.

Mukherjee, K., Sharma, M., Urlaub, H., Bourenkov, G. P., Jahn, R., Südhof, T. C., et al. (2008). CASK functions as a Mg2+−independent neurexin kinase. *Cell*, *133*, 328–339.

Müller, S., Kursula, I., Zou, P., & Wilmanns, M. (2006). Crystal structure of the PB1 domain of NBR1. *FEBS Letters*, *580*, 341–344.

Porto, D., Matsunaga, Y., Franke, B., Williams, R. M., Qadota, H., Mayans, O., et al. (2021). Conformational changes in twitchin kinase in vivo revealed by FRET imaging of freely moving *C. elegans*. *eLife*, *10*, e66862.

Puchner, E. M., Alexandrovich, A., Kho, A. L., Hensen, U., Schäfer, L. V., Brandmeier, B., et al. (2008). Mechanoenzymatics of titin kinase. *Proceedings of the National Academy of Sciences USA*, *105*, 13385–13390.

Ristic, M., Rosa, N., Seabrook, S. A., & Newman, J. (2015). Formulation screening by differential scanning fluorimetry: How often does it work? *Acta Crystallographica F: Structural Biology Communications*, *71*, 1359–1364.

Rosa, N., Ristic, M., Seabrook, S. A., Lovell, D., Lucent, D., & Newman, J. (2015). Meltdown: A tool to help in the interpretation of thermal melt curves acquired by differential scanning fluorimetry. *Journal of Biomolecular Screening*, 7, 898–905.

Sanavia, T., Birolo, G., Montanucci, L., Turina, P., Capriotti, E., & Fariselli, P. (2020). Limitations and challenges in protein stability prediction upon genome variations: Towards future applications in precision medicine. *Computational and Structural Biotechnology Journal*, *18*, 1968–1979.

Scheeff, E. D., Eswaran, J., Bunkoczi, G., Knapp, S., & Manning, G. (2009). Structure of the pseudokinase VRK3 reveals a degraded catalytic site, a highly conserved kinase fold, and a putative regulatory binding site. *Structure*, *17*, 128–138.

Schulz, M. N., Landström, J., & Hubbard, R. E. (2013). MTSA—A Matlab program to fit thermal shift data. *Analytical Biochemistry*, *433*, 43–47.

Schymkowitz, J., Borg, J., Stricher, F., Nys, R., Rousseau, F., & Serrano, L. (2005). The FoldX web server: An online force field. *Nucleic Acids Research*, *33*(SUPPL. 2), W382.
Shah, K., & Kim, H. (2019). The significant others: Global search for direct kinase substrates using chemical approaches. *IUBMB Life*, *71*, 721–737.
Sun, C., Li, Y., Yates, E. A., & Fernig, D. G. (2020). SimpleDSFviewer: A tool to analyze and view differential scanning fluorimetry data for characterizing protein thermal stability and interactions. *Protein Science*, *29*, 19–27.
von Castelmur, E., Strümpfer, J., Franke, B., Bogomolovas, J., Barbieri, S., Qadota, H., et al. (2012). Identification of an N-terminal inhibitory extension as the primary mechanosensory regulator of twitchin kinase. *Proceedings of the National Academy of Sciences USA*, *109*, 13608–13613.
Williams, R. M., Franke, B., Wilkinson, M., Fleming, J. R., Rigden, D. J., Benian, G. M., et al. (2018). Autophosphorylation is a mechanism of inhibition in twitchin kinase. *Journal of Molecular Biology*, *430*, 793–805.
Zou, P., Gautel, M., Geerlof, A., Wilmanns, M., Koch, M. H., & Svergun, D. I. (2003). Solution scattering suggests cross-linking function of telethonin in the complex with titin. *Journal of Biological Chemistry*, *278*, 2636–2644.
Zou, P., Pinotsis, N., Lange, S., Song, Y. H., Popov, A., Mavridis, I., et al. (2006). Palindromic assembly of the giant muscle protein titin in the sarcomeric Z-disk. *Nature*, *439*, 229–233.

CHAPTER EIGHT

Co-expression of recombinant RIPK3:MLKL complexes using the baculovirus-insect cell system

Cheree Fitzgibbon[a,b,†], Yanxiang Meng[a,b,†], and James M. Murphy[a,b,*]

[a]Walter and Eliza Hall Institute of Medical Research, Parkville, VIC, Australia
[b]Department of Medical Biology, University of Melbourne, Parkville, VIC, Australia
[*]Corresponding author: e-mail address: jamesm@wehi.edu.au

Contents

Abstract

Pseudokinase domains are found throughout the kingdoms of life and serve myriad roles in cell signaling. These domains, which resemble protein kinases but are catalytically-deficient, have been described principally as protein interaction domains. Broadly, pseudokinases have been reported to function as: allosteric regulators of conventional enzymes; scaffolds to nucleate assembly and/or localization of signaling complexes; molecular switches; or competitors of signaling complex assembly. From detailed structural and biochemical studies of individual pseudokinases, a picture of how they mediate protein interactions is beginning to emerge. Many such studies have

† These authors contributed equally and are listed alphabetically.

Methods in Enzymology, Volume 667
ISSN 0076-6879
https://doi.org/10.1016/bs.mie.2022.03.029

relied on recombinant protein production in insect cells, where endogenous chaperones and modifying enzymes favor bona fide folding of pseudokinases. Here, we describe methods for co-expression of pseudokinases and their interactors in insect cells, as exemplified by the MLKL pseudokinase, which is the terminal effector in the necroptosis cell death pathway, and its upstream regulator kinase RIPK3. These methods are broadly applicable to co-expression of other pseudokinases with their interaction partners from bacmids using the baculovirus-insect cell expression system.

1. Introduction

Pseudokinase domains are defined by the absence of one or more of the three critical catalytic residues known to mediate phosphoryl transfer by their catalytically-active cousins, the kinases (Manning, Whyte, Martinez, Hunter, & Sudarsanam, 2002). Sequence deviations in one or more of the VAIK, DFG and HRD motifs in pseudokinases compromise the enzymatic activity typical of protein kinases (Boudeau, Miranda-Saavedra, Barton, & Alessi, 2006; Jacobsen & Murphy, 2017; Murphy, Zhang, et al., 2014), with rare exceptions occurring in some proteins that have evolved to modify non-canonical substrates (Sreelatha et al., 2018; Zhu et al., 2016) or phosphorylate substrates via compensatory mechanisms with lower efficiency than conventional kinases (Min, Lee, Cobb, & Goldsmith, 2004). The VAIK, DFG and HRD catalytic motifs occur at the cleft between the smaller N-lobe and the larger C-lobe within the kinase fold where substrate ATP binds and is hydrolyzed in an active kinase. The VAIK lysine is contributed by the β3 strand in the N-lobe and binds the α- and β-phosphates of ATP to position the γ-phosphate for phosphoryl transfer; the DFG aspartate resides at the start of the activation loop and binds the Mg^{2+} cofactor required for efficient phosphoryl transfer; and the HRD aspartate occurs in the activation loop and serves as the catalytic residue. The prevalence of pseudokinases in, for example, mammals and apicomplexans (~10% and 30% of their respective kinase complements) reflects their intrinsic importance to signal transduction across phyla (Kwon et al., 2019).

Instead of performing catalytic functions like conventional enzymes, pseudokinases have been attributed other critical functions in signal transduction that reflect their roles as protein interaction domains (Eyers & Murphy, 2013, 2016; Mace & Murphy, 2021; Murphy, Farhan, & Eyers, 2017; Ribeiro et al., 2019). Broadly, four major functions have been ascribed to pseudokinases: (i) as allosteric regulators of bona fide enzymes; (ii) as protein hubs that nucleate formation and localization of signaling complexes; (iii) as signal receivers/integrators and molecular switches; and (iv) as substrate trapping or competitor (decoy) proteins. These functions rely on pseudokinase domains mediating protein–protein interactions, the precise nature of which have been revealed through detailed mechanistic studies over the past decade.

Structural studies have been integral to enhancing understanding of how pseudokinases can act as hubs that nucleate assembly of signaling complexes, such as those anchored by the PEAK family (Ha & Boggon, 2018; Lecointre et al., 2018; Patel et al., 2017; Patel, Roy, Murphy, & Lucet, 2020) or Tribbles family (Durzynska et al., 2017; Jamieson et al., 2018; Kung & Jura, 2019; Murphy et al., 2015) pseudokinases. However, one of the best characterized functions of pseudokinases is their interaction with conventional protein kinases to allosterically promote the catalytic activities of the latter, as described for the regulation of FAM20C, EGFR and JAK2 kinase domain catalytic activity by their respective pseudokinase partners, FAM20A (Cui et al., 2017; Zhang et al., 2018), HER3 (Littlefield et al., 2014) and the JAK2 pseudokinase domain (Babon, Lucet, Murphy, Nicola, & Varghese, 2014; James et al., 2005; Saharinen, Vihinen, & Silvennoinen, 2003). Relatedly, pseudokinases can bind conventional protein kinases to block their catalytic functions in cells, as exemplified by a virally-encoded ortholog of the Mixed lineage kinase domain-like protein (vMLKL) (Palmer, Chappidi, Pinkham, & Hancks, 2021; Petrie, Sandow, et al., 2019), which can compete with host cell MLKL for binding to the key upstream regulator protein kinase, RIPK3, to block execution of the necroptosis pathway.

In contrast to vMLKL, the cellular ortholog of the MLKL pseudokinase is itself the terminal executioner of the necroptosis programmed cell death pathway (Murphy, 2020; Murphy et al., 2013; Petrie et al., 2018; Sun et al., 2012; Wu et al., 2013; Zhao et al., 2012). Following phosphorylation of the MLKL pseudokinase domain by the upstream kinase, RIPK3 (Cook et al., 2014; Meng, Sandow, Czabotar, & Murphy, 2021; Murphy et al., 2013; Sun et al., 2012; Tanzer et al., 2015), MLKL's N-terminal four-helix bundle domain is unleashed and can execute cell death by mediating MLKL translocation to, and permeabilization of, the plasma membrane (Fig. 1) (Davies et al., 2018; Dondelinger et al., 2014; Garnish et al., 2021; Hildebrand et al., 2014; Petrie et al., 2020; Petrie, Hildebrand, & Murphy, 2017; Samson, Garnish, Hildebrand, & Murphy, 2021; Samson et al., 2020; Tanzer et al., 2016; Wang et al., 2014). Because phosphorylation of MLKL by RIPK3 leads to a structural interconversion (Davies et al., 2020; Garnish

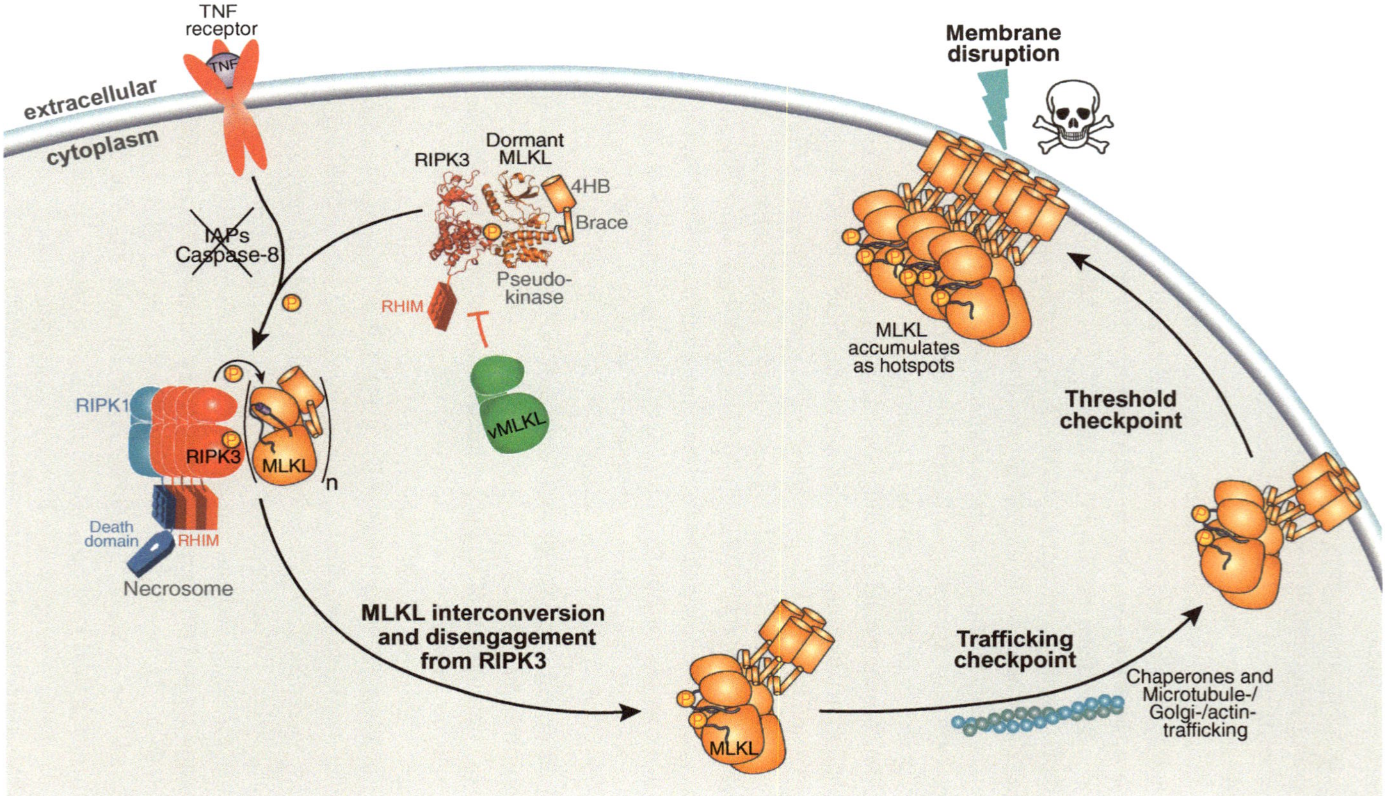

Fig. 1 See figure legend on opposite page.

et al., 2021; Murphy et al., 2013; Petrie, Czabotar, & Murphy, 2019; Petrie et al., 2018), MLKL serves as an exemplar of the molecular switch function that can be mediated by pseudokinases. Our current knowledge of this molecular switch function is derived from structural studies of MLKL alone and in complex with distinct protein ligands (Davies et al., 2020; Garnish et al., 2021; Meng, Davies, et al., 2021; Murphy et al., 2013; Murphy, Lucet, et al., 2014; Petrie, Czabotar, & Murphy, 2019; Pierotti et al., 2020; Xie et al., 2013).

As pseudokinases lack catalytic activities (Jacobsen & Murphy, 2017; Murphy, Zhang, et al., 2014), verification of proper folding is not possible using enzymatic assays but can be inferred from biophysical analyses, such as thermal stability of the recombinant proteins (Lucet & Murphy, 2017; Murphy, Zhang, et al., 2014). Consequently, insect cells are often the preferred host for the expression of recombinant pseudokinases owing to the presence of endogenous protein folding chaperones and enzymes that perform post-translational modifications, which provide some certainty that the recombinant pseudokinase of interest adopts a bona fide native fold. Here, we describe methods for the co-expression of MLKL with binding partners using the baculovirus expression vector system (BEVS), to generate

Fig. 1 Schematic of the necroptotic signaling pathway. The terminal effector of the necroptotic cell death pathway, the MLKL pseudokinase, is maintained in complex with its upstream regulator, the RIPK3 kinase, in a dormant form until the cell receives a necroptotic cue (Garnish et al., 2021; Meng, Davies, et al., 2021). Ligation of death receptors, such as TNF receptor 1, or pathogen sensors, such as Toll-like receptor 3 or 4, can induce necroptotic signaling in cellular contexts where the IAP E3 ligase family and the proteolytic enzyme, Caspase-8, are depleted or disabled. Upon receipt of a necroptotic stimulus, the RIPK3:MLKL subcomplex can be recruited to the necrosome—a large cytoplasmic platform built on the core kinases RIPK1 and RIPK3—where RIPK3 is activated and phosphorylates MLKL (Garnish et al., 2021; Murphy et al., 2013; Rodriguez et al., 2016; Sun et al., 2012; Tanzer et al., 2015; Wang et al., 2014). MLKL phosphorylation promotes conformational interconversion and disengagement from the necrosome (Davies et al., 2020; Garnish et al., 2021; Hildebrand et al., 2014; Murphy et al., 2014; Petrie et al., 2018). MLKL oligomers are then trafficked to the plasma membrane where they accumulate into higher order assemblies termed hotspots, which mediate permeabilization of the plasma membrane, and ultimately cell death (Dondelinger et al., 2014; Garnish et al., 2021; Liu et al., 2021; Murai et al., 2018; Quarato et al., 2016; Samson et al., 2021, 2020; Sethi et al., 2022). Pathogens have evolved mechanisms to subvert necroptotic signaling, such as via expression of RIPK3-inhibiting proteins encoded by some poxviruses, termed viral MLKL (vMLKL) (Petrie et al., 2019). The skull and crossbones image (Mycomorphbox_Deadly.png; by Sven Manguard) was used under a Creative Commons Attribution-Share Alike 4.0 license.

recombinant protein complexes that can be used for structural characterization of MLKL conformations. The methods described herein for the generation of bacmids and protein co-expression in insect cells can be broadly applied to produce recombinant pseudokinases in complex with their binding partners for biochemical and biophysical characterization.

2. Molecular cloning of baculovirus co-expression constructs

Because RIPK3 and MLKL can only form a stable complex when both proteins are co-expressed in insect cells (Garnish et al., 2021; Meng, Davies, et al., 2021; Xie et al., 2013), we must deliver both genes into the expression cell line simultaneously to make this protein complex recombinantly. We accomplished this by inserting the human RIPK3 kinase domain (residues 1–316; untagged) expression cassette from a pAceBac2 plasmid (pAceBac2 *RIPK3*) into a pFastBac HT B plasmid encoding a TEV protease-cleavable, N-terminally His-tagged MLKL pseudokinase domain (residues 190–471) (pFastBac HT B *MLKL*) to generate a co-expression plasmid referred to as pFastBac HT B *MLKL+RIPK3* herein. This bicistronic construct enabled the subsequent generation of a co-expression bacmid containing genes for both proteins of interest to ensure their co-expression in every infected insect cell. A purification tag was only added to MLKL but not RIPK3 to ensure that all of the co-purified RIPK3 was in complex with MLKL. In this example, RIPK3 is abundantly expressed in insect cells and co-expression with MLKL yields a 1:1 complex with negligible apo MLKL enriched following Ni^{2+} chromatography. While usually barely detectable, any apo MLKL is readily separated from the 1:1 MLKL:RIPK3 complex in the subsequent size exclusion chromatography step. This cloning strategy can be applied to co-express other pairs of proteins of interest, exemplified by other MLKL:RIPK3 constructs (Davies et al., 2020; Petrie, Sandow, et al., 2019), MLKL:Monobody complexes (Garnish et al., 2021), and other interacting kinase:pseudokinase pairs (Lange, Nelen, Cohen, & Kulathu, 2021; Oliver et al., 2021; Patel et al., 2020; Zhang et al., 2018).

In this section of the protocol, we describe the molecular cloning of the pFastBac HT B *MLKL+RIPK3* plasmid and the subsequent generation of the bacmid for RIPK3:MLKL co-expression. First, we assemble expression constructs for the individual proteins of interest. In this example, *MLKL* gene is cloned into the multiple cloning site (MCS) of pFastBac HT B vector with an in-frame, TEV protease-cleavable, N-terminal His-tag (Fig. 2A),

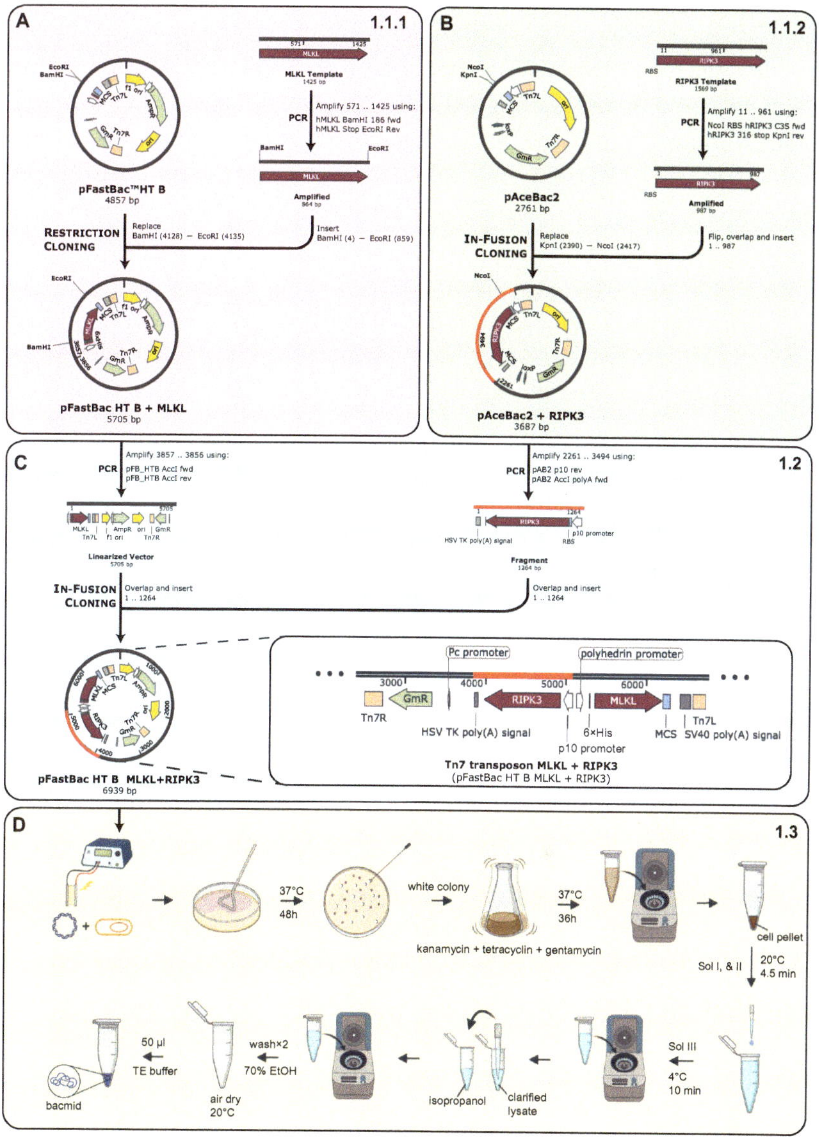

Fig. 2 Overall cloning strategy of the pFastBac HT B *MLKL + RIPK3* recombinant bacmid for co-expressing MLKL and RIPK3 in *Sf*21 cells. (A) Construction of pFastBac HT B *MLKL* plasmid using restriction cloning. (B) Construction of pAceBac2 *RIPK3* plasmid using In-Fusion cloning. (C) Construction of pFastBac HT B *MLKL + RIPK3* plasmid using In-Fusion cloning. pFastBac HT B *MLKL* is linearized by inverse PCR. The *RIPK3* expression cassette (red) from a pAceBac2 construct is amplified using PCR, and In-Fusion cloned into the linearized pFastBac HT B *MLKL* construct. Zoomed panel: the Tn7 transposon of pFastBac HT B *MLKL + RIPK3* co-expression construct. The RIPK3 expression cassette is inserted into a non-coding region between the polyhedrin promoter (upstream of the *MLKL* gene) and the Pc promoter (upstream of the gentamycin resistance gene). (D) A schematic of the workflow for recombinant bacmid preparation from *E. coli* DH10MultiBac cells. Illustrations were created with BioRender.com.

while the *RIPK3* gene is cloned into the MCS of pAceBac2 (Fig. 2B). In the second step, we inserted a cassette from the pAceBac2 *RIPK3* plasmid encoding the p10 promoter, RIPK3 residues 1–316 and the HSV TK poly(A) terminator into a non-coding region adjacent to the Tn7 transposon of pFastBac HT B *MLKL* plasmid (Fig. 2C). Because of the lack of suitable sites in this region of pFastBac HT B vector for restriction cloning, we used In-Fusion cloning, which allows ligation-independent cloning of PCR products into any site using custom oligonucleotides (Table 1). The advantages of In-Fusion cloning and primer design strategies are discussed below. Subsequently, we used the Bac-to-Bac system and DH10MultiBac *Escherichia coli* cells to generate bacmids for baculovirus production in *Sf*21 insect cells (Fig. 2D).

2.1 Cloning of individual expression constructs

In this section of the protocol, we describe the cloning of the individual *MLKL* and *RIPK3* genes into the MCS of pFastBac HT B and pAceBac2, respectively. Here, we present two different cloning methods, one of which was used for MLKL (Fig. 2A) and the other for the RIPK3 (Fig. 2B) construct. There are many equally valid cloning methods that achieve comparable results, which is simply to ensure the inserted gene of interest is cloned in-frame downstream of the N-terminal tag encoded by the destination vector. The more important consideration here is the selection of vector pairs that use different promoters. The expression cassette of pFastBac HT B uses a polyhedrin promoter, while pAceBac2 uses the p10 very late promoter, to initiate transcription of the genes of interest. RIPK3 expression via the p10 promoter will be induced a few hours earlier than polyhedrin promoter-driven MLKL expression, which initiates transcription ~18 h post-infection (Roelvink et al., 1992), as recently reviewed elsewhere (Grose, Putman, & Esposito, 2021). In the case of RIPK3:MLKL co-expression, RIPK3 expression is induced first, which appears to benefit the protein complex yield. We presume this occurs because accumulation of RIPK3 before induction of His_6-MLKL expression favors saturation of MLKL with RIPK3 and consequent stoichiometric complex formation. In support of this strategy comes our previous experience with co-expression of the MLKL and RIPK3 constructs from pFastBac HT B and pAceBac1 (Fig. 3A), which both use polyhedrin promoters. Owing to their simultaneous onset of expression, and perhaps competition for the transcription and translational machinery, we obtained only very low MLKL yields and undetectable RIPK3 expression.

Table 1 List of PCR reactions and primer sequences.

Protocol	Purpose	Primer 1	Primer 2
1.1.1	Amplify *MLKL* pseudokinase domain gene insert	>hMLKL BamHI 186 fwd CGCGGATCCCAAGAGCAAATCAAGGAGATCAAG	>hMLKL Stop *Eco*RI Rev. GAATTCCTACTTAGAAAGGTGGAGAG
1.1.1	Colony screen pFastBac HT B *MLKL*	>hMLKL K230M rev GAGTTTTTGAATACcaTTATGGCCACTGGAGCTCTG	>pBac-up CCGGATTATTCATACCGTC
1.1.2	Amplify RIPK3 gene insert	> *Nco*I RBS hRIPK3 C3S fwd GGATCTCGAGCCATGGAAACCATGTCGTCCGTCAAGTTATGGCCC	>hRIPK3 316 stop *Kpn*I rev CTCCCCCATCTCCCGTTAAGAAAATCTCCTATTGCTGCTCCTGAG
1.2	Linearize pFastBac HT B *MLKL*	>pFB_HTB *Acc*I fwd AGCTATAGTTCTAGTGGTTGGCTACGT	>pFB_HTB AccI rev AGAGTCCTGGGCGAACAAACG
1.2	Amplify RIPK3 expression cassette	>pAB2 p10 rev ACTAGAACTATAGCTCGGACCTTTAATTCAACCCAACACAA	>pAB2 AccI polyA fwd TTCGCCCAGGACTCTCCCCGCGTTTATGAACAAACG
1.3	Verify *MLKL* + *RIPK3* co-expression bacmid	>hMLKL K230M rev GAGTTTTTGAATACcaTTATGGCCACTGGAGCTCTG	>hRIPK3 S227A rev GTACACGAGTGCTGGTTCGGTTGG
1.3	Verify *MLKL* + *RIPK3* co-expression bacmid	>hMLKL K230M Fwd CAGAGCTCCAGTGGCCATAAtgGTATTCAAAAAACTC	>M13Reverse AGCGGATAACAATTTCACACAGG
1.3	Verify *MLKL* + *RIPK3* co-expression bacmid	>hRIPK3 S227A fwd CCAACCGAACCAGCACTCGTGTACG	>M13Forward CCCAGTCACGACGTTGTAAAACG

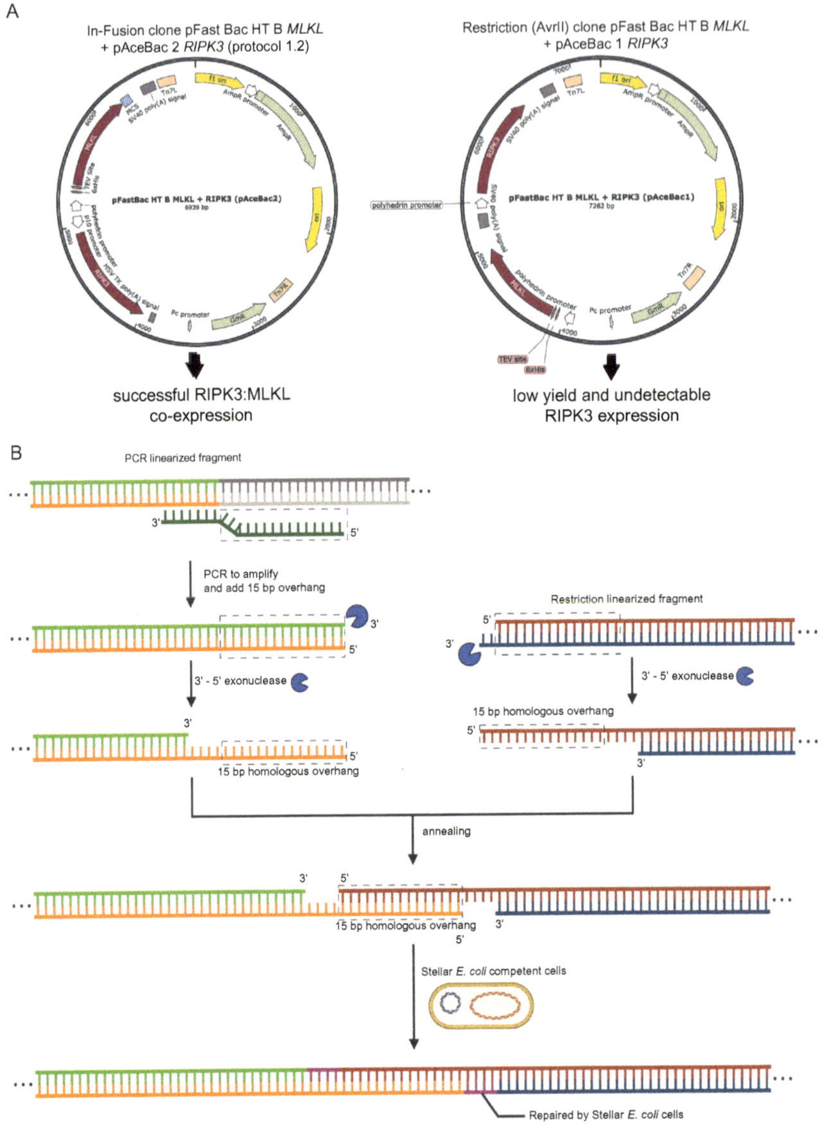

Fig. 3 Overview of In-Fusion cloning of the pFastBac HT B *MLKL* + *RIPK3* co-expression plasmid. (A) Comparison between the successful In-Fusion cloned construct and the failed restriction cloned construct. Left, plasmid map for the In-Fusion cloned pFastBac HT B *MLKL* + *RIPK3* from pAceBac2. Right, plasmid map for the restriction cloned construct made from pAceBac1, which failed to co-express RIPK3 and MLKL. Plasmid maps were generated in SnapGene. (B) The principles of In-Fusion cloning and primer design. In-Fusion cloning can be used to ligate any DNA fragment to any site by designing custom primers (dark green) with 15 bp overhang (dashed box) that

2.1.1 Cloning pFastBac HT B **MLKL** (Fig. 2A)

2.1.1.1 Equipment

- Sterile bacterial hood or Bunsen burner
- Shaker incubator (37 °C)
- Stationary incubator (37 °C)
- Heat block (37 °C)
- Heat block (50 °C)
- Thermocycler
- DarkReader® transilluminator (Clare Chemical Research) for preparative agarose gels
- UV transilluminator for photographing ethidium bromide-stained gels (e.g., Bio-Rad ChemiDoc)
- Scalpel or razor blades
- Agarose gel electrophoresis tank and power supply
- MicroPulser Electroporation Cuvettes (Bio-Rad)
- MicroPulser Electroporator (Bio-Rad)
- Benchtop microcentrifuge
- PCR tubes
- Eppendorf tubes (1.7 mL)
- Sterile round bottom polypropylene tubes (15 mL)
- Sterile pipette tips
- P2, P20, P200, P1000 pipettors

2.1.1.2 Materials and reagents

- CloneAmp HiFi PCR Premix (2×) (Takara; supplied in In-Fusion cloning kit #638951)
- Primers (Table 1) purchased from a vendor, such as IDT, who synthesize custom sequences. Synthesis on the 20 nmol scale, desalted, is sufficient for sequences <40 bases, but HPLC purification should be considered for longer primers

---

is homologous to the end of the target fragment. Either a PCR-linearized DNA fragment (green and yellow; gray illustrates the part of the DNA template that was not amplified by PCR) or restriction linearized DNA fragment (blue and red) can be used as the destination vector. In-Fusion enzyme premix contains a 3′-5′ exonuclease, which creates sticky ends by hydrolyzing the 3′ end of dsDNA. The 15 bp homologous overhangs allow the yellow strand and red strand to anneal. After transforming this annealed plasmid into Stellar™ competent *E. coli* cells, the cells will repair any gap that the 3′-5′ exonuclease created on the green and blue strands (newly incorporated nucleotides are colored magenta) and ligate the phosphodiester backbone.

- Restriction enzymes: *Bam*HI HF (NEB #R3136S) and *Eco*RI HF (NEB R3101S)
- CutSmart Buffer (NEB; supplied with restriction enzymes)
- pFastBac HT B plasmid (ThermoFisher #10584027)
- *MLKL* DNA template (synthesized by ATUM, CA)
- Antarctic phosphatase (NEB #M0289S)
- Antarctic phosphatase reaction buffer (NEB; supplied with enzyme)
- Agarose for molecular biology (Bioline #BIO-41025)
- Ethidium bromide (10 mg/mL; Sigma-Aldrich #E1510)
- Gel Loading Dye, Purple (6 ×) (NEB #B7024A)
- TAE buffer (Sigma-Aldrich #T9650-4 L; 1:10 dilution of 10 × concentrate in reverse osmosis water)
- 1 kb Plus DNA Ladder GeneRuler (ThermoFisher #10787018); diluted to 50 ng/μL in milliQ water and 1 × gel loading dye
- QIAquick Gel Extraction Kit (Qiagen #28706)
- T4 DNA ligase (Promega #M1804)
- T4 ligase buffer (Promega; supplied with T4 ligase)
- *E. coli* DH10B electrocompetent cells (sourced from ThermoFisher #12331013; prepared in-house)
- SOC media (Takara; supplied in In-Fusion cloning kit #638951)
- LG agar (1% *w/v* tryptone, 0.5% w/v yeast extract, 0.5% w/v NaCl, 0.2% w/v D-glucose, 10 mM Tris pH 7.4, 1 mM $MgCl_2$, 1.5% w/v bacto agar, mixed in deionized water; autoclaved to sterilize; prepared in-house)
- LG agar to which 100 μg/mL ampicillin was added to molten agar (when cooled to around 37 °C) before pouring 20 mL into each of 10 cm diameter Petri dish
- Phire Hot Start II DNA polymerase (ThermoFisher Scientific #F122L)
- 5 × Phire II buffer (ThermoFisher Scientific; supplied with polymerase)
- dNTPs (Bioline #BIO-39044; supplied as a 10 mM mixed stock of dATP, dCTP, dGTP and dTTP)
- Ampicillin (Roche #10835269001) prepared as a 100 mg/mL stock in sterile milliQ water
- Super broth (3.5% *w/v* tryptone, 2% w/v yeast extract, 0.5% w/v NaCl, 1 mL of 5 M NaOH per liter of medium, deionized water; autoclaved to sterilize; prepared in-house)
- Super broth supplemented with 100 μg/mL ampicillin
- QIAprep Spin Miniprep Kit (Qiagen; catalog number 27106)

2.1.1.3 Step-by-step protocol

1. Set up overhang PCR to amplify MLKL gene insert and introduce the desired oligonucleotide-encoded restriction sites (Table 1). In a PCR tube, set up the reaction containing the following final quantities or concentrations:
   - CloneAmp HiFi PCR Premix (2×): 12.5 μL
   - Primer 1: 0.3 μM
   - Primer 2: 0.3 μM
   - Template DNA: 10 ng
   - MilliQ water: to 25 μL final volume

Perform the following PCR steps in a thermal cycler for 35 cycles:
- 98 °C for 10 s
- 55 °C for 5 s
- 72 °C for 10 s (recommended 5 s/kb for <100 ng template DNA)

2. Restriction digest using *Bam*HI and *Eco*RI to linearize the pFastBac HT B plasmid at the MCS and create sticky ends on the insert fragment containing *MLKL* gene. Set up the reaction in an Eppendorf tube containing the following final concentrations:
   - DNA: 3 μg pFastBac HT B plasmid; or 25 μL PCR reaction of *MLKL* gene insert
   - NEB Cutsmart buffer: 3.5 μL
   - *Bam*HI HF: 1 μL
   - *Eco*RI HF: 1 μL
   - Deionized water: up to 35 μL

Incubate the restriction digestion in a 37 °C heat block overnight. Oligonucleotide-encoded overhang beyond the restriction site and subsequent restriction digestion of the PCR product are important to ensure the presence of a 5′ phosphate group for T4 ligase-mediated ligation into the destination vector.

3. Treat the restriction linearized pFastBac HT B vector with Antarctic phosphatase to prevent self-annealing during the ligation reaction. To the 35 μL restriction digest mixture, add 4 μL Antarctic phosphatase reaction buffer and 0.5 μL Antarctic phosphatase. Incubate the reaction in a 37 °C heat block for 45 min, then add another 0.5 μL Antarctic phosphatase. Incubate reaction at 37 °C for another 45 min
4. Gel purify the restriction digested *MLKL* insert fragment and Antarctic phosphatase treated linearized pFastBac HT B vector. Add appropriate amount of gel loading dye to the DNA mixture and load the whole mixture onto a 1% agarose gel in TAE buffer with 1 μg/mL ethidium

bromide. Load 7 μL 1 kb Plus DNA Ladder GeneRuler in one well to serve as a molecular marker and the entire restriction digest in another well. It may be necessary to load over two wells if the volume of reaction supersedes the capacity of a single well. Run the agarose gel at 110 V for 30–60 min. Detect the DNA bands using a low energy/long wavelength UV transilluminator, cut the band of appropriate size using a clean scalpel blade and put gel slice into a clean Eppendorf tube. The choice of UV transilluminator is important; a preparative instrument with long wavelength, low energy UV light is crucial to avoid damage to the DNA fragments during excision of the DNA band from the gel. If a specialist low energy transilluminator is not available, it may be possible to excise the bands using the preparative mode on an imaging device, such as a Bio-Rad ChemiDoc. It is important to excise bands under low energy UV light before photographing, so as to avoid damage to the DNA. Excise gel bands then photograph the gel with an instrument such as a Bio-Rad ChemiDoc to obtain an image of the gel for your records. Extract DNA from the excised gel band(s) using a QIAquick Gel Extraction Kit following manufacturer's instructions, which requires a heat block at 50 °C

5. Ligate the purified *MLKL* gene insert with the restriction linearized pFastBac HT B vector using T4 DNA ligase. In a clean Eppendorf tube, set up a 10 μL reaction mixture that contains the following:
   - 10 × Ligase Buffer: 1 μL
   - pFastBac HT B linearized vector: 0.8 μL
   - Purified *MLKL* gene insert: 2.5 μL
   - T4 DNA ligase: 0.8 μL
   - Deionized water: 4.9 μL

Incubate the ligation reaction at room temperature for at least 1 h.

6. Transform the ligation reaction into *E. coli* DH10B electrocompetent cells. All procedures involving cells are to be performed under sterile conditions. Mix 2 μL ligation reaction with 25 μL electrocompetent cells in a 0.2 cm MicroPulser Electroporation Cuvette prechilled on ice
7. Select EC2 on a MicroPulser Electroporator, and electroporate the cuvette containing the cells and DNA. Return the cuvette back to ice
8. To recover the electroporated cells, mix the cells in the cuvette with 500 μL SOC media and incubate at 37 °C with shaking at 200 rpm for 30 min

9. Remove excess SOC media following centrifugation at 10000 × *g* for 1 min at room temperature, then resuspend the cell pellet in 50 μL media and spread on an LG agar plate supplemented with 100 μg/mL ampicillin
10. Incubate the LG agar plate at 37 °C overnight (~16 h) in a stationary incubator, with the plate inverted to prevent condensation collecting on the agar surface, to develop the bacteria colonies
11. (Optional) Perform a colony PCR screen to verify that the genes of interest were correctly incorporated into the plasmid. Pipette 20 μL sterile MilliQ water into sterile 1.5 mL Eppendorf tubes. The precise number of colonies you will need to screen for a positive will depend on your cloning efficiency, which can be dictated by many factors, including freshness of your restriction enzymes and the choice of enzymes themselves. A reasonable starting point is 5–10 colonies per ligation. Using a sterile pipette tip, touch a single colony on the LG-Agar transformation plate surface and transfer the tip (with associated bacteria from the colony) to the 20 μL sterile MilliQ water in an Eppendorf tube. Allow the tip to sit in the MilliQ as you proceed. Then attach your pipettor and pipette up and down mix the bacteria in the water. Remove 1 μL to an LG-Agar plate supplemented with Amp marked with a numbered 1 cm × 1 cm grid to use an archive colony, and incubate as per Step 10. Ensure the tube and the grid numbers match. Boil the Eppendorf tube containing the remaining bacteria in water at 100 °C for 10 min to lyse the cells and thus generate a crude DNA sample for colony PCR analysis. Centrifuge the lysate at 13,000 × *g* for 10 min to clarify. To set up the PCR reactions, in a PCR tube mix the following quantities or concentrations per reaction, preparing a master mix for your chosen number of colonies (plus 1 reaction, such that you make a mix for 6 reactions for 5 colonies, and 11 reactions for 10 colonies, for example):
    - 5 × Phire II buffer: 4 μL
    - dNTPs: 0.1 mM
    - Primer 1: 0.5 μM
    - Primer 2: 0.5 μM
    - Phire Hot Start II DNA polymerase: 0.05 μL
    - Deionized (milliQ) water: to 17 μL

Aliquot 17 μL per reaction to PCR tubes and mix with 3 μL of the cell lysate, ensuring the tubes are numbered correspondingly.

Perform the following PCR steps in a thermal cycler:

- Step 1: 98 °C for 3 min
- Step 2: 30 cycles:
  - 98 °C for 10 s
  - 55 °C for 10 s
  - 72 °C for 5 s/kb
- Step 3: 72 °C for 30 s

Analyze the PCR products by 1% agarose gel electrophoresis. Load 3–5 μL of 50 ng/μL 1 kb Plus DNA ladder to serve as a molecular marker in a control lane to estimate PCR transcript length. Depending on the number of colonies in your screen, you may need to scale from a minigel (10 or 15 wells per row; 3 rows maximum) to a larger agarose gel that will accommodate dozens to hundreds of PCR reactions. In cases where a large number of colonies are being screened, performing PCR reactions in 96-well plates and using a multichannel pipette for liquid handling should be considered to minimize repetitive strain.

12. Inoculate a single colony into ~3 mL Super broth supplemented with 100 μg/mL ampicillin. Incubate the bacteria culture at 37 °C overnight with shaking at 200 rpm. If not possible to pick a single colony, an additional dilution step where the colony is streaked on a fresh LG-Agar + Amp plate with a further overnight stationary 37 °C incubation may be required. Where a colony PCR screen was performed, it is possible to inoculate your miniprep culture directly from the archive plate for colonies that yielded a band of the expected size in the PCR screen
13. Harvest 3 mL of liquid culture by centrifugation at 11,000 × *g* for 1 min. Remove supernatant by pipette or using an aspirator. Extract plasmid DNA using QIAprep Spin Miniprep Kit following manufacturer's instructions
14. Verify the sequence of the miniprepped plasmid by Sanger sequencing. Minimally, sequencing primers that bind specifically to regions within the vector that flank the insertion site should be used to verify the integrity of the amplified and ligated gene and the integration site

2.1.2 Cloning pAceBac2 RIPK3

2.1.2.1 Equipment

- Sterile bacterial hood or Bunsen burner
- Shaker incubator (37 °C)
- Stationary incubator (37 °C)
- Thermocycler

- Water bath (42 °C)
- Sterile inoculating loop
- Benchtop microcentrifuge
- PCR tubes
- Eppendorf tubes (1.5 mL)
- Sterile round bottom polypropylene tubes (15 mL)
- Sterile pipette tips
- P2, P20, P200, P1000 pipettors

2.1.2.2 Materials and reagents

- pAceBac2 plasmid (now supplied by Geneva Biotech)
- RIPK3 DNA template (synthesized by ATUM, CA)
- Primers (Table 1) purchased from a vendor, such as IDT, who synthesize custom sequences. Synthesis on the 20 nmol scale, desalted, is sufficient for sequences <40 bases, but HPLC purification should be considered for longer primers
- CloneAmp HiFi PCR Premix (2 ×) (Takara; supplied in In-Fusion cloning kit #638951)
- Cloning Enhancer (Takara; supplied in In-Fusion cloning kit #638951)
- QIAquick PCR Purification Kit (Optional, Qiagen; #28006)
- In-Fusion HD Enzyme Premix (5 ×) (Takara; #638951)
- Stellar™ chemically competent *E. coli* cells (Takara; supplied in In-Fusion cloning kit #638951)
- SOC media (Takara; supplied in In-Fusion cloning kit #638951)
- LG agar (1% *w/v* tryptone, 0.5% *w/v* yeast extract, 0.5% w/v NaCl, 0.2% w/v D-glucose, 10 mM Tris pH 7.4, 1 mM $MgCl_2$, 1.5% w/v bacto agar, mixed in deionized water; autoclaved to sterilize; prepared in-house)
- LG agar to which 100 μg/mL ampicillin was added to molten agar (when cooled to around 37°C) before pouring into, and setting in, 10 cm diameter Petri dishes
- Super broth (3.5% w/v tryptone, 2% w/v yeast extract, 0.5% w/v NaCl, 1 mL of 5 M NaOH per liter of medium, deionized water; autoclaved to sterilize; prepared in-house)
- Super broth supplemented with 100 μg/mL ampicillin
- QIAprep Spin Miniprep Kit (Qiagen; #27106)
- Ampicillin (Roche #10835269001) prepared as a 100 mg/mL stock in sterile milliQ water
- Sterile milliQ water

2.1.2.3 Step-by-step protocol

1. While conventional restriction enzyme cloning could be used to introduce the *RIPK3* gene into pAceBac2, we describe here a method using In-Fusion cloning (Takara). The principles of In-Fusion cloning and its associated advantages are detailed below in Section 2.2. To start, set up PCR reactions to linearize pAceBac2 vector (inverse PCR) and amplify the *RIPK3* gene insert with appropriate 15 bp homologous overhanging sequences for In-Fusion cloning (overhang PCR). Primer design can be performed using the manufacturer's online tool or with the aid of computer software, such as SnapGene, which contains modules for In-Fusion overhang design (as detailed in Section 2.2). In a PCR tube, set up a reaction mixture containing the following final volumes or concentrations:
   - CloneAmp HiFi PCR Premix (2 ×): 12.5 μL
   - Primer 1: 0.3 μM
   - Primer 2: 0.3 μM
   - Template DNA: 10 ng
   - milliQ water to 25 μL

Perform the following PCR steps in a thermal cycler for 35 cycles:

- 98 °C for 10 s
- 55 °C for 5 s
- 72 °C for 10 s (recommended 5 s/kb for <100 ng template DNA)

2. Clean up the PCR products using Takara Cloning Enhancer treatment. A portion of the PCR reaction in the previous step, 5 μL is mixed with 2 μL Cloning Enhancer and incubated in a thermal cycler:
   - 37 °C for 15 min
   - 80 °C for 15 min

Because we use only a small amount of template DNA which is of high purity, cloning enhancer treatment is typically sufficient. If the template DNA is impure or the PCR reaction results in multiple products that can be observed as several bands by agarose gel electrophoresis, a PCR cleanup step using a QIAquick PCR purification Kit (following manufacturer's instructions) can improve cloning efficiency markedly, and this is especially recommended for longer PCR products (e.g., vectors linearized by inverse PCR).

3. Set up the In-Fusion cloning reactions in PCR tubes. The final reaction solution will contain the following:
   - In-Fusion HD Enzyme Premix (5 ×): 2 μL
   - Enhancer treated linearized vector (pAceBac2): 2 μL

- Enhancer treated insert (*RIPK3* gene with 15 bp homologous overhang): 2 μL
- Deionized water: 4 μL

Incubate the reaction at 50°C for 15 min using a thermal cycler.

4. Transform the In-Fusion cloning products into Stellar™ chemically competent *E. coli* cells. Two microliters of In-Fusion cloning reaction is mixed with 25 μL competent cells thawed on ice and incubated on ice for 30 min. Then, the cells are heat shocked for 45 s at 42 °C in a water bath, before being returned to ice for 2 min
5. Recover transformed competent cells in 500 μL SOC media for 30 min at 37 °C with shaking at 200 rpm
6. Remove excess SOC media by centrifugation at 11,000 × *g* for 1 min at room temperature, before resuspending the cells in 50 μL media and plating on an LG agar plate supplemented with 100 μg/mL ampicillin. Incubate at 37 °C overnight (~16 h) to develop colonies
7. For the extraction of miniprep DNA, an individual colony should be picked using a sterile loop or pipette tip to inoculate 5 mL Super broth containing 100 μg/mL ampicillin in a 15 mL round bottom tube and the culture incubated at 37 °C with shaking at 200 rpm overnight (14–16 h). Multiple colonies may be picked from the same agar plate into distinct Super broth and ampicillin cultures in parallel to increase the chance of identifying successfully cloned plasmids. Typically, with this method colony PCR screening is not required. However, if there are problematic templates or long destination vectors, it may be prudent to consider screening colonies for successful cloning events using a primer targeting a sequence within the insert and another within the flanking region of the vector that you would predict to amplify a 500–1500 bp PCR product
8. Harvest cells from 3 mL of liquid culture by centrifugation at 11,000 × *g*. Aspirate supernatant using a P1000 pipette. Cell pellets may be frozen at −20 °C for long-term storage
9. Extract the plasmids from the cell pellet using a QIAprep Spin Miniprep Kit (Qiagen) following the manufacturer's instructions
10. Verify the sequence of the miniprepped plasmid by Sanger sequencing. Minimally, sequencing primers that bind specifically to regions within the vector and flank the insertion site should be used to verify the integrity of the amplified and introduced gene, as well as its correct integration into the vector

2.2 Making pFastBac HT B *MLKL*+*RIPK3* co-expression plasmid using In-Fusion cloning

After obtaining the individual MLKL and RIPK3 expression constructs, we made a co-expression construct by inserting the expression cassette from pAceBac2 *RIPK3* (Fig. 2C) into pFastBac HT B *MLKL*. We selected the non-coding region between the polyhedrin promoter of *MLKL* and the Pc promoter of gentamycin resistance gene (GmR) (Fig. 2C, zoomed panel) as a site that could be disrupted without compromising pFastBac HT B function. This region is flanked by the Tn7 transposon sites of pFastBac HT B vector that are essential for transposition into a bacmid backbone in the next step. Therefore, insertion of the RIPK3 expression cassette here will allow it to be transposed into the bacmid along with *MLKL* gene and consequently introduced into the baculovirus downstream. This is also the longest non-coding region flanked by the Tn7 transposon, so we anticipated any modifications in this region would minimize any impact to the functions of other genetic elements. In this section of the protocol, we describe how the RIPK3 expression cassette from pAceBac2 can be inserted into pFastBac HT B (*MLKL*) using In-Fusion cloning.

To accomplish such manipulation of the plasmids, we previously attempted conventional restriction cloning by inserting a *RIPK3* expression cassette (excised with *Spe*I and *Avr*II) from pAceBac1 into the AvrII site of pFastBac HT B *MLKL* (Fig. 3A, right). However, due to the lack of unique restriction sites on either vector at the desired position—because these vectors were not originally designed to be cut in this manner—it proved challenging. Furthermore, we could not control the direction of insertion, resulting in both expression cassettes running head to tail in the same direction. This may have been unfavorable, perhaps contributing to the low protein yields arising from the aforementioned pFastBac HT B+pAceBac1 co-expression construct.

To overcome the limitations of restriction cloning in performing such DNA manipulations, we employed In-Fusion cloning. Importantly for our purposes, this technique enables ligation-independent cloning of PCR products into any vector, at any site of linearization, because desired sticky ends can be generated from the 5′ ends of the PCR primers. In addition, In-Fusion cloning offers further advantages over conventional cloning, owing to its rapidity, specificity, directionality, efficiency, and high-throughput compatibility.

The principle for In-Fusion cloning relies on 3′-5′ exonuclease activity to generate the sticky ends for annealing and ligation. First, the DNA fragments to be joined need to have homologous ends (Fig. 3B). As a rule of thumb, it is recommended by the manufacturer, Takara, that the homologous ends are 15 bp in length. Overhangs can be created either by restriction digestion or PCR. Restriction digestion is more suitable for manipulating larger DNA fragments, such as the linearization of a very large plasmid, for which inverse PCR may be difficult owing to limited yields and fidelity that may occur from PCR amplification of large (>10 kb) vectors. PCR can be used to linearize smaller plasmids or amplify the insert and add the 15 bp homologous ends at the same time by introducing them as the overhang of the PCR primers. During the In-Fusion enzyme reaction, 3′-5′ exonuclease activity will hydrolyze nucleotides from the 3′ end of dsDNA, creating sticky ends. The 15 bp homology regions encoded by PCR primers allow these sticky ends to anneal into a linearized vector non-covalently. The In-Fusion reaction can be transformed into competent Stellar™ *E. coli* cells where it assembles into one covalently linked dsDNA plasmid. Stellar™ *E. coli* cells (Takara) or similar strains are preferred for these transformations, because other commonly used cloning strains lacking *recA1* and *endA* mutations may compromise the required nick repairs. Because the 15 bp homologous ends can be easily introduced as overhangs of PCR primers (Table 1), any DNA fragment can be directed to any site with high specificity by simply designing custom oligonucleotides to match the ends of the cloning site of interest. The website of the manufacturer provides a free primer design tool (https://www.takarabio.com/learning-centers/cloning/primer-design-and-other-tools), and design can be performed using some widely used computer software, such as SnapGene. Gibson assembly is another technique that works based on similar principle but relies on a different set of enzymes. Most notably Gibson assembly uses a 5′-3′ exonuclease, which creates sticky ends by removing nucleotides from the opposite strands to the In-Fusion enzymes.

2.2.1 Equipment

- Sterile bacterial hood or Bunsen burner
- Shaker incubator (37 °C)
- Stationary incubator (37 °C)
- Thermocycler
- Water bath (42 °C)
- Sterile inoculating loop
- Benchtop microcentrifuge
- PCR tubes
- Eppendorf tubes (1.5 mL)

- Sterile round bottom polypropylene tubes (15 mL)
- Sterile pipette tips
- P2, P20, P200, P1000 pipettors

2.2.2 Materials and reagents

- pFastBac HT B *MLKL* plasmid (from 1.1.1)
- pAceBac2 *RIPK3* plasmid (from 1.1.2)
- CloneAmp HiFi PCR Premix (2 ×) (Takara; supplied in In-Fusion cloning kit #638951)
- Cloning Enhancer (Takara; supplied in In-Fusion cloning kit #638951)
- QIAquick PCR Purification Kit (Optional, Qiagen #28006)
- In-Fusion HD Enzyme Premix (5 ×) (Takara; supplied in In-Fusion cloning kit #638951)
- Primers (Table 1)
- Stellar™ chemically competent *E. coli* cells (Takara; supplied in In-Fusion cloning kit #638951)
- SOC media (Takara; supplied in In-Fusion cloning kit #638951)
- LG agar (1% *w/v* tryptone, 0.5% w/v yeast extract, 0.5% w/v NaCl, 0.2% w/v D-glucose, 10 mM Tris pH 7.4, 1 mM $MgCl_2$, 1.5% w/v bacto agar, mixed in deionized water; autoclaved to sterilize; prepared in-house)
- LG agar to which 100 μg/mL ampicillin was added to molten agar (when cooled to around 37 °C) before pouring into, and setting in, 10 cm diameter Petri dishes
- Super broth (3.5% w/v tryptone, 2% w/v yeast extract, 0.5% w/v NaCl, 1 mL of 5 M NaOH per liter of medium, deionized water; autoclaved to sterilize; prepared in-house)
- Super broth supplemented with 100 μg/mL ampicillin
- QIAprep Spin Miniprep Kit (Qiagen; #27106)

2.2.3 Step-by-step protocol

11. Begin by linearizing the pFastBac HT B *MLKL* plasmid and amplifying the insert containing the expression cassette of RIPK3 from pAceBac2 *RIPK3* using PCR. In PCR tubes, set up PCR reactions that contain the following final quantities or concentrations:
    - CloneAmp HiFi PCR Premix (2 ×): 12.5 μL
    - Primer 1: 0.3 μM
    - Primer 2: 0.3 μM
    - Template DNA: 10 ng
    - Deionized water: to 25 μL

Perform the following PCR steps in a thermal cycler for 35 cycles:

- 98 °C for 10 s
- 55 °C for 5 s
- 72 °C for 30 s (recommended 5 s/kb for <100 ng template DNA)

We recommend using <100 ng of template DNA per reaction (about 10 ng is sufficient). This not only allows one to use a much shorter extension step during PCR (5 s/kb, according to manufacturer's instructions), but it also allows a much faster cleanup step using Cloning Enhancer (Takara).

12. Cleanup the PCR products using Cloning Enhancer (Takara) treatment. Five microliters of PCR reaction from the previous step is mixed with 2 μl Cloning Enhancer and incubated in a thermal cycler:

- 37 °C for 15 min
- 80 °C for 15 min

Alternatively, PCR cleanup using a QIAquick PCR purification kit (following manufacturer's instructions) may be performed to improve cloning efficiency.

13. Set up In-Fusion cloning reactions in PCR tubes. Final reaction solutions contain the following quantities:

- In-Fusion HD Enzyme Premix (5 ×): 2 μL
- Enhancer treated linearized vector (pFastBac HT B *MLKL*): 2 μL
- Enhancer treated insert (expression cassette from pAceBac2 *RIPK3*): 2 μL
- Deionized water: 4 μL

Incubate the reaction at 50 °C for 15 min in a thermal cycler.

14. Transform the In-Fusion cloned plasmids into Stellar™ chemically competent *E. coli* cells. Two microliters of In-Fusion cloning reaction are mixed with 25 μL ice-cold competent cells and incubated on ice for 30 min. Then, the cells are heat shocked for 45 s at 42 °C in a water bath, before being returned to ice for 2 min

15. The transformed competent cells are recovered in SOC media for 30 min at 37 °C with shaking at 200 rpm

16. Remove excess SOC media by centrifugation at 11,000 × *g* for 1 min at room temperature, then resuspend the cell pellet in 50 μL media before plating on LG agar plate supplemented with 100 μg/mL ampicillin. Incubate at 37 °C overnight to develop the colonies

17. For the extraction of miniprep DNA, an individual colony should be picked using a sterile loop, inoculation needle, or pipette tip to inoculate 5 mL Super broth containing 100 μg/mL ampicillin in a sterile

15 mL round bottom tube, and the culture should be incubated at 37 °C with shaking at 200 rpm overnight (14–16 h). Multiple colonies may be picked from the same agar plate into distinct Super broth + ampicillin cultures in parallel to increase the chance of identifying successfully cloned plasmids

18. Harvest cells from 3 mL of liquid culture by centrifugation at 11,000 × *g*. The supernatant is aspirated off using a pipette. The cell pellets may be frozen at −20 °C for long-term storage
19. Extract the plasmids from the cell pellet using a QIAprep Spin Miniprep Kit (Qiagen) following the manufacturer's instructions
20. Verify the sequence of the miniprepped plasmid by Sanger sequencing. Minimally, sequencing primers that bind specifically to regions within the vector and flank the insertion site should be used to verify the integrity of the amplified and introduced gene

2.3 Generation of bacmids

To deliver the genes of interest along with the rest of the baculoviral genome into insect cells, we generated bacmids through recombination with a parental bacmid harbored within DH10MultiBac *E. coli* cells (MultiBac system; ATG Biosynthetics; Fig. 2D). Alternative *E. coli* strains containing suitable parental bacmids can similarly be used to generate bacmids via Tn7 transposon recombination, such as the widely used DH10Bac strain (Bac-to-Bac System; ThermoFisher). Additional benefits are attainable using Kinase Factory 2.0 or EmBacY *E. coli* cells (Geneva Biosciences), where each parental bacmid encodes yellow fluorescence protein (YFP) to permit real time spectrophotometric and/or microscopy monitoring of baculoviral infection and protein production in infected insect cells downstream. In the pFastBac HT B *MLKL*+*RIPK3* plasmid that we constructed, expression cassettes of both genes of interests, *RIPK3* and *MLKL*, as well as gentamicin resistance gene, are flanked by Tn7 transposon sites (Fig. 2C). This allows them to be transposed into the bacmid upon introduction in the host *E. coli* cells, where their retention in the cells can be selected for with Gentamicin in addition to the Kanamycin and Tetracycline resistance encoded by the parental bacmid. Colonies in which successful transposition has occurred can be selected by blue/white screening. IPTG induction of a *lacZα* gene in the parental bacmid leads to processing of the indicator that we include in the LG agar plates, BluoGal. However, successful recombination of the shuttle vector encoding our genes of interest into the parental vector will disrupt the *lacZα* gene and lead to the occurrence of white colonies.

Bacmids can then be purified from the positive (white) colonies using a modified miniprep procedure that seeks to minimize possible fractures that might occur in bacmid DNA owing to their large size (>100 kb) (Fig. 2D). Following quality control by PCR to confirm the presence of the gene(s) of interest in the purified bacmids, these bacmids can be used for baculovirus production in *Sf*21 insect cells.

2.3.1 Equipment

- Sterile bacterial hood or Bunsen burner
- MicroPulser Electroporation Cuvette (Bio-Rad)
- MicroPulser Electroporator (Bio-Rad)
- Shaker incubator (37 °C)
- Stationary incubator (37 °C)
- Thermocycler
- Water bath (42 °C)
- Sterile inoculating loop
- Benchtop microcentrifuge
- UV transilluminator for photographing ethidium bromide (or similar)-stained gels (e.g., Bio-rad ChemiDoc)
- Agarose gel electrophoresis tank and power supply
- Microdrop Spectrophotometer (e.g., Nanodrop or Denovix instrument)
- Eppendorf tubes (1.5 mL)
- Sterile round bottom polypropylene tubes (15 mL)
- Sterile pipette tips
- P2, P20, P200, P1000 pipettors

2.3.2 Materials and reagents

- pFastBac HT B *MLKL* + *RIPK3* plasmid (from 1.2)
- *E. coli* DH10MultiBac electrocompetent cells (prepared in-house; strain from ATG Biosynthetics, now supplied by Geneva Biotech)
- SOC media (Takara; supplied in In-Fusion cloning kit #638951)
- Kanamycin (ThermoFisher; #11815-032); 50 mg/mL (1000 ×) stock in milliQ water
- Tetracycline (Sigma; #T3383-25G); 10 mg/mL (1000 ×) stock in milliQ water
- Gentamicin (Sigma; #G1397-10ML); supplied as 50 mg/ml (7100 ×) aqueous stock
- BluoGal (ThermoFisher #15519028); 0.04 g dissolved in 2 mL dimethyl formamide (Sigma #D4551-250ML) in a polypropylene or glass tube for 400 mL of LG agar

- IPTG (UBP Bio #P1010-100); 0.2 g/mL stock dissolved in milliQ water and filter sterilized
- LG agar (1% *w/v* tryptone, 0.5% w/v yeast extract, 0.5% w/v NaCl, 0.2% w/v D-glucose, 10 mM Tris pH 7.4, 1 mM $MgCl_2$, 1.5% w/v bacto agar, mixed in deionized water; autoclaved to sterilize; prepared in-house)
- LG agar plates with additives. To 400 mL molten agar (at ~37 °C), add: 400 μL kanamycin stock (final 50 μg/mL), 400 μL tetracycline stock (final 10 μg/mL), 56 μL gentamicin stock (final 7 μg/mL), 2 mL BluoGal (final 100 μg/mL) and 80 μL IPTG stock (40 μg/mL final) prior to pouring ~20 mL of molten agar into, and setting in, 10 cm diameter Petri dishes
- Super broth (3.5% w/v tryptone, 2% w/v yeast extract, 0.5% w/v NaCl, 1 mL of 5 M NaOH per liter of medium, deionized water; autoclaved to sterilize; prepared in-house)
- Super broth supplemented with 50 μg/mL kanamycin, 10 μg/mL tetracycline, and 7 μg/mL gentamycin
- Solution I (15 mM Tris, 10 mM EDTA, pH 8.0; sterile filtered and stored at 4 °C) supplemented with 1:100 RNase A stock (Qiagen; part number 1031301; 10 mg/mL)
- Solution II (200 mM NaOH, 1% SDS, sterile filtered)
- Solution III (3 M potassium acetate, pH 5.4, sterile filtered)
- TE buffer (ThermoFisher Scientific; #12090-015)
- Isopropanol, molecular biology grade (pre-chilled to 4 °C)
- 70% *v*/v ethanol, prepared from milliQ water and molecular biology grade ethanol (pre-chilled to 4 °C)
- Phire Hot Start II DNA polymerase (ThermoFisher Scientific)
- 5 × Phire II buffer (ThermoFisher Scientific)
- dNTPs (Bioline #BIO-39044; 10 mM mixed stock of dATP, dCTP, dGTP, dTTP;)
- Primers (Table 1) purchased from a vendor, such as IDT, who synthesize custom sequences. Synthesis on the 20 nmol scale, desalted, is sufficient for sequences <40 bases, but HPLC purification should be considered for longer primers
- Agarose for molecular biology (Bioline #BIO-41025)
- Ethidium bromide (10 mg/mL; Sigma-Aldrich #E1510)
- TAE buffer (Sigma-Aldrich #T9650-4 L; 1:10 dilution of 10 × concentrate in reverse osmosis water)
- 1 kb Plus DNA Ladder GeneRuler (ThermoFisher #10787018); diluted to 50 ng/μL in milliQ water and 1 × gel loading dye

2.3.3 Step-by-step protocol

21. Transform pFastBac HT B *MLKL + RIPK3* DNA miniprep into *E. coli* DH10MultiBac electrocompetent cells. First, dilute the plasmid miniprep to approximately 1 ng/μL
22. Then, mix 1 μL of the diluted miniprep with 25 μL DH10MultiBac electrocompetent cells (thawed and maintained on ice) in a 0.2 cm MicroPulser Electroporation Cuvette on ice
23. Select EC2 on BioRad MicroPulser Electroporator, and electroporate the cuvette containing the cells and DNA. Return the cuvette to ice
24. To recover the electroporated cells, mix the cells in the cuvette with 500 μL SOC media and incubate at 37 °C with shaking for 5 hours
25. Remove excess SOC media by centrifugation at 11,000 × *g* for 1 min at room temperature, then resuspend the cell pellet in 50 μL media and plate them on a LG agar plate supplemented with 50 μg/mL kanamycin, 10 μg/mL tetracycline, 7 μg/mL gentamicin, 100 μg/mL BluoGal, and 40 μg/mL IPTG. Depending on the number, competency and whether chemically- or electro-competent cells are used at this step, it may be advantageous to plate different volumes of the recovered cells on several plates in parallel.
26. Incubate the agar plate inverted in a stationary 37 °C incubator for 36–48 h and perform blue/white screening. White colonies indicate successful recombination events; the 36–48 h incubation is necessary for BluoGal to be metabolized for the blue pigment to develop in negative colonies
27. Re-streak individual white colonies on to a fresh LG-Agar plate containing 50 μg/mL kanamycin, 10 μg/mL tetracycline, 7 μg/mL gentamicin, 100 μg/mL BluoGal and 40 μg/mL IPTG. Incubate the agar plate inverted in a stationary incubator at 37 °C for 36–48 h
28. Using an inoculation loop or needle, or a sterile pipette tip, inoculate an isolated white colony into 4 mL Super broth supplemented with 50 μg/mL kanamycin, 10 μg/mL tetracycline, and 7 μg/mL gentamicin. Incubate the bacteria culture overnight (~16 h) at 37 °C with shaking at 200 rpm. Several colonies can be inoculated into separate cultures in parallel as backup preparations
29. Harvest 2.5 mL of the by serial centrifugation at 11,000 × *g* for 1 min at room temperature in a 1.5 mL Eppendorf tube. Aspirate and discard the supernatant
30. Resuspend the cell pellet in 0.3 mL ice cold Solution I (supplemented with RNase A) by pipetting the cells gently up and down. Do not vortex the mixture during preparation, because this may cause bacmids to fracture

31. Add 0.3 mL Solution II to the tube. Gently mix the solution by inverting several times and incubate the mixture at room temperature for <5 min
32. Add 0.3 mL pre-chilled Solution III to the mixture, and gently mix the solutions by inverting several times. Incubate the mixture on ice for 10 min, then centrifuge it at 13,000 × *g* for 10 min at 4 °C to remove precipitate
33. Pipette 0.75 mL clarified lysate into a clean Eppendorf tube and add 0.75 mL cold isopropanol on ice. Gently mix the solution by inverting several times and incubate on ice for 10 min
34. Centrifuge the tube at 13,000 × *g* at 4 °C for 20 min to pellet the DNA. Remove the supernatant by pipetting carefully, ensuring that the DNA pellet is not disturbed or aspirated
35. Wash the pellet by adding 0.5 mL ice-cold 70% ethanol, followed by 13,000 × *g* at 4 °C for 2 min. On addition of 70% ethanol, the pellet is usually visible as a small white disc. Next, aspirate the ethanol using a pipette. It is not essential to remove absolutely all of the ethanol at this point, because a further washing step will be performed
36. Repeat step 35
37. Carefully remove as much liquid ethanol as possible using a pipette; a brief spin may be required to collect any ethanol in the bottom of the tube for more complete removal. Leave the DNA pellet to dry at room temperature for approximately 30 min
38. After the pellet is dried, initially add 100 μL TE buffer to re-dissolve the bacmid DNA. Gently agitate the tube to mix and incubate the solution at room temperature for several hours to allow all DNA to fully dissolve. Measure the DNA concentration using a microdrop spectrophotometer at 260 nm wavelength. Depending on the DNA yield, it may be necessary to add further TE buffer to dilute to ~1 μg/μL final concentration
39. Perform PCR to verify that the genes of interest were correctly incorporated into the bacmid using Phire Hot Start II DNA polymerase (ThermoFisher Scientific). Each 20 μL reaction is set up in a PCR tube that contains final concentrations of the following:

- 5 × Phire II buffer: 4 μL
- dNTPs: 0.1 mM
- Primer 1: 0.5 μM
- Primer 2: 0.5 μM
- Phire Hot Start II DNA polymerase: 0.05 μL
- Bacmid: 0.5 μL
- Deionized water: to 20 μL

Perform the following PCR steps in a thermal cycler:

- Step 1: 98 °C for 3 min
- Step 2: 30 cycles:
  - o 98 °C for 10 s
  - o 55 °C for 10 s
  - o 72 °C for 5 s/kb
- Step 3: 72 °C for 30 s

Analyze the PCR products by gel electrophoresis using a minigel of TAE buffer containing 1% *w/v* agarose with 1 μg/mL ethidium bromide. Ensure a suitable molecular marker, such as 1 kb Plus DNA Ladder (3 μL of 50 ng/μL solution), is loaded in one lane of the gel to allow estimation of the size of the PCR products. If bands with correct sizes are observed, proceed with bacmid transfection into *Sf*21 cells to generate baculovirus, as detailed below. Bacmids should be stored at 4 °C to prevent possible DNA fracturing that can accompany freeze-thaw steps. Bacmids are typically stable for several years at 4 °C when EDTA is present in the storage buffer.

3. Baculovirus production, amplification, and recombinant protein expression

Each protein construct or complex will have its own optimal expression conditions. The main point of difference in optimizing yields of hard-to-produce vs robustly-expressed proteins lies in amplification of the baculovirus itself, especially during the second (P2) amplification step. The protocol outlined below for robustly-expressed proteins has been successful for constructs encoding single globular domains, such as the MLKL pseudokinase domain and RIPK3 kinase domain, alone and in combination. Conversely, yields of hard-to-produce proteins, such as full-length (e.g., human MLKL) or unstable (e.g., some human RIPK3) constructs, benefit greatly from the gentler P2 amplification and the expression procedure outlined in the hard-to-produce protein method. Both methods take approximately the same amount of time to complete, with the hard-to-produce protein method requiring modest additional monitoring, but with the benefit of additional yields. Furthermore, a small-scale test expression can be readily conducted following on from the P2 amplification steps of the hard-to-produce protein method, allowing for easy transition into large scale expressions. Both methods commence with the same procedure of bacmid transfection into insect cells to generate P1 baculovirus.

3.1 Preparation of baculovirus from bacmid (Fig. 4)

3.1.1 Equipment

- Biosafety cabinet
- Compound microscope
- Stationary humidified incubator (27 °C)
- Benchtop centrifuge for pelleting cells in 15 and 50 mL Falcon tubes

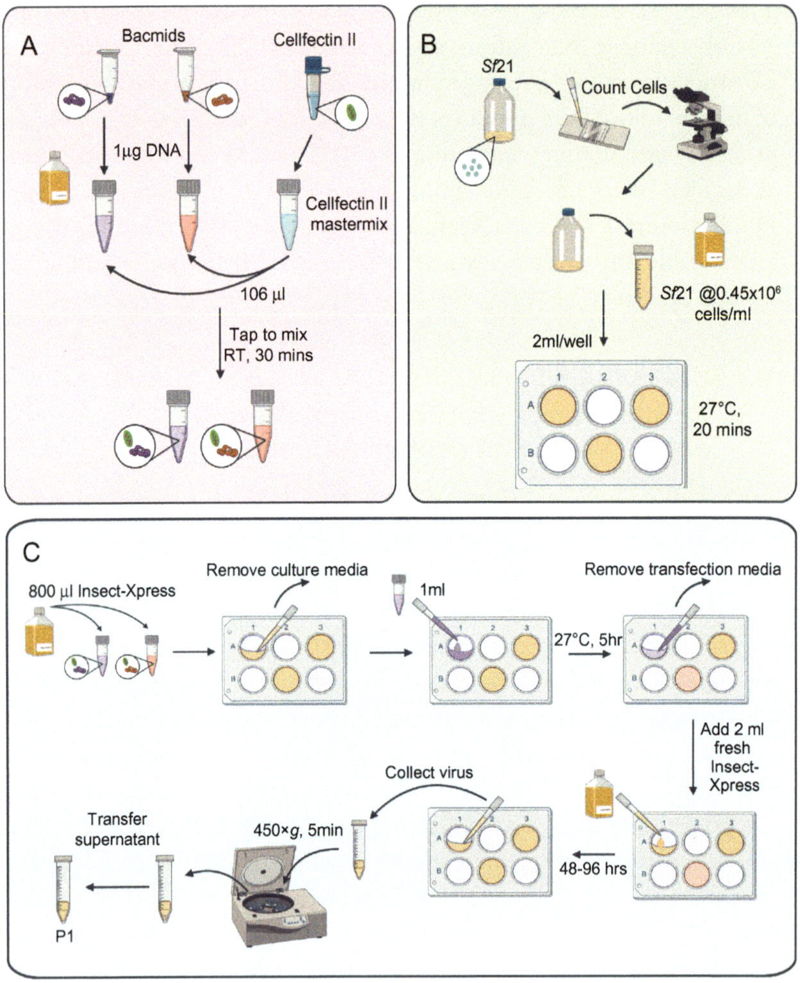

Fig. 4 Baculovirus production. (A) Preparation of DNA:lipid complex transfection mix. (B) Preparation of 0.9×10^6 *Sf*21 cells/well monolayer. (C) Transfecting *Sf*21 cells; culture media is removed from cells and replaced with 1 ml transfection mixture and incubated for 5 h before replacing with fresh media. Incubate for a further 48–96 h before harvesting P1 virus. Illustrations were created with BioRender.com.

3.1.2 Materials and reagents

- Insect-Xpress™ Protein-free insect cell media (Lonza #12-730Q)
- CellFectin™ II (ThermoFisher Scientific #10362100)
- Sterile 5 mL polystyrene tubes
- Sterile 50 mL tubes
- Sterile 6-well plates
- P2, P20, P200, P1000 pipettors
- Sterile Pipette tips
- PipetteAid
- Individually wrapped, sterile 5, 10 and 25 mL graduated pipettes
- Bacmids
- *Sf*21 (or *Sf*9) insect cells (ThermoFisher Scientific #11497013 (or #11496015))
- Hemocytometer slide and cover slip
- Aluminum foil

3.1.3 Preparation of DNA:Lipid complexes (Fig. 4A)

40. All insect cell protocols are carried out in a biosafety cabinet under sterile conditions because no antibiotics or antifungals are typically used in insect cell culture. You will need one 5 mL polystyrene tube per transfection. Add 100 μL Insect-Xpress medium to each before adding 1 μg bacmid DNA. Tap to mix. While larger tubes, such as 15 mL Falcon tubes, can be used, they are often narrow in diameter. As a result, additional care must be taken to avoid contact of the pipette barrel with the inside walls of the tube to prevent contamination between transfections

41. Make a master mix of CellFectin II by adding 100 μL Insect-Xpress media for every transfection to a fresh 5 mL polystyrene tube. Invert CellFectin II several times to insure it is mixed before adding 6 μL CellFectin II reagent per transfection to media. It is advisable to add an extra half aliquot to such a master mix to assist pipetting. So, for 3 transfections: 350 μL Insect-Xpress +21 μL (3.5 × 6 μL) CellFectin II. Gently tap the base of the tube ~8 times to mix

42. Add 106 μL of the CellFectin II master mix to each of your DNA containing tubes, avoiding cross contamination by using filter tips where possible, changing tips between operations, and through avoiding contact between the barrel of the pipettor and the destination tube. Tap the DNA:CellFectin mix gently to ensure homogeneity. Incubate at room temperature (in the biosafety cabinet) for 30–45 min, while preparing cell monolayer

3.1.4 *Preparation of Sf21 cell monolayers* (Fig. 4B)

43. While DNA:CellFectin liposomes are incubating, count *Sf*21 cells using a hemocytometer slide viewed using a compound microscope. *Sf*21 cells can be cultured in sterile 1 L Schott bottles at 27 °C in an incubator shaking at 130 rpm. Caps should be loosened to allow air exchange, and the cap should be covered with aluminum foil. Foil can be pre-sterilized by spraying with 80% ethanol and drying in the biosafety cabinet. Make a master mix of cells at 0.45×10^6 cells/mL with Insect-Xpress medium, allowing 2 mL per transfection. Mix sufficient cells at this dilution to seed an additional "untransfected" control well
44. Seed 6-well plates with 2 mL of cell master mix. Spread by tilting side to side and front to back, avoiding a circular motion that would favor cells pooling in the center of the well. Place in a stationary 27 °C incubator for approximately 20 min to allow cells to adhere. As plates are not sealed, it is important to humidify the stationary incubator, such as by including reservoirs of deionized water in the base of the incubator, to ensure wells do not dry out
45. Check that most cells have adhered before proceeding by examining under a compound microscope and agitating the plate when the cell monolayer is in focus. If there are dead cells that have not adhered, these will be aspirated in the next steps. However, if there are extensive dead cells visible, there may be concerns that your seed cultures are overgrown or have lost viability and are therefore unlikely to be readily transfected with bacmid

3.1.5 *Transfection* (Fig. 4C)

46. Add 800 μL fresh Insect-Xpress medium to each DNA:CellFectin mix. One well at a time, remove and discard seeding media using a P1000 pipette, before transferring the entire ~1 mL of DNA:CellFectin mix to the well slowly in a dropwise manner. This needs to be done slowly and close to the surface of the monolayer to avoid dislodging adhered cells or creating aerosols that may contaminate neighboring wells. It may be advantageous to leave vacant wells between transfections to avoid possible cross-contamination if different bacmids are being transfected in different wells on the same 6-well plate
47. Repeat for remaining wells one-by-one, to ensure monolayers do not dry out. Leave untransfected well untouched
48. Incubate the plate at 27 °C in stationary incubator for 5 h minimum to allow DNA:lipid mixtures to enter the adhered *Sf*21 cells

49. After >5 h, one well at a time, remove transfection mix using a P1000 pipette and discard. Immediately, to avoid monolayer drying out, gently drop 2 mL fresh Insect-Xpress medium onto cell monolayer using P1000 pipette. At the very least, the first 1 mL should be dropped slowly to avoid aerosols and dislodging adhered cells. Ensure care is taken to avoid contaminating media stock with residues from infected wells, such as by decanting a portion of the medium into a 50 mL Falcon tube rather than drawing directly from the stock bottle, and by changing pipette tip between each operation
50. Repeat with remaining wells, one well at a time, to ensure monolayers do not dry out
51. Return to 27 °C stationary incubator
52. 2 and 4 days later visually inspect each well under microscope to confirm no bacterial or fungal contamination is evident. Harvest ~2 mL supernatant, which is the P1 baculovirus, from each well into labeled 15 mL Falcon tubes using P1000. Spin down supernatant at 450 × *g* for 5 min at room temperature to eliminate dead cells. Transfer supernatant into fresh labeled tubes using P1000 and store at 4 °C, protected from light with aluminum foil, until needed for P2 amplification

3.1.6 Tips for baculovirus production

- Aerosol contamination is a major concern throughout virus production. All measures should be taken to avoid cross contamination of wells and stock media/cell cultures. To minimize risk, decant required amount of cell media and stock *Sf*21 culture into appropriate containers in the biosafety cabinet before beginning, so that stock bottles are never in the hood at the same time as any virus
- When using a non-humidified incubator, you may lose P1 virus volume due to evaporation. To avoid media levels dropping too low, add an extra 500 μL Insect-Xpress media at step 49
- Visual inspection of transfected vs untransfected control wells at earlier time points (48–60 h) should provide a good indication of infection, with swollen granular cells present in transfected wells compared to smaller healthy looking control cells. As the incubation time extends to 90–96 h, visual inspection proves futile as cells in untransfected control wells tend to be overconfluent and thus may show similar morphological characteristics to infected cells

3.2 P2 amplification and expression of hard-to-express proteins (Fig. 5)

3.2.1 Equipment

- Biosafety cabinet
- Compound microscope
- Shaking incubator (27 °C)
- Centrifuge for spinning 15 and 50 mL Falcon tubes
- Bench top centrifuge for spinning 1.5 mL Eppendorf tubes
- Sonicator (Bandelin, with MS 73 probe)
- BioRad ChemiDoc
- Optional: CLARIOstar Microplate reader (BMG LabTech)

3.2.2 Materials and reagents

- PBS buffer
- P1 virus
- Sterile 250 or 500 mL Schott bottles/disposable culture bottles
- Sterile 2.8 L Fernbach flasks with foam stoppers/disposable culture bottles
- Insect-Xpress™ Protein-free insect cell media (Lonza #12-730Q)
- *Sf*21 cells
- P1000
- PipetteAid
- Individually wrapped graduated pipettes
- 1.5 mL Eppendorf tubes
- Sterile 50 mL tubes
- Hemocytometer slide and cover slip
- Aluminum foil
- Bio-Rad TGX stain free gel 4–20% (#4568096)
- Bio-Rad Stain free molecular weight ladder (#161-0363)
- Bio-Rad gel running tank and power pack
- 1 × SDS-PAGE gel running buffer (25 mM Tris, 192 mM Glycine, 0.1% SDS)
- Liquid Nitrogen and appropriate cryogenic container for snap freezing

3.2.3 P2 amplification (Fig. 5A)

53. In a sterile 250 or 500 ml Schott bottle, infect 25 mL *Sf*21 culture at a density of 0.5–0.8 × 10^6 cells/mL (as determined by hemocytometer slide counting) with 1 mL P1 baculovirus. Place in shaking incubator (27 °C, 130 rpm).

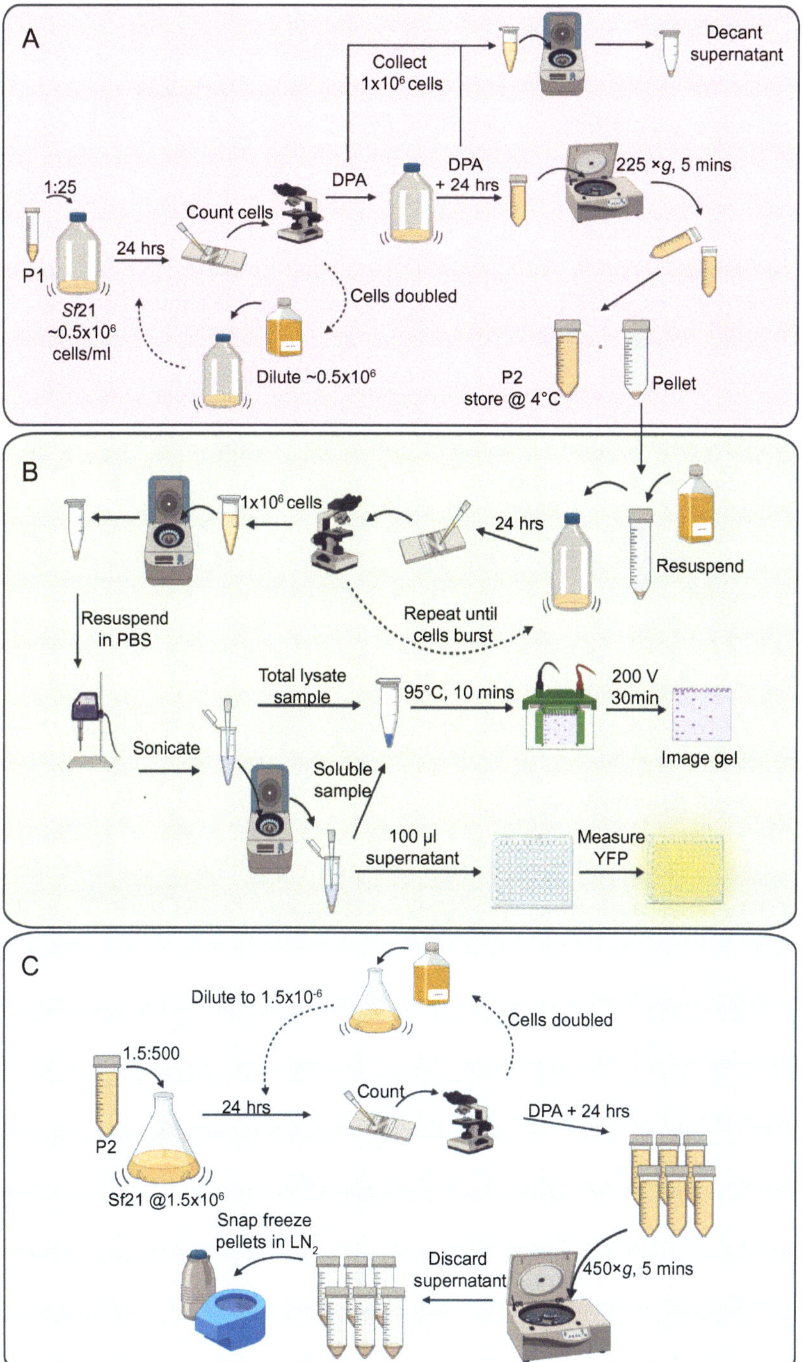

Fig. 5 See figure legend on next page.

54. 24 h later, perform cell count using a hemocytometer slide with a compound microscope. If cells have doubled, dilute the culture to 0.5–0.8 × 10^6 cells/mL with fresh Insect-Xpress media. Return the culture to shaking incubator overnight

55. Repeat step 54, expanding the culture, until the cells cease to double over the period of 24 h. When two approximately comparable cell densities are measured 24 h apart, the time of the second measurement is termed "Day post arrest" (DPA)—indicating that the majority of cells were infected 24 h prior. Do not dilute/expand the culture past this point. Take a 1 × 10^6 cell sample; spin down, remove supernatant and freeze cell pellet as a test expression sample for an SDS-PAGE gel analysis. Return the culture to the incubator

56. 24 h later (48 h post infection), spin down culture at 225 × *g* for 5 min. Transfer supernatant, which is P2 virus, to fresh tubes and store at 4 °C, protected from light by wrapping in aluminum foil, until needed for expression. Resuspend cells with fresh Insect-Xpress media and return to same culture flask, if performing test expression

3.2.4 Test expression (Fig. 5B)

57. Following the centrifugation of the P2 culture and transfer of the P2 supernatant to fresh tubes, resuspend the P2 cells in fresh Insect-Xpress media to a density ~1.5 × 10^6 cells/mL in the same culture flask P2 amplification was performed. Collect 1 × 10^6 cell sample, spin down at 450 × *g* for 5 min at room temperature, discard supernatant and snap freeze pellet for test expression on an SDS-PAGE gel

58. Continue to culture cells shaking at 27 °C at 130 rpm, collecting 1 × 10^6 cell pellets every 24 h until cell count drops and majority of cells appear swollen and grainy, or have burst, under microscope

Fig. 5 P2 amplification and expression of low yield/low stability proteins. (A) P2 virus amplification. Cells are counted each day to monitor cell count, P2 virus is harvested 48 h after cells arrest (day post arrest (DPA) +24 h). (B) Small scale test expression using cell pellet from P2 amplification step. Resuspended cells are monitored daily, taking 1 × 10^6 cell samples. Cell samples are resuspended in PBS, sonicated and used to examine total and soluble protein content (and YFP if applicable) to determine optimized expression timeframe. (C) Upscaling expression using P2. Cells are infected at a density of 1.5 × 10^6 cells/mL and incubated shaking at 27 °C for time determined by test expression. Cultured cells are harvested by centrifugation and cell pellets are snap frozen in liquid nitrogen. Store at −80 °C until ready to purify. Illustrations were created with BioRender.com.

examination. Note when ~85–90% of culture is swollen and only few cells have burst, harvest culture at this stage if you want to perform a test purification that most accurately reflects expression conditions. 1×10^6 cell pellets collected each day should be snap frozen in liquid nitrogen after each collection and stored at −80 °C

59. To determine best expression protocol, thaw pellets of 1×10^6 collected throughout test expression, being sure to include a 1×10^6 pellet of untransfected cells as a control. Resuspend in 200 μL PBS and sonicate on a ~40% amplitude strength setting (Bandelin, with MS 73 probe) for 10 s. Collect 50 μL sample as "total" lysate. Spin down remaining lysate 5 min at 20,000 × *g* and collect 50 μL supernatant as "soluble" fraction. Use remaining 100 μL supernatant to measure yellow fluorescence protein (YFP) in cases where bacmids encode YFP, such as when bacmids were generated using the Kinase Factory or EmBacY *E. coli* strains

60. Add 50 μL 2 × Laemmli buffer to each "total" and "soluble" lysate samples across all time points collected and untreated control. Heat for 10 min at 95 °C and load 10 μL onto Bio-Rad TGX stain free 4–20% SDS-PAGE gel. Include Bio-Rad stain free molecular weight marker. Run a gel in 1 × SDS-PAGE buffer at 200 V for 30 min

61. Image the gel using Bio-Rad ChemiDoc MP on stain free gel setting, 45 s activation and auto exposure. Determine which band is protein of interest based on molecular weight estimated from protein sequence (such as using the ExPasy server) and at what time point the band is strongest in the soluble fraction. This is the best starting point for upscaling an expression

62. If measuring YFP, place 100 μL lysate supernatant into 96-well plate. Include untreated and PBS controls. Measure fluorescence using a CLARIOstar plate reader with default excitation and emission wavelengths of 497-15/540-20, respectively. Peak YFP measurement indicates best time for harvest to ensure highest yields; once overshot, YFP will begin to drop off along with yield

3.2.5 Expression and harvest of non-secreted proteins (Fig. 5C)

63. *Sf*21 cells should be at a density of ~1.5×10^6 cells/mL in 500 mL total volume within an unbaffled 2.8 L Fernbach flask. Seed *Sf*21 cells at appropriate lower density ahead of time, so that the culture has had at least overnight to adjust to the flask and culture conditions

64. To each 500 mL *Sf*21 expression culture, add 1.5 mL P2. Incubate at 27 °C shaking for 24 h at 90 rpm

65. Examine cells under microscope and count all cells using a hemocytometer slide. If they have doubled, expand culture by diluting to ~1.5×10^6 cells/mL with fresh Insect-Xpress media. Continue to culture, checking every 24 h and diluting to ~1.5×10^6 cells/mL with fresh Insect-Xpress media until proliferation has arrested. Once arrested, determine the best harvest time from test expression compared to infection time. If the cells have been completely obliterated by lysis, start again adding less P2 virus. Ideally, at time of harvest, ~85–90% of cells should appear swollen and granular under the microscope, with only few or nil burst cells
66. When ready to harvest, decant culture into 50 ml Falcon tubes and pellet cells at $450 \times g$ using a benchtop centrifuge. Discard supernatant. Snap freeze pellets in liquid nitrogen and store at −80 °C until you are ready to perform purifications

3.2.6 Notes on P2 amplification, expression and harvest of hard-to-express proteins

- During P2 amplification, if expression is expanded into a fresh, larger bottle to maintain culture volume at <20% vessel capacity for optimal oxygenation, cell count can be thrown off while the culture adjusts to the new bottle. Keep this in mind when determining best time to harvest P2 virus
- Whether using an automated cell counter or trypan blue exclusion, it is important to perform total cell counts to monitor proliferation arrest each day, rather than just counting viable cells. The later stages of baculovirus infection can make the insect cells trypan blue permeable, and thus underestimate cell viability. This is an important consideration to avoid harvesting cells too early, which may prevent optimal yields
- YFP is not incorporated into all bacmid systems, so cannot be used universally to monitor infection and expression from all bacmids. Kinase Factory and EmBacY derived bacmids encode YFP in addition to the introduced protein(s) of interest. In contrast, the MultiBac system lacks YFP, so SDS-PAGE-based monitoring is best-suited to this system
- YFP, when present, can also be used to determine best time for expression and harvest in large scale expressions. Simply, collect 1×10^6 cells each day during an expression and measure YFP. Once the YFP measurements plateau from the previous day, it is time to harvest cells
- Some hard-to-express proteins often do not show up clearly on the test expression gel without enrichment. Rather, they can be lost in bands from chaperone proteins (if present) or other endogenous protein bands. To determine whether the protein of interest is expressed, we suggest:

(i) reducing the amount of sample loaded onto gel to reduce background; (ii) performing a Western blot for the target protein before continuing to upscale; or (iii) performing an enrichment/small scale purification with an affinity resin on the test expression culture before proceeding to upscale

- Instead of Bio-Rad TGX stain free gels, any SDS-PAGE gels are suitable. However, use of a stain such as Coomassie or SafeStain will be needed, along with appropriate destaining, to visualize gel bands. Similarly, a stained protein ladder can replace Bio-Rad unstained protein ladder if not using stain free imaging

3.3 P2 amplification and expression of robustly-expressed proteins (Fig. 6)

3.3.1 *Equipment*

- Biosafety cabinet
- Compound microscope
- Shaking incubator (27 °C)
- Benchtop Centrifuge suitable for 50 mL Falcon tubes

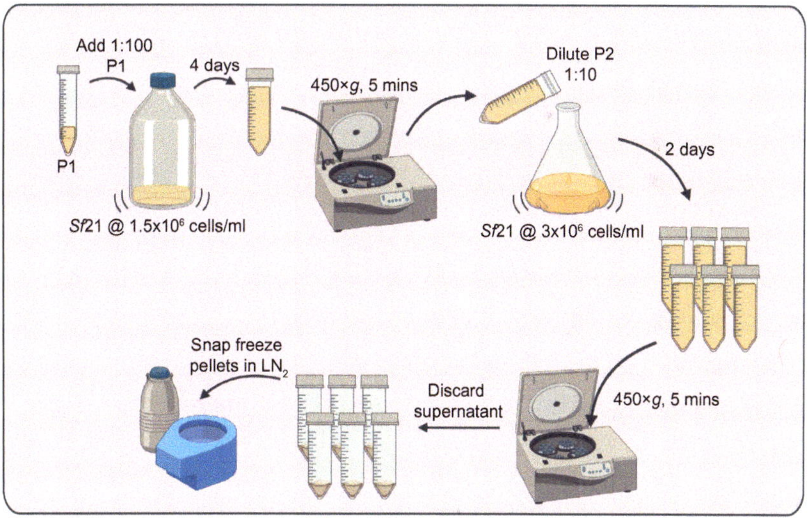

Fig. 6 P2 amplification and expression of high yield, stable proteins. P1 virus is diluted 1:100 into *Sf*21 culture at a density of 1.5×10^6 cells/mL and incubated for 4 days to generate P2. P2 is diluted 1:10 into *Sf*21 cultured to a density of 3×10^6 cells/mL and incubated for 46–50 h. Cultured cells are harvested by centrifugation and cell pellets snap frozen in liquid nitrogen. Store at −80 °C until ready to purify. Illustrations were created with BioRender.com.

3.3.2 Materials and reagents

- P1 virus
- Sterile 1 L Schott bottles/disposable culture bottles
- Sterile unbaffled 2.8 L Fernbach flasks with foam stoppers/disposable culture bottles
- Insect-Xpress™ Protein-free insect cell media (Lonza #12-730Q)
- *Sf*21 cells
- P1000 pipettor
- PipetteAid
- Individually wrapped graduated pipettes
- Aluminum foil
- Sterile 50 mL tubes
- Hemocytometer slide and cover slip
- Compound microscope
- Liquid Nitrogen and appropriate cryogenic container for snap freezing
- −80 °C freezer

3.3.3 P2 amplification

67. In a 1 L sterile Schott bottle, add 1 mL P1 to 100 mL Sf21 cells at 1.5×10^6 cells/mL density. Note, 100 mL P2 will be used per 1 L expression using this method. To ensure optimal oxygenation of the culture, it is advisable to limit the volume to 10% of the vessel size

68. Incubate for 4 days in shaking incubator at 27 °C and 130 rpm

69. After 4 days, harvest P2 virus under sterile conditions by transferring P2 culture to 50 mL tubes and spinning for 5 min at 450 × *g* using a bench-top centrifuge. Transfer supernatant containing virus into fresh clearly labeled tubes and store at 4 °C wrapped in foil until needed for expression. Aim to use within 1 week of harvest, although shelf life will vary depending on the protein being expressed

3.3.4 Expression and harvest of non-secreted proteins

70. *Sf*21 cells should be grown to a density of $3\text{–}4 \times 10^6$ cells/mL with 500 mL per 2.8 L Fernbach flask. Prepare *Sf*21 cells ahead of time, seeding at an appropriate lower density so that the culture has had at least overnight to adjust to the flask and culture conditions. Care should be taken to not overgrow seed cultures beyond 3–4 × 10 cells/mL because overgrowth can compromise baculovirus production and protein yield

71. Add 50 mL P2 to each 500 mL *Sf*21 expression culture. Incubate in 27 °C shaking incubator for 46–50 h at 90 rpm
72. Spin down cells at 450 × *g* in 50 mL Falcon tubes using a benchtop centrifuge, discard supernatants, and snap freeze pellets in liquid nitrogen. Store at −80 °C until ready to purify

3.4 Notes on overall bacmid virus production, amplification and protein expression

- To allow for appropriate oxygenation, culture volume should not exceed more than 20% of the capacity of 2.8 L Fernbach (wide bottomed) flasks. In Schott bottles, cultures to which baculovirus is added, should occupy only ~10% of the vessel volume, while seed cultures can be grown at 20% of the vessel capacity
- Expression and yields can be further optimized for some constructs by performing the same procedures in other insect cell lines, such as High Five or *Sf*9 insect cell lines. Other cell lines will require acclimatization to Insect-Xpress media or use of different media, and may require optimization of conditions, such as growth temperature and incubation times
- Suitable serum-free expression media is available from other vendors, including Expression Systems, Merck and ThermoFisher. Cell lines need to be first acclimatized to different media before using for virus production and expression, and protein yields may vary depending on the selected media
- The method outlined here has been optimized for non-secreted (cytoplasmic) proteins and, as such, test expressions and the harvest of secreted proteins will require distinct protocols using the supernatants rather than the cell pellets

4. Conclusions

Insect cells are extensively used for the production of recombinant pseudokinases owing to the presence of endogenous chaperones and modifying enzymes that enable proper folding of these interesting proteins. Here, we have detailed methods that can be broadly applied to the co-expression of pseudokinases and their protein binding partners. While we have described methods for the co-expression of MLKL and another protein interactor, the methods we have presented are modular and, using ligation-independent cloning methods, constructs can be assembled to co-encode many interactors. This approach is comparable to established methods, such as serial ligations or recombination events using transfer vectors in the MultiBac system.

Acknowledgments

Illustrations were created with BioRender.com. Plasmid maps were created with SnapGene. Y.M. is supported by Melbourne Research Scholarship and AINSE PGRA scholarship. J.M.M. is grateful to the NHMRC for grant (1172929) and infrastructure (9000719) support, and Anaxis Pharma Pty Ltd. for providing funding support. We gratefully acknowledge the Victorian Operational Infrastructure Support Scheme.

Competing interests

The authors contribute a project developing necroptosis inhibitors in collaboration with Anaxis Pharma Pty Ltd.

References

Babon, J. J., Lucet, I. S., Murphy, J. M., Nicola, N. A., & Varghese, L. N. (2014). The molecular regulation of Janus kinase (JAK) activation. *The Biochemical Journal, 462*(1), 1–13.

Boudeau, J., Miranda-Saavedra, D., Barton, G. J., & Alessi, D. R. (2006). Emerging roles of pseudokinases. *Trends in Cell Biology, 16*(9), 443–452.

Cook, W. D., Moujalled, D. M., Ralph, T. J., Lock, P., Young, S. N., Murphy, J. M., et al. (2014). RIPK1- and RIPK3-induced cell death mode is determined by target availability. *Cell Death and Differentiation, 21*(10), 1600–1612.

Cui, J., Zhu, Q., Zhang, H., Cianfrocco, M. A., Leschziner, A. E., Dixon, J. E., et al. (2017). Structure of Fam20A reveals a pseudokinase featuring a unique disulfide pattern and inverted ATP-binding. *eLife, 6*, e23990.

Davies, K. A., Fitzgibbon, C., Young, S. N., Garnish, S. E., Yeung, W., Coursier, D., et al. (2020). Distinct pseudokinase domain conformations underlie divergent activation mechanisms among vertebrate MLKL orthologues. *Nature Communications, 11*(1), 3060.

Davies, K. A., Tanzer, M. C., Griffin, M. D. W., Mok, Y. F., Young, S. N., Qin, R., et al. (2018). The brace helices of MLKL mediate interdomain communication and oligomerisation to regulate cell death by necroptosis. *Cell Death and Differentiation, 25*(9), 1567–1580.

Dondelinger, Y., Declercq, W., Montessuit, S., Roelandt, R., Goncalves, A., Bruggeman, I., et al. (2014). MLKL compromises plasma membrane integrity by binding to phosphatidylinositol phosphates. *Cell Reports*, 7(4), 971–981.

Durzynska, I., Xu, X., Adelmant, G., Ficarro, S. B., Marto, J. A., Sliz, P., et al. (2017). STK40 is a Pseudokinase that binds the E3 ubiquitin ligase COP1. *Structure, 25*(2), 287–294.

Eyers, P. A., & Murphy, J. M. (2013). Dawn of the dead: Protein pseudokinases signal new adventures in cell biology. *Biochemical Society Transactions, 41*(4), 969–974.

Eyers, P. A., & Murphy, J. M. (2016). The evolving world of pseudoenzymes: Proteins, prejudice and zombies. *BMC Biology, 14*(1), 98.

Garnish, S. E., Meng, Y., Koide, A., Sandow, J. J., Denbaum, E., Jacobsen, A. V., et al. (2021). Conformational interconversion of MLKL and disengagement from RIPK3 precede cell death by necroptosis. *Nature Communications, 12*(1), 2211.

Grose, C., Putman, Z., & Esposito, D. (2021). A review of alternative promoters for optimal recombinant protein expression in baculovirus-infected insect cells. *Protein Expression and Purification, 186*, 105924.

Ha, B. H., & Boggon, T. J. (2018). The crystal structure of pseudokinase PEAK1 (Sugen kinase 269) reveals an unusual catalytic cleft and a novel mode of kinase fold dimerization. *The Journal of Biological Chemistry, 293*(5), 1642–1650.

Hildebrand, J. M., Tanzer, M. C., Lucet, I. S., Young, S. N., Spall, S. K., Sharma, P., et al. (2014). Activation of the pseudokinase MLKL unleashes the four-helix bundle domain to induce membrane localization and necroptotic cell death. *Proceedings of the National Academy of Sciences of the United States of America, 111*(42), 15072–15077.

Jacobsen, A. V., & Murphy, J. M. (2017). The secret life of kinases: Insights into non-catalytic signalling functions from pseudokinases. *Biochemical Society Transactions, 45*(3), 665–681.

James, C., Ugo, V., Le Couedic, J. P., Staerk, J., Delhommeau, F., Lacout, C., et al. (2005). A unique clonal JAK2 mutation leading to constitutive signalling causes polycythaemia vera. *Nature, 434*(7037), 1144–1148.

Jamieson, S. A., Ruan, Z., Burgess, A. E., Curry, J. R., McMillan, H. D., Brewster, J. L., et al. (2018). Substrate binding allosterically relieves autoinhibition of the pseudokinase TRIB1. *Science Signaling, 11*(549), eaau0597.

Kung, J. E., & Jura, N. (2019). The pseudokinase TRIB1 toggles an intramolecular switch to regulate COP1 nuclear export. *The EMBO Journal, 38*(4), e99708.

Kwon, A., Scott, S., Taujale, R., Yeung, W., Kochut, K. J., Eyers, P. A., et al. (2019). Tracing the origin and evolution of pseudokinases across the tree of life. *Science Signaling, 12*(578), eaav3810.

Lange, S. M., Nelen, M. I., Cohen, P., & Kulathu, Y. (2021). Dimeric structure of the Pseudokinase IRAK3 suggests an allosteric mechanism for negative regulation. *Structure, 29*(3), 238–251.e4.

Lecointre, C., Simon, V., Kerneur, C., Allemand, F., Fournet, A., Montarras, I., et al. (2018). Dimerization of the Pragmin pseudo-kinase regulates protein tyrosine phosphorylation. *Structure, 26*(4), 545–554.

Littlefield, P., Liu, L., Mysore, V., Shan, Y., Shaw, D. E., & Jura, N. (2014). Structural analysis of the EGFR/HER3 heterodimer reveals the molecular basis for activating HER3 mutations. *Science Signaling*, 7(354), ra114.

Liu, Z., Dagley, L. F., Shield-Artin, K., Young, S. N., Bankovacki, A., Wang, X., et al. (2021). Oligomerization-driven MLKL ubiquitylation antagonizes necroptosis. *The EMBO Journal, 40*(23), e103718.

Lucet, I. S., & Murphy, J. M. (2017). Characterization of ligand binding to Pseudokinases using a thermal shift assay. *Methods in Molecular Biology, 1636*, 91–104.

Mace, P. D., & Murphy, J. M. (2021). There's more to death than life: Noncatalytic functions in kinase and pseudokinase signaling. *The Journal of Biological Chemistry, 296*, 100705.

Manning, G., Whyte, D. B., Martinez, R., Hunter, T., & Sudarsanam, S. (2002). The protein kinase complement of the human genome. *Science, 298*(5600), 1912–1934.

Meng, Y., Davies, K. A., Fitzgibbon, C., Young, S. N., Garnish, S. E., Horne, C. R., et al. (2021). Human RIPK3 maintains MLKL in an inactive conformation prior to cell death by necroptosis. *Nature Communications, 12*(1), 6783.

Meng, Y., Sandow, J. J., Czabotar, P. E., & Murphy, J. M. (2021). The regulation of necroptosis by post-translational modifications. *Cell Death and Differentiation, 28*(3), 861–883.

Min, X., Lee, B. H., Cobb, M. H., & Goldsmith, E. J. (2004). Crystal structure of the kinase domain of WNK1, a kinase that causes a hereditary form of hypertension. *Structure, 12*(7), 1303–1311.

Murai, S., Yamaguchi, Y., Shirasaki, Y., Yamagishi, M., Shindo, R., Hildebrand, J. M., et al. (2018). A FRET biosensor for necroptosis uncovers two different modes of the release of DAMPs. *Nature Communications, 9*(1), 4457.

Murphy, J. M. (2020). The killer Pseudokinase mixed lineage kinase domain-like protein (MLKL). *Cold Spring Harbor Perspectives in Biology*, *12*(8), a036376.

Murphy, J. M., Czabotar, P. E., Hildebrand, J. M., Lucet, I. S., Zhang, J. G., Alvarez-Diaz, S., et al. (2013). The pseudokinase MLKL mediates necroptosis via a molecular switch mechanism. *Immunity*, *39*(3), 443–453.

Murphy, J. M., Farhan, H., & Eyers, P. A. (2017). Bio-zombie: The rise of pseudoenzymes in biology. *Biochemical Society Transactions*, *45*(2), 537–544.

Murphy, J. M., Lucet, I. S., Hildebrand, J. M., Tanzer, M. C., Young, S. N., Sharma, P., et al. (2014). Insights into the evolution of divergent nucleotide-binding mechanisms among pseudokinases revealed by crystal structures of human and mouse MLKL. *The Biochemical Journal*, *457*(3), 369–377.

Murphy, J. M., Nakatani, Y., Jamieson, S. A., Dai, W., Lucet, I. S., & Mace, P. D. (2015). Molecular mechanism of CCAAT-enhancer binding protein recruitment by the TRIB1 Pseudokinase. *Structure*, *23*(11), 2111–2121.

Murphy, J. M., Zhang, Q., Young, S. N., Reese, M. L., Bailey, F. P., Eyers, P. A., et al. (2014). A robust methodology to subclassify pseudokinases based on their nucleotide binding properties. *Biochemical Journal*, *457*(2), 323–334.

Oliver, M. R., Horne, C. R., Shrestha, S., Keown, J. R., Liang, L. Y., Young, S. N., et al. (2021). Granulovirus PK-1 kinase activity relies on a side-to-side dimerization mode centered on the regulatory alphaC helix. *Nature Communications*, *12*(1), 1002.

Palmer, S., Chappidi, S., Pinkham, C., & Hancks, D. C. (2021). Evolutionary profile for (host and viral) MLKL indicates its activities as a battlefront for extensive counter-adaptation. *Molecular Biology and Evolution*, *38*(12), 5405–5422.

Patel, O., Griffin, M. D. W., Panjikar, S., Dai, W., Ma, X., Chan, H., et al. (2017). Structure of SgK223 pseudokinase reveals novel mechanisms of homotypic and heterotypic association. *Nature Communications*, *8*(1), 1157.

Patel, O., Roy, M. J., Murphy, J. M., & Lucet, I. S. (2020). The PEAK family of pseudokinases, their role in cell signalling and cancer. *The FEBS Journal*, *287*(19), 4183–4197.

Petrie, E. J., Birkinshaw, R. W., Koide, A., Denbaum, E., Hildebrand, J. M., Garnish, S. E., et al. (2020). Identification of MLKL membrane translocation as a checkpoint in necroptotic cell death using Monobodies. *Proceedings of the National Academy of Sciences of the United States of America*, *117*(15), 8468–8475.

Petrie, E. J., Czabotar, P. E., & Murphy, J. M. (2019). The structural basis of necroptotic cell death signaling. *Trends in Biochemical Sciences*, *44*(1), 53–63.

Petrie, E. J., Hildebrand, J. M., & Murphy, J. M. (2017). Insane in the membrane: A structural perspective of MLKL function in necroptosis. *Immunology and Cell Biology*, *95*(2), 152–159.

Petrie, E. J., Sandow, J. J., Jacobsen, A. V., Smith, B. J., Griffin, M. D. W., Lucet, I. S., et al. (2018). Conformational switching of the pseudokinase domain promotes human MLKL tetramerization and cell death by necroptosis. *Nature Communications*, *9*(1), 2422.

Petrie, E. J., Sandow, J. J., Lehmann, W. I. L., Liang, L. Y., Coursier, D., Young, S. N., et al. (2019). Viral MLKL homologs subvert necroptotic cell death by sequestering cellular RIPK3. *Cell Reports*, *28*(13), 3309–3319.

Pierotti, C. L., Tanzer, M. C., Jacobsen, A. V., Hildebrand, J. M., Garnier, J. M., Sharma, P., et al. (2020). Potent inhibition of necroptosis by simultaneously targeting multiple effectors of the pathway. *ACS Chemical Biology*, *15*, 2702–2713.

Quarato, G., Guy, C. S., Grace, C. R., Llambi, F., Nourse, A., Rodriguez, D. A., et al. (2016). Sequential engagement of distinct MLKL phosphatidylinositol-binding sites executes necroptosis. *Molecular Cell*, *61*(4), 589–601.

Ribeiro, A. J. M., Das, S., Dawson, N., Zaru, R., Orchard, S., Thornton, J. M., et al. (2019). Emerging concepts in pseudoenzyme classification, evolution, and signaling. *Science Signaling*, *12*(594), eaat9797.

Rodriguez, D. A., Weinlich, R., Brown, S., Guy, C., Fitzgerald, P., Dillon, C. P., et al. (2016). Characterization of RIPK3-mediated phosphorylation of the activation loop of MLKL during necroptosis. *Cell Death and Differentiation*, *23*(1), 76–88.

Roelvink, P. W., van Meer, M. M., de Kort, C. A., Possee, R. D., Hammock, B. D., & Vlak, J. M. (1992). Dissimilar expression of Autographa californica multiple nucleocapsid nuclear polyhedrosis virus polyhedrin and p10 genes. *The Journal of General Virology, 73*(Pt 6), 1481–1489.

Saharinen, P., Vihinen, M., & Silvennoinen, O. (2003). Autoinhibition of Jak2 tyrosine kinase is dependent on specific regions in its pseudokinase domain. *Molecular Biology of the Cell, 14*(4), 1448–1459.

Samson, A. L., Fitzgibbon, C., Patel, K. M., Hildebrand, J. M., Whitehead, L. W., Rimes, J. S., et al. (2021). A toolbox for imaging RIPK1, RIPK3, and MLKL in mouse and human cells. *Cell Death and Differentiation, 28*(7), 2126–2144.

Samson, A. L., Garnish, S. E., Hildebrand, J. M., & Murphy, J. M. (2021). Location, location, location: A compartmentalized view of necroptotic signaling. *Science Signaling, 14*(668), eabc6178.

Samson, A. L., Zhang, Y., Geoghegan, N. D., Gavin, X. J., Davies, K. A., Mlodzianoski, M. J., et al. (2020). MLKL trafficking and accumulation at the plasma membrane control the kinetics and threshold for necroptosis. *Nature Communications, 11*(1), 3151.

Sethi, A., Horne, C. R., Fitzgibbon, C., Wilde, K., Davies, K. A., Garnish, S. E., et al. (2022). Membrane permeabilization is mediated by distinct epitopes in mouse and human orthologs of the necroptosis effector, MLKL. *Cell Death and Differentiation*. https://doi.org/10.1038/s41418-022-00965-6.

Sreelatha, A., Yee, S. S., Lopez, V. A., Park, B. C., Kinch, L. N., Pilch, S., et al. (2018). Protein AMPylation by an evolutionarily conserved Pseudokinase. *Cell, 175*(3), 809–821.

Sun, L., Wang, H., Wang, Z., He, S., Chen, S., Liao, D., et al. (2012). Mixed lineage kinase domain-like protein mediates necrosis signaling downstream of RIP3 kinase. *Cell, 148*(1–2), 213–227.

Tanzer, M. C., Matti, I., Hildebrand, J. M., Young, S. N., Wardak, A., Tripaydonis, A., et al. (2016). Evolutionary divergence of the necroptosis effector MLKL. *Cell Death and Differentiation, 23*(7), 1185–1197.

Tanzer, M. C., Tripaydonis, A., Webb, A. I., Young, S. N., Varghese, L. N., Hall, C., et al. (2015). Necroptosis signalling is tuned by phosphorylation of MLKL residues outside the pseudokinase domain activation loop. *The Biochemical Journal, 471*(2), 255–265.

Wang, H., Sun, L., Su, L., Rizo, J., Liu, L., Wang, L. F., et al. (2014). Mixed lineage kinase domain-like protein MLKL causes necrotic membrane disruption upon phosphorylation by RIP3. *Molecular Cell, 54*(1), 133–146.

Wu, J., Huang, Z., Ren, J., Zhang, Z., He, P., Li, Y., et al. (2013). Mlkl knockout mice demonstrate the indispensable role of Mlkl in necroptosis. *Cell Research, 23*(8), 994–1006.

Xie, T., Peng, W., Yan, C., Wu, J., Gong, X., & Shi, Y. (2013). Structural insights into RIP3-mediated necroptotic signaling. *Cell Reports, 5*(1), 70–78.

Zhang, H., Zhu, Q., Cui, J., Wang, Y., Chen, M. J., Guo, X., et al. (2018). Structure and evolution of the Fam20 kinases. *Nature Communications, 9*(1), 1218.

Zhao, J., Jitkaew, S., Cai, Z., Choksi, S., Li, Q., Luo, J., et al. (2012). Mixed lineage kinase domain-like is a key receptor interacting protein 3 downstream component of TNF-induced necrosis. *Proceedings of the National Academy of Sciences of the United States of America, 109*(14), 5322–5327.

Zhu, Q., Venzke, D., Walimbe, A. S., Anderson, M. E., Fu, Q., Kinch, L. N., et al. (2016). Structure of protein O-mannose kinase reveals a unique active site architecture. *eLife, 5*, e22238.

CHAPTER NINE

CRISPR deletions in cell lines for reconstitution studies of pseudokinase function

Annette V. Jacobsen[a,b] and James M. Murphy[a,b,*]
[a]Walter and Eliza Hall Institute of Medical Research, Parkville, VIC, Australia
[b]Department of Medical Biology, University of Melbourne, Parkville, VIC, Australia
*Corresponding author: e-mail address: jamesm@wehi.edu.au

Contents

Abstract

The non-catalytic cousins of protein kinases, the pseudokinases, have grown to prominence as indispensable signaling entities over the past decade, despite their lack of catalytic activity. Because their importance has only been fully embraced recently, many of the 10% of the human kinome categorized as pseudokinases are yet to be attributed biological functions. The advent of CRISPR-Cas9 editing to genetically delete pseudokinases in a cell line of interest has proven invaluable to dissecting many functions and remains the method of choice for gene knockout. Here, using the terminal effector pseudokinase in the necroptosis cell death pathway, MLKL, as an exemplar, we describe a method for genetic knockout of pseudokinases in cultured cells. This method does not retain the CRISPR guide sequence in the edited cells, which eliminates possible interference in subsequent reconstitution studies where mutant forms of the pseudokinase can be reintroduced into cells exogenously for detailed mechanistic characterization.

Methods in Enzymology, Volume 667
ISSN 0076-6879
https://doi.org/10.1016/bs.mie.2022.03.054

1. Introduction

In cell biology, one of the most powerful tools we have for understanding protein kinase function is knockout cell lines that are deficient in the kinase of interest. This is a particularly important tool for pseudokinases, which are members of the protein kinase family that have absent (or negligible) catalytic activity (Boudeau, Miranda-Saavedra, Barton, & Alessi, 2006; Eyers & Murphy, 2013, 2016; Jacobsen & Murphy, 2017; Mace & Murphy, 2021; Manning, Whyte, Martinez, Hunter, & Sudarsanam, 2002; Murphy et al., 2014; Murphy, Farhan, & Eyers, 2017; Murphy, Mace, & Eyers, 2017; Reiterer, Eyers, & Farhan, 2014; Ribeiro et al., 2019), making interrogation of function and regulatory mechanisms with kinase inhibitors especially challenging. To date, much of our knowledge surrounding the biology of pseudokinases using knockout cell lines has come from the study of the murine proteins, in large part due to long-established techniques to generate knockout mice and, subsequently, cell lines. It is now increasingly recognized that observations made in one species cannot necessarily be applied to another (Davies et al., 2018, 2020; Meng et al., 2021; Petrie et al., 2018; Tanzer et al., 2016). Thus, being able to compare and validate findings from rodents and cells of rodent origin with human cell lines is of critical importance, particularly when investigating pseudokinases that have a role in the pathogenesis of human diseases. In some extreme cases, some pseudokinases, like PSKH2 (Shrestha, Byrne, Harris, Kannan, & Eyers, 2020), are present only in primates and are not encoded in rodent genomes, which underscores the importance of access to knockout human cell lines for detailed interrogation of their function.

Although genome editing in human cells has been possible for over 20 years, through introduction of zinc finger nucleases (ZFNs), transcription-activator-like effector nucleases (TALENs), or meganucleases (Gaj, Gersbach, & Barbas 3rd., 2013; Wright, Li, Yang, & Spalding, 2014), all of these techniques are limited by the requirement to create a new engineered enzyme for each target editing site. However, discoveries in the last decade showing that the Cas9 endonuclease, a component of the clustered regularly interspersed short palindromic repeats (CRISPR) system, could be targeted to different sites in the genome by modification of the sequence of a single guide RNA (sgRNA) has made the task of genome editing in human cells a much simpler process (Knott & Doudna, 2018), with respect to both the practical scientific processes involved and the financial outlay required.

1.1 A brief history of CRISPR

The term CRISPR was first coined by Francisco Mojica and Ruud Jansen, who both identified clusters of short palindromic repeats in a range of archaeal and bacterial species between 1993 and 2005 (Jansen, Embden, Gaastra, & Schouls, 2002; Mojica, Diez-Villasenor, Garcia-Martinez, & Soria, 2005). The recognition that these sequences matched the genomes of bacteriophage allowed these groups to hypothesize that the CRISPR system was used by bacteria as a form of adaptive immunity. Between 2005 and 2010, a number of key discoveries expanded our understanding how of the CRISPR system worked, including: the identification of CRISPR-associated (cas) genes, including the now well-known Cas9 nuclease; validation that DNA itself was the target of the CRISPR system; and, that this targeting required both the presence of a protospacer adjacent motif (PAM) within the target DNA and short RNA sequences transcribed from the host CRISPR loci that guide Cas9 to the target sequence, the latter known as CRISPR RNAs (crRNAs) (Doudna & Charpentier, 2014). In 2011, Emmanuelle Charpentier's group showed that a second short RNA, termed a trans-activating CRISPR RNA (tracrRNA), was also required for targeting a specific DNA sequence by forming a duplex with the crRNA (Deltcheva et al., 2011).

One of the first indications that the CRISPR-Cas9 system could be used as a molecular biology tool came from the group of Virginijus Siksnys, who demonstrated that an exogenous CRISPR system could be introduced into, and expressed in, *Escherichia coli* (Sapranauskas et al., 2011). As well as showing this system could be used to generate plasmid resistance in *E. coli*, they showed that the crRNA sequence could be trimmed to 20 nucleotides while retaining efficient cleavage, and that this sequence could be edited to direct Cas9 to a target site of their choosing. These results were validated in parallel by a collaboration between Emmanuelle Charpentier and Jennifer Doudna, who additionally showed it was possible to fuse the crRNA and the tracrRNA into a single sequence (Jinek et al., 2012), simplifying delivery. In 2013, two independent research groups, led by Feng Zhang and George Church, demonstrated the CRISPR system could also be used for gene targeting in mammalian cells (Cong et al., 2013; Mali et al., 2013). As a result of these findings, the CRISPR-Cas9 system has become established as a core molecular biology tool around the globe, and has simplified the gene editing of various mammalian cell types in a range of applications, including targeted gene knockout, introduction of specific mutations and tags within genes, and as a mechanism to direct proteins to specific places within the genome.

1.2 Selection of the CRISPR delivery mechanism

Utilization of CRISPR in mammalian cells requires the introduction of two distinct components: an exogenous sgRNA containing the crRNA and the tracrRNA sequences, and a paired RNA-guided endonuclease that facilitates the DNA cleavage, most commonly being the Cas9 enzyme of *Streptococcus pyogenes*. When CRISPR was developed, cell biologists already had a range of techniques available to facilitate introduction of proteins and RNAs into mammalian cells, and these methods were quickly adopted by the pioneers of CRISPR-Cas9. Successful introduction of Cas9 and sgRNA was achieved via delivery of both the sgRNA and Cas9 as DNA encoded into plasmids, introducing both in RNA form, or introducing the Cas9 enzyme with the guide RNA as a ribonucleoprotein complex. These components were able to be introduced into cells successfully via physical (e.g., microinjection, electroporation) (Zhang, Shen, Li, & Cheng, 2021), viral (e.g., lentivirus, adenovirus) (Kueh & Herold, 2016), or non-viral (e.g., liposomes, nanoparticles) (Li, Hu, & Chen, 2018) delivery methods.

When deciding to adopt CRISPR technology in our lab, we decided to use a lentiviral delivery method for several reasons, including (i) established high-efficiency methods for transduction of host cells, (ii) broad availability of suitable vectors, (iii) ease of use and flexibility to adapt to a wide range of cell lines, including non-dividing cells, and (iv) relatively low cost. While we found that generation of knockout cell lines was straightforward, when we tried to reintroduce our pseudokinase of interest back into cells, either as wildtype or a modified version, we found we were unable to express the reintroduced protein at sufficient levels to reestablish the phenotype. We attempted to circumvent these problems using several strategies, including (a) using a doxycycline-inducible guide vector in combination with a different driver for our reintroduced protein and (b) silently mutating the PAM motif in the construct encoding our reintroduced protein. However, these methods were not sufficient to overcome the problem of low expression of our reintroduced protein of interest, which we presumed was related to continual expression of the sgRNA within the cell, even at very low levels, preventing the same protein from being expressed when introduced exogenously.

To test this hypothesis, we decided to adapt our technique for introduction of the CRISPR sgRNA component so that it was not retained in the cells after knockout of the target gene. For this, we decided on integrase-deficient lentiviral technology as this retained the aforementioned high-efficiency, amenability to a broad range of cell lines and low cost (Wanisch & Yanez-Munoz, 2009), while enabling transient expression of the sgRNA within cells. As its name suggests, an integrase-deficient lentiviral system allows movement of DNA of interest into the nucleus of the target cell. However, due to either disabling mutations in the viral integrase or mutations within the 5′ and/or 3′ long terminal repeats (LTRs), the introduced DNA is not efficiently integrated into the target cells (Fig. 1). All target cells receive the introduced DNA initially, in which the sgRNA can be expressed episomally in the nucleus, however as cells divide this DNA is gradually diluted out of the cell lines. As CRISPR editing occurs relatively quickly, this method proved to be an appropriate and reliable way to generate stable knockout cell lines using transient gene editing in dividing cell lines.

1.3 MLKL as a model pseudokinase

Over the past decade, pseudokinases have emerged as crucial signaling proteins whose biological functions and underpinning molecular mechanisms of regulation have attracted widespread interest. Our focus has been on the mixed lineage kinase domain-like (MLKL) pseudokinase, which is the most downstream known effector protein in a programmed cell death pathway called necroptosis (Murphy, 2020; Murphy et al., 2013; Sun et al., 2012). MLKL can be activated by engagement of death receptors and other pathogen detectors in conditions where caspase activity is compromised. Our work has primarily focused on necroptosis through tumor necrosis factor receptor 1 (TNFR1), where engagement with the receptor leads to the activation and phosphorylation of receptor interacting protein kinases 1 and 3 (RIPK1/RIPK3). This allows RIPK3 to associate with, and phosphorylate, MLKL, triggering a conformational change and translocation to the plasma membrane (Cook et al., 2014; Garnish et al., 2021; Hildebrand et al., 2014; Meng et al., 2021; Murphy et al., 2013, 2014; Petrie et al., 2018, 2020; Petrie, Czabotar, & Murphy, 2019; Samson et al., 2020; Samson, Garnish, Hildebrand, & Murphy, 2021). Here, MLKL accumulates, leading to membrane disruption, the swelling of the cell and its organelles and, ultimately, rupture of the cell and cell death (Dondelinger et al., 2014; Samson et al., 2021; Sethi et al., 2022; Su et al., 2014; Tanzer et al., 2016; Wang et al., 2014). This mechanism classifies necroptosis as an inflammatory form of cell death and, since its discovery, it has been linked to the pathogenesis of a range of human diseases (Hildebrand et al., 2020; Hildebrand et al., 2021; Karunakaran et al., 2016; Lu et al., 2021; Muller et al., 2017; Newton et al., 2016; Pefanis, Ierino, Murphy, & Cowan, 2019; Rickard et al., 2014, 2014; Tovey Crutchfield, Garnish, & Hildebrand, 2021).

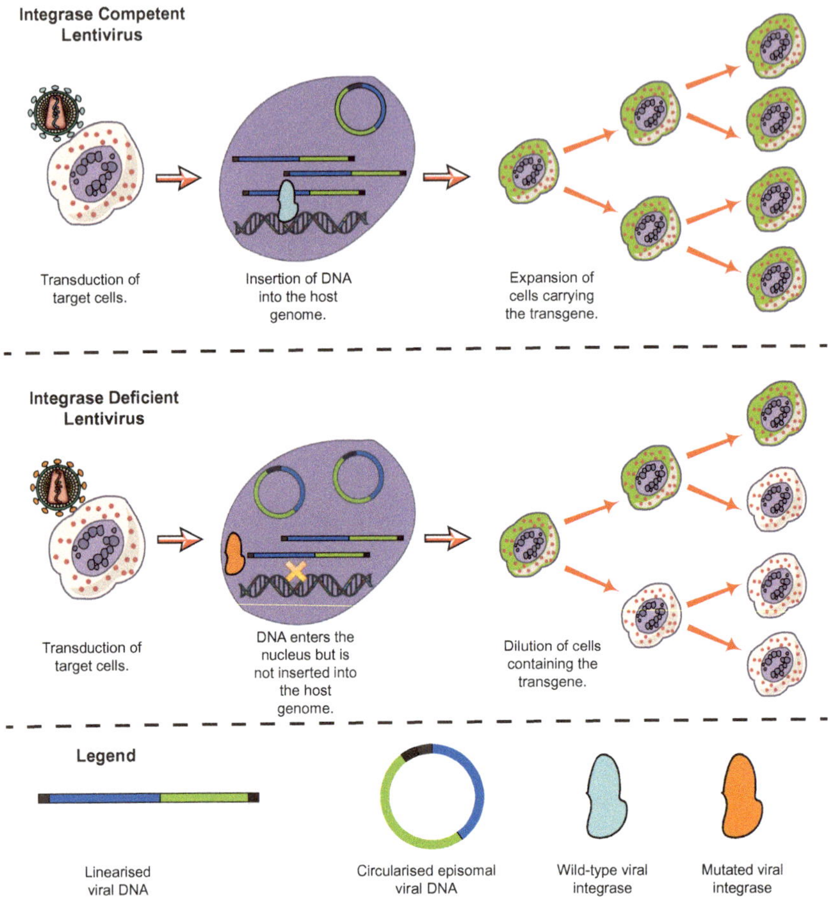

Fig. 1 Comparison of integrase-competent and integrase-deficient lentiviral systems. When introducing a transgene into cells with integrase-competent or integrase-deficient lentiviral particles, the initial steps of the process are the same in both scenarios: the viral particles are transduced across the cellular membrane, then uncoated and dissembled in the cytoplasm, allowing the RNA genome representing the transgene to be reverse-transcribed into double-stranded DNA and subsequently transported into the nucleus. In a system with a competent integrase, the DNA containing the transgene is predominantly present in linear form and subsequently inserted into the host genome. A proportion of the transgene DNA is circularized and retained in the nucleus as metabolically stable episomal DNA molecules capable of being transcribed by the host machinery, albeit at a lower efficiency than transgene DNA that has been integrated into the nucleus. When integration is disabled, this episomal DNA accounts for the bulk of transgene expression within the cells. However, as it cannot be replicated with the host DNA on cell division the proportion of cells carrying the transgene reduces over time.

Like other members of the protein kinase superfamily, pseudokinases are structurally dynamic, enabling them to regulate signaling pathways through a range of mechanisms. Broadly, these mechanisms include: (i) allosteric regulation of a kinase's catalytic activity; (ii) serving as a signaling hub; (iii) functioning as a conformational switch and/or integrator of signals, such as post-translational modifications; and (iv) as a competitor or decoy to prevent kinase engagement in the pathway or access to substrates (Mace & Murphy, 2021; Murphy, Mace, & Eyers, 2017). MLKL is an exemplar of the idea that some pseudokinases can serve as conformational switches, whereby phosphorylation by its upstream regulator, the RIPK3 kinase, induces a conformational change to regulate MLKL's function (Garnish et al., 2021; Meng et al., 2021). Phosphorylation of MLKL's pseudokinase domain leads to conformational toggling and the unveiling of an N-terminal domain four-helical bundle, now recognized as the 'killer' domain of the MLKL protein (Dondelinger et al., 2014; Hildebrand et al., 2014; Su et al., 2014; Tanzer et al., 2016; Wang et al., 2014). Although this mechanism of action is generally consistent across a range of species, we quickly identified key differences to MLKL regulation and activation across species, particularly between mouse and human cells (Davies et al., 2018, 2020; Meng et al., 2021; Sethi et al., 2022; Tanzer et al., 2016).

While we had generated a mouse knockout of MLKL and had access to *Mlkl*-deficient cells derived from these mice (Murphy et al., 2013), further interrogation of the human MLKL mechanism of activation and how substitutions identified in patients could lead to dysregulated necroptotic signaling relied on access to *MLKL*$^{-/-}$ human cell lines. These interests led us to employ CRISPR editing to generate *MLKL*$^{-/-}$ human cell lines, which could be reconstituted (sometimes termed "complemented") upon inducible expression of lentivirally-encoded *MLKL* exogenes (Garnish et al., 2021; Meng et al., 2021; Petrie et al., 2018). Because MLKL is a killer protein, and some mutations can provoke this activity (Jacobsen & Murphy, 2017; Murphy et al., 2013), it was imperative that we used a lentiviral vector with minimal leakiness from which we could induce exogene expression in the target cell lines. Our approach—using a doxycycline-inducible lentiviral vector to introduce MLKL variants—was, however, at odds with established CRISPR editing methods, where a doxycycline-inducible system was similarly used to induce the CRISPR gRNA required for editing. Consequently, we employed the protocol described below in which an integrase-deficient lentiviral system could be used to generate *MLKL*$^{-/-}$ human cell lines using CRISPR-Cas9 via transient editing where the gRNA-encoding vector is not retained in these cells. Subsequently, we were able to reconstitute the necroptosis pathway by stably introducing MLKL variants into these cells encoded via inducible lentiviral vectors, and tightly control expression of these killer proteins using doxycycline induction, to interrogate human MLKL mechanism of action and regulation. While the methods we describe pertain to transiently CRISPR editing human cell lines to knockout the MLKL pseudokinase, these methods can be readily applied to other cell types and pseudokinase (or kinase) targets to enable interrogation of their functions and the impacts of pathogenic mutations on signal transduction.

2. Selection and preparation of CRISPR and lentiviral delivery vectors

This section outlines the preparatory molecular biology techniques necessary to generate your lentiviral and CRISPR delivery vectors for transduction into your target cells (Section 3) and is split into two subsections. The first subsection provides an example protocol for introduction of a disabling mutation into the viral integrase of a standard (integrase-competent) lentiviral delivery vector using overlap PCR. This section has been marked as 'optional' as there are integrase-deficient lentiviral vectors available through plasmid repositories such as Addgene; one we have used with success is the psPAX2 D64V lentiviral delivery vector. However, if, like us, you have a preferred lentiviral delivery vector and would like to make comparisons between the integrase-competent and integrase-deficient versions of the vector, this protocol may be of value.

The second section outlines the preparation of your CRISPR delivery vector. Our example protocol involves introduction of MLKL guides into the pF GH1t UTG CRISPR sgRNA delivery vector, however the general flow of the method will apply no matter which CRISPR delivery vector you are using taking into consideration the following basic requirements: (1) the vector must be compatible with lentiviral delivery, and (2) the vector should have a fluorescent reporter for optimum results. Although the protocol has been set out for introduction of a non-integrating sgRNA into cell lines stably expressing Cas9, it can be adapted for use with a lentiviral vector that dually expresses an sgRNA and Cas9. An example of a suitable dual vector that we have employed successfully is the LentiCRISPRv2-mCherry (available from Addgene).

2.1 Mutating the pCMV ∂R8.2 lentiviral packaging vector using overlap PCR (optional)

When debating the use of integrase-deficient lentiviral technology for delivery of CRISPR, the first consideration was whether to do this by mutation of the viral integrase or the LTR regions. Both strategies have different advantages and disadvantages: LTR mutants generally have lower random integration rates, but integrase mutants tend to have higher expression levels within cells. A further consideration related to the setup of lentiviral delivery systems, as the integrase sequence is part of the Gag-Pol sequence found in packaging vectors and the LTRs are in the delivery vector component. Because the loss of efficiency in both viral transduction and transgene expression within cells is a key challenge of using integrase-deficient lentiviral technology, we decided the integrase mutant would be the better option; particularly as random integration could be eliminated during the FACS sorting steps (Section 3). Additionally, as packaging vectors can be used with a wide range of delivery vectors, this would give the system greater utility, as mutation of the LTRs must be done for each delivery vector.

The integrase mutation we chose to use was the well-characterized mutation of aspartic acid to valine at position 64 (D64V), one of the critical residues of the HIV-1 integrase catalytic triad. We chose to introduce this D64V mutation into the pCMV ∂R8.2 vector, a 2nd generation lentiviral packaging vector that has been used by our group extensively for lentiviral transduction (Davies et al., 2018, 2020; Garnish et al., 2021; Hildebrand et al., 2014; Jacobsen et al., 2022; Meng et al., 2021; Murphy et al., 2013; Petrie et al., 2018, 2019, 2020; Tanzer et al., 2015, 2016), and reliably allows for high-titer generation of lentiviral particles when transfected into HEK293T cells with appropriate delivery and envelope vectors. However, the protocol below can be adapted to suit any lentiviral delivery vector of your choice, as long as you know the sequence and can identify suitable restriction sites. Note, if using a 3rd generation lentiviral packing setup, it is important to note that one of the trade-offs of the increased safety profile is a lower viral titer, due to an additional plasmid being required. Thus, the viral production steps in Section 3 may need to be performed on a larger scale to achieve the same results.

Overlap extension PCR was chosen to introduce the mutation into the vector as it is a simple and reliable method to introduce discrete mutations, requiring only standard molecular biology reagents and equipment. There are a few variations of this method in the literature; our strategy is outlined in Fig. 2 and requires the presence of unique restriction sites at reasonable

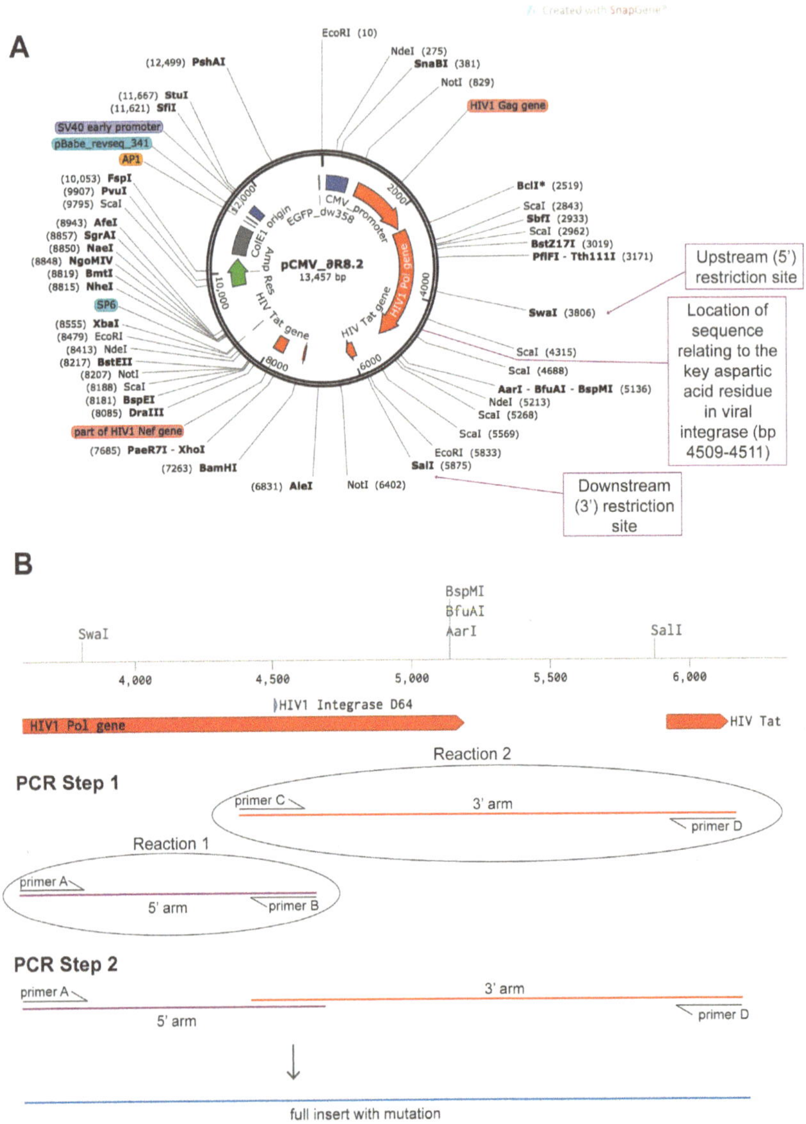

Fig. 2 Introduction of a D64V mutation into the pCMV ∂R8.2 lentiviral packaging vector. (A) Schematic of the pCMV ∂R8.2 lentiviral packaging vector showing key features and restriction enzyme recognition sites—unique sites are indicated bold. The location of the catalytic aspartate in viral integrase, and the upstream and downstream restriction sites selected as boundaries for the overlap extension PCR are marked with purple lines/boxes. (B) Overview of the mutagenesis overlap extension PCR protocol used in Section 2. PCR Step 1 contains two separate reactions: in the first, Primer A (upstream from the *SwaI* restriction site) and Primer B (5′ mutagenesis primer) amplify a section of the pCMV ∂R8.2 vector to create the 5′ arm; in the second, Primer C (3′ mutagenesis primer) and Primer D (upstream from the *SalI* restriction site) amplify a section of the pCMV ∂R8.2 vector to create the 3′ arm. The 5′ and 3′ arms are then purified and combined with Primers A and D to generate the full-length insert segment containing the desired mutation in PCR Step 2. For full details on the primers used refer to Table 1.

Table 1 Primers for overlap extension PCR.

Primer	Description	Sequence	T_m (for Phusion)
Primer A *also used as a sequencing primer	Integrase *Swa*I Fwd—upstream from restriction site	5′-GAA AGC ATA GTA ATA TGG GGA AAG AC-3′	63 °C
Primer B	Integrase D64V Rvs—contains base changes for desired mutation	5′-CTA AAT GTG TAC A**GA C**TA GCT GCC-3′	61 °C
Primer C	Integrase D64V Fwd—contains base changes for desired mutation	5′-GGC AGC TA**G TC**T GTA CAC ATT TAG-3′	61 °C
Primer D	Integrase *Sal*I Rvs—upstream from restriction site	5′-GTA ACG CCT ATT CTG CTA TGT CG-3′	64 °C

For overlap extension PCR, Primers A and D should flank, or overlap, the restriction site. In this case, flanking sites were chosen due to being unable to get a suitable primer profile (regarding T_m, secondary structure, etc.) at the overlap site. Primers B and C must overlap the bases that need to be changed—in this case it involved (in forward orientation, refer Primer C) adjusting the bases corresponding to the codon for aspartic acid at position 64 (D64; GAT) to a codon corresponding to valine (V; GTC)—the bases corresponding to the valine codon are indicated with bold text in Primers B and C. Ideally, this mutation site should be near to the middle of the primer, but should minimally be at least 6 bp from either end. All primers should have a T_m within 2–3 °C of each other, if possible.

distances upstream and downstream from the desired mutation site. Once these sites are identified, and suitable primers have been designed (Table 1), the protocol involves two PCR steps, followed by a restriction digest step and ligation of the insert generated via PCR into the target vector. The reassembled vector can then be replicated in your *E. coli* cloning strain of choice (we use DH10B electrocompetent cells) and purified for validation/sequencing.

2.1.1 Introduction of a D64V mutation into pCMV ∂R8.2 using overlap extension PCR

2.1.1.1 Equipment

- PCR tubes (100 μL)
- P2, P20, P200, P1000 pipettors and associated pipette tips
- Thermocycler (e.g., Bio-Rad T100)
- Agarose gel electrophoresis tank and power supply
- Razor blades or scalpel
- Eppendorf microfuge tubes (1.5 mL)

- DarkReader® transilluminator (Clare Chemical Research) for gel band excision
- UV transilluminator for photographing ethidium bromide-stained gels (e.g., Bio-Rad ChemiDoc)
- Heat block/s, or equivalent, to cover the following temps: 16, 25, 37, 50 °C
- Vortex
- Benchtop microcentrifuge
- Sterile pipette tips
- MicroPulser Electroporation Cuvettes (Bio-Rad)
- MicroPulser Electroporator (Bio-Rad)
- Shaker incubator (37 °C)
- Stationary incubator (37 °C)
- Sterile round bottom polypropylene tubes (15 mL)
- Denovix DS-11 Spectrophotometer (or equivalent, e.g., Nanodrop)

2.1.1.2 Materials and reagents

- Phusion polymerase (ThermoFisher #F530L)
- 5 × Phusion high-fidelity (HF) buffer (supplied with polymerase)
- dNTPs (Bioline #BIO-39044; supplied as a 10 mM mixed stock of dATP, dCTP, dGTP and dTTP)
- pCMV ∂R8.2 lentiviral packaging vector (available from Addgene; #12263); maxiprep of pCMV ∂R8.2 plasmid diluted to 50 ng/μL
- Primers (Table 1) diluted to 50 ng/μL; purchased from a vendor, such as IDT, who synthesize custom sequences. Synthesis on the 20 nmol scale, desalted, is sufficient for sequences <40 bases, but HPLC purification should be considered for longer primers
- MilliQ (or equivalent quality) water
- Agarose for molecular biology (Bioline #BIO-41025)
- Ethidium bromide (10 mg/mL; Sigma-Aldrich #E1510)
- Gel Loading Dye, Purple (6 ×) (NEB #B7024A)
- TAE buffer (Sigma-Aldrich #T9650-4L; 1:10 dilution of 10 × concentrate in reverse osmosis water)
- 1 kb Plus DNA Ladder GeneRuler (ThermoFisher #10787018); diluted to 50 ng/μL in milliQ water and 1 × gel loading dye
- QIAquick Gel Extraction Kit (Qiagen; catalog number 28706)
- Qiagen MinElute PCR Purification kit (Qiagen; catalog number 28006)

- Restriction enzymes: *Swa*I (NEB #R0604S), *Sal*I-HF (NEB #R3138S), *Bam*HI-HF (NEB #R3136S), *Eco*RI-HF (NEB #R3101S), *Nde*I (NEB #R0111S), and *Not*I (NEB #R3189S)
- rCutSmart™ Buffer (NEB; supplied with restriction enzymes)
- NEBuffer™ r3.1 (NEB; supplied with SwaI)
- T4 DNA ligase (Promega #M1804)
- 10 × T4 ligase buffer (Promega; supplied with T4 ligase)
- *E. coli* DH10B electrocompetent cells (sourced from ThermoFisher #12331013; prepared in-house)
- Super broth (3.5% w/v tryptone, 2% w/v yeast extract, 0.5% w/v NaCl, 1 mL of 5 M NaOH per liter of medium, deionized water; autoclaved to sterilize)
- LG agar (1% w/v tryptone, 0.5% w/v yeast extract, 0.5% w/v NaCl, 0.2% w/v D-glucose, 10 mM Tris pH 7.4, 1 mM $MgCl_2$, 1.5% w/v agar, mixed in deionized water; autoclaved to sterilize)
- Ampicillin (Roche #10835269001) prepared as a 100 mg/mL stock in sterile milliQ water
- LG agar to which 100 μg/mL ampicillin was added to molten agar (when cooled to around 37 °C) before pouring 20 mL into each of 20 × 10 cm diameter Petri dish
- Super broth supplemented with 100 μg/mL ampicillin
- Phire Hot Start II DNA polymerase (ThermoFisher Scientific #F124L)
- 5 × Phire II Green buffer (ThermoFisher Scientific; supplied with polymerase)

2.1.1.3 Step-by-step protocol

1. Set up 2 × 50 μL PCR reactions to create the 3′ and 5′ arms, using the strategy in Fig. 3, as follows:
   (a) 10 μL 5 × HF buffer
   (b) 1.5 μL dNTPs (10 mM)
   (c) 2 μL each of both required primers (50 ng/mL)
   (d) 1 μL template DNA (50 ng/mL)
   (e) 33 μL MilliQ water
   (f) 0.5 μL Phusion polymerase

   Perform PCR for 30 cycles using conditions optimized for your primer sets and arm lengths—below are example conditions optimized and used for this specific example:
   - 98 °C for 3 min (initial denaturation step—perform once only)
   - 98 °C for 15 s

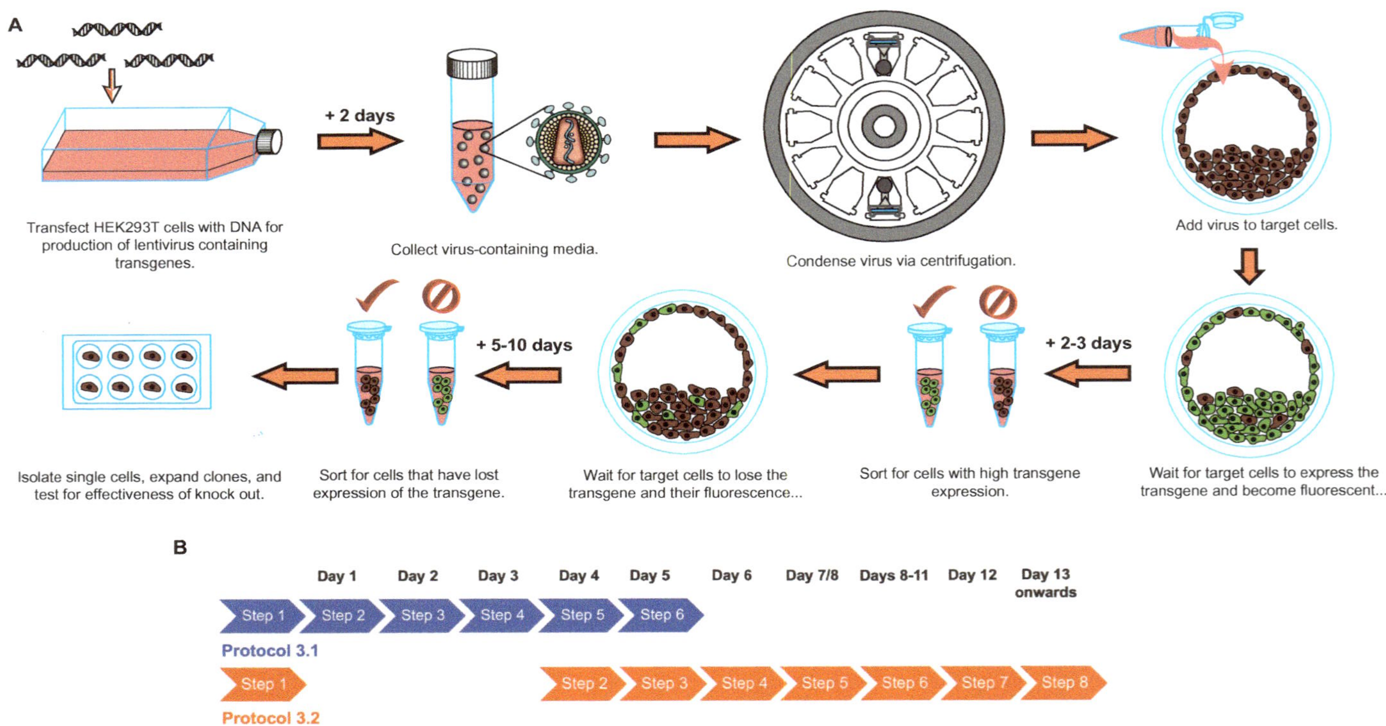

Fig. 3 Overview of cell biology techniques used in Section 3 and recommended timeline. (A) Diagrammatic summary of Protocols 3.1 and 3.2 combined into a continuous flow showing cell biology methods used. (B) Recommended timeline for Protocols 3.1 and 3.2 illustrating overlap points. This timeline is suggested to give maximal efficiency of transduction. Note: Step 1 is the preparatory step involving defrosting and recovery of cells and should be done 3–7 days before starting Protocol 3.1 – we recommend target cells are defrosted at the same time, as U937 and HT29 cells (used in Protocol 3.2) tend to require longer recovery than HEK293T cells (used in Protocol 3.1).

- 61 °C for 15 s (adjust as appropriate based on primer T_m)
- 72 °C for 30 s (adjust as appropriate based on arm length—enzyme speed 15 s/kb)
- 72 °C for 1 min (final extension—perform once only at end)

2. Use agarose gel electrophoresis to resolve the 3′ and 5′ arms created above by combining each reaction with a 1/5 volume of 6 × loading dye and loading into an agarose gel (1% w/v in TAE buffer; containing 1 μg/mL ethidium bromide for UV visualization) alongside 7 μL 1 kb Plus DNA Ladder GeneRuler. We typically pour 50 mL mini-gels and use 8 well combs. Note: depending on well depth you may need to spread the PCR reactions across multiple wells. Run the gel at 100–110 V for 30–60 min, until dye front has migrated halfway down gel.
3. Excise gel band/s with clean razor or scalpel using a **low-energy** UV transilluminator and place fragments in 1.5 mL Eppendorf tubes. Do not photograph the gel until after excision to avoid damaging DNA with high-energy UV light. We use separate transilluminators for excision and photography but if you are able to adjust energy settings for both applications this is also appropriate.
4. Extract DNA separately from the 3′ and 5′ arm gel fragments using the QIAquick Gel Extraction Kit as per manufacturer's instructions and elute in 30 μL Qiagen Elution Buffer (or similar, e.g., 10 mM Tris pH 8). This will yield a concentration appropriate for use directly in the next PCR step. Note: equipment required for this step includes a heat block (set to 50 °C), a vortex (optional but recommended), and a benchtop microcentrifuge.
5. Set up a further 50 μL PCR to create the full vector insert, using the strategy in Fig. 3, as follows:
   (a) 10 μL 5 × HF buffer
   (b) 1.5 μL dNTPs (10 mM)
   (c) 2 μL each of both required primers (50 ng/mL)
   (d) 1 μL of each of the 3′ and 5′ arms prepared in step 4
   (e) 32 μL MilliQ water
   (f) 0.5 μL Phusion polymerase

   Perform PCR for 30 cycles using conditions optimized for your primer sets and insert length—below are example conditions optimized and used for this specific example:
   - 98 °C for 3 min (initial denaturation step—perform once only)
   - 98 °C for 15 s

- 61 °C for 15 s (adjust as appropriate based on primer T_m)
- 72 °C for 45 s (adjust as appropriate based on insert length—enzyme speed 15 s/kb)
- 72 °C for 1 min (final extension—perform once only at end)

6. Purify DNA using a Qiagen PCR purification kit as per manufacturer's instructions and elute in 30 μL Qiagen Elution Buffer (or similar). This will yield an eluted volume of approximately 28 μL. Note, a benchtop microcentrifuge is required for this step.
7. Digest eluted PCR product with appropriate restriction enzymes. An example for setup of the *Swa*I/*Sal*I digest used to create this vector is shown below—please note buffers and incubation temperatures will need to be adjusted specific to the enzymes used. Furthermore, if you are using two enzymes that cannot be incubated in the same buffer (as per manufacturer's instructions), two separate digests will be required and an additional gel purification step may need to be added in between each digest.
   (a) ~28 μL PCR product
   (b) 3.3 μL NEBuffer™ r3.1
   (c) 0.7 μL each of *Swa*I and *Sal*I

   Incubate at 25 °C for 1.5 h in a heat block or incubator, immediately followed by a 1.5 h incubation at 37 °C. Note, the two-step incubation is because, in this case, the enzymes have different optimal cutting temperatures—if this is the case, start with the lowest temperature first. If both enzymes have the same optimal temperature, a single 1.5 h incubation is sufficient.

   In parallel, digest the target vector with the same enzymes, using the same incubation conditions. An example of the reaction setup to digest pCMV ∂R8.2 is below:
   - 2–3 μg pCMV ∂R8.2 (volume dependent on concentration)
   - 3.3 μL NEBuffer™ r3.1
   - 2 μL each of *Swa*I and *Sal*I
   - MilliQ water to total volume of 50 μL
8. Resolve and gel purify both vector and insert fragments post digest as per steps 2–4. Note: while not required in this example, if you are using restriction enzymes that create complementary overhangs in the vector, a dephosphorylation step using a phosphatase, such as Antarctic Phosphatase (NEB), should be included after restriction digestion, prior to gel purification

9. Set up two ligations reaction as follows:
   - 1 μL 10 × T4 ligase buffer
   - 2.5 μL gel-purified insert from step 8 (or 2.5 μL MilliQ for control ligation)
   - 0.8 μL gel-purified vector from step 9
   - 2 μL T4 DNA ligase (additional enzyme used for blunt end ligation)
   - 3.7 μL MilliQ water

   Incubate reactions overnight at 16 °C—we have a heat block set up in a cold room for cool ligations; alternatively, a thermocycler could be used. Note: long cold incubation used here was for a blunt end ligation—if setting up early in the day, a long room temp incubation (≥2 h) could be used for blunt ends, and if ligating cohesive/sticky ends a shorter room temp incubation (≥15 min) is usually sufficient.

10. Transform the ligation reaction into *E. coli* DH10B electrocompetent cells under sterile conditions as follows. Note: all pipetting steps should be done gently to avoid damaging the cells.

   (a) Set up two transformations by mixing 2.5 μL of each ligation reaction (vector + insert and vector only control) with 50 μL electrocompetent cells in a 0.2 cm MicroPulser Electroporation Cuvette prechilled on ice. Gently tap cuvette to ensure mixture is equally distributed across cuvette electrodes and leave on ice for 2–5 min.

   (b) Electroporate the cuvettes containing the cells and DNA using the EC2 setting on a MicroPulser Electroporator. Immediately add 200 μL Super Broth (without ampicillin) and recover, in a shaking incubator, for ~1 h at 37 °C and 200 rpm. If you are doing multiple transformations and addition of medium may be delayed (>30 s), return the cuvette to ice between electroporation and addition of medium.

   (c) Spread cells onto an LG agar plate supplemented with 100 μg/mL ampicillin and incubate in a stationary incubator overnight (16–24 h) at 30 °C. If you are plating the full amount of the transformation, we would recommend incubating the plate agar side down for 30–60 min before inverting to agar side up to prevent condensation. The amount of cell mixture you spread on the plate will depend on the efficiency of both the electrocompetent cells and the transformation/ligation steps—a dilution series onto plates can be used if you desire. Note the slightly cooler temperature used here was optimized for this specific plasmid and ligation to get the highest possible success rate; 37 °C is an appropriate starting point to generate plasmid DNA in *E. coli* in most cases.

11. Check plates and confirm the plate(s) with inserts have markedly more colonies than the plate(s) with the control ligation. Select colonies from the insert plate/s and inoculate each colony into ~2.5 mL Super Broth supplemented with 100 μg/mL ampicillin in a 15 mL round bottom polypropylene tube with a loose-fitting lid to allow aeration and incubate in a shaking incubator overnight (16–24 h) at 30 °C and 220 rpm. Note, the number of colonies expanded for further analysis will depend on the difference between the numbers of colonies on the insert vs control plates—we recommend a minimum of three colonies if the control plate has much fewer colonies than the relative insert plate, but more may be required if the difference is not as pronounced.
12. Transfer culture to 1.5 mL Eppendorf tubes and pellet cells in a benchtop microcentrifuge at 13,000 × *g* for 1 min.
13. Isolate plasmid DNA using a QIAprep Spin Miniprep Kit as per manufacturer's instructions. Assess and quantify DNA using from the absorbance at 280 nm using a spectrophotometer.
14. (Optional) Screen DNA to confirm the presence of an insert of the correct length. This can be done using the same primer set as used in the second overlap PCR reaction (Step 5). Each 20 μL PCR reaction should be set up to contain the following final volumes—note due to the small amount of enzyme required, preparation of a master mix without the template DNA (isolated in step 12) is highly recommended:
    - 4 μL Phire Green buffer
    - 0.2 μL dNTPs (10 mM)
    - 1.5 μL each of forward and reverse primers (50 ng/mL)
    - 1 μL template DNA (for concentration between 0.5 and 1.5 μg/mL)
    - 11.75 μL MilliQ water
    - 0.05 μL Phire polymerase

    Perform PCR for 30 cycles using conditions optimized for your primer sets and insert length—below are example conditions optimized and used for this specific example:
    - 98 °C for 1 min (initial denaturation step—perform once only)
    - 98 °C for 5 s

- 62 °C for 5 s (adjust as appropriate based on primer T_m)
- 72 °C for 15 s (adjust as appropriate based on insert length—enzyme speed 10s/kb)
- 72 °C for 1 min (final extension—perform once only at end)

Resolve samples by electrophoresis on a 1% w/v agarose gel as described in step 2 and image using a high-energy UV transilluminator to confirm there is a single band at the correct molecular weight of the insert. Note: if using the Phire Green buffer, the reaction can be loaded directly into the gel without further addition of loading dye.

15. Confirm plasmids isolated in Step 13 contain the desired mutation using Sanger sequencing. The primer (or primers) should be designed to cover the mutation site at minimum but ideally would cover the whole insertion site—particularly if no other validation methods are being used. An example of a secondary validation method which we used to verify the integrity of the plasmids is a diagnostic digest—for the pCMV ∂R8.2 we performed single digests with *Bam*HI, *Eco*RI, *Nde*I, and *Not*I in their appropriate buffers under conditions recommended by the manufacturer and examined the products by TAE-agarose gel electrophoresis

2.2 Preparing your CRISPR delivery vector

Although there can be variation, the vast majority of CRISPR delivery vectors have an sgRNA insertion site that is created by using paired Type IIS restriction enzyme recognition sites due to two important features. Firstly, these enzymes cut outside their recognition site enabling the creation of non-complementary sticky ends with a single digest. Secondly, Type IIS enzymes recognize asymmetric regions of DNA, enabling directional control of the cut site itself so that the recognition sites can be cut out during vector preparation. These two features substantially improve the efficiency of both the cloning and screening steps, leading to them being widely used for a range of techniques that involve DNA manipulation, so if your vector of choice does not use Type IIS enzymes (e.g., *Bbs*I, *Bsa*I, *Bsm*BI, etc.) you may need to make adjustments to the below protocol and/or add additional steps.

The below protocol does not include any instructions on how to design the sgRNA sequences themselves. Although our knowledge of how to most efficiently direct CRISPR-Cas systems to specific sites in the genome while limiting off-target effects has grown exponentially over the last decade, our understanding of this is still incomplete, with improved predictive algorithms and design tools still being developed at rapid rates. We do not strongly recommend one specific tool or method, although in our case, we successfully designed sgRNA sequences using the crispr.mit.edu server. There are some general considerations that are applicable regardless of the design tool being used to help ensure the generated knockouts provide an accurate representative phenotype. This includes careful consideration of which exon you want to target, which is especially relevant if your pseudokinase of interest has multiple active isoforms in which one or more exons are spliced out. We would also strongly recommend using a minimum of two different sgRNAs per gene to limit the probability of any changes resulting from off-target effects; if possible, these should be targeted to different exons or, if this is not possible/practical, to distinct regions of the same exon. Finally, although it is possible to generate a knockout using intronic recognition sites, the likelihood of this working is much lower than when using an sgRNA that targets an exon. If a suitable Cas9 exonic sgRNA cannot be generated from your pseudokinase of interest, investigation of alternate CRISPR-Cas systems using distinct nucleases/PAMs should be considered.

Once you have sgRNA sequences that you are satisfied with, the final step required (prior to beginning this protocol) is to have single-strand complementary DNA oligonucleotides that can be annealed and inserted into your CRISPR delivery vector synthesized (e.g. via IDT). When doing this, ensure that additional bases have been added to the appropriate ends of your complementary oligonucleotide pairs to create an overhang compatible with the insertion site of your vector of choice (refer to Table 2 for oligonucleotides used to generate the MLKL sgRNAs used in the pF GH1t UTG vector).

2.2.1 Insertion of MLKL sgRNAs into the pF GH1t UTG CRISPR delivery vector

2.2.1.1 Equipment

- Eppendorf tubes (1.5 mL) or PCR tubes (100 μL) if using a thermocycler to anneal
- P2, P20, P200, P1000 pipettors and associated pipette tips
- Heat block/s, or equivalent, to cover the following temperatures: 37, 50, 55, 65, 95 °C
- P2, P20, P200, P1000 pipettors and associated pipette tips

Table 2 Oligonucleotides used for cloning MLKL sgRNA into pF GH1t UTG.

Oligonucleotide	Description	Sequence	T_m (as per IDT)
MLKL1 sense	Exon 2, + oriented, BP 5275–5294, AA 24–31	5′-**TCC C**CA AGA AAC AGT GCC GGC GCC-3'	68.1 °C
MLKL1 antisense	Complementary pair annealed to MLKL1 sense	5′-**AAA C**GG CGC CGG CAC TGT TTC TTG-3′	64.8 °C
MLKL3 sense	Exon 2, + oriented, BP 5533–5552, AA 110–117	5′-**TCC C**GG AGC TCT CGC TGT TAC TTC-3′	62.2 °C
MLKL3 antisense	Complementary pair annealed to MLKL3 sense	5′-**AAA C**GA AGT AAC AGC GAG AGC TCC-3′	59 °C
Sequencing Primer	Validated 5′ sequencing primer for pF GH1t UTG (listed on Addgene)	5′- CAG ACA TAC AAA CTA AAG AAT-3′	45.6 °C

When selecting sgRNA for cloning, distinct exons, or at least distinct positions in the same exon, should be chosen. When designing MLKL guides, we selected distinct positions in exon 2 of the primary isoform for a number of reasons: (1) exon 2 is the first coding exon, and the largest, spanning 462 bp; (2) it codes for the entire N-terminal domain of MLKL, a domain critical for MLKL-mediated cell death; and (3) it is one of only four exons that are common in their entirety to both human isoforms of MLKL, the others being (relative to the primary isoform) exons 3, 10, and 11. Although exon 3 could have been a potential target, it contains only two sgRNA sites, both of which score poorly in design tools. Note, the terms 'sense' or 'antisense' denote orientation to the PAM, and hence the direction they need to be inserted into pF GH1t UTG. The additional bases added to the sense (TCCC) and antisense (AAAC) to create an overhang cloning into pF GH1t UTG are shown in bold text. Information in the description is provided to help locate the guides in the MLKL gene and include orientation (+ or − strand), and the position in the gene (BP) and protein (AA) sequence. Note that although both MLKL1 and MLKL3 are satisfactory guides and produce knockouts, the efficiency of MLKL3 was far superior, and yielded the majority of validated clones. (MLKL2 was designed, but not listed here as was not effective, even when integrated, highlighting the importance of the design of multiple sgRNAs).

- Agarose gel electrophoresis tank and power supply
- Razor blades or scalpel
- DarkReader® transilluminator (Clare Chemical Research) for gel band excision
- UV transilluminator for photographing ethidium bromide-stained gels (e.g., Bio-Rad ChemiDoc)
- Vortex
- Benchtop microcentrifuge
- Sterile pipette tips

- MicroPulser Electroporation Cuvettes (Bio-Rad)
- MicroPulser Electroporator (Bio-Rad)
- Shaker incubator (37 °C)
- Stationary incubator (37 °C)
- Sterile round bottom polypropylene tubes (15 mL)
- Denovix DS-11 Spectrophotometer (or similar, e.g., Nanodrop)

2.2.1.2 Materials and reagents

- 1 M Sodium chloride
- 50 mM Magnesium chloride
- Single strand DNA oligonucleotides corresponding to the sense and antisense of MLKL sgRNA sequences diluted to 100 μM (from IDT; see Table 2)
- MilliQ (or equivalent quality) water
- pF GH1t UTG CRISPR sgRNA delivery vector (available from Addgene #70183; (Aubrey et al., 2015))
- Restriction enzymes: *Bsm*BI (NEB #R0739S) and *Bam*HI-HF (NEB #R3136S)
- NEBuffer™ r3.1 (NEB; supplied with *Bsm*BI) and rCutSmartTM buffer (NEB; supplied with *Bam*HI-HF)
- Agarose for molecular biology (Bioline #BIO-41025)
- Ethidium bromide (10 mg/mL; Sigma-Aldrich #E1510)
- TAE buffer (Sigma-Aldrich #T9650-4L; 1:10 dilution of 10 × concentrate in reverse osmosis water)
- 1 kb Plus DNA Ladder GeneRuler (ThermoFisher #10787018); diluted to 50 ng/μL in milliQ water and 1 × gel loading dye
- Gel Loading Dye, Purple (6 ×) (NEB #B7024A)
- QIAquick Gel Extraction Kit (Qiagen; catalog number 28706)
- T4 DNA ligase (Promega #M1804)
- 10 × T4 ligase buffer (Promega; supplied with T4 ligase)
- *E. coli* DH10B electrocompetent cells (sourced from ThermoFisher #12331013; prepared in-house)
- Super broth (3.5% w/v tryptone, 2% w/v yeast extract, 0.5% w/v NaCl, 1 mL of 5 M NaOH per liter of medium, deionized water; autoclaved to sterilize)
- LG agar (1% w/v tryptone, 0.5% w/v yeast extract, 0.5% w/v NaCl, 0.2% w/v D-glucose, 10 mM Tris pH 7.4, 1 mM $MgCl_2$, 1.5% w/v agar agar, mixed in deionized water; autoclaved to sterilize)

- Ampicillin (Roche #10835269001) prepared as a 100 mg/mL stock in sterile milliQ water
- LG agar to which 100 μg/mL ampicillin was added to molten agar (when cooled to around 37 °C) before pouring 20 mL into each of 20 × 10 cm diameter Petri dish
- Super broth supplemented with 100 μg/mL ampicillin
- STET buffer (8% sucrose, 0.1% Triton X-100, 50 mM EDTA [pH 8.0], 50 mM Tris [pH 8.0])
- Lysozyme (such as Roche # 10837059001; 50 mg/mL in milliQ water)
- Sodium chloride (NaCl; at 0.5 M and 1.2 M in milliQ water)
- Cetyltrimethylammonium bromide (CTAB; such as Sigma-Aldrich #H6269; 5% w/v in 0.5 M NaCl)
- Ethanol (EtOH; molecular biology grade; at 100% and diluted to 70% in milliQ water)
- TE buffer (10 mM Tris [pH 8.0], 1 mM EDTA [pH 8.0] supplemented with 100 μg/mL RNase A (available from Qiagen kits; part number 1031301; 10 mg/mL)

2.2.1.3 Step-by-step protocol

1. Set up 50 μL reactions to anneal each set of sgRNAs as follows:
   (a) 8 μL each of the sense and antisense DNA oligonucleotide pairs (100 μM)
   (b) 2 μL NaCl (1 M)
   (c) 8 μL $MgCl_2$ (50 mM)
   (d) 24 μL MilliQ water
   Assemble in a 1.5 mL Eppendorf tube and place in a heat block set at 65 °C for 5 min then switch off the heat block and allow to cool for approximately 1 h or to room temperature. We have also assembled in 100 μL PCR tubes and used a thermocycler programmed to cool slowly (≥30 min from 100 °C to room temp) as an alternative to using a heat block. Note: if your DNA oligonucleotides are predicted by the manufacturer or design software to form secondary structure with a T_m greater than 65 °C you should adjust the starting temperature for this step so that it is above this T_m to disrupt the intrinsic secondary structure and all successful annealing of oligonucleotide pairs.
2. Digest the CRISPR vector with the appropriate enzyme/s. An example of the reaction setup to digest pF GH1t UTG is below:
   - 2–3 μg pCMV ∂R8.2 (volume dependent on concentration)
   - 3.3 μL NEBuffer™ r3.1

- 2 μL of *Bsm*BI
- MilliQ water to total volume of 50 μL

Incubate at 55 °C for 1.5 h. Note that *Bsm*BI, which is commonly used as the enzyme to prepare CRISPR delivery vectors, cuts outside its recognition site creating non-complementary sticky ends. This limits the chance of re-annealing so a dephosphorylation step is not required after digestion.

3. Use electrophoresis to separate the digested vector from step 2 by combining the reaction with an appropriate amount of loading dye, then into an agarose gel (1% w/v in TAE buffer containing 1 μg/mL ethidium bromide) alongside 7 μL 1 kb Plus DNA Ladder GeneRuler. We typically prepare 50 mL minigels and use 8 well combs. Note: depending on well depth you may need to spread the PCR reactions across multiple wells. Run the gel at 100–110 V for 30–60 min, until dye front has migrated halfway down gel.
4. Excise the gel band with clean razor or scalpel using a **low-energy** UV transilluminator and place fragments in 1.5 mL Eppendorf tubes. Do not photograph the gel until after excision to avoid damaging DNA with high-energy UV light. We use separate transilluminators for excision and photography but, if you are able to adjust energy settings for both applications, this is also appropriate.
5. Extract DNA from the gel fragments using the QIAquick Gel Extraction Kit as per manufacturer's instructions and elute in 30 μL Qiagen Elution Buffer (or similar, e.g., 10 mM Tris pH 8). Note: equipment required for this step includes a heat block (set to 50 °C), a vortex (optional but recommended), and a benchtop microcentrifuge.
6. Set up ligation reactions for all annealed sgRNAs created in Step 1 as follows:
   - 1 μL 10 × T4 ligase buffer
   - 2.5 μL gel-purified insert from Step 1 diluted 1:100 in MilliQ water (or 2.5 μL MilliQ alone for control ligation/s)
   - 0.8 μL digested and gel-purified vector from step 5
   - 0.8 μL T4 DNA ligase
   - 4.9 μL MilliQ water

   Incubate reactions at room temperature for at least 15 min—we typically incubate for approximately 1 h, but shorter is sufficient if using fresh enzyme.

7. Transform the ligation reactions into *E. coli* DH10B electrocompetent cells under sterile conditions as follows. Note all pipetting steps should be done gently to avoid damaging the cells

(a) Set up transformations by mixing 2.5 μL of each ligation reaction (plus a vector + water (no insert) control reaction) with 50 μL electrocompetent cells in a 0.2 cm MicroPulser Electroporation Cuvette prechilled on ice. Gently tap cuvette to ensure mixture is equally distributed across cuvette electrodes and leave on ice for 2–5 min.

(b) Electroporate the cuvettes containing the cells and DNA using the EC2 setting on a MicroPulser Electroporator. Immediately add 200 μL Super Broth (without ampicillin) and recover, in a shaking incubator, for ~1 h at 37 °C at 200 rpm. If you are doing multiple transformations and addition of medium may be delayed (>30 s), return the cuvette to ice between electroporation and addition of medium.

(c) Spread cells onto an LG agar plate supplemented with 100 μg/mL ampicillin and incubate in a stationary incubator overnight (16–24 h) at 37 °C. If you are plating the full amount of the transformation, we would recommend incubating the plate agar side down for 30–60 min before inverting to agar side up to prevent condensation. The amount of cell mixture you spread on the plate will depend on the effectiveness of both the electrocompetent cells and the transformation/ligation steps—a dilution series onto plates can be used if you desire.

8. Check plates and confirm the plate(s) with inserts have markedly more colonies than the plate(s) with the control ligation. Select isolated colonies from the insert plate(s) and inoculate each colony into ~2.5 mL Super Broth supplemented with 100 μg/mL ampicillin in a 15 mL round bottom polypropylene tube with a loose-fitting lid to allow aeration and incubate in a shaking incubator overnight (16–24 h) at 37 °C at 220 rpm. Note, the number of colonies expanded for further analysis will depend on the difference between the numbers of colonies on the insert vs control plates—we recommend a minimum of three colonies per ligation if the control plate has much fewer colonies than the corresponding insert plate(s), but more may be required if the difference is not as clear

9. Transfer culture to 1.5 mL Eppendorf tubes and pellet cells in a benchtop microcentrifuge at 10,000 × *g* for 3 min.
10. Isolate plasmid DNA using your preferred method. If you have many preparations, or do not have access to commercial kits, we have found the following method to produce high yield DNA of suitable quality for downstream applications.
    (a) Aspirate supernatant from pelleted cells (Step 8) and resuspend in 500 μL STET buffer.
    (b) Add 5 μL lysozyme (50 mg/mL) to each cell suspension and incubate at room temperature for 5 min. Note, if doing multiple preparations, lysozyme can be added to an aliquot of STET buffer immediately prior to step (a) to save time.
    (c) Incubate for 1 min at 95–100 °C, then clarify in a benchtop microcentrifuge at 10,000 × g for 10 min at room temperature. Remove pellet with a micropipette tip (or similar) and discard.
    (d) Add 20 μL CTAB (5% w/v in 0.5 M NaCl) to supernatant from step (c) and centrifuge as above for 5 min.
    (e) Aspirate supernatant and resuspend pellet in 400 μL NaCl (1.2 M). Resuspend by vortexing.
    (f) Add 800 μL of 100% EtOH, vortex briefly to combine, then centrifuge for 5 min as above.
    (g) Aspirate supernatant, wash the pellet with 70% EtOH, then spin for 5 min as above. (Optional: this wash step can be repeated if desired).
    (h) Remove the supernatant and allow to air dry for at least 5 min to allow evaporation of residual ethanol. Resuspend in 100 μL TE buffer supplemented with 100 μg/mL RNase.

    Assess and quantify DNA from the absorbance at 260 nm using a spectrophotometer. We find this protocol typically gives a DNA concentration of between 0.5 and 1.5 μg/μL.
11. Digest isolated DNA to confirm the presence of an insert. Note, the enzyme(s) used and digestion pattern may differ depending on your choice of CRISPR vector—the below example describes the protocol for the pF GH1t UTG CRISPR delivery vector:
    - 0.5–1.5 μg DNA isolated in step 10 (volume dependent on concentration)
    - 2 μL NEBuffer™ r3.1
    - 0.4 μL of *Bsm*BI
    - 0.4 μL of *Bam*HI
    - MilliQ water to total volume of 20 μL

Incubate at 37 °C for 1.5 h (for *Bam*HI), followed by a second digestion at 55 °C for 1.5 h (for *Bsm*BI). Note, if using NEB high fidelity (HF) enzymes and buffers, incubation times can be shortened considerably.

12. Resolve samples by electrophoresis on a 1% w/v agarose gel (containing 1 μg/mL ethidium bromide) as described in step 3 and image using a high-energy UV transilluminator. A linearized vector indicates successful integration of the sgRNA; if you see an additional band at ~1.3 kb this indicates the sgRNA is not present in the vector.

13. Confirm plasmids verified in Step 11 contain the desired mutation using Sanger sequencing. If your CRISPR vector has been sourced through Addgene, or similar, the suggested sequencing primer is generally placed at an optimal distance from the insertion site to get a clean read (see Table 1 for primer used for the pF GH1t UTG vector).

3. Preparation of virus and introduction into target cells

This section outlines the cell biology techniques used to effect the knockout of your pseudokinase of interest in cultured mammalian cell lines, using the vectors that were cloned in Section 2. The first subsection details the preparation of the integrase-deficient lentiviral particles using a producer cell line, such as HEK293T, and the second details a protocol for transducing target cell lines and the mechanisms of selection used to maximize the chance of obtaining clonal knockout cells. The example protocols below focus on CRISPR-mediated knockout in human cell lines but can easily be adapted to suit other species compatible with CRISPR.

Unlike the molecular biology protocols, the cell biology protocols do not have any 'pause points' within and should be done continuously; thus, the steps have been arranged to correspond to days to assist with experiment scheduling. Our recommendation for this section is that the two protocols are done in an overlapping fashion, so that fresh lentiviral particles are used to transduce the target cells as, due to the fragility of viral particles, any method of storage will result in loss of titer, leading to a decrease in the proportion of target cells that are transduced. This is particularly important when using integrase-deficient lentiviral particles, as they already have impaired infectivity compared to integrase-competent lentiviral particles. However, because it is possible to decouple the two protocols if needed, we have presented them individually. The overlap points are indicated in the protocols and can also be seen in Fig. 3, along with a diagrammatic summary of the major steps involved.

3.1 Preparation of integrase-deficient lentiviral particles

The preparation of integrase-deficient lentiviral particles is essentially like preparation of integrase-competent viral particles, with two key points of difference. Firstly, production must be done on a much larger scale to generate sufficient infectious particles. Secondly, viral media cannot be used without concentration to infect target cell lines and requires an additional centrifugation step to concentrate the viral media (Al Yacoub, Romanowska, Haritonova, & Foerster, 2007). Taking these two points into account, the below protocol can be adapted to suit preferred methods for lentiviral particle production, regarding aspects such as choice of producer cells and/or transfection reagents.

For the centrifugation step, we have selected a method (originally developed to concentrate lentiviral particles packaging very large delivery vectors) that retains high viral titer and gives superior results compared to ultracentrifugation. However, it should be noted that if you do have access to a centrifuge capable of reaching 20,000 × g, a lower centrifugation speed should still give comparable results.

3.1.1 Production of non-integrating lentiviral particles targeting human MLKL

3.1.1.1 Equipment

- Stationary CO_2 incubator (37 °C; 5–10% CO_2)
- Biosafety cabinet (for all cell culture steps)
- Large sterile cell culture dishes: 10 cm or 15 cm plates
- Pipette boy and suitable sterile graduated pipettes (5, 10, 25 mL)
- Cell counter (any type, manual hemocytometer is sufficient)
- P2, P20, P200, P1000 pipettors and associated pipette tips (sterile for cell culture work)
- Eppendorf tubes (1.5 mL; sterile for cell culture work)
- Vortex (optional)
- Standard fluorescence microscope (optional)
- Falcon tubes (15, 50 mL)
- Sterile filters (0.2, 0.45 μm)
- Sterile syringe (10, 20, or 50 mL, as appropriate)

- Floor centrifuge (with rotor, such as SS34 or equivalent, capable of running at 20,000 × g at +4 °C)
- Sterile 50 mL polycarbonate centrifuge tubes
- Lab scales (sensitive to 0.01 g)

3.1.1.2 Materials and reagents

- HEK293T cell line (available from ATCC and other cell repositories)
- Foetal calf serum (FCS; Sigma-Aldrich)
- Dulbecco's Modified Eagle Medium (DMEM; Gibco) supplemented with 8% v/v FCS
- Trypsin (sterile, for cell culture)
- Phosphate Buffered Saline (PBS; sterile, for cell culture)
- Effectene transfection reagent (Qiagen #301425)
- pCMV ∂R8.2 lentiviral packaging vector with integrase D64V mutation (created in protocol 2.1) or similar integrase deficient lentiviral packing vector, such as the psPAX2-D64V (available from Addgene #63586), maxiprep DNA
- pCMV-VSV-G lentiviral envelope plasmid (available from Addgene #8454), maxiprep DNA
- pF GH1t UTG CRISPR sgRNA delivery vectors with MLKL sgRNA inserts (created in protocol 2.2), maxiprep DNA
- Sucrose buffer, sterile filtered and chilled to +4 °C (20% w/v sucrose in 50 mM Tris-HCl, pH 7.4, 100 mM NaCl, 0.5 mM EDTA)

3.1.1.3 Step-by-step protocol

1. (Cell line preparation—3-7 days prior to starting experiment) Defrost HEK293T and culture in DMEM with 8% FCS in a CO_2 incubator set at 37 °C for at least 3 days, until cells are doubling well and look healthy and well attached.
   If transducing target cells immediately after virus generation, defrost these now also (see Section 3.2.1.3 step 1).
2. (Day 1) Detach HEK293T cells, by either forceful pipetting or short exposure to trypsin (~30 s), and seed for transfection the next day, based on the number of cell lines you want to transduce with the CRISPR lentivirus particles. As a guide, one 10 cm plate typically generates enough virus for one infection, and one 15 cm plate generates enough virus for three transfections. Cells should be ~70–80% confluent on the day of transfection; seeding at a density of $1–1.2 \times 10^5$ cells/cm^2 18–24 h before should achieve this, but you may need to adjust this specific to the doubling rate

of your cells. The precise volume of media is not crucial, but the cells should be covered by at least 1 cm of media at the middle of the plate. This is typically 8–10 mL of media for a 10 cm plate.

3. (Day 2) The following day, transfect HEK293T cells with lentiviral components using Effectene transfection reagent. The below instructions are sufficient to transfect 1 × 15 cm plate; if you are working with 10 cm plates, divide by amounts by 2.5 (surface area of a 15 cm plate is ~2.5 × that of a 10 cm plate).

   For each transfection, you will need:
   - 2.5 μg lentiviral packaging vector (pCMV ∂R8.2 with integrase D64V mutation)
   - 2.5 μg lentiviral envelope plasmid (pCMV-VSV-G)
   - 1.1 mL of EC buffer (supplied in Effectene transfection kit)
   - 5 μg CRISPR sgRNA delivery vector (pF GH1t UTG with MLKL sgRNA inserts)
   - 40–80 μL of Enhancer (supplied in Effectene transfection kit; amount dependent on how easy your cell lines are to transfect)
   - 80 μL of Effectene

   Working in a biosafety cabinet under sterile conditions, set up transfection as follows:

   i. Combine the lentiviral packaging vector and envelope plasmid (2.5 μg of each) with 1.1 mL of EC buffer in a 15 mL Falcon tube and mix to combine (tap or briefly vortex). Note, if preparing multiple transfections, multiply amounts here to cover the number of transfections.

   ii. Add 1 mL of the mix created to 5 μg of your prepared CRISPR sgRNA delivery vector. Repeat this in a clean 15 mL Falcon tube for each sgRNA being used. Tap, or briefly vortex to combine.

   iii. Add 40–80 μL of Enhancer to each tube and mix well by tapping or vortexing for 10 s. Allow to stand at room temperature for 2–5 min.

   iv. Add 80 μL of Effectene to each tube, mix briefly as above, and allow to stand at room temperature for 15–30 min.

   v. Remove media from the HEK293T cells plated on Day 1 and, gently and dropwise, add the transfection mix, distributing around the plate(s). Swirl plate(s) gently to ensure transfection mix is well distributed and return cells to the incubator.

4. (Day 3) Carefully remove media from HEK293T plates, discard media and replace with fresh DMEM (8% FCS), remembering to add media slowly to prevent dislodging the cells. Return the plate/s to the incubator. If desired, cells can be observed briefly under a standard fluorescence microscope prior to removal of media; if the transfection has been successful, you should see a weak GFP positive signal.
5. (Day 4) Take up the media (the viral supernatant) from HEK293T plates into a sterile syringe before attaching a 0.45 μm sterile filter and passing the media through into a 50 mL Falcon tube. Filters with smaller pores should not be used as they can damage lentiviral particles and decrease titer. Carefully replace the media as above and return HEK293T cells to the incubator. Note: cells should be highly confluent and starting to detach by now. Store collected and filtered viral media at +4 °C.
   Prepare sucrose buffer and pass through a 0.2 μm pore sterile filter. Chill overnight at +4 °C. This can be prepared earlier, but if prepping more than 1 week in advance, freeze aliquots in Falcon tubes for long term storage—if you have done this, place required aliquot/s at +4 °C overnight now to defrost.
   If transducing target cells immediately after virus generation, and are using an adherent cell line, plate these in preparation for transduction now (see Section 3.2.1.3 step 2).
6. (Day 5) Remove and filter the viral supernatant from the HEK293T cells returned to the incubator, as described above, and combine with Day 4's collection. Chill viral supernatant on ice while setting up for centrifugation. Discard HEK293T cells. Prechill floor centrifuge and rotor to +4 °C.
   Place 5 mL of prechilled sucrose buffer into a sterile 50 mL centrifuge tube. Very gently top with ~20 mL of the cold, filtered viral medium, so that it sits above the sucrose buffer in a separate layer. Place tubes on ice.
   Weigh tubes, taking care to not disrupt layering (i.e., place in a small beaker to keep upright). Note: tubes in opposite positions in the centrifuge rotor should be within 0.1 g of each other—if this is not the case, top up with cold media. Likewise, if you have an uneven number of tubes, make a mock tube at matched weight, preferably with a sucrose buffer layer at the bottom. Place tubes carefully into the pre-chilled rotor and centrifuge at 20,000 × *g* for 4 h at +4 °C.
   Carefully decant fluid from centrifuge tubes and resuspend each pellet in 2 mL of cell culture medium (supplemented with FCS) appropriate for

the target cell line being transduced. Note, the pellet itself may not be visible, especially if using plastic centrifuge tubes, so ensure that media is washed right around the bottom of the centrifuge tube when resuspending.

Resuspended viral particles containing the same sgRNA can be pooled at this point if desired and if using a compatible medium. To keep things simple, when transducing HT29 and U937 cells in parallel, as described in Section 3.2, we typically resuspend in DMEM as U937 grow well in this medium also and can be fully transitioned back to RPMI in 2–3 days.

If not using immediately to transduce cells (see Section 3.2), condensed viral medium can be stored at +4 °C for up to a week, or snap-frozen in 1 mL aliquots and stored at −80 °C longer term.

3.2 Transduction of target cells and isolation of CRISPR-edited clones

This section details the transduction of target cells and subsequent selection, isolation, and validation of CRISPR-edited clones. We have provided examples for 1 × adherent cell line (HT29) and 1 × suspension cell line (U937) in the below protocol, but due to the broad tropism of lentiviral delivery systems, this protocol can be adapted to suit most cell types of interest. That said, it is important to note that, if working with non-dividing (terminally differentiated) cells, then there is no advantage to using this method over standard integrase-competent lentiviral transduction, as the episomal transgene has no mechanism to be diluted out of the nucleus.

The protocol, as detailed below, requires access to a cell sorter, either in house or via a flow cytometry facility. However, if access to this equipment is not possible, clonal selection can be done by using a limiting dilution method into 96-well plates in place of the first tube sort (Step 5), eliminating the second sort, and moving straight on to the maintenance and validation phase (Step 8). Once clones have grown out, colonies that do not express GFP (which can be visualized using a standard fluorescence microscope) represent cells that have not integrated the transgene, which should be the majority of the colonies, whereas colonies that still express GFP are those that have retained the transgene through random integration and can be used as validation controls. If you choose to select via this method, note that, as the two-stage sorting process is not possible, efficiency will be greatly reduced. Thus, we would strongly recommend that you spend additional time validating the efficacy of your sgRNAs by testing pooled cells transduced with standard integrase-competent lentivirus before attempting transduction with an integrase-deficient lentivirus.

Finally, this protocol requires target cells that constitutively express Cas9, however if you are using a dual sgRNA/Cas9 CRISPR delivery vector these cells are not required and can be replaced with the wild-type variant. In our case, we used HT29 and U937 stably transduced with FUCas9Cherry, as we predominantly reintroduce MLKL back into cells using vectors that utilize puromycin selection. However, for the purpose of the protocol, the type of vector used to introduce Cas9 is not critical and should be selected based on the primary concerns of your lab; for example, a Cas9 vector that utilizes antibiotic selection may be more appropriate if your downstream applications involve microscopy. The protocol for generation of these cells (using standard integrase-competent lentiviral transduction) is not included below, so ensure that you have these cells preprepared before starting.

3.2.1 *Introduction of MLKL sgRNAs into cells using lentiviral transduction and isolation of MLKL$^{-/-}$ clonal cells*

3.2.1.1 Equipment

- Stationary CO_2 incubator (37 °C; 5–10% CO_2)
- Biosafety cabinet (for all cell culture steps)
- Sterile cell culture dishes and plates, including: 96-well plates (both U-bottom and flat-bottom for single cell sorting), 48-, 24-, 12-, and 6-well plates for cell line expansion and general culture work, and 10 cm dishes for larger scale freezing/defrosting of adherent cell lines, if desired
- Pipette boy and suitable sterile graduated pipettes (5, 10, 25 mL)
- Cell counter (any type, manual hemocytometer is sufficient)
- P2, P20, P200, P1000 pipettors and associated pipette tips (sterile for cell culture work)
- Falcon tubes (10, 50 mL)
- Swinging-bucket benchtop centrifuge with adaptors for Falcon tubes and rectangular cell culture plates
- Sterile FACS collection tubes (5 mL; Sarstedt)
- Sterile filters (0.2 μm)
- Sterile syringe (10, 20, or 50 mL, as appropriate)
- Cell strainer (0.7 μm pore)
- Access to a FACS sorting facility (or your own FACS sorter)

3.2.1.2 Materials and reagents

- HT29 and U937 cell lines (available from ATCC and other cell repositories)

- HT29 and U937 cell lines stably transduced with FUCas9Cherry (plasmid available from Addgene #70182) or any other plasmid that will give expression of the Cas9 protein
- Foetal calf serum (FCS; Sigma-Aldrich)
- Dulbecco's Modified Eagle Medium (DMEM; Gibco) supplemented with 8% FCS
- RPMI medium (we use in-house preparations; commercially supplied medium from Gibco is also suitable) supplemented with 8% *v*/v FCS
- Trypsin (sterile, for cell culture)
- Phosphate Buffered Saline (PBS; sterile, for cell culture)
- Polybrene (such as Sigma-Aldrich #107689)
- Doxycycline (Sigma-Aldrich #D9891)
- Sorting buffer (PBS, 10mM HEPES pH7.5–8.0, 5mM EDTA pH8.0, 1% FCS)

3.2.1.3 Step-by-step protocol

1. (Cell line preparation – 3-7days prior to starting experiment) Defrost HT29 and culture in DMEM with 8% FCS in a 5–10% CO_2 incubator at 37 °C for at least 3 days, until cells are doubling well and look healthy and well attached. In parallel, defrost U937 cells and culture into RPMI medium with 8% FCS, and culture as above until doubling well. Note, if cell line aliquots have been frozen for extended periods of time it may be necessary to extend the preparation time beyond 7 days.
2. (Day 1) Seed adherent cells into 6-well plates for transduction the next day, based on the number of cell lines you want to transduce with the CRISPR lentivirus particles. For HT29 cells, we detach cells using trypsin (2–3min in the incubator at 37 °C) and typically seed around 3×10^5 cells/well, which gives both efficient transduction and ensures $>1 \times 10^6$ cells for sorting in 3–4 days' time.
   If transducing target cells immediately after virus generation, this step should be done on the same day as first media change of transfected HEK293T (see Section 3.1.1.3 step 5).
3. (Day 2) Transduce cells with condensed viral media collected in protocol 3.1.
   Prior to adding media to cells, add 4 μg/mL polybrene to increase transduction efficiency. This may need to be optimized if using other cell lines as some are more sensitive to polybrene than others.

Remove media from HT29 cells adhered within a 6-well plate and replace the media in each well with 1 mL viral medium (with polybrene) and return to the incubator.

For U937, harvest, count and aliquot 5×10^5 cells per transduction into 1.5 mL Eppendorf or 15 mL Falcon tubes, depending on density. Spin down cells at 300 × g for 5 min at room temperature in a microfuge or benchtop centrifuge, respectively. Aspirate medium using a pipette or vacuum aspirator, and resuspend in 1 mL viral medium (with polybrene). Transfer cells into a 24-well plate and centrifuge at 200 × g for 1–2 h at 32 °C in a benchtop centrifuge, then return cells to incubator. Note, this final centrifugation step can be done in tubes if preferred or if you do not have access to a temperature-controlled plate centrifuge.

Add 0.5–1 mL of normal DMEM (8% FCS) to each well of HT29 prior at the end of the working day.

4. (Day 3) Add 1–2 mL of RPMI (8% FCS) to each well of U937—transfer to 12-well or 6-well plates as appropriate; small flasks can also be used if preferred.

   Harvest HT29 and transfer into 10 cm plates for expansion. Note, if cells are <50% confluent or appear stressed, do a media change only instead. To all cell lines, add 1 μg/mL doxycycline to induce sgRNA expression.

5. (Late Day 4 or early Day 5) The timing of when to do this step depends on when viral transduction was initiated on Day 2; ideally there should be 2–3 days between this step and step 3. As we prefer transducing cells immediately after the harvest of virus, meaning that cells are transduced late on Day 2, our preference is to prepare cells for the first sort early Day 5 to allow a little extra time for recovery and expansion. If you need to book an instrument at a sorting facility, make sure this is done in advance and takes into account the following requirements:

   (a) The instrument needs to be able to perform sterile-condition two-way tube sorting and detect all the fluorophores present in your target cell lines—in this case GFP and mCherry.

   (b) To promote viability, a large sorter nozzle should be used—typically we use a 100 μm diameter nozzle. Although cells can potentially fit through smaller nozzles, the increase in speed and force can be damaging to cell walls, which is particularly relevant for the HT29 cells as their cell walls will already be weakened due to being trypsinized prior to the sorting process

(c) Allow 30 min for every cell line to be sorted—for example, if you have two cell lines each transduced with two sgRNAs, you should book a 2 h slot. Usually, the process is a little quicker than this, but 30 min will allow a buffer to account for set-up and any potential issues during the sort (e.g., dealing with a nozzle blockage).

For this step, you will require the following cell lines:

- Cas9-Cherry HT29 transduced with MLKL sgRNA
- Wild-type HT29 (as a control for sorter set-up)
- Cas9-Cherry U937 transduced with MLKL sgRNA
- Wild-type U937 (as a control for sorter set-up)

For each of the cell lines transduced with sgRNA prepare a sterile FACS collection tube with ~1 mL of the appropriate culturing media. Place on ice.

If not already prepared, make up sorting buffer, pass through a sterile 0.2 μm filter, and chill on ice.

For all cells (including control cell lines) perform the following steps:

- Harvest cells (typically $1–2 \times 10^6$) and wash 3 × with 1–2 mL sterile PBS, collecting cells in between by centrifugation at 300 × g in a benchtop centrifuge. This washing step is required for any cells that have been exposed to lentivirus within 3 days of sorting as per local biosafety regulations
- Pass cell suspension through a sterile 70 μm pore cell strainer fitted atop a 50 mL Falcon tube to remove any clumps that could potentially block the sorter nozzle.
- Count cells and resuspend in sorting buffer at a density of $3–5 \times 10^6$ cells/mL in a 15 mL Falcon tube. If you have less than 1.5×10^6 cells, resuspend in 500 μL, no matter how many cells you have, so there is sufficient volume for analysis of the cell population, and set-up of appropriate gating, before the sort itself begins. Note: it is not unusual to have less than 1.5×10^6 cells for this initial sort. If desired, you can add a live-dead counterstain (e.g., propidium iodide at 1 μg/mL), although this is not critical as healthy populations are relatively easy to identify without this.
- Transfer each prepared cell line (including controls) into separate sterile FACS tubes, label, and place on ice for transfer to your sorting facility (remembering to include the collection tubes prepared above).

For this first sort, collect the **GFP-positive** population, and sorting parameters should be relatively generous to include the top 40–50%

of the GFP-positive population (refer to Fig. 4 for examples). This is to account for the fact that, even though integrase is disabled, there will still be a proportion of cells that have integrated the DNA into their genome, and these will express GFP more strongly due to increased transcription efficiency for integrated DNA.

Post sort, spin down the collected cells and place into a cell culture plate appropriate for the number of cells you have collected (typical range is $2–4 \times 10^5$ cells). Return cells to the incubator.

6. (Days 5–8) Maintain and passage cell lines as necessary. This period is to allow time for cells that have not integrated the sgRNA transgene to dilute this out. We would recommend freezing a portion of these cells when recovered/expanded sufficiently to act as a back-up in case of any problems.
7. (Day 9) This second sort should be done at least 1 week after cells were transduced, but the timing of this is not critical, so can be scheduled up to a week later than this if necessary.

   Prepare previously sorted cells, plus controls, for the second round of sorting as per step 5. Although the purpose of this step is to identify the cells that have lost the transgene, there is also value to retaining the cells that have managed to integrate the transgene as a comparison, particularly if you have not done any other validation for these sgRNAs at this point. Additionally, you will be collecting single cell clones at this point. As such, please note the following adjustments to step 5.

   (a) The sorter should have adjustments to do both single-cell plate sorting and four-way tube sorting. For the latter, three-way is sufficient (GFP-high, GFP-negative, waste), but for many instruments the adjustments come in multiples of two. Note: two-way sorting is sufficient if you chose not to collect the cells that have integrated the transgene.

   (b) Prepare 2× collection tubes per cell line (if also collecting the GFP-high cells).

   (c) Prepare 96-well plates for single cell clones (see below).

   (d) The PBS wash step is not required, although it can be retained.

   Regarding plate preparation for single cell clones, it is important to note that different cell types cope better than others when isolated like this, so if you are using other cell types, please be sure to optimize this prior to beginning this experiment. For HT29 cells, we find that around one third of the clones will proliferate post-sorting, whereas U937 are much

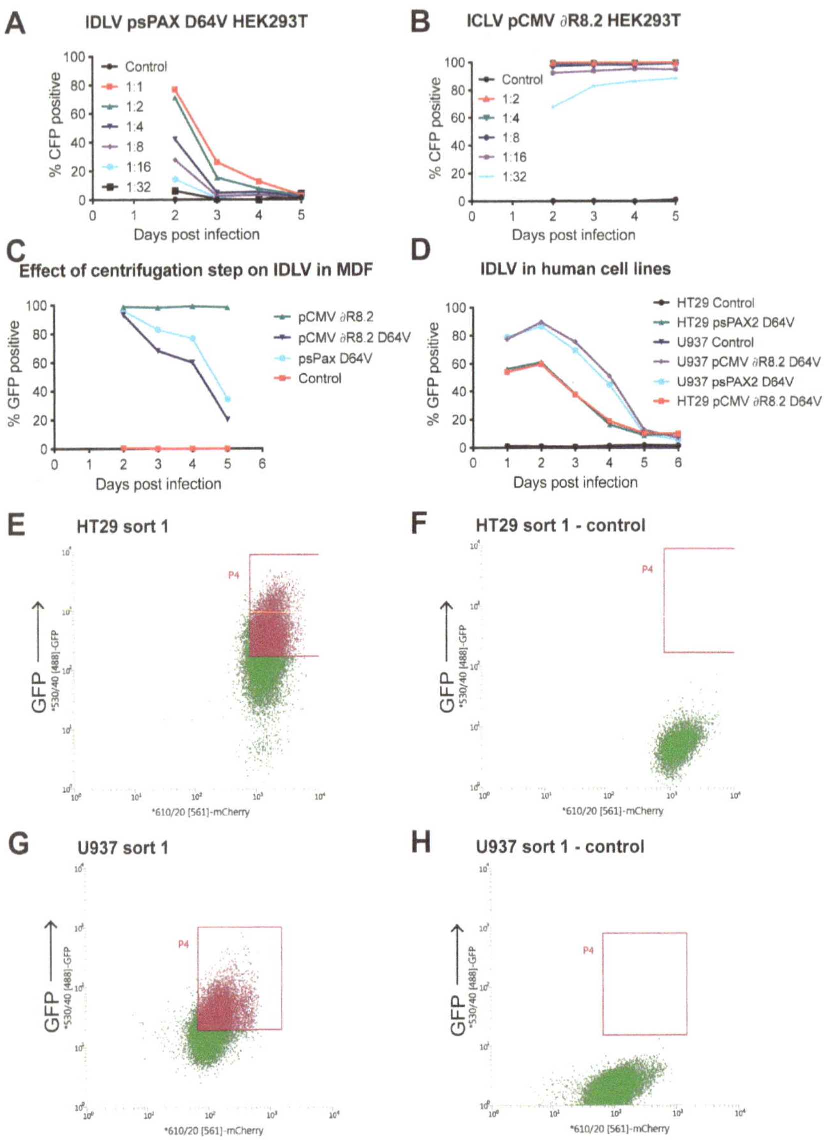

Fig. 4 Integrase-deficient lentivirus effectiveness in various cell types and sorting rationale of HT29 and U937. (A, B) Raw viral media containing either integrase-deficient lentiviral (IDLV; A) or integrase-competent lentiviral (ICLV; B) particles were harvested and used to transduce HEK293T cells. Cyan fluorescent protein signal was measured using flow cytometry daily between 2 and 5 days after transduction, counting a minimum of 10,000 cells to evaluate the difference in infectivity, and the rate of expulsion of the transgene with IDLV transduction. HEK293T treated with IDLV have a lower proportion of cells taking up the transgene (compared to those treated with ICLV) and expel the transgene rapidly. (C) Murine dermal fibroblasts (MDFs) were transduced with ICLV (pCMV ∂R8.2) diluted 1:2 or IDLV (pCMV ∂R8.2 D64V or psPAX2 D64V) that had been

more robust, with between half to two thirds of the clones able to expand. As such, the number of plates required will vary for each cell line.

- For HT29: Prepare 3 × flat-bottomed 96-well plates **per cell line** by adding 120 μL DMEM with FCS per well for non-integrating clones. If you want to collect single cell clones from the integrated population, one further plate per cell line is sufficient.
- For U937: Prepare 2 × U-bottomed 96-well plates **per cell line** by adding 120 μL RPMI with FCS per well for non-integrating clones. If you want to collect single cell clones from the integrated population, a further half a plate per cell line is sufficient.

For the second sort, you primarily want to collect the **GFP-negative** population, which represent your cells that have not integrated the sgRNA, although you may also be collecting the high GFP-positive population. A diagrammatic explanation of this collection procedure, and some examples of gating strategies, can be found in Fig. 5.

Post sort, cells singly sorted into plates can be returned immediately to the incubator. Cells collected into tubes as pools should be spun down and replated in fresh media as per step 5 before returning to the incubator.

---

concentrated via centrifugation as per Protocol 3.1 or left untreated. Cells were analyzed for GFP signal using flow cytometry daily between 2 and 5 days after transduction, counting a minimum of 10,000 cells to evaluate the difference in infectivity, and the rate of expulsion of the transgene with IDLV transduction. After adding the spin protocol, MDFs transduced with either IDLV or ICLV took up the transgene at equivalent rates; additionally, the rate of expulsion of the transgene slowed at early timepoints, suggesting cells were taking up more viral particles initially. (D) HT29 and U937 were transduced with IDLV (pCMV ∂R8.2 D64V or psPAX2 D64V) concentrated by centrifugation as above or left untreated. Cells were analyzed for GFP signal using flow cytometry daily between 1 and 6 days after transduction, counting a minimum of 10,000 cells to evaluate the optimal time for sorting cells. Based on maximal fluorescence at Day 2, and return to near baseline at Day 6, we determined the first sort should be approximately 48 h after transduction, and the second sort should be at least a week after transduction. (E, F, G, H) Example FACS plots showing Cas9-mCherry HT29 and U937 transduced with pF GH1t UTG (E, G) and respective parental controls (F, H) during Sort 1. Although the majority of both cell lines took up the transgene (shift in GFP signal compared to control cells), there is a broad range of signal, indicating variability in how many viral particles were initially taken up by each cell across the population. The cells collected from this sort are indicated by the pink gating (P4) and represent the brightest 40–50% of the GFP positive population, ensuring there is a clear gap between this and the negative population.

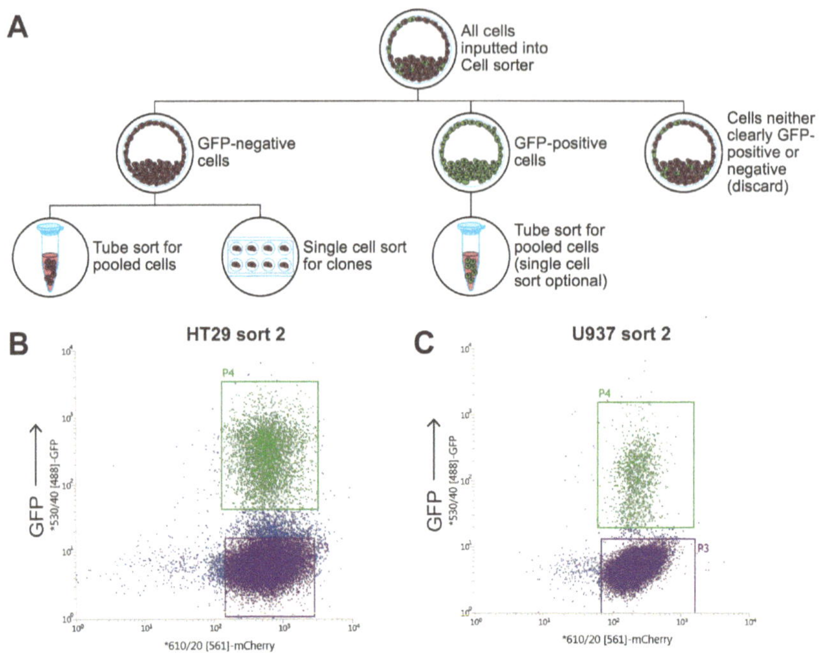

Fig. 5 Overview of strategy and flow cytometry gating for sort 2. (A) For the second sort, there are three key populations to be aware of. GFP-negative cells represent those that have expelled the transgene and are the population of cells that should be screened for potential knockout clonal cells that no longer express the CRISPR sgRNA and are thus able to be reconstituted with doxycycline-inducible exogene vectors. From this population, you should collect single cell clones (into 96 well plates) and a tube sort of the pooled population for testing knockout efficacy. GFP-positive cells indicate cells that have managed to retain the transgene via random integration. We recommend you collect a tube sort of pooled cells from this population, as these cells will have increased likelihood of an effective knockout and can be used to assess sgRNA efficiency, particularly if this has not been previously tested. Single cell clones can also be collected from this population if desired. Cells that sit between the two populations and cannot clearly be classified as GFP positive or negative should be discarded as waste. (B, C) Example FACS plots showing Cas9-mCherry HT29 (B) and U937 (C) transduced with pF GH1t UTG during Sort 2. Purple (P3) gating indicates GFP-negative cells, and green (P4) gating indicates the GFP-positive cells. The top of the P3 gate should not exceed the height of the control population and there should be a clear gap between the top of the P3 gate and the bottom of the P4 gate.

8. (Day 10 onward) Maintain and expand cells using standard cell culture procedures. As a guide:
   - Pooled cells can be frozen as backup, and tested for knockout efficacy, a week after this final sort.

- U937 clones usually require 2 weeks in the 96-well plate (with addition of media or full/partial media changes every 3–4 days), before expanding. Testing of clones can usually begin 4 weeks after sorting, although not all will grow out at the same rate.
- HT29 clones typically require at least 3 weeks in the 96-well plate before expanding, also with media changes every 3–4 days. Clones are usually ready for testing 5–6 weeks after sorting.

The methods used for validation will vary depending on the tools available for your pseudokinase of interest. For MLKL, we used three validation methods:

- Phenotype testing: As MLKL is a critical protein in necroptosis, samples of each colony that grew out could be interrogated with a necroptotic stimulus, with potential knockout clonal populations being resistant to cell death.
- Western blot analysis: We have access to a sensitive and specific antibody against MLKL that can be reliably used to detect presence or absence of the protein. Clonal populations that passed phenotype testing were further analyzed in this way.
- Next Generation Sequencing (NGS): For clonal populations that passed both screening methods above, final confirmation of knockout was validated using NGS.

We would recommend that at least one screening method is used for validation in combination with NGS, but if there is not an appropriate phenotype test or western blot antibody available, then alternate methods such as semi-quantitative PCR can be used to determine whether gene expression is occurring in clones.

Once validated, clonal populations can be maintained and frozen individually for downstream applications or combined to make a mixed pool of clones, if desired.

4. Conclusion

Here we describe methodology for genetic knockout of pseudokinases in cultured cells via non-integrating (transient) transduction with lentiviruses encoding a gRNA that targets the gene encoding a pseudokinase of interest. We have illustrated the utility of this approach using MLKL, the terminal effector in the necroptosis cell death pathway, but have successfully employed this method to knockout other genes, such as that encoding MLKL's upstream regulatory kinase, RIPK3 (Meng et al., 2021). The power of this approach is that mutant forms of the protein of interest can be introduced into the edited cells and robustly expressed in place of the endogenous protein via stable transduction with a lentivirus encoding the protein of interest. We have successfully defined key residues in the MLKL pseudokinase (Davies et al., 2020; Garnish et al., 2021; Meng et al., 2021; Petrie et al., 2018, 2020), which underpin its activation and regulation, by exogenously expressing truncated forms, mutants and orthologs in human cell lines in which we knocked out endogenous MLKL using the method described herein. While the focus of this chapter is on MLKL, the methods we describe herein can be employed generally to knockout other pseudokinases of interest in cultured cells, which provides a powerful platform for re-expression of mutant forms, including those observed in human disease, to dissect a pseudokinase's functions and mechanisms of regulation and activation.

Acknowledgments

We thank the Walter and Eliza Hall Institute's graphics team for providing stock images for use in Fig. 3 and in parts of Figs. 1 and 5. We are grateful for funding provided by the Australian National Health and Medical Research Council (NHMRC) through grant (1124735), fellowship (1172929 to J.M.M.) and infrastructure (IRIISS grant 9000719) support. We thank the Australian Government for Research Training Program Stipend Scholarship support (to A.V.J.).

References

Al Yacoub, N., Romanowska, M., Haritonova, N., & Foerster, J. (2007). Optimized production and concentration of lentiviral vectors containing large inserts. *The Journal of Gene Medicine*, *9*(7), 579–584.

Aubrey, B. J., Kelly, G. L., Kueh, A. J., Brennan, M. S., O'Connor, L., Milla, L., et al. (2015). An inducible lentiviral guide RNA platform enables the identification of tumor-essential genes and tumor-promoting mutations in vivo. *Cell Reports*, *10*(8), 1422–1432.

Boudeau, J., Miranda-Saavedra, D., Barton, G. J., & Alessi, D. R. (2006). Emerging roles of pseudokinases. *Trends in Cell Biology*, *16*(9), 443–452.

Cong, L., Ran, F. A., Cox, D., Lin, S., Barretto, R., Habib, N., et al. (2013). Multiplex genome engineering using CRISPR/Cas systems. *Science*, *339*(6121), 819–823.

Cook, W. D., Moujalled, D. M., Ralph, T. J., Lock, P., Young, S. N., Murphy, J. M., et al. (2014). RIPK1- and RIPK3-induced cell death mode is determined by target availability. *Cell Death and Differentiation*, *21*(10), 1600–1612.

Davies, K. A., Fitzgibbon, C., Young, S. N., Garnish, S. E., Yeung, W., Coursier, D., et al. (2020). Distinct pseudokinase domain conformations underlie divergent activation mechanisms among vertebrate MLKL orthologues. *Nature Communications*, *11*(1), 3060.

Davies, K. A., Tanzer, M. C., Griffin, M. D. W., Mok, Y. F., Young, S. N., Qin, R., et al. (2018). The brace helices of MLKL mediate interdomain communication and oligomerisation to regulate cell death by necroptosis. *Cell Death and Differentiation*, *25*(9), 1567–1580.

Deltcheva, E., Chylinski, K., Sharma, C. M., Gonzales, K., Chao, Y., Pirzada, Z. A., et al. (2011). CRISPR RNA maturation by trans-encoded small RNA and host factor RNase III. *Nature*, *471*(7340), 602–607.

Dondelinger, Y., Declercq, W., Montessuit, S., Roelandt, R., Goncalves, A., Bruggeman, I., et al. (2014). MLKL compromises plasma membrane integrity by binding to phosphatidylinositol phosphates. *Cell Reports*, 7(4), 971–981.

Doudna, J. A., & Charpentier, E. (2014). Genome editing. The new frontier of genome engineering with CRISPR-Cas9. *Science*, *346*(6213), 1258096.

Eyers, P. A., & Murphy, J. M. (2013). Dawn of the dead: Protein pseudokinases signal new adventures in cell biology. *Biochemical Society Transactions*, *41*(4), 969–974.

Eyers, P. A., & Murphy, J. M. (2016). The evolving world of pseudoenzymes: Proteins, prejudice and zombies. *BMC Biology*, *14*(1), 98.

Gaj, T., Gersbach, C. A., & Barbas, C. F., 3rd. (2013). ZFN, TALEN, and CRISPR/Cas-based methods for genome engineering. *Trends in Biotechnology*, *31*(7), 397–405.

Garnish, S. E., Meng, Y., Koide, A., Sandow, J. J., Denbaum, E., Jacobsen, A. V., et al. (2021). Conformational interconversion of MLKL and disengagement from RIPK3 precede cell death by necroptosis. *Nature Communications*, *12*(1), 2211.

Hildebrand, J. M., Kauppi, M., Majewski, I. J., Liu, Z., Cox, A. J., Miyake, S., et al. (2020). A missense mutation in the MLKL brace region promotes lethal neonatal inflammation and hematopoietic dysfunction. *Nature Communications*, *11*(1), 3150.

Hildebrand, J. M., Lo, B., Tomei, S., Mattei, V., Young, S. N., Fitzgibbon, C., et al. (2021). A family harboring an MLKL loss of function variant implicates impaired necroptosis in diabetes. *Cell Death and Disease*, *12*(4), 345.

Hildebrand, J. M., Tanzer, M. C., Lucet, I. S., Young, S. N., Spall, S. K., Sharma, P., et al. (2014). Activation of the pseudokinase MLKL unleashes the four-helix bundle domain to induce membrane localization and necroptotic cell death. *Proceedings of the National Academy of Sciences of the United States of America*, *111*(42), 15072–15077.

Jacobsen, A. V., & Murphy, J. M. (2017). The secret life of kinases: Insights into non-catalytic signalling functions from pseudokinases. *Biochemical Society Transactions*, *45*(3), 665–681.

Jacobsen, A. V., Pierotti, C. L., Lowes, K. N., Au, A. E., Zhang, Y., Etemadi, N., et al. (2022). The Lck inhibitor, AMG-47a, blocks necroptosis and implicates RIPK1 in signalling downstream of MLKL. *Cell Death and Disease*, *13*, 291.

Jansen, R., Embden, J. D., Gaastra, W., & Schouls, L. M. (2002). Identification of genes that are associated with DNA repeats in prokaryotes. *Molecular Microbiology*, *43*(6), 1565–1575.

Jinek, M., Chylinski, K., Fonfara, I., Hauer, M., Doudna, J. A., & Charpentier, E. (2012). A programmable dual-RNA-guided DNA endonuclease in adaptive bacterial immunity. *Science*, *337*(6096), 816–821.

Karunakaran, D., Geoffrion, M., Wei, L., Gan, W., Richards, L., Shangari, P., et al. (2016). Targeting macrophage necroptosis for therapeutic and diagnostic interventions in atherosclerosis. *Science Advances*, *2*(7), e1600224.

Knott, G. J., & Doudna, J. A. (2018). CRISPR-Cas guides the future of genetic engineering. *Science*, *361*(6405), 866–869.

Kueh, A. J., & Herold, M. J. (2016). Using CRISPR/Cas9 technology for manipulating cell death regulators. *Methods in Molecular Biology*, *1419*, 253–264.

Li, L., Hu, S., & Chen, X. (2018). Non-viral delivery systems for CRISPR/Cas9-based genome editing: Challenges and opportunities. *Biomaterials*, *171*, 207–218.

Lu, Z., Van Eeckhoutte, H. P., Liu, G., Nair, P. M., Jones, B., Gillis, C. M., et al. (2021). Necroptosis signalling promotes inflammation, airway remodelling and emphysema in COPD. *American Journal of Respiratory and Critical Care Medicine*, *204*(6), 667–681.

Mace, P. D., & Murphy, J. M. (2021). There's more to death than life: Noncatalytic functions in kinase and pseudokinase signaling. *The Journal of Biological Chemistry*, *296*, 100705.

Mali, P., Yang, L., Esvelt, K. M., Aach, J., Guell, M., DiCarlo, J. E., et al. (2013). RNA-guided human genome engineering via Cas9. *Science*, *339*(6121), 823–826.

Manning, G., Whyte, D. B., Martinez, R., Hunter, T., & Sudarsanam, S. (2002). The protein kinase complement of the human genome. *Science*, *298*(5600), 1912–1934.

Meng, Y., Davies, K. A., Fitzgibbon, C., Young, S. N., Garnish, S. E., Horne, C. R., et al. (2021). Human RIPK3 maintains MLKL in an inactive conformation prior to cell death by necroptosis. *Nature Communications*, *12*(1), 6783.

Mojica, F. J., Diez-Villasenor, C., Garcia-Martinez, J., & Soria, E. (2005). Intervening sequences of regularly spaced prokaryotic repeats derive from foreign genetic elements. *Journal of Molecular Evolution*, *60*(2), 174–182.

Muller, T., Dewitz, C., Schmitz, J., Schroder, A. S., Brasen, J. H., Stockwell, B. R., et al. (2017). Necroptosis and ferroptosis are alternative cell death pathways that operate in acute kidney failure. *Cellular and Molecular Life Sciences*, *74*(19), 3631–3645.

Murphy, J. M. (2020). The killer pseudokinase mixed lineage kinase domain-like protein (MLKL). *Cold Spring Harbor Perspectives in Biology*, *12*(8), a036376.

Murphy, J. M., Czabotar, P. E., Hildebrand, J. M., Lucet, I. S., Zhang, J. G., Alvarez-Diaz, S., et al. (2013). The pseudokinase MLKL mediates necroptosis via a molecular switch mechanism. *Immunity*, *39*(3), 443–453.

Murphy, J. M., Farhan, H., & Eyers, P. A. (2017). Bio-zombie: The rise of pseudoenzymes in biology. *Biochemical Society Transactions*, *45*(2), 537–544.

Murphy, J. M., Lucet, I. S., Hildebrand, J. M., Tanzer, M. C., Young, S. N., Sharma, P., et al. (2014). Insights into the evolution of divergent nucleotide-binding mechanisms among pseudokinases revealed by crystal structures of human and mouse MLKL. *The Biochemical Journal*, *457*(3), 369–377.

Murphy, J. M., Mace, P. D., & Eyers, P. A. (2017). Live and let die: Insights into pseudoenzyme mechanisms from structure. *Current Opinion in Structural Biology*, *47*, 95–104.

Murphy, J. M., Zhang, Q., Young, S. N., Reese, M. L., Bailey, F. P., Eyers, P. A., et al. (2014). A robust methodology to subclassify pseudokinases based on their nucleotide-binding properties. *The Biochemical Journal*, *457*(2), 323–334.

Newton, K., Dugger, D. L., Maltzman, A., Greve, J. M., Hedehus, M., Martin-McNulty, B., et al. (2016). RIPK3 deficiency or catalytically inactive RIPK1 provides greater benefit than MLKL deficiency in mouse models of inflammation and tissue injury. *Cell Death and Differentiation*, *23*(9), 1565–1576.

Pefanis, A., Ierino, F. L., Murphy, J. M., & Cowan, P. J. (2019). Regulated necrosis in kidney ischemia-reperfusion injury. *Kidney International*, *96*(2), 291–301.

Petrie, E. J., Birkinshaw, R. W., Koide, A., Denbaum, E., Hildebrand, J. M., Garnish, S. E., et al. (2020). Identification of MLKL membrane translocation as a checkpoint in necroptotic cell death using monobodies. *Proceedings of the National Academy of Sciences of the United States of America*, *117*(15), 8468–8475.

Petrie, E. J., Czabotar, P. E., & Murphy, J. M. (2019). The structural basis of necroptotic cell death signaling. *Trends in Biochemical Sciences*, *44*(1), 53–63.

Petrie, E. J., Sandow, J. J., Jacobsen, A. V., Smith, B. J., Griffin, M. D. W., Lucet, I. S., et al. (2018). Conformational switching of the pseudokinase domain promotes human MLKL tetramerization and cell death by necroptosis. *Nature Communications*, *9*(1), 2422.

Petrie, E. J., Sandow, J. J., Lehmann, W. I. L., Liang, L. Y., Coursier, D., Young, S. N., et al. (2019). Viral MLKL homologs subvert necroptotic cell death by sequestering cellular RIPK3. *Cell Reports*, *28*(13), 3309–3319.e3305.

Reiterer, V., Eyers, P. A., & Farhan, H. (2014). Day of the dead: Pseudokinases and pseudophosphatases in physiology and disease. *Trends in Cell Biology*, *24*(9), 489–505.

Ribeiro, A. J. M., Das, S., Dawson, N., Zaru, R., Orchard, S., Thornton, J. M., et al. (2019). Emerging concepts in pseudoenzyme classification, evolution, and signaling. *Science Signaling*, *12*(594), eaat9797.

Rickard, J. A., Anderton, H., Etemadi, N., Nachbur, U., Darding, M., Peltzer, N., et al. (2014). TNFR1-dependent cell death drives inflammation in Sharpin-deficient mice. *eLife*, *3*, e03464.

Rickard, J. A., O'Donnell, J. A., Evans, J. M., Lalaoui, N., Poh, A. R., Rogers, T. W., et al. (2014). RIPK1 regulates RIPK3-MLKL driven systemic inflammation and emergency hematopoiesis. *Cell, 157*(2), 1175–1188.

Samson, A. L., Fitzgibbon, C., Patel, K. M., Hildebrand, J. M., Whitehead, L. W., Rimes, J. S., et al. (2021). A toolbox for imaging RIPK1, RIPK3, and MLKL in mouse and human cells. *Cell Death and Differentiation, 28*(7), 2126–2144.

Samson, A. L., Garnish, S. E., Hildebrand, J. M., & Murphy, J. M. (2021). Location, location, location: A compartmentalized view of necroptotic signaling. *Science Signaling, 14*(668), eabc6178.

Samson, A. L., Zhang, Y., Geoghegan, N. D., Gavin, X. J., Davies, K. A., Mlodzianoski, M. J., et al. (2020). MLKL trafficking and accumulation at the plasma membrane control the kinetics and threshold for necroptosis. *Nature Communications, 11*(1), 3151.

Sapranauskas, R., Gasiunas, G., Fremaux, C., Barrangou, R., Horvath, P., & Siksnys, V. (2011). The Streptococcus thermophilus CRISPR/Cas system provides immunity in *Escherichia coli*. *Nucleic Acids Research, 39*(21), 9275–9282.

Sethi, A., Horne, C. R., Fitzgibbon, C., Wilde, K., Davies, K. A., Garnish, S. E., et al. (2022). Membrane permeabilization is mediated by distinct epitopes in mouse and human orthologs of the necroptosis effector, MLKL. *Cell Death & Differentiation*. https://doi.org/10.1038/s41418-41022-00965-41416.

Shrestha, S., Byrne, D. P., Harris, J. A., Kannan, N., & Eyers, P. A. (2020). Cataloguing the dead: Breathing new life into pseudokinase research. *The FEBS Journal, 287*(19), 4150–4169.

Su, L., Quade, B., Wang, H., Sun, L., Wang, X., & Rizo, J. (2014). A plug release mechanism for membrane permeation by MLKL. *Structure, 22*(10), 1489–1500.

Sun, L., Wang, H., Wang, Z., He, S., Chen, S., Liao, D., et al. (2012). Mixed lineage kinase domain-like protein mediates necrosis signaling downstream of RIP3 kinase. *Cell, 148*(1–2), 213–227.

Tanzer, M. C., Matti, I., Hildebrand, J. M., Young, S. N., Wardak, A., Tripaydonis, A., et al. (2016). Evolutionary divergence of the necroptosis effector MLKL. *Cell Death and Differentiation, 23*(7), 1185–1197.

Tanzer, M. C., Tripaydonis, A., Webb, A. I., Young, S. N., Varghese, L. N., Hall, C., et al. (2015). Necroptosis signalling is tuned by phosphorylation of MLKL residues outside the pseudokinase domain activation loop. *The Biochemical Journal, 471*(2), 255–265.

Tovey Crutchfield, E. C., Garnish, S. E., & Hildebrand, J. M. (2021). The role of the key effector of necroptotic cell death, MLKL, in mouse models of disease. *Biomolecules, 11*(6), 803.

Wang, H., Sun, L., Su, L., Rizo, J., Liu, L., Wang, L. F., et al. (2014). Mixed lineage kinase domain-like protein MLKL causes necrotic membrane disruption upon phosphorylation by RIP3. *Molecular Cell, 54*(1), 133–146.

Wanisch, K., & Yanez-Munoz, R. J. (2009). Integration-deficient lentiviral vectors: A slow coming of age. *Molecular Therapy, 17*(8), 1316–1332.

Wright, D. A., Li, T., Yang, B., & Spalding, M. H. (2014). TALEN-mediated genome editing: Prospects and perspectives. *The Biochemical Journal, 462*(1), 15–24.

Zhang, S., Shen, J., Li, D., & Cheng, Y. (2021). Strategies in the delivery of Cas9 ribonucleoprotein for CRISPR/Cas9 genome editing. *Theranostics, 11*(2), 614–648.

CHAPTER TEN

Protein engineering strategies to stimulate the functions of bacterial pseudokinases

Xiaole Yang, Kimberly A. Kowallis, and W. Seth Childers*
Department of Chemistry, University of Pittsburgh, Pittsburgh, PA, United States
*Corresponding author: e-mail address: wschild@pitt.edu

Contents

Abstract

Enzymes orchestrate an array of concerted functions that often culminate in the chemical conversion of substrates into products. In the bacterial kingdom, histidine kinases autophosphorylate, then transfer that phosphate to a second protein called a response regulator. Bacterial genomes can encode large numbers of histidine kinases that provide surveillance of environmental and cytosolic stresses through signal stimulation of histidine kinase activity. Pseudokinases lack these hallmark catalytic functions but

Methods in Enzymology, Volume 667
ISSN 0076-6879
https://doi.org/10.1016/bs.mie.2022.03.030

often retain binding interactions and allostery. Characterization of bacterial pseudokinases then takes a fundamentally different approach than their enzymatic counterparts. Here we discuss models for how bacterial pseudokinases can utilize protein–protein interactions and allostery to serve as crucial signaling pathway regulators. Then we describe a protein engineering strategy to interrogate these models, emphasizing how signals flow within bacterial pseudokinases. This description includes design considerations, cloning strategies, and the purification of leucine zippers fused to pseudokinases. We then describe two assays to interrogate this approach. First is a *C. crescentus* swarm plate assay to track motility phenotypes related to a bacterial pseudokinase. Second is an in vitro coupled-enzyme assay that can be applied to test if and how a pseudokinase regulates an active kinase. Together these approaches provide a blueprint for dissecting the mechanisms of cryptic bacterial pseudokinases.

1. Introduction

Pseudokinases mimic the structural folds of kinases but lack kinase activity (Eyers & Murphy, 2016; Ribeiro et al., 2019). Instead of canonical enzymatic functions, these pseudokinases often serve as critical regulators and scaffolds of active signaling networks. Within the bacterial kingdom, active histidine kinase and pseudokinases function as a part of two-component systems (Rong Gao & Stock, 2009). Two-component systems are a common bacterial signaling pathway architectures that regulate cell division (Wu, Ohta, Zhao, & Newton, 1999), quorum sensing (Neiditch et al., 2006), and pathogenesis (Gao & Lynn, 2005; Wang, Zhao, Novick, & Muir, 2014). These systems are composed of a sensor histidine kinase that senses environmental or cytosolic signals. Sensing of these signals then stimulates autophosphorylation and phosphotransfer to its associated response regulator. Then in a phosphorylation-dependent manner, the response regulators exert control over gene transcription, chemotaxis, or proteolysis (Gao, Bouillet, & Stock, 2019). This chapter will consider methods for a protein engineering strategy to interrogate bacterial pseudokinase mechanisms. As illustrative examples, these methods will be applied toward the model bacterial pseudokinase DivL. DivL regulates the CckA-CtrA signaling pathway that regulates asymmetric cell division in the bacterium *Caulobacter crescentus*.

1.1 The function of active bacterial histidine kinases

Histidine kinases are bifunctional enzymes that contain both autokinase and phosphatase activity toward its associated response regulator. The namesake autokinase activity involves phosphorylation of a conserved histidine within their own dimerization and histidine phosphotransferase (DHp) domain.

Phosphorylation is mediated by the catalytic ATP (CA) binding domain. This domain includes an ATP binding site consisting of the N, G1, G2, and F boxes named based upon conserved residues within each motif (Gao & Stock, 2009). A third critical motif of the catalytic ATP domain is the gripper helix (Bhate, Molnar, Goulian, & DeGrado, 2015; Huynh & Stewart, 2011), which mediates conformational changes between the catalytic ATP and the DHp domains (Bhate et al., 2015). Bioinformatic inspection of these domains may point toward the loss of typical catalytic functions (Childers et al., 2014).

The DHp domain contains three distinct functional motifs. First is the highly conserved histidine phosphorylation site. The second structural feature within the DHp domain is the response regulator binding site. This comprises a set of protein-protein interaction residues that mediate specific binding to its cognate response regulator (Chen, Tsokos, Biondi, Perchuk, & Laub, 2009). The third critical structural feature is a four-residue motif that mediates phosphatase activity. This phosphatase motif is positioned between the histidine phosphorylation site and the response regulator binding site. These residues chelate, organize and activate a water molecule to hydrolyze the phosphoaspartate within its cognate response regulator (Huynh & Stewart, 2011).

Together the DHp and CA domains can exist in conformations that promote kinase, phosphatase, and inactive functions (Bhate et al., 2015). These functional states are allosterically regulated by an N-terminal sensory domain that binds small molecules or protein signals. The C-terminus of the sensory domain commonly directly connects to a coiled-coil linker that feeds into a dimeric four-helix bundle that makes up the DHp domain. Therefore, signal binding to the N-terminal sensory domain regulates the downstream helical bundle through twisting or bending conformational changes within the coiled-coil. These conformational changes either result in the engagement of the CA and DHp domains to promote kinase functions or the engagement of the DHp domain and the response regulator to promote phosphatase functions (Bhate et al., 2015).

1.2 The potential mechanisms of bacterial pseudokinases

As recently reviewed (Collins & Childers, 2021), pseudokinases lack one or more of these histidine kinase enzymatic functions. The ability of pseudokinases to engage response regulators with their DHp domain or to engage proteins with their N-terminal sensor domains lies at the heart of their function. Through protein-protein interactions, bacterial pseudokinases capture many of the same functions as their eukaryotic pseudokinase counterparts (Eyers & Murphy, 2016; Murphy, Mace, & Eyers, 2017; Ribeiro et al., 2019). For example, bacterial kinases can act as subcellular localization anchors, allosteric modulators of active enzymes, and competitors for substrate binding (Fig. 1) (Collins & Childers, 2021).

When investigating the mechanism of a bacterial pseudokinase, three functions should be considered: response regulator binding, ligand or protein binding to the N-terminal sensor, and sensor-kinase allostery. Without sensor-kinase allostery, the pseudokinase domain may exhibit functions constitutively and rely on expression and proteolysis for regulation. In contrast, bacterial pseudokinases that retain the sensor-kinase helix may act as a protein switch. For example, signal binding to the N-terminal sensor may regulate response regulator binding and sequestration via the C-terminal pseudokinase domains (Fig. 1). We describe this typical signaling flow from the sensor to kinase as forward signaling. With very few exceptions, the functions of active histidine kinases have been characterized in this "forward signaling" manner (Fig. 2).

Forward signaling may also be leveraged with pseudokinases to regulate a second active histidine kinase (Fig. 2). This will depend on whether the pseudokinase interacts with the active kinase via the DHp or N-terminal sensing domains. The DHp domain typically mediates histidine kinase homodimerization. However, this domain may also promote heterodimerization (Ashenberg, Keating, & Laub, 2013). Thus, active histidine kinases and inactive pseudokinases may heterodimerize via the DHp domain. Within a pseudokinase-kinase heterodimer, the N-terminal sensory domain may regulate kinase-pseudokinase heterodimerization. Alternatively, the sensory domains of the pseudokinase-kinase heterodimer may allosterically regulate the function of the active kinase (Fig. 2). While this mode of pseudokinase-kinase regulation has not yet been reported, it would provide a unique mechanism to integrate signals from two sensory domains to regulate a single active histidine kinase.

In contrast, studies of the *C. crescentus* pseudokinase DivL have revealed that the DivL engages an active histidine kinase CckA through sensory domain protein-protein interactions (Childers et al., 2014; Kowallis et al., 2020; Tsokos, Perchuk, & Laub, 2011). Meanwhile, DivL uses its DHp domain to interact with its cognate response regulator DivK. Response regulator binding to DivL's histidine kinase domain stimulates changes in its N-terminal sensory domain (Fig. 2). Therefore, the DivK-DivL-CckA protein complex leverages the same sensor-kinase allostery in reverse.

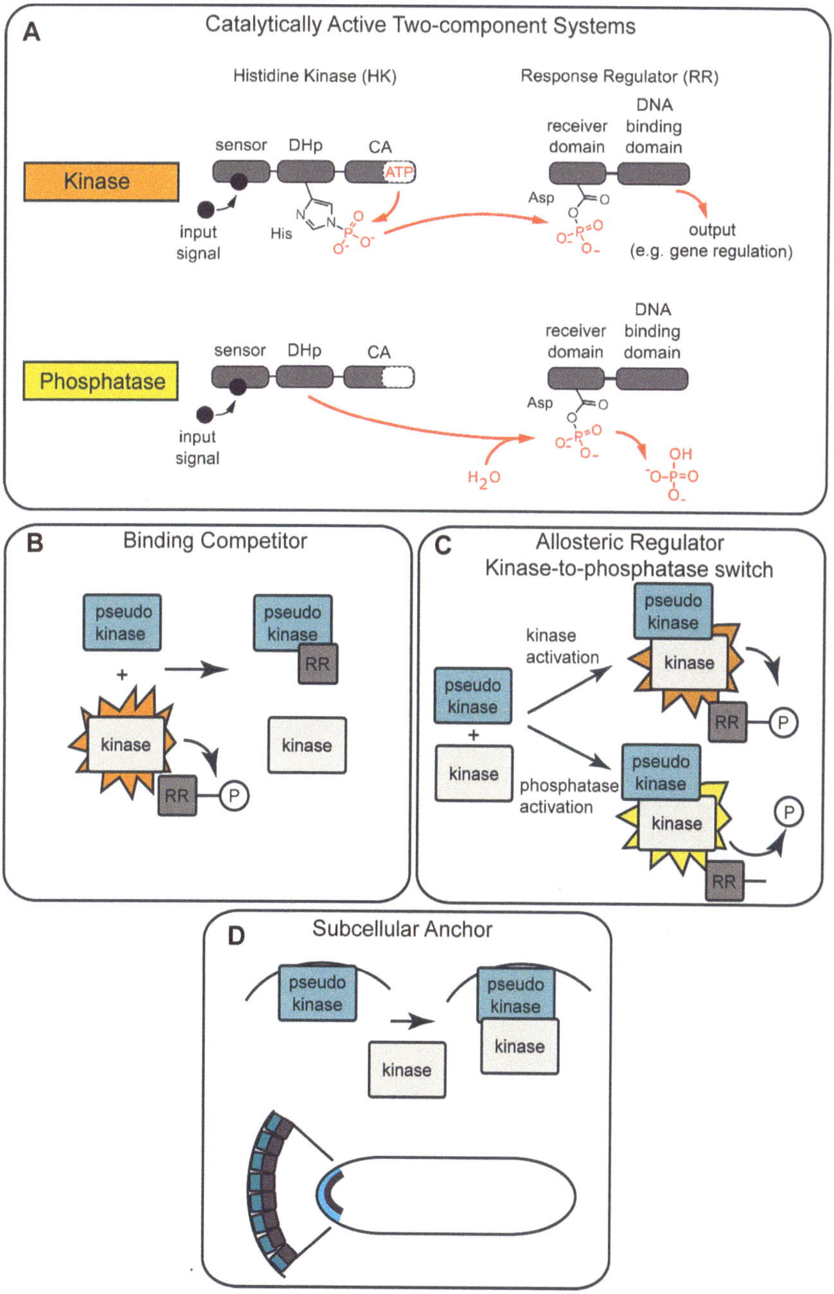

Fig. 1 Bacterial pseudokinases exhibit properties similar to their eukaryotic counterparts. (A) Two-component systems are composed of histidine kinase (HK) and response regulator (RR). Histidine kinases contain a dimerization and histidine phosphotransferase (DHp) domain and a catalytic ATP binding domain (CA). Histidine kinases are bifunctional and can function as a kinase or a phosphatase. (B–D) Bacterial pseudokinases lack kinase and phosphatase functions. Instead, these pseudokinases can exhibit allosteric regulation of kinase or phosphatase activity, competitive binding, and anchoring an active kinase to a subcellular location.

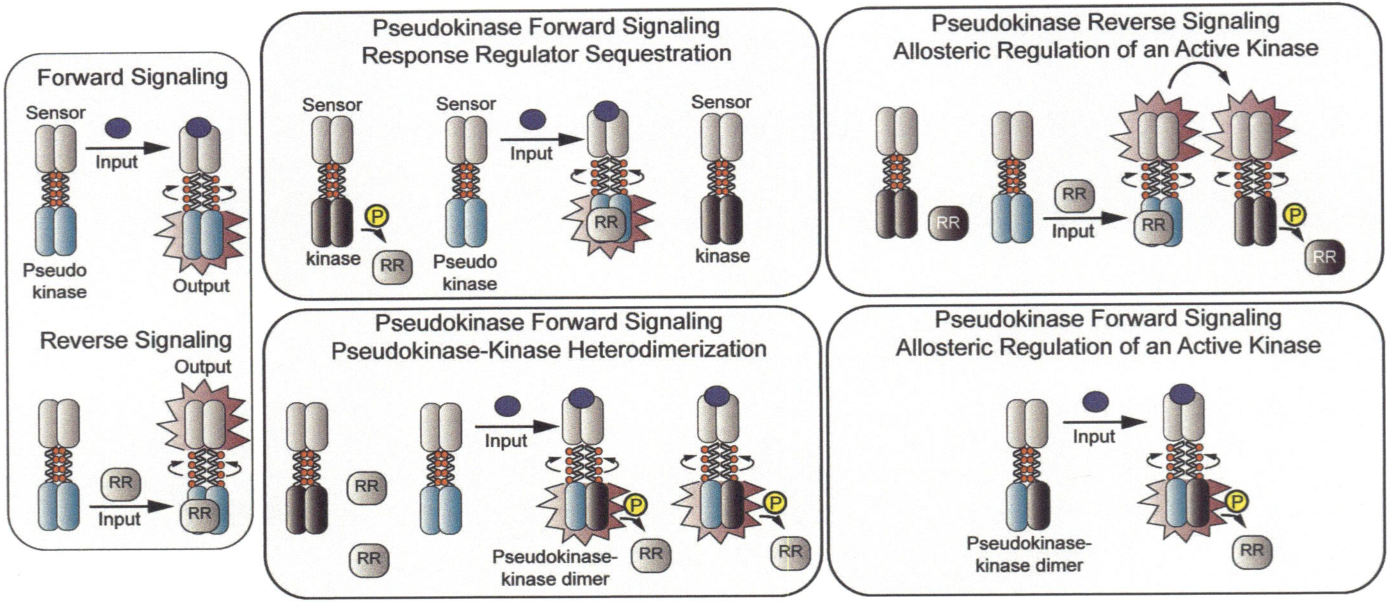

Fig. 2 Models of how pseudokinases can function as a protein switch through forward and reverse signaling flow. This includes the sequestration of response regulators and the allosteric regulation of active histidine kinases. Kinase domain (black), pseudokinase domain (cyan), signals that bind to the sensory domain (blue), and response regulators (RR) are shown in black or gray. The coiled-coil linker residues along one heptad position are colored red to emphasize conformational change within the coiled-coil that occurs upon stimulation.

Since signals flow from the pseudokinase domain to the sensory domain, we term this "reverse signaling" flow (Fig. 2). Critically, both forward and reverse signaling utilizes twisting or bending conformational changes within the coiled-coil linker that connects the sensor and kinase domains.

1.3 Synthetic stimulation of pseudokinase functions

The natural coiled-coil linkers between the sensor and kinase domains adopt two or more distinct conformations that are coupled to the structural state of the attached sensor and kinase domains (Bhate et al., 2015). Thus, protein engineering strategies can bias and lock the coiled-coil in one conformation, subsequently altering the attached domains' conformation (Lin, Gao, Binns, & Lynn, 2008; Wang et al., 2014; Wang, Gao, & Lynn, 2002). Here we describe a method to lock the interconnecting sensor-kinase coiled coils into one state by fusing a high-affinity leucine zipper directly to the coiled-coil region. This fusion event results in continuous coiled coils that force the pseudokinase coiled-coil to adopt conformations compatible with the heptad registry of the high-affinity GCN4 leucine protein (Fig. 3). This approach has been successfully applied toward active kinases (Wang et al., 2002, 2014) to stimulate forward signaling. This involves the fusion of the leucine zipper to the N-terminus of the coiled-coil linker to lock the kinase domain into a distinct structural state.

In contrast, we also showed that the leucine zipper could be fused to the C-terminus of the coiled-coil to lock the sensor domain into a distinct structural state (Fig. 3) (Kowallis et al., 2020). This fusion event subsequently stimulates reverse signaling and impacts downstream functions of the N-terminal sensory domain (Fig. 2). In this chapter, we will describe the design and implementation of this leucine zipper fusion strategy to interrogate the mechanism of bacterial pseudokinases.

1.4 Experimental design considerations

(1) Hypothesize how the pseudokinase domain and sensor of interest are being leveraged (Figs. 1 and 2). Consider what is likely the final output of the pseudokinase. Sequestration of a response regulator? Regulation of an active kinase?

(2) If available, evaluate data for protein-protein interactions with the pseudokinase. Identify kinases and response regulators that interact with the DHp domain. In addition, identify proteins and signals that interact with the N-terminal sensory domain.

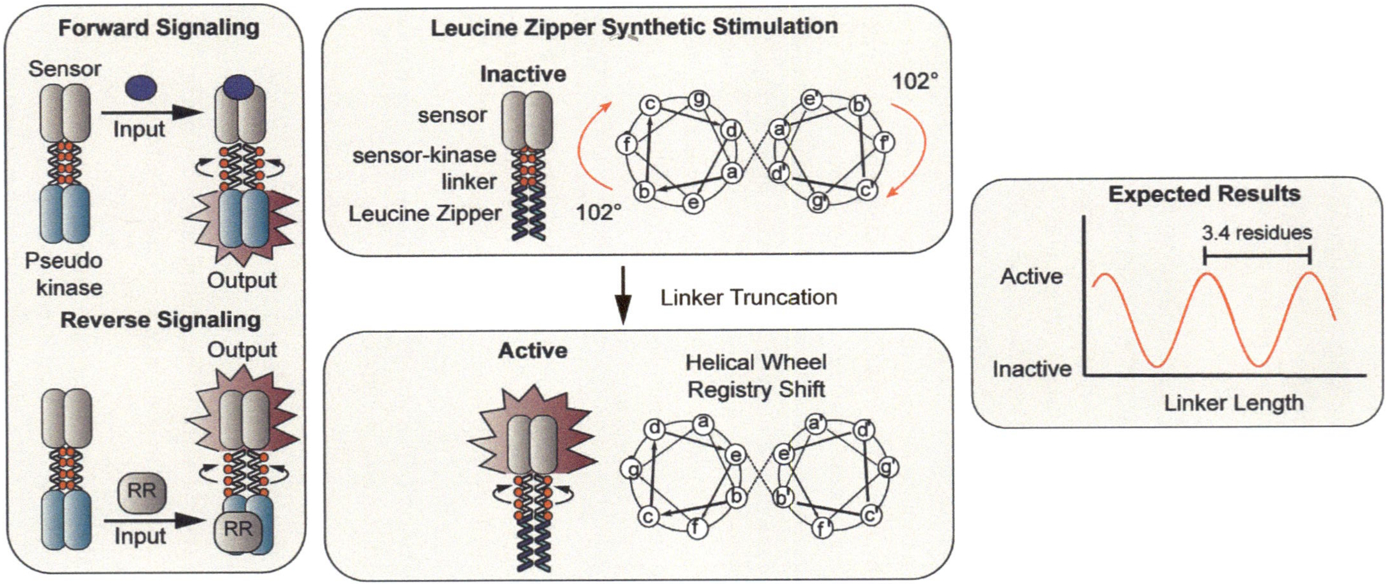

Fig. 3 Leucine Zipper Synthetic Simulation Strategy to detect reverse signaling mediated by a pseudokinase. A comparison of forward and reverse signaling flow. In each case, signals flow from one domain to the next through a set of interconnecting coiled coils. Fusion of a GCN4 leucine zipper (blue) to the sensor-kinase coiled-coil linker locks the coiled-coil into a distinct conformation compatible with the GCN4 leucine zipper structure. Truncation of the sensor-HK linker will thus result in a registry shift of the helical wheel. This may place the attached domains into a distinct structural conformation. Systematic deletion of residues is anticipated to oscillate the state of the pseudokinase from active to inactive with a periodicity that matches the coiled-coil at 3.6 residues per turn.

(3) Consider all possible pseudokinase regulation models from the observed protein-protein interactions (Figs. 1 and 2). Consider if forward or reverse signaling is likely utilized (Fig. 2). Or consider testing both modes of regulation (Fig. 3).

(4) If available, evaluate data from domain deletion experiments. Deletion of an output domain will often lead to a loss of function of a pseudo histidine kinase. In contrast, the removal of an input domain may retain unregulated functions of the output domain. If the pseudokinase domain is essential, that may point toward forward signaling. In contrast, if the sensory domain is essential, that may point toward reverse signaling. For example, in the case of the pseudo histidine kinase DivL, nearly all of the pseudokinase domains could be removed and this truncation could support cell viability (Christen et al., 2010). This result suggested that the N-terminal sensory domain carried out essential functions. Based on these findings, we hypothesized that DivL utilized reverse signaling in which stimulation of the pseudokinase domain altered the N-terminal sensor function.

(5) If typical "forward" signaling is conserved, the N-terminal sensory domain may regulate pseudokinase domain functions (e.g., binding to a response regulator or another active kinase—see Fig. 2). To test forward signaling the N-terminal sensory domain can be swapped for a high-affinity leucine zipper.

(6) In contrast, in reverse signaling, the binding of the response regulator may regulate sensory domain functions. To test reverse signaling the C-terminal pseudokinase can be swapped for a high-affinity leucine zipper.

(7) Note: If phenotypes of the pseudokinase are well characterized but protein interaction partners are unknown, you can still interrogate if "forward signaling" or "reverse signaling" is utilized.

(8) Apply bioinformatics tools to identify the end of the sensory domain and the start of the DHp domain.

(9) Apply bioinformatics tools, such as the PCOILS method, to confirm the linker is predicted to utilize a coiled-coil structural motif.

(10) Many linkers between folded domains are commonly composed of flexible intrinsically disordered proteins. Therefore, apply bioinformatics programs to predict disorder. The leucine zipper approach will not be suitable to fuse against disordered mechanisms, as the allosteric coupling between the leucine zipper and intrinsically disordered segments is unclear.

(11) To test reverse signaling, select a fusion site at the C-terminus of the coiled-coil linker at the junction of the coiled-coil and the DHp domain.

(12) Next, design a library of 7–14 leucine zipper chimeras in which residues are deleted incrementally. Two truncation strategies have been successfully applied in other work. First is the incremental truncation of the sensor-HK linker (Stabel, Stüven, Ohlendorf, & Möglich, 2017). Then the second strategy is the incremental deletion of the GCN4 leucine zipper (Lin et al., 2014). An example of a truncation series used to test reverse signaling with the pseudokinase DivL is shown below in Fig. 4.

(13) Consider the epitope tag to include within the chimeric protein to confirm protein expression. These could include fluorescent proteins, FLAG tags or hexahistidine tags. Consider the optimal placement of tags to retain the desired function (Kowallis et al., 2020).

(14) A final consideration is the selection of biological or enzymatic assays. Phenotypes regulated by the pseudokinase could be applied. For example, two-component systems commonly regulate genes that control motility, biofilm formation, cell division, chromosome replication and other bacterial development. In the swarm assay bacterial cultures are inoculated on low percent agar plates that permit radially growth of bacterial cultures that exhibit functional motility. Misregulation of two-component systems such as the *C. crescentus* CckA-CtrA two-component system leads to a loss of cell motility, which can be detected through a swarm plate assay (see the Swarm Plate Assay Protocol in Section 3.3).

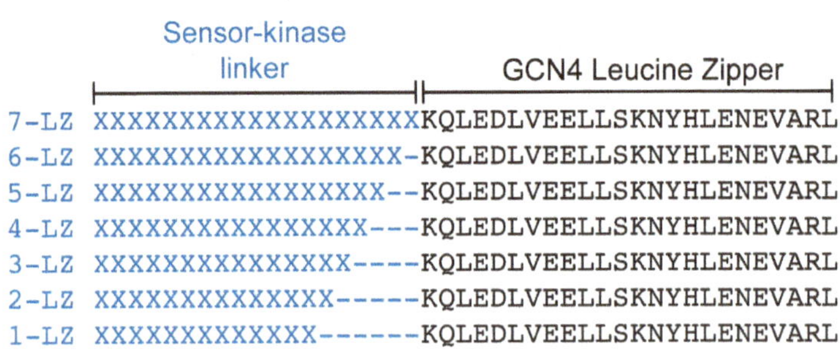

Fig. 4 An example library of pseudokinase-leucine chimeras designed and tested to interrogate the reverse signaling. The leucine zipper sequence is appended to the C-terminus pseudokinase's sensor-HK linker. For each variant, amino acid residues are incrementally removed (Kowallis et al., 2020).

(15) If the pseudokinase is a known regulator of an active histidine kinase, then an array of kinases could be employed, such as the coupled enzyme assay could be applied.

2. Equipment and materials

2.1 Cloning DivL-leucine zipper (DivL-LZ) chimeras

Materials

1. *Sac* I (ER1131, Thermo Scientific)
2. DMSO (180778,Thermo fisher)
3. Betaine (AC204240050, Fisher Scientific)
4. Primers ordered from IDT
5. dNTPs (R0181,Thermo Fisher)
6. Phusion High-Fidelity DNA Polymerase (F530L, Thermo Scientific)
7. T5 Exonuclease (M0663S, New England Biolabs)
8. Taq DNA Ligase (M0208S, New England Biolabs)
9. Agarose (207569, Fisher Scientific)
10. Ethidium Bromide (BP1302-10, Fisher Scientific)
11. Peptone Yeast Extract (212750, Bacto™)
12. GeneJET Plasmid Miniprep Kit (K0502, Thermo Scientific)
13. GeneJET Gel Extraction Kit (K0692, Thermo Scientific)
14. Electroporation Cuvettes (FB101, Fisherbrand)
15. *C. crescentus* NA1000 (CB15N)
16. *Escherichia coli* DH5α (Invitrogen)
17. BL21 (Novagen)
18. DNA Molecular Weight Marker, 1Kb Ladder, (500–10,000 bp) (J67225.4EQ, Thermo Scientific)
19. 6 × DNA Gel Loading Dye (R0611,Thermo Scientific)
20. Dream Taq PCR Master Mix (K1071,Thermo Scientific)

Equipment

1. PCR SimpliAmp Thermal Cycler (A24811,Thermo Scientific)
2. Heating block (Thermo Scientific)
3. Bio-Rad micropulse
4. Thermo Scientific Easy Cast B1A Mini Gel Electrophoresis Systems
5. Incubator (Thermo Fisher)
6. Small Benchtop Centrifuge (Thermal Scientific)
7. Nanodrop (Thermo Fisher)
8. ChemiDoc XRS+ system (Bio-Rad)
9. pH meter (METTLER TOLEDO)

2.2 Western blot confirmation of expression

Materials

1. Peptone Yeast Extract (212750, Bacto™)
2. Xylose
3. Non-fat dry Milk (AmericanBio)
4. Tris Base (BP152-500, Fisher Scientific)
5. Glycine (VWR)
6. TEMED (BP150-20, Fisher Scientific)
7. Acrylamide/Bisacrylamide 37.5:1, 40% solution (BP1410-1, Fisher Bioreagents)
8. Ammonium Persulfate (1610700, Bio-RAD)
9. 2-Mercaptoethanol (BME) (444203, Sigma)
10. Bromophenol Blue (A18469.18, Thermo Scientific)
11. Methanol (A452-SK, Fisher Scientific)
12. Tween-20 (BP337-500, Fisher Scientific)
13. Sodium Chloride (BP-358, Fisher Scientific)
14. Anti-M2 (FLAG) antibody (F3165, Sigma-Aldrich)
15. Goat anti-rabbit IgG-peroxidase secondary antibody (AP132P, Sigma-Aldrich)
16. PVDF membrane (GE Healthcare)
17. Pierce ™ Fast Western Blot Kit ECL Substrate (35055, Thermo Scientific)

Western blot buffers preparations

10 × Transfer Buffer (1 L)
- 30.3 g Tris Base
- 144 g Glycine
- Fill to 1 L with Distilled Water

1 × Transfer Buffer (1 L)
- 100 mL 10 × Transfer Buffer
- 50 mL Methanol
- 850 mL Distilled Water

10 × Tris Buffered Saline (TBS) pH 7.5 (1 L)
- 12.114 g Tris Base
- 87.66 g NaCl
- 1000 mL Distilled Water
- pH to 7.5

1 × TBS (1 L)
- 100 mL 10 × TBS
- 900 mL Distilled Water

1 × TBS + 0.1% Tween (TBST) (500 mL)
500 μL Tween 20 for 1%, 250 μL Tween 20 for 0.05%
5% Blocking Buffer (25 mL)
25 mL 1 × TBST
1.25 g Nonfat Dry Milk

Equipment

1. Heating block (Thermo Fisher)
2. mini gel tank (Thermo Fisher)
3. SDS-PAGE glass plates and Empty Mini Gel Cassette Combs, 12-well Invitrogen ™
4. Mini Blot module (Fisher Scientific)
5. Container with lid
6. Rocker (Fisher Scientific)
7. Power supply for mini gel tank

2.3 *C. crescentus* swarm plate assay

Materials

1. 70% Ethanol
2. Peptone Yeast Extract (212750, Bacto™)
3. Agar (BP1423-500, Fisher Scientific)
4. 15 cm Petri Dish Plate
5. Ampicillin (AC611770250, Fisher Scientific)
6. Kanamycin Sulfate (AC611290050, Fisher Scientific)

Equipment

1. Bunsen burner
2. Boekel Scientific replicator
3. Tecan Infinite 200 Pro plate reader
4. Incubator (Thermo scientific)
5. Clear flat-bottom 96-well plate (Greiner)

2.4 DivL-LZ protein expression and purification

Materials

1. LB Broth (Miller) (BP1426, Fisher Scientific)
2. Agar (BP1423-500, Fisher Scientific)
3. Ampicillin (AC611770250, Fisher Scientific)
4. Isopropyl-β-D-thiogalactopyranoside (IPTG) (BP1755100, Fisher Scientific)
5. HEPES (AC172570010, Fisher Scientific)

6. Potassium Hydroxide (P250-1, Fisher Scientific)
7. Potassium Chloride (P217-3, Fisher Scientific)
8. DTT, Dithiothreitol (AC165680250, Fisher Scientific)
9. Imidazole (I5513, Sigma)
10. Glycerol (BP229-4, Fisher Scientific)
11. Benzonase Nuclease (E8263, Sigma)
12. SIGMAFAST protease inhibitor tablets (S8820, Sigma)
13. Liquid Nitrogen
14. NanoPure Water
15. 10% SDS-PAGE gel
16. PageRuler™ Plus Prestained Protein ladder (26619,Thermo Scientific)

Equipment

1. Incubator (Thermo Scientific)
2. Stackable Shaker (Thermo Scientific)
3. 2800 mL Erlenmeyer Flasks
4. UV/VIS Spectrophotometer
5. Legend ™ Centrifuge (Thermo Scientific)
6. Centrifuge Bottles
7. Emulsiflex
8. Protein purification spin column (Thermo Scientific)
9. pH meter (METTLER TOLEDO)
10. Sterile culture tubes
11. Sterile culture flask 50 mL
12. Amicon Ultra Centrifugal Filter Units, MWCO 30 kDa
13. PCR tubes (VWR)
14. 50 mL Falcon Tubes

2.5 Coupled enzyme assay

Materials

1. Phosphoenolpyruvate (PEP) (10228209, Alfa Asesar)
2. Nicotinamide adenine dinucleotide (NADH) (N05E014, Alfa Asesar)
3. Adenosine triphosphate (ATP) (A2383, Sigma)
4. HEPES (AC172570010, Fisher Scientific)
5. Potassium chloride (P217-3, Fisher Scientific)
6. Potassium hydroxide (P250-1,Fisher Scientific)
7. DivL variants proteins
8. Magnesium chloride (BP214, Fisher Scientific)

9. Pyruvate kinase/Lactate dehydrogenase from rabbit muscle (P0294, Sigma)
10. Nanopure water
11. Calcium chloride (SC10-1, Fisher Scientific)

Equipment

1. Tecan Infinite 200 Pro plate reader
2. Flat clear bottom 96 well plate (Greiner Bio-One)
3. Nano Drop (Thermo Scientific)

3. Protocol

3.1 Cloning DivL-leucine zipper chimeras into a *C. crescentus* high-copy plasmid

The small size of the GCN4 leucine zipper (72 bp) allows for many different strategies for incorporating them into DNA plasmids. Here we describe the embedment of the leucine zipper coding sequence into the PCR primers as an approach.

1. For reverse signaling stimulation, design the reverse primer to include the 3′ end of the sensory domain sequence adjacent to the coding sequence for the GCN4 leucine zipper followed by the necessary overhangs for the adjacent Gibson assembly insert or vector. The design of the plasmid can be facilitated using online software such as the J5 device editor software (Hillson, Rosengarten, & Keasling, 2012). An example of this primer design is shown in Fig. 5 for the DivL-LZ chimera.

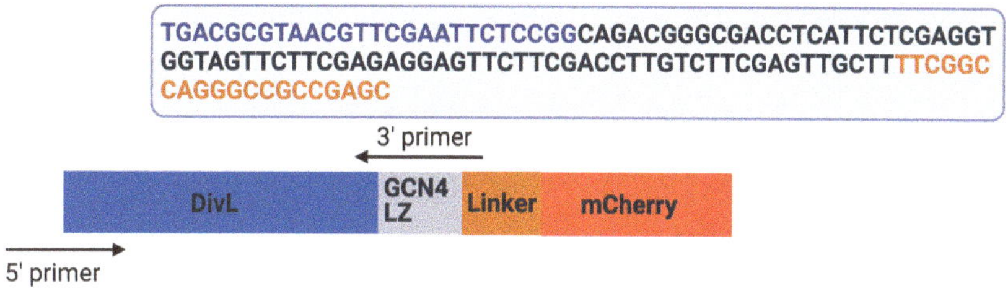

Fig. 5 The cloning strategy of embedding the leucine zipper into the PCR reverse primer. An example of the leucine zipper sequence designed into the reverse primer with overhangs overlapping the DivL truncation and the linker or backbone. The reverse primer contained blue coded sequence partly overlapping with the DivL sequence, the black coded sequence representing the GCN4 leucine zipper sequence, orange coded sequence partly overlapping the linker sequence. *Created with BioRender.com.*

2. Prepare PCR reactions at the 50 μL scale containing 3% (v/v) DMSO, 1.3 M betaine, 0.3 μM forward and reverse primer respectively, and 0.2 mM each dNTP, and 1 U Phusion High-Fidelity DNA polymerase (Thermo Scientific).

   Note: (1) The addition of DMSO and betaine facilitates PCR of gene fragments with high GC content, such as the *C. crescentus* genome that displays high GC content. (2). If the PCR reaction is still not successful, KOD polymerase can be considered.
3. Setup the PCR thermocycler using standard protocols specified for Phusion DNA polymerase, including a 55 °C annealing temperature and 25 cycles.
4. Linearize the destination *C. crescentus* integrating plasmid pXCHYC-6 through digestion with *Sac* I (Thermo Scientific), and linearize the *C. crescentus* high copy plasmid pBXSPA-2 backbone through PCR.
5. Apply DNA gel electrophoresis to purify the Gibson assembly inserts and linearized backbones. Confirm the size of the DNA fragments by running 1% agarose gel.
6. Purify the resulting DNA bands using standard manufacturer's protocols provided with the GeneJET gel extraction kit (Thermo Fisher Scientific).
7. Perform Gibson assembly reactions at 50 °C for 1 h (Gibson et al., 2009) at the 20 μL scale with 100 ng backbone and DNA inserts at a molar ratio of backbone: insert (1:5), 0.08 U T5 Exonuclease (New England Biolabs), 1 U Phusion High-Fidelity DNA Polymerase (Thermo Fisher Scientific), and 80 U Taq DNA Ligase (New England Biolabs).
8. Incubate 10 μL Gibson assembly reaction mixture and 100 μL *E. coli* DH5 alpha competent cells) with 20 μL 5 × KCM buffer containing 0.5 M KCl, 0.25 M $MgCl_2$, 0.15 M $CaCl_2$, 70 μL sterile water on ice for 20 min.
9. Place the mixture mentioned above in a preheated heating block by heat shock at 42 °C for 45 s.
10. Transfer the heat-shocked cells to 2 mL of fresh super broth, allowing cells to grow and recover at 37 °C for 45 min.
11. Plate a 100 μL aliquot of these recovered cells on LB agar plates supplemented with a final concentration of 50 μg/mL of kanamycin.
12. Place the inverted plates into a temperature-controlled incubated at 37 °C overnight.
13. Select 10–15 colonies for analysis colony PCR using corresponding DNA primers to detect the DivL-LZ inserted sequences by running 1% agarose gel.

14. Inoculate isolated colonies containing plasmids with inserts of the proper size into 5 mL of fresh super broth media at 37 °C overnight.
15. Isolate plasmids from saturated cultures using the manufacturer protocol provided with the GeneJET Plasmid Miniprep Kit (Thermo Fisher Scientific).
16. Submit plasmids containing inserts of the proper size for DNA Sanger sequencing (e.g., Genewiz (South Plainfield, NJ)).
17. Confirm the DNA sequence of the isolated DNA plasmids to ensure they contain the divL-LZ chimeric sequence in frame.
18. Prepare competent cells for electroporation by growing *C. crescentus* cells in PYE (Peptone Yeast Extract) medium overnight at 28 °C.
19. Wash the saturated cultures three times with ice-cold sterile water.
20. Resuspend the rinsed cells in sterile water using 80 μL of cells and 10 μL of the sequenced plasmids. Electroporate the resulting cell-plasmid suspension using a time constant of 3.5–5.0 ms.
21. Incubate the electroporated cells in fresh PYE media for 3 h to allow them to grow and recover.
22. Plate the recovered *C. crescentus* cells on PYE agar solid media supplemented with antibiotics.
23. Screen selected colonies for protein expression using the western blot assay described in Section 3.2.

3.2 Western blot confirmation of expression

3.2.1 Methods

In the subsequent swarm plate assay, bacterial cultures are inoculated on a low percent agar and motile cultures increase swarm size radially in diameter. Given that DivL regulation controls motility via the CckA-CtrA signaling pathway, expression of functional DivL-LZ variants will lead to a reduction in swarm size. In contrast, expression of non-functional DivL-LZ variants will result in no changes in swarm size. Therefore, the observation of a swarm size that is unchanged could be due to the expression of a DivL-LZ chimera in an inactive conformational state. In contrast, it could be the result of poor DivL-LZ chimera protein expression. Therefore, Western blot analysis should be performed to confirm the expression of DivL-LZ chimeric proteins.

3.2.2 Prepare 10% SDS-PAGE gel

1. Set up gel cast apparatus on an even bench. Then add a small amount of water to test for potential leaks.

2. Prepare a 10% running gel buffer in a 50 mL falcon tube by mixing 4.8 mL Nano pure water, 2.5 mL Acrylamide/Bisacrylamide 37.5:1, 40% solution, 2.5 mL 1.5 M Tris pH 8.8, 100 μL SDS 10% (w/v), 10 μL TEMED (N,N,N′,N′-tetramethyl ethylenediamine), Ammonium Persulfate (APS) 10% (w/v).

   Note: Add TEMED and Ammonium Persulfate 10% (w/v) lastly and mix the running gel buffer well by pipetting up and down.
3. Transfer the 10% running gel mixture into the glass gel cast cassette until 1.5 cm from the top of the glass plates.
4. Overflow the glass plates gap using Nanopure Water or isopropanol to ensure the running gel is horizontal and straight.
5. Prepare a 4% stacking gel buffer in another 50 mL FALCON tube by mixing 6.40 mL Distilled Water, 1.00 mL Acrylamide/Bisacrylamide 37.5:1, 40% solution, 2.5 mL 1 M Tris pH 6.8, 100 μL SDS 10% (w/v), 5 μL TEMED, 100 μL APS 10% (w/v).
6. Allow approximately 35 min for the running gel to solidify.
7. Remove the Nanopure water, then add the 4% stacking gel mixture to the gel cast apparatus with an overflow to remove any potential air bubbles.
8. Insert the desired well volume comb between the cover plate and glass plate gap.
9. Prepare a 10 × SDS-PAGE running buffer (25 mM Tris, 192 mM glycine, 3.5 mM SDS) weighing 30.3 g Tris Base, 144.0 g Glycine, and 10.0 g SDS and mix with 1 L of nanopure water. 1 L, 1 × SDS-PAGE running buffer could be diluted by adding 900 mL water to 100 mL 10 X stock running buffer.
10. Prepare 10 mL 4 × Sample buffer containing 2.4 mL 1 M Tris HCl pH 6.8, 0.8 g SDS, 4 mL glycerol, 0.5 mL beta mercaptoethanol (BME) and Bromophenol Blue (around 3 mL).

3.2.3 Prepare samples of strains that express the DivL-LZ chimeric proteins

1. Inoculate 2 mL PYE medium with *C. crescentus* strains containing the DivL-LZ expression plasmids. Allow cells to grow overnight until the early log phase at 28 °C.
2. Supplement the PYE media with 0.3% xylose to induce expression of the DivL-LZ chimeric proteins.
3. Harvest cells from 1 mL of culture by centrifugation. Then, resuspend the cell pellet in 100 μL of 4 × sample loading buffer.

4. Heat samples at 95 °C for 5 min.

   Note: Cooled cell samples with SDS-loading dye down are very sticky. Preheat samples before addition to the SDS-PAGE gel.
5. Normalize samples to the lowest OD600 upon collection to ensure samples are loaded at similar concentrations.
6. Load samples and pre-stained protein ladder (26619,Thermo Scientific) onto a 10% SDS-PAGE with a constant voltage at 150 V for 50 min, at 4 °C, until the loading dyes are near the bottom of the gel.
7. Transfer the proteins from the SDS-PAGE gel onto a PVDF membrane (GE Healthcare). Note: (1) Presoak sponges, filter paper, PVDF with 1 × transfer buffer

When working on protein transferring procedure with mini blot assembly core contained two sponges, two filter paper, SDS-PAGE protein gel, and PVDF membrane shown in Fig. 6 a roller could be used to gently remove the air bubble with the addition of 1 × transfer.

8. Submerge the mini blot sandwich assembly with transfer buffer in a mini gel tank and run at 20 V for 80 min, at 4 °C.

The assembly order is from anode to cathode, namely, a sponge, filter paper, SDS-PAGE gel, PVDF membrane, filter paper, and sponge.

9. Wash the PVDF membrane with 25 mL 1 × TBS for 5 min at room temperature and block in freshly prepared 5% (m/v) milk (AmericanBio) overnight at 4 °C.
10. Incubate membrane with anti-M2 (FLAG) antibody (F3165, Sigma-Aldrich) in TBST buffer for 1 h at 4 °C with 1:5000 (v/v). Note: the antibody dilution could be optimized based on the experimental conditions. Additionally, batch-to-batch variation in antibody quality should be considered.

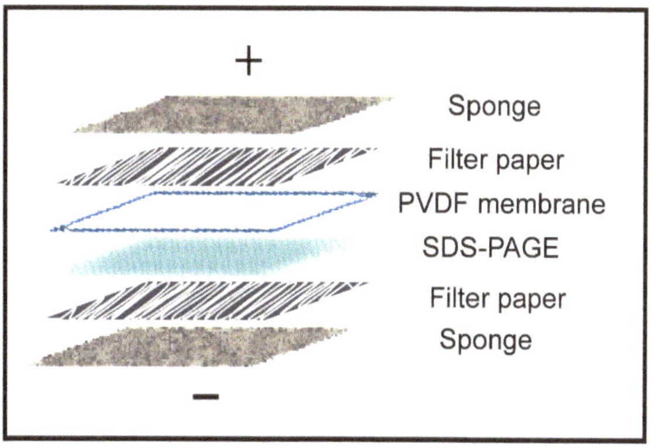

Fig. 6 A diagram of Western blot protein transfer.

11. Wash 3 times for 5 min each in 15 mL of TBST buffer.
12. Incubate anti-rabbit IgG-peroxidase secondary antibody (AP132P, Sigma-Aldrich) at 1:10,000 dilution for 1 h at room temperature.
13. Decant the TBST buffer with anti-rabbit IgG-peroxidase secondary antibody after 1 h incubation.
14. Rinse the PVDF membrane with ~10 mL TBST via rocking for 10 min and decant the TBST buffer. Repeat the same rinse procedure twice.
15. Treat PVDF membranes with an ECL Western blotting kit (Thermo Scientific). Add 1 mL SuperSignal West Femto Stable Peroxide Buffer together with 1 ml of SuperSignal West Femto Lumino/Enhancer Solution, allowing 15 min development.

    Note: The signal duration ranges from 1 to 2 h.
16. Visualize the treated PVDF membranes using a ChemiDoc XRS+ system (Bio-Rad).
17. Visualize the protein ladder using colorimetric imaging, then align the blots to confirm the molecular weight of proteins.

3.3 *C. crescentus* swarm plate assay

Given that pseudokinases commonly lack catalytic functions, phenotypic assays provide an indirect way to evaluate their functions. Here we describe the application of a swarm plate assay as the pseudokinase DivL regulates the CckA-CtrA pathway (Childers et al., 2014; Tsokos & Laub, 2012). The CtrA pathway regulates cell motility, division, and replication, all of which impact swarm size. Thus, the swarm plate assay provides a straightforward way to compare DivL variants (Fig. 7).

Methods

1. Streak out *C. crescentus* strains expressing the DivL-LZ chimeric proteins cell stock with sterile plastic pipette tip onto sterile PYE plate and grow at 28 °C for 2 days. Note: a strain with an empty vector will be used as a control.
2. Prepare the desired number of 0.3% PYE agar plates supplemented with antibiotics and 0.3% xylose inducer in 15 cm diameter Petri plates.
3. Inoculate the *C. crescentus* colonies into PYE medium and allow cultures to grow to mid-log phase at 28 °C overnight.
4. Normalize individual cell cultures to the same OD600 by adding a PYE medium.
5. Load 200 μL of each sample into a 96-well plate. Samples should be spaced out every other well to provide adequate space for the cultures to grow on the swarm plate.

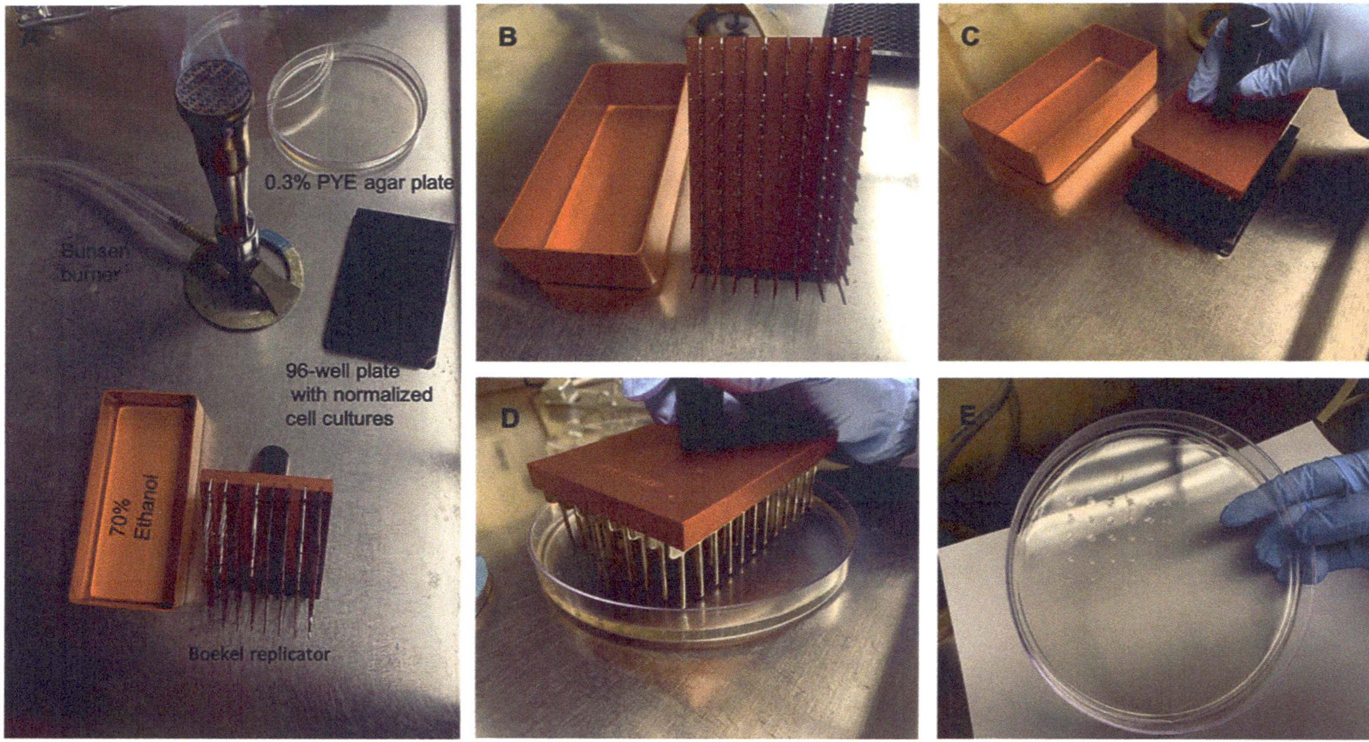

Fig. 7 Workflow for *C. crescentus* Swarm Plate Assay. (A). An overview of swarm assay components including Boekel Scientific replicator with a metal container for sterilizing, 96-well flat clear bottom plate, Bunsen burner, 15 cm perish plate contained 0.3% PYE agar plate. (B). Before inoculation, the Boekel replicator was sterilized with 70% Ethanol. (C). The sterile replicator was gently transferred to a 96-well plate to obtain normalized *C. crescentus* cell cultures. (D). The replicator with corresponding cell cultures was gently stabbed into a 0.3% PYE agar plate. (E). The stabbed 0.3% PYE agar plate with small holes, and the plate was placed upward for incubation.

6. Sterilize the Boekel Scientific replicator 96 pins in ethanol, then flame the pins to remove the residual ethanol and allow the pins to cool in the air near the flame.
7. Submerge the 96 pins of the Boekel Scientific replicator into the 96-well plate containing the *C. crescentus* cultures.
8. Gently inoculate the cultures into the prepared 0.3% PYE agar plates with the appropriate antibiotic, and 0.3% xylose in 15 cm diameter perish plates using the Boekel replicator into triplicates.

   Note: when the PYE agar concentration is above 1% might prohibit swarming as *C. crescentus* swarmer cells do not exhibit swarming motility on solid surfaces (Kearns, 2010). Due to the potential large swarm size, the cell culture replicating could be spaced with sufficient room by dispensing normalized cell cultures to 96 well plates in a staggering arrangement.
9. Incubate swarm plates at 28 °C for 3 days.

   Note: Temperature and humidity should be closely monitored. In addition, the petri dish should be placed with the media on the bottom plate, as the low percent agar is too fragile to incubate inverted.
10. Inspect *C. crescentus* swarms for deviations from circular swarm patterns, as this may indicate a genetic suppressor event during the swarm plate assay.
11. Visualize swarm plates in triplicates by ChemiDoc XRS+ (Bio-Rad).
12. Normalize the swarm areas to the empty vector control for each plate.
13. Calculate the standard deviation of the area, then apply the Dunnett's multiple comparisons test to compare the mean area of each variant strain to the mean area of DivL-WT (ns: $P > 0.05$, $*P \leq 0.05$, $**P \leq 0.01$, $***P \leq 0.001$) (Fig. 8).

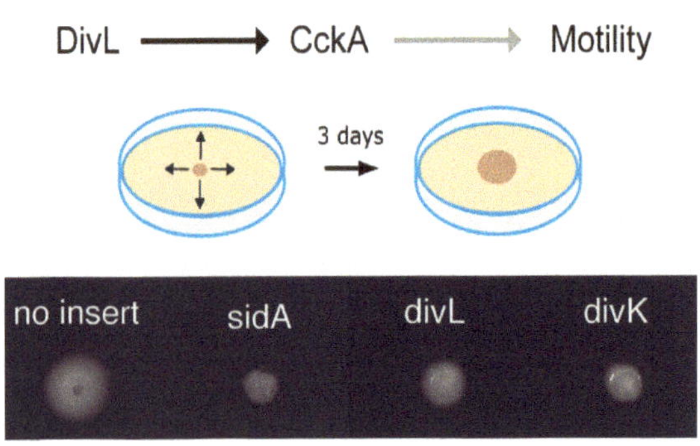

Fig. 8 Application of the swarm plate assay to study the DivL-CckA signaling pathway in *C. crescentus*. Cartoon schematic illustrating the expected swarm size growth over 3 days. Example swarms generated from *C. crescentus* strains that carry an empty vector negative control versus *C. crescentus* strains overexpression the DivL pseudokinase, and sidA and divK as positive controls.

3.4 Protein overexpression and purification

3.4.1 Heterologous protein expression of DivL-LZ chimera in E. coli

1. Transform expression plasmids into the *E. coli* BL21 cells, then plate onto LB agar plates containing appropriate antibiotics and incubate overnight at 37 °C.
2. Pick and inoculate a single colony into a 10 mL sterile LB medium with appropriate antibiotics and grow overnight to saturation.
3. Inoculate 10 mL of saturated overnight cultures into two 1 L LB cultures and grow to mid-log phase between 0.4 and 0.6 (OD600) at 37 °C, 275 rpm.

   Induce the expression of the DivL-LZ fusions proteins with 333 μM isopropyl-β-D-thiogalactopyranoside (IPTG) for 4 h at 25 °C.
4. Centrifuge the overexpressed cells at 3700 × *g*, 4 °C for 20 min, then decant the supernatant and collect the pellet.
5. Resuspend the cell pellet with 50 mL of lysis buffer containing 50 mM HEPES pH 8.0, 500 mM KCl, 1 mM DTT, 25 mM imidazole, 10% glycerol, and 200 U of Benzonase Nuclease (Sigma) supplemented with SIGMAFAST protease inhibitor tablets (S8820,Sigma).
6. Lyse the cell suspension with three passes through the Emulsiflex at 20,000 psi.

   Centrifuge the cell lysate at 30,000 × *g* at 4 °C for 50 min. Collect the supernatant and discard the cell debris.
7. Incubate the combined supernatants with 1 mL of a 50% slurry of HisPur™ Ni-NTA agarose resin was (Catalog number: 88221, Thermo Fisher Scientific) at 4 °C for 1 h.

   Note: the slurry HisPur Ni-NTA volume should be optimized, and too much Ni-NTA will result in high levels of impurities that bind non-specifically to the Ni-NTA resin.
8. Using a low centrifugation speed at 270 × *g* for 5 min, pellet the Ni-NTA agarose containing DivL-LZ chimera, then wash with 30 mL of wash buffer a total of three times (50 mM HEPES KOH pH 8.0, 500 mM KCl, 1 mM DTT, 25 mM imidazole, and 10% glycerol).
9. Elute the DivL-LZ fusion protein from the Ni-NTA agarose resin using 15 mL of elution buffer (50 mM HEPES KOH pH 8.0, 500 mM KCl, 1 mM DTT, 250 mM imidazole, and 10% glycerol).
10. Per the manufacturer's suggested protocol, concentrate the elution fractions via Amicon Centrifugal Filter Units (30 kDa cutoff).
11. Aliquot and freeze the concentrated DivL-LZ Fusion protein samples at −80 °C.

12. Determine the DivL-LZ fusion protein concentration using the predicted molecular weight of 58 kDa and 78,630 M^{-1} cm^{-1}.
13. Confirm protein purity by running an SDS-PAGE gel.

3.5 Coupled enzyme assay

3.5.1 Methods

Coupled enzyme assay is a robust, non-radioactive, continuous assay utilizing kinase or ATPase hydrolysis product ADP accumulation coupled with NADH oxidation reactions to quantify inorganic phosphates at a wavelength of 340 nm real-time and further characterize ATPase or kinase enzymatic kinetics (Fig. 9).

1. Develop the design of your experiment to test the role of the chimeric leucine zipper protein. At a minimum, this should include the reaction buffer with no kinase or pseudokinase, kinase only, pseudokinase only and the kinase and pseudokinase combination. Additionally, select pseudokinase-leucine zipper chimeras or the entire truncation series to test for synthetic stimulation of the pseudokinase. Here we will describe assays that utilize DivL-LZ as the pseudokinase-LZ chimera and CckA as the active kinase.
2. Dilute 10 μL concentrated protein stock solution 200 μM DivL-LZ and 10 μL of 200 μM CckA protein within kinase buffer containing 50 mM HEPES KOH, pH 8.0, 50 mM KCl to the final concentration of 10 μM. Note: the desired final concentration of enzymes range is between 2 and 10 μM.

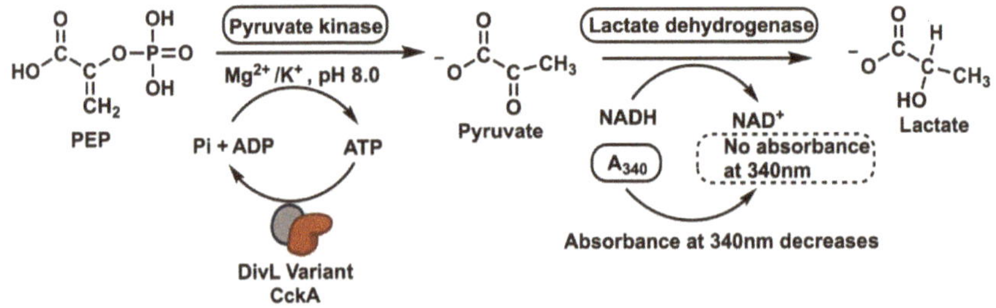

Fig. 9 The Principle of Coupled Enzyme Assay. DivL variant and CckA hydrolyze ATP to ADP and phosphate with the presence of Mg^{2+} at pH 8.0. ADP and PEP (phosphoenolpyruvate) serve as substrates for pyruvate kinase. This enzymatic reaction product pyruvate is further catalyzed by lactate dehydrogenase in the presence of the NADH cofactor to generate lactate. The decreased NADH absorbance can be monitored continuously by a plate reader at a wavelength of 340 nm, 60 s interval for 1 h at room temperature.

3. Supplement the enzymes with 20 μL of 100 mM $MgCl_2$ with final concentration of 10 mM $MgCl_2$, 6 μL of 100 mM phosphoenolpyruvate to make final concentration at 3 mM phosphoenolpyruvate, 2 μL of 20 mM NADH diluting to final concentration of 0.2 mM NADH, 2 units of pyruvate kinase, and 6.6 units of lactate dehydrogenase (P0294, Sigma). The ATP substrate will be added lastly to initiate the reaction. The total enzymatic reaction volume is 200 μL.

   Note: NADH concentration should be low to reduce UV exposure-induced NADH decomposition (Kiianitsa, Solinger, & Heyer, 2003).
4. Run side-by-side control samples without kinases and ATP.

   Note: It is imperative to know ATP background hydrolysis and NADH oxidation to subtract these background rates from the observed ATP hydrolysis rates.
5. Dispense all the reaction and control samples as triplicates into 96 well transparent flat-bottom microplates.

   Note: there might be some air bubbles that occurred while dispensing samples. The loaded plate could be centrifuged at 700 × *g*, 2 min to remove the air bubbles.
6. Initiate kinase reactions upon the addition of 10 μL of 10 mM ATP to make the final concentration at 0.1 mM ATP using a multi-channel pipette to distribute substrates at the same time evenly.

   Analyze the loaded 96-well microplate using a Tecan Infinite 200 Pro plate reader to record absorbance at 340 nm every 60 s for 60 min.

   Note: to reduce UV exposure causing NADH decomposition, plate reader intervals between readings could be extended (Kiianitsa et al., 2003).
7. Calculate the ATP hydrolysis rate by determining the slope of a stable, linear absorbance decay with NADH molar extinction coefficient of $6220 M^{-1} S^{-1}$ (Vo et al., 2017).

   Note: after the pathlength correction, identify a time range that is relatively stable to calculate the rate of change of A_{340}. Subtract the ATP background hydrolysis and NADH oxidation from the observed ATP hydrolysis rates.

$$\text{ATP hydrolysis rate} = \frac{\Delta[\text{ADP}]}{\Delta\text{time}} = \frac{(-1)\,(\Delta[\text{A340}]/\Delta\text{time})}{(\text{pathlength})(6220\ \text{M}^{-1}\text{cm}^{-1})}$$

4. Data analysis

A critical assumption in this experiment is that the high affinity of the GCN4 leucine zipper will alter the conformation and registry of the pseudokinase's associated coiled-coils directly fused to it. Therefore, an incremental truncation series analysis is expected to observe periodic changes in activity that correspond to the periodicity of an alpha helix at approximately 3.6 residues per turn. This observation would support the model that the coiled-coil linker is allosterically coupled to its attached domain.

In contrast, alternative results may suggest different functions of the coiled-coil. For example, patterns with increased periodicity could point to multiple active registries of the alpha helix within the coiled-coil or alternative mechanisms. Suppose no changes are observed within the leucine zipper fusion series. Such results may indicate that the coiled-coil is not allosterically coupled to the attached domain or requires another input for regulation. It may also suggest that the predicted linker's structure adopts a conformation other than the predicted coiled-coil, such as a disordered linker. In contrast, if changes in activity occur for every residue beyond the truncation of a specific residue, it may point toward alternative models such as the removal of a binding site within the coiled-coil.

Upon identifying functional and non-functional leucine zipper chimera, generate helical wheel diagrams to interrogate the results. In these helical wheels, models consider the following two criteria. First, is the positioning of hydrophobic, charged, and polar residues consistent with coiled-coil structure? Second, what is the position of highly conserved residues? For example, designs that place charged residues at the a and d leucine-rich interface may be conformations less likely to occur in the wild-type protein. In contrast, observations of hydrophobic leucines and valines at the a and d positions are expected. Secondly, one may expect conserved residues to be positioned at the hydrophobic interface in the a and d positions or residues that make salt-bridge interactions at the g and e positions. In this manner, conservation would point toward conserved roles in maintaining coiled-coil structure.

5. Summary

This chapter discussed a protein engineering experimental strategy to interrogate the mechanism of action of bacterial pseudokinases. Three functions should be considered: response regulator binding, ligand or protein binding to the N-terminal sensor, and sensor-kinase allostery. If a coiled-coil linker is conserved between the sensor and pseudokinase domains, one should consider mechanisms in which the pseudokinase functions as a protein switch (Figs. 1–3). For example. If typical "forward" signaling is conserved, the N-terminal sensory domain may regulate pseudokinase domain functions (e.g., binding to a response regulator or another active kinase). To test forward signaling the N-terminal sensory domain can be swapped for a high-affinity leucine zipper. Fusion of the leucine zipper then locks the pseudokinase domain into one well-defined conformational state. In contrast, in reverse signaling, the binding of the response regulator may regulate sensory domain functions. To test reverse signaling, the C-terminal pseudokinase can be swapped for a high-affinity leucine zipper. Using this approach, variations in pseudokinase function should vary with the expected periodicity of a coiled-coil of 3.6 residues per turn. Implementation of this protocol will reveal the functional significance of coiled coils attached to pseudokinases and reveal if they function as protein switches.

Acknowledgment

This work was supported by start-up funds to W.S.C. from the University of Pittsburgh.

Reference

Ashenberg, O., Keating, A. E., & Laub, M. T. (2013). Helix bundle loops determine whether histidine kinases autophosphorylate in cis or in trans. *Journal of Molecular Biology*, *425*(7), 1198–1209.

Bhate, M. P., Molnar, K. S., Goulian, M., & DeGrado, W. F. (2015). Signal transduction in histidine kinases: Insights from new structures. *Structure*, *23*(6), 981–994.

Chen, Y. E., Tsokos, C. G., Biondi, E. G., Perchuk, B. S., & Laub, M. T. (2009). Dynamics of two phosphorelays controlling cell cycle progression in *Caulobacter crescentus*. *Journal of Bacteriology*, *191*(24), 7417–7429.

Childers, W. S., Xu, Q., Mann, T. H., Mathews, I. I., Blair, J. A., Deacon, A. M., et al. (2014). Cell fate regulation governed by a repurposed bacterial histidine kinase. *PLoS Biology*, *12*(10), e1001979.

Christen, B., Fero, M. J., Hillson, N. J., Bowman, G., Hong, S. H., Shapiro, L., et al. (2010). High-throughput identification of protein localization dependency networks. *Proceedings of the National Academy of Sciences of the United States of America*, *107*(10), 4681–4686.

Collins, M. J., & Childers, W. S. (2021). The upcycled roles of pseudoenzymes in two-component signal transduction. *Current Opinion in Microbiology*, *61*, 82–90.

Eyers, P. A., & Murphy, J. M. (2016). The evolving world of pseudoenzymes: Proteins, prejudice and zombies. *BMC Biology*, *14*(1), 98.

Gao, R., Bouillet, S., & Stock, A. M. (2019). Structural basis of response regulator function. *Annual Review of Microbiology*, *73*, 175–197.

Gao, R., & Lynn, D. G. (2005). Environmental pH sensing: Resolving the VirA/VirG two-component system inputs for Agrobacterium pathogenesis. *Journal of Bacteriology*, *187*(6), 2182–2189.

Gao, R., & Stock, A. M. (2009). Biological insights from structures of two-component proteins. *Annual Review of Microbiology, 63*(1), 133–154.
Gibson, D. G., Young, L., Chuang, R.-Y., Venter, J. C., Hutchison, C. A., & Smith, H. O. (2009). Enzymatic assembly of DNA molecules up to several hundred kilobases. *Nature Methods, 6*(5), 343–345.
Hillson, N. J., Rosengarten, R. D., & Keasling, J. D. (2012). j5 DNA assembly design automation software. *ACS Synthetic Biology, 1*(1), 14–21.
Huynh, T. N., & Stewart, V. (2011). Negative control in two-component signal transduction by transmitter phosphatase activity. *Molecular Microbiology, 82*(2), 275–286.
Kearns, D. B. (2010). A field guide to bacterial swarming motility. *Nature Reviews Microbiology, 8*(9), 634–644.
Kiianitsa, K., Solinger, J. A., & Heyer, W.-D. (2003). NADH-coupled microplate photometric assay for kinetic studies of ATP-hydrolyzing enzymes with low and high specific activities. *Analytical Biochemistry, 321*(2), 266–271.
Kowallis, K. A., Silfani, E. M., Kasumu, A., Rong, G., So, V., & Childers, W. S. (2020). Synthetic control of signal flow within a bacterial multi-kinase network. *ACS Synthetic Biology, 9*(7), 1705–1713.
Lin, Y. H., Gao, R., Binns, A. N., & Lynn, D. G. (2008). Capturing the VirA/VirG TCS of agrobacterium tumefaciens. In *Vol. 631. Bacterial signal transduction: Networks and drug targets* (pp. 161–177). New York: Springer.
Lin, Y.-H., Pierce, B. D., Fang, F., Wise, A., Binns, A. N., & Lynn, D. G. (2014). Role of the VirA histidine autokinase of *Agrobacterium tumefaciens* in the initial steps of pathogenesis. *Frontiers in Plant Science, 5*(195).
Murphy, J. M., Mace, P. D., & Eyers, P. A. (2017). Live and let die: Insights into pseudoenzyme mechanisms from structure. *Current Opinion in Structural Biology, 47*, 95–104.
Neiditch, M. B., Federle, M. J., Pompeani, A. J., Kelly, R. C., Swem, D. L., Jeffrey, P. D., et al. (2006). Ligand-induced asymmetry in histidine sensor kinase complex regulates quorum sensing. *Cell, 126*(6), 1095–1108.
Ribeiro, A. J. M., Das, S., Dawson, N., Zaru, R., Orchard, S., Thornton, J. M., et al. (2019). Emerging concepts in pseudoenzyme classification, evolution, and signaling. *Science Signaling, 12*(594), eaat9797.
Stabel, R., Stüven, B., Ohlendorf, R., & Möglich, A. (2017). Primer-aided truncation for the creation of hybrid proteins. In V. Stein (Ed.), *Synthetic protein switches: Methods and protocols* (pp. 287–304). New York, NY: Springer New York.
Tsokos, C. G., & Laub, M. T. (2012). Polarity and cell fate asymmetry in *Caulobacter crescentus*. *Current Opinion in Microbiology, 15*(6), 744–750.
Tsokos, C. G., Perchuk, B. S., & Laub, M. T. (2011). A dynamic complex of signaling proteins uses polar localization to regulate cell-fate asymmetry in *Caulobacter crescentus*. *Developmental Cell, 20*(3), 329–341.
Vo, C. D., Shebert, H. L., Zikovich, S., Dryer, R. A., Huang, T. P., Moran, L. J., et al. (2017). Repurposing Hsp90 inhibitors as antibiotics targeting histidine kinases. *Bioorganic & Medicinal Chemistry Letters, 27*(23), 5235–5244.
Wang, Y., Gao, R., & Lynn, D. G. (2002). Ratcheting up vir gene expression in *Agrobacterium tumefaciens*: Coiled coils in histidine kinase signal transduction. *Chembiochem, 3*(4), 311–317.
Wang, B., Zhao, A., Novick, R. P., & Muir, T. W. (2014). Activation and inhibition of the receptor histidine kinase AgrC occurs through opposite helical transduction motions. *Molecular Cell, 53*(6), 929–940.
Wu, J. G., Ohta, N., Zhao, J. L., & Newton, A. (1999). A novel bacterial tyrosine kinase essential for cell division and differentiation. *Proceedings of the National Academy of Sciences of the United States of America, 96*(23), 13068–13073.

CHAPTER ELEVEN

Dynamics of protein kinases and pseudokinases by HDX-MS

Joshua B. Sheetz, Mark A. Lemmon*, and Yuko Tsutsui*
Department of Pharmacology and Yale Cancer Biology Institute, Yale University School of Medicine, West Haven, CT, United States
*Corresponding authors: e-mail address: mark.lemmon@yale.edu; yuko.tsutsui@yale.edu

Contents

Abstract

Dynamics of the protein kinase fold are deeply intertwined with its structure. The past three decades of kinase biophysical studies revealed key dynamic features of the kinase domain and, more recently, how these features may endow catalytically impaired kinases—or pseudokinases—with signaling properties. Hydrogen-deuterium exchange coupled with mass spectrometry (HDX-MS) is proving to be a valuable approach for studies of kinase and pseudokinase domain dynamics. Here, we briefly discuss the methods that have provided insights into protein kinase dynamics, describe how HDX-MS is being used to answer questions in the kinase/pseudokinase field, and provide a detailed protocol for collecting an HDX-MS dataset to study the impacts of small molecule binding to a pseudokinase domain. As more small molecules are discovered that can disrupt pseudokinase conformations, HDX-MS is likely to be a powerful approach for exploring drug-induced changes in pseudokinase dynamics and structure.

Methods in Enzymology, Volume 667
ISSN 0076-6879
https://doi.org/10.1016/bs.mie.2022.03.037

1. Introduction

In eukaryotes, the protein kinase fold is remarkably well conserved and yet wonderfully diverse (Manning, Plowman, Hunter, & Sudarsanam, 2002). From vastly different substrate specificities (Miller & Turk, 2018), to distinct regulatory mechanisms (Endicott, Noble, & Johnson, 2012), to non-catalytic signaling modes (Kung & Jura, 2016), the kinase domain is a molecular marvel of signal transduction. Since the first structural elucidation of a protein kinase domain over three decades ago (Knighton et al., 1991), a key question has been: *how does the fold achieve such broad biological functions*? To understand how this common architecture is repurposed in cell signaling with altered specificity and functionality, the field has turned to a variety of biophysical techniques (Endicott et al., 2012; Lorenzen & Pawson, 2014; Xiao, Liddle, Pardi, & Ahn, 2015). The answer appears to partially lie in the dynamic nature of the domain.

The first hints of kinase conformational plasticity came from static snapshots of kinase domains (Huse & Kuriyan, 2002; Johnson, Noble, & Owen, 1996). X-ray crystal structures of the cAMP-dependent protein kinase, or PKA, catalytic subunit revealed a canonical protein kinase domain that is bi-lobal, comprising a primarily β-stranded N-lobe and a largely α-helical C-lobe (Knighton et al., 1991). This architecture allows ATP, divalent metal ions, and peptide substrates (containing serine, threonine, or tyrosine) to dock precisely between the two lobes in an orientation suitable for the phosphotransferase reaction (Adams, 2001). To allow for efficient association and dissociation of substrates, the N-lobe is relatively dynamic (Kornev & Taylor, 2015). Comparing the crystal structure of a cyclin dependent kinase 2 (CDK2) monomer with a breakthrough structure of cyclin A-bound CDK2 first demonstrated that a single helix within the N-lobe—helix αC—can adopt a range of conformations, forming a critical salt bridge in the inward, active state (De Bondt et al., 1993; Jeffrey et al., 1995). In addition, flexibility of the glycine-rich loop above the ATP-binding site is crucial for facilitating nucleotide binding and catalysis, as was visualized in early structures of PKA and of phosphorylase kinase bound to different nucleotides (Narayana, Cox, Nguyen-huu, Ten Eyck, & Taylor, 1997; Owen, Noble, Garman, Papageorgiou, & Johnson, 1995). A third important feature found in nearly all protein kinases is the 'activation loop' region, the conformational landscape of which is frequently controlled by post-translational modification (such as phosphorylation). This was seen vividly in structural comparisons of the inactive (unphosphorylated) and active (phosphorylated) forms of the insulin receptor tyrosine kinase domain(IRK) (Hubbard, 1997; Hubbard, Wei, Ellis, & Hendrickson, 1994). The dynamics of these three regions, highlighted in Fig. 1, work in concert with the rest of the domain to facilitate catalysis, but they are also exploited by numerous regulatory mechanisms (Huse & Kuriyan, 2002). Many kinases are controlled via the αC helix, which can be allosterically toggled inward and outward by intra- and/or intermolecular interactions (*e.g.*, cyclin binding to CDKs) and/or the flexible autoinhibitory activation loop, which is often phosphorylated to relieve pseudosubstrate autoinhibition (*e.g.*, IRK).

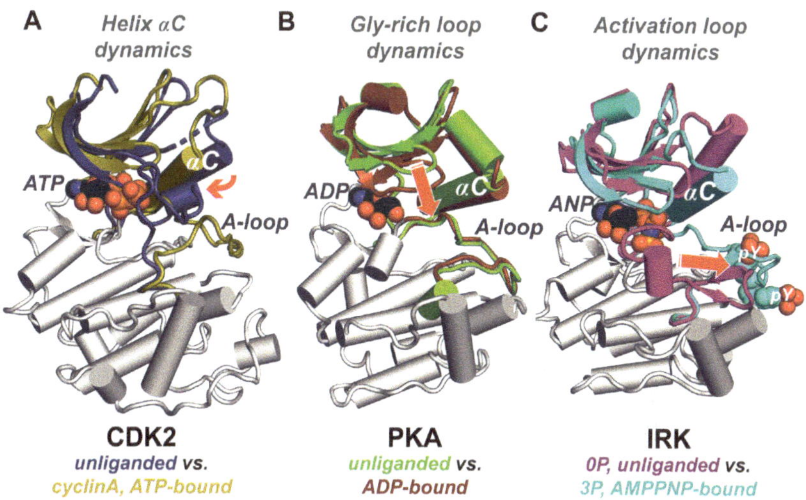

Fig. 1 Dynamic features of the eukaryotic protein kinase domain revealed by early crystal structures. (A) Overlay of inactive, unliganded CDK2 (dark blue) and active, cyclin A- and ATP-bound CDK2 (gold) illustrates movement of the αC helix (see red arrow), as well as the activation- and glycine-rich loops, upon allosteric activation of CDK2 by cyclin A. (B) Overlay of the PKA catalytic domain in its unliganded state (green) and ADP-bound state (red), revealing closure of the glycine-rich loop upon nucleotide binding (as depicted by the red arrow). (C) Overlay of inactive (unphosphorylated) and unliganded IRK (purple) and active (phosphorylated) AMPPNP-bound IRK (cyan), illustrating rearrangement of the activation loop (A-loop) as depicted by the red arrow, as well as closure of the αC helix and glycine-rich loops upon phosphorylation and nucleotide binding. In each case, the kinase C-lobes are colored white. PDB IDs: 1HCL (Schulze-Gahmen, De Bondt, & Kim, 1996), 1FIN (Jeffrey et al., 1995), 1CMK (Zheng et al., 1993), 1BKX (Narayana et al., 1997), 1IRK (Hubbard et al., 1994), 1IR3 (Hubbard, 1997).

Although crystal structures have provided invaluable clues into kinase mechanisms, many questions of how protein motions achieve catalysis still remained (Xiao et al., 2015). After the initial set of protein kinase structures in the 1990s, a battery of in-solution biophysical methods advanced our understanding of kinase dynamics in the decade that followed. Studies using nuclear magnetic resonance (NMR) or hydrogen-deuterium exchange coupled to mass spectrometry (HDX-MS) provided global insights into kinase dynamics (Lorenzen & Pawson, 2014; Xiao et al., 2015). The use of two-dimensional NMR to study PKA, for instance, revealed that the binding of nucleotide to this kinase domain results in large perturbations that propagate to the peptide-binding site of the enzyme (Masterson, Mascioni, Traaseth, Taylor, & Veglia, 2008). Early HDX-MS studies of extracellular regulated protein kinase 2 (ERK2) shed light on the changes in conformational dynamics that occur upon phosphorylation of the activation loop, demonstrating how activation loop movements allosterically impact other key regions (Hoofnagle, Resing, Goldsmith, & Ahn, 2001). In more recent years, advances in NMR instrumentation have allowed detailed observations of protein kinase energy landscapes and inactive/active state transitions (Hadzipasic et al., 2020; Pitsawong et al., 2018; Xie, Saleh, Rossi, & Kalodimos, 2020). Progress with HDX-MS is also providing key insights into dynamics of ligand binding and complex formation (described below). In addition, spectroscopic methods, such as infrared (IR) spectroscopy and electron paramagnetic resonance (EPR) spectroscopy, employed for a more focused analysis of particular regions within the kinase have enabled more detailed assessment of the dynamics of the activation loop of kinases like Aurora A (Ruff et al., 2018). Given all of these approaches, along with advances in computational modeling, members of the kinase community now have many powerful tools at their fingertips to investigate and understand dynamics.

We focus here on HDX-MS as an approach that is well-equipped to address a variety of questions that are particularly crucial for pseudokinases—where phosphotransfer has been dispensed with, but dynamic conformational transitions provide the primary window on function (Kung & Jura, 2019). HDX-MS reveals information on intrinsic dynamics as well as local conformational changes induced upon perturbation, making it complementary to other in-solution methods, as well as to structure determination approaches like crystallography and cryo-electron microscopy (Engen, 2009; Engen & Komives, 2020; Hoofnagle, Resing, & Ahn, 2003). Although NMR is currently more capable of providing residue-specific information on conformation states at fast time scales, the information obtained is generally limited to residues that can be labeled and detected in the NMR experiment. HDX-MS, by contrast, can measure protein movements on a wide range of time scales and is amenable to large proteins and protein complexes without the need for isotopic labeling during cell culture or access to strong, expensive magnets (Peacock & Komives, 2021). Below, we present an overview of HDX-MS, followed by some discussion of how the technique has provided valuable insights into the dynamics of the kinase fold and its regulation. We then review how HDX-MS has helped resolve questions about how the catalytically impaired pseudokinases may function by virtue of dynamic conformational changes related to those that normally control the activation state of a *bona fide* protein kinase.

2. Overview of HDX-MS

HDX-MS has emerged in recent years as a core method in structural biology. The principle behind the method is the propensity of labile protons bound to heteroatoms within a polypeptide chain to exchange with protons present in aqueous solvent (Englander, Mayne, Bai, & Sosnick, 1997; Hvidt & Linderstrom-Lang, 1954, 1955). Protons within simple, unstructured peptides exchange rapidly, whereas amide protons engaged in hydrogen bonds (or otherwise shielded from solvent) do so at a measurably slower rate. Biophysicists can take advantage of this property to measure relative exchange rates for local regions within a protein, thereby providing insight into the global conformational dynamics of a given protein or protein complex in a region-specific manner (Engen, 2009). In practice, this is achieved by replacing the $^{1}H_2O$-based solvent with a $^{2}H_2O$ (deuterium oxide, herein referred to as D_2O)-based solvent via dilution, allowing for isotopic exchange of protium atoms with deuterium. The presence of these deuterated amide protons is then monitored over time using a sensitive method like mass spectrometry.

To prevent deuterated amide protons from undergoing back-exchange when in contact with excess H_2O-based buffer (such as during chromatographic separation), the reaction must be quenched by lowering the pH and temperature. Rosa and Richards demonstrated that acidic conditions are optimal for minimizing the rate of amide hydrogen exchange ($pH_{min} \sim 2.5$), and are thus often used, along with low temperatures (*e.g.*, close to 0 °C), as quench conditions in standard HDX-MS workflows today (Jensen & Rand, 2016; Rosa & Richards, 1979). Although hydrogens bound to heteroatoms in peptide side chains or at N- and C-termini also undergo rapid exchange, their much higher pH_{min} values result in much higher back exchange rates compared to amide protons, allowing standard HDX-MS to specifically probe protein backbone exchange dynamics. That is, with the exception of proline residues, which lack an amide proton. For more details on fundamental hydrogen exchange theory, we refer the reader to a helpful review (Jensen & Rand, 2016).

Whereas unstructured peptides exchange at millisecond-second time scales, peptides engaged in higher order structure can exchange on significantly slower time scales due to hydrogen bonding and solvent inaccessibility. A general HDX reaction mechanism, derived from the Linderstrøm-Lang model of peptide exchange kinetics, can be expressed:

$$N-H_{cl} \underset{k_{cl}}{\overset{k_{op}}{\rightleftharpoons}} N-H_{op} \underset{k_{int}}{\overset{D_2O}{\rightarrow}} N-D_{op} \underset{k_{op}}{\overset{k_{cl}}{\rightleftharpoons}} N-D_{cl}$$

where the observed rate of exchange is dependent upon the intrinsic exchange rate (k_{int}), or the rate at which an amide proton N–H is converted into N–D, and the equilibrium of the proton between open (H_{op}) and closed (H_{cl}) states, given that it must be in the open state to undergo exchange (Hoofnagle et al., 2003; Jensen & Rand, 2016). As protein refolding is much faster than deuterium exchange (i.e. $k_{cl} >> k_{int}$) under most conditions for structured proteins, a so-called EX2 kinetic exchange regime with a single transition from the undeuterated state to the fully deuterated state is observed, where the observed exchange rate (k_{ex}) can be expressed:

$$k_{ex} = \frac{k_{op}}{k_{cl}} * k_{int}$$

An alternative regime, called EX1, is possible for cases with multiple conformation states present when amide proton exchange is faster than refolding (i.e. $k_{cl} << k_{int}$), and thus $k_{ex} = k_{op}$. For globular proteins like protein kinase domains in their native state, peptides largely follow the EX2 kinetic regime, making HDX-MS a useful measure of differential solvent accessibility within local regions of the protein.

A typical experimental workflow for HDX-MS is quite simple (Fig. 2), although high-resolution LC-MS instrumentation (resolution at least 15,000 full width at half maximum) is generally preferred. The experiment begins with a purified protein sample; for example, we generally aim for a

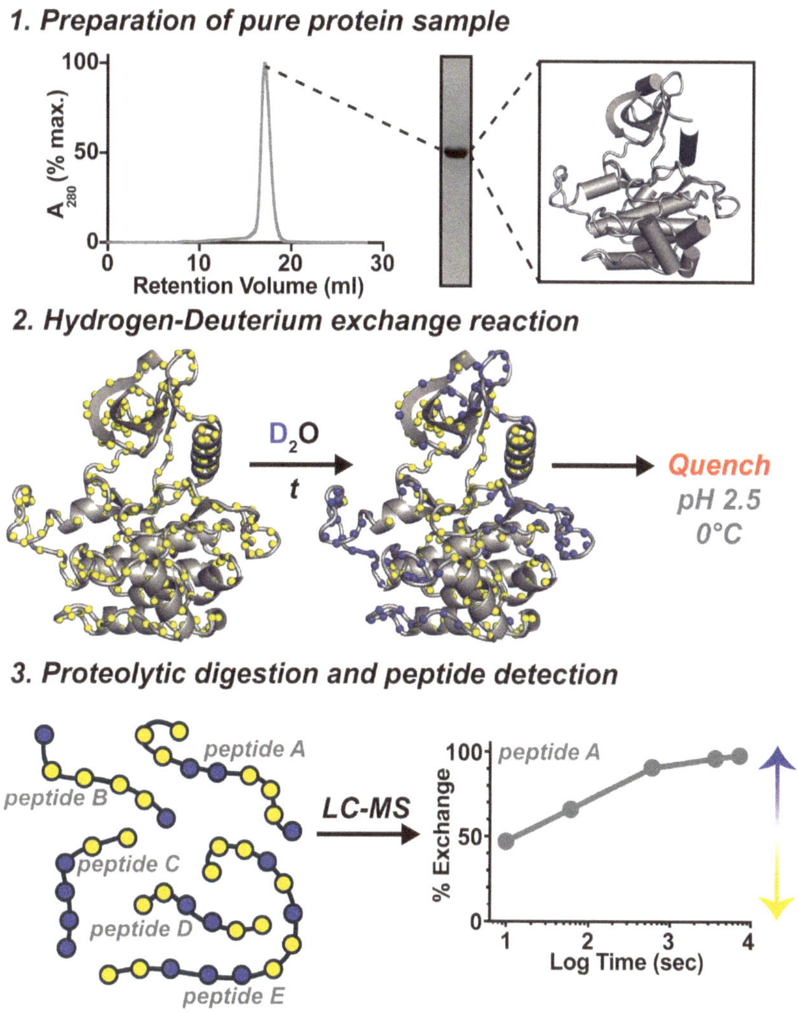

Fig. 2 Overview of a typical HDX-MS workflow. The protein sample is first purified and its purity is assessed using standard approaches such as gel filtration chromatography and SDS-PAGE (discussed in Section 5.1). Hydrogen-deuterium exchange samples are then prepared by diluting the protein sample into D_2O-based buffer for a given time, *t*, and the reaction is quenched at low pH and low temperature (discussed in Section 5.2). During this reaction, amide protons (shown in yellow) exchange with deuterium (shown in purple) at different rates that depend on their structural context. The deuterated samples are then digested using an acid-stable protease such as pepsin, and the resulting peptides are separated and analyzed by liquid chromatography-mass spectrometry (LC-MS) (discussed in Section 5.3). The deuterium uptake levels for a given peptide can then be presented as a function of time in a classic uptake plot shown at bottom right, or by other useful data representations that include structural context (discussed in Section 6).

protein concentration ranging from 0.3 mg/mL to 0.6 mg/mL (thus injecting 1–3 μg protein) with >95% estimated purity by Coomassie-stained SDS-PAGE, although the amount of required protein will depend on instrumentation. The sample is diluted 10- to 20-fold into D_2O-based buffer (25 °C, pD 7.4) for a given time, followed by 1:1 dilution into cold H_2O-based quench buffer at low pH (4 °C, pH 2.3) to minimize amide proton back-exchange. Typically, protein buffers with minimal temperature sensitivity are preferred, such as HEPES or sodium phosphate buffers, and buffer concentration should be at least 10 times lower than that of quench buffer so that pH will be lowered to the acidic pH upon quenching. Samples are then rapidly digested using an acid-stable protease such as pepsin, followed by a short (~7 min) linear acetonitrile gradient for peptide separation and subsequent mass spectrometry analysis for peptide detection and identification. The deuterium uptake profiles of individual peptides are then compared across various labeling time points to elucidate local exchange behavior. Datasets can be presented in a number of ways to connect region-specific deuterium uptake values to protein structure. In a later section, we provide a detailed protocol for collecting an HDX-MS dataset, as well as suggestions for presenting HDX-MS data.

3. HDX-MS to probe dynamics of the kinase fold

HDX-MS had an early impact on the kinase field (Lorenzen & Pawson, 2014). The laboratory of Natalie Ahn paved the way with HDX-MS studies of ERK2 that shed light on the effects of phosphorylation (as mentioned above) and nucleotide binding (Hoofnagle et al., 2001; Lee, Hoofnagle, Resing, & Ahn, 2005). While binding of the nucleotide probe AMP-PNP impacted the N-lobe of ERK2 similarly whether the activation loop was phosphorylated or not, a key regulatory region known as the DFG motif was specifically protected from exchange upon AMP-PNP binding only to the phosphorylated form of ERK2 (Lee et al., 2005). This result suggested that ERK2 activation loop phosphorylation and nucleotide binding cooperate to properly orient the active site (which contains the DFG motif) for catalysis. In a separate study, this group further utilized HDX-MS to discover the location of a docking site within the ERK2 kinase domain that can be bound by regions of substrates distal to the site of phosphorylation (Lee et al., 2004). These docking sites are key determinants of substrate specificity in mitogen-activated protein kinases (Miller & Turk, 2018).

More recent studies have benefitted from the advances in both LC-MS instrumentation and data analysis software to enable greater peptide coverage and more complex ligand binding studies. For example, a recent set of studies utilized HDX-MS to uncover regulatory inter-domain interactions within the kinase AKT1. AKT1 has an N-terminal pleckstrin homology (PH) domain that is known to bind phosphatidylinositol-3,4,5-trisphosphate (PtdIns(3,4,5)P_3) in the plasma membrane to promote AKT1 kinase activity (Siess & Leonard, 2019). HDX-MS experiments of full length AKT1 revealed that its binding to liposomes containing PtdIns(3,4,5)P_3 retained protection of the PH domain from exchange, but increased exchange in the C-terminal kinase domain, supporting a model of auto-inhibitory intramolecular PH domain/kinase domain interactions that are disrupted when the PH domain binds phosphoinositides (Lucic et al., 2018). Subsequent crystallographic studies showed that the PH domain docks onto the C-lobe of the kinase domain in a manner reminiscent of its engagement with PtdIns(3,4,5)P_3 (Truebestein et al., 2021). These studies nicely illustrate both the capability of HDX-MS to address or refine structural hypotheses and the power of combining HDX-MS with complementary approaches.

In its current state, HDX-MS is most useful for qualitative biophysical inquiry, making it suitable for comparative studies across different peptide-, ligand-, drug-, or lipid-bound conditions or across differently mutated protein variants. In addition, HDX-MS is exceptionally well suited for identifying intrinsically disordered regions within proteins. This can be particularly valuable in designing constructs for protein crystallization, for example, employing HDX-MS to identify unstructured regions likely to be a barrier to crystal packing (Fowler et al., 2016). In the coming years, we predict that the application of HDX-MS will increasingly impact our understanding of drug binding on kinase domain dynamics. Although HDX-MS is already an important component in drug discovery (Masson, Jenkins, & Burke, 2017), it will be a key technique for understanding how current FDA-approved kinase inhibitors (such as the numerous and structurally diverse inhibitors of the epidermal growth factor receptor, or EGFR) bias kinase conformational dynamics in a way that could modulate kinase non-catalytic functions (Kung & Jura, 2016, 2019). Indeed, the Ahn lab recently used HDX-MS to reveal differences in activation loop dynamics upon binding to ERK2 of two inhibitors that induce distinct conformations (Pegram et al., 2019). As chromatography, mass spectrometry, and HDX-MS data processing systems all improve, we also predict more insight into allostery within larger macromolecular complexes. For example, the binding site of the stress response kinase GCN2 (general control nonderepressible 2) within purified ribosomes was mapped to the ribosomal P-stalk region, and GCN2 was shown to be allosterically activated by this interaction (Inglis et al., 2019) in a *tour de force* study enabled by HDX-MS. Despite these examples, studies of the protein kinases within higher order complexes have largely lagged behind those of the lipid kinases, which includes impressive work by the Burke, Hurley, and Williams laboratories, and from which useful insights give us a glimpse of what the protein kinase field can expect in the future (Burke et al., 2011; Dornan et al., 2020; Rostislavleva et al., 2015; Vadas & Burke, 2015; Young, Cho, Lawrence, Zoncu, & Hurley, 2016).

4. HDX-MS as a tool to study pseudokinase dynamics

HDX-MS has only recently been applied to studies of pseudokinases. Lacking key residues thought necessary for catalytic activity, pseudokinases were initially considered remnants of evolution that may serve primarily as scaffolds for assembly of signaling complexes (Zeqiraj & van Aalten, 2010). As the field has progressed, more dynamic roles for pseudokinases have been discovered, including roles as signaling switches and allosteric activators (Boudeau, Miranda-Saavedra, Barton, & Alessi, 2006; Mace & Murphy, 2021). Key to this shift in perspective has been the appreciation that pseudokinases are not simply rigid molecules as the term scaffold implies, but are just as dynamic as their catalytically active relatives. Two key studies highlight the power of HDX-MS for investigating the conformational dynamics of pseudokinase domains and for helping to settle the debate over rigidity versus dynamics in pseudokinases.

In one study on the highly divergent mixed lineage kinase domain-like (MLKL) pseudokinase, Petrie et al. used HDX-MS to understand its "molecular switch"-like function (Petrie et al., 2018). MLKL is the terminal effector protein in the necroptosis pathway that mediates pro-inflammatory cell death (Murphy, 2020). MLKL contains an N-terminal four-helix bundle (4HB) domain that is thought to engage lipids and permeabilize the plasma membrane to execute the final step in necroptosis (Petrie, Czabotar, & Murphy, 2019). Release/exposure of the 4HB domain and MLKL oligomerization (also crucial) are tightly regulated by the C-terminal pseudokinase domain of MLKL. In several structural and functional studies, Murphy and colleagues have discovered key steps in this process, including the importance of conformational switching of the pseudokinase domain

(Murphy, 2020; Murphy et al., 2013; Petrie et al., 2018). Although MLKL lacks kinase activity, its conserved ability to bind ATP appears to be a crucial feature for MLKL tetramerization, which is required for necroptosis. In an effort to understand allosteric communication between the 4HB and pseudokinase domains of MLKL, Petrie et al. turned to HDX-MS. One key experiment examined the impact of ATP binding to a monomeric variant of MLKL and revealed allosteric effects on both helix αC and the adjacent region of the 4HB domain—both of which increased in hydrogen exchange in the presence of ATP, suggesting inter-domain communication resulting from ATP-site ligation. HDX-MS studies like those of Petrie et al. can help illuminate how pseudokinase domains contribute in larger signaling complexes and how conformational dynamics, likely influenced by nucleotide binding, direct and control pseudokinase domain signaling.

Not all pseudokinases have retained the ability to bind ATP (Murphy et al., 2014). In our own work, we leveraged HDX-MS to address a longstanding question of whether pseudokinases incapable of ATP binding are rigid or dynamic molecules (Sheetz et al., 2020). An unusual group of receptor tyrosine kinases (RTKs) with intracellular pseudokinase domains, namely ROR1, ROR2, RYK, and PTK7, all play important roles in WNT signaling. Through X-ray crystallography, we showed that the pseudokinase domains of these receptors all adopt an inactive conformation highly reminiscent of IRK, with their ATP-sites fully occluded by side-chains (Artim, Mendrola, & Lemmon, 2012; Mendrola, Shi, Park, & Lemmon, 2013; Sheetz et al., 2020). Given suggestions in the field that pseudokinases that do not bind ATP may be 'locked' in a more rigid conformation state (Kornev & Taylor, 2009; Patel et al., 2017; Scheeff, Eswaran, Bunkoczi, Knapp, & Manning, 2009) to function as rigid scaffolds, we turned to HDX-MS to compare the intrinsic dynamics of these four pseudokinases with those of the *bona fide* kinase IRK. We found that the intrinsic dynamics of the RTK pseudokinase domains are largely similar to those of IRK, particularly in functionally dynamic regions such as the activation loop and the signature αC helix (Fig. 3A). To understand whether the conformation of the pseudokinase domains can be modulated though binding of small molecules—which might be useful in exploiting them as therapeutic targets—we screened for compounds that bind these four pseudokinases. HDX-MS served as a valuable tool in determining that small molecules that bind one of these pseudokinases (ROR1) can disrupt the conformation, impacting the dynamics of the ROR1 pseudokinase in a manner consistent with a transition towards a more active-like state. In the next section, we

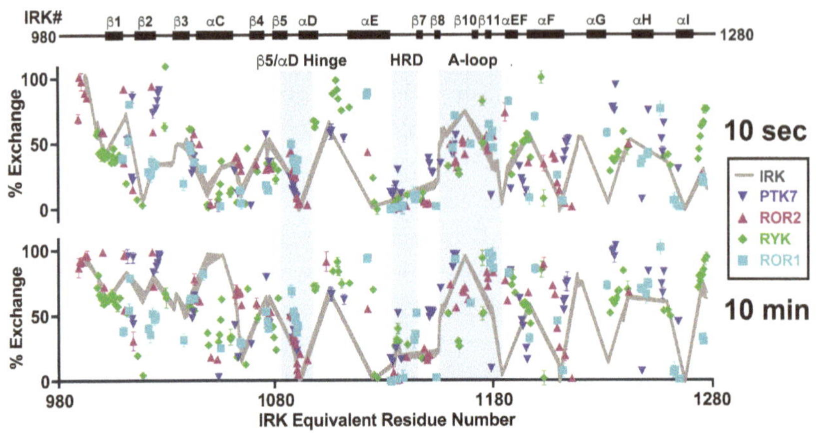

Fig. 3 HDX-MS comparisons of IRK with the pseudokinase domains of ROR1, ROR2, RYK, and PTK7 demonstrate that IRK-like dynamics are largely conserved in pseudokinases that do not bind nucleotide (Sheetz et al., 2020). HDX data (mean ± SD for three independent labeling experiments) at 10 s and 10 min are shown for the PTK7 (slate blue), ROR2 (magenta), RYK (green), and ROR1 (cyan) pseudokinase domains. Data for IRK (gray) are depicted as the range for each point. The x axis represents IRK-equivalent median residue number. The % exchange value for the pseudokinase domains rises and falls across the sequence with a similar pattern to that seen in IRK, suggesting similar conformational dynamics (Sheetz et al., 2020).

present a detailed protocol for collecting an HDX-MS dataset to study drug-induced changes in pseudokinase dynamics, using our ROR1 experiments as an example.

5. Protocol: Bottom-up HDX-MS to probe ligand-induced conformational dynamics of a pseudokinase domain

5.1 Purification of the ROR1 pseudokinase domain from insect cells

5.1.1 Equipment

1. Standard static incubator for insect cell culture, maintained at 27 °C (*e.g.* Fisherbrand Isotemp BOD Refrigerated Incubator, 566 L, Porcelain Steel, Fisher Scientific, catalog # 3720A)
2. Standard shaking incubators for insect cell culture, maintained at 27 °C (*e.g.* Multitron, INFORS-HT)
3. Standard shaking bacterial incubator set to 37 °C (*e.g.* Eppendorf 1-24 Incubating Shaker, catalog # 14-285-852)

4. Standard refrigerated tabletop centrifuge (*e.g.* Sorvall ST 40R, ThermoFisher Scientific, catalog # 75-004-525, with a BIOLiner swinging bucket rotor, ThermoFisher Scientific, catalog # 75-003-667)
5. Standard refrigerated benchtop microcentrifuge (*e.g.* Eppendorf, catalog # 5406000640)
6. Standard refrigerated high speed floor centrifuge for large volumes (*e.g.* Avanti JXN-26, Beckman Coulter, catalog # B38619, with a 6 x 1 l fixed-angle rotor JLA-8.1000, Beckman Coulter, catalog # 969328)
7. Standard refrigerated high speed floor centrifuge for medium volumes (*e.g.* Sorvall Lynx 6000, catalog # 75006590, with a T29-8 x 50 fixed angle rotor, catalog # 75003009)
8. Water bath maintained at 42 °C
9. Sterile standard 1.6 mL microcentrifuge tubes (*e.g.* VWR, catalog # 490004-436)
10. Sterile standard 6-well tissue culture plates (*e.g.* Corning, catalog # 3471)
11. Sterile standard T25 and T75 flasks (*e.g.* Corning, catalog # 430639 and # 3290)
12. Sterile standard glass unbaffled 125 and 300 mL DeLong flasks for insect cell propagation (Bellco Glass Inc., catalog # 2510-00125 and # 2510-00300)
13. Sterile standard unbaffled 2.8-liter Fernbach flasks (*e.g.* Chemglass Life Sciences, catalog # CLS-2020-11)
14. Sterile Falcon 15 mL conical centrifuge tubes (ThermoFisher Scientific, catalog # 14-959-53A)
15. Nanodrop Microvolume UV-Vis Spectrophotometer (*e.g.* Nanodrop One, ThermoFisher Scientific, catalog #ND-ONE-W)
16. QSonica Q125 ultrasonic processor with a 1/8-inch mini-probe (catalog # 1204P07)
17. Cole-Parmer Ultrasonic Processor with a ¼-inch (6.5-mm) tapered microtip (catalog # EW-04711-70)
18. KONTES FLEX-COLUMN preparative chromatography column, 98 mL, 25 x 200 mm (VWR, catalog # KT420401-2520)
19. Tricorn 5/50 column (Cytiva, catalog # 28406409) packed with Fractogel trimethylammoniumethyl (TMAE) anion exchange resin (Millipore Sigma, catalog # 1168810100)
20. Tricorn 5/50 column (Cytiva, catalog # 28406409) packed with Fractogel Sulfoisobutyl (SO_3^-) cation exchange resin (Millipore Sigma, catalog # 1.16890.0100)

21. Centrifugal spin concentrating unit (*e.g.*, Amicon Ultra, 10,000 Dalton molecular weight cutoff, Millipore Sigma, catalog # UFC801024)
22. Fast protein liquid chromatography (FPLC) system (*e.g.*, AKTA Pure, Cytiva)

5.1.2 Reagents

1. pFastbac1 6xHis-ROR1$^{457-752}$ plasmid DNA (or pseudokinase of choice)
2. Antibiotic stocks: 50 mg/mL kanamycin, 7 mg/mL gentamicin, 10 mg/mL tetracycline
3. Bluo-Gal (*e.g.*, Invitrogen, catalog # 15519028)
4. Isopropyl-β-D-thiogalactoside (IPTG; Sigma-Aldrich catalog # 367-93-1)
5. Ethanol, molecular biology grade
6. Isopropanol, molecular biology grade
7. Phosphate-buffered saline (PBS)
8. RNase A from bovine pancreas (ThermoFisher Scientific, catalog # J61996.MC)
9. Tris(2-carboxyethyl)phosphine (TCEP; Hampton Research, catalog # HR2-801)
10. Luria Broth (LB) media: 1% w/v tryptone, 0.5% w/v yeast extract, 1% w/v NaCl
11. Super Optimal broth with Catabolite repression (SOC) media: 2% w/v tryptone, 0.5% w/v yeast extract, 10 mM NaCl, 2.5 mM KCl, 10 mM $MgCl_2$, 10 mM $MgSO_4$, and 20 mM glucose
12. MAX Efficiency DH10Bac Competent Cells (ThermoFisher Scientific, catalog # 10361012)
13. Solution I: 15 mM Tris–HCl pH 8, 10 mM ethylenediaminetetraacetic acid (EDTA)
14. Solution II: 0.2 M NaOH, 1% Sodium dodecyl sulfate (SDS)
15. Solution III: 3 M potassium acetate, pH 5.5
16. *Spodoptera frugiperda* Sf9 cells (Expression Systems, catalog # 94-001S)
17. Serum-free Sf9 medium such as ESF921 (Expression Systems catalog # 96-001-01)
18. Cationic lipid transfection reagent (*e.g.*, Gibco Cellfectin II, catalog #10362100)
19. TE buffer: 10 mM Tris, pH 8.0, 1 mM EDTA
20. Lysis buffer: 50 mM Tris–HCl, pH 8.5, 300 mM NaCl, 10 mM 2-mercaptoethanol, 10 mM imidazole, 1 mM phenylmethylsulfonyl fluoride (PMSF)

21. cOmplete protease inhibitor cocktail, EDTA-free (Millipore Sigma, catalog # 11873580001)
22. Benzonase nuclease (Millipore Sigma, catalog # 70746-4)
23. Nickel nitrilotriacetic acid (NiNTA) agarose beads (Qiagen, catalog # 30210)
24. Mouse monoclonal anti-his tag antibody, clone 6AT18 (Sigma-Aldrich, catalog # SAB1305538)
25. AEX buffer: 50 mM Tris–HCl, pH 8.0, 2 mM dithiothreitol (DTT)
26. CEX buffer: 50 mM MES, pH 6.0, 2 mM DTT
27. SEC buffer: 20 mM HEPES, pH 7.0, 150 mM NaCl, and 100 μM TCEP
28. Laemmli sample buffer (4X): 125 mM Tris, pH 6.8, 4% SDS, 40% glycerol, 20% 2-mercaptoethanol, 0.02% Bromophenol Blue
29. Tris-Glycine SDS-PAGE running buffer (10X): 250 mM Tris, 1.92 M glycine, 1% SDS, pH 8.3
30. Coomassie staining solution: 50% methanol, 10 glacial acetic acid, 0.1% Coomassie Brilliant Blue R-250 (Bio-Rad, catalog # 1610400)
31. Coomassie destaining solution: 40% methanol, 10 % glacial acetic acid
32. Gel filtration standards (Bio-Rad, catalog # 1511901)
33. Precision Plus Protein Dual Color standards (Bio-Rad, catalog # 1610374)

5.1.3 Expression and purification of the ROR1 pseudokinase domain

5.1.3.1 Generation of baculovirus

1. Prior to start, ensure that the pseudokinase or kinase domain of interest (in this case, ROR1 pseudokinase domain, residues 457-752) is cloned into the pFastbac1 vector (or similar). This protocol uses nickel ion affinity purification, so just upstream of the pseudokinase open reading frame the construct contains an N-terminal hexahistidine tag. Verify cloning by Sanger sequencing.
2. Prepare fresh LB/agar bacterial plates for the generation and isolation of recombinant bacmid DNA. Plates should contain 50 μg/mL kanamycin, 7 μg/mL gentamicin 10 μg/mL tetracycline, 100 μg/mL Bluo-gal, 40 μg/mL IPTG. Air dry plates overnight, then store at 4 °C wrapped in both plastic and aluminum foil for a maximum of 6 weeks.
3. Transform 20–50 μL of DH10Bac chemically component cells by adding 0.5 μg purified pFastbac1 6xHis-ROR1$^{457\text{-}752}$, incubating for 30 min on ice, heat shocking at 42 °C for 30 s in a pre-warmed water bath, adding 500 μL SOC media, and incubating for 4 h at 37 °C with

shaking at 225 RPM. Plate 50 μL and 150 μL onto two separate plates and incubate for 48 h at 37 °C.

4. Check plates after 48 h. Pick a white colony from a bacmid plate and re-streak onto a fresh bacmid plate, and incubate for another 48 h at 37 °C.
5. Using a sterile pipet tip, pick a single white colony from a bacmid plate, and inoculate a 5 mL LB culture containing 50 μg/mL kanamycin, 7 μg/mL gentamicin, and 10 μg/mL tetracycline. Incubate with shaking at 37 °C for 12–24 h. It is recommended to also pick a blue colony as a negative control for bacmid PCR described below.
6. Isolate the bacmid DNA. Spin down 1.5 mL culture at 14,000 x *g* for 1 min to pellet cells. Remove supernatant and resuspend pellet in 0.3 mL of Solution I (15 mM Tris–HCl pH 8, 10 mM EDTA, with freshly added 100 μg/mL RNase A). Add 0.3 mL of freshly made Solution II (0.2 M NaOH, 1% SDS) and gently mix. Incubate for 5 min at room temperature. Add 0.4 mL of Solution III dropwise (3 M potassium acetate, pH 5.5), adding slowly, and mixing gently during addition by flicking tube. A thick white precipitate (containing protein and genomic DNA) should form. Incubate sample on ice for 5–10 min. Centrifuge for 10 min at 14,000 x *g* at 4 °C. Transfer supernatant to a fresh tube containing 0.8 mL of 100% isopropanol. Avoid transferring white precipitate. Mix gently by inverting, and incubate on ice for 5–10 min. Centrifuge sample for 15 min at 14,000 x *g* at room temperature. Remove supernatant and add 0.5 mL of 70% ethanol. Invert to wash pellet. Centrifuge 5 min at 14,000 x *g* at room temperature. Carefully remove supernatant with a pipette. Allow pellet to air-dry in an open tube for ~30 min at room temperature. Add 40 μL of TE buffer (10 mM Tris, pH 8.0, 1 mM EDTA) to the pellet and incubate at room temperature for 10 min. Gently tap the tube to mix and check the DNA concentration using a Nanodrop spectrophotometer; typical yield is 10–100 μg bacmid DNA. Check for proper insertion of ROR1 (or other pseudokinase/kinase domain insert of interest) by PCR and DNA gel electrophoresis. Use bacmid DNA for transfection as soon as possible. For long-term storage, make aliquots and store at −20 °C to avoid freeze/thaw cycles.
7. Transfect insect cells with recombinant bacmid DNA. For ROR1 production, we use Sf9 cells from the fall armyworm *Spodoptera frugiperda*. Healthy Sf9 cells are typically maintained in ESF921 serum-free medium from Expression Systems at a density ranging from 0.7

to 8.0×10^6 cells/mL by shaking at 110 RPM at 27 °C, and cells are split every 3–4 days. On the day prior to transfection, 35 mL propagation culture is seeded at a final density of 1.0×10^6 cells/mL in an unbaffled 125 mL glass DeLong shaker flask. On the day of transfection, seed 2 mL of cells at a final density of 0.8×10^6 into wells of a six-well plate. Prepare a transfection mixture by adding 5 μL of bacmid (at 400 ng/μL for a total of 2 μg bacmid DNA) + 100 μL cell culture media to 8 μL Cellfectin II + 100 μL medium and incubate at room temperature for 30 min. Add this entire mixture to one of the wells containing 2 mL seeded cells. Include a negative control condition by transfecting cells with bacmid lacking the ROR1 insert for comparison during test expressions. After incubation at 27 °C for 4–5 days in a static incubator, agitate the cells by pipetting, collect the entire contents, and centrifuge at 1000 x *g* for 10 min in a standard tabletop centrifuge. Store the supernatant in 2 mL tubes at 4 °C.

8. Scale up the P0 transfection virus to a P1 virus stock. In a T75 flask, seed 15 mL of Sf9 cells at a final density of 0.6×10^6 cells/mL and allow them to adhere for 30 min. Gently add 1 mL P0 virus to the flask and allow for viral amplification over 4 days in a static incubator maintained at 27 °C. Collect the P1 supernatant by transferring to a Falcon 15 mL conical tube followed by gentle centrifugation at 1000 x *g* for 10 min. Store the viral stock at 4 °C.

5.1.3.2 Small-scale protein test expression

1. Before scaling up for large-scale expression, it is prudent to ensure efficient viral production and soluble protein expression with a smaller scale test expression. Begin by seeding 15 mL of healthy Sf9 cells into a T75 flask at a density of $1.5–2.0 \times 10^6$ cells/mL. After allowing cells to adhere for 30 min, infect the cells with 1 mL of P1 virus. Alongside, prepare control conditions of uninfected cells and cells infected with a non-ROR1 control. Incubate cells at 27 °C in a static incubator for 3 days.
2. Harvest the cells by transferring cell suspension into Falcon 15 mL conical tubes and spinning down at 1000 x *g* for 10 min. Pour off supernatant and gently resuspend cells in 10 mL ice cold 1X PBS buffer, and repeat the centrifugation step. Again, pour off supernatant and resuspend cells in 1 mL ice cold lysis buffer (50 mM Tris–HCl, pH 8.5, 300 mM NaCl, 10 mM 2-mercaptoethanol, 10 mM Imidazole, 1 mM PMSF, protease inhibitor cocktail) and transfer to a 1.6 mL microcentrifuge tube on ice.

3. Sonicate to lyse cells using a mini-sonicator probe. For 1 mL volumes, we typically use a QSonica Q125 ultrasonic processor with a 1/8-inch mini-probe. We sonicate test expressions using two rounds of 30 s sonication at 50% amplitude with 1 s–1 s on-off cycles and brief end-over-end mixing of lysate between rounds. Sonication is performed at in a 4 °C cold room with cell suspensions kept on ice during sonication.
4. Spin down the lysate at 4 °C in a refrigerated microcentrifuge at 14,000 x *g* for 30 min. Transfer the supernatant to a fresh 1.6 mL microcentrifuge tube and add 50 μL of pre-equilibrated 50% NiNTA-agarose slurry. Rock samples at 4 °C for 30 min. Wash the beads three times with lysis buffer containing 40 mM imidazole using a gentle centrifugation speed of 500 x *g*.
5. Remove final wash supernatant and elute protein by adding 25 μL of lysis buffer plus 5 mM EDTA and 15 μL of 4X Laemmli sample buffer.
6. Assess expression levels by running 10 μL of sample on two SDS-polyacrylamide gel electrophoresis (PAGE) gels containing 12% acrylamide (37.5:1 acrylamide:bis-acrylamide), staining one gel with Coomassie blue stain and using the other for a Western blot to probe for the 6xHis tag (we use a mouse anti-His antibody from Sigma-Aldrich, clone 6AT18). Compare ROR1 expression to control conditions and to a molecular weight standard, such as Precision Plus Dual Color Protein Standard (Bio-Rad). If little-to-no protein is detectable by Western blotting, add an additional viral amplification step by repeating step 8 in the previous section using the P1 stock to generate a P2 stock, and repeat the test expression. Ensure that cells are >98% viable and growing at a healthy rate (~24–40 h doubling time).

5.1.3.3 Large-scale protein expression

1. Once expression of ROR1 in a small-scale test is confirmed, one can confidently proceed with large-scale protein expression. Begin by generating fresh baculovirus for large-scale expression by seeding 100 mL of healthy Sf9 cells into a 300 mL DeLong flask at a density of 1.0×10^6 cells/mL and incubate at 27 °C shaking at 110 RPM. The next day, cells should have a density of ~$1.5–2.0 \times 10^6$ cells/mL. Add 1 mL of P1 virus stock and incubate for 4–5 days at 27 °C shaking at 110 RPM. Harvest the P2 virus by spinning down the cells at 1000 x *g* for 10 min. Transfer the supernatant to fresh tubes and store at 4 °C until ready for large-scale infection.

2. Prior to large-scale protein purification, we recommend performing a pilot experiment at intermediate scale (15–50 mL media), to ensure that protein expression conditions are optimal.
3. To proceed with large-scale expression, scale healthy cells up to 4 L in four 2.8-L unbaffled Fernbach flasks (1 L culture/flask) from initial starting culture using standard conditions: 27 °C shaking at 110 RPM.
4. Once cells reach 4 L at 1.5–2.0 x 10^6 cells/mL cell density, infect the cells with fresh P2 virus generated in step one. Use ~24 mL virus per liter of culture. Allow cells to incubate at standard shaking conditions for 3 days.

5.1.3.4 Purification of the ROR1 pseudokinase domain

1. To harvest cells for purification, pour each liter culture into a 1-L centrifuge bottle and spin down the cells at a slow speed (so as not to lyse cells) of 2000 x *g* for 10 min at 4 °C. We use an Avanti JXN-26 refrigerated floor centrifuge with a 6 x 1 L fixed-angle rotor (JLA-8.1000 from Beckman Coulter) to perform this step. Slowly pour off media and gently wash the pellets with 0.5–1 L ice-cold PBS and repeat centrifugation step.
2. Slowly pour off PBS and gently resuspend pellets in 40 mL of ice-cold lysis buffer (50 mM Tris–HCl, pH 8.5, 300 mM NaCl, 10 mM 2-mercaptoethanol, 10 mM Imidazole, 1 mM PMSF, protease inhibitor cocktail) and transfer to 50 mL conical tubes. Add 250 units benzonase per resuspended pellet to digest cellular DNA.
3. Sonicate to lyse cells. We use a Cole-Parmer 500-Watt Ultrasonic Processor with a ¼-inch (6.5-mm) tapered microtip to lyse 40 mL of cell suspension at a time. Though conditions may vary depending on sonicator and probe, we use a gentle sonication program of two rounds of 30 s sonication at 35% amplitude with 1 s–1 s on-off cycles and brief end-over-end mixing of lysate between rounds. Sonication is performed in a 4 °C cold room with cell suspensions kept on ice during sonication. Clarify lysate by centrifugation for 30 min at 30,000 x *g* at 4 °C.
4. Load clarified lysate onto a clean KONTES column containing 3 mL pre-equilibrated NiNTA resin. Wash beads with 50 column volumes of lysis buffer.
5. Elute hexahistidine-tagged $ROR1^{457-752}$ with step-wise elutions containing increasing concentrations of imidazole up to 400 mM, taking one fraction per column volume. We typically elute with 3 column volumes (CVs) of lysis buffer containing 40 mM, 100 mM, 200 mM, and finally 400 mM imidazole. Assess purity and yield at this stage by

SDS-PAGE using Tris-Glycine based running buffer and Coomassie staining. We typically prepare SDS-PAGE samples by adding 9 μL of each fraction to 3 μL of 4X Laemmli sample buffer and run each sample side-by-side in a 15-well, reducing SDS-PAGE gel containing 12% acrylamide (37.5:1 acrylamide/bis-acrylamide), running the gel for ~1 h at a constant 150 V, or until dye front has reached the bottom of the gel. We run samples alongside a Precision Plus Protein Dual Color Standard (Bio-Rad), expecting the $ROR1^{457\text{-}752}$ protein product to run at a molecular weight of approximately 35 kDa.

6. Combine desired fractions and dilute with freshly made, ice-cold anion exchange (AEX) buffer (50 mM Tris–HCl, pH 8.0, 2 mM DTT) until the final NaCl concentration is 20 mM. Using an FPLC instrument such as an AKTA Pure (Cytiva) maintained at 4 °C, load protein onto a TMAE anion-exchange column pre-equilibrated with AEX buffer plus 20 mM NaCl. We typically use Tricorn 5/50 columns (Cytiva) packed in-house with Fractogel TMAE strong anion exchanger resin (Millipore Sigma) with a standard flow rate of 3 mL/min. Elute ROR1 protein with a linear NaCl gradient up to 1 M. We typically elute over 60 CVs (60 mL for a 1 mL column) and collect the eluate in 2 mL fractions. Assess purity and yield by SDS-PAGE and Coomassie staining using the same gel and running conditions as above. For this step, we typically run 15 μL fraction sample + 5 μL (4X) Laemmli sample buffer per well and compare samples with Precision Plus molecular weight standards as before. If necessary, repeat this step using cation exchange (CEX) in freshly made, ice-cold CEX buffer (50 mM MES, pH 6.0, 2 mM DTT) with an NaCl gradient of 50 mM to 1 M using an SO_3^- cation exchange column (*e.g.*, a Tricorn 5/50 column packed with Fractogel SO_3^- strong cation exchanger resin) using otherwise equivalent chromatography conditions as above for AEX.

7. Combine desired fractions and concentrate using a pre-washed centrifugal spin concentrating unit with a 10,000 Dalton molecular weight cut-off at 3000 x *g* until the sample volume is reduced to 0.5–1 mL. Using an FPLC, inject up to 0.5 mL concentrated protein at a time onto a Superdex 75 10/300 gel filtration column pre-equilibrated with freshly made, ice-cold 20 mM HEPES, pH 7.0, 150 mM NaCl, and 100 μM TCEP. Following gel filtration of ROR1, run fresh gel filtration standards (Bio-Rad) to directly compare with elution conditions of ROR1. $ROR1^{457\text{-}752}$ should elute as a globular monomer of approximately 35 kDa, and a single symmetrical peak should indicate high purity.

Further assess purity and yield by SDS-PAGE and Coomassie staining similarly to step 6, looking for the absence of non-ROR1$^{457\text{-}752}$ bands. Additionally, protein purity can be assessed using intact mass spectrometry.

5.2 Hydrogen/deuterium exchange reactions of ROR1 +/- ponatinib

5.2.1 Equipment

1. Standard set of single-channel and/or multi-channel benchtop pipettes
2. 2.0 mL dolphin microcentrifuge tubes, not autoclaved (*e.g.* VWR, catalog # 490010-618)
3. Ice bucket or cold room
4. Liquid nitrogen

5.2.2 Reagents

1. Purified ROR1$^{457\text{-}752}$ protein (from Section 5.1 above) at a concentration of ~0.5 mg/mL
2. Deuterium oxide (99.9%), low paramagnetic (Cambridge Isotope Laboratories, Inc. catalog # DLM-11)
3. Exchange buffer, H_2O-based: 20 mM HEPES, pH 7.4, 200 mM NaCl, 100 μM TCEP
4. Exchange buffer, D_2O-based: 20 mM HEPES, pD 7.4, 200 mM NaCl, 100 μM TCEP
   (Note: pD = pH + 0.4. Buffer should be adjusted to a desirable pD by adding DCl (Cambridge Isotope Laboratories, Inc. catalog # DLM-2) or NaOD (Cambridge Isotope Laboratories, Inc. catalog # DLM-45)
5. 1 mM Ponatinib (Selleckchem, catalog # S1490), or another drug/ligand of interest in 100% DMSO
6. Quench buffer: 100 mM sodium phosphate buffer, pH 2.4, 2 M optical-grade guanidine hydrochloride (ThermoFisher), 1% formic acid (FA)
7. Urea-D4 (Cambridge Isotope Laboratories, Inc. catalog # DLM-1269) dissolved in the D_2O exchange buffer, 7 M solution at pD 7.4

5.2.3 HDX sample preparation

1. Preparation of the HDX samples can be done well in advance of data collection, as samples can be stored at −80 °C for up to one month. Alternatively, robotic sample preparation systems can be utilized to improve reproducibility and increase throughput. Robotic HDX sample preparation typically takes longer than 8–10 h for an entire labeling time course series, including blank wash steps between runs. Thus, for protein

samples that may be unstable over the course of several hours at room temperature (while awaiting processing by the robot), manual sample preparation should be used instead. The use of multi-channel pipettors for manually preparing manual HDX-MS samples with technical replicates can save a significant amount of time.

2. For manual sample preparation, begin with a ~0.5 mg/mL stock solution of purified ROR1 pseudokinase domain in size exclusion buffer. To prepare undeuterated standards, pipette 5 μL protein into a fresh non-autoclaved 2.0 mL dolphin microcentrifuge tube. Add 95 μL of H_2O-based buffer containing 20 mM HEPES, pH 7.4, 200 mM NaCl, 100 μM TCEP. Perform a mock quench step by transferring 90 μL of diluted ROR1 into a fresh tube on ice containing 90 μL of ice-cold quench buffer (100 mM sodium phosphate buffer, pH 2.4, 2 M guanidine hydrochloride, 1% FA). Immediately snap-freeze in liquid nitrogen and store at −80 °C.
3. Repeat step 2 at least three times for technical replicates, or use a multi-channel pipettor to make technical replicates. We recommend preparing several additional undeuterated samples if LC-MS optimization is required.
4. To prepare deuterium-labeled samples, repeat step 2 but replace H_2O-based buffer with D_2O-based buffer containing 20 mM HEPES, pD 7.4, 200 mM NaCl, 100 μM TCEP at 25 °C. Incubate the samples at 25 °C for the desired amount of time and quench using ice-cold quench buffer. Immediately following the cold quench step, sample tubes should be immersed in liquid nitrogen to snap freeze and then stored at −80 °C until data collection. For kinase/pseudokinase samples such as ROR1, we generally recommend the following time points as a starting point: 10 s, 1 min, 10 min, 1 h, and 2 h. This range is a "broad sweep" found to be appropriate for monitoring uptake kinetics of peptides within typical protein kinase domains and was selected based on scouting experiments.
5. Again, repeat step 4 for a minimum of three technical replicates and for the desired number of time points. Samples should be stored at −80 °C prior to data collection. With adequate LC-MS systems, it should not be necessary to collect replicate data in series and can be carried out over multiple days if preferred (Houde, Berkowitz, & Engen, 2011).
6. In order to assess the impact of small molecules on the dynamics of a pseudokinase (or kinase) domain, this procedure can be performed in the presence of the compound for comparison with data obtained with

the compound. For example, to measure changes in ROR1 dynamics when bound to ponatinib, the initial protein stock should be prepared by diluting 1 mM ponatinib (in 100% DMSO) 20-fold into ~0.5 mg/mL ROR1 in size exclusion buffer, to give a final 50 μM ponatinib concentration. Additionally, the D_2O-based HDX buffer should also include 50 μM ponatinib (and 5% DMSO) final concentration to match. The remaining sample preparation conditions are unchanged. Here, it is important to add sufficient drug such that >85% protein is drug-bound after the 10- to 20-fold D_2O dilution (Kochert, Iacob, Wales, Makriyannis, & Engen, 2018), so knowledge of the affinity of the compound for the pseudokinase being studied is important.

7. Finally, fully deuterated standards should be prepared to correct the deuterium-labeled dataset for back exchange. This can be done by repeating step 2, but replacing the H_2O-based buffer with a D_2O-based denaturant such as 7 M urea. We typically incubate protein samples in 7 M-urea-D_2O (pD 7.4) for 1–10 min to exchange all amide hydrogen atoms in the protein.

5.3 In-line pepsin digestion, reversed-phase chromatography, and mass spectrometry

5.3.1 Equipment

1. UPLC (*e.g.*, Waters ACQUITY UPLC M-Class System with HDX Technology)
2. Mass spectrometer (*e.g.*, Waters Synapt G2-Si HDMS Q-TOF Mass Spectrometer)
3. UPLC C18 peptide trap column (*e.g.*, Waters ACQUITY UPLC BEH C18 VanGuard Pre-column, 2.1 mm x 5 mm, catalog # 186003975)
4. UPLC C18 separation column (*e.g.*, Waters AQUITY UPLC BEH C18 column, 1 mm x 100 mm, catalog # 186002346)
5. Pepsin column (*e.g.*, Waters Enzymate BEH pepsin column catalog #186007233)
6. 250 μL Hamilton glass syringe, or similar, for sample injection (*e.g.*, Waters, catalog # 430000865)

5.3.2 Reagents

1. Mobile phase A: mass spectrometry grade water + 0.1% Formic Acid (FA) (*e.g.*, Burdick and Jackson catalog # 39253 and # 56302, respectively)

2. Mobile phase B: mass spectrometry grade acetonitrile (ACN) + 0.1% FA (*e.g.*, Burdick and Jackson catalog # LC015)

5.3.3 LC-MS data acquisition

1. For data collection, the UPLC-MS system must be fitted with the appropriate components for HDX-MS. We use a Waters HDX nanoAcquity UPLC-M with in-line protein digestion capability equipped with a Waters Enzymate BEH pepsin column a Waters ACQUITY UPLC BEH C18 VanGuard pre-column peptide trap, and a Waters ACQUITY UPLC BEH C18 separation column. All columns are housed in a Waters HDX Manager at 2 °C. Upon pepsin digestion, desalting, and peptide separation, the separated peptides are analyzed using a Waters Synapt G2-Si mass spectrometer for determining the mass of each peptic peptide by a time-of-flight (TOF) analyzer. We utilize data-independent analysis (DIA) involving MS^E data collection performed with a 15 V to 30 V ramp collision-induced dissociation (CID) for determining the amino acid sequence of each peptic peptide.
2. Begin data collection by determining the optimal LC-MS conditions for the pseudokinase/kinase domain being studied. To do this, we recommend running undeuterated protein samples (50 μL) with the following starting LC conditions: a 3 min desalting step at 5% acetonitrile, followed by a 7 min 5–35% mobile phase B linear gradient (Fig. 4).
3. We generally aim for a sequence coverage above 90%. Sub-optimal LC separation of peptides often results in poor sequence coverage, so altering the acetonitrile gradient or using an alternative reversed-phase column (such as a charge-surface-modified one, *e.g.*, Waters ACQUITY UPLC CSH C18 separation column) for the protein under investigation should be considered when necessary. Additionally, mass spectrometers capable of ion mobility mode (such as the Waters Synapt G2-Si HDMS Q-TOF mass spectrometer) may provide a more optimal peptide coverage than the standard TOF mode alone, especially when studying large protein complexes.
4. Once optimal LC-MS conditions are determined, frozen HDX samples may then be individually thawed and 50 μL injected immediately (to minimize back-exchange) onto the pepsin column for MS^E data acquisition. Wash syringe between injections with mass spectrometry grade water. Follow each injection with a blank run by injecting water + 0.1% FA to ensure sample carryover between runs is not a concern.
5. To ensure reproducibility and to obtain sufficient information for statistical analysis, a minimum of three technical replications per biological

sample for each labeling timepoint should be obtained. When conclusions are to be drawn about specific structural elements, we recommend a minimum of three biological repeats be performed, using separate protein preparations. Biological repeats are especially important when complex protein sample preparation procedures are used, such as the purification of post-translationally modified proteins. Based on our experience, inconsistent deuterium uptake for the same protein sample

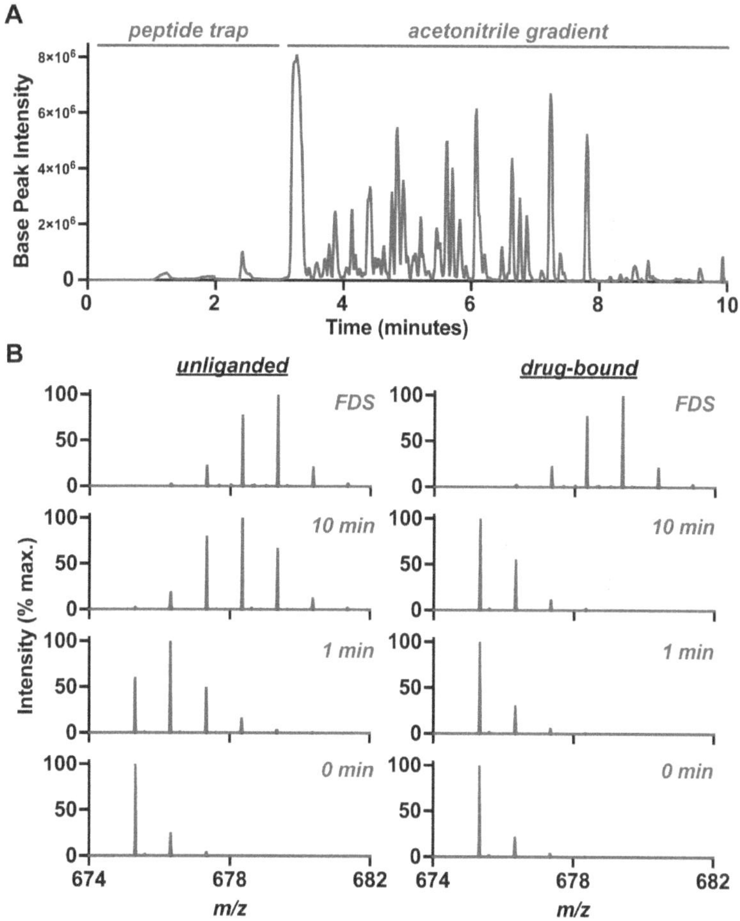

Fig. 4 Example LC-MS raw data. (A) Sample UPLC chromatogram of separated peptides prior to MS analysis. (B) Sample stacked spectral plots from a ROR1 peptide that is protected from deuterium exchange in the presence of the drug ponatinib. Centroids of mass envelopes from each time point from these spectra are used to generate uptake plots. FDS, fully-deuterated standard.

occurs when a protein is not freshly purified. Avoid using protein samples that have repeatedly been frozen and thawed, since these will undoubtedly yield variable results as protein integrity is compromised. The quality of the protein sample is often the most important (and overlooked) element for reproducibility and interpretability of results. In particular, if the protein under study has any tendency to aggregate, this can be manifest as increased protection from exchange, and can cause results to be misleading as well as irreproducible.

6. Data analysis and presentation

6.1 Analyzing HDX-MS data

6.1.1 Initial peptide identification

To initially assess the efficiency of the LC-MS protocol and ultimately to analyze complete HDX-MS datasets, the analysis must start with peptide identification from the MS^E spectra of peptic peptides using data obtained from the undeuterated protein sample. An MS^E experiment involves linking MS with LC information for each peptide: the parent ion m/z of each peptic peptide is determined, followed by CID for detecting the b- and y ion-series of each peptic peptide—and this information is linked with the retention time of each parent ion. The analysis software uses this mass information to determine the amino acid sequence of each peptic peptide. Several software packages are available for peptide identification; in our studies we use ProteinLynx Global SERVER 3.0.3 (PGLS) from Waters Corporation. The undeuterated MS^E dataset is used to search specifically against the sequence of the recombinant protein of interest (*e.g.*, 6xHis-ROR1$^{457\text{-}752}$) as well as porcine pepsin to monitor the health of the pepsin column. This initial search generates ion accounting files (IA files) that include information on the amino acid sequence, the monoisotopic as well as centroid mass, the retention time, the charge state, and the peptide intensity count of each peptic peptide.

As for peptide identification, several software packages have been developed for determining the relative deuterium uptake of each peptic peptide, including DynamX (Waters Corporation), ExMS (Kan, Mayne, Chetty, & Englander, 2011; Kan, Ye, Skinner, Mayne, & Englander, 2019), and HDExaminer (Sierra Analytics). Using DynamX, IA files are imported in order to determine the deuterium uptake of each peptic peptide relative to the corresponding undeuterated peptic peptide—

based on the centroid mass difference of each peptide derived from the unlabeled- and labeled-protein. In our studies of pseudokinase dynamics, the following stringent MS^E data filtering are applied to minimize peptide misidentification and to increase the accuracy of peptide identification: a minimum sequence length of 5, a maximum peptide length of 20 residues, minimum products per amino acid of 0.3, and a maximum MH^+ error of 5 ppm. Raw MS files for deuterated protein samples are uploaded for determining the centroid mass of each peptic peptide. DynamX then performs an automatic search for peptide peaks in MS spectra for deuterium-unlabeled and -labeled samples and generates relative deuterium uptake plots for each condition at different labeling timepoints, based on the centroid mass difference between the unlabeled and the labeled peptic peptide. Once the relative deuterium uptake for each peptic peptide is calculated by DynamX, it is imperative to inspect the assigned MS spectra manually in DynamX in order to discard any misassigned peptides. The deuterium uptake of a peptide at different charge states ought to be consistent; otherwise, a high standard deviation of deuterium uptake for the corresponding peptide is observed. Higher charge states tend to undergo greater back-exchange during LC-MS analysis compared to lower charge states of the same peptide. In addition, significant broadening of mass envelopes is often observed with higher charge state peptides. Therefore, it is acceptable to exclude +4 or higher charge states from the analysis, as long as all experiment conditions are processed equivalently.

It is also important to note that mass informatics software generally uses the monoisotopic mass (MH^+) of peptides to determine their amino acid sequence. Longer peptides have a greater tendency to deviate from the expected MH^+, which can result in peptide misidentifications. For this reason, in addition to greater back-exchange in longer peptides, we set the peptide maximum cutoff length of 20–25 residues in our analysis. The extent of back-exchange will vary depending on instrumentation, so we recommend that readers assess the extent of the back exchange under their conditions by first comparing the deuterium uptake of an unlabeled and fully-deuterated protein before proceeding to labeling time course experiments.

6.1.2 Downstream data analysis

Upon initial HDX-MS data processing, key information such as relative deuterium uptake ± standard deviation for each peptide at a given time point can be extracted for further processing and/or data visualization. We highly recommend correcting the relative deuterium uptake of each peptide for back exchange using the fully-deuterated standard sample. Calculating the relative deuterium uptake using an experimentally determined sample accounts for variability in back-exchange among different peptide lengths and sequence compositions. In order to obtain this normalized percentage of deuterium uptake ($\%D_t$) at incubation time t for a given peptide, the following equation can be used:

$$\%D_t = 100 * \frac{(m_t - m_0)}{(m_f - m_0)}$$

where m_t is the centroid mass at time t, m_0 is the centroid mass of the undeuterated control, and m_f is the centroid mass of the fully deuterated control. When datasets are generated comparing two separate conditions of the same protein sample, for instance in the analysis of exchange differences upon drug binding, a percent deuteration difference value ($\Delta\%D_{exchange}$) can be calculated. This is expressed by the equation:

$$\Delta\%D_{exchange} = \%D_{x,t}(bound) - \%D_{x,t}(unbound)$$

where x is a given peptide, and t is the D_2O incubation time. Several post-processing software packages have been developed by HDX-MS users to facilitate analysis and visualization, such as DECA (Lumpkin & Komives, 2019), Deuteros (Lau, Claesen, Hansen, & Politis, 2021; Lau, Ahdash, Martens, & Politis, 2019), and HaDeX (Puchala et al., 2020).

6.2 Presenting HDX-MS data

As with most other methodologies, HDX-MS datasets can be presented a number of different ways. A few useful representations (visualized in Fig. 5) that we find particularly informative for our studies include:

(A) *Individual peptide uptake plots* that display the level of deuterium uptake for a particular peptide over time (Fig. 5A). Such uptake plots are very useful for highlighting exchange differences (or similarities) in key regions of interest.

(B) *Structural heat maps* that display uptake levels or differences in deuterium exchange for a given time point projected onto a structural model of the protein (Fig. 5B). Such structure heat maps are invaluable for connecting dynamic HDX-MS information with structural information, for both monomeric proteins as well as protein complexes.

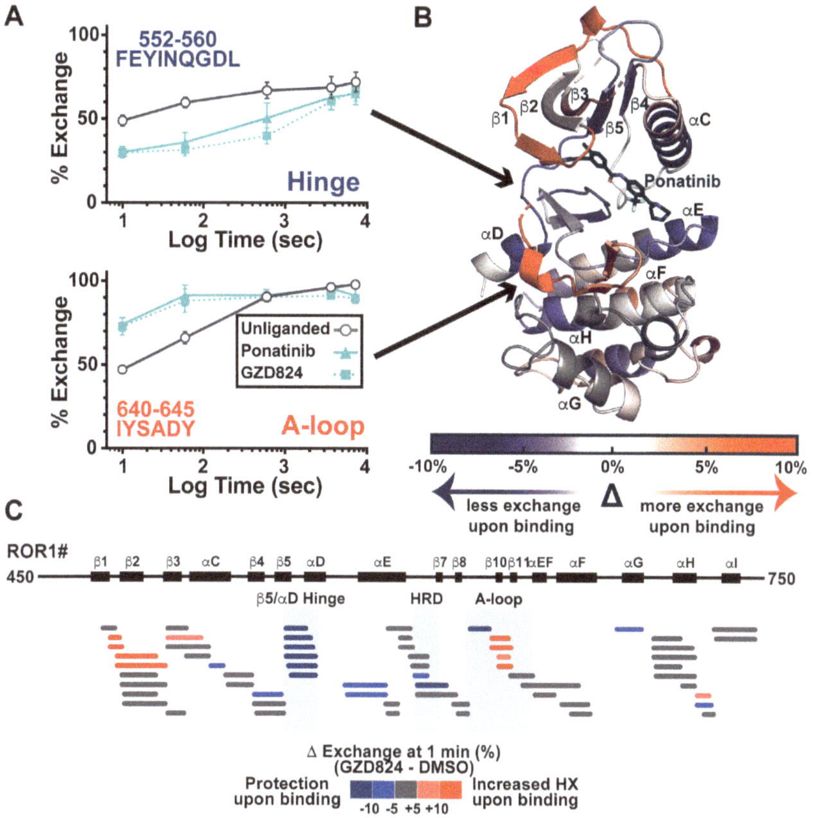

Fig. 5 Example representations of pseudokinase domain HDX-MS data. (A) Peptide uptake plots for differentially exchanging regions of ROR1 upon drug binding. (B) Structural heat map for the difference in deuterium exchange upon ponatinib binding to the ROR1 pseudokinase domain. Data from the 1 min timepoint are projected onto the structure of ponatinib-bound ROR1, color coded as described in the associated legend. (C) Coverage heat map for the difference in deuterium exchange upon ponatinib binding to the ROR1 pseudokinase domain. At the top of the panel, a secondary structure diagram for the ROR1 pseudokinase domain is shown in black. Below, short lines represent peptides from ROR1, colored according to data from the 1 min timepoint in the HDX-MS experiment. All data shown represent biological triplicates, each with technical triplicates as described (Sheetz et al., 2020). PDB ID: 6TU9 (Sheetz et al., 2020).

(C) *Peptide coverage heat maps* that display both uptake information and the level of peptide coverage for the protein being investigated (Fig. 5C). Generally, a linear diagram indicating secondary structural elements and/or domain boundaries is depicted horizontally at the top of the figure, with adjacent lines corresponding to peptide residue boundaries from peptides analyzed in the experiment. Each peptide can be colored by level of deuterium uptake or levels of differential exchange.

(D) *Whole dataset uptake plots* that display deuterium uptake for all peptides for a given labelling experiment, with median residue number of the peptide on the *x*-axis (*e.g.*, as shown in Fig. 3). Datasets across timepoints, protein variants, or different experimental conditions can be combined with this representation to provide a comprehensive view of deuterium uptake differences or similarities.

Finally, it is worth considering the following tips when presenting HDX-MS data at conferences or in publications:

1. Choose a visual representation that best fits the question being investigated. For example, structural heat maps are very useful for conveying how the presence of a drug alters structural dynamics in the context of the protein's structure. When exchange differences in key regions of interest are to be highlighted, peptide uptake plots comparing different experimental conditions are often an excellent choice for conveying specific peptide information.
2. Show more data than a few selected uptake plots to avoid cherry-picking. In addition to presenting uptake plots, provide data for the entire polypeptide chain (for example using a peptide coverage heat map) so that the peptide of interest can be understood within the context of the larger protein being examined. When possible, uptake plots of multiple, overlapping peptides can provide greater confidence when interpreting exchange differences. The goal is both to provide a complete view of the dataset and to highlight key areas of interest.
3. Follow data reporting standards. Resulting from discussions at the first International Conference on Hydrogen-Deuterium Exchange in 2017, a set of community recommendations serves as useful guidelines for HDX-MS users. These guidelines provide specifications on data reporting and include useful supplemental data templates to facilitate the evaluation of HDX-MS datasets during and after peer review. We refer the reader to these important guidelines for more details (Masson et al., 2019), and encourage their adoption.

7. Summary

Like protein kinases, pseudokinase domains are dynamic and competent to adopt (and switch between) distinct conformation states. Although X-ray crystallography has led the way in providing high resolution structural information on kinases and pseudokinases, the resulting structural pictures are static, and the field also needs rigorous methods for assessing the dynamics of the kinase fold. HDX-MS is one technique that is well suited to probe pseudokinase dynamics in solution and to validate changes in conformation and in conformational exploration induced by binding of small molecule pseudokinase binders, protein binding partners, post-translational modifications, and mutations. In this chapter, we provide a detailed protocol for applying HDX-MS to a pseudokinase domain to assess intrinsic dynamics as well as drug-induced changes in conformational dynamics. Discovery of pseudokinase-targeted small molecules (and protein binding partners) is in its infancy, but we envision that HDX-MS will be a powerful methodology for connecting structure to mechanism and function in the coming years.

Acknowledgments

We dedicate this chapter to the many excellent researchers who contributed to the development of HDX-MS, making it the accessible structural biology tool that it is today. We thank members of the Lemmon and Ferguson labs for helpful discussion and comments. This work was supported by NIGMS grant R35-GM122485 (to M.A.L.), NCI grant R03-CA259881 (to Y.T.), and an NSF Graduate Research Fellowship (DGE1122492 to J.B.S.).

References

Adams, J. A. (2001). Kinetic and catalytic mechanisms of protein kinases. *Chemical Reviews*, *101*(8), 2271–2290. https://doi.org/10.1021/cr000230w.

Artim, S. C., Mendrola, J. M., & Lemmon, M. A. (2012). Assessing the range of kinase autoinhibition mechanisms in the insulin receptor family. *The Biochemical Journal*, *448*(2), 213–220. https://doi.org/10.1042/BJ20121365.

Boudeau, J., Miranda-Saavedra, D., Barton, G. J., & Alessi, D. R. (2006). Emerging roles of pseudokinases. *Trends in Cell Biology*, *16*(9), 443–452. https://doi.org/10.1016/j.tcb.2006.07.003.

Burke, J. E., Vadas, O., Berndt, A., Finegan, T., Perisic, O., & Williams, R. L. (2011). Dynamics of the phosphoinositide 3-kinase p110delta interaction with p85alpha and membranes reveals aspects of regulation distinct from p110alpha. *Structure*, *19*(8), 1127–1137. https://doi.org/10.1016/j.str.2011.06.003.

De Bondt, H. L., Rosenblatt, J., Jancarik, J., Jones, H. D., Morgan, D. O., & Kim, S. H. (1993). Crystal structure of cyclin-dependent kinase 2. *Nature*, *363*(6430), 595–602. https://doi.org/10.1038/363595a0.

Dornan, G. L., Stariha, J. T. B., Rathinaswamy, M. K., Powell, C. J., Boulanger, M. J., & Burke, J. E. (2020). Defining how oncogenic and developmental mutations of PIK3R1 alter the regulation of class IA phosphoinositide 3-kinases. *Structure*, *28*(2), 145–156 e145. https://doi.org/10.1016/j.str.2019.11.013.

Endicott, J. A., Noble, M. E., & Johnson, L. N. (2012). The structural basis for control of eukaryotic protein kinases. *Annual Review of Biochemistry*, *81*, 587–613. https://doi.org/10.1146/annurev-biochem-052410-090317.

Engen, J. R. (2009). Analysis of protein conformation and dynamics by hydrogen/deuterium exchange MS. *Analytical Chemistry, 81*(19), 7870–7875. https://doi.org/10.1021/ac901154s.

Engen, J. R., & Komives, E. A. (2020). Complementarity of hydrogen/deuterium exchange mass spectrometry and cryo-electron microscopy. *Trends in Biochemical Sciences, 45*(10), 906–918. https://doi.org/10.1016/j.tibs.2020.05.005.

Englander, S. W., Mayne, L., Bai, Y., & Sosnick, T. R. (1997). Hydrogen exchange: The modern legacy of Linderstrom-Lang. *Protein Science, 6*(5), 1101–1109. https://doi.org/10.1002/pro.5560060517.

Fowler, M. L., McPhail, J. A., Jenkins, M. L., Masson, G. R., Rutaganira, F. U., Shokat, K. M., et al. (2016). Using hydrogen deuterium exchange mass spectrometry to engineer optimized constructs for crystallization of protein complexes: Case study of PI4KIIIbeta with Rab11. *Protein Science, 25*(4), 826–839. https://doi.org/10.1002/pro.2879.

Hadzipasic, A., Wilson, C., Nguyen, V., Kern, N., Kim, C., Pitsawong, W., et al. (2020). Ancient origins of allosteric activation in a Ser-Thr kinase. *Science, 367*(6480), 912–917. https://doi.org/10.1126/science.aay9959.

Hoofnagle, A. N., Resing, K. A., & Ahn, N. G. (2003). Protein analysis by hydrogen exchange mass spectrometry. *Annual Review of Biophysics and Biomolecular Structure, 32*, 1–25. https://doi.org/10.1146/annurev.biophys.32.110601.142417.

Hoofnagle, A. N., Resing, K. A., Goldsmith, E. J., & Ahn, N. G. (2001). Changes in protein conformational mobility upon activation of extracellular regulated protein kinase-2 as detected by hydrogen exchange. *Proceedings of the National Academy of Sciences of the United States of America, 98*(3), 956–961. https://doi.org/10.1073/pnas.98.3.956.

Houde, D., Berkowitz, S. A., & Engen, J. R. (2011). The utility of hydrogen/deuterium exchange mass spectrometry in biopharmaceutical comparability studies. *Journal of Pharmaceutical Sciences, 100*(6), 2071–2086. https://doi.org/10.1002/jps.22432.

Hubbard, S. R. (1997). Crystal structure of the activated insulin receptor tyrosine kinase in complex with peptide substrate and ATP analog. *The EMBO Journal, 16*(18), 5572–5581. https://doi.org/10.1093/emboj/16.18.5572.

Hubbard, S. R., Wei, L., Ellis, L., & Hendrickson, W. A. (1994). Crystal structure of the tyrosine kinase domain of the human insulin receptor. *Nature, 372*(6508), 746–754. https://doi.org/10.1038/372746a0.

Huse, M., & Kuriyan, J. (2002). The conformational plasticity of protein kinases. *Cell, 109*(3), 275–282. https://doi.org/10.1016/s0092-8674(02)00741-9.

Hvidt, A., & Linderstrom-Lang, K. (1954). Exchange of hydrogen atoms in insulin with deuterium atoms in aqueous solutions. *Biochimica et Biophysica Acta, 14*(4), 574–575. https://doi.org/10.1016/0006-3002(54)90241-3.

Hvidt, A., & Linderstrom-Lang, K. (1955). Exchange of deuterium and 18O between water and other substances. III. Deuterium exchange of short peptides, Sanger's A-chain and insulin. *Comptes Rendus des Travaux du Laboratoire Carlsberg. Série Chimique, 29*(22–23), 385–402. https://www.ncbi.nlm.nih.gov/pubmed/13305138.

Inglis, A. J., Masson, G. R., Shao, S., Perisic, O., McLaughlin, S. H., Hegde, R. S., et al. (2019). Activation of GCN2 by the ribosomal P-stalk. *Proceedings of the National Academy of Sciences of the United States of America, 116*(11), 4946–4954. https://doi.org/10.1073/pnas.1813352116.

Jeffrey, P. D., Russo, A. A., Polyak, K., Gibbs, E., Hurwitz, J., Massague, J., et al. (1995). Mechanism of CDK activation revealed by the structure of a cyclinA-CDK2 complex. *Nature, 376*(6538), 313–320. https://doi.org/10.1038/376313a0.

Jensen, P. F., & Rand, K. D. (2016). Hydrogen exchange: A sensitive analytical window into protein conformation and dynamics. In D. D. Weis (Ed.), *Hydrogen exchange mass spectrometry of proteins: Fundamentals, methods, and applications* John Wiley & Sons, Ltd.

Johnson, L. N., Noble, M. E., & Owen, D. J. (1996). Active and inactive protein kinases: Structural basis for regulation. *Cell*, *85*(2), 149–158. https://doi.org/10.1016/s0092-8674(00)81092-2.

Kan, Z. Y., Mayne, L., Chetty, P. S., & Englander, S. W. (2011). ExMS: Data analysis for HX-MS experiments. *Journal of the American Society for Mass Spectrometry*, *22*(11), 1906–1915. https://doi.org/10.1007/s13361-011-0236-3.

Kan, Z. Y., Ye, X., Skinner, J. J., Mayne, L., & Englander, S. W. (2019). ExMS2: An integrated solution for hydrogen-deuterium exchange mass spectrometry data analysis. *Analytical Chemistry*, *91*(11), 7474–7481. https://doi.org/10.1021/acs.analchem.9b01682.

Knighton, D. R., Zheng, J. H., Ten Eyck, L. F., Ashford, V. A., Xuong, N. H., Taylor, S. S., et al. (1991). Crystal structure of the catalytic subunit of cyclic adenosine monophosphate-dependent protein kinase. *Science*, *253*(5018), 407–414. https://doi.org/10.1126/science.1862342.

Kochert, B. A., Iacob, R. E., Wales, T. E., Makriyannis, A., & Engen, J. R. (2018). Hydrogen-deuterium exchange mass spectrometry to study protein complexes. *Methods in Molecular Biology*, *1764*, 153–171. https://doi.org/10.1007/978-1-4939-7759-8_10.

Kornev, A. P., & Taylor, S. S. (2009). Pseudokinases: Functional insights gleaned from structure. *Structure*, *17*(1), 5–7. https://doi.org/10.1016/j.str.2008.12.005.

Kornev, A. P., & Taylor, S. S. (2015). Dynamics-driven allostery in protein kinases. *Trends in Biochemical Sciences*, *40*(11), 628–647. https://doi.org/10.1016/j.tibs.2015.09.002.

Kung, J. E., & Jura, N. (2016). Structural basis for the non-catalytic functions of protein kinases. *Structure*, *24*(1), 7–24. https://doi.org/10.1016/j.str.2015.10.020.

Kung, J. E., & Jura, N. (2019). Prospects for pharmacological targeting of pseudokinases. *Nature Reviews. Drug Discovery*, *18*(7), 501–526. https://doi.org/10.1038/s41573-019-0018-3.

Lau, A. M. C., Ahdash, Z., Martens, C., & Politis, A. (2019). Deuteros: Software for rapid analysis and visualization of data from differential hydrogen deuterium exchange-mass spectrometry. *Bioinformatics*, *35*(17), 3171–3173. https://doi.org/10.1093/bioinformatics/btz022.

Lau, A. M., Claesen, J., Hansen, K., & Politis, A. (2021). Deuteros 2.0: Peptide-level significance testing of data from hydrogen deuterium exchange mass spectrometry. *Bioinformatics*, *37*(2), 270–272. https://doi.org/10.1093/bioinformatics/btaa677.

Lee, T., Hoofnagle, A. N., Kabuyama, Y., Stroud, J., Min, X., Goldsmith, E. J., et al. (2004). Docking motif interactions in MAP kinases revealed by hydrogen exchange mass spectrometry. *Molecular Cell*, *14*(1), 43–55. https://doi.org/10.1016/s1097-2765(04)00161-3.

Lee, T., Hoofnagle, A. N., Resing, K. A., & Ahn, N. G. (2005). Hydrogen exchange solvent protection by an ATP analogue reveals conformational changes in ERK2 upon activation. *Journal of Molecular Biology*, *353*(3), 600–612. https://doi.org/10.1016/j.jmb.2005.08.029.

Lorenzen, K., & Pawson, T. (2014). HDX-MS takes centre stage at unravelling kinase dynamics. *Biochemical Society Transactions*, *42*(1), 145–150. https://doi.org/10.1042/BST20130250.

Lucic, I., Rathinaswamy, M. K., Truebestein, L., Hamelin, D. J., Burke, J. E., & Leonard, T. A. (2018). Conformational sampling of membranes by Akt controls its activation and inactivation. *Proceedings of the National Academy of Sciences of the United States of America*, *115*(17), E3940–E3949. https://doi.org/10.1073/pnas.1716109115.

Lumpkin, R. J., & Komives, E. A. (2019). Deca, a comprehensive, automatic post-processing program for HDX-MS data. *Molecular & Cellular Proteomics*, *18*(12), 2516–2523. https://doi.org/10.1074/mcp.TIR119.001731.

Mace, P. D., & Murphy, J. M. (2021). There's more to death than life: Noncatalytic functions in kinase and pseudokinase signaling. *The Journal of Biological Chemistry*, *296*, 100705. https://doi.org/10.1016/j.jbc.2021.100705.

Manning, G., Plowman, G. D., Hunter, T., & Sudarsanam, S. (2002). Evolution of protein kinase signaling from yeast to man. *Trends in Biochemical Sciences*, *27*(10), 514–520. https://doi.org/10.1016/s0968-0004(02)02179-5.

Masson, G. R., Burke, J. E., Ahn, N. G., Anand, G. S., Borchers, C., Brier, S., et al. (2019). Recommendations for performing, interpreting and reporting hydrogen deuterium exchange mass spectrometry (HDX-MS) experiments. *Nature Methods*, *16*(7), 595–602. https://doi.org/10.1038/s41592-019-0459-y.

Masson, G. R., Jenkins, M. L., & Burke, J. E. (2017). An overview of hydrogen deuterium exchange mass spectrometry (HDX-MS) in drug discovery. *Expert Opinion on Drug Discovery*, *12*(10), 981–994. https://doi.org/10.1080/17460441.2017.1363734.

Masterson, L. R., Mascioni, A., Traaseth, N. J., Taylor, S. S., & Veglia, G. (2008). Allosteric cooperativity in protein kinase A. *Proceedings of the National Academy of Sciences of the United States of America*, *105*(2), 506–511. https://doi.org/10.1073/pnas.0709214104.

Mendrola, J. M., Shi, F., Park, J. H., & Lemmon, M. A. (2013). Receptor tyrosine kinases with intracellular pseudokinase domains. *Biochemical Society Transactions*, *41*(4), 1029–1036. https://doi.org/10.1042/BST20130104.

Miller, C. J., & Turk, B. E. (2018). Homing in: Mechanisms of substrate targeting by protein kinases. *Trends in Biochemical Sciences*, *43*(5), 380–394. https://doi.org/10.1016/j.tibs.2018.02.009.

Murphy, J. M. (2020). The killer pseudokinase mixed lineage kinase domain-like protein (MLKL). *Cold Spring Harbor Perspectives in Biology*, *12*(8). https://doi.org/10.1101/cshperspect.a036376.

Murphy, J. M., Czabotar, P. E., Hildebrand, J. M., Lucet, I. S., Zhang, J. G., Alvarez-Diaz, S., et al. (2013). The pseudokinase MLKL mediates necroptosis via a molecular switch mechanism. *Immunity*, *39*(3), 443–453. https://doi.org/10.1016/j.immuni.2013.06.018.

Murphy, J. M., Zhang, Q., Young, S. N., Reese, M. L., Bailey, F. P., Eyers, P. A., et al. (2014). A robust methodology to subclassify pseudokinases based on their nucleotide-binding properties. *The Biochemical Journal*, *457*(2), 323–334. https://doi.org/10.1042/BJ20131174.

Narayana, N., Cox, S., Nguyen-huu, X., Ten Eyck, L. F., & Taylor, S. S. (1997). A binary complex of the catalytic subunit of cAMP-dependent protein kinase and adenosine further defines conformational flexibility. *Structure*, *5*(7), 921–935. https://doi.org/10.1016/s0969-2126(97)00246-3.

Owen, D. J., Noble, M. E., Garman, E. F., Papageorgiou, A. C., & Johnson, L. N. (1995). Two structures of the catalytic domain of phosphorylase kinase: An active protein kinase complexed with substrate analogue and product. *Structure*, *3*(5), 467–482. https://doi.org/10.1016/s0969-2126(01)00180-0.

Patel, O., Griffin, M. D. W., Panjikar, S., Dai, W., Ma, X., Chan, H., et al. (2017). Structure of SgK223 pseudokinase reveals novel mechanisms of homotypic and heterotypic association. *Nature Communications*, *8*(1), 1157. https://doi.org/10.1038/s41467-017-01279-9.

Peacock, R. B., & Komives, E. A. (2021). Hydrogen/deuterium exchange and nuclear magnetic resonance spectroscopy reveal dynamic allostery on multiple time scales in the serine protease thrombin. *Biochemistry*. https://doi.org/10.1021/acs.biochem.1c00277.

Pegram, L. M., Liddle, J. C., Xiao, Y., Hoh, M., Rudolph, J., Iverson, D. B., et al. (2019, Jul 30). Activation loop dynamics are controlled by conformation-selective inhibitors of ERK2. *Proceedings of the National Academy of Sciences of the United States of America*, *116*(31), 15463–15468. https://doi.org/10.1073/pnas.1906824116.

Petrie, E. J., Czabotar, P. E., & Murphy, J. M. (2019). The structural basis of necroptotic cell death signaling. *Trends in Biochemical Sciences*, *44*(1), 53–63. https://doi.org/10.1016/j.tibs.2018.11.002.

Petrie, E. J., Sandow, J. J., Jacobsen, A. V., Smith, B. J., Griffin, M. D. W., Lucet, I. S., et al. (2018). Conformational switching of the pseudokinase domain promotes human MLKL tetramerization and cell death by necroptosis. *Nature Communications*, *9*(1), 2422. https://doi.org/10.1038/s41467-018-04714-7.

Pitsawong, W., Buosi, V., Otten, R., Agafonov, R. V., Zorba, A., Kern, N., et al. (2018). Dynamics of human protein kinase Aurora A linked to drug selectivity. *eLife*, 7. https://doi.org/10.7554/eLife.36656.

Puchala, W., Burdukiewicz, M., Kistowski, M., Dabrowska, K. A., Badaczewska-Dawid, A. E., Cysewski, D., et al. (2020). HaDeX: An R package and web-server for analysis of data from hydrogen-deuterium exchange mass spectrometry experiments. *Bioinformatics*, *36*(16), 4516–4518. https://doi.org/10.1093/bioinformatics/btaa587.

Rosa, J. J., & Richards, F. M. (1979). An experimental procedure for increasing the structural resolution of chemical hydrogen-exchange measurements on proteins: Application to ribonuclease S peptide. *Journal of Molecular Biology*, *133*(3), 399–416. https://doi.org/10.1016/0022-2836(79)90400-5.

Rostislavleva, K., Soler, N., Ohashi, Y., Zhang, L., Pardon, E., Burke, J. E., et al. (2015). Structure and flexibility of the endosomal Vps34 complex reveals the basis of its function on membranes. *Science*, *350*(6257), aac7365. https://doi.org/10.1126/science.aac7365.

Ruff, E. F., Muretta, J. M., Thompson, A. R., Lake, E. W., Cyphers, S., Albanese, S. K., et al. (2018). A dynamic mechanism for allosteric activation of Aurora kinase A by activation loop phosphorylation. *eLife*, 7. https://doi.org/10.7554/eLife.32766.

Scheeff, E. D., Eswaran, J., Bunkoczi, G., Knapp, S., & Manning, G. (2009). Structure of the pseudokinase VRK3 reveals a degraded catalytic site, a highly conserved kinase fold, and a putative regulatory binding site. *Structure*, *17*(1), 128–138. https://doi.org/10.1016/j.str.2008.10.018.

Schulze-Gahmen, U., De Bondt, H. L., & Kim, S. H. (1996). High-resolution crystal structures of human cyclin-dependent kinase 2 with and without ATP: Bound waters and natural ligand as guides for inhibitor design. *Journal of Medicinal Chemistry*, *39*(23), 4540–4546. https://doi.org/10.1021/jm960402a.

Sheetz, J. B., Mathea, S., Karvonen, H., Malhotra, K., Chatterjee, D., Niininen, W., et al. (2020). Structural insights into pseudokinase domains of receptor tyrosine kinases. *Molecular Cell*, *79*(3), 390–405e397. https://doi.org/10.1016/j.molcel.2020.06.018.

Siess, K. M., & Leonard, T. A. (2019). Lipid-dependent Akt-ivity: Where, when, and how. *Biochemical Society Transactions*, *47*(3), 897–908. https://doi.org/10.1042/BST20190013.

Truebestein, L., Hornegger, H., Anrather, D., Hartl, M., Fleming, K. D., Stariha, J. T. B., et al. (2021). Structure of autoinhibited Akt1 reveals mechanism of PIP3-mediated activation. *Proceedings of the National Academy of Sciences of the United States of America*, *118*(33). https://doi.org/10.1073/pnas.2101496118.

Vadas, O., & Burke, J. E. (2015). Probing the dynamic regulation of peripheral membrane proteins using hydrogen deuterium exchange-MS (HDX-MS). *Biochemical Society Transactions*, *43*(5), 773–786. https://doi.org/10.1042/BST20150065.

Xiao, Y., Liddle, J. C., Pardi, A., & Ahn, N. G. (2015). Dynamics of protein kinases: Insights from nuclear magnetic resonance. *Accounts of Chemical Research*, *48*(4), 1106–1114. https://doi.org/10.1021/acs.accounts.5b00001.

Xie, T., Saleh, T., Rossi, P., & Kalodimos, C. G. (2020). Conformational states dynamically populated by a kinase determine its function. *Science*, *370*(6513). https://doi.org/10.1126/science.abc2754.

Young, L. N., Cho, K., Lawrence, R., Zoncu, R., & Hurley, J. H. (2016). Dynamics and architecture of the NRBF2-containing phosphatidylinositol 3-kinase complex I of autophagy. *Proceedings of the National Academy of Sciences of the United States of America*, *113*(29), 8224–8229. https://doi.org/10.1073/pnas.1603650113.

Zeqiraj, E., & van Aalten, D. M. (2010). Pseudokinases-remnants of evolution or key allosteric regulators? *Current Opinion in Structural Biology*, *20*(6), 772–781. https://doi.org/10.1016/j.sbi.2010.10.001.

Zheng, J., Knighton, D. R., Xuong, N. H., Taylor, S. S., Sowadski, J. M., & Ten Eyck, L. F. (1993). Crystal structures of the myristylated catalytic subunit of cAMP-dependent protein kinase reveal open and closed conformations. *Protein Science*, *2*(10), 1559–1573. https://doi.org/10.1002/pro.5560021003.

CHAPTER TWELVE

Cellular thermal shift assay (CETSA) for determining the drug binding affinity using Ba/F3 clones stably expressing receptor pseudokinases

Hanna Karvonen[a,†], Juuli Raivola[b], and Daniela Ungureanu[a,b,c,*]
[a]Cancer Signaling, Faculty of Medicine and Health Technology, Tampere University, Tampere, Finland
[b]Applied Tumor Genomics, Faculty of Medicine, University of Helsinki, Helsinki, Finland
[c]Faculty of Biochemistry and Molecular Medicine, University of Oulu, Oulu, Finland
*Corresponding author: e-mail address: daniela.ungureanu@helsinki.fi

Contents

† Current address: Herbert Irving Comprehensive Cancer Center, Columbia University Irving Medical Center, New York, NY 10032, USA

Methods in Enzymology, Volume 667
ISSN 0076-6879
https://doi.org/10.1016/bs.mie.2022.03.038

Abstract

The majority of drug screening approaches are performed using recombinant proteins, however, drug binding to its target(s) in cells should be also assessed, especially for drugs aimed at modulating intracellular signaling pathways. As a result, the development of a cellular thermal shift assay (CETSA) has become an important tool for determining the binding affinity of drugs to their intracellular targets. Cell lines, such as Ba/F3, are an excellent model system to stably express and study a target protein when this protein is not endogenously expressed or only present at low levels. Together with CETSA, Ba/F3 clones allow study of the transforming properties of the protein in question, its downstream intracellular signaling activation pathways, as well as its drug binding kinetics. This chapter describes in detail the establishment of Ba/F3 clones stably expressing receptor pseudokinases, such as receptor tyrosine kinase-like orphan receptors (ROR1, ROR2) and protein tyrosine kinase 7 (PTK7), and the use thereof to evaluate binding of small molecule inhibitors to their intracellular (pseudo)kinase domain by CETSA.

1. Introduction

1.1 Ba/F3 cells

Ba/F3 cells are murine pro-B cells widely used in cell-based assays, especially while studying oncogenic kinases and their potential inhibitors. The survival and growth of Ba/F3 cells are dependent on the presence of interleukin-3 (IL-3) in their growth media. IL-3 is an essential growth factor for a subpopulation of bone marrow-derived early murine lymphoid cells committed to B-cell lineage, but not for mature B-cells (Palacios, Henson, Steinmetz, & McKearn, 1984; Palacios & Steinmetz, 1985). The effects of IL-3 are mediated through IL-3 receptor that transduces downstream signaling through JAK/STAT, PI3K/AKT and MEK/ERK pathways (Reddy, Korapati, Chaturvedi, & Rane, 2000). As fast-growing suspension cells that can be easily transfected by electroporation or infected by commonly used retro- and lentiviral expression systems, Ba/F3 cells have become a useful cellular model especially for screening drug-like inhibitors to determine their affinity and activity toward kinases and other intracellular targets. Several kinase inhibitors targeting activated downstream IL-3 signaling have been validated using Ba/F3 cells, such as JAK1/JAK2 and Src/Abl inhibitors (Gwanmesia et al., 2009; Hornakova et al., 2011; Liu et al., 2009; Warmuth, Kim, Gu, Xia, & Adrián, 2007).

Previous studies have shown that the ectopic expression of constitutively active oncogenic kinases (receptor or non-receptor) in Ba/F3 cells can compensate for the IL-3 dependent growth and can induce growth factor independence. This process was first shown with BCR-ABL fusion gene (Daley & Baltimore, 1988), which is the main driver of chronic myeloid leukemia (CML) and in some cases of acute lymphoid leukemia (ALL). Since then, several other kinases and their mutated oncogenic forms have been shown to induce IL-3 independent growth of Ba/F3 cells, including TEL-JAK2 fusion, mutated JAK2, FLT3, EGFR, and HER2 (Warmuth et al., 2007). Furthermore, we have also shown that the catalytically inactive receptor pseudokinase ROR1 is able to transform Ba/F3 cells to proliferate in the absence of IL-3 (Sheetz et al., 2020).

In addition to transforming assays, Ba/F3 clones can also be used to assess the activation of intracellular signaling pathways following the expression of an oncogenic protein in the absence of IL-3. Moreover, Ba/F3 cells are also used for kinase drug discovery and to study the effect of mutations on drug sensitivity and resistance (Warmuth et al., 2007). For example, the effects of BCR-ABL mutations on imatinib, dasatinib, and nilotinib sensitivity and resistance have been assessed using Ba/F3 cells (Azam, Latek, & Daley, 2003; Shah et al., 2007; Weisberg et al., 2007). Ba/F3 cells expressing mutant EGFR have also been used to test the efficacy of EGFR-targeting protein degraders, a more recent approach for targeted therapies (Jang et al., 2020). Efforts have also been made to develop high-throughput drug screening approaches using Ba/F3 cells, such as by testing a panel of 1400 kinase inhibitors against 35 clones of TEL-tyrosine kinase fusion proteins using an automated compound profiling system (Melnick et al., 2006).

Although Ba/F3 cells are a very useful tool for kinase research and drug discovery, several limitations should be taken into consideration, such as their murine and B-cell lineage specific origin, which determines their tissue specific gene expression profile and signaling pathways. These factors should be carefully considered when the molecule of interest that is ectopically introduced into Ba/F3 cells is not murine and/or is not usually expressed in hematological cells, as it might influence the results and their interpretation. Furthermore, it has been observed that especially with weak transforming genes, additional acquired mutations can arise within the transgene of interest in Ba/F3 cells (Watanabe-Smith, Godil, Agarwal, Tognon, & Druker, 2017).

These additional mutations might confound results from the Ba/F3 transforming assay and affect downstream signaling, which would be reflected, for example, in drug sensitivity studies.

Section 1 of this chapter describes the generation of Ba/F3 clones whereas Section 2 details the use of CETSA to evaluate drug binding to ROR1 along with methods suitable for calculation of the binding affinity.

1.2 Thermal shift assay (TSA)

Thermal shift assays provide information about a protein's stability that is highly dependent on the environment or the state of the protein itself. Thus, the TSA is widely applied for evaluating the stability of recombinant proteins, for buffer optimization in crystallization studies, and for screening and analyzing novel ligands together with their mode of action (Niesen, Berglund, & Vedadi, 2007).

TSA measures thermal denaturation of a molecule, which is related to the Gibbs free energy (ΔG) and is temperature-dependent (Niesen et al., 2007). As the temperature increases, the ΔG decreases until a state of equilibrium is reached where half of the protein is unfolded ($\Delta G = 0$). This is the melting temperature (T_m) of the protein. Typically, but not always, binding of a ligand stabilizes the protein, indicated by higher T_m-values compared with the unbound (apo) forms. The stabilizing effect is proportional to the concentration and affinity of the ligand, and hence the change in the melting temperatures (ΔT_m) can be used to profile potential binders (Niesen et al., 2007). TSA values can be used to estimate the dissociation constant (K_d), which is a typical parameter used to evaluate the binding of the ligand (drug) to its substrate. K_d is related to ΔG and to temperature via the equation:

$$\Delta G = -RT\ln\frac{1}{Kd}$$

(where R is the molar gas constant) (Du et al., 2016). Consequently, K_d can be determined by titrating the ligand against a constant protein concentration (Vivoli, Novak, Littlechild, & Harmer, 2014).

Differential scanning fluorimetry (DSF) is a type of TSA where changes in the protein stability are observed using a fluorescent dye, such as SYPRO Orange protein gel stain that is highly active in non-polar environment, i.e., when in contact with the hydrophobic sites of the unfolded protein (Niesen et al., 2007; Semisotnov et al., 1991). The reaction volumes in DSF are small and only micromolar amounts of purified protein are required. Thus, in comparison to label free methods such as isothermal titration calorimetry or differential scanning calorimetry, DSF requires less protein per reaction and can be parallelized for high throughput, which is crucial for drug screening (Wu et al., 2020). Thermal unfolding is monitored using real-time PCR instrument that increases the temperature stepwise, after which the detected fluorescence values are plotted against the temperature values to obtain a sigmoidal curve that represents the two-state transition (Fig. 1). From these data, the T_m can be determined as the maximum value of the first derivative of the curve.

In addition to DSF, light scattering has been used to measure thermal stability of proteins (D'Arcy, 1994; Kopec & Schneider, 2011; Senisterra

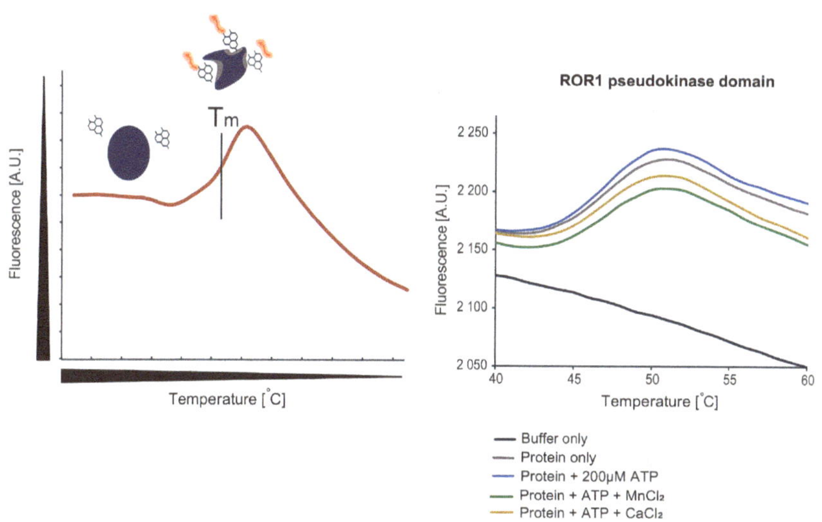

Fig. 1 Schematic representation of the principle behind thermal protein unfolding, and an example of the melting curves. (A) An example of unfolding curve of fluorescence intensity vs temperature in the presence of a fluorescent dye. The dye is depicted as an arbitrary aromatic molecule structure. In the presence of a folded, globular protein (at the baseline of the curve), a basic fluorescence intensity is excited by light (depicted as orange curved arrows). As the protein unfolds, hydrophobic patches (in gray) become exposed, and strong fluorescent light is emitted by the dye molecules bound to it. Following the peak intensity, a gradual decrease is observed, which is mainly due to the protein being removed from solution due to precipitation and aggregation. The melting temperature (T_m) can be evaluated as the midpoint between the baseline and the peak (and calculated as the first derivative). (B) Example of a TSA data set showing various melting curves of ROR1 pseudokinase domain (purified recombinant ROR1 pseudokinase domain in buffer containing 20 mM Tris-HCl pH 8.0, 150 mM NaCl, 20% glycerol and 0.5 mM TCEP) in the presence or absence of ATP and salt ($MnCl_2$ and $CaCl_2$) as well as the negative control (no protein).

et al., 2006). Dynamic light scattering (DLS) evaluates the stabilization and interactions of the protein based on the aggregates formed by denatured proteins. The method is label free and enables the measurement of multiple samples in parallel. This offers an advantage compared to DSF where the fluorescent dye may interact with the potential ligands (such as probes or drugs). However, DLS is not applicable for proteins that do not form aggregates and the method is not as sensitive as DSF. Alternatives to DSF as methods to measure ligand binding include time-resolved fluorescence resonance energy transfer (TR-FRET) (Patel et al., 2008) and high-affinity radiolabeled ligand binding (Hulme & Trevethick, 2010), among others. However, due to its simple set-up, low reaction volume and protein usage, DSF is today a highly used method in drug discovery.

The DSF assay is especially valuable for defining the enzymatic potential of kinases and pseudokinases based on their ability to bind ATP and/or divalent cations, which in active kinases are required for proper catalysis (Murphy et al., 2014). This method has been applied to identify specific domains and amino acid residues that are crucial for the catalytic function. This is especially important in the case of pseudokinases that have more functional variation compared to conventional kinases and can act as a dynamic scaffolds, modulators of protein–protein interactions, or competitors to signaling complex assembly (Mace & Murphy, 2021). The mechanism of action can be studied by comparing the stability of the protein in the apo- vs ligand-bound form, using wild-type and mutated isoforms. For example, DSF studies revealed specific regions in SgK223 that are required for dimerization of the pseudokinase (Patel et al., 2017). Moreover, the method has been used for mechanistic analysis of the pseudokinase TRIB2, assigning two cysteine residues that are crucial for both binding of ATP and covalent inhibitors (Foulkes et al., 2018). Lastly, DSF is a useful method for finding small molecule modulators that could alter the function of (pseudo)kinases, which is important for the development of new therapeutic treatments. It allows the discovery of pseudokinase modulators, which is difficult via other methods than direct bindings (i.e., DSF), due to their diverse functional mechanisms and lack of catalytic activity (Kung & Jura, 2019).

1.3 Cellular thermal shift assay (CETSA)

Similar to TSA discussed above, cellular thermal shift assay (CETSA) is based on the biophysical principle that ligand binding can induce thermal stabilization of target proteins. Instead of using purified proteins, the assay can evaluate binding of drug(s) to its target proteins in cells and tissue samples (Martinez Molina et al., 2013). After its initial description in 2013 (Martinez Molina et al., 2013), CETSA has been used in a wide range of model systems to study thermal stability or compound binding to target proteins (Seashore-Ludlow, Axelsson, & Lundbäck, 2020). For a basic CETSA assay (Fig. 2), cells or cell lysates are treated with the compound of interest and divided into aliquots and then heated at different temperatures followed by isolation of the soluble protein fraction and the quantification of target protein levels in each sample by Western blotting (Martinez Molina et al., 2013).

In addition to obtaining protein melting curves, CETSA can also be used to investigate the effect of drug concentration on the target engagement in so-called isothermal dose-response fingerprint CETSA ($ITDRF_{CETSA,}$ also referred as $ITDR_{CETSA}$ in later publications), where heating time and temperature are the same between samples, but the drug concentration varies (Martinez Molina et al., 2013). It is noted that in $ITDRF_{CETSA}$, the half-saturation point is not directly linked to the binding affinity but it still gives a characteristic fingerprint of target engagement in relation to drug concentration, making it possible to evaluate relative differences in binding affinities as long as measurement parameters, such as cell culture conditions, temperature and time, are kept as constant as possible (Martinez Molina et al., 2013).

Most of the published CETSA experiments rely on Western blotting as a final readout of protein levels, but several other detection methods have also been tested (Seashore-Ludlow et al., 2020). These alternative detection methods are mainly focused on adapting CETSA for a high-throughput format. These high-throughput detection methods include enzyme fragment complementation technology (McNulty et al., 2018), AlphaScreen® (Shaw et al., 2018), and SplitLuc reporter assay (Martinez et al., 2018). Many of these adaptations require specialized reagents and equipment as well as tagged proteins in some cases. One readout approach for high-throughput that has gained momentum in recent years is combining CETSA with mass spectrometry (called MS-CETSA or thermal proteome profiling (TPP)), which enables simultaneous assessments of melting curves for several target proteins (Dai et al., 2019).

The advantage of using CETSA is that the protein–ligand interactions are assessed with proteins in their cellular context, in comparison to traditional TSA where purified proteins and different buffer conditions are used. Although the basic CETSA protocol with Western blot detection can be

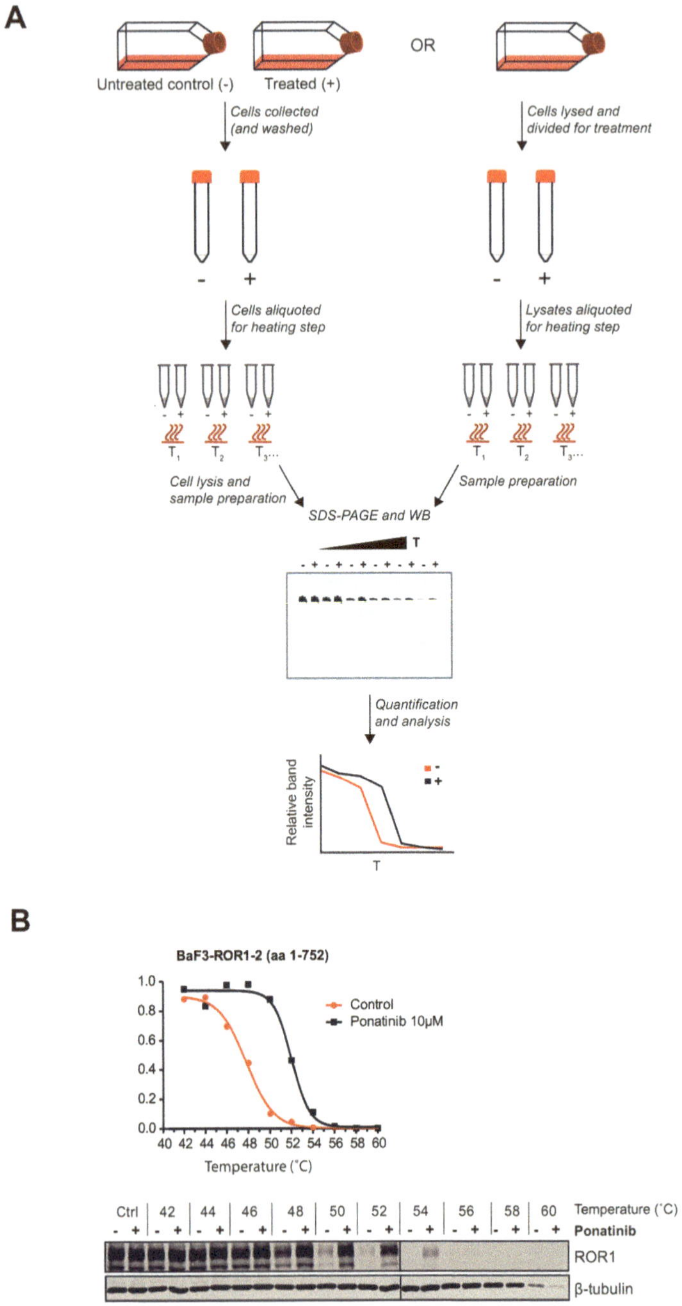

Fig. 2 See figure legend on opposite page.

somewhat labor intensive, one of its advantages is that it does not necessarily require very specific laboratory equipment, which makes it easy to adapt under many circumstances. On the other hand, a major limitation of the Western blot-based detection in CETSA is the requirement for multiple samples comprising large number of cells, which might not be feasible with some cell lines.

Thus far, there are few examples for the utilization of CETSA with pseudokinases, probably linked to the fact that there are not many small molecule inhibitors known to bind to pseudokinases. The binding of afatinib, lapatinib, and erlotinib to TRIB2 has been studied with CETSA, and it was found that afatinib binding destabilized TRIB2, whereas lapatinib partially stabilized TRIB2 (Foulkes et al., 2018). Additionally, CETSA has been used to study bosutinib binding to HER3 and its mutant forms HER3^{KGG} and HER3^{T787M} (Claus et al., 2018). Finally, we have used CETSA to study the binding of ponatinib and GZD824 to ROR1 (Sheetz et al., 2020).

2. Generation of stable Ba/F3 clones expressing receptor pseudokinases using a nucleofection protocol

Here, we describe the protocol for establishing stable Ba/F3 clones expressing receptor pseudokinases such as ROR1, ROR2 and PTK7 using

Fig. 2 Schematic representation of a basic CETSA protocol with Western blot detection. (A) Work flow for CETSA: drug treatment can be done either with intact cells (left) or, alternatively, cells can be lysed and cell lysates treated with drugs (right). Following drug treatment, cells or lysates are aliquoted and heated pairwise (control- and drug-treated samples). Finally, cells are lysed, and the lysates are run on an SDS-PAGE gel followed by Western blotting with antibodies targeting the protein of interest. The signal intensity of each band is quantified and the CETSA melt curve is calculated from the band intensity for each condition as temperature is increased relative to the intensity of the sample heated at the lowest temperature. (B) An example of CETSA showing ponatinib binding to the pseudokinase domain of ROR1. Ba/F3-ROR1 (aa 1–752, with a C-terminal HA-tag) stably expressing cells were treated with vehicle (DMSO, Ctr.) or ponatinib (10 μM) for 2 h before cells were collected and divided into pairs for heating at the indicated time-points. After lysis, samples were run onto SDS-PAGE gel and ROR1 protein levels were visualized by Western blotting using HA-tag. β-Tubulin was used as a loading control. Band intensities were quantified first using β-tubulin as a loading control and then normalized using the value of the non-heated sample.

nucleofection, which is a specialized version of electroporation that introduces nucleic acids into host cell's cytoplasm and/or nucleus. Nucleofection requires specific equipment and reagents and is an efficient way to introduce DNA into cells that are hard to transfect by standard approaches, which includes many cell lines and primary cells.

2.1 Before you begin

- Clone the protein of interest into an appropriate expression plasmid with a suitable selection marker such as antibiotic-based marker, (e.g., we have used pEFIRES-P vector (Hobbs, Jitrapakdee, & Wallace, 1998; Sheetz et al., 2020) with puromycin selection). Ba/F3 cells are usually sensitive to several antibiotics that can be used as selection marker; however, each batch of cells could behave differently, therefore, the antibiotic sensitivity should be tested beforehand by culturing cells with different concentrations of antibiotic for 3–7 days and then monitoring cell survival. For some antibiotics, longer treatment times up to 14 days might be needed. We have found that puromycin has a fast killing activity for sensitive cells, which takes place within 3–4 days. The lowest antibiotic concentration that results in complete cell death should be used as a starting concentration for the antibiotic-based selection

  Note: A tag (e.g., HA- or His-tag) could be added to the protein of interest to facilitate detection via Western blotting.
- Prepare WEHI-3B (Merck, #86013003) conditioned media as described below in Section 2.2
- Culture Ba/F3 cells following thawing to let the cells to recover from freezing for at least a week or 2–3 cell passages before proceeding with nucleofection

  Ba/F3 culture conditions: Ba/F3 cells are cultured in RPMI 1640 media supplemented with IL-3 source (for details see Section 2.2) at 37 °C in a humidified incubator with 5% CO_2. Cells should be split when they reach the density of over 1×10^6 cells/mL and can be split down to $1–3 \times 10^5$ cells/mL. Usually, Ba/F3 cells need to be split every 2–3 days and highest cell density for Ba/F3 cells should be around 2.5×10^6 cells/mL. For each nucleofection 3×10^6 cells are needed, so for 1–2 nucleofections a T25 flask (with 8–10 mL culture) is sufficient, but for three or more, expansion of the culture to a T75 flask (with 15–20 mL culture media) is advisable.

2.2 Materials and equipment

Materials

- Ba/F3 cells (DSMZ #ACC 300)
- WEHI-3B cells (Merck, #86013003)
- Ba/F3 media: RPMI 1640 supplemented with 10% FBS, 2mM L-glutamine, 50U/mL Penicillin, 50μg/mL Streptomycin, and 10% WEHI-3B conditioned media as IL-3 source
- WEHI-3B conditioned media is prepared by culturing WEHI-3B cells in RPMI 1640 supplemented with 10% FBS, 2mM L-glutamine, 50U/mL Penicillin, 50μg/mL Streptomycin at concentration of 1×10^6/mL for 96h, after which the supernatant is collected in sterile conditions and stored frozen at −80°C in 20mL aliquots until further use. For using the conditioned media, the aliquots are thawed at +37°C water bath and added to the Ba/F3 growth media at 10% final concentration

  Alternative: mouse recombinant IL-3 can be used as IL-3 source at concentration 10ng/mL.
- T25 cell culture flasks (e.g., Greiner #690175) and possibly T75 flasks (Greiner #658175)
- SF Cell line 4D-Nucleofector X Kit L X (#V4XC-2012, Lonza), including

o SF Cell Line Nucleofector™ Solution
o Supplement 1
o pmaxGFP™ Vector (for control nucleofections)
o 12 × 100μL Nucleocuvette™ Vessel
o 12 × Single use pipettes (for transferring cells from cuvettes)

Alternative: SG Cell line 4D-Nucleofector X Kit L X (#V4XC-3012, recommended for Ba/F3 cells on Lonza transfection database).

- Puromycin (Merck, #P9620) (or any other antibiotic for cell selection depending on the expression vector that is used)
- antibiotic-free RPMI 1640 (Lonza): supplemented with 10% FBS and 2mM L-glutamine

  Note: RPMI 1640 or other low-calcium/calcium-free media is recommended to be used right after nucleofection regardless of base media used to culture cells. The use of media with high-calcium can lead to increased cell death, possibly because calcium can enter into cells through the pores generated during the nucleofection and activate calcium-sensitive pathways.

Equipment

- 4D-Nucleofector™ Core Unit (#AAF-1003B) and X Unit (#AAF-1003X) (Lonza)
- Cell counter and counting chambers, e.g., Countess II and Countess™ Cell Counting Chamber Slides (#C10228) (ThermoFisher Scientific)
  Alternative: microscope and cell counting chamber
- Centrifuge, e.g., Eppendorf 5810R, swing bucket rotor S-4-104

2.3 Step-by-step method details

Nucleofection

Duration: around 60 min

1. Count the Ba/F3 cells and take 3×10^6 cells/nucleofection
2. Collect cells by centrifugation at $90 \times g$ for 10 min and remove media
3. While cells are in centrifuge, prepare the SF solution by combining for each nucleofection 82 μL of SF Cell Line Nucleofector™ Solution and 18 μL of Supplement 1 that is included in the nucleofection kit. Also, it is advisable to turn on and set up Nucleofector ready for the pulse at this point
   Alternative: Combine one tube of Supplement 1 to one tube of SF Cell Line Nucleofector™ Solution, this can be stored for 3 months.
4. Resuspend cells in 100 μL of SF solution
   Critical: After this point, work fast as it is not recommended to keep cells in SF solution for extended periods. Therefore, it is also best to turn on and set up Nucleofector during the previous collection step.
5. Add 1–3 μg of the purified plasmid to the cells
   Note: The concentration of plasmid should be at least 100 ng/mL as the volume of added plasmid should not exceed 10 μL. In most cases, plasmid purification with Miniprep kit (e.g., Macherey-Nagel NucleoSpin Miniprep kit # 740588.50) is sufficient.
6. Transfer cell suspension into the Nucleocuvette™ Vessel
7. Perform nucleofection with Nucleofector 4D using pulse code DS-137
8. Incubate the Nucleocuvette™ Vessel for 10 min at room temperature (RT).
9. Add 500 μL of antibiotic-free RPMI 1640 media to the Nucleocuvette™ Vessel and incubate for 10 min at 37 °C (in the cell culture incubator with 5% CO_2).
10. Using a small pipette provided in the nucleofection kit, transfer cells from the Nucleocuvette™ Vessel to a T25 cell culture flask with 5.5 mL of Ba/F3 media

Selection with puromycin

Duration: 1–2 weeks

1. 2–3 days after nucleofection add 0.5 mg/mL puromycin to the cell culture flask
2. After an additional 2–3 days, change media and increase puromycin concentration to 1 μg/mL

   Note: 1 μg/mL puromycin usually kills most of the untransfected cells.
3. Continue refreshing puromycin and checking cells every 2–3 days. Split cells as needed and if they are growing well increase puromycin concentration to 2 μg/mL for the last week of the selection to ensure that all non-transfected cells are killed

   After about 2 weeks from the addition of puromycin, the selection of positive Ba/F3 clones should be ready. The expression of the protein of interest can be confirmed by Western blotting and, for cell surface proteins, flow cytometry is also a good option.

2.4 Expected outcomes

This protocol should generate stable Ba/F3 cell lines with high expression of the transfected protein. A high final puromycin concentration ensures that only transfected cells will survive and grow, in a relatively short period of time (typically 2–3 weeks). For instance, ectopic expression of receptor pseudokinases, such as ROR1, ROR2 and PTK7, showed high and uniform cell-surface expression levels of these proteins in Ba/F3 cells based on flow cytometry staining, and these clones were generated on average within 1 month.

2.5 Advantages

Puromycin selection of positive Ba/F3 clones is relatively fast, only 2–3 weeks are required. Moreover, since Ba/F3 cells are fast growing suspension cells, it is possible to grow sufficient numbers of cells for downstream applications in a reasonable timeframe. Additionally, the expression of proteins of interest in this system is quite robust.

Other commonly used method for the introduction of exogenes into Ba/F3 cells is the viral vector-mediated transduction (lenti- or retroviruses). Compared to viral transduction, nucleofection is a much faster method, as the nucleofection protocol takes only about 1 h whereas the viral transduction requires lengthy generation of viruses with careful attention to the viral titer. Furthermore, nucleofection is associated with less biosafety issues due to the absence of viruses that integrate the transgene into the genomic DNA.

Electroporation has also been used to transduce Ba/F3 cells and, in principle, it is very similar to nucleofection but the advantage of nucleofection is that it is able to transfer nucleic acids directly into nucleus, which results in faster expression of the exogenous DNA.

2.6 Limitations

If the transfection efficiency (discussed further below) is not very high, then a large proportion of cells could be lost during the puromycin selection. However, puromycin selection should be done with high enough concentration (up to at least 2 μg/mL) and in proper timeframe in order to avoid spontaneous puromycin resistance in non-transfected cells, while giving enough time for positive cells to adapt and grow during puromycin selection.

It is also important to note that Ba/F3 cells are murine cells, which might be a limiting factor for the efficiency of expression of human proteins. Furthermore, because the described Ba/F3 cell clones involve exogene overexpression, higher expression of the target protein than what would be observed in endogenous context could affect downstream signaling and should be taken into account when interpreting the results.

2.7 Optimization and troubleshooting

Problem: Too many cells are dead after nucleofection

It is normal that some of the cells will die during nucleofection but based on our experience with this protocol there should be clearly alive cells (20–30%) after nucleofection. However, there are differences in Ba/F3 cell batches and culture conditions between different labs that could reduce cell viability during the nucleofection.

Potential solution: Adjusting the pulse code

Lonza provides a table of pulse codes which could be used to adjust nucleofection for either higher viability of cells or higher transfection efficiency.

Problem: All cells die during the puromycin selection/puromycin-selected cells do not express the protein of interest

If all cells die during the puromycin selection it might be that the transfection did not work or the puromycin selection was started too soon and the cells did not have time to manifest resistance before subjected to antibiotic selection. It is highly advisable that each lab should test the sensitivity of their parental Ba/F3 cells to the antibiotic of interest, in our case puromycin. A dose-response curve for 3–7 days should be generated to identify whether the cells are sensitive to the antibiotic and what is the lowest concentration that could be used to start the selection. For Ba/F3 cells, a typical puromycin concentration range tested would be around 0.25–10 μg/mL. In the case where some cells survive the selection, but the target protein is not expressed, the problem might be intrinsic to the plasmid or because the antibiotic selection has been done with too low concentrations of puromycin and/or over too long period of time, causing cells to develop spontaneous puromycin resistance.

Potential solution: Check the expression vector/adjust selection protocol

If the problem seems to originate from the plasmid, the sequence of the plasmid should be checked for errors by sequencing, and if possible, the expression of the transgene should be tested by transient transfection in an orthogonal cell line, e.g., 293T (DSMZ #ACC 635) or Cos-7 (DSMZ #ACC 60). If the problem seems to be rooted in the nucleofection protocol, the GFP-expressing plasmid included in the nucleofection kit can be used to test the nucleofection efficiency. If there are no issues with the plasmid or the nucleofection, then the problem is most likely in the antibiotic selection protocol. It is not recommended to keep cells for long periods of time in low puromycin concentrations following the nucleofection protocol as this might lead to the development of spontaneous antibiotic resistance. In contrast, if after adding 1 μg/mL puromycin the vast majority of cells die, for example, it might be better to remove puromycin completely for few days and let cells to recover before transitioning to a higher puromycin concentration.

3. CETSA with stable Ba/F3 cell lines

The protocol below describes a basic CETSA protocol using Western blot as a readout. In this assay, Ba/F3 clones are treated with the inhibitor and lysed after the heating step. Although many protocols use quick freeze-thaw cycles to lyse cells (Ishii et al., 2017; Jafari et al., 2014; Martinez Molina et al., 2013); here, we have used Triton X100-containing lysis buffer to lyse cells as it has been shown that the presence of detergent improves CETSA results for membrane proteins (Kawatkar et al., 2019; Reinhard et al., 2015), like our proteins of interest.

3.1 Before you begin

- Make sure you have sufficient Ba/F3 cells with over 90% viability for CETSA. You would need around 15×10^6 cells for each control-drug pair of samples
- Prepare drug/inhibitor with appropriate concentration for an optimal dilution

3.2 Materials and equipment

Materials

- Ba/F3 clones (minimum 15 million cells, with over 90% viability)
- Ba/F3 growth media: RPMI 1640 supplemented with 10% FBS, 2 mM L-glutamine, 50 U/mL Penicillin, 50 μg/mL Streptomycin, and 10% WEHI-3B conditioned media as IL-3 source

  Alternative: mouse recombinant IL-3 can be used as IL-3 source at concentration 10 ng/mL.
- PBS
- Drug/inhibitor
- Triton X lysis buffer: 50 mM Tris-HCl pH 7.5, 10% glycerol, 150 mM NaCl, 1 mM EDTA, 1% Triton-X100, 50 mM NaF
- Protein and phosphatase inhibitor cocktails (Bimake, #B14001 and B15001)
- 4 × Laemmli sample buffer (Bio-Rad, #1610747)

Equipment

- Cell counter or microscope and counting chamber
- Cell culture facilities (incubator, laminar hood, etc.)
- Centrifuge for 15 mL conical tubes, e.g., Eppendorf 5810R, swing bucket rotor S-4-104 and for 1.5 mL microcentrifuge tubes, e.g., Eppendorf 5424R
- PCR machine, e.g., Bio-Rad T100 Thermal Cycler #1861096

  Note: Also, other heating devices like heat blocks or water baths could be used, but PCR machine would be preferable as it can reach the target temperature fast and is expected to hold the sample at the precise temperature for the duration of the incubation better than a heat block or water bath
- SDS-PAGE and Western blotting equipment

3.3 Step-by-step method details

CETSA

Duration: 6–7 h

This protocol is adapted from intact cells protocol from Martinez Molina et al. (2013).

1. Count Ba/F3 cells and collect 15×10^6 cells. All these steps are done using regular Ba/F3 growth media. Cells should have good viability, over 90%
2. Spin down cells by centrifugation at $300 \times g$ for 5 min and remove the media
3. Resuspend cells in 20 mL of Ba/F3 media and divide cell suspension equally into two cell culture T25 flasks (10 mL/flask)
4. Add the drug at high concentration, e.g., 10 μM to one of the flasks and DMSO to the other flask

   Note: The concentration should be optimized for each drug. Using a high drug concentration (5–10 μM) would ensure that there is sufficient amount of drug to bind the target protein in cells. To determine what is the best drug concentration that would efficiently bind to the target in cells, a log-fold drug dilution (from nM to mM concentrations) could be carried out by incubating cells with various drug concentrations and monitoring the drug effect on cells (e.g., inhibition of kinase activity for kinase inhibitors, cell cytotoxic effect, depending on the drug mode of action) for each dilution. For kinase inhibitors such as ponatinib, a 10 μM concentration is sufficient for a strong target binding in cells. As a control, use the solvent in which the drug has been dissolved.
5. Incubate cells 2 h at 37 °C in a humidified cell culture incubator with 5% CO_2
6. After the incubation, collect cells by centrifugation at $300 \times g$ for 5 min and remove media
7. Wash cells once with 10 mL of PBS, collect cells by centrifugation at $300 \times g$ for 5 min and remove PBS
8. Resuspend cells in 375 μL of PBS
9. Divide both drug-treated and control cells into 13 of 25 μL aliquots in 0.2 mL PCR tubes

   Note: Each aliquot should have approximately 0.5×10^6 cells at this point.
10. Sample heating:

a) Heat one aliquot per condition (control and drug treated) for 3 min at a different temperature, ranging from 42 °C to 64 °C with 2 °C increments. Leave one aliquot as unheated control

Note: The temperature range used here is expected to work with most proteins. If higher temperatures are needed, it is important to test that the cells remain intact at those temperatures. It has been shown that cells might start to lose their membrane integrity at temperatures above 65 °C (Martinez Molina et al., 2013). After the initial experiment, the temperature range can be narrowed.

b) After heating, let each sample pair to cool down at room temperature for 3 min and then transfer the tubes to ice.

11. Transfer samples to new 1.5 mL microcentrifuge tubes and collect cells by centrifugation at 500 × *g* for 10 min at 4 °C. Aspirate to remove supernatant.

Note: Transferring samples to new tubes can be avoided if there is a centrifuge that can accommodate 0.2 mL PCR tubes and can reach the centrifugation speed required to clear the lysates in the next step.

12. Sample lysis

a) Add 20 μL of Triton X lysis buffer with protease and phosphatase inhibitors (Bimake #B14001 and #B15001, Biotool, Houston, TX, USA) to each tube

b) Incubate samples for 15 min on ice

c) Clear lysates by centrifugation at 20,000 × *g* for 20 min at 4 °C

d) Transfer supernatant (18 μL) to new tubes and add 6 μL of 4 × Laemmli sample buffer

Pause point: Samples can be stored at −20 °C.

13. Run samples on an SDS-PAGE gel and continue with standard Western Blotting protocol. Briefly, first transfer proteins from a gel to a membrane using wet or semi-dry transfer followed by blocking of membrane using 4% *w*/*v* BSA (bovine serum albumin) in TBS (Tris Buffer Saline) with 0.5% Tween-20 (Merck, # 8221840050). After blocking, probe the membrane with primary antibody that recognizes the target protein (or the tag used in construct), typically diluted 1:1000 in TBS 0.05% Tw20 (Tween 20) for overnight with gentle shacking at 4 °C. Of note, each antibody should be diluted according to provider's instruction, and if required, further optimized. The next day, wash the membrane 3 × with TBS 0.1% Tw20 and add the secondary antibody such as IRDye® 680RD Donkey anti-Goat IgG or IRDye® 800CW Donkey anti-Mouse IgG or IRDye® 680RD Donkey anti-Rabbit IgG (LI-COR, Lincoln, NE, USA) at 1:10,000 dilution in TBS 0.05% Tw20 for 1.5 h with gentle shacking at room temperature. Finally, wash the membrane 3 × TBS 0.1% Tw20 with and scan with Odyssey® CLx Imaging System (LI-COR). Images are processed with Image Studio Lite (LI-COR).

Note: Protein detection could be done using an antibody specific for the protein of interest or using an antibody against the tag, if available.

3.4 Expected outcomes

If the drug binds to the protein of interest, it leads to its thermal stabilization that would be reflected in an increased protein level compared to the control treated sample heated at the same temperature. Target proteins without the drug bound to them would unfold more readily upon heating and aggregate, eventually precipitating from the soluble fraction (Martinez Molina et al., 2013). As an example, binding of ponatinib to the ROR1 pseudokinase domain stabilizes ROR1 protein in comparison with the non-drug treated sample that is degraded during heat-induced denaturation (Fig. 2B). In some cases, like afatinib with TRIB2, drug binding can also result in thermal destabilization that would appear as decreased protein levels compared to control treated sample (Foulkes et al., 2018).

3.5 Quantification and statistical analysis

Analysis is conducted by quantifying band intensities for the target protein at different temperature conditions on the Western blots, using suitable software (e.g., we have used Image Studio Lite; and IRDye secondary antibodies, e.g., IRDye® 800CW Donkey anti-Mouse IgG #926-32212 from LI-COR). Quantified band intensities are normalized using the value of the sample heated at the lowest temperature. To obtain the melting curve, relative band intensities are plotted against the heating temperature in GraphPad PRISM (or equivalent software). For a proper melting curve, you would need at least three independent replicate experiments.

3.6 Advantages

In CETSA, drug-target binding can be rapidly detected in cells and the optimization of drug concentrations and incubation times is relatively easy. If tagged proteins are used, even rare targets lacking good specific antibodies can be studied. The use of suspension cells, like Ba/F3 cells, makes cell handling easy and avoids the cell detachment step that can be cumbersome and can also affect cell viability.

3.7 Limitations

The protocol presented here only enables analysis of the drug-target interactions that affect thermal stability of the target protein. Western blotting is not a very sensitive detection method if good antibodies are not available for the target protein. Also, the drug concentration needs to be relatively high to ensure sufficient drug-target binding for the detection of changes in thermal stability. Finally, some drugs might be too toxic for the cells to enable the analysis or might have poor membrane permeability and will be unable to cross the cell membrane to react with the target.

3.8 Optimization and troubleshooting

Problem: Cells die during the 2h drug incubation period

Depending on drugs used, it is possible that with high drug concentration cells are already dying during the 2h incubation period used here for drug treatment. This could be the case for drugs that have multiple targets, like many kinase inhibitors, and thus they have very strong cytotoxic effect.

Potential solution: Decreasing drug concentration/incubation time or adjust the cell number after drug treatment

The first alternative is to further optimize the drug to be used at lower concentrations or reduce the incubation time. The other alternative to overcome this problem is to count the cells after the drug incubation step and take an equal number of cells from drug- and control-treated cells for next steps of the protocol.

4. Summary

A major challenge associated with working with pseudokinases that are catalytically impaired, or fully inactive, is finding appropriate methods to measure their functional states and their drug-binding properties. Likewise, it is challenging to evaluate whether post-translational modifications are important for regulation of these properties. Wnt ligand binding receptor pseudokinases such as ROR1, ROR2 and PTK7, have been shown to be catalytically inactive (Murphy et al., 2014) and to likely function more like scaffolding proteins that modulate intracellular signaling. Their biochemical and structural properties have only recently been investigated (Sheetz et al., 2020). Using an approach frequently employed in studies of receptor tyrosine kinases (Collins et al., 1988; Warmuth et al., 2007), we expressed ROR1, ROR2 and PTK7 in the interleukin-3 (IL-3) dependent Ba/F3 murine pro-B cell line to obtain stable Ba/F3 clones via puromycin selection. Whereas parental Ba/F3 cells die following IL-3 withdrawal, a significant number of those expressing ROR1, ROR2 and PTK7 survived in the absence of IL-3. In addition, ROR1-expressing cells ultimately achieved IL-3 independent proliferation (Sheetz et al., 2020)—consistent with the contribution of ROR1 to chronic lymphocytic leukemia (CLL) and some other B-cell lineage hematological malignancies (Karvonen, Perttilä, Niininen, Barker, & Ungureanu, 2018).

Given the status of Wnt-binding receptor pseudokinases as therapeutic targets, we sought to identify inhibitors that could bind to and modulate the signaling properties of their pseudokinase domains. To identify small molecules that bind the ROR1, ROR2, PTK7 and RYK pseudokinase domains, we used DSF to screen a total of 1486 compounds from the published protein kinase inhibitor set and the FDA-approved drug set from Selleckchem (Sheetz et al., 2020). Despite the relatively large and kinase inhibitor-focused screening set, we failed to identify any hits for PTK7, ROR2, or RYK. For ROR1, however, we identified two closely related inhibitors that significantly increased T_m—indicating strong binding. These inhibitors included ponatinib, a multitargeted kinase inhibitor approved by the FDA for chronic myeloid leukemia (CML) and Philadelphia chromosome-positive acute lymphoblastic leukemia (ALL), and GZD824—which is structurally related to ponatinib (O'Hare et al., 2009; Ren et al., 2013). Using our stable Ba/F3-ROR1 clones and CETSA, we further showed that both ponatinib and GZD824 bind to and stabilize full-length ROR1 in Ba/F3 cells when added at 10 μM concentration.

Thus, as we had shown for the ROR1 pseudokinase, application of the stable expressing Ba/F3 clones and CETSA (Sheetz et al., 2020), provides a model system where the functional consequences of pathogenic mutations in ROR1 on its signaling and sensitivity to drugs can now be assessed.

5. Key resource table

Reagent or resource	Source	Identifier
Chemicals, Peptides, and Recombinant Proteins		
Puromycin	Sigma	P8833-10MG
RPMI	Lonza	12-167F
Penicillin-Streptomycin	Lonza	17-602E

Continued

—cont'd

Reagent or resource	Source	Identifier
FBS	ThermoFisher Scientific	10270-106
BSA	ThermoFisher Scientific	BP9702-100
Tween 20	Merck	8221840050
Protease and Phosphatase Inhibitor Cocktails	Bimake, Biotool, Houston, TX, USA	B14001 B15001
Critical Commercial Assays		
SF Cell line 4D-Nucleofector X Kit L	Lonza	V4XC-2012
Experimental Models: Cell Lines		
Mouse: *M. musculus* Ba/F3 pro-B cells	DSMZ	ACC-300,
Mouse *M. musculus* WEHI-3B cells	Merck	86013003
Recombinant DNA		
pEFIRES-puromycin	(Hobbs et al., 1998)	N/A
Software and Algorithms		
Image Studio Lite 5.0	LI-COR	
GraphPad PRISM	GraphPad Software	
Other		
Nucleofector 4D (Core unit + X unit)	Lonza	

References

Azam, M., Latek, R. R., & Daley, G. Q. (2003). Mechanisms of autoinhibition and STI-571/imatinib resistance revealed by mutagenesis of BCR-ABL. *Cell*, *112*, 831–843.

Claus, J., Patel, G., Autore, F., Colomba, A., Weitsman, G., Soliman, T. N., et al. (2018). Inhibitor-induced HER2-HER3 heterodimerisation promotes proliferation through a novel dimer interface. *eLife*, 7, e32271.

Collins, M. K., Downward, J., Miyajima, A., Maruyama, K., Arai, K., & Mulligan, R. C. (1988). Transfer of functional EGF receptors to an IL3-dependent cell line. *Journal of Cellular Physiology*, *137*, 293–298.

Dai, L., Prabhu, N., Yu, L. Y., Bacanu, S., Ramos, A. D., & Nordlund, P. (2019). Horizontal cell biology: Monitoring global changes of protein interaction states with the proteome-wide cellular thermal shift assay (CETSA). *Annual Review of Biochemistry*, *88*, 383–408.

Daley, G. Q., & Baltimore, D. (1988). Transformation of an interleukin 3-dependent hematopoietic cell line by the chronic myelogenous leukemia-specific P210bcr/abl protein. *Proceedings of the National Academy of Sciences of the United States of America*, *85*, 9312–9316.

D'Arcy, A. (1994). Crystallizing proteins – A rational approach? *Acta Crystallographica Section D*, *50*, 469–471.

Du, X., Li, Y., Xia, Y.-L., Ai, S.-M., Liang, J., Sang, P., et al. (2016). Insights into protein–ligand interactions: Mechanisms, models, and methods. *International Journal of Molecular Sciences*, *17*, 144.

Foulkes, D. M., Byrne, D. P., Yeung, W., Shrestha, S., Bailey, F. P., Ferries, S., et al. (2018). Covalent inhibitors of EGFR family protein kinases induce degradation of human tribbles 2 (TRIB2) pseudokinase in cancer cells. *Science Signaling*, *11*. https://doi.org/10.1126/scisignal.aat7951.

Gwanmesia, P. M., Romanski, A., Schwarz, K., Bacic, B., Ruthardt, M., & Ottmann, O. G. (2009). The effect of the dual Src/Abl kinase inhibitor AZD0530 on Philadelphia positive leukaemia cell lines. *BMC Cancer*, *9*, 53.

Hobbs, S., Jitrapakdee, S., & Wallace, J. C. (1998). Development of a bicistronic vector driven by the human polypeptide chain elongation factor 1alpha promoter for creation of stable mammalian cell lines that express very high levels of recombinant proteins. *Biochemical and Biophysical Research Communications*, *252*, 368–372.

Hornakova, T., Springuel, L., Devreux, J., Dusa, A., Constantinescu, S. N., Knoops, L., et al. (2011). Oncogenic JAK1 and JAK2-activating mutations resistant to ATP-competitive inhibitors. *Haematologica*, *96*, 845–853.

Hulme, E. C., & Trevethick, M. A. (2010). Ligand binding assays at equilibrium: Validation and interpretation. *British Journal of Pharmacology*, *161*, 1219–1237.

Ishii, T., Okai, T., Iwatani-Yoshihara, M., Mochizuki, M., Unno, S., Kuno, M., et al. (2017). CETSA quantitatively verifies in vivo target engagement of novel RIPK1 inhibitors in various biospecimens. *Scientific Reports*, *7*, 13000.

Jafari, R., Almqvist, H., Axelsson, H., Ignatushchenko, M., Lundbäck, T., Nordlund, P., et al. (2014). The cellular thermal shift assay for evaluating drug target interactions in cells. *Nature Protocols*, *9*, 2100–2122.

Jang, J., To, C., De Clercq, D. J. H., Park, E., Ponthier, C. M., Shin, B. H., et al. (2020). Mutant-selective allosteric EGFR degraders are effective against a broad range of drug-resistant mutations. *Angewandte Chemie (International Ed. in English)*, *59*, 14481–14489.

Karvonen, H., Perttilä, R., Niininen, W., Barker, H., & Ungureanu, D. (2018). Targeting Wnt signaling pseudokinases in hematological cancers. *European Journal of Haematology*, *101*, 457–465.

Kawatkar, A., Schefter, M., Hermansson, N.-O., Snijder, A., Dekker, N., Brown, D. G., et al. (2019). CETSA beyond soluble targets: A broad application to multipass transmembrane proteins. *ACS Chemical Biology*, *14*, 1913–1920.

Kopec, J., & Schneider, G. (2011). Comparison of fluorescence and light scattering based methods to assess formation and stability of protein-protein complexes. *Journal of Structural Biology*, *175*, 216–223.

Kung, J. E., & Jura, N. (2019). Prospects for pharmacological targeting of pseudokinases. *Nature Reviews. Drug Discovery*, *18*, 501–526.

Liu, P. C. C., Caulder, E., Li, J., Waeltz, P., Margulis, A., Wynn, R., et al. (2009). Combined inhibition of Janus kinase 1/2 for the treatment of JAK2V617F-driven neoplasms: Selective effects on mutant cells and improvements in measures of disease severity. *Clinical Cancer Research*, *15*, 6891–6900.

Mace, P. D., & Murphy, J. M. (2021). There's more to death than life: Noncatalytic functions in kinase and pseudokinase signaling. *Journal of Biological Chemistry*, *296*, 100705.

Martinez, N. J., Asawa, R. R., Cyr, M. G., Zakharov, A., Urban, D. J., Roth, J. S., et al. (2018). A widely-applicable high-throughput cellular thermal shift assay (CETSA) using split Nano luciferase. *Scientific Reports, 8*, 9472.

Martinez Molina, D., Jafari, R., Ignatushchenko, M., Seki, T., Larsson, E. A., Dan, C., et al. (2013). Monitoring drug target engagement in cells and tissues using the cellular thermal shift assay. *Science (New York, N.Y.), 341*, 84–87.

McNulty, D. E., Bonnette, W. G., Qi, H., Wang, L., Ho, T. F., Waszkiewicz, A., et al. (2018). A high-throughput dose-response cellular thermal shift assay for rapid screening of drug target engagement in living cells, exemplified using SMYD3 and IDO1. *SLAS DISCOVERY: Advancing the Science of Drug Discovery, 23*, 34–46.

Melnick, J. S., Janes, J., Kim, S., Chang, J. Y., Sipes, D. G., Gunderson, D., et al. (2006). An efficient rapid system for profiling the cellular activities of molecular libraries. *Proceedings of the National Academy of Sciences, 103*, 3153–3158.

Murphy, J. M., Zhang, Q., Young, S. N., Reese, M. L., Bailey, F. P., Eyers, P. A., et al. (2014). A robust methodology to subclassify pseudokinases based on their nucleotide-binding properties. *The Biochemical Journal, 457*, 323–334.

Niesen, F. H., Berglund, H., & Vedadi, M. (2007). The use of differential scanning fluorimetry to detect ligand interactions that promote protein stability. *Nature Protocols, 2*, 2212–2221.

O'Hare, T., Shakespeare, W. C., Zhu, X., Eide, C. A., Rivera, V. M., Wang, F., et al. (2009). AP24534, a pan-BCR-ABL inhibitor for chronic myeloid leukemia, potently inhibits the T315I mutant and overcomes mutation-based resistance. *Cancer Cell, 16*, 401–412.

Palacios, R., Henson, G., Steinmetz, M., & McKearn, J. P. (1984). Interleukin-3 supports growth of mouse pre-B-cell clones in vitro. *Nature, 309*, 126–131.

Palacios, R., & Steinmetz, M. (1985). Il-3-dependent mouse clones that express B-220 surface antigen, contain Ig genes in germ-line configuration, and generate B lymphocytes in vivo. *Cell, 41*, 727–734.

Patel, O., Griffin, M. D. W., Panjikar, S., Dai, W., Ma, X., Chan, H., et al. (2017). Structure of SgK223 pseudokinase reveals novel mechanisms of homotypic and heterotypic association. *Nature Communications, 8*, 1157.

Patel, R., Lebrun, L. A., Wang, S., Howett, L. J., Thompson, P. A., Appleman, J. R., et al. (2008). ATLAS- -a high-throughput affinity-based screening technology for soluble proteins: Technology application using p38 MAP kinase. *Assay and Drug Development Technologies, 6*, 55–68.

Reddy, E. P., Korapati, A., Chaturvedi, P., & Rane, S. (2000). IL-3 signaling and the role of Src kinases, JAKs and STATs: A covert liaison unveiled. *Oncogene, 19*, 2532–2547.

Reinhard, F. B. M., Eberhard, D., Werner, T., Franken, H., Childs, D., Doce, C., et al. (2015). Thermal proteome profiling monitors ligand interactions with cellular membrane proteins. *Nature Methods, 12*, 1129–1131.

Ren, X., Pan, X., Zhang, Z., Wang, D., Lu, X., Li, Y., et al. (2013). Identification of GZD824 as an orally bioavailable inhibitor that targets phosphorylated and nonphosphorylated breakpoint cluster region-Abelson (Bcr-Abl) kinase and overcomes clinically acquired mutation-induced resistance against imatinib. *Journal of Medicinal Chemistry, 56*, 879–894.

Seashore-Ludlow, B., Axelsson, H., & Lundbäck, T. (2020). Perspective on CETSA literature: Toward more quantitative data interpretation. *SLAS Discovery: Advancing Life Sciences R & D, 25*, 118–126.

Semisotnov, G. V., Rodionova, N. A., Razgulyaev, O. I., Uversky, V. N., Gripas', A. F., & Gilmanshin, R. I. (1991). Study of the "molten globule" intermediate state in protein folding by a hydrophobic fluorescent probe. *Biopolymers, 31*, 119–128.

Senisterra, G. A., Markin, E., Yamazaki, K., Hui, R., Vedadi, M., & Awrey, D. E. (2006). Screening for ligands using a generic and high-throughput light-scattering-based assay. *Journal of Biomolecular Screening*, *11*, 940–948.

Shah, N. P., Skaggs, B. J., Branford, S., Hughes, T. P., Nicoll, J. M., Paquette, R. L., et al. (2007). Sequential ABL kinase inhibitor therapy selects for compound drug-resistant BCR-ABL mutations with altered oncogenic potency. *The Journal of Clinical Investigation*, *117*, 2562–2569.

Shaw, J., Leveridge, M., Norling, C., Karén, J., Molina, D. M., O'Neill, D., et al. (2018). Determining direct binders of the androgen receptor using a high-throughput cellular thermal shift assay. *Scientific Reports*, *8*, 163.

Sheetz, J. B., Mathea, S., Karvonen, H., Malhotra, K., Chatterjee, D., Niininen, W., et al. (2020). Structural insights into Pseudokinase domains of receptor tyrosine kinases. *Molecular Cell*, *79*, 390–405. e7.

Vivoli, M., Novak, H. R., Littlechild, J. A., & Harmer, N. J. (2014). Determination of protein-ligand interactions using differential scanning fluorimetry. *Journal of Visualized Experiments*, *51809*.

Warmuth, M., Kim, S., Gu, X., Xia, G., & Adrián, F. (2007). Ba/F3 cells and their use in kinase drug discovery. *Current Opinion in Oncology*, *19*, 55–60.

Watanabe-Smith, K., Godil, J., Agarwal, A., Tognon, C., & Druker, B. (2017). Analysis of acquired mutations in transgenes arising in Ba/F3 transformation assays: Findings and recommendations. *Oncotarget*, *8*, 12596–12606.

Weisberg, E., Catley, L., Wright, R. D., Moreno, D., Banerji, L., Ray, A., et al. (2007). Beneficial effects of combining nilotinib and imatinib in preclinical models of BCR-ABL+ leukemias. *Blood*, *109*, 2112–2120.

Wu, T., Yu, J., Gale-Day, Z., Woo, A., Suresh, A., Hornsby, M., et al. (2020). Three essential resources to improve differential scanning fluorimetry (DSF) experiments. *BioRxiv*. 2020.03.22.002543.

CHAPTER THIRTEEN

Conformational control and regulation of the pseudokinase KSR via small molecule binding interactions

Arthur Chow[†], Zaigham M. Khan[†], William M. Marsiglia[†], and Arvin C. Dar*

Department of Oncological Sciences, Department of Pharmacological Sciences, The Tisch Cancer Institute, Mount Sinai Centre for Therapeutic Discovery, Icahn School of Medicine at Mount Sinai, New York, NY, United States

*Corresponding author: e-mail address: arvin.dar@mssm.edu

Contents

[†] Co-first, listed alphabetically.

Methods in Enzymology, Volume 667
ISSN 0076-6879
https://doi.org/10.1016/bs.mie.2022.03.039

Abstract

Pseudokinases often operate through functionally related enzymes and receptors. A prime example is the pseudokinase KSR (Kinase Suppressor of RAS), which can act as both an amplifier and inhibitor of members in the RAS-MAPK (Mitogen Activated Protein Kinase) signaling pathway. KSR is structurally related to the active RAF kinases over multiple domains; moreover, the pseudokinase domain of KSR forms physical and regulatory complexes with both RAF and MEK through distinct interfaces. Characterization of small molecule interactions on KSR has been used to uncover novel chemical tools and understand the mechanism of action of clinical drugs. Here, we elaborate on assays and structural methods for measuring binding at orthosteric and interfacial binding sites on KSR. These distinct small molecule pockets provide therapeutic paths for targeting KSR1 and KSR2 pseudokinases in disease, including in RAS and RAF mutant cancers.

1. KSR identification

Kinase Suppressor of RAS (KSR) was originally identified in forward genetic screens conducted in the fly and worm as a modulator of RAS signaling (Kornfeld, Hom, & Horvitz, 1995; Sundaram & Han, 1995; Therrien et al., 1995). Specifically, KSR was identified by virtue of mutations that suppress phenotypes associated with oncogenic RAS signaling. Epistatic analysis further suggested that KSR functioned downstream of RAS but upstream or parallel to RAF (Karim et al., 1996; Sundaram & Han, 1995). We now know that KSR and RAF are highly related, with KSR characterized as the pseudokinase homolog of the RAF kinases (Lavoie & Therrien, 2015). Indeed, all metazoan species that we have analyzed possess at least a single KSR and RAF paralog (Khan et al., 2020; Kondo et al., 2019; Lavoie et al., 2018; Liau et al., 2020; Mysore et al., 2021; Park et al., 2019), demonstrating broad coexistence and potentially conserved pseudokinase-kinase functions within the RAS-mitogen-activated protein kinase/extracellular signal-regulated kinase (MAPK/ERK) signaling system.

2. KSR domain structure

Humans possess two KSRs (KSR1 and KSR2) and three RAFs (ARAF, BRAF, and CRAF). Below we focus on KSR1 and BRAF as exemplary, but also highlight some general features within each subfamily. KSR and RAF proteins share several elements including C-terminal kinase domains (Fig. 1). The N-terminal regions differ whereby the RAF kinases possess the prototypical RAS-binding domain (RBD), which participates in

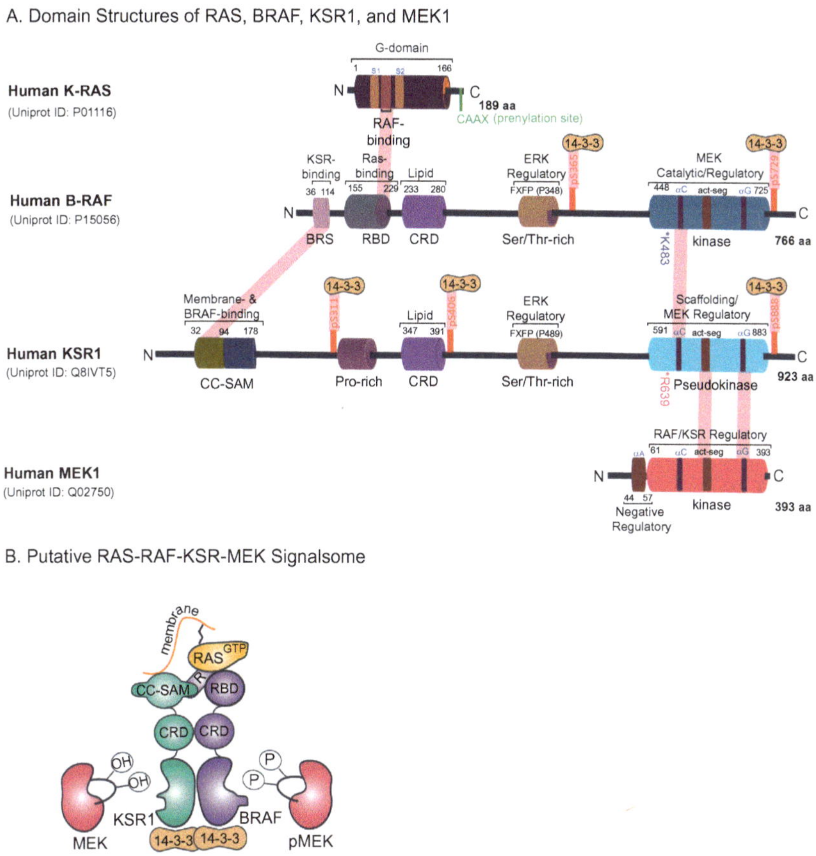

Fig. 1 Domain structure, regulatory motifs, and interaction sites between RAS/RAF/KSR/MEK/14-3-3. (A) The domain structures of human KRAS, BRAF, KSR1, and MEK1 are shown and numbered according to the listed Uniprot accession numbers. Domain boundaries are marked based on structural data available in the PDB or are based on structure-based sequence alignments. 14-3-3 binding sites are based on published studies (Müller, Ory, Copeland, Piwnica-Worms, & Morrison, 2001; Xing, Kornfeld, & Muslin, 1997); in BRAF, these binding sites have been recently confirmed via structural studies (Kondo et al., 2019; Liau, Wendorff, et al., 2020; Park et al., 2019). In KSR1, the 14-3-3 binding sites are based on experimental data or predicted based on structural alignments (Cacace et al., 1999). (B) A cartoon depiction of the putative signalsome including KSR1 highlighting expected interactions sites based on known structures of fragments, as well as functional data (Brennan et al., 2011; Dhawan, Scopton, & Dar, 2016; Khan et al., 2020; Kondo et al., 2019; Lavoie et al., 2018; Liau, Wendorff, et al., 2020; Park et al., 2019).

direct interactions with active GTP-bound RAS (Herrmann, Martin, & Wittinghofer, 1995; Nassar et al., 1995; Van Aelst, Barr, Marcus, Polverino, & Wigler, 1993; Weber, Slupsky, Kalmes, & Rapp, 2001; Zhang et al., 1993). Among the RAF homologs, BRAF also contains a specific N-terminal region (BRS) that is involved in homotypic and heterotypic dimerization interactions, including with KSR1 or KSR2 (Lavoie et al., 2018). In contrast, the KSR proteins possess an N-terminal coiled-coiled sterile alpha motif (CC-SAM) that has been shown to participate in targeting KSR1 to the plasma membrane (Koveal et al., 2012). Both KSR and RAF kinases possess a cysteine-rich domain (CRD), which has been shown to be required for interactions and also activation of RAF1 by KSR1 (Michaud et al., 1997) as well as to participate directly in binding to RAS (Brtva et al., 1995; Cutler, Stephens, Saracino, & Morrison, 1998; Tran et al., 2021). Recent structures have further demonstrated that the CRD folds back onto the kinase domain in the auto-inhibited structures of BRAF via interactions that are reinforced through a 14-3-3 dimer (Heppner & Eck, 2021; Park et al., 2019). 14-3-3 has also been demonstrated to directly associate with KSR1 through several phosphorylated residues, including Ser297 and Ser392 within the areas flanking the CRD (Cacace et al., 1999). Although not experimentally established, but by analogy to BRAF, these and other phospho-regulatory sites may be important in maintaining an auto-inhibited and putative monomeric conformation of full-length KSR.

3. Higher-order KSR–RAF–MEK complexes: Heterodimers and heterotetramers

At the moment, experimental structures for the "KSR signalosome" are lacking, however, insights from fragments reveal the complexity of binding between distinct components of the RAS-MAPK cascade and imply the potential for overlapping and possible competing interactions between KSR, RAF and MEK to regulate RAS–MAPK pathway output (Khan et al., 2020; Kondo et al., 2019; Lavoie et al., 2018; Liau, Wendorff, et al., 2020; Mysore et al., 2021; Park et al., 2019). Current models suggest that the signalosome occurs when KSR bridges to RAS through heterodimerization with RAF kinases and thereby indirectly through Ras-binding domain (RBD) interactions. Within the larger assembly, KSR directly interacts with both RAF and MEK through the pseudokinase domain (Brennan et al., 2011; Khan et al., 2020; Lavoie et al., 2018; Rajakulendran, Sahmi, Lefrançois, Sicheri, & Therrien, 2009).

Crystal structures have revealed that the pseudokinase domain of KSR1/2 binds to MEK via the lower C-lobe centered on interactions mediated by helix αG and the activation loop (Fig. 2; Khan et al., 2020; Dhawan

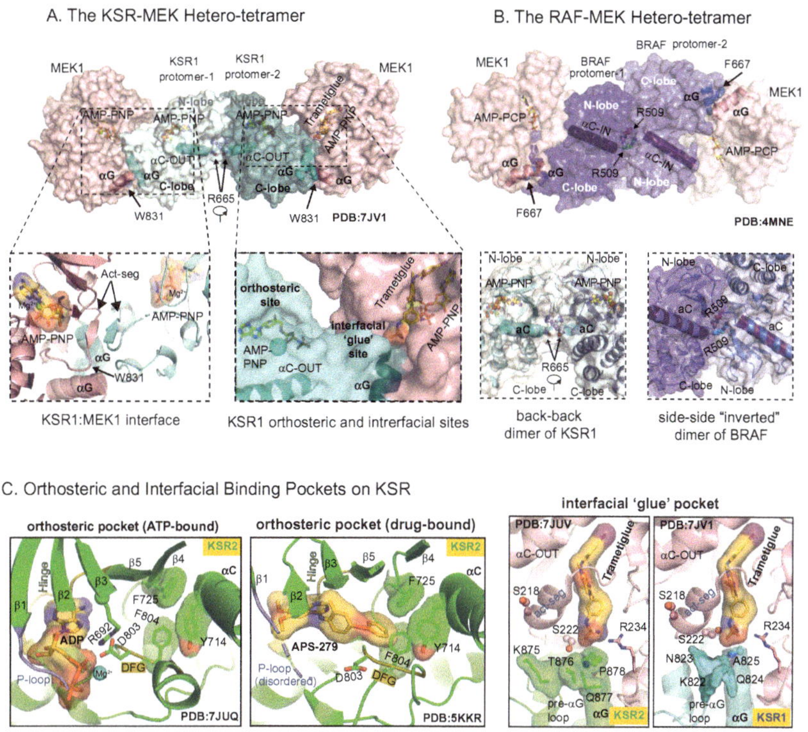

Fig. 2 Structural features of KSR:MEK and RAF–MEK complexes. Tetrameric assembly of KSR1:MEK (A) and BRAF:MEK (B). The available crystal structures of KSR1-bound MEK (PDB ID: 7JV1; Khan et al., 2020) provide a snap-shot of KSR:KSR homodimerization centered around R665. MEK binds KSR on the flanking sides of the KSR homodimer. The left inset represents the heterodimerization interface between KSR1 and MEK1 centered around the respective helix ⊠G (W831 of KSR1 is shown as spheres) and the activation loops, which form a 2-stranded antiparallel β-sheet. The right inset highlights the location of the KSR1 orthosteric pocket with AMP-PNP bound and interfacial glue pocket. (B) A crystal structure of BRAF:MEK (PDB ID: 4MNE; Haling et al., 2014) highlighting BRAF homodimerization centered around R509. The lower insets highlight the homodimerization interface of KSR1 and BRAF. Whereas KSR1 homodimerization involves an extensive interaction of residues at the N-lobe of the two KSR1 molecules that results in a back-back dimer, two molecules of BRAF are assembled in an inverted fashion from N- to C-lobe. (C) Orthosteric pockets (left; bound to ADP, PDB ID 7JUQ; Khan et al., 2020) and interfacial binding pockets (middle left; bound to APS-2-79, PDB ID 5KKR; Dhawan et al., 2016). In KSR2, APS-2-79 directly projects toward a back pocket that is lined by hydrophobic residues Y714 (helix ⊠C), F725 (strand β4) and F825 (DFG motif). The binding of APS-2-79 results in the restructuring of the activation loop, downstream of the DFG motif (not shown) and disordering of P-loop (Dhawan et al., 2016). Right two panels show interfacial binding pockets of KSR1-MEK (PDB ID 7JV1) and KSR2-MEK (PDB ID 7JUV) occupied with trametiglue. The terminal sulfamide motif of trametiglue occupies the interfacial space between KSR and MEK, with complementarity between trametiglue and KSR spanning the pre-helix ⊠G loop and the tip of the helix ⊠G (Khan et al., 2020).

et al., 2016), whereas binding to RAF occurs through the dimer interface centered on interactions in the N-lobe and helix αC (Rajakulendran et al., 2009). In the inactive state conformation of KSR1 and KSR2, helix αC adopts an OUT configuration (Khan et al., 2020; Brennan et al., 2011). Differences between inhibitor and ATP/AMP-PNP structures have revealed additional variations in the P-loop and N-terminal portion of the activation segment that have highlighted potential intermediate-state conformations of KSR (Dhawan et al., 2016). For example, in previous work we proposed that the ATP-bound state permits the transition from the inactive to active state of KSR through alterations in the conformation of the P-loop and an "induced" locked within the KSR activation segment that maintains the Ser218–Ser222 sites on bound MEK within a protected arrangement (Dhawan et al., 2016). The first characterized inhibitor pocket on KSR is referred to as the orthosteric site as this pocket overlaps directly with the ATP-binding site.

A secondary pocket on KSR exists directly at the interface for MEK binding and is mediated through the region directly preceding helix αG. We refer to this pocket as the interfacial "glue" site. This site enables binding and direct interactions with allosteric MEK inhibitors, including the clinical anti-cancer drug trametinib and is exemplified by our structures of the KSR1–MEK1 and KSR2–MEK1 complexes bound to a tool compound that we created called trametiglue (Khan et al., 2020). Trametiglue is distinct from trametinib in that the former can form high affinity and glue-like interactions with both KSR–MEK and, as we have speculated, RAF–MEK complexes. Moreover trametiglue is markedly enhanced in terms of potency of growth inhibition in KRAS mutant cell lines relative to trametinib and other MEKi including CH5126766. The enhanced activity of trametiglue has been correlated to reduced "rebound" signaling. This form of adaptive drug resistance has emerged as a major limitation of current MEKi including trametinib (Yaeger & Corcoran, 2019). More recently, structures of BRAF-MEK1 bound to a series of MEKi have provided further evidence for binding of MEKi within RAF-MEK complexes (Pino et al., 2021), and have demonstrated functional agreement of differential binding of MEKi to isolated MEK and BRAF-MEK (Ishii et al., 2013; Lito et al., 2014).

Both KSR and RAF form homotypic dimers through similar interfaces on the N-lobe and helix αC, but via distinct quaternary arrangements. For example, in the inactive state conformations of KSR1 and KSR2, we found that KSR-KSR dimerization is centered on Arg665 and Arg718 in KSR1 and KSR2, respectively (Khan et al., 2020; Brennan et al., 2011). The quaternary arrangement of the KSR-KSR dimer is related to the RAF-RAF side-to-side dimer in that the conserved Arg at the base of helix αC forms part of the dimer interface (Haling et al., 2014; Rajakulendran et al., 2009; Thevakumaran et al., 2015; Wan et al., 2004). However, the RAF-RAF arrangement is rotated relative to the KSR-KSR dimer (Haling et al., 2014; Khan et al., 2020). We previously suggested that KSR and RAF most likely would dimerize in a RAF-RAF-like side-to-side arrangement (Brennan et al., 2011; Dhawan et al., 2016); in this configuration helix αC of KSR would be expected to rotate toward the IN state, which we further predicted would disengage inhibitory interactions of the activation segments between KSR and MEK and thereby release the Ser218–Ser222 sites. We predicted that this release mechanism induced through KSR-RAF heterodimerization would enable phosphorylation on the Ser218–Ser222 sites in MEK by separate RAF protomers; data to support this model was generated through biochemical reconstitution and chemical genetic experiments (Brennan et al., 2011). More recent data using a cellular reconstitution system and point mutations have suggested that catalytic RAF kinases prefer to phosphorylate unbound MEK over KSR-bound MEK, implying that KSR may not physically bridge the RAF → MEK phosphorylation step directly (Lavoie et al., 2018). However, given that RAF also forms similar inhibitory interactions on bound MEK, it is likely that both KSR-bound MEK and RAF-bound MEK can serve as adaptors and scaffolds for separate catalytic RAF protomers within large macromolecular assemblies. Further structural data on RAF–KSR complexes will be needed to better understand how exactly heterodimerization enables phosphorylation on MEK.

4. Is KSR an active kinase?

While KSR possesses a conserved kinase domain, like many pseudokinases, it lacks a physiological substrate and several lines of evidence suggest that its role in RAS-MAPK signaling is predominantly mediated via scaffolding, allostery, and higher order interactions. For example, the mammalian homologs of KSR, including human KSR1 and KSR2, include an arginine residue (R639 in human KSR1 and R692 in human KSR2) at the invariant lysine of subdomain II of the kinase domain; this residue is critical for orienting ATP in the active site of canonical kinases and mutations to arginine have been shown to severely compromise phospho-transferase activity of canonical kinases (e.g., Carrera, Alexandrov, & Roberts, 1993; Iyer, Garrod, Woods, & Taylor, 2005; Robinson et al., 1996). That the mammalian homologs of KSR naturally possess a lysine-to-arginine substitution at this critical position suggested that KSR may function independent of catalytic activity, and this notion was in fact supported in early studies. Stewart et al. were able to demonstrate that several catalytic site mutants of KSR, including at the conserved aspartic acid residue in the DFG motif, were able to rescue KSR loss-of-function in animals as evidenced by RAS-dependent vulval induction in *C. elegans* (Stewart et al., 1999). Notably, a KSR allele within the KSR-RAF heterodimerization interface mutated could not rescue, supporting the specificity of the tested catalytic site mutants and further suggesting that KSR phospho-transferase activity is either weak or unnecessary to support RAS signaling. However, while some studies suggested that KSR does not utilize catalytic function, other studies in fact argued toward the importance of KSR catalytic activity at least in some contexts (e.g., Therrien, Michaud, Rubin, & Morrison, 1996; Zhang et al., 1997).

In support of catalytic function, we previously demonstrated that KSR can phosphorylate MEK1 in vitro using purified KSR2:MEK1 complexes (Brennan et al., 2011). Intriguingly, while the efficiency and stoichiometry of KSR2 → MEK1 phosphorylation was low, we identified that heterodimerization with BRAF could accelerate KSR2 activity, again in vitro using reconstitution assays. However, further studies testing a quadruple site mutant of MEK1 (S18A + T23A + S24A + S72A) that removed all of the identified KSR phosphorylation sites found that the mutant could be activated and signal downstream to ERK at levels that were essentially identical to wild-type MEK1, leading us to surmise that phosphorylation on KSR-mediated sites within MEK1 were dispensable for MAPK signaling (Dhawan et al., 2016). Moreover, we conducted experiments using two distinct types of inhibitors (ASC24 and APS-2-79) and demonstrated that inhibition of KSR2 catalytic activity alone was insufficient to antagonize KSR2-accelerated BRAF → MEK phosphorylation in vitro (Dhawan et al., 2016). Shaw, Taylor and colleagues identified an ATP-binding site mutant (A587F in mKSR1) that has been modeled to fill the adenosine pocket in KSR so as to stabilize an active and closed conformation of the pseudokinase domain (Filbert, Taylor, & Shaw, 2011). This mutant has been studied in both KSR and RAF family kinases (Filbert et al., 2011; Hu et al., 2013), and suggested to mimic the ATP-bound state through stabilization of the hydrophobic spine; this mutant would also be expected to impede conformational changes within the pseudokinase domain of KSR, including by disfavoring switching between the open and closed states of the pseudokinase domain. Intriguingly, the A587F mutant in KSR1 could form complexes with both RAF and MEK but not support MAPK/ERK signaling as assessed by cell transformation and colony formation (Filbert et al., 2011).

In summary, no one experiment has clearly answered the question of whether KSR is a physiologically relevant active kinase or phosphotransferase in the RAS-MAPK pathway. However, collective data from many studies suggest that the role of the KSR pseudokinase is primarily linked to its conformational changes that mirror changes pervasive across many active kinases. One such change includes an orientation of helix αC relative to the ATP-binding pocket. Data to support that KSR does not simply serve as a static scaffold for the recruitment of MEK to RAF have been provided through elegant structural and mutagenesis studies suggesting that heterodimerization of KSR and RAF first requires the formation of KSR–MEK complexes to enable RAF-mediated phosphorylation of KSR-free MEK on the essential activating residues Ser218 and Ser222 (Lavoie et al., 2018). ATP binding and/or hydrolysis may play a permissive role in the mechanism and formation of KSR–RAF–MEK complexes, which could explain some of the confounding results testing the importance of KSR catalytic function. Indeed, one possibility is that conformational switching of the KSR pseudokinase domain, and thereby binding to MEK and RAF, is regulated through ATP binding and release cycles that are coupled to ATP hydrolysis but not through phospho-transferase activity per se. Consistent with the importance of ATP in the RAS-MAPK pathway, nucleotide binding has been demonstrated to modulate side-to-side dimerization of BRAF in the absence of bound 14-3-3 (Liau et al., 2020; Liau, Wendorff, et al., 2020); and further a C-terminal regulatory segment of BRAF has been suggested to also regulate ATP binding (Kondo et al., 2019; Kondo, Paul, Subramaniam, & Kuriyan, 2021). Likewise ATP, through its binding and hydrolysis, could play an important role in KSR-dependent MAPK signaling by acting as an endogenous cofactor to regulate conformational changes within the active site that are linked to protein–protein interactions or possibly heterodimerization toward RAF kinases. Such a model would suggest that the pseudokinase domain in KSR shares regulatory principles with enzymes such as GTPases, like K-RAS, or other ATPases. Furthermore, such a model would suggest that small molecules that lock and thereby disrupt conformational cycles in KSR may block specific steps in the RAS-MAPK pathway. We have in fact found two classes of molecules—one targeting the orthosteric site on KSR and a second at the interface of KSR–MEK complexes—that can modulate RAS signaling (Dhawan et al., 2016; Khan et al., 2020). We next describe purification methods to isolate KSR1:MEK1 and KSR2:MEK1 complexes for structural studies. We further describe assays and methods to characterize small molecule binding interactions at the orthosteric and interfacial bindings sites on KSR both in vitro and in live cells.

5. Purification of KSR1–MEK1 and KSR2–MEK1 for biochemical assays and cocrystallography

5.1 Production of KSR:MEK in insect cells

We routinely use insect cell expression systems to produce KSR1:MEK1 and KSR2:MEK1 complexes, and have been successful in isolating full-length MEK1 bound to the pseudokinase domains of KSR1 [residues 591–899] or KSR2 [634–950]. However, for cocrystallography, we most often use a construct that lacks the first 34 residues at the N-terminus of MEK1. We have also utilized mutations that improve the stability of MEK1, including Ser298Asn/Ser299Lys/Tyr300Phe (Fischmann et al., 2009). We have reported the isolation of both sets of complexes using our purification strategies in detail previously (Brennan et al., 2011; Dhawan, Scopton, & Dar, 2016; Khan et al., 2020), which we outline in Fig. 3. Here, we provide additional practical guidelines.

To generate baculovirus and determine viral titres, we follow standard protocols, detailed in Jarvis (2014). Virus titres can be determined using plaque assays or alternatively using other approaches (Mulvania, Hayes, & Hedin, 2004). Several contract research organizations, including Expression Systems, offer titering as a service. We also test for a functional multiplicity of infection (MOI) of 1, which should be an amount of virus solution that when added to insect cells allows for a single doubling before growth reaches a plateau.

In terms of insect cell lines, we most often use the *Sf*21 line; however, it is useful to test new constructs (e.g., mutant variants) in additional lines, including *Sf*9 and Hi5, as the yield of soluble complex from each can vary significantly. Our typical growth format for large scale infection includes 4 L of culture, with each liter split into a 2.4 L baffled and aerated flask that is grown at 27 °C and at 125 rpm shaking. We also find that the highest yields of protein occur after 3–4 passages of cells. Each passage is conducted when cells reach $3–4 \times 10^6$ cells/mL.

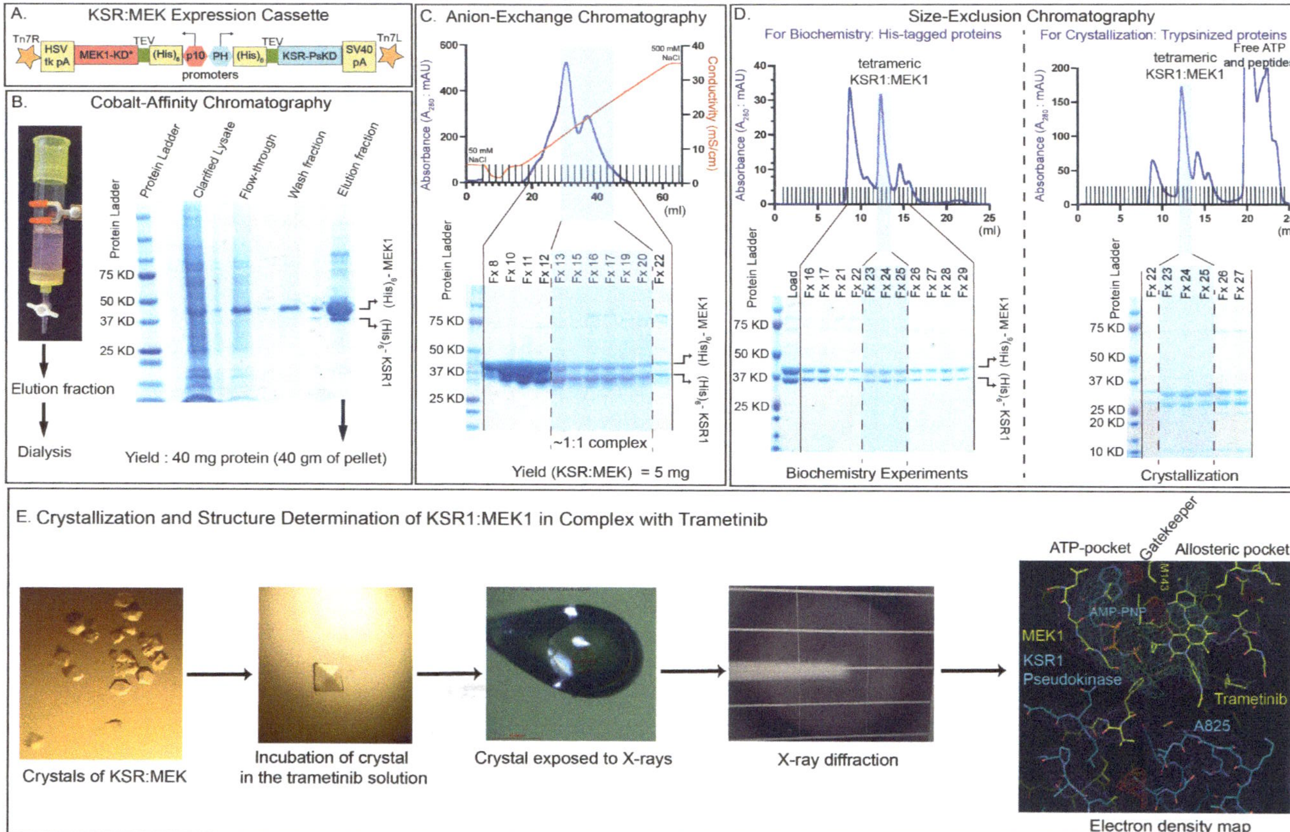

Fig. 3 See figure legend on next page.

Equipment:

- Biosafety cabinet (Labconco, Cat# 302420001)
- Countess III FL automated cell counter (ThermoFisher, Cat# AMQAF2000)
- P2, P20, P200, P1000 pipettes (Gilson, Cat# F144054M, F144056M, F144058M, F144059M, respectively)
- Serological Pipet-Aid (ThermoFisher, Cat# 03–840-313)
- Sorvall LYNX 6000 centrifuge (ThermoFisher, Cat# 75006590)
- Nalgene 1L Super-Speed Centrifuge Bottles with Sealing Closure (ThermoFisher, Cat# 05-564-25)
- Sonicator (Branson 450 Digital Sonifier)
- Benchtop microcentrifuge (ThermoFisher, Cat# 75002447)
- Benchtop centrifuge (ThermoFisher, Cat# 75008801)

Fig. 3 Purification of the KSR1:MEK1 complex for biochemical and structural analysis. (A) Copurification of the KSR1 pseudokinase domain and MEK1 kinase domain is shown as an example (Khan et al., 2020). KSR1 [residues 591–899] and MEK1 [residues 33–393] were coexpressed as His-tagged proteins using the *Sf21* insect cell system. (B) Cobalt-affinity purification of KSR1-MEK1 complex using the gravity-flow column (left); the corresponding 8–16% SDS-gel is shown on the right. (C) Partially-purified KSR1:MEK1 complex is further purified using anion exchange chromatography to remove excessive MEK. Several impurities are detected on the 8–16% SDS gel (below). (D) KSR1-MEK1 (1:1) complex is obtained after size-exclusion chromatography. Either the KSR1:MEK1 complex can be isolated as His-tagged protein (left) or KSR1-MEK1 can be treated with trypsin (1:1000, incubated overnight at 4 °C) to remove partially unfolded proteins and flexible loops, and hence highly homogenous proteins are obtained for crystallization. SDS gels are shown below, and the protein fractions corresponding to the tetrameric peak (150 kDa—see Fig. 2 for a tetrameric assembly) are pooled, and concentrated for subsequent analysis. (E) The flow-chart is provided to illustrate the procedure for determining structures of the KSR1:MEK1 complexes bound to interfacial binders, including trametinib (Khan et al., 2020). Similar procedures can be used for solving structures of other MEKi. Cocrystallization of the KSR:MEK1 complex in the presence of orthosteric binders can also be used to solve structures (Brennan et al., 2011; Dhawan, Scopton, & Dar, 2016). (left to right) Crystals of purified KSR1:MEK1 or KSR2:MEK1 complexes are first grown in the presence of 5 mM AMP-PNP, and typically first appear within 24 h at room temperature. Crystals of KSR2-MEK1 are shown as an example. Next, individual crystals (72–96 h old) are transferred into a fresh drop containing crystallization solution (20% PEG2000, 200 mM Mg acetate, 100 mM MES pH 6.5) supplemented with 5 mM AMP-PNP and 1 mM trametinib. The drop is re-sealed and further incubated for 48–96 h at room temperature. Next, crystals are harvested and flash-frozen in liquid nitrogen. A representative X-ray diffraction image is shown, followed by a model of the inhibitor binding pocket. The electron density map of KSR1:MEK1 in complex with AMP-PNP and trametinib is displayed (PDB ID: 7JUX).

- NanoDrop UV–vis Spectrophotometer (ThermoFisher, Cat# ND-ONEC-W)
- Superloop (Cytiva, Cat# 18102385)
- HiTrap Q anion-exchange column (Cytiva, Cat#17505301)
- HiTrap SP cation exchange column (Cytiva, Cat# 17515701)
- Size exclusion chromatography column: SD200 10/30 GL (Cytiva, Cat# 28990944)
- ÄKTA pure protein purification system (Cytiva, Cat# 29018224)

Consumables:

- Celltreat Scientific Erlenmeyer Flasks with baffled bottom, sterile, various sizes (e.g., ThermoFisher, Cat# 50-202-096)
- Pierce Protease Inhibitor Tablets, EDTA-free (Thermo Fisher, Cat# A32965)
- Cobalt agarose beads (Gold Bio, Cat# H-310-100)
- Gravity flow columns (Bio-rad, Cat# 7372512)
- 2-way Stopcock valve (Bio-rad, Cat# 7328102)
- Slide-A-Lyzer™ Dialysis Cassettes (ThermoFisher, Cat# 66030)
- Amicon® Ultra Centrifugal Filter Units: various sizes (Millipor, Cat# UFC903008)
- 8–16% Criterion™ TGX™ Precast Midi Protein Gel (Bio-rad, Cat# 5671105)
- Pellet-wash buffer: 20 mM Tris-HCl, pH 7.5, 200 mM NaCl, 5% Glycerol
- Lysis buffer: 50 mM Tris-HCl, pH 7.5, 200 mM NaCl, 10% Glycerol, 10 mM Imidazole; optional 0.25% Triton X-100, 0.25% Tween 20. To add immediately prior to lysis: 5 mM beta mercaptoethanol, and 1 tablet of EDTA-free protease inhibitor
- Immobilized metal affinity chromatography (IMAC) wash buffer: 20 mM Tris-HCl pH 7.5, 200 mM NaCl, 10% glycerol, 10 mM Imidazole
- IMAC elution buffer: 20 mM Tris-HCl pH 7.5, 200 mM NaCl, 10% glycerol, 500 mM Imidazole
- Dialysis buffer: 20 mM Tris-HCl pH 7.5, 200 mM NaCl, 10% glycerol, 5 mM DTT
- Ion-exchange (IEX) buffer A: 20 mM Tris-HCl, pH 7.5, 10% Glycerol
- IEX buffer B: 20 mM Tris-HCl, pH 7.5, 10% Glycerol, 500 mM NaCl
- Size exclusion chromatography (SEC buffer: 20 mM Tris-HCl pH 7.5, 200 mM NaCl, 5 mM DTT, 1 mM TCEP
- Adenylyl-imidodiphosphate (AMP-PNP) (Sigma, Cat# 10102547001)
- Trametinib 10 mM solution (Selleckchem, Cat# S2673)

- Crystallization condition C9 from JBS Kinase Screen (Jena Biosciences, Cat# CS-204 L)
- TPCK treated trypsin (ThermoScientific, Cat# 20233)

Here, we are detailing the steps for production of KSR1:MEK1 and KSR2: MEK1 complexes:

1. Infection: On the day before an infection, for 1 L of cell culture per growth flask, inoculate at 0.5–1 × 10^6 cells/mL (counted using trypan blue dye). Next day, when cells are at 1–2 × 10^6 cells/mL, add the virus at a MOI of ~1. (Note: At the beginning of an infection, cells are usually 90–95% viable. After 72 h, when cells are harvested, we can measure viability, as low as ~50%. This decrease in viability is often a marker of a strong infection and can be associated with larger yields of recombinant protein).
2. Harvest the insect cells using low-speed centrifugation (1000 × *g*), and wash the pellets using TBS buffer. (Note: typically, a 10–20 g pellet is achieved from 1 L insect cell culture.) For both the KSR1:MEK1 and KSR2:MEK1 complexes, we have developed expression vectors based on the pFASTBac-Dual cassette. In this system 6 ×-His-MEK1 is produced under the influence of the early p10 promoter, whereas 6 ×-His-KSR1 or 6 ×-His-KSR2 are expressed from the later PH promoters. Tags on both proteins to isolate the complex are not essential. However, we find the dual-tagging strategy advantageous as this allows for capture of KSR-bound MEK and also excess free MEK. Moreover, due to the timing of the promoters, and the relative expression levels, the amount of free His-MEK1 can exceed the levels of His-KSR1/2 by several fold, as can be observed on SDS-PAGE
3. Protein extraction: (Note: Optimum yield from infected pellets is the highest without extensive or long-term cold storage). All the purification steps are performed at 4 °C. Lyse the cell pellets in 30 mL buffer per 10 g pellet in Lysis buffer on ice by sonication using a Branson 450 Digital Sonifier, installed with 3 mm diameter horn, using 2 rounds of 60 pulses with 0.3 s burst of 45% output and 0.7 s resting time
4. Centrifugation and incubation of the cleared lysate with cobalt-agarose beads for IMAC: an initial high-speed centrifugation step (45,000 × *g*, 20 min) separates soluble protein from cellular debris. However, downstream processing at this stage can be difficult so one can conduct batch purification in two different ways: either (A) protamine sulfate (final 1 mg/mL) is added to the supernatant after the first step and the protein is precipitated via a second high-speed centrifugation step

before the supernatant is decanted onto cobalt agarose beads (5 mL of beads per 20 g of starting pellets) in a gravity flow-column; or (B) the highly-viscous supernatant is directly incubated for 2 h with cobalt agarose beads to sequester the histidine-tagged proteins in a 50 mL falcon tube, which are then sedimented by a low-speed centrifugation step (100 × *g*) and the viscous supernatant can be poured off and discarded. The cobalt beads with the bound proteins are then suspended in wash buffer, and poured onto the gravity columns attached to a valve to control the flow of buffer

5. Elution of His-KSR1/2:His-MEK1 complexes from cobalt beads. After washing the beads with 3 rounds of 50 mL wash buffer, the bound proteins are eluted using 25 mL elution buffer. Immediately after the elution, add EDTA to a final concentration of 1 mM to chelate any leached cobalt in the eluate. After an incubation of 5 min on ice with EDTA, add DTT to a final concentration of 5 mM
6. Dialysis: Perform using dialysis cassette for minimally 4 h against 2 L of dialysis buffer
7. SDS-PAGE analysis: Use 8–16% gradient SDS-PAGE to assess supernatant, pellets, washes, elutions, and beads following elution for expression and solubility of KSR–MEK complexes over the initial purification steps (see Fig. 3B).
8. Ion-exchange chromatography: this step is required for additional purification and to separate excess free MEK from KSR-bound MEK. For KSR1:MEK1, use Q anion exchange chromatography, and for KSR2:MEK1, use an SP cation exchange column
9. Sample loading onto ion-exchange column: Remove the dialysis retentate from the dialysis cassette and dilute the sample in buffer A to reduce the concentration of sodium chloride to 50 mM. Load the sample onto the IEX column at a flow-rate of 1 mL/min using a 150 mL superloop attached to an FPLC (e.g., AKTA) purification system
10. Washing and elution step of IEX: wash the column with 10 CV of buffer containing 50 mM NaCl. Elute the bound protein using a gradient of buffer A and buffer B over a volume of 40 mL. Collect 2 mL fractions in tubes or deep well plates. Note: In the case of KSR1:MEK1, the complex will elute at ~250 mM NaCl concentration. In the case of KSR2:MEK1, the complex will elute at ~250 mM NaCl, with free MEK1 in the flow through or low salt washes (i.e., $<$100 mM NaCl).
11. SDS-PAGE analysis: Use 8–16% gradient SDS-PAGE to confirm which fractions contain stoichiometric KSR:MEK complex

12. Concentration of KSR:MEK complex: Pool the fractions including 1:1 KSR:MEK complex and add fresh DTT at a final concentration of 5 mM. Pour the sample into a 15 mL Amicon Ultra centrifugal device (with 20 kDa MWCO), and centrifuge the sample at 4000 rpm for 10 min. (Note: pipette the sample inside the centrifugal device up and down between each spin to mix the sample evenly). Repeat the centrifugation to achieve protein at a concentration of 4–5 mg/mL
13. Concentration estimation: use nanodrop to measure the concentration of KSR:MEK complex with buffer as a blank. For both KSR1:MEK1 and KSR2:MEK1, 1.0 Absorbance units at 280 nm equates to 1 mg/mL of complex
14. Size exclusion chromatography (SD200 10/30 GL) to isolate the KSR1:MEK1 or KSR2:MEK1 heterotetramer. The isolation of the larger molecular weight heterotetrametric complex ensures reproducible biochemical results and crystal formation. Collect the sample into 0.5 mL fractions. Estimate purity by SDS-PAGE analysis. Concentrate the fractions containing 1:1 KSR-MEK complex to 5–7 mg/mL. (Note: Following gel filtration, the purified complexes can either be flash frozen in liquid nitrogen or used immediately for crystallization trials). Typical yields are 0.2 mg of tetrameric complex from 50 g of pellet for KSR1:MEK1 and 0.8 mg from 50 g of pellet for KSR2:MEK1

5.2 Crystallization of KSR:MEK complexes to map orthosteric and interfacial binders

Crystals of purified KSR1:MEK1 and KSR2:MEK1 complexes can be generated using the hanging drop vapor diffusion method (Bergfors, 2011), with a 1:1 μL mix ratio of protein and crystallization mother liquor above crystallization well-solutions.

1. For crystallization, KSR:MEK complexes typically in the range of 5–7 mg/mL are supplemented with 10 mM $MgCl_2$ and 5 mM ATP or AMP-PNP
2. Conditions for optimal crystals can be found from focused screens; for example, slight variations on the concentration of PEG and pH around the following: 20% PEG 2000, 200 mM magnesium acetate, 100 mM MES, pH 6.5. In our hands, the best crystals have come from the condition C9 in the Jena Biosciences Kinase screen
3. We also found serendipitously that proteases can cleave the KSR:MEK complex in crystallization drops, and the crystals that subsequently

appeared from these drops were of better quality in terms of size and X-ray diffraction properties (Khan et al., 2020). Testing a series of proteases, we determined that trypsin is ideal to control this process; we assume the protease removes flexible loops and/or partially folded proteins. Eventually, the trypsinization step can be included in the purification prior to size exclusion chromatography. Trypsinization is done at 1:1000 ratio of TPCK treated trypsin to protein (IEX-purified) overnight at 4 °C. Subsequent SEC allows a separation of hetero-tetrameric KSR:MEK (150 kDa) from trypsin (27 kDa). Intriguingly, from our determined crystal structures, we have not identified an obvious cleavage site suggesting that limiting amounts of trypsin does not cleave at crystallographically-visible lysine or arginine residues in the determined structures of KSR–MEK complexes (Khan et al., 2020).

4. Once crystals have appeared (first within 24 h, and grown for 3–5 days) we then transfer them into soaking solutions that include compounds (typically ~1 mM, with 5–10% DMSO final) for 24–96 h at 20 °C to allow the diffusion of the compounds
5. After soaking, crystals are harvested in crystallization well solutions (20% PEG2000, 200 mM magnesium acetate, 100 mM MES, pH 6.5) supplemented with 25% ethylene glycol as cryoprotectant. One can also attempt 20–25% glycerol as a cryoprotectant
6. Crystals are then flash frozen in liquid nitrogen and preserved for data collection
7. We find that we need to screen on average 5 crystals to find a crystal that provides diffraction data to a resolution of ~3.5 Å or better. However, we have screened in some cases up to 20–30 crystals to find a single strong diffractor. High solvent content in the crystal form that we most often obtain for KSR1:MEK1 and KSR2:MEK1, as well as marked sensitivity to radiation damage, likely contribute to the variability in crystal quality and data collection that commonly occurs. However, given that we repeatedly obtain crystals with P6 symmetry, we can collect relatively small data wedges. Moreover, techniques such as low-dose of X-rays, reduced degree of oscillation (0.2 deg.) per frame, exposure time of 1 s/deg., vector data collection and grid screening while employing the software-driven strategy of data collection (McCoy et al., 2007) at modern beamlines, such as AMX-NSLS2, FMX-NSLS2 and LS-CAT-APS, where we have collected most of our data, have helped overcome some of the challenges in solving KSR1:MEK1 and KSR2:MEK1 structures bound to ligands

Standard software packages are employed for data processing (XDS; Kabsch, 2010), phase determination (Phaser; McCoy et al., 2007) and refinement (Phenix; Afonine et al., 2012).

6. Assays to measure binding of ATP-competitive compounds within the orthosteric site on KSR1 and KSR2

The original forward genetic screens conducted in the fly and worm, which led to the identification of KSR, found point mutations in KSR that could suppress signaling specifically by oncogenic mutants of RAS. Several years ago, we mapped the mutations from these screens onto the predicted structures of the RAF:KSR:MEK complex (Fig. 4A) and found that a number of the alleles spanned the RAF and MEK interfaces extending to the ATP-binding pocket (Dhawan et al., 2016). This allowed us to speculate that we may identify ATP-competitive binders of KSR that could function equivalently to genetic suppressor alleles or, as we hypothesized, "chemical suppressors of RAS." To identify such compounds, we used a probe for labeling the ATP binding pocket of purified KSR2:MEK1 complexes. This probe, referred to as ATP-biotin (Fig. 4B), places an acyl-phosphate within the active site of a kinase to label proximal lysine residues (Dhawan et al., 2016; Patricelli et al., 2007). The probe is available commercially (Pierce 88,311) or can be generated based on published procedures (Patricelli et al., 2011). When using ATP-biotin we find it important to calibrate labeling based on varying time and concentration of the probe as well as the target; this can also be a good quality control step for new batches of purified proteins. For example, in Fig. 4C and D, we show labeling on isolated MEK1, isolated BRAF, and the KSR1:MEK1, and KSR2:MEK1 complexes.

Small molecule binding at the orthosteric site on KSR can be detected in the competition format with free inhibitor or nucleotide. Moreover, an initial assessment of inhibitor specificity can be conducted using this assay as well; as we demonstrate BRAF-specific inhibition using dabrafenib as an example in Fig. 4E. However, it is important to note that we run ATP-biotin competition assays without physiological concentrations of free ATP, which can lead to potentially false positive measurements of small molecule-target interactions. For example, we have found that several of our probes, including APS-2-79, can strongly inhibit labeling of both MEK1 and KSR2 in vitro (Fig. 4E); however, we have only been able to

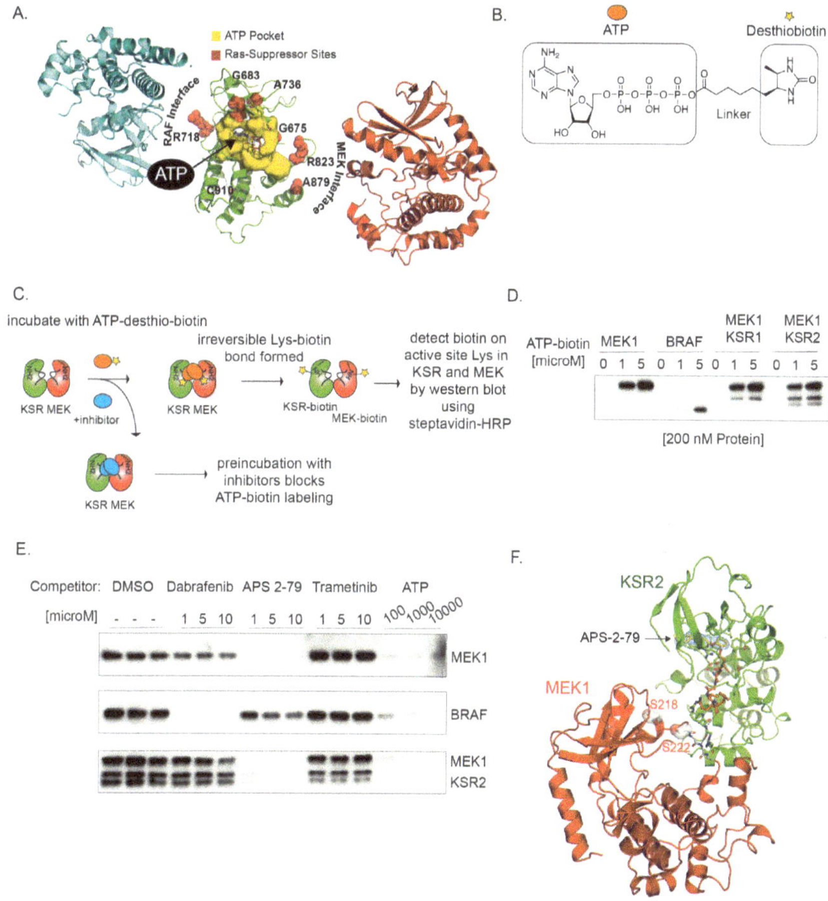

Fig. 4 An assay for orthosteric binders at the ATP-binding pocket of purified KSR-MEK complexes. (A) Putative structure of RAF (blue)–KSR (green)–MEK (red) complex based on structural and functional studies (Brennan et al., 2011; Burack & Shaw, 2000; Dhawan et al., 2016; Khan et al., 2020; Lavoie et al., 2018; Michaud et al., 1997; Morrison, 2001; Nguyen et al., 2002; Rajakulendran et al., 2009; Roy, Laberge, Douziech, Ferland-McCollough, & Therrien, 2002; Therrien et al., 1996). KSR mutations that suppress oncogenic RAS signaling (shown in red) localized around the ATP-binding pocket (yellow) as well as the RAF and MEK interfaces (Dhawan et al., 2016). (B) Structure of desthiobiotin-ATP: ATP (left) linker (middle) desthiobiotin (right). (C) Schematic for labeling of the orthosteric sites of KSR and MEK with ATP-biotin in the presence and absence of competing ligands probe. (D) Representative western blot of recombinant MAPK proteins (MEK1, BRAF, KSR1:MEK, and KSR2:MEK) labeled with increasing concentrations of the ATP-biotin probe. (E) Probe-labeled kinases (MEK, BRAF, KSR1:MEK, and KSR2:MEK) treated with increasing concentrations of inhibitors (5, 10 and 20 μM) were measured via western blotting. 2.5 μM of probe was used to label each protein. DMSO was used as a negative control (No inhibitor) and free ATP (100, 1000, and 10,000 μM) was used as a positive control for blocking labeling. (F) The KSR2-MEK1 complex with APS-2-79 identified the inhibitor bound in the orthosteric site of KSR2 (Dhawan et al., 2016).

validate binding of APS-2-79 to KSR2 in vivo, using orthogonal assays (Dhawan et al., 2016). Moreover, when APS-2-79 was soaked or cocrystallized with KSR2:MEK1, we only found density for APS-2-79 within the KSR2 binding pocket and not MEK1 (Fig. 4F). Again, these data suggest that APS-2-79 binds preferentially to KSR2 within the KSR2: MEK1 complex. Another potential source of false positives for this assay can occur from compounds that possess a free amine (-NH) group that can react with the probe and quench the critical acyl-phosphate moiety; although this is rare and we have not observed reaction with buffer components such as Tris with the probe. False-positives can be assessed by testing for inhibitor + probe adducts using LC-MS. Additionally, competition experiments conducted using control kinases can rule out non-specific reactions between the probe and test compounds.

6.1 Equipment

- P2, P20, P200, P1000 pipettes (Gilson, Cat# F144054M, F144056M, F144058M, F144059M, respectively)
- Isotemp Heat Block (Fisher Scientific, Cat# 11-71806)
- Criterion™ Cell and PowerPac™ Basic Power Supply (BioRad, Cat# 1656019)
- Criterion™ Blotter with Plate Electrodes (BioRad, Cat# 1704070)
- PowerPac™ HC High-Current Power Supply (Bio-Rad, Cat# 1645052)
- Magnetic Stirrer Plate: 5 L Load Capacity (Velp Scientific, Cat# 56LK76)

6.2 Materials and reagents

- Gel loading pipet tips (Fisherbrand, Cat# 02-707-181)
- PCR tube strips (ThermoFisher, Cat# AB2000)
- 1.5 mL microcentrifuge tubes
- Pipette tips
- Assay Buffer (200 mM NaCl, 50 mM Tris, 5% Glycerol, 1 mM DTT, and 1 mM EDTA)
- Proteins: MEK1 and BRAF (Dhawan et al., 2016; Khan et al., 2020), MEK1:KSR2 (Refer to Section 5)
- Magnesium chloride hexahydrate (Sigma-Aldrich, Cat# M9272-500G)
- Pierce Kinase Enrichment Kit with ATP Probe (Thermo Scientific, Cat# 88310)
- Dimethyl Sulfoxide (DMSO) (Fisher Scientific, Cat# BP231-100)
- Dabrafenib (GSK2118436) (Selleckchem, Cat# S2807)

- APS-2-79
- Adenosine 5′-triphosphate disodium salt hydrate (Sigma-Aldrich, Cat# A2383)
- Molecular biology grade water (Corning, Cat# 46-000-Cl)
- 6 × Laemmli Buffer (Alfa Aesar, Cat# J61337)
- 4–15% Bis-tris glycine 26-well gel (Bio-Rad, Cat# 340125)
- 20 × MOPS SDS Running Buffer (ThermoFisher Scientific, Cat# NP0001)
- Nitrocellulose blotting membrane (GE Healthcare, Cat# 10600002)
- Western blotting filter paper (Thermo Scientific, Cat# 84784)
- Tris-Glycine Transfer Buffer (25 mM Tris, 192 mM Glycine, and 20% methanol)
- Tris-Buffered Saline (200 mM Tris and 1500 mM NaCl) and Tween 20 (TBS-T)
- Bovine Serum Albumin Fatty Acid-free Powder (Fisher Scientific, Cat# BP9704-100)
- Magnetic stir bar
- Streptavidin-horseradish peroxidase (HRP) (Cell Signaling, Cat# 3999S)

6.3 Step-by-step protocol

1. Prepare the assay buffer fresh before each experiment. Dilute each purified protein to a concentration of 200 nM (e.g., 8.4 ng/L for MEK1, 6.4 ng/μL for BRAF kinase domain, and 15.4 ng/μL for KSR2: MEK1) and keep on ice
2. Aliquot 27.3 μL of each protein mix per each condition desired
3. Prepare a 1 M stock solution of $MgCl_2$ and add 0.3 μL to each condition
4. Prepare a 30 × stock of each inhibitor in 100% DMSO (e.g., 30, 150, and 300 μM) and add 1 μL to the appropriate condition for a final concentration of 1, 5, and 10 μM. Incubate the reaction mixture for 15 min
5. Prepare 30 × free ATP stock concentrations in molecular biology grade water (e.g., 3, 30, and 300 mM) and add 1 μL to the appropriate conditions for a final concentration of 100, 1000, and 10,000 μM. Incubate the reaction mixture for 15 min
6. During the incubation, prepare a 166 μM working stock of the ATP-desthiobiotin probe in 60 μL of molecular biology grade water. Add 1.6 μL of the 166 μM stock solution to each reaction mixture at a final concentration of 2.5 μM and incubate for 10 min

7. After the incubation, add 6 μL of 6 × Laemmli SDS sample buffer to each reaction mixture and heat each sample mix to 80 °C for 5 min
8. Load each prepared sample into SDS-PAGE wells of a 4–15% bis-tris glycine 26-well gel and run the gel in 1 × MOPS running buffer for 60 min at 125 V at room temperature
9. Next, transfer the gel onto a nitrocellulose membrane in 20% methanol in tris-glycine transfer buffer for 1 h at 90 V at 4 °C. Add a magnetic stir bar at the bottom of the transfer tank (~400 RPM) to help distribute heat and keep the buffer homogenous
10. After confirmation of a successful transfer via Ponceau Red, extensively wash the membrane until the Ponceau Red is completely removed
11. Block the membrane in 5% BSA TBS-T for 45 min at room temperature on a rocker
12. While the membrane is blocking, prepare the Streptavidin-HRP antibody cocktail in 5% BSA TBS-T (1:7500 dilution).
13. Add the antibody cocktail to the membrane and incubate overnight at 4 °C on a rocker
14. The following day, extensively wash the membrane 3 × with TBS-T on a rocker at room temperature
15. Prepare a 1:1 mixture of SuperSignal West Pico Chemiluminescent Substrate and directly add it onto the membrane to detect Streptavidin-HRP specifically bound to the protein of interest
16. Image the membrane by film or using a ChemDoc XRS+ Imaging System
17. Export the desired image exposure using the Image Lab program
18. Integrate band intensities for inhibitor and DMSO treated samples to calculate percent inhibition

7. Assays to measure binding of interfacial compounds of KSR-MEK complexes

Protein–protein interactions are critical to facilitating normal and oncogenic signal transduction. Biochemical characterization of inhibitors is generally focused on measuring interactions of small molecules with single targets. This type of analysis does not capture the physiological factors, such as endogenous competitors or the influence of protein–protein interactions in the cell. Moreover, inhibitor influence on protein–protein interactions is usually assessed using lytic techniques such as coimmunoprecipitation and cellular thermal shift assays. While robust, these approaches could potentially alter pre-existing or dynamic protein–protein complexes that occur within live cells.

NanoBRET is a technique that overcomes the limitations of in vitro or lytic assays of target engagement by allowing for pharmacological measurements in living cells (Robers et al., 2019; Vasta et al., 2018). In this technique, a protein of interest is transfected into mammalian cell lines (usually HEK293T) as an N- or C-terminal NanoLuciferase (NL) fusion and a small molecule called a tracer is applied to the cell. A tracer is a bifunctional molecule that can participate in bioluminescence resonance energy transfer (BRET) with NL-tagged proteins and consists of a drug moiety targeting a protein(s) of interest, and a Bodipy dye (576/589 nm) connected by a linker (usually PEG or alkyl). Once the tracer is bound to its target, the NL substrate, furimazine, is applied to generate blue light (460 nm). If the Bodipy moiety of the tracer is within $\sim$100 Å from the NL (donor) it will undergo BRET and emit red light (acceptor-610 nm). The red and blue light signal can be quantified separately on a plate reader (GloMax Discover plate reader using the BRET 618 protocol) and divided to generate a BRET ratio (acceptor/donor). The emission profiles of light emitted from the Nanoluciferase and the Bodipy moiety have minimal overlap, which minimizes bleed-through of signal between the donor and acceptor channels, and thus, false-positive BRET.

Two examples of NanoBRET assays are steady-state and washout experiments (Fig. 5). The readout of a steady-state experiment is a dose-response curve where varying concentrations of a drug and a constant tracer concentration are introduced to cells transfected with the protein of interest. These dose-response curves can be fitted to generate apparent IC_{50} values. In a washout experiment, transfected cells are incubated with a drug (at a saturating concentration $\sim 15\times$ higher than the apparent IC_{50} determined from steady-state experiments) for 2 h, spun down, washed with Opti-MEM to remove the drug, and introduced to a constant tracer concentration. The readout of this experiment is the buildup of BRET signal over time as the tracer occupies more open targets as they become available due to release of pre-bound drugs. It is important to include DMSO as a negative control and to establish the baseline kinetics of tracer binding in the absence of a drug. Step-by-step instructions for both NanoBRET assays have been previously described (Robers et al., 2019).

When developing a NanoBRET assay to measure pharmacology on a single target and/or protein–protein interaction, it is important to consider

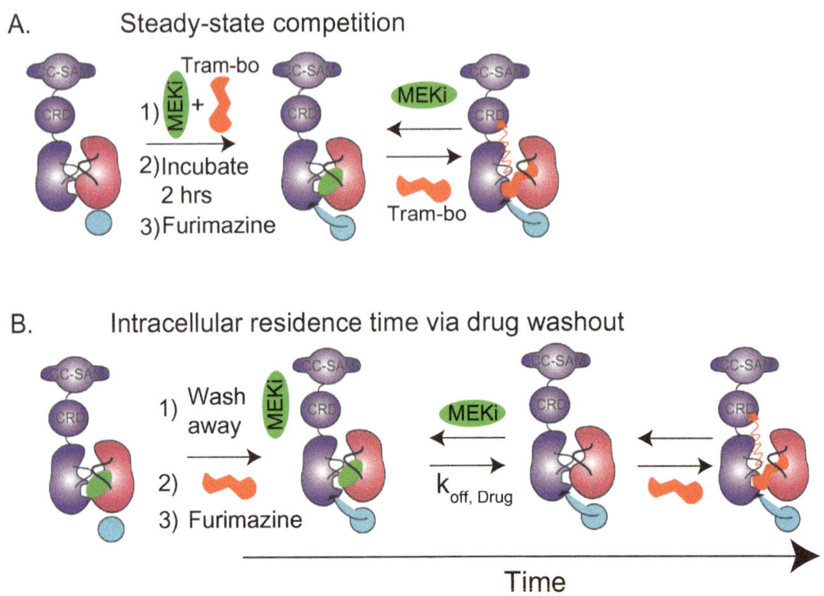

Fig. 5 Live cell assay schematic for allosteric and interfacial binders on KSR–MEK complexes. Schematic examples of steady-state (A) and washout (B) experiments performed on a MEK-NL (pink) construct coexpressed with unlabeled KSR (purple) (Note: other assay configurations to measure pharmacology on the MEK–KSR complex can be found in Fig. 6B). In both assays, BRET can only be observed when the donor (e.g., Nanoluciferase or NL) is in relatively close proximity (<50 Å) to an acceptor (e.g., the BODIPY moiety of Tram-bo). Under these conditions, the blue light produced from activity of Nanoluciferase on its substrate furimizine produces high-intensity luminescence (measured at 460 nm) and thereby energy transfer on Tram-bo, which subsequently emits a second signal as red light (measured at 610 nm). Competition with free ligands (e.g., MEKi) that overlap with the Tram-bo binding site block the energy transfer and therefore can be recorded as a lack of BRET relative to vehicle control. In the residence time experiments, drugs are added to cells first and then washed out. Build-up of BRET upon addition of Tram-bo to the washed cells then serves as a proxy for the dissociation of free MEKi and thereby k_{off} MEKi, as binding sites become available for tracer binding on the complex (far right cartoon). Step-by-step instructions to estimate k_{off} and drug residence time from wash-out experiments can be found in Robers et al. (2019).

(1) the potency and specificity of drug that becomes the basis of a tracer, and (2) NL placement in the context of a fusion construct. MEK inhibitors (MEKi) pharmacology on free MEK and MEK in complex with KSR can be measured using a potent Trametinib-based tracer, which we termed Tram-bo (Fig. 6A). Tram-bo is composed of Trametinib, a two-PEG unit linker, and a Bodipy dye (576/589 nm), and can be used to measure pharmacology of MEKi on free MEK1 using a MEK1-NL construct, and MEK1

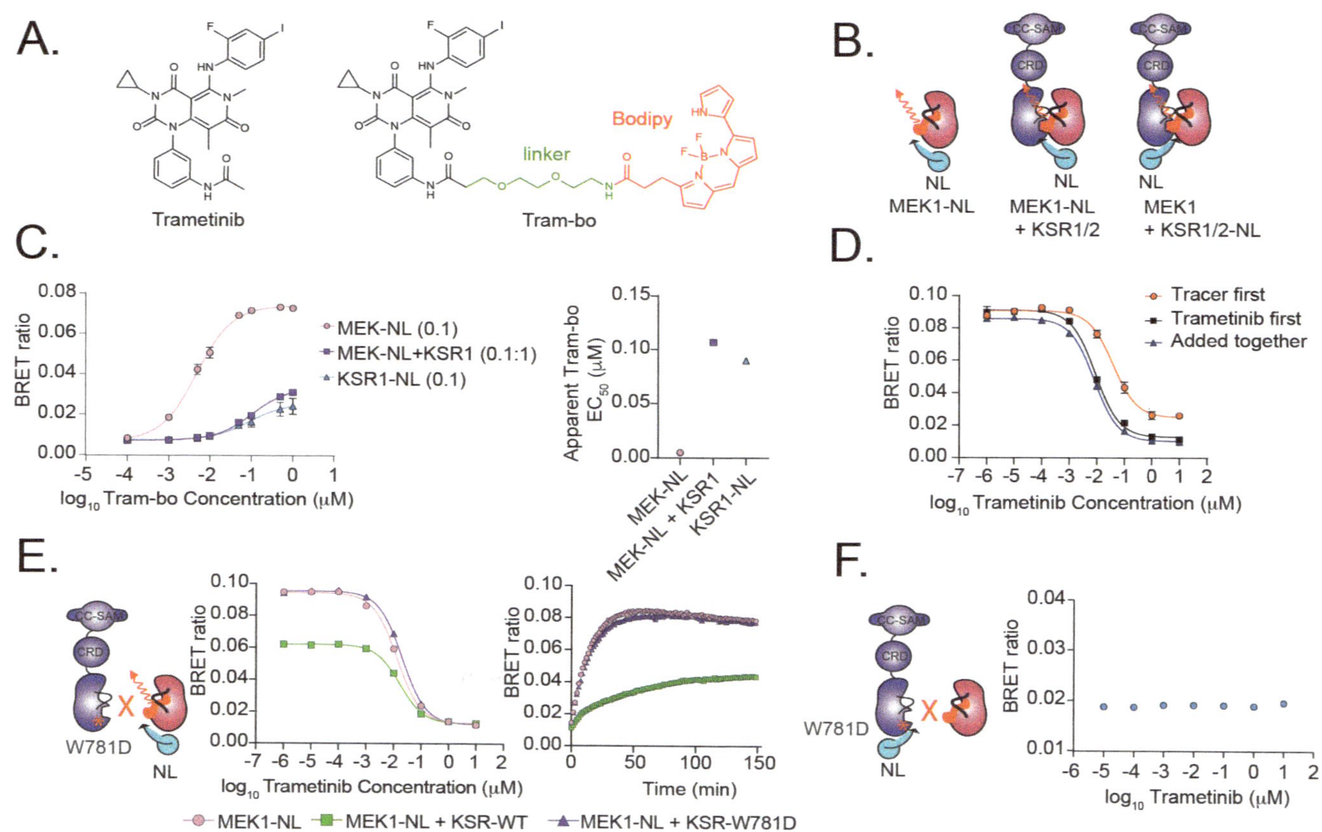

Fig. 6 See figure legend on next page.

in complex with mKSR1 using a mKSR1-NL construct, or MEK1-NL coexpressed with excess unlabeled mKSR1 (Fig. 6B). To compare steady state and washout experiments across multiple complexes, it is important to determine a saturating concentration of a tracer (i.e., a concentration where Tram-bo gives the highest signal and is bound to the same extent across all proteins and protein complexes). Note that this may not be achievable for all tracers since the solubility of tracers only extends generally into the low single-digit micromolar range. For Tram-bo, this was accomplished by generating an EC_{50} curve for each scenario where the concentration of Tram-bo was varied to determine its binding affinity and highest BRET signal (Fig. 6C). From these assays, we determined a working concentration of Tram-bo above the EC_{50} on MEK1-NL plus/minus mKSR1 coexpression or in mKSR1-NL assays. Although the EC_{50} of Tram-bo on MEK1-NL is ~20 times more potent than on the other two MEK1–mKSR1 complex scenarios, using a saturating constant concentration (1 μM) is important to compare MEKi pharmacology on free-MEK1 and the MEK1–mKSR1 complex. Further, since the EC_{50} of Tram-bo is in the nanomolar range, its tight binding can cause reproducibility issues in steady-state experiments that are dependent on the order of addition of Tram-bo and a drug. This is illustrated by the apparent IC_{50} and curve height differences resulting from the order of addition of a drug and Tram-bo on MEK1-NL (Fig. 6D).

Fig. 6 Developing NanoBRET to measure the pharmacology of MEK inhibitors on KSR–MEK complexes in live cells. (A) Structures of Trametinib and the NanoBRET probe we call Tram-bo. Within Tram-bo, the parent compound is shown in black, the linker portion is green, and Bodipy is red. (B) Nanoluciferase (-NL) placement on MEK or KSR allows pharmacology to be measured on MEK and the MEK-KSR complex in model systems (e.g., HEK293 cells). Shown on the left is a cartoon for measuring Tram-bo bound to MEK1-NL. Also shown are putative complexes of Tram-bo bound to MEK-KSR formed through MEK1-NL coexpressed with either KSR1 or KSR2 (middle) or from the expression of KSR1-NL or KSR2-NL, which is expected to form complexes with endogenous MEK (far right). (C) Build-up curves to measure the apparent EC50 of Tram-bo on MEK and the MEK–KSR complex. (D) Order of addition matters for tracers that have nanomolar apparent EC50 values as seen for Trametinib on MEK-Nanoluciferase. (E) Steady-state dose-response and washout experiments depend on specific interactions at the interface between MEK and KSR1 as determined by the use of a KSR1 mutant, W781D. (F) Specificity of KSR1-NL assays are evidenced by the lack of a Trametinib dose-response using the W781D mutant. Please see Fig. 2 for the location of the W781 residue at helix αG and at the interface of the KSR1–MEK1 complex. *Data in panels (E and F) reproduced from Khan, Zaigham M., Alexander M. Real, William M. Marsiglia, Arthur Chow, Mary E. Duffy, Jayasudhan R. Yerabolu, Alex P. Scopton, and Arvin C. Dar. 2020. "Structural basis for the action of the drug trametinib at KSR-bound MEK." Nature 588 (7838): 509–14.*

Either adding the drug first, or in combination with a tracer provides the most reproducible results with the largest BRET window.

When considering protein–protein interactions it is important to use mutants at the interface to test specificity. In the scenario where NL is on MEK1 and is coexpressed with mKSR1, one of the most noticeable observables is a decreased maximum BRET ratio in steady-state and washout experiments that is most likely due to reorientation of the tracer in the complex. Mutation of W781D on mKSR1 generates similar steady-state and washout experiment results to MEK-NL alone indicating the specificity of the complex when testing MEK1-NL in combination with wild-type mKSR1 (Fig. 6E). Harmoniously, specificity of the MEK1:mKSR1 complex can also be tested using the KSR-NL assay configuration where W781D results in the loss of a Trametinib dose-response (Fig. 6F). These data support that Tram-bo based BRET in the context of the mKSR1-bound complex occurs through interactions with endogenous MEK1.

Following tracer optimization and confirmation of complex specificity, inhibitors can be characterized to develop cellular apparent IC_{50} and residence time values on free MEK1 and complexes (e.g., mKSR1-bound MEK1). When comparing free MEK1 and the mKSR1:MEK1complex, MEKi such as trametinib have a similar apparent IC_{50} value for both free protein and the complex, whereas a related clinical MEKi cobimetinib displays more potent binding to the complex (Fig. 7A). This change in potency may be related to the sterics that hinder the dissociation of cobimetinib in the MEK1–mKSR1 complex as compared to free MEK1. These differences become less pronounced in washout experiments (Fig. 7B) because most of the MEKi in Fig. 7A display nanomolar apparent binding affinities with long k_{off}. Even though a drug can have a similar apparent IC_{50} for free protein vs a complex, differences can be observed with respect to the lifetime of the drug–receptor complex as measured by k_{off} and residence time (Copeland, 2015; Lu & Tonge, 2010). In NanoBRET experiments, residence time can be approximated and compared between compounds through wash-out assays (Fig. 5B). For highly potent compounds, residence time can be approximated by titrating drugs at subsaturating concentrations. In the case of trametinib, longer residence time (i.e., slower Tram-bo buildup) is observed for the mKSR1:MEK1-NL complex as evidenced by the lack of Tram-bo buildup and recovery and the small changes from titrations of free drug from base-line signal (Fig. 7C). For MEK1-NL alone, equivalent free drug concentrations produced a more progressive movement toward the DMSO control, which is consistent with

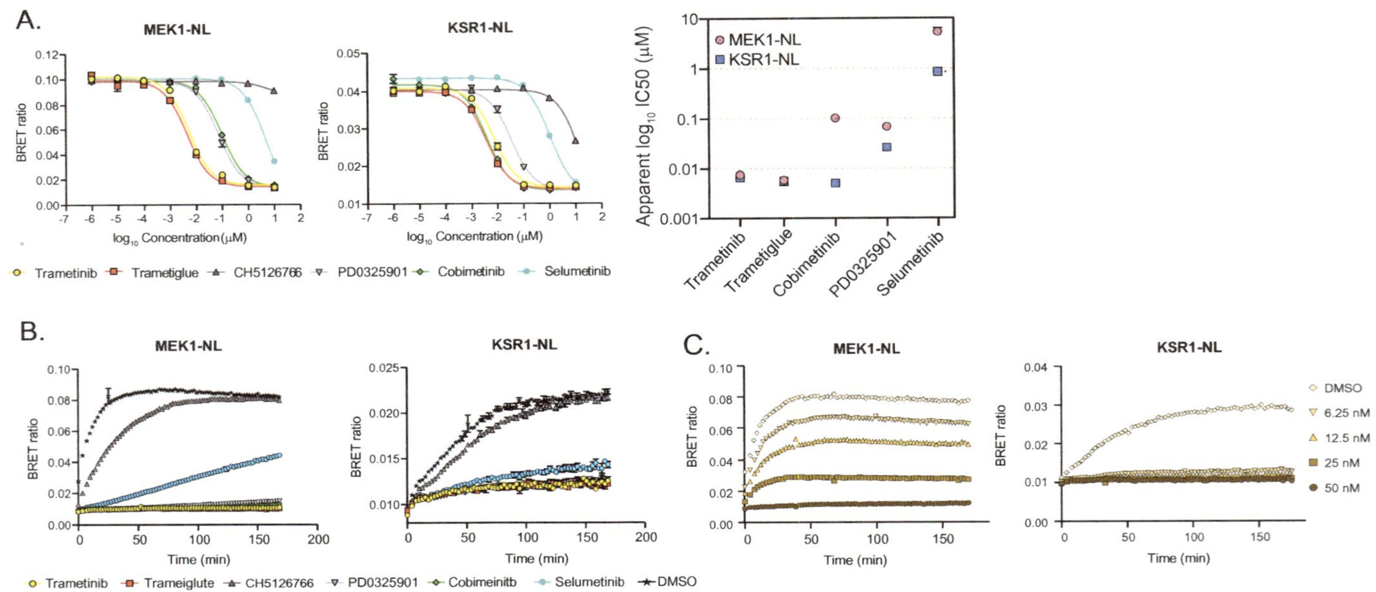

Fig. 7 Steady-state and kinetic measures of MEKi on MEK and the KSR–MEK complex in live cells. (A) Dose-response curves and fitted apparent IC50 values of MEKi on free MEK and the KSR–MEK complex. (B) Characterization of MEKi residence time on MEK and the KSR–MEK complex. (C) Residence time differences of Trametinib (i.e., drugs with long k_{off} values) on MEK and the KSR–MEK complex can be observed using subsaturating drug concentrations in washout experiments (Khan et al., 2020). *All panels reproduced from Khan, Zaigham M., Alexander M. Real, William M. Marsiglia, Arthur Chow, Mary E. Duffy, Jayasudhan R. Yerabolu, Alex P. Scopton, and Arvin C. Dar. 2020. "Structural basis for the action of the drug trametinib at KSR-bound MEK." Nature 588 (7838): 509–14.*

shorter residence time of the analyzed MEKi on free MEK, supporting the notion that, once bound, mKSR1 serves as a barrier to dissociation.

Taken together, NanoBRET is a useful tool to interrogate pharmacology of pseudokinase protein–protein interactions, as well as on-target pharmacology. This assay provided a physiological measure of KSR-bound MEK in living cells and demonstrated the unique pharmacology of MEK inhibitors on the complex versus free forms of the enzyme. The judicious use of mutations at the interface allows one to interrogate in vivo binding signals and interpret the most likely forms of the complex and the binding interface requirements that may strongly influence structure–activity relationships (SAR) between compounds.

7.1 Equipment

- Bioluminescent plate reader (GloMax Discover, Promega Cat# GM3000 or PerkinElmer EnVision Cat# 2105-0010)
- Biosafety cabinet (Labconco, Cat# 302420001)
- Bench top microcentrifuge (ThermoFisher, Cat# 75002447)
- Benchtop centrifuge (ThermoFisher, Cat# 75008801)
- Countess III FL automated cell counter (ThermoFisher, Cat# AMQAF2000)
- P2, P20, P200, P1000 pipettes (Gilson, Cat# F144054M, F144056M, F144058M, F144059M, respectively)
- P20 and P200 12-channel pipettes (Rainin, Cat# 17013808 and 170123810, respectively)
- Serological Pipet-Aid (ThermoFisher, Cat# 03-840-313)

7.2 Material and reagents

- HEK293T cells (ATCC CRL-3216)
- Opti-MEM without Phenol Red (Gibco, Cat# 11058021)
- DMEM (Gibco, Cat# 11965092)
- PBS (Gibco, Cat# 20012027)
- 25 mL reservoirs (ThermoFisher, Cat# 95128093)
- Trypsin-EDTA (0.25%) with phenol red (Gibco, Cat# 25200114)
- White 96-well plates (Corning, Cat# 3990)
- Clear plastic 96-well plate (Celltreat, Cat# 229196))
- Serological pipette tips (2 mL aspirating, 2, 5, 10 mL)
- Fugene HD Transfection Reagent (Promega, Cat# 2311)
- Carrier DNA (Promega, Cat# E4882)

- MEK inhibitors (Selleckchem)
- DMSO (Sigma, Cat# 275855-100 mL)
- Countess slides (ThermoFisher, Cat# C10228)
- Promega NL substate and Inhibitor kit (Promega, Cat# N2161)
- Tracer Dilution Buffer (Promega, Cat# N2191)
- Trami-bo (Khan et al., 2020)
- CMV-mKSR1, CMV-mKSR1-NL, and CMV-MEK1-NL DNA (Khan et al., 2020)
- Stackable pipette tips (Rainin, Cat# 17005873 and 17005875, respectively)
- PCR tube strips (ThermoFisher, Cat# AB2000)
- 1.5 mL microcentrifuge tubes (ThermoFisher, Cat# 05-4080130)
- Trypan Blue (ThermoFisher, Cat# C10282)
- Tissue culture dishes (Celltreat, Cat# 229660)
- 15 mL Falcon tubes (Falcon, Cat# 352196)

7.3 Step-by-step protocol

Day 1: Transfection

1. For MEK1-NL or mKSR1-NL single transfections: prepare a mixture of Carrier DNA and with MEK1-NL or mKSR1-NL DNA in Opti-MEM. For MEK1-NL coexpressed with mKSR1: prepare a mixture of Carrier, MEK1-NL and mKSR1 DNA in Opti-MEM. Add Fugene HD and leave at room temperature for 20 min. If using Fugene HD, the final concentration of total DNA should be 10 μg/mL. For single transfections, use 0.1 μg/mL + 9.9 μg/mL for MEK-NL or KSR-NL and Carrier DNA, respectively. For coexpressions use 0.1 μg/mL + 1:8.9 μg/mL MEK1-NL:mKSR1:Carrier DNA. The total amount of Opti-MEM should be 1/20th the volume of the entire transfection. For example, if 1.1 million cells in 5.28 mL of DMEM (200,000 cells/mL according to Fugene HD manufacturer's protocol) need to be transfected to have enough cells on Day 2 for half of a 96-well plate, 0.264 mL of Opti-MEM should be used to generate DNA complexes
2. During the incubation, remove HEK293T cells from incubator, check that they are ~70% confluent, remove DMEM, wash with PBS, trypsinize, add DMEM, and spin down at 300 × *g* for 5 min at room temperature. Remove the supernatant, adjust the cell density to 200,000 cells/mL, and plate in a tissue culture treated plate
3. Add the DNA complex mixture, mix well, and incubate at 37 °C for at least 20 h

Day 2: Steady-state assay setup

1. Thaw 400 × Tram-bo (solution in DMSO), tracer dilution buffer, and 1000 × MEK inhibitors in DMSO. Note that it is good practice to make a large 1000 × stock of Tram-bo and 1000 × stock of each inhibitor (10 mM is a good place to start), aliquot into single use PCR tubes to store at −80 °C
2. Warm up DMEM, PBS, Trypsin-EDTA, and Opti-MEM in a 37 °C water bath
3. Prepare a 1000 × dose-response curve in a strip of eight PCR tubes. Pipette 18 μL of DMSO into tubes 2–8, and 5 μL of the 1000 × MEK inhibitor into the first tube. Then take 2 μL from the first tube, and pipette up and down in the second tube. Take 2 μL from the second tube and then pipette up and down in the third tube. Repeat until the end of the strip, and the entire procedure for each MEK inhibitor
4. In a clear plastic 96-well plate, pipette 198 μL of warm Opti-MEM into each well of a column. Using a multichannel pipette, transfer 2 μL of each PCR tube in the strip to the column of Opti-MEM. Pipette up and down to mix. The concentration of the dose-response will be 10 × and enough volume for 20 columns of a 96-well plate
5. Prepare a 20 × stock of Tram-bo aliquoted into a strip of 12 PCR tubes. To prepare enough to add to one 96-well plate mix 28.8 μL of 400 × Tram-bo with 86.4 μL of DMSO and 460.8 μL of Tracer dilution buffer. Mix well and aliquot. Keep in a dark place until use
6. Remove transfected HEK293T cells from the incubator, remove DMEM, wash with PBS, trypsinize, add DMEM, and spin down at 300 × *g* for 5 min at room temperature. Remove supernatant and resuspend in Opti-MEM
7. Adjust cell density to 235,000 cells/mL using Opti-MEM and use a multichannel pipette to dispense 85 μL of cells into each well of a white NBS Corning 96-well plate from a reservoir
8. Using a multichannel pipette, transfer 10 μL of each 10 × drug dose-response from the clear 96-well plate, to the 96-well plate with cells
9. Next, transfer 5 μL of Tram-bo to each well using a multichannel pipette and pipette up and down
10. Incubate for 2 h at 37 °C
11. Prepare a 3 × solution of NL substrate and inhibitor in Opti-MEM. For one 96-well plate add 33 μL of the NL-substrate and 11 μL of the inhibitor to 5.24 mL of Opti-MEM

12. Take plate out of incubator and let it cool to room temperature for 10 min. Then using a multichannel pipette, add 50 μL of Opti-MEM solution to each well
13. Read on bioluminescent plate reader (GloMax Discover) using the stand NanoBRET 618 protocol
14. Export the data as an Excel document, and copy and paste into Prism. The data can be fitted using the non-linear regression option followed by the [Inhibitor] vs response (three parameter) equation

Day 2: Washout assay setup

1. Thaw 1000 × solutions of MEK inhibitors in DMSO and add 2 μL into 198 μL of Opti-MEM in a well of a clear 96-well plate to make a 10 × solution. Transfer 30 μL of each solution into a clean microcentrifuge tube
2. Remove transfected HEK293T cells from the incubator, remove DMEM, wash with PBS, trypsinize, add DMEM, and spin down at 300 × *g* for 5 min at room temperature. Remove supernatant and resuspend in Opti-MEM
3. Adjust cell density to 1.67 million cells/mL using Opti-MEM and transfer 270 μL to each microcentrifuge tube containing MEK inhibitors for a total volume of 300 μL. There will be 450,000 cells in each tube
4. Place tube at 37 °C for 2 h
5. Spin tubes down at 300 × *g* and remove supernatant
6. Add 300 μL of Opti-MEM and spin down again
7. Remove the supernatant, resuspend in 300 μL of Opti-MEM, and transfer 95 μL into three wells of an NBS corning 96-well plate
8. Add 5 μL of a 20 × Tram-bo solution to each well, followed immediately by 50 μL of a 33.3 × solution of NL-substrate and inhibitor
9. Read immediately on a bioluminescent plate reader (GloMax Discover). Adjust the number of reads and time between reads so the length of the experiment will be 2.5 h
10. Export the data into Excel and plot with Prism

8. Final conclusions

KSR-MEK-RAF has proven itself as a model to study signal transduction that occurs via changes in the assembly and regulation of higher order complexes (e.g., Brennan et al., 2011; Burack & Shaw, 2000; Dhawan et al., 2016; Khan et al., 2020; Lavoie et al., 2018; Michaud et al., 1997; Morrison, 2001; Nguyen et al., 2002; Rajakulendran et al., 2009;

Roy et al., 2002; Therrien et al., 1996). Moreover, subtle but important structural differences between KSR and RAF have enabled unique pharmacological mechanisms to modulate the RAS-MAPK pathway. For example, our structural and mechanistic data suggests that KSR can serve as a coreceptor for potent and long-lived binding of several clinical MEKi (Khan et al., 2020). The model that emerges from these studies is that certain compounds (e.g., trametinib) favor binding of MEK to KSR while disfavoring binding of MEK to RAF. This proposed mechanism suggests that clinically important compounds can be developed to exploit the defining features of highly conserved enzyme-pseudoenzyme signaling systems. These data also suggest that specific pseudokinases and related complexes (KSR:MEK and BRAF:MEK) may serve as distinct and physiologically relevant targets for therapeutics. Pseudokinase:kinase complexes are pervasive, and the methods described in this chapter that aim to find both ATP-competitive orthosteric inhibitors or interfacial binders could similarly be applied to other related systems. Indeed, we recently used the ATP-biotin assay, combined with additional biophysical and computational approaches, to identify the first type II binders of the pseudokinase STRAD (Smith et al., 2021).

We have highlighted the identification of small molecule binders to exploit or harness specific aspects of KSR function. Such chemical tools, in particular molecular glues for KSR, may be effective at limiting common resistance and drug escape mechanisms from MAPK-targeted therapies. Other pharmacological approaches, such as small molecule degraders and PROTACs, could also be developed. Recent efforts described for BRAF and MEK have demonstrated that degraders can be achieved on both wild-type and mutant forms of these proteins (Alabi et al., 2021; Posternak et al., 2020; Wei et al., 2019). Small molecules to eliminate KSR, or molecular glues and orthosteric inhibitors to lock particular functional states, may yield useful chemical probes and therapeutic leads. Several studies have given somewhat mixed results with respect to the impact of KSR depletion across cancer and disease models (e.g., (Costanzo-Garvey et al., 2009; Germino et al., 2018; Kortum et al., 2006; Lozano et al., 2003; Nguyen et al., 2002; Pearce et al., 2013; Therrien et al., 1996). Future efforts using chemical biology approaches should help deconvolute how KSR may be best exploited for therapeutic applications. This will continue to be an important goal as KSR functions in the pathway and via association with targets for two of the most frequently mutated oncogenes (i.e., BRAF and KRAS) across human cancers.

Acknowledgments

The Dar laboratory gratefully receives funding from the NIH (RO1s CA227636, CA258736, CA256480, and R56 AG066712) and the Mark Foundation for Cancer Research (20-030-ASP, 21-039-ASP). A.C.D. also thanks the Pershing-Square Sohn Cancer Research Alliance and Alex's Lemonade Stand Foundation for Childhood Cancer for support. A.C. and W.M.M. are recipients of NIH 5T32CA078207 and F99/K00 CA212474 awards, respectively. The authors are also supported by NCI grant P30 CA196521 to the Tisch Cancer Institute. A.C.D. is a founder, shareholder, advisory board member, and consultant to Nested Therapeutics.

References

Afonine, P. V., Grosse-Kunstleve, R. W., Echols, N., Headd, J. J., Moriarty, N. W., Mustyakimov, M., et al. (2012). Towards automated crystallographic structure refinement with phenix.refine. *Acta Crystallographica. Section D: Biological Crystallography, 68*(Pt 4), 352–367.

Alabi, S., Jaime-Figueroa, S., Yao, Z., Gao, Y., Hines, J., Samarasinghe, K. T. G., et al. (2021). Mutant-selective degradation by BRAF-targeting PROTACs. *Nature Communications, 12*(1), 920.

Bergfors, T. (2011). Macromolecular crystallization and crystal perfection, by Naomi E. Chayen, John R. Helliwell, and Edward H. Snell. *Crystallography Reviews, 17*, 225–226. https://doi.org/10.1080/0889311x.2010.544114.

Brennan, D. F., Dar, A. C., Hertz, N. T., Chao, W. C. H., Burlingame, A. L., Shokat, K. M., et al. (2011). A Raf-induced allosteric transition of KSR stimulates phosphorylation of MEK. *Nature, 472*(7343), 366–369.

Brtva, T. R., Drugan, J. K., Ghosh, S., Terrell, R. S., Campbell-Burk, S., Bell, R. M., et al. (1995). Two distinct Raf domains mediate interaction with Ras. *The Journal of Biological Chemistry, 270*(17), 9809–9812.

Burack, W. R., & Shaw, A. S. (2000). Signal transduction: Hanging on a scaffold. *Current Opinion in Cell Biology, 12*(2), 211–216.

Cacace, A. M., Michaud, N. R., Therrien, M., Mathes, K., Copeland, T., Rubin, G. M., et al. (1999). Identification of constitutive and Ras-inducible phosphorylation sites of KSR: Implications for 14-3-3 binding, mitogen-activated protein kinase binding, and KSR overexpression. *Molecular and Cellular Biology, 19*(1), 229–240.

Carrera, A. C., Alexandrov, K., & Roberts, T. M. (1993). The conserved lysine of the catalytic domain of protein kinases is actively involved in the Phosphotransfer reaction and not required for anchoring ATP. *Proceedings of the National Academy of Sciences of the United States of America, 90*(2), 442–446.

Copeland, R. A. (2015). The drug–target residence time model: A 10-year retrospective. *Nature Reviews. Drug Discovery, 15*(December), 87.

Costanzo-Garvey, D. L., Pfluger, P. T., Dougherty, M. K., Stock, J. L., Boehm, M., Chaika, O., et al. (2009). KSR2 is an essential regulator of AMP kinase, energy expenditure, and insulin sensitivity. *Cell Metabolism, 10*(5), 366–378.

Cutler, R. E., Jr., Stephens, R. M., Saracino, M. R., & Morrison, D. K. (1998). Autoregulation of the Raf-1 serine/threonine kinase. *Proceedings of the National Academy of Sciences of the United States of America, 95*(16), 9214–9219.

Dhawan, N. S., Scopton, A. P., & Dar, A. C. (2016). Small molecule stabilization of the KSR inactive state antagonizes oncogenic Ras Signalling. *Nature, 537*(7618), 112–116.

Filbert, E. L., Taylor, S. S., & Shaw, A. S. (2011). Mutation that blocks ATP binding creates a Pseudokinase stabilizing the scaffolding function of kinase suppressor of Ras, CRAF and BRAF. *PNAS*. https://www.pnas.org/content/108/15/6067.short.

Fischmann, T. O., Smith, C. K., Mayhood, T. W., Myers, J. E., Reichert, P., Mannarino, A., et al. (2009). Crystal structures of MEK1 binary and ternary complexes with nucleotides and inhibitors. *Biochemistry, 48*(12), 2661–2674.

Germino, E. A., Miller, J. P., Diehl, L., Swanson, C. J., Durinck, S., Modrusan, Z., et al. (2018). Homozygous KSR1 deletion attenuates morbidity but does not prevent tumor development in a mouse model of RAS-driven pancreatic cancer. *PLoS One, 13*(3), e0194998.

Haling, J. R., Sudhamsu, J., Yen, I., Sideris, S., Sandoval, W., Phung, W., et al. (2014). Structure of the BRAF-MEK complex reveals a kinase activity independent role for BRAF in MAPK signaling. *Cancer Cell, 26*(3), 402–413.

Heppner, D. E., & Eck, M. J. (2021). A structural perspective on targeting the RTK/Ras/MAP kinase pathway in cancer. *Protein Science: A Publication of the Protein Society, 30*(8), 1535–1553.

Herrmann, C., Martin, G. A., & Wittinghofer, A. (1995). Quantitative analysis of the complex between p21ras and the Ras-binding domain of the human Raf-1 protein kinase. *The Journal of Biological Chemistry, 270*(7), 2901–2905.

Hu, J., Stites, E. C., Haiyang, Y., Germino, E. A., Meharena, H. S., Stork, P. J. S., et al. (2013). Allosteric activation of functionally asymmetric RAF kinase dimers. *Cell, 154*(5), 1036–1046.

Ishii, N., Harada, N., Joseph, E. W., Ohara, K., Miura, T., Sakamoto, H., et al. (2013). Enhanced inhibition of ERK signaling by a novel allosteric MEK inhibitor, CH5126766, that suppresses feedback reactivation of RAF activity. *Cancer Research*. https://doi.org/10.1158/0008-5472.can-12-3937.

Iyer, G. H., Garrod, S., Woods Jr, V. L., & Taylor, S. S. (2005). Catalytic independent functions of a protein kinase as revealed by a kinase-dead mutant: Study of the Lys72His mutant of cAMP-dependent kinase. *Journal of Molecular Biology, 351*(5), 1110–1122.

Jarvis, D. L. (2014). Recombinant protein expression in baculovirus-infected insect cells. *Methods in Enzymology, 536*, 149–163.

Kabsch, W. (2010). XDS. *Acta Crystallographica. Section D: Biological Crystallography, 66*(Pt 2), 125–132.

Karim, F. D., Chang, H. C., Therrien, M., Wassarman, D. A., Laverty, T., & Rubin, G. M. (1996). A screen for genes that function downstream of Ras1 during drosophila eye development. *Genetics, 143*(1), 315–329.

Khan, Z. M., Real, A. M., Marsiglia, W. M., Chow, A., Duffy, M. E., Yerabolu, J. R., et al. (2020). Structural basis for the action of the drug trametinib at KSR-bound MEK. *Nature, 588*(7838), 509–514.

Kondo, Y., Ognjenović, J., Banerjee, S., Karandur, D., Merk, A., Kulhanek, K., et al. (2019). Cryo-EM structure of a dimeric B-Raf:14-3-3 complex reveals asymmetry in the active sites of B-Raf kinases. *Science, 366*(6461), 109–115.

Kondo, Y., Paul, J. W., Subramaniam, S., & Kuriyan, J. (2021). New insights into Raf regulation from structural analyses. *Current Opinion in Structural Biology*. https://doi.org/10.1016/j.sbi.2021.07.005.

Kornfeld, K., Hom, D. B., & Horvitz, H. R. (1995). The Ksr-1 gene encodes a novel protein kinase involved in Ras-mediated signaling in C. Elegans. *Cell, 83*(6), 903–913.

Kortum, R. L., Johnson, H. J., Costanzo, D. L., Volle, D. J., Razidlo, G. L., Fusello, A. M., et al. (2006). The molecular scaffold kinase suppressor of Ras 1 is a modifier of RasV12-induced and replicative senescence. *Molecular and Cellular Biology, 26*(6), 2202–2214.

Koveal, D., Schuh-Nuhfer, N., Ritt, D., Page, R., Morrison, D. K., & Peti, W. (2012). A CC-SAM, for coiled coil-sterile α motif, domain targets the scaffold KSR-1 to specific sites in the plasma membrane. *Science Signaling, 5*(255), ra94.

Lavoie, H., Sahmi, M., Maisonneuve, P., Marullo, S. A., Thevakumaran, N., Jin, T., et al. (2018). MEK drives BRAF activation through allosteric control of KSR proteins. *Nature*, *554*(7693), 549–553.

Lavoie, H., & Therrien, M. (2015). Regulation of RAF protein kinases in ERK Signalling. *Nature Reviews. Molecular Cell Biology*, *16*(5), 281–298.

Liau, N. P. D., Venkatanarayan, A., Quinn, J. G., Phung, W., Malek, S., Hymowitz, S. G., et al. (2020). Dimerization induced by C-terminal 14-3-3 binding is sufficient for BRAF kinase activation. *Biochemistry*, *59*(41), 3982–3992.

Liau, N. P. D., Wendorff, T. J., Quinn, J. G., Steffek, M., Phung, W., Liu, P., et al. (2020). Negative regulation of RAF kinase activity by ATP is overcome by 14-3-3-induced dimerization. *Nature Structural & Molecular Biology*, *27*(2), 134–141.

Lito, P., Saborowski, A., Yue, J., Solomon, M., Joseph, E., Gadal, S., et al. (2014). Disruption of CRAF-mediated MEK activation is required for effective MEK inhibition in KRAS mutant tumors. *Cancer Cell*, *25*(5), 697–710.

Lozano, J., Xing, R., Cai, Z., Jensen, H. L., Trempus, C., Mark, W., et al. (2003). Deficiency of kinase suppressor of Ras1 prevents oncogenic Ras signaling in mice. *Cancer Research*, *63*(14), 4232–4238.

Lu, H., & Tonge, P. J. (2010). Drug–target residence time: Critical information for Lead optimization. *Current Opinion in Chemical Biology*. https://doi.org/10.1016/j.cbpa.2010.06.176.

McCoy, A. J., Grosse-Kunstleve, R. W., Adams, P. D., Winn, M. D., Storoni, L. C., & Read, R. J. (2007). Phaser crystallographic software. *Journal of Applied Crystallography*, *40*(Pt 4), 658–674.

Michaud, N. R., Therrien, M., Cacace, A., Edsall, L. C., Spiegel, S., Rubin, G. M., et al. (1997). KSR stimulates Raf-1 activity in a kinase-independent manner. *Proceedings of the National Academy of Sciences of the United States of America*, *94*(24), 12792–12796.

Morrison, D. K. (2001). KSR: A MAPK scaffold of the Ras pathway? *Journal of Cell Science*, *114*(Pt 9), 1609–1612.

Müller, J., Ory, S., Copeland, T., Piwnica-Worms, H., & Morrison, D. K. (2001). C-TAK1 regulates Ras signaling by phosphorylating the MAPK scaffold, KSR1. *Molecular Cell*, *8*(5), 983–993.

Mulvania, T., Hayes, B., & Hedin, D. (2004). A flow cytometric assay for rapid, accurate determination of baculovirus titers. *Bioprocessing Journal*, *3*(3), 47.

Mysore, V. P., Zhou, Z.-W., Ambrogio, C., Li, L., Kapp, J. N., Chunya, L., et al. (2021). A structural model of a Ras-Raf signalosome. *Nature Structural & Molecular Biology*, *28*(10), 847–857.

Nassar, N., Horn, G., Herrmann, C., Scherer, A., McCormick, F., & Wittinghofer, A. (1995). The 2.2 a crystal structure of the Ras-binding domain of the serine/threonine kinase c-Raf1 in complex with Rap1A and a GTP analogue. *Nature*, *375*(6532), 554–560.

Nguyen, A., Richard Burack, W., Stock, J. L., Kortum, R., Chaika, O. V., Afkarian, M., et al. (2002). Kinase suppressor of Ras (KSR) is a scaffold which facilitates mitogen-activated protein kinase activation in vivo. *Molecular and Cellular Biology*, *22*(9), 3035–3045.

Park, E., Rawson, S., Li, K., Kim, B.-W., Ficarro, S. B., Pino, G. G.-D., et al. (2019). Architecture of autoinhibited and active BRAF–MEK1–14-3-3 complexes. *Nature*, *575*(7783), 545–550.

Patricelli, M. P., Katrin Szardenings, A., Liyanage, M., Nomanbhoy, T. K., Min, W., Weissig, H., et al. (2007). Functional interrogation of the Kinome using nucleotide acyl phosphates. *Biochemistry*, *46*(2), 350–358.

Patricelli, M. P., Nomanbhoy, T. K., Jiangyue, W., Brown, H., Zhou, D., Zhang, J., et al. (2011). In situ kinase profiling reveals functionally relevant properties of native kinases. *Chemistry & Biology*, *18*(6), 699–710.

Pearce, L. R., Atanassova, N., Banton, M. C., Bottomley, B., van der Klaauw, A. A., Revelli, J.-P., et al. (2013). KSR2 mutations are associated with obesity, insulin resistance, and impaired cellular fuel oxidation. *Cell, 155*(4), 765–777.

Pino, G.-D., Gonzalo, L., Li, K., Park, E., Schmoker, A. M., Ha, B. H., et al. (2021). Allosteric MEK inhibitors act on BRAF/MEK complexes to block MEK activation. *Proceedings of the National Academy of Sciences of the United States of America, 118*(36). https://doi.org/10.1073/pnas.2107207118.

Posternak, G., Tang, X., Maisonneuve, P., Jin, T., Lavoie, H., Daou, S., et al. (2020). Functional characterization of a PROTAC directed against BRAF mutant V600E. *Nature Chemical Biology, 16*(11), 1170–1178.

Rajakulendran, T., Sahmi, M., Lefrançois, M., Sicheri, F., & Therrien, M. (2009). A dimerization-dependent mechanism drives RAF catalytic activation. *Nature, 461*(7263), 542–545.

Robers, M. B., Vasta, J. D., Corona, C. R., Ohana, R. F., Hurst, R., Jhala, M. A., et al. (2019). Quantitative, Real-time measurements of intracellular target engagement using energy transfer. *Methods in Molecular Biology, 1888*, 45–71.

Robinson, M. J., Harkins, P. C., Zhang, J., Baer, R., Haycock, J. W., Cobb, M. H., et al. (1996). Mutation of position 52 in ERK2 creates a nonproductive binding mode for adenosine 5′-triphosphate. *Biochemistry, 35*(18), 5641–5646.

Roy, F., Laberge, G., Douziech, M., Ferland-McCollough, D., & Therrien, M. (2002). KSR is a scaffold required for activation of the ERK/MAPK module. *Genes & Development, 16*(4), 427–438.

Smith, R. H. B., Khan, Z. M., Ung, P. M.-U., Scopton, A. P., Silber, L., Mack, S. M., et al. (2021). Type II binders targeting the 'GLR-out' conformation of the Pseudokinase STRADα. *Biochemistry, 60*(4), 289–302.

Stewart, S., Sundaram, M., Zhang, Y., Lee, J., Han, M., & Guan, K. L. (1999). Kinase suppressor of Ras forms a multiprotein signaling complex and modulates MEK localization. *Molecular and Cellular Biology, 19*(8), 5523–5534.

Sundaram, M., & Han, M. (1995). The C. Elegans Ksr-1 gene encodes a novel Raf-related kinase involved in Ras-mediated signal transduction. *Cell, 83*(6), 889–901.

Therrien, M., Chang, H. C., Solomon, N. M., Karim, F. D., Wassarman, D. A., & Rubin, G. M. (1995). KSR, a novel protein kinase required for RAS signal transduction. *Cell, 83*(6), 879–888.

Therrien, M., Michaud, N. R., Rubin, G. M., & Morrison, D. K. (1996). KSR modulates signal propagation within the MAPK Cascade. *Genes & Development, 10*(21), 2684–2695.

Thevakumaran, N., Lavoie, H., Critton, D. A., Tebben, A., Marinier, A., Sicheri, F., et al. (2015). Crystal structure of a BRAF kinase domain monomer explains basis for allosteric regulation. *Nature Structural & Molecular Biology, 22*(1), 37–43.

Tran, T. H., Chan, A. H., Young, L. C., Bindu, L., Neale, C., Messing, S., et al. (2021). KRAS interaction with RAF1 RAS-binding domain and cysteine-rich domain provides insights into RAS-mediated RAF activation. *Nature Communications, 12*(1), 1176.

Van Aelst, L., Barr, M., Marcus, S., Polverino, A., & Wigler, M. (1993). Complex formation between RAS and RAF and other protein kinases. *Proceedings of the National Academy of Sciences of the United States of America, 90*(13), 6213–6217.

Vasta, J. D., Corona, C. R., Wilkinson, J., Zimprich, C. A., Hartnett, J. R., Ingold, M. R., et al. (2018). Quantitative, wide-Spectrum kinase profiling in live cells for assessing the effect of cellular ATP on target engagement. *Cell Chemical Biology, 25*(2), 206–214. e11.

Wan, P. T. C., Garnett, M. J., Mark Roe, S., Lee, S., Niculescu-Duvaz, D., Good, V. M., et al. (2004). Mechanism of activation of the RAF-ERK signaling pathway by oncogenic mutations of B-RAF. *Cell, 116*(6), 855–867.

Weber, C. K., Slupsky, J. R., Kalmes, H. A., & Rapp, U. R. (2001). Active Ras induces heterodimerization of cRaf and BRaf. *Cancer Research, 61*(9), 3595–3598.

Wei, J., Jianping, H., Wang, L., Xie, L., Jin, M. S., Chen, X., et al. (2019). Discovery of a first-in-class mitogen-activated protein kinase kinase 1/2 degrader. *Journal of Medicinal Chemistry*, *62*(23), 10897–10911.

Xing, H., Kornfeld, K., & Muslin, A. J. (1997). The protein kinase KSR interacts with 14-3-3 protein and Raf. *Current Biology: CB*, 7(5), 294–300.

Yaeger, R., & Corcoran, R. B. (2019). Targeting alterations in the RAF-MEK pathway. *Cancer Discovery*, *9*(3), 329–341.

Zhang, X. F., Settleman, J., Kyriakis, J. M., Takeuchi-Suzuki, E., Elledge, S. J., Marshall, M. S., et al. (1993). Normal and oncogenic p21ras proteins bind to the amino-terminal regulatory domain of c-Raf-1. *Nature*, *364*(6435), 308–313.

Zhang, Y., Yao, B., Delikat, S., Bayoumy, S., Lin, X. H., Basu, S., et al. (1997). Kinase suppressor of Ras is ceramide-activated protein kinase. *Cell*, *89*(1), 63–72.

CHAPTER FOURTEEN

Computational tools and resources for pseudokinase research

Brady O'Boyle[a], Safal Shrestha[b], Krzysztof Kochut[c], Patrick A. Eyers[d], and Natarajan Kannan[a,b,*]

[a]Department of Biochemistry & Molecular Biology, University of Georgia, Athens, GA, United States
[b]Institute of Bioinformatics, University of Georgia, Athens, GA, United States
[c]Department of Computer Science, University of Georgia, Athens, GA, United States
[d]Department of Biochemistry and Systems Biology, Institute of Systems, Molecular and Integrative Biology, University of Liverpool, Liverpool, United Kingdom
*Corresponding author: e-mail address: nkannan@uga.edu

Contents

Abstract

Pseudokinases regulate diverse cellular processes associated with normal cellular functions and disease. They are defined bioinformatically based on the absence of one or more catalytic residues that are required for canonical protein kinase functions. The ability to define pseudokinases based on primary sequence comparison has enabled the systematic mapping and cataloging of pseudokinase orthologs across the tree of life. While these sequences contain critical information regarding pseudokinase evolution and functional specialization, extracting this information and generating testable hypotheses based on integrative mining of sequence and structural data requires specialized computational tools and resources. In this chapter, we review recent advances in the

Methods in Enzymology, Volume 667
ISSN 0076-6879
https://doi.org/10.1016/bs.mie.2022.03.040

development and application of open-source tools and resources for pseudokinase research. Specifically, we describe the application of an interactive data analytics framework, KinView, for visualizing the patterns of conservation and variation in the catalytic domain motifs of pseudokinases and evolutionarily related canonical kinases using a consistent set of curated alignments organized based on the widely used kinome evolutionary hierarchy. We also demonstrate the application of an integrated Protein Kinase Ontology (ProKinO) and an interactive viewer, ProtVista, for mapping and analyzing primary sequence motifs and annotations in the context of 3D structures and AlphaFold2 models. We provide examples and protocols for generating testable hypotheses on pseudokinase functions both for bench biologists and advanced users.

Abbreviations

BPPS	Bayesian partitioning with pattern selection
DFG-Asp	metal-binding motif-Aspartate
HRD-Asp	catalytic motif-Aspartate
MSA	multiple sequence alignment
PK	protein kinase
ProKinO	Protein Kinase Ontology
PSKH2	Protein Serine Kinase H2
ULK4	Unc-51-Like Kinase 4
VRK3	Vaccinia-Related Kinase 3
β3-Lys	ATP-binding motif-Lysine

1. Introduction

Protein phosphorylation is one of the most widespread mechanisms for regulating cellular functions in both eukaryotes and prokaryotes. The enzymes that catalyze protein phosphorylation, protein kinases, are ubiquitous across the tree of life and constitute nearly 2% of most eukaryotic genomes. Protein kinases are also prevalent in prokaryotes, viruses, and pathogens (Kannan, Taylor, Zhai, Venter, & Manning, 2007; Kwon et al., 2019; Manning et al., 2011; Olson, Wang, Rico, & Wiebe, 2019; Talevich & Kannan, 2013; Wang, Zhao, Savas, Zhang, & Feng, 2020) and are defined at the sequence level based on the presence of key hallmark motifs that define three-dimensional structure and catalytic function (Hanks & Hunter, 1995; Knighton et al., 1991; Manning, Whyte, Martinez, Hunter, & Sudarsanam, 2002). These hallmark motifs include: the glycine rich loop, the ATP-binding β3-Lys, the catalytic aspartate and the cation-coordinating asparagine within the catalytic HRDXXXN loop motif, and the metal binding aspartate of the DFG motif at the beginning of the activation loop (Eyers & Murphy, 2013).

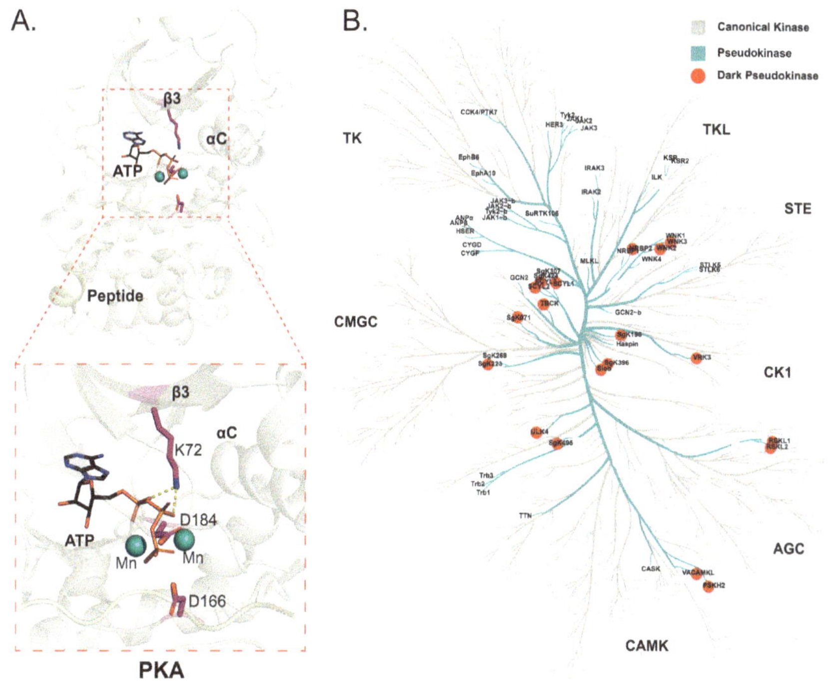

Fig. 1 Pseudokinases within the human kinome. (A) Key catalytic motifs in the Protein Kinase (PK) fold enzymes. (Top) Crystal structure of the prototypical protein kinase PKA (PDB ID: 1ATP; Zheng et al., 1993) is shown as cartoon and colored in gray. (Bottom) Zoomed in view of the catalytic pocket with β3-Lys (K72), catalytic HRD-Asp (D166), and metal binding DFG-Asp (D184) shown as sticks and colored in magenta. Manganese ions (Mn) are shown as spheres and colored in teal. ATP is shown as sticks and colored in black. (B) Pseudokinases are present within the human kinome. Pseudokinases present within the human kinome are shown in cyan. Dark pseudokinases are highlighted with a red circle (Berginski et al., 2020). The tree figure was generated using Coral (Metz et al., 2018).

Pseudokinases are bioinformatically defined based on the absence of one or more of the canonical active site residues (β3-Lys, HRD-Asp, DFG-Asp) (Fig. 1A), and the ability to define pseudokinases based on primary sequence comparison has enabled their systematic mapping across the tree of life (Kwon et al., 2019). The original cataloging of the human kinome in 2002 revealed that about 10% of human kinases are pseudokinases (Manning et al., 2002). The pseudokinases identified within the human genome were not confined to a single kinase family, but rather spread across all major kinase groups. Recent efforts to identify and annotate pseudokinases using profile-based methods have revealed not only the extent of the prevalence of pseudokinases but also taxonomic specific expansions across the tree of life. A survey of the more than 10,000 reference proteomes from the tree of life suggested that around 10% of the protein kinase (PK) members in each genome are pseudokinases (Kwon et al., 2019). Moreover, the study also identified several family-specific expansions such as interleukin-1 receptor–associated kinase (IRAK)–like pseudokinases in plants and tyrosine kinase–like (TKL) pseudokinases in fungi. Within prokaryote kinomes, pseudokinases are not ubiquitously present as seen in eukaryotes, however, a major expansion of *Mycobacterium tuberculosis* protein kinase B (pknB)-like pseudokinases was identified within many bacteria. In terms of pathogens, an expansion of pseudokinases was revealed within eukaryotic pathogens, including *Plasmodium falciparum* (Kwon et al., 2019; Manning et al., 2011) and *Giardia lamblia* (Kwon et al., 2019; Ward, Equinet, Packer, & Doerig, 2004). While there have not been any substantial efforts into studying the pseudokinomes of viruses, the presence of viral pseudokinases was established with the discovery of the B12 pseudokinase in vaccinia virus (Olson et al., 2019; Petrie et al., 2019; Wang et al., 2020). Indeed, the widespread presence of pseudokinases across the tree of life suggests multiple independent emergences of pseudokinases throughout kinase evolution.

Identification of pseudokinase orthologs provides a valuable resource for understanding pseudokinase evolution; however, a common issue during ortholog identification is the level of sequence divergence that is present when working with large sequence datasets. Sequence dissimilarities because of insertions, deletions, and other sequence restructuring events can make identification and classification of pseudokinase orthologs difficult without curated alignments (Neuwald, 2009). However, alignment curation is a difficult, time-consuming, and error-prone process that can result in inconsistencies in ortholog classification and subsequently inaccurate comparative and structural analyses. More recently, a human kinome centric ortholog detection method termed KinOrtho (Huang et al., 2021) was developed, which employs a graph-clustering approach using both primary sequence similarity and domain organization to help define human kinase orthologs across fully-sequenced proteomes. Recent application of this method across 17,000 proteomes identified 75 million overlapping kinase orthologous relationships. In comparison, other orthology inference methods only identified 35–60% of the orthologs defined by KinOrtho. KinOrtho's ability to identify weakly-related orthologs makes it ideal for studying pseudokinases, particularly the many understudied "dark" pseudokinases (Fig. 1B), where orthologous relationships from other organisms can be used as features in machine learning models for function prediction (Huang et al., 2021).

Integrative mining of the evolutionary information encoded in primary sequences with structural and functional data has provided new insights into the allosteric and non-catalytic functions of various pseudokinases (Eyers, Keeshan, & Kannan, 2017; Foulkes et al., 2018; Kwon et al., 2019; Murphy et al., 2015; Preuss et al., 2020; Shrestha, Byrne, Harris, Kannan, & Eyers, 2020). This approach was used to identify functionally constrained residues within the activation loop of the TRIB pseudokinase family as well as constraints within the MEK1 and COP1 binding sites of the C-tail (Eyers et al., 2017). Mapping of these constraints to the TRIB1 crystal structure (Jamieson et al., 2018; Murphy et al., 2015) revealed potential allosteric coupling that drives conformational changes regulating substrate binding to the C-lobe (Eyers et al., 2017). Consistently, C-tail tethering to the pseudokinase domain renders the protein incapable of substrate binding (Eyers, 2015; Murphy et al., 2015). Additionally, functional constraints within a leucine-rich repeat receptor-like plant pseudokinase family, termed LRRVI-2, and a receptor-like cytoplasmic kinase family, termed RLCKXII-1, were mapped to *Zea mays* LRRVI-2 (PDB ID: 6CPY; Aquino et al., 2018) and BSK8 (PDB ID: 4I92; Grütter, Sreeramulu, Sessa, & Rauh, 2013) crystal structures, respectively (Kwon et al., 2019). Integration of constraints and three-dimensional (3D) structure revealed surface exposed residues in LRRVI-2 involved in the unique tethering of the C-terminal tail to the kinase domain. Likewise, RLCKXII-1 specific constraints stabilize a unique inactive conformation involving β3-αC loop and the αC-helix glutamate residue.

Several new tools are available for the broader scientific community to access, visualize, and mine the evolutionary, structural, and functional information on pseudokinases. KinOrtho defined orthologs and their corresponding alignments for the core kinase domains are available through KinView (McSkimming et al., 2016), a sequence logo-based visualization tool (Crooks, Hon, Chandonia, & Brenner, 2004; Schneider & Stephens, 1990). In addition, UniProtKB, a comprehensive database for protein sequences and their functional annotations (Zaru, Magrane, Orchard, & UniProt Consortium, 2020), in conjunction with Mechanism and Catalytic Site Atlas (M-CSA), a database for enzyme active site and mechanism (Ribeiro et al., 2018), can be used as an orthogonal approach to identify enzymatically inactive kinase orthologs (Ribeiro, Tyzack, Borkakoti, & Thornton, 2020). These resources can serve as a conceptual starting point for generating new testable hypotheses on unique catalytic and non-catalytic mechanisms through integration of sequence data with 3D models and structures. Recent advancements in deep learning-based protein structure prediction, namely AlphaFold2 (Jumper et al., 2021), have generated a vast number of structural models for full-length kinases and pseudokinases, which are available for open-access analysis (https://alphafold.ebi.ac.uk/). Valuable information from these models can only be extracted by mapping important sequence, structural, and functional annotations to 3D models. ProKinO (Gosal, Kochut, & Kannan, 2011; McSkimming et al., 2015), a protein kinase-specific ontology framework, was previously developed to unify various protein kinase annotations, including data on kinase mutations and pathways. Recently, ProtVista viewer (Watkins, Garcia, Pundir, Martin, & UniProt Consortium, 2017), a graphical representation to interactively map sequence annotations on 3D structures, has been integrated into the ProKinO browser. In this chapter, we describe these computational tools (KinView and ProtVista) and describe their contribution to previously published pseudokinase research pertaining to two understudied pseudokinases, Unc-51-Like Kinase 4 (ULK4) and Protein Serine Kinase H2 (PSKH2). In addition, we also provide step-by-step instruction to use these to study and generate testable hypotheses for the dark pseudokinase, Vaccinia-Related Kinase 3 (VRK3).

2. Investigating pseudokinase functional specialization using comparative kinomics

2.1 KinView

Accurate multiple sequence alignments (MSAs) of protein kinase sequences from diverse organisms are valuable resources for investigating the relationships connecting pseudokinase sequence, structure, function, and regulation (Eyers et al., 2017; Foulkes et al., 2018; Jamieson et al., 2018; Preuss et al., 2020). MSAs can be used to identify functionally constrained residues in pseudokinases of interest, especially when compared to an evolutionarily similar protein, such as an active canonical counterpart (Chatzou et al., 2016; Morrison, 2006). However, taking a manual approach to identify constrained residues in an alignment is an error-prone, time-consuming endeavor. Furthermore, inconsistencies in MSAs based on the alignment method of choice, input sequences, and parameters pose additional challenges in data interpretation, reproducibility, and sharing (Pervez et al., 2014). Therefore, KinView was developed to resolve such issues and prevent inconsistencies. KinView is a visual comparative sequence analysis tool that provides sequence logo-based graphical representations of large multiple sequence alignments for protein kinases (McSkimming et al., 2016).

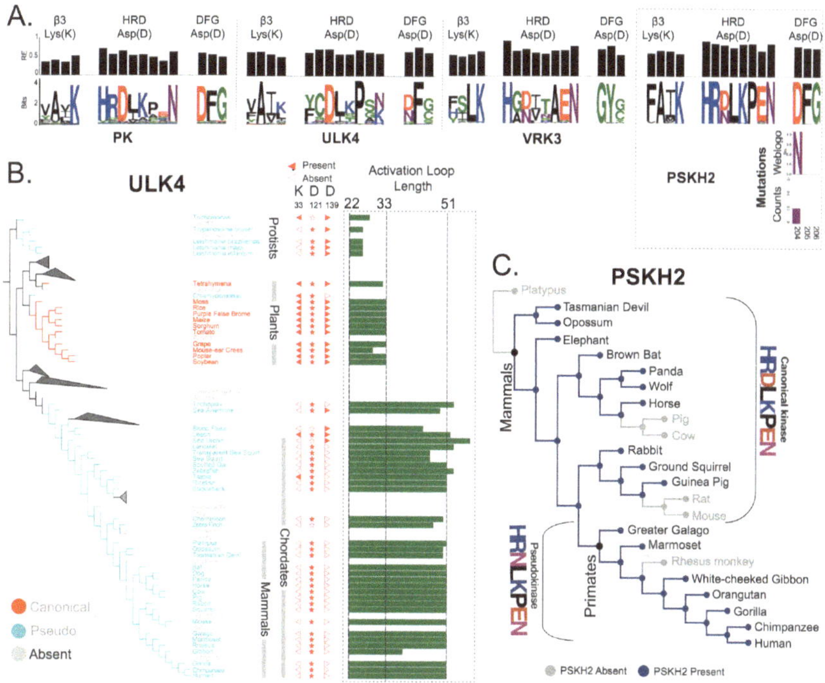

Fig. 2 Comparative kinomics using KinView identifies key amino acid differences in several pseudokinases. (A) Snapshot of sequence logos of key catalytic motifs in various dark pseudokinases in KinView (https://prokino.uga.edu/kinview/). Relative entropy (RE) was calculated using a BLOSUM62 as the background and matrix with a window size of 3 and sequence weighting enabled (Capra & Singh, 2007). Kinase domain sequence alignments were obtained from KinOrtho. PSKH2 mutations of the DFG motif found in cancer and their frequency are shown. (B) Presence of pseudo/canonical ULK4 orthologs mapped to SwissTree species tree (https://swisstree.sib.swiss/cgi-bin/swisst?page=species_tree). Key catalytic residue losses are shown as unfilled shapes. The lengths of the activation loop as compared to the core kinase domain are shown as bar charts and colored in green. (C) PSKH2 orthologs mapped to the SwissTree species tree. Absence and presence of PSKH2 orthologs are colored gray and blue, respectively.

It enables comparative analysis and visualization of the patterns of conservation and variation at each aligned position in the kinase domain using sequence logos, which provide a graphical representation of the amino acid frequencies at each aligned position (Fig. 2A) (Capra & Singh, 2007; Kullback & Leibler, 1951). For more details on KinView features, see Protocol Section 4.1.

An example showing the sequence logos for the ATP-binding motif (β3-Lys), the catalytic motif (HRD-Asp), and the metal binding motif (DFG-Asp) in an alignment of 95,591 annotated protein kinases (PK) is shown in Fig. 2A. This PK sequence logo can serve as a reference for investigating pseudokinase divergence at catalytic positions. Sequence logos for some of the dark pseudokinases (ULK4, VRK3, and PSKH2) are shown next to the PK sequence logo (Fig. 2A). Differences in the distribution of different amino-acid residues in pseudokinases are immediately apparent. For ULK4 (Eyers, 2020), the β3-Lys and DFG-Asp are less constrained, whereas the HRD-Asp remains highly conserved in ULK4 across diverse organisms (Fig. 2A, B). Although most of the chordate ULK4 (including humans) are missing canonical residues at β3-Lys and DFG-Asp positions, all ULK4 orthologs in plants possess all three canonical residues (Fig. 2B). Moreover, the increase in the length of the activation loop seems to correlate with the emergence of ULK4 pseudokinases in higher organisms (Fig. 2B). Despite the lack of β3-Lys (L33) and DFG-Asp (N139), human ULK4 has been convincingly shown to bind ATP in a metal-independent manner (Khamrui, Ung, Secor, Schlessinger, & Lazarus, 2020; Murphy et al., 2014; Preuss et al., 2020). The sequence basis for these adaptations can be further probed by using the advanced statistical methods described below (Section 4.3) (Preuss et al., 2020).

In the case of PSKH2, an asparagine is observed at the catalytic aspartate (HRD-Asp) position (Fig. 2A). This substitution is selectively observed in primates (N) (Fig. 2C) (Reiterer, Eyers, & Farhan, 2014; Shrestha et al., 2020), suggesting primate-specific emergence of PSKH2 pseudokinases. Another feature of KinView is a visual mapping of cancer mutations and post-translational modifications (PTMs). Interestingly, in a few sequenced cancer samples, the DFG-Asp is substituted to an asparagine (N) (Fig. 2A), although whether this a driver or bystander mutation requires further investigation. The presence of asparagine in place of DFG-Asp in pseudokinases nearly always abolishes cation binding (Murphy et al., 2014), and consequently, experimentally mutating DFG-Asp to an asparagine in canonical kinases also prevents metal-ion binding (Haydon et al., 2003; Tsuchiya et al., 2020) generating questions as to the specific nature of the cancer variant and whether it modifies the ability of PSKH2 to bind to metal ions.

In VRK3 pseudokinases, the metal binding aspartate (DFG-Asp) is replaced by a glycine (G) whereas the catalytic aspartate (HRD-Asp) is present but only partially conserved (Fig. 2A). In humans and a subset of VRK3 orthologs, the catalytic aspartate is substituted to an asparagine (N) (Fig. 2A), resulting in a degraded nucleotide binding site. This suggested that VRK3 is incapable of binding and hydrolyzing ATP in a metal dependent manner, which was supported by biochemical and biophysical studies (Murphy et al., 2014; Nichols & Traktman, 2004; Scheeff, Eswaran, Bunkoczi, Knapp, &

Manning, 2009). However, more recently, recombinant GST-VRK3 or FLAG-VRK3 were reported to phosphorylate the nuclear envelope protein barrier-to-autointegration factor (BAF) *in vitro* (Park et al., 2015). The N-terminal flanking region of VRK3, which was absent in recombinant VRK3 when the kinase domain was crystallized in an inactive form containing an occluded ATP site, was required for phosphorylation of BAF. Thus, a potential for a compensatory catalytic residue and an undescribed flanking region interaction with the kinase domain warrants more thorough investigation in VRK3 and many other pseudokinases. In this context, it is worth noting that purified VRK3 can be destabilized by non-specific kinase inhibitors such as DAP and VI16832 *in vitro* (Murphy et al., 2014), suggesting atypical effects of small molecules that might involve the highly degraded "nucleotide-binding" site.

2.2 Bayesian partitioning with pattern selection (BPPS)

While active site variations can be visually examined using KinView, identifying allosteric networks that couple the active site to distal regulatory sites requires more advanced statistical analysis of sequence datasets. Bayesian Partitioning with Pattern Selection (BPPS) is one such approach that has been successfully employed to investigate pseudokinase functional specialization (Eyers et al., 2017; Kwon et al., 2019; Preuss et al., 2020; Shrestha et al., 2020). Conceptually, BPPS clusters aligned sequences into functionally divergent subgroups based on correlated residue patterns that are highly conserved within subgroup sequences but divergent in sequences outside of the subgroup (Neuwald, 2011, 2014). The pattern-based classification method identified a conserved surface patch in the N-terminal ATP-binding lobe of ULK4 pseudokinase that distinguishes it from other canonical ULK members (Preuss et al., 2020). Likewise, BPPS analysis revealed a subset of co-conserved residues distinguishing the pseudokinase PSKH2 (His 291, Leu303, His307, and Asp316) from closely related canonical PSKH1. These co-conserved residues map to the substrate-binding C-lobe of PSKH2 kinase domain model, presumably conferring specificity in protein-protein interactions (Shrestha et al., 2020). For a step-by-step protocol to perform constraint-based analysis using BPPS, see Protocol Section 4.3.

3. Structural informatics and visualization tools for pseudokinase research

3D structures of proteins are critical for investigating the unique non-catalytic functions of pseudokinases. Indeed, analyzing the active site variations and co-variations considering available crystal structures, homology models, and molecular dynamics (MD) simulations have provided unique insights into non-catalytic functions of various pseudokinases (Eyers et al., 2017; Foulkes et al., 2018; Jamieson et al., 2018; Kwon et al., 2019; Preuss et al., 2020; Shrestha et al., 2020; Talevich & Kannan, 2013). However, the lack of structural modeling data for the vast majority of pseudokinases has hindered functional insights. Recent advancement in the field of protein structure prediction using Artificial Intelligence (AI) is now partially alleviating this issue. AlphaFold2 (Jumper et al., 2021) is an AI-based protein structure prediction tool developed by DeepMind. Based on Critical Assessment of protein Structure Prediction 14 (CASP14) measures, AlphaFold2 can predict protein structures to near experimental accuracy (median backbone accuracy of 0.96 Å r.m.s.d. as compared to 2.8 Å r.m.s.d. for the next best performing method). The method employs novel neural network based on evolutionary, physical, and geometric constraints. It takes the protein sequence and multiple sequence alignment of related sequences as inputs and directly predicts the 3D structure. Through a collaboration between DeepMind and EMBL-EBI, structures predicted by AlphaFold2 for the human proteome and 20 other model organisms have been made available online (https://alphafold.ebi.ac.uk/).

The Protein Kinase Ontology (ProKinO) is an integrated knowledge graph capturing the relationships connecting sequence, structure, function, mutation, and pathways on protein kinases. We have recently integrated the ProtVista viewer within ProKinO to provide 3D context for sequence annotations. We demonstrate the utility of this viewer in analyzing 3D structures and predicted structural models of pseudokinases using human ULK4 full length AlphaFold2 model (UniProt ID: Q96C45) as an example (Fig. 3A). ULK4 lacks the conserved salt bridge between the lysine ($K72^{PKA}$) in the β3 strand and glutamate ($E92^{PKA}$) in the αC-helix. Identifying these variations in the active site typically requires structural and sequence alignment of ULK4 with canonical protein kinase such as Protein Kinase A (PKA). However, using the ProtVista viewer, such variations can be rapidly identified simply by selecting the residues annotated as "KE Salt bridge" under the "Structural Motifs" tab in the sequence viewer. Selection of these residues highlights L33 and W46 in the ULK4 structural model, providing insights into ULK4's unique mode of ATP binding. Similarly, the DFG-Asp is substituted to an asparagine (N139) (Fig. 3B) in ULK4, visualized in the context of 3D models or crystal structures (PDB ID: 6U5L (Khamrui et al., 2020) and 6TSZ (Preuss et al., 2020)) co-crystallized with an inhibitor and

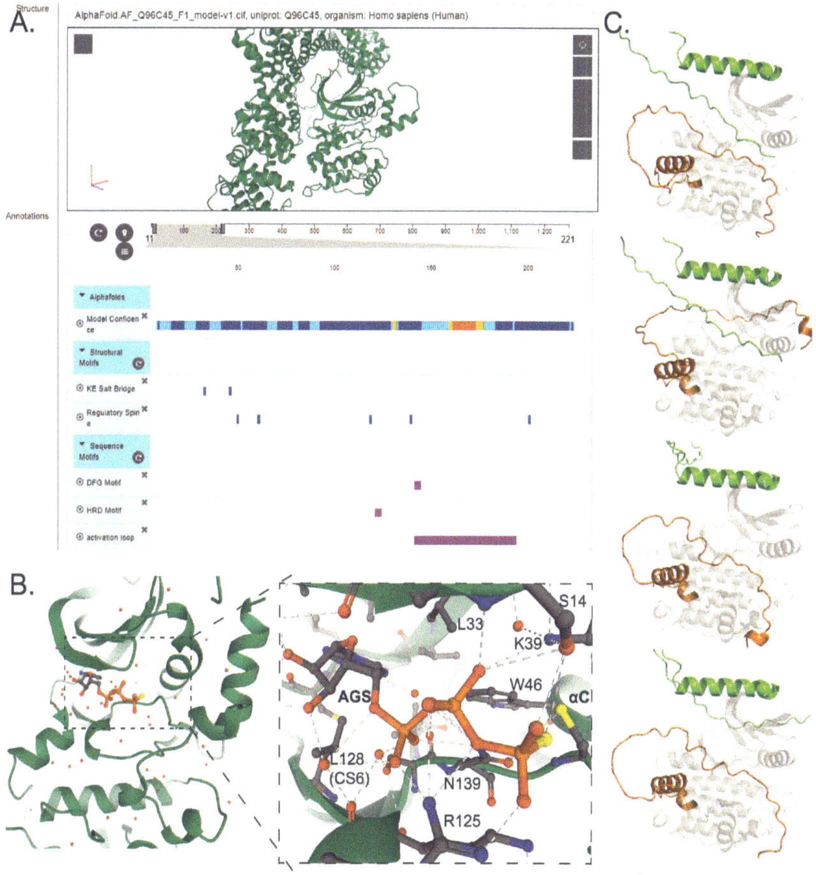

Fig. 3 Integration of AlphaFold2 models and ProtVista viewer in ProKinO for pseudokinase research. (A) Snapshot of the ProKinO page for human ULK4 (UniProt ID: Q96C45). Various sequence and structural annotations can be accessed and easily mapped to the AlphaFold2 model or crystal structures in the ProKinO browser. (B) Cartoon representation of the kinase domain of human ULK4 co-crystallized with AGS, an ATP analog (PDB ID: 6TSZ; Preuss et al., 2020), as shown using the ProtVista viewer. (Right) Zoomed in view of the interactions made by AGS with the kinase domain. (C) Multiple AlphaFold2 predicted models of full length human PSKH2. The models are shown as cartoons with N- (1–62) and C- (321–385) terminal regions colored in green and orange, respectively. PyMOL version 2.3.2 was used to generate the figure.

ATP-analog (AGS). The phosphates of the AGS molecules interact with positively charged residues such as K39 and R125 (Fig. 3B). Moreover, L128, which is annotated as catalytic spine 6 (CS6) residue, forms van der Waals interaction with the adenine ring. This interactive mining of the sequence motifs and annotations in the context of 3D models allows one to further explore the binding modes. The ProtVista viewer can also be used to define domain boundaries of the catalytic domain and visualize interactions between the kinase domain and flanking regulatory segments, as predicted by AlphaFold2 models. This is illustrated for the pseudokinase PSKH2, showing predicted interactions between the kinase domain and flanking N- (1–62) and C- (321–385) terminal flanking regions (shown in green and orange respectively) (Fig. 3C). For detailed instruction on how to perform similar analysis, refer to Protocol Section 4.2.

4. Protocols

4.1 Comparative kinomics of VRK family members using KinView (Fig. 4)

1.1. Open a web browser (Chrome, Firefox, Safari, Edge) on a computer connected to the internet.

1.2. Open the URL https://prokino.uga.edu/. KinView can be accessed through the ProKinO browser (Fig. 4A). To access the KinView app on the web, click "Launch" button under KinView (**1**). For convenience, users can download an offline version by clicking on "Launch" button under "Downloads." KinView can be installed on any standard operating system (Linux, Mac, Windows 10) and requires an installation of "node js" (https://nodejs.org/en/). For offline use, it is recommended to have at least 16 GB of RAM and 10 GB of free disk space.

(**2**). Alternatively, KinView can be accessed through this link: https://prokino.uga.edu/kinview/.

1.3. Once the KinView app is launched, toggle selection for the kinase families of interest by clicking on the boxes near the names (Fig. 4B). Here, pseudokinase VRK3 as well as its paralogs, VRK1 and VRK2, are selected (**3**). Protein Kinase (PK) has been selected as reference (**4**). The search bar can be used to directly select the family of interest (**5**).

1.4. Turn on the kinase-specific sequence motifs by clicking on the "Motif" button under "Settings" (**6**). The motifs such as HRD, DFG, APE and so on are shown in graphical format (**7**). To get to these C-lobe motifs, the users will need to scroll to the right using the scroll bar at the bottom of the page.

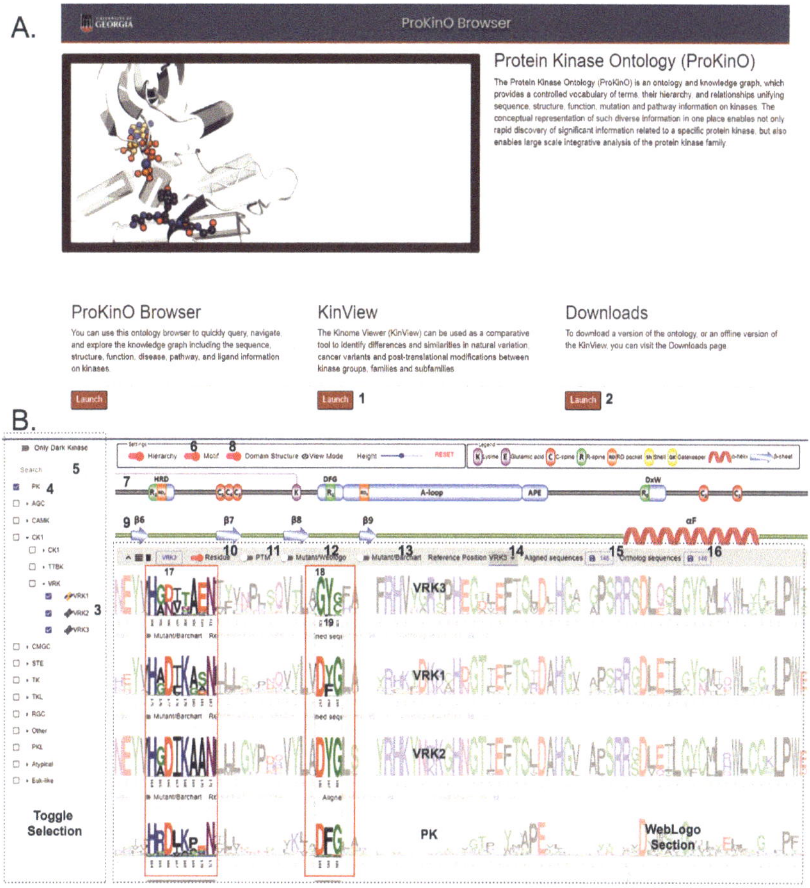

Fig. 4 The KinView web interface. (A) The ProKinO Browser. Buttons to go to the online version or to download the offline version of KinView are indicated by numbers 1 and 2, respectively. (B) Comparison of VRK paralogs on KinView. The kinase family of interests can be toggled on/off using the selection on the left. The multiple sequence alignments for VRK1–3 are represented as sequence logos. The sequence and structural annotations are at the top. HRD and DFG motifs are highlighted with Protein Kinase (PK) as reference.

1.5. Turn on the secondary structure representation by clicking on the "Domain Structure" button under "Settings" (**8**). The secondary structures, such as β6–9 and αF, are shown in graphical format (**9**).

1.6. Sequence logos for the selected families (3,4) are shown in the sequence logo section (Fig. 4B). By default, the "Residue" option is turned on (**10**). Turning off the "Residue" option will hide the sequence logos.

1.7. Post Translational Modifications (PTM) type and counts can be toggled on using the "PTM" button (**11**). Cancer mutations in the form of a sequence logo (**12**) or mutation counts grouped by amino acid properties can also be toggled on to aid in visualizing documented mutations (**13**).

1.8. The residue numbering below the sequence logos for each family can be changed by selecting the options under "Reference Position" (**14**). Alignment numbering can be set to "VRK3," which is the residue numbering within VRK3, or to "PKA," which changes the residue numbering in relation to the alignment numbering of PKA. Setting the numbering to "PKA" allows for position specific comparisons across kinases. In this example, VRK3 numbering is used for the VRK3 sequence logo.

1.9. To download the alignment in the FASTA format, click on the floppy disk icon (**15**). Cancer mutations and PTM data are also downloadable options (**15**). Kinase domain or full-length domains can be downloaded by clicking on the button next to "Ortholog sequences" (**16**).

1.10. The sequence logos can then be used to compare key sequence motifs. HRD and DFG motifs are highlighted (Fig. 4B, Red Boxes). Only a subset of VRK3 members conserve the HRD-Asp. In some VRK3 sequences it is replaced by an asparagine (N) (**17**), whereas VRK1 and VRK2 sequences strictly conserve an Asp at this position (**17**). Another key catalytic residue, the DFG-Asp is a strictly a glycine (G, Gly) in VRK3 sequences (**18**). Moreover, the DFG-Phe is strictly conserved as a tyrosine (Y, Tyr) residue (**19**). One can see that the same change is seen in a subset of VRK1 sequences and all VRK2 sequences.

4.2 Structural insights into VRK3 active site using ProtVista viewer in ProKinO browser (Fig. 5)

2.1. Open a web browser (Chrome, Firefox, Safari, Edge) on a computer connected to the internet.

2.2. Open the URL https://prokino.uga.edu/browser/. ProKinO homepage with the search bar is shown in Fig. 5A.

2.3. Type "VRK3" in the search bar (**1**) as indicated Fig. 5A to search for VRK3. You will see multiple results. Click on the first link "prokino: Human_VRK3."

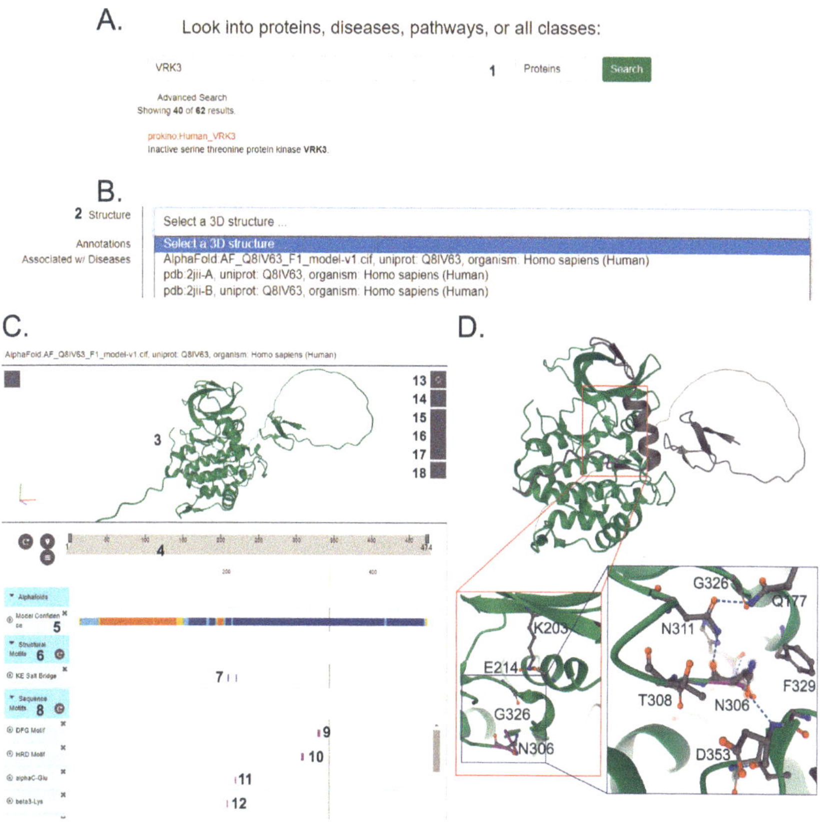

Fig. 5 The ProtVista Viewer on ProKinO. (A) Home page of the ProKinO Browser (B) ProKinO page for the Human_VRK3 showing the Structure section. (C) ProtVista viewer view of the AlphaFold2 model for human VRK3 full length protein. The different annotations are also shown. (D) Snapshot of the human VRK3 AlphaFold2 model. Kinase domain is colored green with N-terminal region colored in gray. The residues are shown as Ball & Stick with carbon atoms colored in gray.

2.4. You will get to the ProKinO page for Human VRK3 (UniProt Id: Q8IV63). Here, you will find multiple information about the protein such as cellular localization, functional domains, sequence, disease annotations, and so on. You can access both crystal structures and AlphaFold2 models under "Structure" (**2**) (Fig. 5B). For this example, click on the AlphaFold2 model "AlphaFold: AF_Q8IV63_F1_model-v1.cif, uniprot: Q8IV63, organism: *Homo sapiens* (Human)."

2.5. The ProtVista viewer is loaded, and you will see the AlphaFold2 model of full length human VRK3 (**3**) (Fig. 5C).

2.6. The sequence view is shown (**4**). Human VRK3 is 474 residues long. You can zoom to residue level resolution by hovering over it and using the scroll wheel on the mouse.

2.7. Under the sequence view, you will find "Model Confidence" (**5**). It is an AlphaFold2 calculated per-residue confidence score (pLDDT). Blue represents "Very High" confidence while orange represents "Very Low" confidence.

2.8. To map the KE salt bridge, go to "Structural Motifs" (**6**). The residues in the salt bridge can then be directly selected and shown as Ball & Stick representation by clicking on the vertical bars (**7**).

2.9. To map sequence motifs to the AlphaFold2 model, go to "Sequence Motifs" (**8**). The motifs DFG, HRD, αC-Glu, and β3-Lys can be viewed in the structure by clicking on the vertical bars **9–12** respectively.

2.10. To make publication quality figures (Fig. 5D), use the different ProtVista viewer tools (**13–18**). The functions of the tools are as follows:

13. Reset Camera.

14. Screenshot: HD (1280 × 720), Full HD (1920 × 1080), Ultra HD (3840 × 2160) and custom resolutions available for download. Ultra HD resolution was used for snapshots in Fig. 5D.

15. Structure Tools: You can make measurements as well as create new selections and representations.

16. Expanded Viewport: Allows user to toggle on/off full screen mode.

17. Settings: You can change the background color, lighting, clipping and so on.

18. Selection: By toggling on this tool, you can select residues of interest, create selection objects, and color them.

2.11. The snapshots in Fig. 5D were generated using ProtVista tools. The selection tool (18) was used to select the N-terminal region colored in gray (Fig. 5D, Top).

2.12. Using the selection tool (18) and the annotations (9–12), the key catalytic residues as well as αC-Glu were mapped and shown as Ball & Stick representation (Fig. 5D, Bottom Left, Red box).

2.13. The viewer also provides an interaction network for residues within 5 Å of the selected residue. N306 (HRD-Asp equivalent residue in human VRK3) is selected and interactions with residues within 5 Å are shown (Fig. 5D, Bottom Right, Blue box). The side-chain carbonyl of N306 forms a hydrogen bond with D353 backbone amide.

4.3 Comparative kinomics of VRK family members using BPPS (Fig. 6)

3.1. Open a web browser (Chrome, Firefox, Safari, Edge) on a computer connected to the internet. A Linux-based operating system, such as RedHat or Ubuntu, is recommended when running BPPS and no non-standard hardware is required.

3.2. Navigate to https://www.igs.umaryland.edu/labs/neuwald/software/bpps/ and left click "bpps1.1.6.tar" to download. Move the downloaded file, bpps1.1.6.tar.gz, to the desired folder. Extract the contents of the bpps1.1.6.tar.gz file.

3.3. Download the VRK kinase domain ortholog sequences from KinView using the protocol described above (Section 4.1). As well, download CSNK1A1 kinase domain ortholog sequences from KinView to serve as an outgroup. Merge all the FASTA files together so that the sequences are all in FASTA format and contained within a single file.

```
cat *.fasta >> vrk_kd_unaligned.fasta
```

3.4. Align the sequences using MUSCLE (https://github.com/rcedgar/muscle/releases/tag/v5.0.1428) to generate FASTA formatted alignments of ortholog sequences

```
muscle -in vrk_kd_unaligned.fasta -out vrk_kd_aligned.fasta
```

3.5. The aligned FASTA file must be converted to a CMA file for BPPS input. Download ConvertMSA from https://www.igs.umaryland.edu/labs/neuwald/software/auxiliary/. Unzip the file to and run the following line to convert the FASTA to CMA format:

```
ConvertMSA fa2cma vrk_kd_aligned.fasta > vrk_kd.cma
```

3.6. Open command prompt or terminal and navigate to the folder containing the BPPS executable file and aligned sequences in CMA format. Run BPPS 1 using the command below. Note that the file extension is not included when running BPPS, just the file name. Output files will be .cmd, .chk, _user.sma, _bst.out, and _aln.rtf.

```
bpps 1 vrk_kd [options]
```

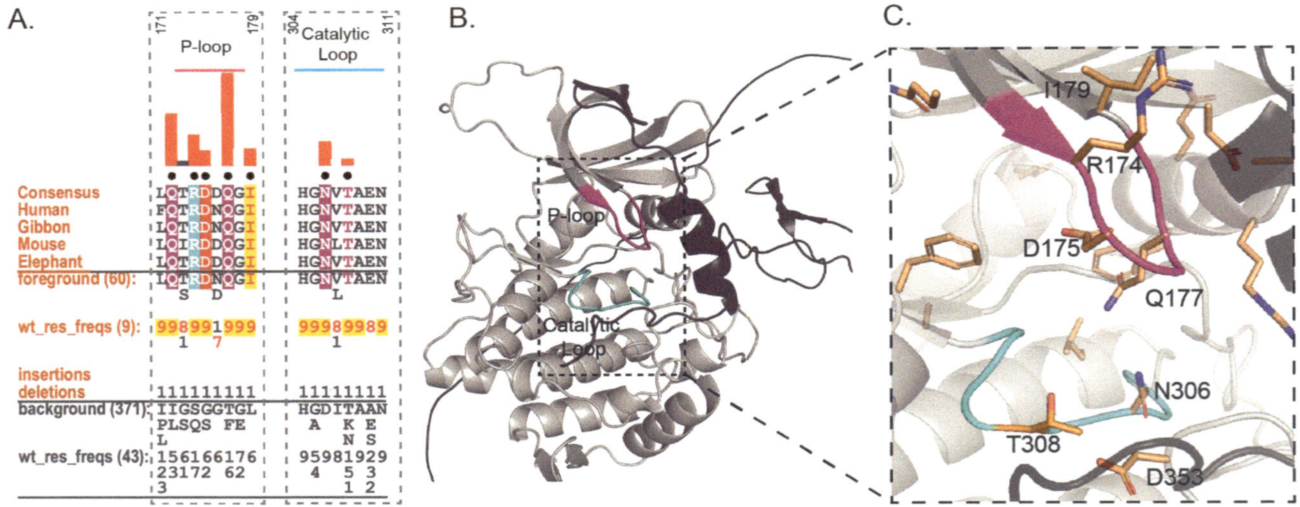

Fig. 6 Mammalian VRK3-specific sequence constraints. (A) Sequence constraints in the glycine-rich loop and catalytic loop of mammalian VRK3 sequences (foreground) when compared to the Set 3 sequences (background). Set 3 contains VRK1, VRK2 and VRK3 sequences that are clustered together based on shared constraints identified by BPPS. The alignment shows mammalian VRK3 sequences from diverse organisms. The histogram in red above the residue indicates the extent to which the foreground alignment diverges from the corresponding position in the background alignment. Black dots mark the alignment positions used by the BPPS procedure when distinguishing the mammalian VRK3 sequences from Set 3 sequences. (B) AlphaFold2 model for full length human VRK3. The glycine-rich loop and the catalytic loop are colored in magenta and cyan, respectively. The N-terminal region is colored darker gray as compared to the kinase domain. (C) Zoomed in view showing the sequence constraints identified in (A). The residues are shown as sticks and colored orange. Glycine-rich loop and the catalytic loop are colored as in (B). A co-conserved aspartate within the activation loop, D353, is also shown.

3.7. Run BPPS 2 using the output of BPPS 1 by using the file name (but not the extension) as the input for BPPS 2. Output files will be _himsa.cma, _himsa.tpl, _himsa.sets, _himsa.dft, _himsa.hpt, and _himsa.ph.

```
bpps 2 vrk_kd [options]
```

3.8. BPPS 3 can be used to generate a PyMOL (Schrödinger, 2019) session file that will overlay constrained residue patterns onto a 3D structure. A pattern set must first be selected from the vrk_kd_himsa.sets file. The set that contains the human VRK3 sequence will be used to generate a PyMOL session file. The set of interest can be identified by searching the vrk_kd_himsa.sets file for the human VRK3 uniProt ID (Q8IV63). First run the following command to separate the sets file into searchable files:

```
tweakcma vrk_kd_himsa -write
```

3.9. Several vrk_kd_himsa_Set files will have been generated. Run the following command to locate the set that the human VRK3 sequence belongs to

```
grep 'Q8IV63' *Set*
```

3.10. The human VRK3 sequence is present in Sets 1, 3, and 7. The number associated with the pattern will be different for different runs, but the patterns identified will generally remain consistent. Set 1 consists mainly of VRK sequences whereas Set 3 consists of VRK3 sequences. Set 7 is a subgroup of Set 3 and contains only mammalian VRK3 sequences.

3.11. Open vrk_kd_himsa.hpt file. This file contains hierarchal grouping of the sequences, with each group being defined by the constrained residues that are shared by a set of sequences. Identify the hierarchical grouping number associated with Set 7, in this case it is 9. This grouping number will be used to run BPPS 3.

3.12. Search online for an rtf-pdf converter to turn vrk_kd_aln.rtf to a pdf (https://online2pdf.com/convert-rtf-to-pdf is free and easy-to-use). This file indicates the constrained residue patterns for each hierarchal group in comparison with the residues present at the corresponding positions of the parent hierarchal group (Fig. 6A). It also includes the frequency of amino acids in the foreground and the background of each group. The statistical significance of a residue within a pattern is indicated by a red bar above the corresponding position. A larger bar indicates a residue is more highly conserved within the foreground when compared to the residues at the same position within the background sequences.

3.13. Download a 3D structure of human VRK3. This can be obtained from AlphaFold2 (https://alphafold.ebi.ac.uk/) by searching for "VRK3," selecting the "*Homo sapiens*" protein and clicking the "PDB File" button. If a crystal structure is available, then it can be obtained from RCSB (https://www.rcsb.org/) by searching "VRK3," selecting the VRK3 structure, clicking "Download Files" and then selecting PDB format. Save the structure as VRK3_H.pdb.

3.14. Make a text file that indicates the path to the pdb file and save it as path_to_pdb.txt. The path to a file can generally be obtained by right-clicking on the file and opening the file properties. This can then be copied and pasted in the text file and should appear as shown below:

```
this/is/the/path/to/your/file/path_to_pdb.txt
```

3.15. Generate a PyMOL session by running BPPS 3 using the following command

```
bpps 3 vrk_kd_himsa 9 -pdb=path_to_pdb.txt -pml
```

3.16. Download PyMOL (https://pymol.org/2/) and open it on the computer. Click "File" at the top left, select "Open" and select vrk_kd_himsa_6_1_1.pml. This will open the 3D structure file that was provided in BPPS 3 (Fig. 6B) and annotate the constrained residue patterns from Set 7, as well as the patterns from the parent hierarchal groups.

3.17. On the right side of the PyMOL window, there are classes that correspond to each group's constrained residue patterns. Class Y contains the residue patterns of Set 1, Class R contains the residue patterns of Set 3 and Class O contains the residue patterns of Set 7. Patterns can be hidden or shown by clicking the class name on the right side of the screen.

3.18. Side chain interactions can be predicted using the tools available in PyMOL. Some useful plugins available for PyMOL are: APBS for electrostatic surface visualization (Baker, Sept, Joseph, Holst, & McCammon, 2001), Autodock for protein-ligand docking (Trott & Olson, 2010), GROMACS for molecular dynamics simulation (Makarewicz & Kaźmierkiewicz, 2013) and PyMod for homology modeling (Janson & Paiardini, 2021). Distance between atoms and polar/hydrophobic contacts can also be determined using tools available on the right side of the PyMOL window. Residue side chains of the G-loop and those of the catalytic loop are shown as reference in Fig. 6C.

5. Summary

Advances in open-source computational tools for pseudokinase analysis allow for tracing of robust evolutionary histories and are paving the way for a more thorough understanding of their structural adaptations. KinOrtho's ability to accurately identify and classify pseudokinase orthologs provides a means to utilize the vast wealth of exponentially increasing sequence data. With these orthologs, tools such as KinView and BPPS can identify constrained residues through comparative sequence analysis. The conservation of these residues will provide the foundation for generating and subsequently testing new hypotheses on pseudokinase functional specialization. Tools such as AlphaFold2 and ProKinO can then be utilized to integrate these evolutionary data with structural models and provide context for the conserved residue patterns. While these tools are valuable resources for the pseudokinase community, additional work is needed to further enhance the utility of these tools. In particular, families that are selectively expanded in plants, prokaryotes, fungi and viruses need to be systematically classified and incorporated in the protein kinase ontology and kinase evolutionary hierarchy to enable evolutionary comparisons. Likewise, models of protein-protein interactions will significantly enhance the value of ProtVista in pseudokinase function prediction and annotation.

Acknowledgments

We acknowledge members of the Kannan lab for valuable comments and suggestions. Funding from NIH (R35GM139656 and U01CA239106) is acknowledged.

References

Aquino, B., Counago, R. M., Godoi, P. H. C., Massirer, K. B., Elkins, J. M., Arruda, P., et al. (2018). *Structure of apo GRMZM2G135359 Pseudokinase*. https://doi.org/10.2210/pdb6cpy/pdb.

Baker, N. A., Sept, D., Joseph, S., Holst, M. J., & McCammon, J. A. (2001). Electrostatics of nanosystems: Application to microtubules and the ribosome. *Proceedings of the National Academy of Sciences of the United States of America*, *98*(18), 10037–10041.

Berginski, M. E., Moret, N., Liu, C., Goldfarb, D., Sorger, P. K., & Gomez, S. M. (2020). The dark kinase knowledgebase: An online compendium of knowledge and experimental results of understudied kinases. *Nucleic Acids Research*, *49*(D1), D529–D535.

Capra, J. A., & Singh, M. (2007). Predicting functionally important residues from sequence conservation. *Bioinformatics*, *23*(15), 1875–1882.

Chatzou, M., Magis, C., Chang, J. M., Kemena, C., Bussotti, G., Erb, I., et al. (2016). Multiple sequence alignment modeling: Methods and applications. *Briefings in Bioinformatics*, *17*(6), 1009–1023.

Crooks, G. E., Hon, G., Chandonia, J.-M., & Brenner, S. E. (2004). WebLogo: A sequence logo generator. *Genome Research*, *14*(6), 1188–1190.

Eyers, P. A. (2015). Tribbles: A twist in the pseudokinase tail. *Structure*, *23*(11), 1974–1976.

Eyers, P. A. (2020). Marveling at the incredible ULK4. *Structure*, *28*(11), 1181–1183.

Eyers, P. A., Keeshan, K., & Kannan, N. (2017). Tribbles in the 21st century: The evolving roles of tribbles pseudokinases in biology and disease. *Trends in Cell Biology*, *27*(4), 284–298.

Eyers, P. A., & Murphy, J. M. (2013). Dawn of the dead: Protein pseudokinases signal new adventures in cell biology. *Biochemical Society Transactions*, *41*(4), 969–974.

Foulkes, D. M., Byrne, D. P., Yeung, W., Shrestha, S., Bailey, F. P., Ferries, S., et al. (2018). Covalent inhibitors of EGFR family protein kinases induce degradation of human tribbles 2 (TRIB2) pseudokinase in cancer cells. *Science Signaling*, *11*(549), eaat7951.

Gosal, G., Kochut, K. J., & Kannan, N. (2011). ProKinO: An ontology for integrative analysis of protein kinases in cancer. *PLoS One*, *6*(12), e28782.

Grütter, C., Sreeramulu, S., Sessa, G., & Rauh, D. (2013). Structural characterization of the RLCK family member BSK8: A pseudokinase with an unprecedented architecture. *Journal of Molecular Biology*, *425*(22), 4455–4467.

Hanks, S. K., & Hunter, T. (1995). The eukaryotic protein kinase superfamily: Kinase (catalytic) domain structure and classification1. *The FASEB Journal*, *9*(8), 576–596.

Haydon, C. E., Eyers, P. A., Aveline-Wolf, L. D., Resing, K. A., Maller, J. L., & Ahn, N. G. (2003). Identification of novel phosphorylation sites on Xenopus laevis Aurora A and analysis of phosphopeptide enrichment by immobilized metal-affinity chromatography. *Molecular & Cellular Proteomics: MCP*, *2*(10), 1055–1067.

Huang, L.-C., Taujale, R., Gravel, N., Venkat, A., Yeung, W., Byrne, D. P., et al. (2021). KinOrtho: A method for mapping human kinase orthologs across the tree of life and illuminating understudied kinases. *BMC Bioinformatics*, *22*(1), 446.

Jamieson, S. A., Ruan, Z., Burgess, A. E., Curry, J. R., McMillan, H. D., Brewster, J. L., et al. (2018). Substrate binding allosterically relieves autoinhibition of the pseudokinase TRIB1. *Science Signaling*, *11*(549), eaau0597.

Janson, G., & Paiardini, A. (2021). PyMod 3: A complete suite for structural bioinformatics in PyMOL. *Bioinformatics*, *37*(10), 1471–1472.

Jumper, J., Evans, R., Pritzel, A., Green, T., Figurnov, M., Ronneberger, O., et al. (2021). Highly accurate protein structure prediction with AlphaFold. *Nature*, *596*(7873), 583–589.

Kannan, N., Taylor, S. S., Zhai, Y., Venter, J. C., & Manning, G. (2007). Structural and functional diversity of the microbial kinome. *PLoS Biology*, *5*(3), e17.

Khamrui, S., Ung, P. M. U., Secor, C., Schlessinger, A., & Lazarus, M. B. (2020). High-resolution structure and inhibition of the schizophrenia-linked Pseudokinase ULK4. *Journal of the American Chemical Society*, *142*(1), 33–37.

Knighton, D. R., Zheng, J., Eyck, L. F. T., Ashford, V. A., Xuong, N.-H., Taylor, S. S., et al. (1991). Crystal structure of the catalytic subunit of cyclic adenosine monophosphate-dependent protein kinase. *Science*, *253*(5018), 407–414.

Kullback, S., & Leibler, R. A. (1951). On information and sufficiency. *The Annals of Mathematical Statistics*, *22*(1), 79–86.

Kwon, A., Scott, S., Taujale, R., Yeung, W., Kochut, K. J., Eyers, P. A., et al. (2019). Tracing the origin and evolution of pseudokinases across the tree of life. *Science Signaling*, *12*(578), eaav3810.

Makarewicz, T., & Kaźmierkiewicz, R. (2013). Molecular dynamics simulation by GROMACS using GUI plugin for PyMOL. *Journal of Chemical Information and Modeling*, *53*(5), 1229–1234.

Manning, G., Reiner, D. S., Lauwaet, T., Dacre, M., Smith, A., Zhai, Y., et al. (2011). The minimal kinome of Giardia lamblia illuminates early kinase evolution and unique parasite biology. *Genome Biology*, *12*(7), 1–20.

Manning, G., Whyte, D. B., Martinez, R., Hunter, T., & Sudarsanam, S. (2002). The protein kinase complement of the human genome. *Science*, *298*(5600), 1912–1934.

McSkimming, D. I., Dastgheib, S., Baffi, T. R., Byrne, D. P., Ferries, S., Scott, S. T., et al. (2016). KinView: A visual comparative sequence analysis tool for integrated kinome research. *Molecular BioSystems*, *12*(12), 3651–3665.

McSkimming, D. I., Dastgheib, S., Talevich, E., Narayanan, A., Katiyar, S., Taylor, S. S., et al. (2015). ProKinO: A unified resource for mining the cancer kinome. *Human Mutation*, *36*(2), 175–186.

Metz, K. S., Deoudes, E. M., Berginski, M. E., Jimenez-Ruiz, I., Aksoy, B. A., Hammerbacher, J., et al. (2018). Coral: Clear and customizable visualization of human Kinome data. *Cell Systems*, 7(3), 347–350.

Morrison, D. A. (2006). Multiple sequence alignment for phylogenetic purposes. *Australian Systematic Botany*, *19*(6), 479–539.

Murphy, J. M., Nakatani, Y., Jamieson, S. A., Dai, W., Lucet, I. S., & Mace, P. D. (2015). Molecular mechanism of CCAAT-enhancer binding protein recruitment by the TRIB1 Pseudokinase. *Structure*, *23*(11), 2111–2121.

Murphy, J. M., Zhang, Q., Young, S. N., Reese, M. L., Bailey, F. P., Eyers, P. A., et al. (2014). A robust methodology to subclassify pseudokinases based on their nucleotide-binding properties. *Biochemical Journal*, *457*(2), 323–334.

Neuwald, A. F. (2009). Rapid detection, classification and accurate alignment of up to a million or more related protein sequences. *Bioinformatics*, *25*(15), 1869–1875.

Neuwald, A. F. (2011). Surveying the manifold divergence of an entire protein class for statistical clues to underlying biochemical mechanisms. *Statistical Applications in Genetics and Molecular Biology*, *10*(1), 36. https://doi.org/10.2202/1544-6115.1666.

Neuwald, A. F. (2014). A Bayesian sampler for optimization of protein domain hierarchies. *Journal of Computational Biology*, *21*(3), 269–286.

Nichols, R. J., & Traktman, P. (2004). Characterization of three paralogous members of the mammalian vaccinia related kinase family. *Journal of Biological Chemistry*, *279*(9), 7934–7946.

Olson, A. T., Wang, Z., Rico, A. B., & Wiebe, M. S. (2019). A poxvirus pseudokinase represses viral DNA replication via a pathway antagonized by its paralog kinase. *PLoS Pathogens*, *15*(2), e1007608.

Park, C.-H., Ryu, H. G., Kim, S.-H., Lee, D., Song, H., & Kim, K.-T. (2015). Presumed pseudokinase VRK3 functions as a BAF kinase. *Biochimica et Biophysica Acta (BBA) - Molecular Cell Research*, *1853*(7), 1738–1748.

Pervez, M. T., Babar, M. E., Nadeem, A., Aslam, M., Awan, A. R., Aslam, N., et al. (2014). Evaluating the accuracy and efficiency of multiple sequence alignment methods. *Evolutionary Bioinformatics Online*, *10*, 205–217.

Petrie, E. J., Sandow, J. J., Lehmann, W. I., Liang, L. Y., Coursier, D., Young, S. N., et al. (2019). Viral MLKL homologs subvert necroptotic cell death by sequestering cellular RIPK3. *Cell Reports*, *28*(13), 3309–3319.

Preuss, F., Chatterjee, D., Mathea, S., Shrestha, S., St-Germain, J., Saha, M., et al. (2020). Nucleotide binding, evolutionary insights, and interaction Partners of the Pseudokinase Unc-51-like kinase 4. *Structure*, *28*(11), 1184–1196.

Reiterer, V., Eyers, P. A., & Farhan, H. (2014). Day of the dead: Pseudokinases and pseudophosphatases in physiology and disease. *Trends in Cell Biology*, *24*(9), 489–505.

Ribeiro, A. J. M., Holliday, G. L., Furnham, N., Tyzack, J. D., Ferris, K., & Thornton, J. M. (2018). Mechanism and catalytic site atlas (M-CSA): A database of enzyme reaction mechanisms and active sites. *Nucleic Acids Research*, *46*(Database issue), D618–D623.

Ribeiro, A. J. M., Tyzack, J. D., Borkakoti, N., & Thornton, J. M. (2020). Identifying pseudoenzymes using functional annotation: Pitfalls of common practice. *The FEBS Journal*, *287*(19), 4128–4140.

Scheeff, E. D., Eswaran, J., Bunkoczi, G., Knapp, S., & Manning, G. (2009). Structure of the Pseudokinase VRK3 reveals a degraded catalytic site, a highly conserved kinase fold, and a putative regulatory binding site. *Structure*, *17*(1), 128–138.

Schneider, T. D., & Stephens, R. M. (1990). Sequence logos: A new way to display consensus sequences. *Nucleic Acids Research*, *18*(20), 6097–6100.

Schrödinger, L. L. C. (2019). *The PyMOL Molecular Graphics System*. Version 2.3.2.

Shrestha, S., Byrne, D. P., Harris, J. A., Kannan, N., & Eyers, P. A. (2020). Cataloguing the dead: Breathing new life into pseudokinase research. *The FEBS Journal*, *287*(19), 4150–4169.

Talevich, E., & Kannan, N. (2013). Structural and evolutionary adaptation of rhoptry kinases and pseudokinases, a family of coccidian virulence factors. *BMC Evolutionary Biology*, *13*(1), 1–17.

Trott, O., & Olson, A. J. (2010). AutoDock Vina: Improving the speed and accuracy of docking with a new scoring function, efficient optimization, and multithreading. *Journal of Computational Chemistry*, *31*(2), 455–461.

Tsuchiya, Y., Byrne, D. P., Burgess, S. G., Bormann, J., Baković, J., Huang, Y., et al. (2020). Covalent Aurora a regulation by the metabolic integrator coenzyme A. *Redox Biology*, *28*, 101318. https://doi.org/10.1016/j.redox.2019.101318.

Wang, T., Zhao, J., Savas, A. C., Zhang, S., & Feng, P. (2020). Viral pseudoenzymes in infection and immunity. *The FEBS Journal*, *287*(19), 4300–4309.

Ward, P., Equinet, L., Packer, J., & Doerig, C. (2004). Protein kinases of the human malaria parasite plasmodium falciparum: The kinome of a divergent eukaryote. *BMC Genomics*, *5*(1), 1–19.

Watkins, X., Garcia, L. J., Pundir, S., Martin, M. J., & UniProt Consortium. (2017). ProtVista: Visualization of protein sequence annotations. *Bioinformatics*, *33*(13), 2040–2041.

Zaru, R., Magrane, M., Orchard, S., & UniProt Consortium. (2020). Challenges in the annotation of pseudoenzymes in databases: The UniProtKB approach. *The FEBS Journal*, *287*(19), 4114–4127.

Zheng, J., Trafny, E. A., Knighton, D. R., Xuong, N., Taylor, S. S., Eyck, L. F. T., et al. (1993). 2.2 Å refined crystal structure of the catalytic subunit of cAMP-dependent protein kinase complexed with MnATP and a peptide inhibitor. *Acta Crystallographica Section D: Biological Crystallography*, *49*(3), 362–365.

CHAPTER FIFTEEN

Methods to assess small molecule allosteric modulators of the STRAD pseudokinase

Tingting Qing[a,b], Jin Liu[a,b], Fen Liu[a,b], Dom C. Mitchell[c], Richard T. Beresis[a,b], and John D. Gordan[c,]*

[a]Chempartner Co, Ltd., Shanghai, China
[b]Chempartner Co, Ltd., South San Francisco, CA, United States
[c]Division of Hematology Oncology and Quantitative Biosciences Institute, University of California, San Francisco (UCSF), San Francisco, CA, United States
*Corresponding author: e-mail address: john.gordan@ucsf.edu

Contents

Abstract

With the increased appreciation of the biological relevance of pseudokinase (PSK) allostery, the broadening of small molecule strategies to target PSK function is of particular importance. We and others have pursued the development of small molecule allosteric

Methods in Enzymology, Volume 667
ISSN 0076-6879
https://doi.org/10.1016/bs.mie.2022.03.041

modulators of the STRAD pseudokinase by targeting its ATP binding pocket. The purpose of this effort is to modulate the function of the LKB1 tumor suppressor kinase, which exists in a trimer with the STRAD PSK and the adaptor protein MO25. Here we provide detailed guidance regarding the different methods we have used for medium throughput screening to identify STRAD ligands and measure their impact on LKB1 kinase activity. Our experience supports preferential use of direct measurements of LKB1 kinase activity, and demonstrates the limitations of indirect assessment methods in the development trans-acting allosteric modulators.

1. Introduction

1.1 Overview of discovery and role of LKB1 in cancer biology

LKB1 (Liver Kinase B1, also called *STK11*) mutation was first described in 1998 in a kindred with Peutz-Jeghers Syndrome (PJS), marked by gastrointestinal hamartomas and cancers (Jenne et al., 1998). Inactivating *LKB1* mutations were soon found in sporadic lung and colon tumors (Resta et al., 1998; Sanchez-Cespedes et al., 2002). Across the Cancer Genome Atlas (TCGA) project, *LKB1* inactivation is found in 16% of lung adenocarcinomas, 6% of cervical squamous carcinomas, and 1–3% of common malignancies such as breast, ovarian, pancreatic, gastro/esophageal and melanoma, with LKB1 inactivation also associated with worse prognosis for both recurrence and overall survival (Bailey et al., 2018; Liu et al., 2018). Mouse models and clinical trials have demonstrated the impact of LKB1 on tumor behavior: *Lkb1* loss accelerates tumor initiation in lung and other cancers, and is associated with higher rates of metastasis (Hollstein et al., 2019; Murray et al., 2019). *Lkb1*-deficient mouse cancer models show resistance to chemotherapy and targeted therapy (Chen et al., 2012), and clinical studies have connected *LKB1* loss with resistance to EGFR inhibition, chemotherapy, radiotherapy and the PD1 inhibitor nivolumab (Moro et al., 2018; Park et al., 2019; Sitthideatphaiboon et al., 2021; Skoulidis et al., 2018).

Initial studies of LKB1 signaling identified it as the kinase upstream of the metabolic sensor AMP-activated kinase (AMPK) (Hawley et al., 2003; Shaw et al., 2004; Woods et al., 2003), and thus a regulator of protein translation (Shaw et al., 2004) and autophagy (Liang et al., 2007) via effects on mTOR, ULK1 and others. 13 other AMPK-related kinases (ARKs) were also identified (Lizcano et al., 2004), and have a diverse set of functions, including effects on DNA damage responses (Sapkota et al., 2002; Wang et al., 2016), epigenetic control and cellular differentiation

(Kottakis et al., 2016; Zhang et al., 2017), mitochondrial quality control (Gurumurthy et al., 2010; Ozturk et al., 2019; Rabinovitch et al., 2017), and the cell-intrinsic immune response (Kitajima et al., 2019). Genetic screens have tied these mechanisms to context- specific effects (Cossa et al., 2020; Goodwin et al., 2014; Zagorska et al., 2010), including a role for the MAP/microtubule affinity-regulating kinases (MARK1-4) regulating the Hippo kinase pathway, which inhibits the YAP/TAZ co-activators to reduce cellular proliferation (Mohseni et al., 2014). Recent studies in cancer cell lines and mouse models revealed that the Salt-inducible kinases (SIK1-3) are the essential effectors of LKB1-mediated tumor suppression *KRAS*-mutant lung cancer, with effects *in vivo* on inflammation and metabolism. Interestingly, their deletion results in a similar rate of tumor initiation but smaller tumors with reduced growth compared to *Lkb1* loss, suggesting that SIK-independent functions of LKB1 also play a role in tumor progression (Hollstein et al., 2019; Murray et al., 2019).

1.2 Structure/function relationships and allosteric signaling in the LKB1/STRAD/MO25 trimer

LKB1 signals in an obligate 1:1:1 trimer that includes the pseudokinase (PSK) STRAD (Ste20-related adaptor) and the adaptor protein MO25. There are two STRAD isoforms (α and β). STRADα (STRADA) is much better studied, but there is no evidence to suggest differences in function between the two isoforms. Here, we will refer to both as STRAD. STRAD contacts are necessary for LKB1 activity *in vitro*, and disease-associated LKB1 mutations that disrupt the LKB1/STRAD interface have been found (Zeqiraj et al., 2009; Zeqiraj, Filippi, Deak, Alessi, & van Aalten, 2009). Multiple reports have also shown STRAD expression also increases LKB1 protein levels, likely by stabilizing its tertiary structure (Gaude et al., 2012; Nony et al., 2003; ten Klooster et al., 2009). These results suggested two potential models for STRAD's support of LKB1 kinase activity: First, complex formation may be sufficient to increase LKB1 protein levels to increase its overall catalysis. However, it has been suggested that STRAD induces specific allosteric changes in LKB1 that support a more active confirmation. STRAD-induced changes in LKB1 appear independent of more typical mechanisms of kinase activation, as suggested in the crystal structure by its adoption of an active kinase conformation despite the lack of autophosphorylation.

Detailed biochemical analysis also supports an allosteric effect of STRAD on LKB1 (Zeqiraj, Filippi, Deak, et al., 2009; Zeqiraj, Filippi, Goldie, et al., 2009). The PSK domain of STRAD interacts with the N- and C-lobes of LKB1, with STRAD's active site oriented towards LKB1 while LKB1's active site is oriented away from STRAD, presumably allowing phosphotransfer to its substrates. PSKs often have divergence from the canonical ATP-binding pockets seen in protein kinases. This is seen in STRAD, which directly binds the phosphates in ATP via H200 and R215 rather than coordinating them through a magnesium necessary for catalysis. However, ATP binding by STRAD increases its affinity for MO25 and is necessary to bind LKB1 (Zeqiraj, Filippi, Goldie, et al., 2009). Furthermore, while multiple mutations engineered at the LKB1/STRAD interface disrupt trimeric complex formation, the $LKB1^{R74A}$, $STRAD^{H231A/Y185F}$ and $MO25^{F243A/R240A}$ interface mutations significantly disrupt LKB1 kinase activity without altering complex formation. While these are engineered mutations rather than those identified in disease, studies have identified a range of cancer-associated STRAD and LKB1 mutations at their binding interface (Bailey et al., 2018; Zeqiraj, Filippi, Deak, et al., 2009), further supporting the hypothesis that allosteric communication between STRAD and LKB1 is essential for LKB1 function.

1.3 Small molecule ATP mimetics as allosteric modulators

While much of the focus in the development of ATP-competitive kinase inhibitors is on the disruption of kinase activity, many such compounds also have activating and inhibitory allosteric effects. This is because of the conformational changes in the kinase domain that accompany binding of different types of inhibitors. Most common inhibitors are classified as having a type I (binding to an active conformation) or type II (binding to an inactive conformation) binding mode. A type II molecule binds to a "DFG-out" state of the active site in which the aspartate within the conserved DFG motif is rotated away from the active site, with the DFG-phenylalanine taking its position. Another subtype of the inactive conformation binders, sometimes referred to as type I ½, bind in a conformation in which helix αC of the kinase is swung away from the active conformation disabling proper positioning of the catalytic lysine for kinase catalysis (Zhang, Yang, & Gray, 2009). Allosteric signal transduction despite catalytic inhibition as a results of inhibitor binding has been described in several cases: The BRAF inhibitor vemurafenib has been shown to paradoxically augment signaling downstream of mutant KRAS by inducing dimerization of RAF-family kinases resulting in activation of the downstream mediator MEK (Hatzivassiliou et al., 2010). Vemurafenib is classified as a type I ½ inhibitor, based on an intermediate perturbation of the position of the kinase helix αC (Hymowitz & Malek, 2018). Similarly, type I small molecules targeting the ATP binding pocket of the integrated stress response (ISR) kinase IRE-1, while inhibiting phosphotransfer, can result in activation of its RNase domain (Papa, Zhang, Shokat, & Walter, 2003). Successor molecules for each target have been shown to reverse this effect, with type II IRE1 inhibitors blocking its RNase activity (Wang et al., 2012). Similarly, "paradox breaking" BRAF inhibitors such as PLX7904 do not cause downstream pathway activation. Curiously, paradox breaking inhibitors still bind BRAF in the type I ½ mode but are suspected to affect αC-helix displacement differently than vemurafenib, likely increasing its dynamics, and consequently RAF dimerization (Hymowitz & Malek, 2018; Zhang et al., 2015).

Interestingly, similar allosteric disruption of MEK activation can be accomplished with small molecules targeting the KSR PSK, providing an example of allosteric modulation of a PSK by targeting its ATP-binding pocket (Dhawan, Scopton, & Dar, 2016). This work has motivated investigation of several other allosteric inhibitors of pro-oncogenic PSK, including the receptor-like PSKs ROR1, ROR2, PTK7 and RYK, discovering enough dynamic properties in these PSKs to allow small molecule binding and thus the possibility to develop conformational disrupters (Sheetz et al., 2020). Covalent binding of epidermal growth factor (EGFR) inhibitors to the PSK TRIB2 has also been described, and results in TRIB2 destabilization and degradation (Foulkes et al., 2018). PSK stabilizers have also been developed, including some in relative advanced stages. A key example here is BMS-986165, a stabilizer of the PSK domain in the non-receptor tyrosine kinase TYK2, which has tandem kinase/PSK domains. This agent is a potent inhibitor of IL-12, IL-23, and type I IFN receptor-mediated activation of TYK2 and blocks multiple autoimmune disorders without causing apparent immunodeficiency (Burke et al., 2019). Also of interest for the methods below, a primarily computational strategy has been used to uncover small molecules resulting in allosteric changes in STRAD (Smith et al., 2021). Thus, PSK-mediated allostery is an established drug target with potential in treatment of cancer and other important diseases (Kung & Jura, 2019), most notably inflammatory syndromes such as systemic lupus erythematosus (SLE), and Crohn's disease as discussed above for TYK2 inhibition.

1.4 Considerations in assays of kinase activation

Allosteric modulators of STRAD could result in either activation or inhibition of LKB1 kinase activity. Although there is extensive methodological expertise in the development of kinase inhibitors, the kinetics of kinase activation create additional complexity to be considered in assay optimization. Specifically, while kinase inhibitors tend to have a rapid on rate, allowing for the use of an endpoint assay, kinase activators may also require some time to increase kinase activity, requiring the evaluation of multiple time points to assess compound function and optimization of assay parameters to allow increased activity as well as sensitive detection of decreased activity. This is exemplified in the report of an allosteric SRC kinase activator, where a rate of acceleration is tested across multiple compound concentrations and compared to conditions with maximal ATP and enzyme levels (Yang et al., 2011). More commonly, though, longer kinase assays are used as an endpoint, and compounds are compared based on the concentration for half-maximal activation, which may be called a $K_{0.5}$ or EC_{50}. A useful example of the use of the EC_{50} to understand activator function may be seen in the work describing small molecule activators of AMPK (Scott et al., 2014) and how their impact is modified by phosphorylation of AMPK at different sites. Such analyses have also been adapted to activity-based protein profiling (Kok et al., 2020).

Typically, an allosteric activators and inhibitors alter V_{max} but not K_m for an enzyme. Thus, unlike a kinase assay that might be used to assess an ATP-competitive inhibitor, considering lower kinase ATP concentrations may improve the ability to measure the impact of an activator compound by putting kinase activity at less than the biochemical V_{max}. Furthermore, a special challenge is posed for LKB1 activators that act via binding to the ATP pocket in STRAD, as their binding to STRAD would be anticipated to be sensitive to ATP concentration. Thus, as it will be detailed below, careful assay optimization across multiple parameters is appropriate.

2. Thermal stabilization assays

2.1 Background and principles

As the majority of PSKs lack catalytic activity, enzymatic assays are not available to determine their ligand-binding capabilities. In addition to the nucleotide displacement assays shown below, *in vitro* and cellular thermal stabilization assays provide an opportunity to assess either the interaction of a small molecule with a PSK of interest *in vitro* or altered protein-protein interactions that a PSK of interest engages in.

Differential scanning fluorimetry (DSF), commonly referred to as a Thermofluor assay, is a tractable, efficient *in vitro* strategy to identify ligand binding to proteins based on alterations in protein thermal stability (Niesen, Berglund, & Vedadi, 2007). DSF is performed by incubating a protein of interest with a dye that binds hydrophobic regions (typically SYPRO orange) and then subjecting the solution to stepwise increases in temperature. Fluorescence is measured at each step, typically in a 96 well plate in a real-time PCR device, and fitted to a sigmoidal curve using the Boltzmann equation to determine a half-maximal melting temperature (T_M). In general, analysis focuses around a ΔT_M, either with or without a small molecule ligand or following the engineering of a mutation in the protein of interest. One key complexity in this analysis (addressed below) is the presence of multiple proteins in the same complex, which are likely to have overlapping melting curves.

DSF has been described specifically as a strategy for assessment across the majority of the pseudokinome, with a focus on identifying nucleotide binding as well as cation dependence (Murphy et al., 2014), and has been used for drug development and other studies of various PSKs including JAK2, IRAK3, TRIB1, TYK2 and VRK3 (Lange, Nelen, Cohen, & Kulathu, 2021; McNally et al., 2019; Murphy et al., 2015; Scheeff, Eswaran, Bunkoczi, Knapp, & Manning, 2009; Tokarski et al., 2015). Interestingly, isothermal dose analysis was used for VRK3 to also identify the K_M of ligand binding (Scheeff et al., 2009). Typically, DSF is used to identify ligands whose binding results in stabilization of the protein target. In principle it can also be used to define destabilizing ligands, although that approach is less common.

The cellular thermal shift assay (CETSA) employs the same principles as DSF but in intact cells and tissues. CETSA was first described as a strategy for assessing the pharmacological properties of advanced compounds in tissue (Jafari et al., 2014; Martinez Molina et al., 2013). Here, cellular lysates are subjected to a range of temperatures specific to the target of interest, and then protein stability at distinct temperatures is determined by evaluating the presence of the protein in the soluble fraction by Western blotting, with the inference being that as proteins denature they are more likely to form insoluble aggregates. CETSA is most commonly readout by Western blot, although high-throughput proximity ligation-based strategies and mass spectrometry-based strategies have both been described (Dziekan et al., 2019; Jafari et al., 2014). CETSA has been applied to the study of the PSKs TRIB2 and ROR1 (Dziekan et al., 2019; Sheetz et al., 2020).

2.2 DSF assay protocol

This was developed following Murphy et al. (2014).

2.2.1 Reagents

10x Thermofluor Buffer: 250 mM Hepes pH 7.5, 500 mM NaCl

Thermofluor dye: SYPRO Orange protein gel stain (Invitrogen Cat# S6650)

2.2.2 Assay solution (per well)

2.5 μL 10x buffer

Thermofluor dye 1:5000 final concentration

STRAD/MO25 dimer, 2 μM final concentration

Compound to 20 μM

H_2O: up to 25 μL

2.2.3 Protein preparation

We prepared recombinant STRAD/MO25 following established protocols (Zeqiraj, Filippi, Goldie, et al., 2009). Briefly, His-STRAD/MO25 were expressed in *E. coli* BL21 using the pOPCH vector. Bacteria were induced for 18h with 250 μM IPTG, pelleted and resuspended in 20 mL of the Lysis Buffer per 1 L of culture (50 mM Tris pH 7.8, 50 mM NaCl, 10% glycerol, 20 mM imidazole, 0.2 mM EGTA, 0.2 mM EDTA, 0.075% β-mercaptoethanol, 0.5 mg/mL lysozyme (ThermoFisher 89833), and 0.3 mg/mL Benzonase (Millipore Sigma E1014)). Cells were lysed using a microfluidizer at 20,000–22,000 PSI for 2 cycles. Lysate was clarified with ultracentrifugation at 30,000 x *g* for 30 min and loaded onto Ni-NTA resin (Qiagen, Cat# 30210). After two washes in the Lysis Buffer and one in the Lysis Buffer with 40 mM imidazole, protein was then eluted in Lysis Buffer with 200 mM imidazole. The sample was then concentrated to 1 mL and loaded onto a Superdex 75 gel filtration column (GE Healthcare Cat# 29148722). Fractions were collected on an FPLC, with the STRAD/MO25 dimer eluting early while STRAD monomer eluted later. Purity was then confirmed by Coomassie to assess contaminants and separate STRAD monomer from STRAD/MO25 dimer-containing fractions. Protein was then concentrated to 95 μM for use.

2.2.4 Assay set up and performance

1. Prepare Master Mix containing all components but ligand on ice.
2. Set aside the volume for 3 wells per ligand in a secondary tube or in a 96 well plate.

3. Add adequate ligand quantity for three wells with final volume up to 80 μL, vortex and centrifuge briefly.
4. Aliquot 25 μL/well x3 into a 96 well plate; may be performed on ice if necessary.
5. Once plate is completed, apply clear adhesive film to cover, and spin, before data acquisition.
6. Place plate on a qPCR device (we have used Stratagene Mx3005P and Bio Rad CFX connect).
7. Perform a temperature curve from 25 -> 95 °C increased at 0.5 °C every 30 s, readout is using the FRET channel.

2.2.5 Data analysis

Once data are acquired (see Fig. 1), results are normalized and analyzed as follows. Importantly, as we had both STRAD and MO25 in our assay,

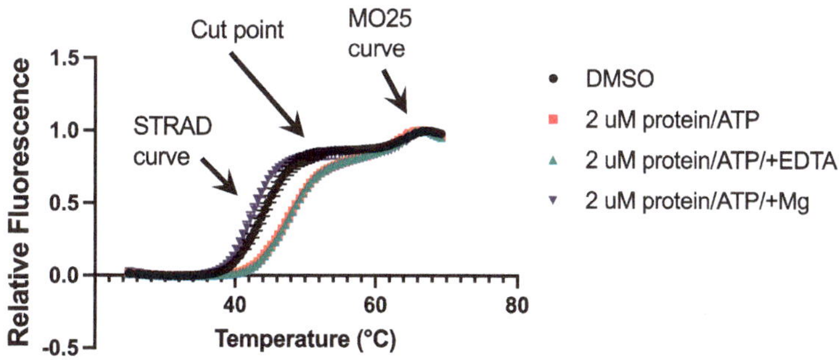

Protein	2 uM	2 uM	2 uM	2 uM
ATP		+	+	+
Additive			EDTA	Mg2+
V50 (n)	44.44	48.81	48.8	42.42
dV50 (n)	0	4.37	4.36	-2.02

Fig. 1 DSF of the STRAD/MO25 complex: We evaluated the stability of the STRAD/MO25 complex purified without nucleotide bound and subsequently treated with a vehicle control (DMSO) or with ATP in DMSO. As a control, we assessed the impact of 2 mM $MgCl_2$ (anticipated to disrupt nucleotide binding by STRAD), and found that it indeed blocked STRAD binding by ATP. Consequently, 0.5 mM EDTA (which is disrupts nucleotide binding to active kinases by chelating metal cations) did not interfere with STRAD stabilization by ATP. Delta V50 (dV50) for normalized fluorescence intensity was calculated for each condition vs vehicle treated protein. Results are from a single experiment done in triplicate; error bars ± standard deviation (SD), but are representative.

we had to determine which curve to measure. Based on published DSF analysis of a STRAD monomer (Murphy et al., 2014), we selected the first of two sigmoidal curves as the STRAD melting curve. Formal analytical methods are available to fit multiple curves, although our typical strategy has been direct assessment.

1. First, data normalization was performed in Microsoft Excel: After visual inspection of the raw data for potential artefacts, normalization was performed with the following overall equation: $(F - F_{start})/(F_{max} - F_{start})$. F_{start} and F_{max} were determined per row to address potential compound interference with the absorption and emission of SYPRO.
2. Normalized values were then pasted into Graphpad Prizm using an XY table with triplicate values for each Y point.
3. Analysis is done by normalizing data, pasting into Graphpad and then performing a non-linear regression with Boltzman Sigmoidal. V50 value is used to estimate thermal shift.

2.2.6 Assay troubleshooting and limitations

We have generally found this assay to be quite robust and to give highly reproducible thermal shifts, although we note that not all PSK are stable enough to start the assay at 25 °C and may need to be started at 4 °C. In a >600 compound screen, we found the majority of compounds to be stabilizing with a narrow subset being destabilizing; approximately 3% of compounds stabilized STRAD by >4 °C, the ΔT_M induced by ATP.

In addition to protein quality and purity, the primary artifact that we encountered was compound interference. In general this could be addressed through normalization to the baseline afforded by the added compound (Shapiro, Walkup, & Keating, 2009); however in rare cases the compound can quench SYPRO orange. In such scenarios, an investigator could consider substituting different dyes such as the nucleophile reactive dye *N*-[4-(7-diethylamino-4-methyl-3-coumarinyl)phenyl]maleimide (CPM) (Alexandrov, Mileni, Chien, Hanson, & Stevens, 2008), which has an excitation and emission spectrum of 387 nM and 463 nM. Thus, the latter option also may address rare cases of compound interference.

2.3 CETSA assay protocol

2.3.1 Reagents

Standard cell culture supplies including 10 cm diameter plates, culture media and serum as appropriate for your application.

Dulbecco's Phosphate Buffered Saline (DPBS; ThermoFisher, Cat# 14190144): 9% NaCl.
Protease inhibitors (Roche cOmplete Mini; ThermoFisher, Cat# 11836153001 or other as preferred).

2.3.2 Assay setup and performance

1. Plate 1–3x10^6 cells of interest in a 10 cm diameter cell culture plate with appropriate media.
2. The next day, add a saturating quantity of the compound of interest (typically 2–3x the *in vitro* EC_{50}).
3. 4 hours later, harvest the cell by rinsing 1x and then scraping in 1 mL room temperature (RT) phosphate buffered saline (PBS).
4. Transfer suspended cells to labeled 1.5 mL tubes and pellet scraped cells by centrifugation at 1000 rpm for 5 min at RT.
5. Aspirate off PBS and resuspend cell pellets in 75 μL in PBS supplemented with protease and phosphatase inhibitors.
6. Distribute each cell suspension to 5 PCR tubes (15 μL/tube).
7. Heat PCR tubes on a gradient in a benchtop thermal cycler for 3 min; we use a gradient from 43.9 to 59.6 °C, but this can be adjusted to suit your cells and proteins of interest.
8. Immediately snap freeze samples in EtOH + dry ice bath for 3 min.
9. Freeze-thaw cells 2x: use thermal cycler set to 25 °C for 3 min for thawing. Vortex tubes after each thaw and use EtOH + dry ice bath for freezing.
10. Transfer to 1.5 mL tubes, vortex and then centrifuge at max speed at 4 °C for 10 min. There will be cell debris and precipitated protein at the bottom of the tube after this step, the supernatant is the lysate.
11. Combine 10 μL lysate with 3 μL of a preferred 5x Western blot loading dye; at this point material may be stored at −20 °C or analyzed by conventional Western blot.
12. Our antibodies are anti-LKB1 (Cell Signaling, Cat# 27D10), anti-STRAD (a.k.a. LYK5; Abcam, Cat# ab192879), anti-MO25 (Cell Signaling, Cat# C49D8) and anti-Vinculin (Cell Signaling, Cat# E1E9V). Vinculin is a highly stable protein which is not impacted by our STRAD ligands and as such serves as a loading control.
13. We have typically performed qualitative analysis using conventional HRP/ECL development techniques. Quantitative analysis using fluorescently labeled secondary antibodies are also reasonable.

2.3.3 *Assay troubleshooting and limitations*

We generally find this assay to be robust, if laborious. Our primary optimization steps have been around generating adequate lysate quantities, and we have frequently scaled out cell culture up to a 15 cm dish if needed. In some experiments, we have also used exogenously expressed 3xFlag-tagged proteins, which ease detection without altering thermal stabilization effects.

3. Nucleotide displacement from the STRAD/MO25 trimer

3.1 Introduction

STRAD, like some but not all PSKs, binds ATP. Thus, assays using the displacement of fluorescent ATP analogs such as ATP- 2′,3′-O-(2,4,6-trinitrophenyl) (TNP), or metabolic labeling with probes such as ATP-biotin may be considered to measure occupancy of the ATP binding site. Of note, while others have published ATP-biotin-based assays conducted with STRAD (Smith et al., 2021), we focused on assays displacing fluorescent ATP from STRAD. Following published work (Zeqiraj, Filippi, Goldie, et al., 2009), we used ATP-TNP (Fig. 2A); however multiple other fluorescent ATP analogs exist and are used for a variety of purposes, most typically (2′-(or-3′)-*O*-(*N*-Methylanthraniloyl) adenosine 5′-triphosphate (MANT-ATP) (Colombo et al., 2021; Hiratsuka, 2003; LaConte, Srivastava, & Mukherjee, 2017). Importantly, ATP-TNP and MANT-ATP both link the fluorophore to ATP at the 3' hydroxyl or the 2' and 3' hydroxyl of the ribose (Fig. 2A).

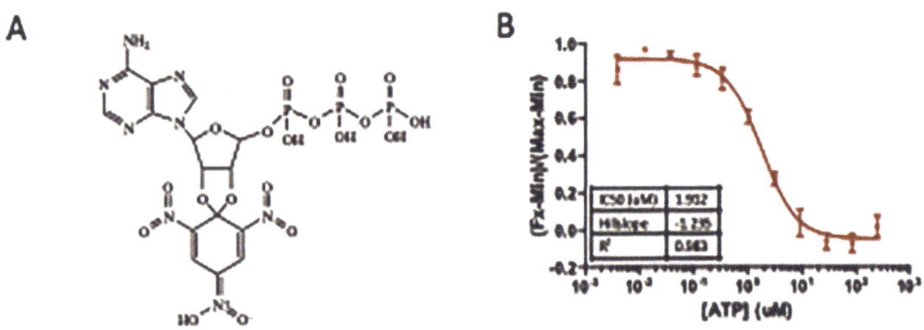

Fig. 2 TNP-ATP assay of STRAD occupancy: (A) Chemical structure of TNP-ATP, showing location of the fluorophore. (B) Example results of TNP-ATP assay, measuring displacement of TNP-ATP by ATP. Results are from a single experiment done in duplicate; error bars ± standard deviation (SD), but are representative.

In the course of our efforts, we found that a subset of compounds of interest had significant fluorescence interference with TNP-ATP. While the absorption/emission spectrum of MANT-ATP is shifted by 55 nM and 90 nM, respectively, shifting to MANT-ATP was not adequately sufficient to address this issue. The tempo of the project and development of successor kinase assays for this project meant that we did not extensively assess other fluorescent ATP analogues. However, for selected compounds we also performed competition experiments displacing STRAD/MO25 from an ATP-coupled bead, also described below.

3.2 ATP-TNP assay optimization and conditions

In the course of assay optimization, we accounted for many parameters. A fixed concentration of 1 μM TNP-ATP was used to avoid any inner filter effects, and then used to determine the appropriate STRAD/MO25 protein concentration. As could be anticipated from the lack of magnesium in the STRAD ATP-binding pocket (Zeqiraj, Filippi, Goldie, et al., 2009), $MgCl_2$ reduced TNP-ATP binding but the addition of Triton X-100 and Casein did not alter binding conditions, resulting in the conditions selected. A standardized dose curve for ATP was developed, with an IC50 of 1.8–2.2 μM (see Fig. 2B).

3.2.1 Assay protocol

This was developed following Zeqiraj, Filippi, Goldie, et al. (2009).

3.2.2 96 well format

3.2.2.1 Reagents

10x TNP-ATP Buffer: 500 mM Tris pH 7.5, 500 mM NaCl
Probe: TNP-ATP (Thermofisher, Cat# T7602)

3.2.2.2 Assay solution (per well)

3 μL 10x buffer
TNP-ATP, 1 μM final concentration
STRAD/MO25 dimer, 5 μM final concentration
H_2O: up to 30 μL

3.2.2.3 Assay set up and performance

1. Prepare Master Mix containing all components but ligand on ice
2. Set aside the volume for 2–3 wells per ligand in a secondary tube or in a 96 well plate, allow to stand at RT for 10 min to bind TNP-ATP

3. Add adequate ligand quantity for 2–3 wells with final volume up to 60–90 μL, mix by manual pipetting
4. Aliquot 30 μL/well x 2 into a 96 well plate, incubate again at RT for a minimum of 10 min
5. Read fluorescence (Ex410 nm/Em538 nm) on a plate reader

3.2.3 384 well format

3.2.3.1 Reagents

10x TNP-ATP Buffer: 500 mM Tris pH 7.5, 1 M NaCl, 10 mM DTT
Probe: TNP-ATP (Thermofisher, Cat# T7602)

3.2.3.2 Assay solution (per well)

0.5 μL 20x buffer
STRAD/MO25 dimer (produced in-house), 5 μM final concentration
H_2O: up to 9 μL

Added to: 1 μL ATP or compound

3.2.3.3 Assay performance

1. Aliquot 1 μL ATP or compound to each well, shake at RT for at least 10 min
2. Prepare assay solution by combining 2x ATP-TNP in buffer with 2x STRAD/MO25 in buffer; mix well and spin down at 4 °C
3. Aliquot 9 μL/well x2 into a 96 well plate, incubate again at RT for a minimum of 10 min
4. Read fluorescence (Ex410 nm/Em538 nm) on a plate reader

3.2.4 Data analysis

As above.

3.2.5 Assay troubleshooting and limitations

We noted two issues with this assay. The first, as discussed above, was significant interference between many of the tested compounds and the TNP-ATP signal. In some cases, normalization could be performed to mitigate this issue. However, it was a persistent enough issue that TNP-ATP was removed from our compound assessment strategy. More importantly, we did not identify a strong correspondence between TNP-ATP displacement and allosteric effects on LKB1 kinase activity by kinase assay either in a large screen of advanced clinical compounds or in the ensuing SAR of lead compounds. Thus, this indirect measure was used for validation of compound mechanism of action but did not prove useful for guiding medicinal chemistry.

One secondary point is that fluorophores themselves may be able to contribute to PSK binding. As part of initial troubleshooting, it can be valuable to determine a binding curve for the labeled nucleotide to the ATP-bound PSK and contrast that to the ability of ATP to displace the labeled nucleotide. In principle, the affinity for ATP and the labeled nucleotide should be quite similar.

3.3 Bead-coupled nucleotide displacement assays

A selected number of compounds were tested for their effects on ATP binding using ATP coupled to sepharose at the γ-phosphate position, as is commonly used for affinity purification of nucleotide binding proteins. This strategy was inspired by the work of Kuster, Johnson and others using ATP-competitive kinase inhibitors, so called Multiplex Inhibitor Beads or Kinobeads, to determine the selectivity of kinase inhibitors *in vitro* and in cellular assays (Bantscheff et al., 2007; Duncan et al., 2012; Sos et al., 2014).

3.3.1 Assay protocol

3.3.1.1 Reagents

1x Binding buffer:
50 mM HEPES (pH 7.5)
0.5% Triton X-100
150 mM NaCl
1 mM EDTA
1 mM EGTA

Probe: γ-linked ATP agarose resin (Novus Biologicals, Cat# 510-0002)

Assay Solution (per microcentrifuge tube):
500 μL Binding buffer
500 nM STRAD/MO25 complex
30 μL ATP bead slurry

3.3.1.2 Assay protocol

1. Pre-incubate STRAD/MO25 complex with vehicle or compound (recommended dose range 80 nM–20 μM) for 30 min at RT in 500 μL Binding buffer.
2. Add to 30 μL ATP bead slurry, washed 1x with 500 μL Binding buffer.
3. Rotate at 4 °C for 2.5 h.
4. Wash three times with Binding buffer by spinning at 100 x *g* for 1 min, aspirating off liquid and resuspending in 1 mL binding buffer.
5. Add 8 μL 5x SDS loading buffer, boil samples and run a gel; blot for STRAD as above.

6. Images acquired using a LiCOR, quantified and then using Graph Pad Prism as above. In this case normalization was not used, but could have been done.

3.3.1.3 Assay troubleshooting and limitations

Similar to TNP-ATP competition, bead displacement assays were not a major part of our drug development. The primary limitations were the laboriousness of the assay itself and the unclear connection between affinity for an immobilized ligand and compound effects on the LKB1 kinase *in vitro* or in cells. That said, we have extensive experience with using ATP-mimetics as affinity and activity probes either multiplexed or individually (Martins et al., 2015; Miller et al., 2016; Saha et al., 2016; Sos et al., 2014). Strategies for optimization in those systems included altered protein concentration, altered NaCl concentration in buffer (150 nM to 1 μM), and altered sequence of adding compound vs binding STRAD/MO25 to beads.

4. *In vitro* kinase assay

4.1 Introduction

In vitro kinase assays are a mainstay of kinase drug development. However, they pose a theoretical challenge in the earliest stages in the development of compounds targeting the ATP-binding pocket of a PSK to modulate kinase behavior. It is possible that a PSK-binding scaffold may have off target effects on activity of an associated kinase, either in the same protein (as for kinases with tandem PSK/kinase domains) or in the same complex as shown here. However, once initial PSK binding scaffolds are found using direct binding strategies, transitioning to a kinase assay as the primary compound assessment seems favorable.

We addressed two key considerations in the application of this assay for the STRAD/Lkb1 complex: reaction conditions and sensitivity of readout. We compared three different systems, summarized in Fig. 3 and detailed below. The first was a Caliper-based Peptide Mobility Shift Assay (MSA) (Perrin, Fremaux, & Scheer, 2006). This system uses capillary electrophoresis of a Fluorescein amidite (FAM)-labeled peptide, the quantitative result a percent conversion (i.e. phosphorylation). We also employed the Promega ADP-Glo system (Vidugiriene, Zegzouti, & Goueli, 2009), which measures ATP consumption directly by determining the quantity of ADP produced during a kinase reaction. Finally, we also piloted a fluorescence resonance

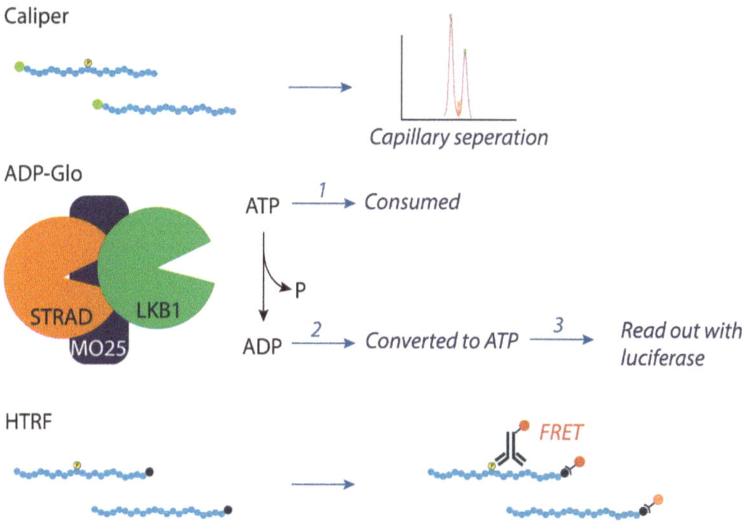

Fig. 3 Kinase assay readouts. Shown here are the three methods described for measuring kinase activity. For Caliper, a FAM-labeled substrate peptide is reacted with the LKB1/STRAD/MO25 trimer, then separated by voltage in a capillary. This optimizes for sensitivity and reproducibility by measuring phosphorylated peptide and comparing to non-phosphorylated peptide. For ADP-Glo, after a kinase reaction is performed, the Promega ADP-Glo kit is used. Here, after some portion of ATP is converted to ADP by LKB1, unreacted ATP is consumed by luciferin, generating AMP and PP. ADP is then converted back to ATP and measured by the luciferase signal as well. This assay is simple and has particular advantages when a conventional substrate peptide is not available. Finally, in HTRF a biotinylated peptide is reacted with the LKB1 trimer, then bound to streptavidin labeled with XL665. It is also incubated with an anti-phospho-Serine/Threonine antibody labeled with Europium Cryptate. Thus, a phosphopeptide can create a FRET pair between Europium Cryptate and streptavidin-XL665, allowing greater sensitivity and specificity in detection.

energy transfer (FRET)-based Homogeneous Time Resolved Fluorescence assay (HTRF; CsBio/Perkin Elmer) (Park et al., 1999). With HTRF, a biotinylated peptide substrate is combined with a streptavidin-coated bead and an anti-phosphoserine/threonine or phospho-tyrosine antibody. Details and background information for each assay is provided below, but each required separate optimizations and had unique strengths and weaknesses, as detailed below.

4.2 High throughput screening (HTS) considerations

As one of our goals was to perform HTS, these assays were variably developed for both low/medium throughput (e.g. 96 well format) and HTS applications, with the latter done in 384 well format. The general considerations we used were:

1. Sensitivity of assay
2. Cost of assay, both in terms of assay reagents and other reagents consumed (e.g. kinase)
3. Robustness/reproducibility of assay, i.e. Z'
4. Stability of assay readout
5. Availability of instrumentation

As Caliper devices are relatively rare, many investigators may wish to consider HTRF or ADP-Glo. In general, we have found the Caliper MSA to be optimal for both sensitivity and reproducibility. HTRF had lower sensitivity but excellent reproducibility with the added advantage of a stable readout over time. Finally, ADP-Glo was the simplest but also least sensitive and least robust.

4.2.1 Optimization strategy

The parameters of the Lkb1 kinase assay needed to be optimized in order to maximize the ability to detect modulation of the kinase activation by the compound, while also considering that an excess of ATP will result in reduced compound binding to STRAD. Our optimization strategies have generally proceeded as follows:

1. Initial enzyme titration at saturated substrate concentration (Fig. 4A)
2. Buffer optimization for additives
3. Peptide substrate concentration optimization
4. DMSO tolerance
5. ATP K_m measurement (Fig. 4B)
6. Enzyme titration with ATP concentration at the K_m (Fig. 4C)

Validation with reference compound sunitinib (Moret et al., 2019) (Fig. 4D)

Further optimization was then undertaken for the readout. Final details are given below.

4.3 Protocols

4.3.1 Caliper assay protocol

This was prepared in 96 well and 384 well format; the details for a 96 well format are provided.

4.3.1.1 Reagents

5x Kinase Assay buffer:

200 mM MOPs-NaOH, pH 7.0

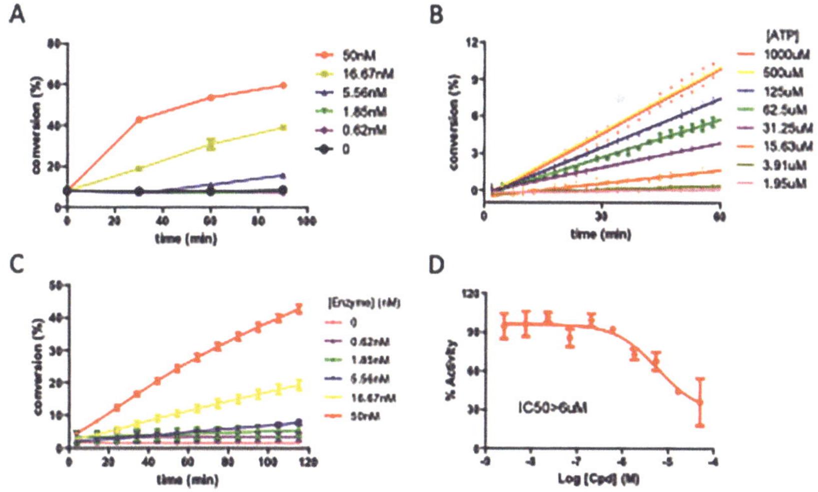

Fig. 4 LKB1 kinase assay optimization. (A) Initial determination of enzyme concentration to be used in the screen for the activator compounds. The STRAD/Mo25/LKB1 trimer was reacted with the LKB1 substrate peptide, as described in the text, at saturating ATP concentrations (500 μM). % conversion from unphosphorylated to phosphorylated peptide is shown on the y axis. Concentration of 6 nM enzyme was initially selected to carry forward as it suggested the greatest ceiling for activation by an activator compound although results below showed that it was not further activated by other optimization steps. (B) After optimizing additives including DTT, glycerol, BSA, EDTA and peptide concentration, K_m of ATP was measured and found to be 60 μM. % conversion from unphosphorylated to phosphorylated peptide is shown on the y axis. (C) Enzyme was then titrated at the ATP K_m, and a concentration of 16 nM was selected for the enzyme to give a greater dynamic range to identify both activation and inhibition. % conversion from unphosphorylated to phosphorylated peptide is shown on the y axis. (D) Assay validation using the known LKB1 inhibitor sunitinib, showing a dose-dependent reduction in kinase activity. Here, we used the optimized conditions from Fig. 4C. % activity is shown, normalized to a vehicle-treated control kinase reaction. In all cases in this figure, results are from a single experiment with each condition run in duplicate, shown ± SD.

5 mM EDTA
10 mM DTT
50 mM $MgCl_2$
0.5% Triton X-100

LKB1/STRAD/MO25 timer (Sigma-Aldrich 14-596)

Fam-labeled LKBtide: (5FAM)-LSNLYHQGKLQTFCGSPLYRRR (COOH) [Threonine 12 is the substrate amino acid], available from Anaspec, Cat# AS-64871

ATP (final concentration 50 μM)

4.3.1.2 Assay solutions (per well of a 384 well plate)

Compound solution (per well):

1 μL 5x assay buffer
DMSO or compound
H_2O: up to 5 μL

Enzyme solution (per well):

2 μL 5x assay buffer
5 nM LKB1/STRAD/MO25 trimer
H_2O: up to 10 μL

Substrate solution (per well):

2 μL 5x assay buffer
2.5 μM Fam-labeled LKBtide
50 μM ATP
H_2O: up to 10 μL

4.3.1.3 Assay protocol

1. Plate 5 μL compound in 5x kinase buffer per well
2. Add 10 μL enzyme, incubate for 15 min at RT
3. Add 10 μL substrate/ATP, incubate for 2 h at RT
4. Add stop buffer (150 mM EDTA)
5. Acquire on Caliper device:
   a. Sample buffer: 100 mM HEPES, pH 7.5, 1 mM EDTA, 0.015% Brij-35, 0.1% CR-3, 5% DMSO
   b. Settings: Upstream Voltage -270 V; Downstream Voltage -1850 V; Pressure -1.2 psi, Sample Sip Time 0.3 s, Buffer Sip Time 90 s
   c. Retention time: 225–230 s for product, 240 s for substrate

4.3.1.4 Analysis

% conversion was calculated using the equation %c = Prod/(Prod + Subst). Data were again plotted using GraphPad Prism. EC_{50} was calculated using the nonlinear fit: log(inhibitor) vs response—Variable slope (four parameters) function, and fold activation determined by comparing %conversion with drug to baseline.

4.4 ADP-Glo

As part of a conversion to 384 well format/high throughput screening, multiple assays were tested including the Promega ADP-Glo system. In addition to the optimized buffer system above, a second round of optimization was performed to improve performance in that system. For ADP-Glo, we developed a second set of assay conditions using a HEPES-based buffer and a 90 min reaction time. However, after extensive testing we found that while these conditions were excellent for direct LKB1 inhibition (e.g. with compounds such as sunitinib), they were consistently less sensitive for STRAD-mediated modulation of LKB1 activation, likely due to different quantities of Triton X-100 in the mix. Accordingly, we have proceeded to using the same reaction conditions for ADP-Glo as for Caliper, described above in Section 4.3.1.

Reactions were prepared for ADP-Glo just as for Caliper, and analyzed using an Envision luminometer, with an integration time of 0.5 s.

4.5 HTRF

Finally, in parallel we developed a 384 well HTRF assay protocol for the LKB1 kinase activity assay. This strategy allowed more direct detection than was possible using ADP-Glo, and obviated the need for a Caliper device for readout. This assay was performed using the Perkin Elmer (CsBio) HTRF KinEASE STK kit. Of note, as HTRF is a modular assay, additional optimization was performed for not only buffer, kinase, ATP and substrate peptide concentration, but also the ratio of the fluorescently labeled streptavidin to substrate peptide quantity. Also, we note that not all plate readers are HTRF compatible, so we include the readout details here.

4.5.1 Reagents

5x Kinase assay buffer

250 mM MOPs-NaOH, pH 7.0
5 mM EDTA
10 mM DTT
50 mM $MgCl_2$
0.5% Triton X-100
LKB1/STRAD/MO25 trimer (Millipore/Sigma)
ATP
KinEASE STK substrate 1
Streptavidin-XL665 (1:8 ratio with substrate)
Antibody-Cryptate

4.5.2 Assay solutions (per well of a 384 well plate)

Compound: variable concentration, final volume 0.5 μL, diluted in H_2O from a 10 mM stock in DMSO

Solution 1

1 μL 5x Kinase Assay buffer
5 nM LKB1/STRAD/MO25 timer
H_2O: up to 2.5 μL

Solution 2

1 μM STK Substrate1-Biotin
50 μM ATP
H_2O: up to 2 μL
Stop/Analysis Buffer
Streptavidin-XL665
H_2O: up to 5 μL

4.5.3 Assay protocol

1. 0.5 μL compound or vehicle was applied directly into each well after dilution.
2. Solution 2 was prepared on ice, and then 2 μL pipetted per well.
3. Solution 1 was prepared on ice, and then 2.5 μL pipetted per well.
4. Plate was covered and allowed to reaction in the dark for 120 min.
5. Fluorescence was acquired using a Molecular Devices Flexstation 3, with the following settings: Excitation: 314 nM, Emission 1: 620 nM, emission 2: 668 nM. Delay between reads was 50 ms, integration/read 400 ms.
6. Signal stability was measured, and was >90% maintained at 24 h.

4.5.4 Analysis

No kinase, no peptide and no ATP controls were performed, although they were not used for the quantitative analysis. Data were initially transformed as 1000x(V668 nM/V620 nM), and then plotted as a %change vs DMSO control. As above, data were plotted using GraphPad Prism. EC_{50} was calculated using the nonlinear fit: log(inhibitor) vs. response—Variable slope (four parameters) function, and fold activation determined by comparing % conversion with drug to baseline.

Determination of Specificity: It is highly likely that our compounds did not bind the LKB1 ATP-binding site as they did not appear to interfere with LKB1 kinase activity but rather to enhance it. However, as an important step in the validation of advanced compounds, we did perform an off-target analysis. This analysis was done through a commercial vendor, Eurofins, which applies the ActivX method to identify kinase binding through blocking the attachment of ATP-biotin (Patricelli et al., 2007).

5. Summary and conclusions

The significance of PSK to human biology has led to their increasing importance as drug targets (Kung & Jura, 2019). We present here our detailed methods for assessing small molecules targeting the PSK STRAD. Our experience highlights the importance of measuring not only compound binding to the PSK target of interest, but utilizing functional readouts of the desired signaling outcome (allosteric or otherwise) of compound binding. As PSK-directed therapies advance further through early stage development, the streamlining of these processes, and addressing the impact of multiple nucleotide binding pockets in target proteins (e.g. kinases with tandem PSK and kinase domains) and target complexes (e.g. the LKB1/STRAD/MO25 trimer, as well as other in-trans PSK complexes such as EGFR/HER3 and HER2/HER3) will be of particular importance.

References

Alexandrov, A. I., Mileni, M., Chien, E. Y., Hanson, M. A., & Stevens, R. C. (2008). Microscale fluorescent thermal stability assay for membrane proteins. *Structure*, *16*, 351–359.

Bailey, M. H., Tokheim, C., Porta-Pardo, E., Sengupta, S., Bertrand, D., Weerasinghe, A., et al. (2018). Comprehensive characterization of cancer driver genes and mutations. *Cell*, *174*, 1034–1035.

Bantscheff, M., Eberhard, D., Abraham, Y., Bastuck, S., Boesche, M., Hobson, S., et al. (2007). Quantitative chemical proteomics reveals mechanisms of action of clinical ABL kinase inhibitors. *Nature Biotechnology*, *25*, 1035–1044.

Burke, J. R., Cheng, L., Gillooly, K. M., Strnad, J., Zupa-Fernandez, A., Catlett, I. M., et al. (2019). Autoimmune pathways in mice and humans are blocked by pharmacological stabilization of the TYK2 pseudokinase domain. *Science Translational Medicine*, *11*.

Chen, Z., Cheng, K., Walton, Z., Wang, Y., Ebi, H., Shimamura, T., et al. (2012). A murine lung cancer co-clinical trial identifies genetic modifiers of therapeutic response. *Nature*, *483*, 613–617.

Colombo, J., Antkowiak, A., Kogan, K., Kotila, T., Elliott, J., Guillotin, A., et al. (2021). A functional family of fluorescent nucleotide analogues to investigate actin dynamics and energetics. *Nature Communications*, *12*, 548.

Cossa, G., Roeschert, I., Prinz, F., Baluapuri, A., Silveira Vidal, R., Schulein-Volk, C., et al. (2020). Localized inhibition of protein phosphatase 1 by NUAK1 promotes spliceosome activity and reveals a MYC-sensitive feedback control of transcription. *Molecular Cell*, 77(1322–1339), e1311.

Dhawan, N. S., Scopton, A. P., & Dar, A. C. (2016). Small molecule stabilization of the KSR inactive state antagonizes oncogenic Ras signalling. *Nature*, *537*, 112–116.

Duncan, J. S., Whittle, M. C., Nakamura, K., Abell, A. N., Midland, A. A., Zawistowski, J. S., et al. (2012). Dynamic reprogramming of the kinome in response to targeted MEK inhibition in triple-negative breast cancer. *Cell*, *149*, 307–321.

Dziekan, J. M., Yu, H., Chen, D., Dai, L., Wirjanata, G., Larsson, A., et al. (2019). Identifying purine nucleoside phosphorylase as the target of quinine using cellular thermal shift assay. *Science Translational Medicine*, *11*.

Foulkes, D. M., Byrne, D. P., Yeung, W., Shrestha, S., Bailey, F. P., Ferries, S., et al. (2018). Covalent inhibitors of EGFR family protein kinases induce degradation of human Tribbles 2 (TRIB2) pseudokinase in cancer cells. *Science Signaling, 11*.

Gaude, H., Aznar, N., Delay, A., Bres, A., Buchet-Poyau, K., Caillat, C., et al. (2012). Molecular chaperone complexes with antagonizing activities regulate stability and activity of the tumor suppressor LKB1. *Oncogene, 31*, 1582–1591.

Goodwin, J. M., Svensson, R. U., Lou, H. J., Winslow, M. M., Turk, B. E., & Shaw, R. J. (2014). An AMPK-independent signaling pathway downstream of the LKB1 tumor suppressor controls Snail1 and metastatic potential. *Molecular Cell, 55*, 436–450.

Gurumurthy, S., Xie, S. Z., Alagesan, B., Kim, J., Yusuf, R. Z., Saez, B., et al. (2010). The Lkb1 metabolic sensor maintains haematopoietic stem cell survival. *Nature, 468*, 659–663.

Hatzivassiliou, G., Song, K., Yen, I., Brandhuber, B. J., Anderson, D. J., Alvarado, R., et al. (2010). RAF inhibitors prime wild-type RAF to activate the MAPK pathway and enhance growth. *Nature, 464*, 431–435.

Hawley, S. A., Boudeau, J., Reid, J. L., Mustard, K. J., Udd, L., Makela, T. P., et al. (2003). Complexes between the LKB1 tumor suppressor, STRAD alpha/beta and MO25 alpha/beta are upstream kinases in the AMP-activated protein kinase cascade. *Journal of Biology, 2*, 28.

Hiratsuka, T. (2003). Fluorescent and colored trinitrophenylated analogs of ATP and GTP. *European Journal of Biochemistry, 270*, 3479–3485.

Hollstein, P. E., Eichner, L. J., Brun, S. N., Kamireddy, A., Svensson, R. U., Vera, L. I., et al. (2019). The AMPK-related kinases SIK1 and SIK3 mediate key tumor-suppressive effects of LKB1 in NSCLC. *Cancer Discovery, 9*, 1606–1627.

Hymowitz, S. G., & Malek, S. (2018). Targeting the MAPK pathway in RAS mutant cancers. *Cold Spring Harbor Perspectives in Medicine, 8*.

Jafari, R., Almqvist, H., Axelsson, H., Ignatushchenko, M., Lundback, T., Nordlund, P., et al. (2014). The cellular thermal shift assay for evaluating drug target interactions in cells. *Nature Protocols, 9*, 2100–2122.

Jenne, D. E., Reimann, H., Nezu, J., Friedel, W., Loff, S., Jeschke, R., et al. (1998). Peutz-Jeghers syndrome is caused by mutations in a novel serine threonine kinase. *Nature Genetics, 18*, 38–43.

Kitajima, S., Ivanova, E., Guo, S., Yoshida, R., Campisi, M., Sundararaman, S. K., et al. (2019). Suppression of STING associated with LKB1 Loss in KRAS-driven lung cancer. *Cancer Discovery, 9*, 34–45.

Kok, B. P., Ghimire, S., Kim, W., Chatterjee, S., Johns, T., Kitamura, S., et al. (2020). Discovery of small-molecule enzyme activators by activity-based protein profiling. *Nature Chemical Biology, 16*, 997–1005.

Kottakis, F., Nicolay, B. N., Roumane, A., Karnik, R., Gu, H., Nagle, J. M., et al. (2016). LKB1 loss links serine metabolism to DNA methylation and tumorigenesis. *Nature, 539*, 390–395.

Kung, J. E., & Jura, N. (2019). Prospects for pharmacological targeting of pseudokinases. *Nature Reviews Drug Discovery, 18*, 501–526.

LaConte, L. E. W., Srivastava, S., & Mukherjee, K. (2017). Probing protein kinase-ATP interactions using a fluorescent ATP analog. *Methods in Molecular Biology, 1647*, 171–183.

Lange, S. M., Nelen, M. I., Cohen, P., & Kulathu, Y. (2021). Dimeric structure of the pseudokinase IRAK3 suggests an allosteric mechanism for negative regulation. *Structure, 29*(238–251), e234.

Liang, J., Shao, S. H., Xu, Z. X., Hennessy, B., Ding, Z., Larrea, M., et al. (2007). The energy sensing LKB1-AMPK pathway regulates p27(kip1) phosphorylation mediating the decision to enter autophagy or apoptosis. *Nature Cell Biology, 9*, 218–224.

Liu, J., Lichtenberg, T., Hoadley, K. A., Poisson, L. M., Lazar, A. J., Cherniack, A. D., et al. (2018). An integrated TCGA pan-cancer clinical data resource to drive high-quality survival outcome analytics. *Cell, 173*(400–416), e411.

Lizcano, J. M., Goransson, O., Toth, R., Deak, M., Morrice, N. A., Boudeau, J., et al. (2004). LKB1 is a master kinase that activates 13 kinases of the AMPK subfamily, including MARK/PAR-1. *The EMBO Journal, 23*, 833–843.

Martinez Molina, D., Jafari, R., Ignatushchenko, M., Seki, T., Larsson, E. A., Dan, C., et al. (2013). Monitoring drug target engagement in cells and tissues using the cellular thermal shift assay. *Science, 341*, 84–87.

Martins, M. M., Zhou, A. Y., Corella, A., Horiuchi, D., Yau, C., Rakshandehroo, T., et al. (2015). Linking tumor mutations to drug responses via a quantitative chemical-genetic interaction map. *Cancer Discovery, 5*, 154–167.

McNally, R., Li, Q., Li, K., Dekker, C., Vangrevelinghe, E., Jones, M., et al. (2019). Discovery and structural characterization of ATP-site ligands for the wild-type and V617F mutant JAK2 pseudokinase domain. *ACS Chemical Biology, 14*, 587–593.

Miller, R. E., Brough, R., Bajrami, I., Williamson, C. T., McDade, S., Campbell, J., et al. (2016). Synthetic lethal targeting of ARID1A-mutant ovarian clear cell tumors with dasatinib. *Molecular Cancer Therapeutics*.

Mohseni, M., Sun, J., Lau, A., Curtis, S., Goldsmith, J., Fox, V. L., et al. (2014). A genetic screen identifies an LKB1-MARK signalling axis controlling the Hippo-YAP pathway. *Nature Cell Biology, 16*, 108–117.

Moret, N., Clark, N. A., Hafner, M., Wang, Y., Lounkine, E., Medvedovic, M., et al. (2019). Cheminformatics tools for analyzing and designing optimized small-molecule collections and libraries. *Cell Chemical Biology, 26*(765–777), e763.

Moro, M., Caiola, E., Ganzinelli, M., Zulato, E., Rulli, E., Marabese, M., et al. (2018). Metformin enhances cisplatin-induced apoptosis and prevents resistance to cisplatin in co-mutated KRAS/LKB1 NSCLC. *Journal of Thoracic Oncology, 13*, 1692–1704.

Murphy, J. M., Nakatani, Y., Jamieson, S. A., Dai, W., Lucet, I. S., & Mace, P. D. (2015). Molecular mechanism of CCAAT-enhancer binding protein recruitment by the TRIB1 pseudokinase. *Structure, 23*, 2111–2121.

Murphy, J. M., Zhang, Q., Young, S. N., Reese, M. L., Bailey, F. P., Eyers, P. A., et al. (2014). A robust methodology to subclassify pseudokinases based on their nucleotide-binding properties. *Biochemical Journal, 457*, 323–334.

Murray, C. W., Brady, J. J., Tsai, M. K., Li, C., Winters, I. P., Tang, R., et al. (2019). An LKB1-SIK axis suppresses lung tumor growth and controls differentiation. *Cancer Discovery, 9*, 1590–1605.

Niesen, F. H., Berglund, H., & Vedadi, M. (2007). The use of differential scanning fluorimetry to detect ligand interactions that promote protein stability. *Nature Protocols, 2*, 2212–2221.

Nony, P., Gaude, H., Rossel, M., Fournier, L., Rouault, J. P., & Billaud, M. (2003). Stability of the Peutz-Jeghers syndrome kinase LKB1 requires its binding to the molecular chaperones Hsp90/Cdc37. *Oncogene, 22*, 9165–9175.

Ozturk, D. G., Kocak, M., Akcay, A., Kinoglu, K., Kara, E., Buyuk, Y., et al. (2019). MITF-MIR211 axis is a novel autophagy amplifier system during cellular stress. *Autophagy, 15*, 375–390.

Papa, F. R., Zhang, C., Shokat, K., & Walter, P. (2003). Bypassing a kinase activity with an ATP-competitive drug. *Science, 302*, 1533–1537.

Park, Y. W., Cummings, R. T., Wu, L., Zheng, S., Cameron, P. M., Woods, A., et al. (1999). Homogeneous proximity tyrosine kinase assays: Scintillation proximity assay versus homogeneous time-resolved fluorescence. *Analytical Biochemistry, 269*, 94–104.

Park, S., Shim, J. H., Lee, B., Cho, I., Park, W. Y., Kim, Y., et al. (2019). Paired genomic analysis of squamous cell carcinoma transformed from EGFR-mutated lung adenocarcinoma. *Lung Cancer, 134*, 7–15.

Patricelli, M. P., Szardenings, A. K., Liyanage, M., Nomanbhoy, T. K., Wu, M., Weissig, H., et al. (2007). Functional interrogation of the kinome using nucleotide acyl phosphates. *Biochemistry, 46*, 350–358.

Perrin, D., Fremaux, C., & Scheer, A. (2006). Assay development and screening of a serine/threonine kinase in an on-chip mode using caliper nanofluidics technology. *Journal of Biomolecular Screening, 11*, 359–368.

Rabinovitch, R. C., Samborska, B., Faubert, B., Ma, E. H., Gravel, S. P., Andrzejewski, S., et al. (2017). AMPK maintains cellular metabolic homeostasis through regulation of mitochondrial reactive oxygen species. *Cell Reports, 21*, 1–9.

Resta, N., Simone, C., Mareni, C., Montera, M., Gentile, M., Susca, F., et al. (1998). STK11 mutations in Peutz-Jeghers syndrome and sporadic colon cancer. *Cancer Research, 58*, 4799–4801.

Saha, S. K., Gordan, J. D., Kleinstiver, B. P., Vu, P., Najem, M. S., Yeo, J. C., et al. (2016). Isocitrate dehydrogenase mutations confer dasatinib hypersensitivity and SRC dependence in intrahepatic cholangiocarcinoma. *Cancer Discovery, 6*, 727–739.

Sanchez-Cespedes, M., Parrella, P., Esteller, M., Nomoto, S., Trink, B., Engles, J. M., et al. (2002). Inactivation of LKB1/STK11 is a common event in adenocarcinomas of the lung. *Cancer Research, 62*, 3659–3662.

Sapkota, G. P., Deak, M., Kieloch, A., Morrice, N., Goodarzi, A. A., Smythe, C., et al. (2002). Ionizing radiation induces Ataxia telangiectasia mutated kinase (ATM)-mediated phosphorylation of LKB1/STK11 at Thr-366. *The Biochemical Journal, 368*, 507–516.

Scheeff, E. D., Eswaran, J., Bunkoczi, G., Knapp, S., & Manning, G. (2009). Structure of the pseudokinase VRK3 reveals a degraded catalytic site, a highly conserved kinase fold, and a putative regulatory binding site. *Structure, 17*, 128–138.

Scott, J. W., Ling, N., Issa, S. M., Dite, T. A., O'Brien, M. T., Chen, Z. P., et al. (2014). Small molecule drug A-769662 and AMP synergistically activate naive AMPK independent of upstream kinase signaling. *Chemistry & Biology, 21*, 619–627.

Shapiro, A. B., Walkup, G. K., & Keating, T. A. (2009). Correction for interference by test samples in high-throughput assays. *Journal of Biomolecular Screening, 14*, 1008–1016.

Shaw, R. J., Bardeesy, N., Manning, B. D., Lopez, L., Kosmatka, M., DePinho, R. A., et al. (2004). The LKB1 tumor suppressor negatively regulates mTOR signaling. *Cancer Cell, 6*, 91–99.

Shaw, R. J., Kosmatka, M., Bardeesy, N., Hurley, R. L., Witters, L. A., DePinho, R. A., et al. (2004). The tumor suppressor LKB1 kinase directly activates AMP-activated kinase and regulates apoptosis in response to energy stress. *Proceedings of the National Academy of Sciences of the United States of America, 101*, 3329–3335.

Sheetz, J. B., Mathea, S., Karvonen, H., Malhotra, K., Chatterjee, D., Niininen, W., et al. (2020). Structural insights into pseudokinase domains of receptor tyrosine kinases. *Molecular Cell, 79*(390–405), e397.

Sitthideatphaiboon, P., Galan-Cobo, A., Negrao, M. V., Qu, X., Poteete, A., Zhang, F., et al. (2021). STK11/LKB1 mutations in NSCLC are associated with KEAP1/NRF2-dependent radiotherapy resistance targetable by glutaminase inhibition. *Clinical Cancer Research, 27*, 1720–1733.

Skoulidis, F., Goldberg, M. E., Greenawalt, D. M., Hellmann, M. D., Awad, M. M., Gainor, J. F., et al. (2018). STK11/LKB1 mutations and PD-1 inhibitor resistance in KRAS-mutant lung adenocarcinoma. *Cancer Discovery, 8*, 822–835.

Smith, R. H. B., Khan, Z. M., Ung, P. M., Scopton, A. P., Silber, L., Mack, S. M., et al. (2021). Type II binders targeting the "GLR-Out" conformation of the pseudokinase STRADalpha. *Biochemistry, 60*, 289–302.

Sos, M. L., Levin, R. S., Gordan, J. D., Oses-Prieto, J. A., Webber, J. T., Salt, M., et al. (2014). Oncogene mimicry as a mechanism of primary resistance to BRAF inhibitors. *Cell Reports*, *8*, 1037–1048.

ten Klooster, J. P., Jansen, M., Yuan, J., Oorschot, V., Begthel, H., Di Giacomo, V., et al. (2009). Mst4 and Ezrin induce brush borders downstream of the Lkb1/Strad/Mo25 polarization complex. *Developmental Cell*, *16*, 551–562.

Tokarski, J. S., Zupa-Fernandez, A., Tredup, J. A., Pike, K., Chang, C., Xie, D., et al. (2015). Tyrosine kinase 2-mediated signal transduction in T Lymphocytes is blocked by pharmacological stabilization of its pseudokinase domain. *The Journal of Biological Chemistry*, *290*, 11061–11074.

Vidugiriene, J., Zegzouti, H., & Goueli, S. A. (2009). Evaluating the utility of a bioluminescent ADP-detecting assay for lipid kinases. *Assay and Drug Development Technologies*, 7, 585–597.

Wang, Y. S., Chen, J., Cui, F., Wang, H., Wang, S., Hang, W., et al. (2016). LKB1 is a DNA damage response protein that regulates cellular sensitivity to PARP inhibitors. *Oncotarget*, 7, 73389–73401.

Wang, L., Perera, B. G., Hari, S. B., Bhhatarai, B., Backes, B. J., Seeliger, M. A., et al. (2012). Divergent allosteric control of the IRE1alpha endoribonuclease using kinase inhibitors. *Nature Chemical Biology*, *8*, 982–989.

Woods, A., Johnstone, S. R., Dickerson, K., Leiper, F. C., Fryer, L. G., Neumann, D., et al. (2003). LKB1 is the upstream kinase in the AMP-activated protein kinase cascade. *Current Biology*, *13*, 2004–2008.

Yang, J., Campobasso, N., Biju, M. P., Fisher, K., Pan, X. Q., Cottom, J., et al. (2011). Discovery and characterization of a cell-permeable, small-molecule c-Abl kinase activator that binds to the myristoyl binding site. *Chemical Biology*, *18*, 177–186.

Zagorska, A., Deak, M., Campbell, D. G., Banerjee, S., Hirano, M., Aizawa, S., et al. (2010). New roles for the LKB1-NUAK pathway in controlling myosin phosphatase complexes and cell adhesion. *Science Signaling*, *3*, ra25.

Zeqiraj, E., Filippi, B. M., Deak, M., Alessi, D. R., & van Aalten, D. M. (2009). Structure of the LKB1-STRAD-MO25 complex reveals an allosteric mechanism of kinase activation. *Science*, *326*, 1707–1711.

Zeqiraj, E., Filippi, B. M., Goldie, S., Navratilova, I., Boudeau, J., Deak, M., et al. (2009). ATP and MO25alpha regulate the conformational state of the STRADalpha pseudokinase and activation of the LKB1 tumour suppressor. *PLoS Biology*, 7, e1000126.

Zhang, H., Fillmore Brainson, C., Koyama, S., Redig, A. J., Chen, T., Li, S., et al. (2017). Lkb1 inactivation drives lung cancer lineage switching governed by polycomb repressive complex 2. *Nature Communications*, *8*, 14922.

Zhang, C., Spevak, W., Zhang, Y., Burton, E. A., Ma, Y., Habets, G., et al. (2015). RAF inhibitors that evade paradoxical MAPK pathway activation. *Nature*, *526*, 583–586.

Zhang, J., Yang, P. L., & Gray, N. S. (2009). Targeting cancer with small molecule kinase inhibitors. *Nature Reviews. Cancer*, *9*, 28–39.

CHAPTER SIXTEEN

Targeting the HER3 pseudokinase domain with small molecule inhibitors

Audrey Colomba[a,†], Jeroen Claus[b,†], Forson Gao[b,c], Roger George[d], Arianna Fornili[e], and Angus J.M. Cameron[c,f,*]

[a]Alzheimer Research UK- UCL Drug Discovery Institute, University College London, London, United Kingdom
[b]Phospho Biomedical Animation, The Greenhouse Studio 6, Unit 17, Millmead Industrial Estate, London, United Kingdom
[c]Section of Structural Biology, Department of Medicine, Imperial College London, London, United Kingdom
[d]Structural Biology Science Technology platform, The Francis Crick Institute, London, United Kingdom
[e]Department of Chemistry, School of Physical and Chemical Sciences, Queen Mary University of London, London, United Kingdom
[f]Kinase Biology Laboratory, Barts Cancer Institute, Queen Mary, University of London, John Vane Science Centre, London, United Kingdom
*Corresponding author: e-mail address: a.cameron@qmul.ac.uk

Contents

† These authors contributed equally.

Methods in Enzymology, Volume 667
ISSN 0076-6879
https://doi.org/10.1016/bs.mie.2022.03.044

Abstract

HER3 is a potent oncogenic growth factor receptor belonging to the human epidermal growth factor (HER/EGFR) family of receptor tyrosine kinases. In contrast to other EGFR family members, HER3 is a pseudokinase, lacking functional kinase activity. As such, efforts to develop small molecule tyrosine kinase inhibitors against this family member have been limited. In response to HER3-specific growth factors such as neuregulin (NRG, also known as heregulin or HRG), HER3 must couple with catalytically active family members, including its preferred partner HER2. Dimerization of the intracellular HER2:HER3 kinase domains is a critical part of the activation mechanism and HER3 plays a specialized role as an allosteric activator of the active HER2 kinase partner. Intriguingly, many pseudokinases retain functionally important nucleotide binding capacity, despite loss of kinase activity. We demonstrated that occupation of the nucleotide pocket of the pseudokinase HER3 retains functional importance for growth factor signaling through oncogenic HER2:HER3 heterodimers. Mutation of the HER3 nucleotide pocket both disrupts signaling and disrupts HER2:HER3 dimerization. Conversely, ATP competitive drugs which bind to HER3, but not HER2, can stabilize HER2:HER3 dimers, induce signaling and promote cell growth in breast cancer models. This indicates a nucleotide-dependent conformational role for the HER3 kinase domain. Critically, our recent proof-of-concept work demonstrated that HER3-directed small molecule inhibitors can also disrupt HER2:HER3 dimerization and signaling, supporting the prospect that HER3 can be a direct drug target despite its lack of intrinsic activity. In this chapter we will describe methods for identifying and validating small molecule inhibitors against the HER3 pseudokinase.

1. Introduction

Targeting the epidermal growth factor receptor (EGFR) family of receptor tyrosine kinases (RTKs) has been a major success story in oncology with numerous small molecule and biologic drugs demonstrating clinical efficacy across a broad range of malignancies (Gschwind, Fischer, & Ullrich, 2004; Lemmon & Schlessinger, 2010; Saraon et al., 2021; Scaltriti & Baselga, 2006). The EGFR family comprises four family members, EGFR (HER1/erbB1), HER2, HER3 and HER4, with each receptor exhibiting distinct growth factor biding properties and dimerization preferences (Yarden, 2001). Targeting strategies are broadly divided into two categories. First biologic drugs, such as monoclonal antibodies have been developed to bind to the cell surface receptor domains, to disrupt function or kill cancer cells through a variety of strategies: ligand/dimer blocking, induction of receptor degradation, antibody-drug conjugate, immune targeting. Second, small molecule tyrosine kinase inhibitors (TKIs) have been developed to inhibit oncogenic receptor signaling (Roskoski, 2019). The majority of licensed drugs focus of EGFR and HER2, where significant success has been achieved in breast, lung, colon, head and neck cancer (Bonner et al., 2010; Cortes et al., 2011; Jonker et al., 2007) and other solid malignancies (Oh & Bang, 2020). As with many targeted therapies, resistance to both biologics and TKIs remains a key problem and HER3 has been shown to play a key role here in a number of settings (Mishra, Patel, Alanazi, Yuan, & Garrett, 2018). Additional strategies to target this family are urgently needed and HER3 likely represents the most underexploited target (Haikala & Janne, 2021; Jacob, James, Hasmann, & Weisser, 2018).

1.1 HER3 is a potent oncogene

HER family receptors are activated by growth factor signaling. EGFR, HER3 and HER4 have a range of known ligands, while for HER2 none have yet been identified. Growth factor binding to the extracellular region of these receptors results in conformational re-organization and asymmetric homo- or heterodimerization of the intracellular kinase domains of two separate receptors. Asymmetric interaction between the kinase domains is essential for activation with one protomer acting as the "activator" subunit, while the partner "receiver" kinase domain becomes activated and initiates signal transduction (Zhang, Gureasko, Shen, Cole, & Kuriyan, 2006). HER2 is the preferred binding target of HER3 and this partnership is unique; HER3 acts as a specialized activator, retaining ligand binding but with no substantive catalytic activity, while HER2 is a specialized receiver, with no ligand but retaining kinase activity (Diwanji et al., 2021; Wallasch et al., 1995). Together the HER2:HER3 heterodimer represents a potent oncogenic unit, most extensively studied and targeted in breast cancer (Alimandi et al., 1995; Bodey et al., 1997; Holbro et al., 2003; Wallasch et al., 1995). Approximately 20% of breast cancers are HER2 positive and this stratifies patients for targeted therapy (Lovekin et al., 1991; Owens, Horten, & Da Silva, 2004; Yaziji et al., 2004), most commonly with HER2 targeting antibody therapies such as Trastuzumab (Herceptin) or Pertuzumab, which directly blocks HER2:HER3 dimerization (Mendes et al., 2015). Certain patients may also receive EGFR/HER2 TKIs such as lapatinib or neratinib to block oncogenic signaling. While often initially effective, a number of resistance mechanisms have been described, including upregulation or HER3 or its growth factor ligands, and a range of HER2/3 mutations (Hanker et al., 2021; Jaiswal et al., 2013). As many mechanisms rely on HER3, considerable focus has fallen on developing additional strategies to target HER3 (Amin et al., 2010).

1.2 HER3-targeting strategies

As HER3 lacks kinase activity the majority of strategies have focused on biological drugs to block dimerization or ligand binding (Claus, Patel, Ng, & Parker, 2014; Gala & Chandarlapaty, 2014; Jacob et al., 2018; Mishra et al., 2018). These include blocking antibodies (Jacob et al., 2018), antibody/aptamer-drug conjugates (Dou et al., 2018; Gandullo-Sanchez et al., 2020; Yonesaka et al., 2019) and T-cell targeting strategies (Kumai et al., 2015; Larcombe-Young, Papa, & Maher, 2020). Despite a lack of appreciable activity, HER3 retains nucleotide binding capacity (Jura, Shan, Cao, Shaw, & Kuriyan, 2009; Murphy, Zhang, et al., 2014; Shi, Telesco, Liu, Radhakrishnan, & Lemmon, 2010) and nucleotide pocket occupation appears to be critical for transactivation of the active receiver kinase binding partner (Claus et al., 2018). Mutations, which prevent nucleotide pocket occupation, can thus disrupt function and impede dimer formation (Claus et al., 2018). The presence of an intact and functionally important nucleotide-binding pocket raises the tantalizing prospect that small molecule inhibitors might be developed to impede HER3 function. Being a pseudokinase, the intent of docking a small molecule inhibitor into HER3 should not be to block any residual low-level catalytic activity (Shi et al., 2010), but rather to conformationally impede the ability of HER3 to occupy the "activator" role within a signaling-competent heterodimer. The challenge here lies in the identification of inhibitors which prevent productive pairing of HER3 with partner RTKs. To do this, a few mechanisms might be considered: (i) inhibitors could allosterically impair HER3 function by impeding active heterodimer formation with other HER receptors; (ii) inhibitors might promote sequestration of HER3 within inactive complexes or to a cellular localization not conducive to productive signal transduction or (iii) inhibitors might promote receptor degradation either through structural destabilization or targeting to the proteasome (Xie et al., 2014). Each of these mechanisms will be considered for HER3-directed inhibitors.

Inhibitor occupation of the kinase nucleotide-binding pocket has been widely reported to allosterically modify kinase domain structure and function. As an early example, we demonstrated that PKC inhibitors can promote adoption of an active PKC conformation and even drive priming of kinase inactive PKC (Cameron, Escribano, Saurin, Kostelecky, & Parker, 2009). Similar mechanisms have been reported for diverse kinases, including Akt, Ire1, RAF and Erk5 (Hatzivassiliou et al., 2010; Lochhead et al., 2020; Okuzumi et al., 2009; Papa, Zhang, Shokat, & Walter, 2003; Poulikakos, Zhang, Bollag, Shokat, & Rosen, 2010). Retention of nucleotide-binding pocket occupation within a significant portion of the pseudokinome indicates that, while having lost catalytic potential, ATP-binding pocket occupation can still retain function in pseudokinases in a way that is analogous to the inhibitor-induced allosteric effects observed in otherwise active kinases (Boudeau, Miranda-Saavedra, Barton, & Alessi, 2006; Cameron, 2011; Claus, Cameron, & Parker, 2013; Claus et al., 2018; Mace & Murphy, 2021; Murphy, Zhang, et al., 2014; Zeqiraj et al., 2009). Two pseudokinases of note in this context are FAM20A and VRK3. The former is still able to bind ATP, but does so in a highly unconventional, inverted ATP orientation without cation present (Cui et al., 2017). This atypical ATP binding in FAM20A provided a significant increase in thermal stability of the pseudokinase itself, and also stabilized the FAM20A/FAM20C complex. On the other hand, VRK3 is unable to bind ATP but instead has substitution residues that occupy the general space that ATP would otherwise take up, with F313 and L262 protruding into the adenine space, and D175 adding a negative charge to the α/β-phosphate space (Scheeff, Eswaran, Bunkoczi, Knapp, & Manning, 2009). As a result, the apo form of VRK3 has a similar thermal stability to the ATP-bound conformers of the related family members VRK1 and VRK2.

The "activator"/"receiver" dichotomy in the HER2:HER3 heterodimer provides an interesting case study in this context, because the "activator" role is structural and performed by the pseudokinase, providing a platform for the "receiver" kinase HER2 to be catalytically active. Critically we demonstrated that nucleotide pocket directed drugs, against either HER2 or HER3 can counterintuitively drive formation of HER2:HER3 dimers by allosterically inducing kinase domain dimerization. In the presence of NRG, low dose lapatinib can induce HER2:HER3 dimerization, signal transduction and breast cancer cell proliferation, despite its ability to inhibit HER2 kinase activity (Claus et al., 2018). This may represent an important drug escape mechanism. Likewise, bosutinib, which binds to HER3 but not HER2, can induce productive HER2:HER3 dimerization to drive oncogenic signaling and cell proliferation, even in the absence of growth factor (Claus et al., 2018). The critical problem here is that the majority of nucleotide pocket induced allosteric changes appear to favor promotion of active conformers.

1.3 Toward small molecule targeting of HER3

Pharmacological studies targeting the HER3 kinase domain to block function remain rare. Xie et al. developed an irreversible ATP pocket-directed HER3 ligand but this failed to block signaling or growth (Xie et al., 2014). Conjugation to a hydrophobic adamantane-moiety to target to the proteasome, resulted in partial HER3 degradation and some evidence of HER2:HER3 dimer disruption (Xie et al., 2014). As a critical proof-of-concept, we recently reported identification of a novel HER3 nucleotide-pocket directed small molecule, AC3573, capable of inhibiting HER3-induced signaling and dimerization with HER2, without binding to HER2 (Colomba et al., 2020). This demonstrates that a small molecule inhibitor can act allosterically to inhibit HER3 function.

Here we will detail the methods used to identify AC3573 as a small molecule inhibitor of HER3 function. While identifying HER3-binding compounds is relatively straightforward, identifying compounds which disrupt function appears to be much more difficult, as the majority of HER3 binders do not disrupt function. The reasons behind this, and strategies to improve hit rates will also be covered. Finally, we will examine the methods required to probe the mechanism of action of AC3573 and other HER3-directed inhibitors, which will be key for the future development of effective pharmacological agents targeting the pseudokinase domain. This will include *in silico* modeling of AC3573 binding to HER3 in efforts to establish why this compound is able to block function.

2. Identifying small molecule binding to HER3 and HER2 by thermal shift assay

Identifying kinase inhibitors typically involves the development of an *in vitro* kinase activity assay; inhibition of activity in high throughput compound screens can then be used to identify hits for further evaluation. As pseudokinases lack intrinsic kinase activity, alternative approaches must be taken to identify compound binding. The binding of small molecules to proteins can increase thermal stability, which can be measured using thermal shift assays (TSAs) (Niesen, Berglund, & Vedadi, 2007). TSAs are cheap, require minimal specialized equipment and are well suited for high throughput screens.

TSA uses specialized fluorogenic dyes to monitor the thermal denaturation of proteins across an increasing temperature gradient, using widely available real-time PCR instruments. Binding of small molecules can enhance protein stability, resulting in an increase of their melting temperature (T_m). This small-molecule induced increase in T_m is then measured as compared to the apo form, to give the ΔT_m.

Here we describe the defined and validated conditions to identify small molecules binding to HER3 or HER2 by TSA.

2.1 Purification of recombinant HER3 and HER2 kinase domains

2.1.1 Materials

2.1.1.1 Reagents

1. Lysis Buffer: 50 mM HEPES pH 8.0, 300 mM NaCl, 1% (v/v) Triton X-100, 10% (v/v) glycerol, 10 mM β-glycerophosphate, 10 mM benzamidine, 0.5 mM EDTA, 1 mM NaF, 1 mM DTT, 5 mM $MgCl_2$
2. Wash Buffer: 50 mM HEPES pH 8.0, 300 mM NaCl, 5% (v/v) glycerol, 25 mM imidazole and 1 mM DTT
3. Elution Buffer: 50 mM HEPES pH 8.0, 300 mM NaCl, 5% (v/v) glycerol, 250 mM imidazole and 1 mM DTT
4. 200 U Basemuncher (Abcam #AB270049)
5. Protease inhibitor tablets (Roche #11836170001)
6. pFastBac1 expression vector
7. DH10Bac *E. coli* cells (ThermoFisher #10361012)
8. Antibiotics (Ampicillin, Kanamycin, Gentamicin and Tetracycline)
9. Sf21 cells (ThermoFisher #12682019)
10. Sf21 culture media (SF900 III, ThermoFisher #12658027)
11. FuGene HD transfection reagent (Promega #E2311)

2.1.1.2 Consumables

1. 1 L Erlenmeyer flasks (Corning #CLS431147)
2. 2 L Roller Bottles (Greiner #680060)
3. Tissue Culture Plates
4. Diverse plastic tubes as required
5. Protein concentrator (Vivaspin #VS0602) with a 20,000 Da cut off

2.1.1.3 Tools and instruments

1. Bacterial 37 °C static incubator for agar plates
2. Shaking Incubator set at 120 rpm and 27 °C
3. Roller Mixer (Stuart SRT6)
4. AKTA Pure protein purification system
5. S200 10/300 size exclusion column (GE Healthcare #GE28-9909-44)
6. Branson Digital Sonifier 450 (400 W) with a ½″ Horn sonicator probe

2.1.2 Method

In order to conduct TSA or other activity-based screens against any protein drug target, a source of high quality, high purity and correctly folded recombinant protein is essential. For our work on HER3 we opted to generate protein using the baculovirus-insect cell system. Many other systems are now available for generating recombinant protein, including mammalian, yeast and bacterial-based approaches. Choice of system will depend on individual protein requirements for post-translational modification, protein folding, required scale and availability. Both HER2 and HER3 kinase domains were purified from Sf21 cells via a His-tag as outlined in Fig. 1. For detailed coverage of His-tagged protein purification please also see Spriestersbach, Kubicek, Schäfer, Block, and Maertens (2015).

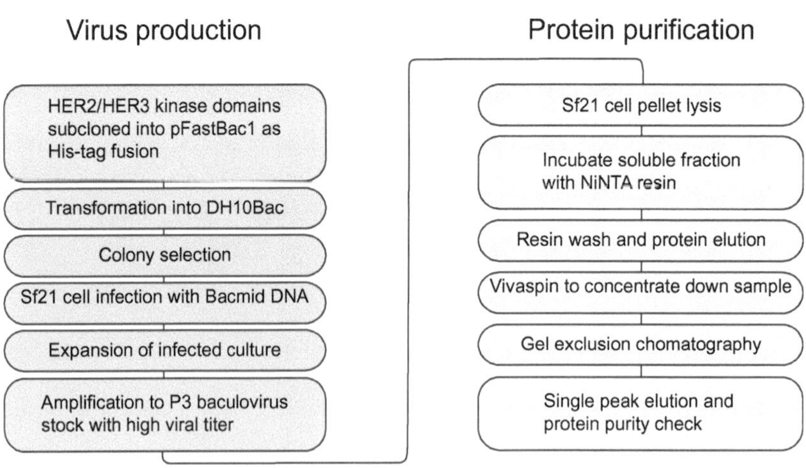

Fig. 1 HER3 compound screen setup. Schematic overview of the HER3 compound screen steps.

2.1.2.1 Virus Production and expression of HER3 kinase domain

1. The HER3 kinase domain (residues G684–D1020) and HER2 kinase domain (residues S703-G1029; M706A, Q711L, M712L) sequences were subcloned into pFastBac1 expression vector as a Hexa-His-tag fusion. The HER2 construct (S703-G1029; M706A, Q711L, M712L) was provided by Dr. Caroline Truman (AstraZeneca) with mutations to stabilize kinase domain and enhance recombinant protein yield (see Colomba et al., 2020).
2. Transform pFastBac1 HER3 or HER2 into DH10Bac *E. coli* cells (ThermoFisher) using strain-specific transformation protocols. DH10Bac cell contain a baculovirus shuttle vector which will combine with the pFastBac1 to generate recombinant Bacmid for infection of insect cell. Correct baculovirus pFastBac transposition will result in Bacmid constructs with kanamycin (25 μg/ml), tetracycline (10 μg/mL) and gentamicin (7 μg/mL) resistance. Transposition positive colonies are selected by blue-white screening with X-Gal (200 μg/mL) and IPTG (40 μg/mL).
3. Pick white colonies to generate high purity Bacmid DNA (Qiagen). For bacmid DNA preparation 1 mL of overnight culture is resuspended with 350 μL P1 buffer. Cells are lysed by addition of 250 μL P2 solution and inverted several times to promote complete lysis. N3 buffer is added and after inverting, incubated on ice for 5 min

   Centrifugation at 16,000 × *g* for 10 min is used to separate insoluble material from the soluble fraction containing bacmid DNA. To precipitate bacmid DNA, 800 μL of ice-cold isopropanol is added, thoroughly mixed and left at −20 °C for 2 h. Precipitated bacmid DNA is then pelleted by centrifugation at 16,000 × *g* for 20 min and washed twice with 1 mL of 70% ethanol. Once dried the DNA is resuspended in 100 μL of distilled, autoclaved water.
4. Transfect 7.5×10^5 of settled Sf21 cells in a single well of a six well plate with 500 ng of bacmid DNA and incubate at 27 °C for 72 h
5. Transfer the resulting P1 baculovirus, including the adhered cells to a 50 mL Sf21 culture at a density of 1×10^6 cells/mL in a 1 L Erlenmeyer flask. Allow the infected culture to grow for a further 72 h at 27 °C with shaking at 120 rpm
6. Further amplify this P2 baculovirus, to generate a P3 baculovirus stock with a high viral titer and use this to infect 500 mL cultures of Sf21 cells at an MOI = 1 in 2 L Roller Bottles (Greiner). Cells are harvested by centrifugation (750 × *g*, for 10 min) after 72 h and stored at −80 °C until required (snap freezing not required).

2.1.2.2 Purification of the HER3 kinase domain

1. Sf21 cell pellet lysis: Thaw cell pellets obtained from 500 mL of Sf21 cultures on ice and resuspend in 20 mL lysis buffer containing 50 mM HEPES pH 8.0, 300 mM NaCl, 1% (v/v) Triton X-100, 10% (v/v) glycerol, 10 mM β-glycerophosphate, 10 mM benzamidine, 0.5 mM EDTA, 1 mM NaF, 1 mM DTT, 5 mM $MgCl_2$, 200 U Basemuncher (Abcam) and an EDTA-free protease inhibitor tablet (Roche). Sonicate on ice to ensure complete lysis (30 s on, 20 s off, at an amplitude of 20% for a total sonication time of 5 min).
2. Remove insoluble material by centrifugation at 100,000 × *g* and incubate the soluble material with 200 μL bead volume of NiNTA resin (Qiagen) for 1 h at 4 °C with constant agitation on a roller mixer (Stuart SRT6).
3. Wash the resin extensively with buffer containing 50 mM HEPES pH 8.0, 300 mM NaCl, 5% (v/v) glycerol, 25 mM imidazole and 1 mM DTT prior to elution in the same buffer supplemented with 250 mM imidazole.
4. Concentrate the eluted protein down to a final volume of 0.5 mL using 6 mL Vivaspin protein concentrators with a 20,000 Da cut off. Careful observation and tracking of protein concentration is important to avoid recombinant protein from crashing out of solution at this stage. Vivaspin concentrators were centrifuged at 4000 × *g* at 4 °C in a swing bucket rotor for 15 min at a time, with mixing between spins, until the volume reached 0.5 mL.
5. Size exclusion chromatography on an AKTA Pure system (or equivalent) is conducted to isolate monomeric HER2 or HER3. The 0.5 mL of concentrated protein is applied to a S200 10/300 size exclusion column (GE Healthcare) equilibrated with buffer containing 50 mM HEPES pH 8.0, 300 mM NaCl, 5% (v/v) glycerol and 1 mM DTT at a flow rate of 0.5 mL/min and 1 mL fractions are collected.
6. The HER2 or HER3 kinase domain should elute as a single peak. Fractions should be assessed by SDS-PAGE to check protein purity. Relevant pure fractions can be pooled and concentrated to 1 mg/mL. 100 μL aliquots are snap frozen in liquid nitrogen and stored at −80 °C

2.2 TSA screen validation

2.2.1 Materials

2.2.1.1 Reagents

1. Assay buffer: 50 mM Hepes pH 7.5, 150 mM NaCl, 1 mM DTT
2. HER3 and HER2 kinase domain recombinant proteins (Section 2.1)
3. $MgCl_2$
4. DMSO

5. ATP (ThermoFisher Scientific, #R0441)
6. Lapatinib (Selleckchem, #S2111)
7. Bosutinib (LC Labs, #B-1788)
8. SYPRO Orange protein gel stain 5000× (ThermoFisher Scientific, #S6650)

2.2.1.2 Consumables

1. MicroAmp Optical 384-well qPCR plates (Applied Biosystems, ThermoFisher Scientific, #4309849)Adhesive aluminum qPCR-compatible plate seal (Starlab)
2. Multichannel pipette reservoirs
3. Diverse plastic tubes as required

2.2.1.3 Tools and instruments

1. 10 μL multichannel pipettes (recommended 16 channels) handling 0.5–10 μL volume
2. Real time PCR instrument. The protocol described below has been developed on a QuantStudio7 Flex Real-Time PCR instrument (Applied Biosystems, ThermoFisher Scientific), but should be easily adapted to similar instruments having adequate filter sets for SYPRO fluorescence monitoring (excitation 490 nm and emission around 600 nm)

2.2.1.4 Software

Data analysis and fitting software, e.g., GraphPad Prism8.

2.2.2 Method

2.2.2.1 Setting up the plate

1. Design the lay out of the plate, so there is, at least, a whole column of 16 wells for each condition. DMSO for neutral controls (ApoHER3 and ApoHER2, without compound), ATP/$MgCl_2$ or Bosutinib for HER3 positive controls, Lapatinib for HER2 positive controls.
2. Thaw the recombinant HER3 and HER2 proteins on ice. Dilute HER3 and HER2 stock solutions to 3 μM in the Assay buffer (the final concentration of HER3 and HER2 for the TSA assay is 1.5 μM/well). Prepare enough volume for the required number of wells (100 μL is sufficient for 16 wells, taking into account dead volume). For HER3 positive controls using ATP as ligand, dilute HER3 in the Assay buffer containing 400 μM ATP and 20 mM $MgCl_2$ (the final ATP and $MgCl_2$ concentrations for the TSA are 200 μM and 10 mM/well, respectively).

3. Dilute Bosutinib and Lapatinib stock solutions to 10 μM in the Assay buffer, so that the DMSO percentage is 10% (1% DMSO final in the TSA assay).
4. Dilute SYPRO Orange 5000 × into assay buffer at a 5 × concentration (the final concentration of SYPRO Orange for the TSA assay is 2.5 ×/well). Prepare enough SYPRO Orange 5 × for the required number of wells to assay (2.5 mL is sufficient for one full 384-well plate, taking into account dead volume).
5. For HER3 TSA: using a multichannel pipette, dispense 1 μL of 10% DMSO, diluted in assay buffer, to the neutral control wells and to the ATP positive control wells (1% DMSO final for the TSA assay). Dispense 1 μL of 10 μM Bosutinib to the Bosutinib positive control wells (the final concentration of Bosutinib for the TSA assay is 1 μM/well). ATP and Bosutinib dose responses can be performed to calculate EC50 values for ligands binding to HER3 as an indirect indicator of binding affinity (Fig. 2).
6. For HER2 TSA: using a multichannel pipette, dispense 1 μL of 10% DMSO, diluted in the Assay buffer, to the neutral control wells (1% DMSO final for the TSA assay) and 1 μL of 10 μM Lapatinib to the positive control wells (the final concentration of Lapatinib for the TSA assay is 1 μM/well). As an alternative, a Lapatinib dose-response can be performed (Fig. 2).
7. For HER3 TSA: dispense 4 μL of the buffered HER3 into the wells for DMSO neutral control and Bosutinib positive controls. Dispense 4 μL of the buffered HER3 + ATP/$MgCl_2$ into the wells for ATP positive controls.
8. For HER2 TSA: dispense 4 μL of the buffered HER2 into the wells for DMSO neutral control and Lapatinib for positive controls
9. Spin the plate briefly (300 × *g*, 30 s).
10. Using a multichannel pipette, dispense 5 μL of SYPRO 5 × into the wells to assay.
11. Seal the plate with the qPCR-compatible adhesive sealing foil and ensure that the plate is well sealed by using a smooth rubber roller.
12. Spin the screening plate briefly (300 × *g*, 1 min at room temperature).

2.2.2.2 TSA assay procedure

1. Place the PCR plate in the QuantStudio7 Flex Real-Time PCR instrument (or equivalent). Open the Real-time PCR instrument software and under "Experiment properties" select the following parameters:
   - 384-well block
   - Experiment type: melt curve

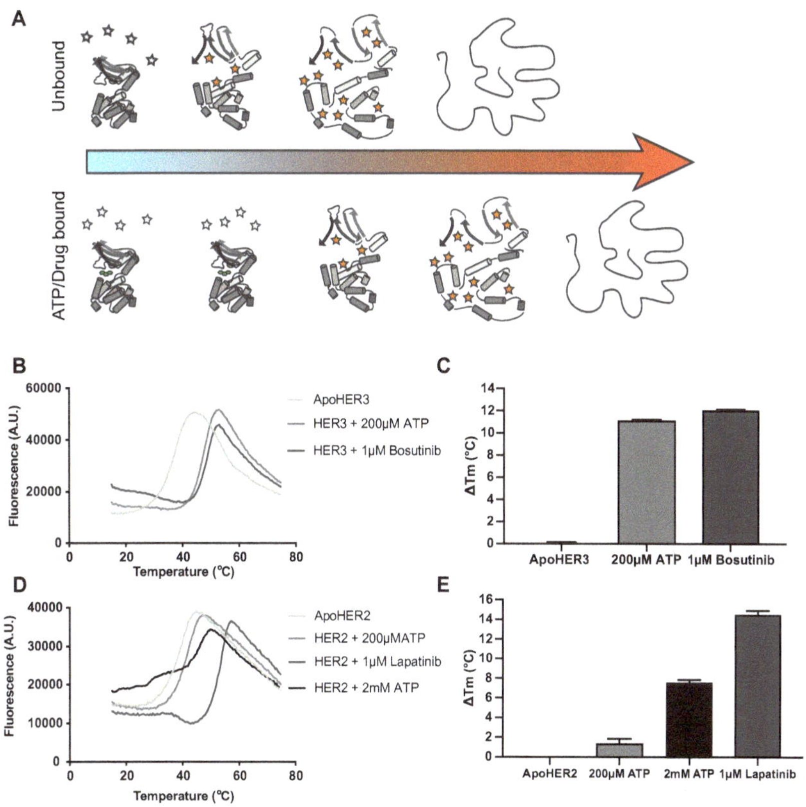

Fig. 2 Effects of different ligands on HER2 and HER3 stability. (A) Schematic overview of the Thermal Shift Assay principle. Unliganded kinase domain has decreased thermal stability and is therefore accessible to SYPRO Orange at a lower temperature compared to liganded kinase domain. (B) The change in HER3 melting temperature value (ΔT_m), in presence of bosutinib or ATP, compared to 1% DMSO ApoHER3 control in absence of any ligand, is reported as mean ± SD from $n=2$ independent experiments. A representative plot of the thermal denaturation profiles of recombinant HER3 kinase domain, in the presence of 1% DMSO (ApoHER3), 1 μM bosutinib or 200 μM ATP is shown. (C) Data shows that 1 μM bosutinib and 200 μM ATP induce a similar shift (around 11 °C) in HER3 melting temperature (T_m). (D) The change in HER2 melting temperature value (ΔT_m), in presence of Lapatinib or ATP, compared to 1% DMSO ApoHER2 control in absence of any ligand, is reported as mean ± SD from $n=2$ independent experiments. A representative plot of the thermal denaturation profiles of recombinant HER2 kinase domain, in the presence of 1% DMSO (ApoHER2), 1 μM Lapatinib or ATP is shown in (E) Data shows a greater increase in HER2 stability in presence of Lapatinib compared to ATP.

- Reagents used to detect the target sequence: other
- Properties for the instrument run: standard

2. Select the "Define" menu on the left and select the following parameters:
   - Target name: Target 1

- Reporter: TAMRA
- Quencher: none
- Passive reference: none

3. In the "Assign" menu on the left, perform the following actions:
   - Highlight the wells to assay in the assay plate
   - Check the box next to "Target 1" on the top left of the plate layout
4. Select the "Run Method" menu on the left, then in the "Graphical view" tab make the following changes to the default Melt Curve profile:
   - Set the "Reaction volume per well" at 10 μL
   - In the "melt curve stage," select "Step and Hold" with a 30 s hold time
   - For the step 1, set the temperature at 15 °C with an initial 1 min hold
   - For the step 2, set temperature ramping up in increments of 0.5 °C (rate of 1 °C/min) to a final temperature of 75 °C with a final 15 s hold
5. In the "Optical Filters" tab, select the following filters:
   - x1-m1: 470/520
   - x2-m2: 520/558
   - x3-m3: 550/586
   - x4-m4: 580/623
6. Click on "Run" in the menu on the left to initiate thermal denaturation. For each plate, measurement time is around 60 min

2.2.3 Data analysis

1. The following steps describe how to manually analyze the data to complete a basic non-linear fitting of thermal denaturation curves to derive the melting temperature (T_m). As an alternative, for more automated data analysis refer to Huynh and Partch (2015) and Fedorov, Niesen, and Knapp (2012). Export the melt curve raw data to Excel. In the Real-Time qPCR software select "Export" on the left menu, then select "File type, xls" and tick "melt curve raw." Data will be exported in column format, indicating well position, and for each temperature the fluorescence recorded.
2. Use a standard fitting software such as GraphPad Prism to enable fitting of the T_m values. In GraphPad Prism, create a new project file and choose "create XY data table." In the Enter/Import data section, check X as "numbers" and Y as "enter and plot a single Y value for each point." Create a data table for each TSA conditions (Apo neutral controls, ATP/$MgCl_2$, Inhibitors).

3. Open the Excel raw data file. Report the data from the melt curve raw data in GraphPad Prism. Copy and paste the temperature values in the X column of the GraphPad XY tables created. For each well paste the fluorescence values recorded for the temperature range. Example melting curves for HER3 with ATP and Bosutinib and HER2 with lapatinib are shown in Fig. 2B and D.
4. Reduce the curves to the actual transition, by truncating post-peak fluorescence data. For each curve, identify the maximum fluorescence intensity (fluorescence peak) and temperature at which the peak is reached. Trim the temperatures above this fluorescence peak temperature.
5. Import the truncated data into a new XY table for analysis. Copy-paste the range of temperatures of the actual transition and the associated fluorescence intensity in this XY truncated data table.
6. Perform a non-linear fitting of the truncated data set to a Boltzmann Sigmoidal curve.
7. In GraphPad Prism, select the "Analyze" tab and choose "XY analyses/ non-linear regression (curve fitting)." Select the data sets to analyze, click "OK." In the "parameters: non-linear regression" window, in the "Model" tab, choose "Boltzmann sigmoidal" and click "OK." This will provide best-fit values, with V_{50} being the T_m of HER3 or HER2 for each condition tested. Standard errors, 95% confidence intervals and the R squared values of the goodness of fit are also displayed.
8. For each condition, average the fit results (i.e., V_{50}) from the 16 wells to determine the T_m values. Perform subsequent analysis to determine the thermal shift ΔT_m (i.e., calculate the change in melting temperature, ΔT_m) obtained in each condition relative to the DMSO Apo neutral control. For each condition calculate ΔT_m as (average T_m condition 1) − (average Tm Apo neutral control DMSO).
9. Plot ΔT_m as a function of each condition (Fig. 2C and E). Dose response experiments can be used to examine the IC_{50} values for ligands or inhibitors (Fig. 3A–C).

2.2.3.1 Assay quality control

The robustness of assay performance should be assessed, making sure a suitable assay window is generated. Check that the raw signals are sufficiently tight and that there is significant separation between the "Max" (ATP/ $MgCl_2$ or Bosutinib and Lapatinib, for HER3 and HER2 positive controls, respectively) and "Min" neutral controls (DMSO ApoHER3 or ApoHER2) signals to conduct screening. A clear separation between the

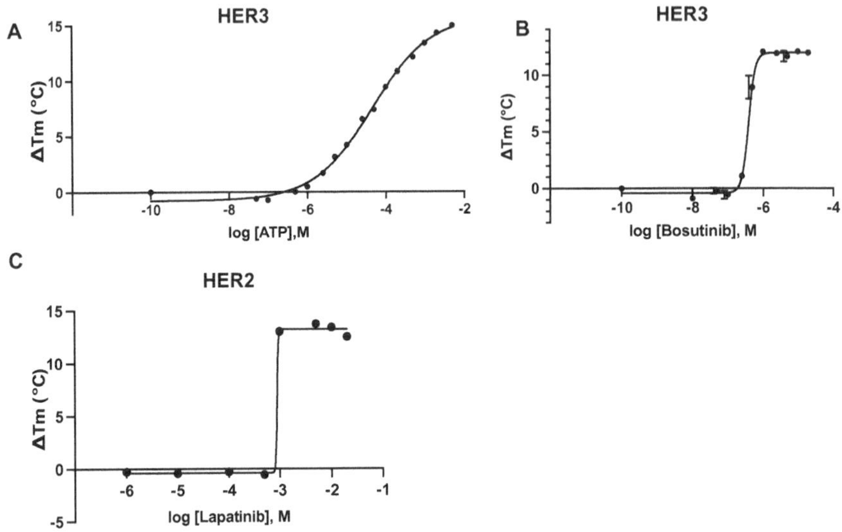

Fig. 3 HER3 differential scanning fluorimetry assay validation. (A) Dose-dependent analysis of thermal shifts induced by ATP tested at the indicated concentrations. The change in HER3 T_m value (ΔT_m), compared to 1% DMSO ApoHER3 control in absence of any ligand, is reported as mean ± SD from $n = 2$ independent experiments. Data shows an ATP dose-dependent increase in HER3 melting temperature (T_m). (B) Dose-dependent analysis of thermal shifts induced by bosutinib tested at the indicated concentrations. The change in HER3 T_m value (ΔT_m), compared to 1% DMSO ApoHER3 control, is reported as mean ± SD from $n = 2$ independent experiments. Data shows the maximum thermal shift in HER3 Tm is achieved with 1 μM bosutinib. (C) Dose-dependent analysis of thermal shifts induced by Lapatinib tested at the indicated concentrations. The change in HER2 Tm value (ΔT_m), compared to 1% DMSO ApoHER2 control, is reported as mean ± SD from $n = 2$ independent experiments. Data shows the maximum thermal shift in HER2 T_m is achieved with 1 μM Lapatinib.

tails of the distributions of the controls and a large signal window dictates how reliable the assay is at detecting active compounds in the presence of typical errors of measurement.

The overall requirement is that the Robust Z factor (RZ′) for each plate is above 0.5, with RZ′ calculated as: $1-(3\times\mathrm{rSD}\ T_m$ positive controls $+3\times\mathrm{rSD}\ T_m$ Apo neutral controls)/(median T_m positive controls − median T_m Apo neutral controls).

$RZ' > 0.5$ provides good statistical estimates from HTS data replication to assess the reliability and performance of the assay.

2.3 High-throughput screen to identify HER3 binders

The small scale and high throughput capacity of TSA screens makes them an excellent platform for the discovery of small molecule ligands.

The following presents a methodological workflow for screening small molecules libraries to identify compounds binding specifically to HER3 using TSA.

2.3.1 Materials

2.3.1.1 Small molecule libraries and reagents

Small molecules libraries of choice in DMSO, ideally at a 10 mM stock concentration.

Depending on the size of the library of interest, perform a primary screen either at one concentration (30 or 50 μM), or as concentration-response curves (CRC; 10, 8 or 4 points) depending on the size of the library and the availability of the recombinant HER3 pseudokinase.

It is also possible to screen multiplex libraries, i.e., where several compounds are screened as mixtures, as we have shown that HER3 binders can be identified from pools of 5 or 10 compounds (Fig. 4) (Colomba et al., 2020). However, it is recommended to screen at a concentration range

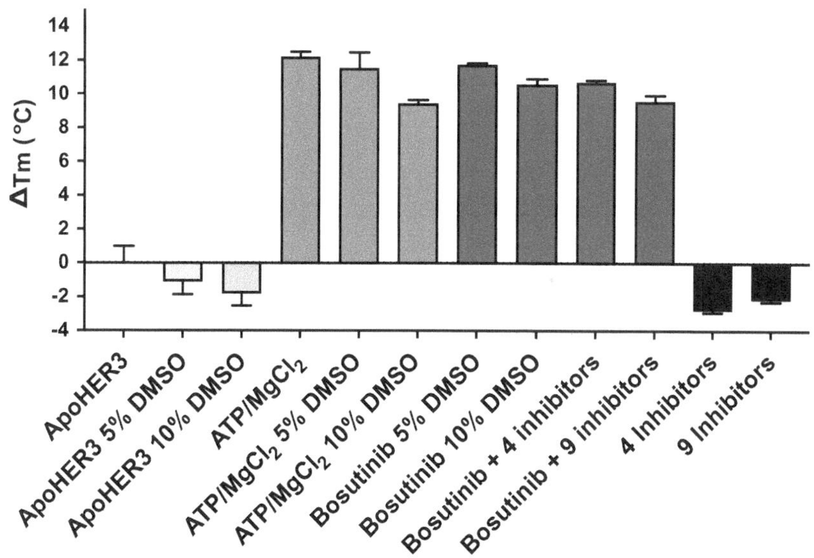

Fig. 4 Small molecule binding to HER3 can be identified from mixtures of several compounds. Recombinant HER3 kinase domain was assayed by TSA in absence of any ligand (ApoHER3), or in presence of 200 μM ATP/10 mM $MgCl_2$ (ATP/$MgCl_2$) or in presence of 1 μM of bosutinib alone or in combination with a mixture of four other inhibitors at 1 μM each (BLU577, MG-132, LY294002, CRT6854) or in combination with a mixture of 9 other inhibitors at 1 μM each (BLU577, BIM1, BX-912, LY294002, staurosporine, MG-132, bortezomib, MG-132, CRT0066854). Data showed that HER3-binding compounds, such as bosutinib, could be identified from pools of compounds, as the shift in HER3 T_m induced by bosutinib was not affected as a result of pooling compounds.

that will keep the final percentage of DMSO below 2% in the assay plates to minimize artifacts. If running a primary screen as a multiplex library first, pools of compounds identified as hits should then be deconvoluted and screened as individual compounds at a single dose. Compounds identified as hits from the single dose primary screen should then be confirmed as HER3 binders in dose-response TSA screens.

2.3.1.2 Tools and instruments

In addition to the tools and instruments described in Section 2.2.1, a liquid handling robot, such as an Echo acoustic dispenser, is recommended to dispense the small molecules to screen, from the stock compound plates to the screening plates, as nL volumes. If a liquid dispensing platform is not available, compounds can be dispensed directly into each well of the assay plate using adjustable volume multichannel micropipettes. In that case, it is recommended to perform intermediate dilutions of compounds from the stock compound plates in assay buffer, to work with more tractable volumes and minimize the final percentage of DMSO.

2.3.2 Method

2.3.2.1 Preparation of the compounds assay plates

The following describes the steps to screen small molecule libraries for HER3 binders on a 384-well platform. However, the assay can be further miniaturized to 1536-well format to allow for automation and increase of screening capacity.

1. Design the layout of the screening plates, as single-dose (either at one compound per well or as mixtures of several compounds per well) or CRC. For each plate, include control positions (DMSO for ApoHER3 neutral controls, with final percentage of DMSO matching the final percentage of DMSO in the compounds wells, and DMSO-ATP/$MgCl_2$ or 1 μM final Bosutinib for positive controls). We included a whole column of 16 wells for each control. For our HER3 screen we pooled 4 inhibitors/well at 30 μM in the initial first pass screen, followed by two rounds of deconvolution as single compound/well at 30 μM. To do this it was important to ensure the assay is robust at high DMSO concentrations, to accommodate multiple inhibitors. Fig. 4 demonstrates that the HER3 assay works equally well up to 10% DMSO and that bosutinib ΔT_m is unaffected by mixing with up to 9 additional inhibitors

2. As a common feature of HTS, edge effects and patterns on plates are generally observed. It is therefore important to keep the same positioning of controls throughout the whole screen. Ideally, on each plate include two full columns of neutral controls and two full columns of positive controls, placed at the edges of the plates. If the HER3 recombinant protein availability and the space on plates are limiting factors, dedicate one whole column for each type of control and place them at the center of the plate
3. Dispense the small molecules to screen from the stock compound plate(s) to the screening 384-well plate(s), ideally using an Echo dispenser. Place a sufficient volume of each compound stock solution in DMSO into the wells of the assay plate(s) to yield the desired final concentration in 10 μL final volume
4. Seal the screening plates and store at −80 °C

2.3.2.2 Setting the screening plates

Follow the steps as described in Section 2.2.2.

1. For screening homogeneity and consistency, it is recommended to perform each round of screening with the same solution of HER3. Buffered HER3 solution will be dispensed into the assay plates containing compounds and plates will be sealed and kept at −80 °C until the TSA runs can be performed. At the time of the TSA run, thaw the assay plates containing compounds and HER3 on ice, spin briefly and add SYPRO Orange 5 × (see step 5).
2. Dilute HER3 stock solution to 3 μM in the Assay buffer and prepare enough volume for the required number of plates to assay (2.5 mL is sufficient to screen one full 384-well plate, taking into account dead volume). If using ATP/$MgCl_2$ as positive control, dilute HER3 stock solutions to 3 μM in the Assay buffer containing 400 μM ATP and 20 mM $MgCl_2$ (the final ATP and $MgCl_2$ concentrations for the TSA are 200 μM and 10 mM/well, respectively. 100 μL is sufficient for 16 wells, taking into account dead volume).
3. Thaw the compound plates, spin briefly (300 × *g*, 1 min at room temperature), then remove the aluminum sealing foil.
4. Using a multichannel pipette, dispense 5 μL of the buffered HER3 into the wells of the screening plates. If compounds have not been dispensed with an Echo acoustic dispenser, adjust the volume of buffered HER3 taking into account the volume of compound plated.

5. Spin the screening plates briefly (300 × g, 1 min at 4 °C) and store at −80 °C until the TSA run can be performed.
6. Dilute the SYPRO Orange 5000 × into the assay buffer at a 5 × concentration (the final concentration of SYPRO Orange for the TSA assay is 2.5 ×) and prepare enough SYPRO Orange 5 × for the required number of plates ready to be run (2.5 mL is sufficient to screen one full 384-well plate, taking into account dead volume). Using a multichannel pipette, dispense 5 μL of SYPRO 5 × into the wells of the screening plates just before performing the TSA run. Do not freeze plates containing buffered HER3 and SYPRO Orange 5 ×, as higher background will be generated when the TSA run is performed.
7. Seal the plates with the qPCR-compatible adhesive plate sealing foil and ensure that the plate is well sealed by using a smooth rubber roller.
8. Spin the screening plate briefly (300 × g, 1 min at room temperature).

2.3.2.3 Experiment procedure

Run the TSA experiment as described in Section 2.2.2.

2.3.3 Data analysis

Melt curve raw data can be processed with data analysis software which is able to analyze TSA data, e.g., Genedata Screener®. As an alternative, melt curve raw data can be analyzed as described in Section 2.2.3 and, for a more automated analysis, by creating a macro or using the Excel-based tool "DSF Analysis" (Fedorov et al., 2012; Huynh & Partch, 2015).

It is recommended to inspect the data and analyses manually for abnormality in curves of fluorescence measurements. Problems can include high intensity of the baseline fluorescence background at low temperature, low fluorescence signals at the transition peak, multiple transitions or lack of sigmoidal transition to the unfolded state. Data should be normalized to the DMSO neutral control ApoHER3, and the thermal shift ΔT_m of each well relative to the DMSO control determined to identify compounds that bind to HER3, i.e., promote a shift in its T_m.

It is vital to ensure that each plate in the screen is valid and that data is comparable across plates and experimental runs. In order to assess the assay reproducibility and reliability, for each plate:

1. Calculate the DMSO ApoHER3 neutral controls and positive controls (HER3 + Bosutinib or HER3 + ATP/$MgCl_2$) T_m medians and rSD. To ensure good screening data quality, neutral and positive controls

T_m should not vary significantly across plates. It is recommended to plot medians of control wells across the plates to look for trends and drifts

2. Determine the RZ′ and disregard any plate which would not pass the QC, i.e., $RZ' > 0.5$

2.3.4 Hit selection

TSA measures protein thermostability which can be influenced by compound binding. This can result either in a destabilizing effect, conferring a decrease in thermostability (lower T_m), or a stabilizing effect, which leads to an increase in thermostability (higher T_m). For a first pass, consider any small molecule that gives rise to a positive ΔT_m over the DMSO ApoHER3 control as a potential hit.

In large scale screens it is challenging to distinguish true hits from the inactive bulk of the data to minimize false positives and false negatives. For practical reasons, the number of compounds to be taken forward for further investigation must also be kept manageable. A common method is to select hits where the value is at least three standard deviations from the mean ($3 \times SD$). Calculate the cut-off value of ΔT_m for compound binding as $3 \times SD$ of the DMSO ApoHER3 neutral controls in each round of screening. In our screen of over 100,000 compounds, we applied a cut-off of 3 °C, corresponding to $3 \times SD$ of the ApoHER3 neutral controls, which was not found to vary significantly across all the different rounds of screening performed. However, it may be necessary for consistency, to apply the same ΔT_m cut-off determined in the primary screen, throughout all the screening stages.

Depending on the size of libraries and the screening strategy chosen, it is recommended to run each screen at least twice with replicates, to get a direct estimate of data variability for each compound. Only compounds showing consistency as HER3 binders are selected as hits of interest. For practical reasons when screening large libraries, compounds may be tested just once at a single concentration in the primary screen. HER3 binders are then triaged and it is recommended to discard any whose structures are annotated as pan-assay interference compounds (frequent hits across multiple screens). A number of studies and resources are available to help identify such compounds (Baell & Holloway, 2010; Baell & Nissink, 2018; Dahlin & Walters, 2016; Jasial, Hu, & Bajorath, 2017). If performing the primary screen as pools of compounds, it is recommended to run the deconvolution screen twice and consider as putative hits compounds which passed the selection criteria in both runs.

For the dose-response confirmatory screens, perform at least two runs with compounds plated in replicates and select as putative hits compounds which passed the selection criteria in both screens with consistent inhibitory characteristics. In the dose-response screen, a binding affinity criterion can also be applied, selecting the most potent compounds as hits for further investigation.

3. Validating the specificity and inhibitory capacity of HER3 binding compounds

Compounds identified as HER3 binders in the TSA screen, should be then further validated as HER3 binders and assessed for their specificity toward HER3 through confirmatory orthogonal assays. As with all ATP competitive inhibitors, generating compounds with sufficient selectivity remains a key challenge. However, as a pseudokinase, binding of ATP to the HER3 binding pocket is not subject to as many evolutionary constraints as with an active kinase, and this is reflected in the atypical arrangement of the active site. Such nucleotide pocket divergence is elegantly illustrated for the MLKL pseudokinase, where distinct non-canonical modes of ATP binding have arisen between mouse and human orthologues, in the absence of a requirement for catalytic activity (Murphy, Lucet, et al., 2014). Atypical nucleotide binding modalities may offer opportunity for development of more selective inhibitors (Byrne, Foulkes, & Eyers, 2017; Kung & Jura, 2019; Murphy, Zhang, et al., 2014).

3.1 Assessment of compound specificity

As HER2 and HER3 are highly homologous EGFR-family members, compounds identified as HER3 binders in the TSA screen should be counterscreened for binding to HER2. As a first approach, TSA is a useful platform to further triage hit compounds and identify those that can also bind to HER2. Hit compounds can be assayed in a dose-response screen and compounds inducing a shift in HER2 T_m, and therefore binding to HER2, will be disregarded. While we acknowledge that compounds binding both HER3 and HER2 might still be effective clinically, cross-inhibition of both HER2 and HER3 complicates validation of whether a HER3-pseudokinase directed therapeutic intervention strategy is a fruitful avenue of investigation.

- Follow the screening method described in Section 2.2, compounds can be counter-screened against HER2 (1.5 μM), using DMSO ApoHER2 as neutral control and 1 μM Lapatinib as positive control.
- Proceed to data analysis and hit selection as described for HER3, applying a cut-off over 3 × SD compared to the DMSO ApoHER2 neutral controls. In our HER2 counter screen of over 500 compounds, we applied a cut-off of 2 °C, corresponding to 3 × SD of the ApoHER2 neutral controls, and select compounds which did not induce a shift in HER2 T_m over 2 °C. The cut-off values applied to select small molecules binding to HER3 and to HER2 should not be significantly different, so the selection criterion for HER2 binding is, neither too stringent (eliminating too many compounds of interest), nor too permissive (selecting too many compounds not specific for HER3 binding).

At this stage, depending on the number of compounds identified, lack of activity toward HER2, and other EGFR family members, can be examined using *in vitro* kinase assays. For a more thorough determination of compounds specificity, broader panels of kinases can be examined. Many resources for testing compound selectivity across the kinome are available through both academic and commercial sources. Development of more selective compounds is a next challenge as we move from proof-of-concept compounds such as AC3573 toward pre-clinical leads.

3.2 Orthogonal assays to confirm HER3 binding

Depending on the number of compounds identified as HER3 binders and HER2 non-binders in the TSA screens, a validation stage, employing other types of assays, can be performed to confirm hit compound binding to HER3. Putative compounds of interest can be further investigated by biophysical methods, such as isothermal titration calorimetry (ITC), Microscale Thermophoresis (MST), or label-free Surface Plasmon Resonance (SPR), which allow low-throughput dose-response affinity screens. As an alternative, Cellular Thermal Shift Assay (CETSA) or NanoBRET assays can provide high-throughput in-cell target engagement assay to support the drug-discovery pipeline.

3.2.1 Biophysical methods

ITC directly measures the thermal characteristics of target-compound binding and can provide detailed assessment of binding parameters in a label free system; this is not readily useable as a screening tool but can be used to validate high-confidence hits (Todd & Gomez, 2001). MST measures temperature dependent changes in fluorescence of a reporter or tracer fluorophore, which can change on binding of a ligand to a target (Jerabek-Willemsen, Wienken, Braun, Baaske, & Duhr, 2011). For kinase inhibitors, the kinase itself can be fluorescently tagged or alternatively a competitive fluorescent tracer—displaced by inhibitor binding—can be used for a higher throughput approach. Surface Plasmon Resonance assays can also be effectively used to assess kinase domain-inhibitor interactions but require immobilization of kinase domain or ligand; once optimized however SPR can provide detailed assessment of binding kinetics and target residence, and has been successfully used to assess TKI binding (Willemsen-Seegers et al., 2017). These techniques should largely be considered for follow up studies once functional HER3 directed inhibitors have been identified.

3.2.2 In-cell target engagement

To confirm target engagement under actual physiological conditions, Cellular Thermal Shift Assay (CETSA) can be used to assess and quantify hit compounds binding to HER3 in cellular environments, providing a more biologically relevant data set. We previously used this approach to confirm HER3 binding of bosutinib in cells under conditions where bosutinib was driving oncogenic signaling and proliferation (Claus et al., 2018). While our previous use of CETSA was performed using Western Blot analysis, recent reports of Mass Spectrometry-based CETSA are promising for the use of this technique on a more high-content proteomic scale (Friman, 2020). The use of CETSA to study pseudokinases is covered in another Chapter "Cellular thermal shift assay" by Karvonen, Raivola and Ungureanu.

NanoBRET assays have emerged as a leading technology for examining in-cell target engagement. Many commercially available assay systems are available for many kinases, including EGFR, HER2 and their oncogenic mutants (Promega). NanoBRET measures bioluminescence resonance energy transfer (BRET) between a NanoLuc luciferase tagged target (e.g., HER3) and a nucleotide pocket binding fluorescent tracer. Competitive displacement of the tracer results in a reduction in BRET signal which can be measured in live cells at high throughput for relatively low cost, using standard high-sensitivity laboratory luminometers. While efforts would be required to establish a robust NanoBRET assay for HER3, the value as an orthogonal binding and in-cell target engagement assay would be of huge value to the discovery pipeline.

3.2.3 *Hydrogen-deuterium exchange mass spectrometry (HDX-MS)*

To gain structural insight on in the absence of a crystal structure, HDX-MS can be used to identify inhibitor contact points with HER3. The use of HDX-MS is covered in detail in Chapter "Dynamics of protein kinases and pseudokinases by HDX-MS" by Sheetz, Lemmon and Tsutsui. The rate of deuterium exchange is reduced in contact residues in the presence of AC3573 or other HER3 binding compounds relative to the ApoHER3. AC3573, bosutinib and the ATP analogue AMP-PNP all protected the 786–795 peptide from deuterium exchange indicating binding to the hinge-region of the ATP binding pocket (Colomba et al., 2020; Jura, Shan, et al., 2009). Additional proteomic techniques such as Limited Proteolysis Mass-Spectrometry (LiP-MS) could give further information about binding and allosteric conformational changes (Feng et al., 2014).

3.3 Assessing HER3 binders as biochemical inhibitors of HER3 function

Compounds identified and confirmed as HER3 specific binders, can next be assessed for their inhibitory effects on HER3 in cell-based assays. Distinct inhibitors are likely to stabilize specific conformers of HER3 and many binders may be expected to promote rather than suppress HER3 interactions with HER2, exemplified by bosutinib induction of oncogenic HER2-HER3 signaling (Claus et al., 2018).

The HER2-HER3 heterodimer is a potent mitogenic and survival unit, in particular through the recruitment of the PI3K-AKT pathway, especially in HER2-dependent cancer cell lines. Hit compound activity can be evaluated for their ability to affect HER2-HER3 downstream signaling in response to NRG-induced HER3 activation in such cell lines (e.g., HER2-dependent breast cancer cell line SK-BR-3). Screening hit compounds for their ability to inhibit NRG-induced HER3 phosphorylation and downstream signaling in a dose dependent manner (i.e., Akt and ERK phosphorylation), will provide further information on compound inhibitory effects and potency. Assessment of Akt and ERK activation, can be performed by Western blot, though other assays, such as enzyme-linked immunosorbent assay (ELISA) or Meso Scale Discovery (MSD), can provide a higher throughput, sensitive and quantitative evaluation.

Hit compounds able to inhibit HER3-dependent signaling in cells can be assessed for their lack of activity on HER2 autonomous activation, in for example engineered HER2-dependent cell lines expressing low or null level of HER3. HER2 constructs (full-length wild-type or oncogenic point mutants, such as G660D and V664E) can be transiently expressed in, for example, CHO or HEK cells, and hit compound effects on HER2 autophosphorylation status (i.e., activation) determined to ensure their mechanism of action on NRG-induced HER2-HER3 signaling is not via HER2. While HER2 is the preferred binding target of HER3, coupling with other TKIs, and notably EGFR, must also be considered when examining the inhibitory effects of HER3 targeting inhibitors on downstream pathways. The development of drug resistant HER3 mutants will provide a valuable additional tool here for target validation, but this will require a structural understanding of drug binding mode. This will be addressed in further detail in Section 5.

4. Methods for assessing inhibitor mode of action

As a pseudokinase, the oncogenic functions of HER3 all require heterodimerization with an active RTK. Small molecule inhibitors binding to HER3 must either suppress expression or modulate interaction with binding partners. In addition to the favored HER2 and EGFR partners, HER3 may interact more broadly with other RTKs including MET (Engelman et al., 2007). Further, while attention has largely focused on dimer pairs, higher order organization of HER family receptors has emerged as an additional layer of complexity (Clayton, Tavarnesi, & Johns, 2007; Clayton et al., 2005; Jura, Shan, et al., 2009; Roberts et al., 2018; van Lengerich, Agnew, Puchner, Huang, & Jura, 2017). In this section, we will provide an overview of methods which can probe distinct modes of action.

4.1 Fluorescence resonance energy transfer (FRET) to assess HER3-HER2 interaction

HER2 is the preferred signaling partner for HER3 and expression of both receptors is necessary for HER2 oncogenic signaling in HER2+ breast cancer (Alimandi et al., 1995; Bodey et al., 1997; Holbro et al., 2003; Wallasch et al., 1995). FRET can be used to quantify direct protein-protein interactions and this is ideally suited for measuring receptor pairing of either endogenous HER2 and HER3 or using overexpressed constructs. FRET occurs between a donor and acceptor fluorophore when they are in close proximity (>10 nm). When FRET occurs between a donor and acceptor, the lifetime of the donor fluorophore is reduced, which can be measured by fluorescence lifetime imaging microscopy (FLIM), to provide a quantitative spatial map of FRET efficiency. For our studies examining HER2-HER3 pairing,

FRET-FLIM was conducted using time-correlated single photon counting (TCSPC) with a multi-photon microscope system as described elsewhere (Barber et al., 2009; Parsons & Ng, 2002; Peter et al., 2005).

4.1.1 Endogenous FRET-FLIM

For endogenous studies, the critical factors are: (i) identifying cell lines expressing sufficient levels of HER2 and HER3 and (ii) identifying antibody pairs against the two receptors compatible with efficient FRET. Antibody pairings must be assessed empirically. SKBR3 breast cancer cells express high levels of both HER2 and HER3 and can be successfully used to probe HER2-HER3 dimerization in response to ligand stimulation or small molecule inhibitor binding. Using this system, bosutinib can be shown to induce HER2-HER3 coupling (Claus et al., 2018) whereas the proof-of-concept HER3 inhibitor AC3573 decreases their interaction (Colomba et al., 2020). This provides a critical indication that AC3573 likely acts through an allosteric mechanism.

Importantly, endogenous FRET-FLIM to examine HER2-HER3 dimerization is also compatible with *in vivo* preclinical studies to establish whether HER3 targeting drugs can disrupt dimer pairs in tumors. Such studies could focus on human orthotopic xenografts of human HER2-dependent cancer cells or genetically engineered mouse models (GEMM) of HER2 positive breast cancer, such as MMTV-c-Neu mice (Muller, Sinn, Pattengale, Wallace, & Leder, 1988). Mouse-compatible antibody pairs would need to be identified for GEMM models. Additionally, FRET-FLIM can used to measure HER2-HER3 dimerization directly from patient tissue (Barber et al., 2020). In this study, HER2-HER3 dimerization was shown to be associated with a subclass of colorectal cancer patients from the phase 3 MRC COIN trial showing better overall survival. This is a promising development, allowing more granular investigation of the specific receptor dimerization behavior at the core of a particular patient's disease. Further, the ability to measure HER2-HER3 dimerization in tumors provides an ideal pharmacodynamic biomarker for assessing target engagement with HER3 targeting agents.

4.1.2 Exogenous FRET-FLIM

While endogenous studies are useful for examining the impact of HER3 inhibitors or activators in physiologically relevant settings, exogenous studies allow better interrogation of mechanism. Expression of distinct receptor pairings, receptor truncations and mutants can differentiate between distinct modes of interaction. A number of crystal structures of the HER3 kinase domain have been resolved, including as part of an asymmetric "active" HER3:EGFR kinase domain pair (PDB ID 4RIW) (Littlefield et al., 2014) and also as a putative symmetric inhibitory HER3:HER3 dimer (PDB ID 3KEX) (Jura, Shan, et al., 2009). FRET-FLIM coupled with site-directed mutagenesis can be used to discriminate between distinct kinase dimer models. Careful consideration of all available structural and modeling data can aid the design of mutants that inhibit a particular interface, but not another. Additionally, HER3 pairing with additional RTKs, such as EGFR or MET can be readily explored using overexpressed constructs.

When designing a FRET-FLIM experiment to investigate RTK interaction behavior with exogenously expressed receptors, two aspects should be not be overlooked. First, the presence of endogenously expressed protein of interest can be a confounding factor. High levels of endogenous receptor can sequester labeled receptors in dimers or high order oligomers, thereby precluding them from forming FRET pairs with their labeled partners. Therefore, in these receptor pairing co-transfection settings, a choice of cell line with low endogenous expression of the receptors under investigation is advantageous to avoid interactions with "dark" endogenous receptors. For HER2-HER3 studies, MCF7 cell have proved a useful platform as these cells express low levels of endogenous HER2 and EGFR (Claus et al., 2018). Second, the specifics of construct design are important. RTKs have multiple interaction interfaces spanning the extracellular, transmembrane, and intracellular domains. When interrogating kinase domain behavior, care should be taken to create plasmids with FRET-paired fluorophores in close proximity to the actual domain interactions. Otherwise, a growth factor-induced extracellular interaction could give a positive FRET-FLIM dimerization signal, even when careful design of exogenously expressed mutants prohibited kinase domain dimerization. Validation of receptor pairs is essential, which ideally will include demonstration of ligand or inhibitor induced dimerization coupled with kinase domain interface mutations that reduce FRET.

4.1.3 Dimer interface site-directed mutagenesis to assess allosteric inhibitor effects

In order to discriminate between potential HER3 dimer conformers, kinase domain dimer interfaces need to be examined to identify key residues essential for interaction. In the case of HER2:HER3 active dimers, HER3 acts as the specialized activator of the HER2 kinase domain, and mutations which

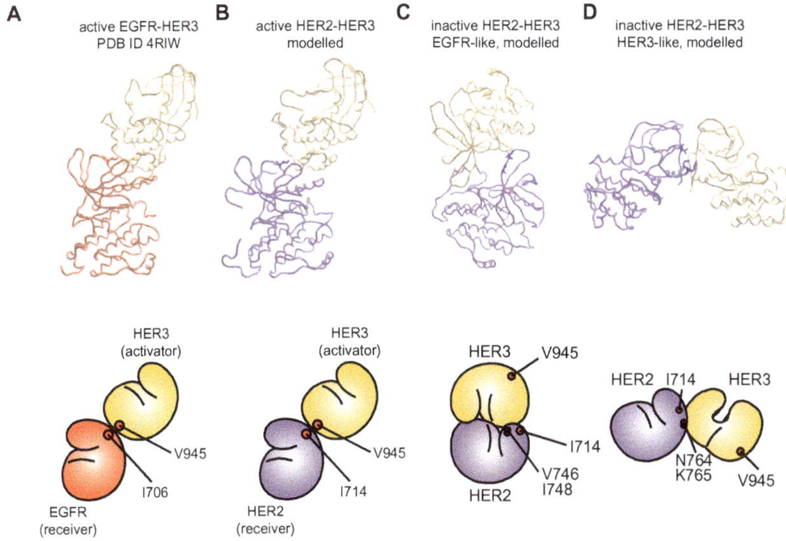

Fig. 5 Dimerization interaction modalities of EGFR family kinase domains. (A) The canonical asymmetric "activator"/"receiver" pairing in which HER3 performs the role of "activator" and EGFR that of the "receiver" kinase. Interface residues highlighted. (B) The same dimerization interface, but modeled onto the HER2:HER3 kinase domain dimer from Claus et al. (2018). (C) The symmetrical association first reported for EGFR homodimerization in Jura et al. (2009), as modeled onto the HER2:HER3 heterodimer (Claus et al., 2018). (D) The symmetrical association observed in the HER3 crystal structure by Jura, Shan, et al. (2009), as modeled onto the HER2:HER3 heterodimer (Claus et al., 2018).

disrupt the asymmetric dimer in both HER3 and HER2 are well-documented (Fig. 5) (Littlefield et al., 2014; van Lengerich et al., 2017). Mutations in the asymmetric dimer interface of either HER2 or HER3 result in reduced FRET efficiency in response to NRG growth factor stimulation (Claus et al., 2018). Interestingly, however lapatinib binding to HER2 induces a distinct dimer which is not disrupted by mutation of HER3 at this asymmetric dimer interface. This is important as it shows that distinct ATP pocket directed inhibitors can promote distinct dimer conformers. Binding of the type 1½ kinase inhibitor lapatinib, sometimes categorized as an atypical type II inhibitor, to HER2, induces an inactive conformation with the αC-helix tilted outwards, inferred from lapatinib binding mode with EGFR and HER4 (Dar & Shokat, 2011; Qiu et al., 2008; Wood et al., 2004). Modeling of lapatinib into the canonical active HER2-HER3 structure also indicated steric clashes with the αC-helix of HER2, which is tilted inwards in its active conformer position

(Claus et al., 2018). Two distinct HER family dimers have been described: a head-to-tail homodimer for EGFR and a head-to-head homodimer for HER3 (Fig. 5) (Jura, Endres, et al., 2009; Jura, Shan, et al., 2009). To test whether lapatinib induced either of these dimers, we designed site-directed mutants to discriminate between these distinct binding modes. FRET-FLIM analyses using these mutants suggested that lapatinib induces a HER2:HER3 head-to-head heterodimer, as described for HER3 homodimers (Claus et al., 2018).

Importantly, these three sets of discriminatory dimer interface mutations can be used to interrogate the effects of any inhibitors arising from HER3 or HER2 directed drug screens (Fig. 5). While bosutinib drives dimerization and activation of HER2:HER3 dimers at concentrations from 41 nM to 1 μM (Claus et al., 2018), we might expect HER3-directed allosteric inhibitors to either disrupt dimer pairing entirely or promote adoption of non-signaling inhibitory dimers or oligomers. In Section 5 we explore this concept further by modeling AC3573 binding to a variety of available HER3 structures and models.

4.2 Orthogonal methods to assess HER3 interaction with RTKs

4.2.1 Split-luciferase assays

HER family interactions have been probed in detail using luciferase fragment complementation. Here, the N-terminal fragment of luciferase (NLuc) or the C-terminal fragment of luciferase (CLuc) are fused—in various configurations—to HER family receptor pairs. This was first successfully used to study EGFR homodimer interactions (Yang, Ilagan, Piwnica-Worms, & Pike, 2009) but has been extended to assess interactions between multiple HER family members, including HER2 and HER3 (Li, Macdonald-Obermann, Westfall, Piwnica-Worms, & Pike, 2012; Macdonald-Obermann, Adak, Landgraf, Piwnica-Worms, & Pike, 2013; Macdonald-Obermann, Piwnica-Worms, & Pike, 2012). The impact of targeting HER family members with small-molecule and antibody monoclonal therapeutics has also revealed much about mechanisms of drug action (Macdonald-Obermann et al., 2013). This study strongly supports a model in which preformed HER3 clusters are disrupted by stimulation with NRG to allow active dimerization with HER2. Both erlotinib and lapatinib are also shown to induce HER2 and EGFR dimerization with HER3, concurring with FRET (Section 4.1) and super-high resolution microscopy techniques (Sections 4.2.2 and 4.2.3). Split-luciferase reporters are highly dynamic and particularly well-suited to tracking receptor pairings over time and for assessing interactions between distinct stimuli and inhibitors.

Although not covered in detail here, a split luciferase assay would act as an excellent complementary screening platform for identifying HER receptor directed inhibitors capable of disrupting, rather than promoting, kinase dimerization (see discussion).

4.2.2 Single-particle tracking

Two-color single particle tracking provides a mechanism to assess the degree of co-localization of two receptors. HER2 and HER3 can be labeled with different fluorophores. Single-molecule images are tracked in live cells by TIRF microscopy (Webb, Needham, Roberts, & Martin-Fernandez, 2006), with simultaneous acquisition of tracks for two distinct fluorophores (HER2 and HER3) (Rolfe et al., 2011; Webb et al., 2006). Comparing the co-localization of HER2 and HER3 tracks over time under different conditions of stimulation or inhibition can provide a measure of inhibitor interaction. This method was successfully used to show that AC3573 decreases pairwise interaction frequency of HER2 with HER3 (Colomba et al., 2020). This method has also been extensively used to track EGFR dynamics (Clarke & Martin-Fernandez, 2019; Zanetti-Domingues et al., 2018).

4.2.3 Stochastic optical reconstruction microscopy (STORM)

While both FRET and split-luciferase techniques are ideal for examining binary interactions, less information is provided about higher order oligomerization. Many studies have now indicated that RTKs can form clusters and that cluster formation is associated with both inhibition and robust induction of signaling (Clayton et al., 2007; Martin-Fernandez, Clarke, Tobin, Jones, & Jones, 2002; van Lengerich et al., 2017). For the HER2-HER3 pairing, studies have suggested the formation of tetramers or higher order oligomers may be required for trans-phosphorylation of the HER2 cytoplasmic tail (Zhang, Park, Kani, & Landgraf, 2012). HER3 coupling with EGFR is also a particularly interesting case in point where dimer-oligomer behavior can vary depending on the growth factor stimulus. Thus, EGF stimulation can incorporate HER3 into higher order oligomers, whereas NRG induces EGFR-HER3 heterodimers (van Lengerich et al., 2017). In a recent study from the Kuriyan lab, the importance of higher order oligomers for effective EGFR signaling was addressed (Huang et al., 2016). Mutants of EGFR which are capable of forming active dimers, but defective in their ability to form oligomers, were severely compromised with only one half of the dimer pair undergoing C-terminal phosphorylation (Huang et al., 2016). These studies reveal the need to also address formation of higher order oligomers in HER receptor signaling. Allosteric inhibitors against HER3 might act to preclude active oligomerization or induce sequestration into higher order inhibitory structures.

The most widely used method for assessing receptor clustering is stochastic optical reconstruction microscopy (STORM). STORM can provide super-resolution by imaging single fluorophore blinking across multiple images, to allow reconstruction of high-resolution images mapping individual fluorophore localizations. For detailed protocols for examining HER2 and HER3 localization using STORM please see Roberts et al. (2018). Treatment with NRG, bosutinib or lapatinib, all induce an increase in cluster size as assessed by STORM although stoichiometry of these higher order oligomers is difficult to assess (Claus et al., 2018). In these cases, ligand or inhibitor binding was associated with the potential to promote active HER2:HER3 complexes (Claus et al., 2018). The STORM data suggest that activation of HER2:HER3 probably involves the formation of higher order oligomers, rather than just heterodimers and based on the results for EGFR homodimers (Huang et al., 2016) and EGFR-HER3 heterodimers (van Lengerich et al., 2017), this is likely to be important for functional output. In the somewhat paradoxical case of lapatinib (as a HER2 inhibitor), combination with NRG was able to promote transition to an active HER2:HER3 signaling state only at low lapatinib concentrations, while higher concentrations remained effective at suppressing HER2 kinase mediated signaling (Claus et al., 2018). AC3573 in contrast does not induce any obvious change in higher order oligomerization with both cluster size and the number of HER3 molecules per cluster remaining constant (Colomba et al., 2020). The distance between near neighbors for HER2 and HER3 was however increased which concurs with the orthogonal FRET and single-particle tracking suggesting that AC3573 causes disruption of dimers. Together, these data suggest a distinct impact of the inhibitory AC3573 compound on HER3 partnering and oligomerization, when compared with interventions which can promote active HER2-HER3 complexes.

5. *In silico* modeling of inhibitor binding to HER3

It has been established that the nucleotide pocket of HER3 is most probably occupied by nucleotide in active complexes, and this occupation appears to be critical for transactivation of the catalytically active "receiver" kinase binding partner. Mutations which prevent nucleotide pocket occupation can thus disrupt function and impede dimer formation (Claus et al., 2018). As many ATP competitive inhibitors stabilize active kinase conformers, this raises the problem that many nucleotide pocket-directed HER3 binding compounds may activate rather than inhibit function.

This is borne out by our HER3 compound screening, with the majority of hits failing to suppress HER3 mediated signaling (Colomba et al., 2020). Indeed we previously reported that bosutinib can allosterically activate HER3 in the absence of a stimulating ligand by driving HER3-HER2 dimerization (Claus et al., 2018). In order for a small molecule to act as an inhibitor of HER3 activity, it must not only bind selectively but must also induce adoption of a HER3 kinase conformer not conducive to productive interaction with its active oncogenic binding partners. AC3573 provides a proof-of-concept that such inhibitors can be developed. Understanding how this inhibitor modulates HER3 interaction with binding partners remains a key next step. Here we will explore methods for elucidating how AC3573 works, which will be applicable to the characterization of other HER3 directed compounds, and can assist in the rational development of more potent and selective drugs. While X-ray crystallography of AC3573 bound to HER3 will be necessary to provide direct experimental evidence of binding pose, *in silico* modeling can provide a dynamic orthogonal approach to explore the implications of distinct inhibitor binding modes.

5.1 Inhibitor docking optimization

To explore potential AC3573 binding modes, we first performed preliminary calculations to determine a docking method and protocol that can reliably predict binding of kinase inhibitors and ATP analogues into HER family kinase domains (Fig. 6A), where co-crystal structures have been resolved (Table 1). Three different docking applications were compared by redocking bosutinib to PDB ID 6OP9 (Campbell et al., 2022). (Fig. 6B). The applications were chosen to represent three different types of docking software: Mcule 1-click (online docking tool), AutoDock Vina (Trott & Olson, 2010) (open-source software) and Glide (docking tool from the commercial Schrödinger suite) (Friesner et al., 2004).

Mcule 1-click was run using the online interface to upload the chemical structure of bosutinib using the crystal structural of the HER3/bosutinib complex: PDB ID 6OP9 (Campbell et al., 2022) (https://mcule.com/apps/1-click-docking) (Fig. 6C). This tool is based on a completely

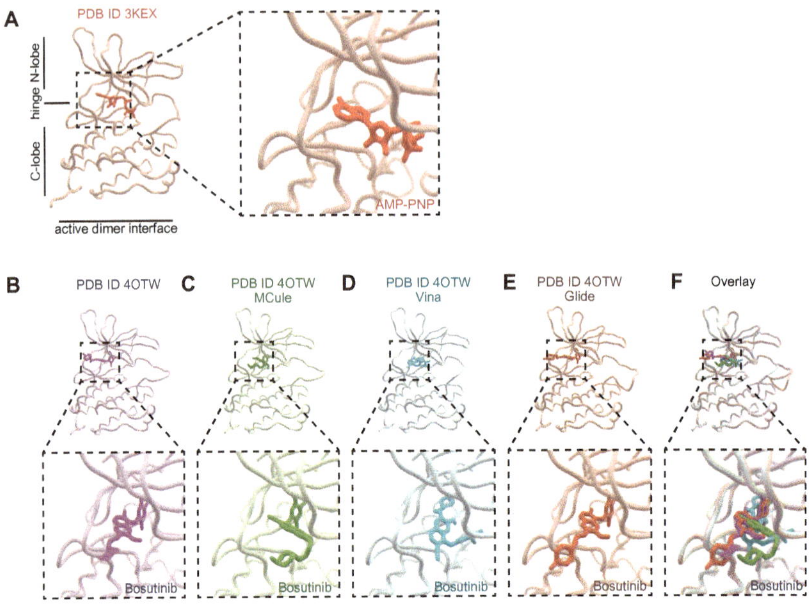

Fig. 6 Assessing *in silico* approaches for HER3 inhibitor docking. (A) Overview of HER3 kinase domain architecture and ATP binding orientation based on PDB ID 3KEX (Jura, Shan, et al., 2009). (B) Crystal structure of HER3 bound to the small molecule inhibitor bosutinib as solved experimentally in PDB ID 6OP9 (Campbell et al., 2022). (C) The HER3 kinase domain with bosutinib docked using MCule 1-click. (D) The HER3 kinase domain with bosutinib docked using Autodock Vina. (E) The HER3 kinase domain with bosutinib docked using Glide. (F) Overlay of all docking methods and the bosutinib binding orientation from crystal structure PDB ID 6OP9 (Campbell et al., 2022).

Table 1 PDB structures used for the redocking calculations.

PDB ID	Ligand	Protein	RMSD[a]	Citation
3KEX	AMP-PNP	HER3	1.33	Jura, Shan, et al. (2009)
3LMG	AMP-PNP	HER3	4.25	Shi et al. (2010)
6OP9	Bosutinib	HER3	1.81 (0.22)	Campbell et al. (2022)
4RIW	AMP-PNP	HER3	2.99	Littlefield et al. (2014)
1XKK	Lapatinib	EGFR	2.39 (1.02)	Wood et al. (2004)
1M17	Erlotinib	EGFR	2.23 (0.24)	Stamos, Sliwkowski, and Eigenbrot (2002)

[a]Root mean square deviation (Å) between the Glide-predicted binding pose and the experimental structure after restrained energy-minimization. The values in parentheses refer to the RMSD calculated without the flexible groups attached to the quinoline/quinazoline group in the ligands.

automated pipeline, where the user only needs to provide the ligand (as chemical structure or different types of ID) and the target (as PDB ID or external file). The docking engine is based on AutoDock Vina with default parameters, which cannot be changed except for the position of the docking grid box. The binding pose with the most negative binding affinity (−7.9 kcal/mol) was selected as the best one.

AutoDock Vina calculations were performed with the 1.1.2 release installed on a local workstation (Fig. 6D); the workstation we used for AutoDock Vina runs Ubuntu and has a 4-core processor (Intel(R) Xeon(R) CPU E5-1630 v3 @ 3.70GHz) and 16GB of RAM. Hydrogen atoms were added to the ligand by using PyMOL, both neutral and charged forms of the piperazine ring were considered. Ligand and protein structures were then prepared and converted to pdbqt files using AutoDock Tools. A $20 \times 16 \times 20\,Å^3$ grid box centered on the bosutinib binding site was used. Different exhaustiveness values were considered, ranging from 10 to 500. Two calculations were run for each exhaustiveness value, where a different random seed was used. Binding affinities and poses were observed to converge at an exhaustiveness value of 100. The final binding pose was selected as the one corresponding to the most negative binding affinity (−9.2 kcal/mol). A binding pose very similar to the Mcule 1-click one and with the same binding affinity (−7.9 kcal/mol) was found when using the lowest exhaustiveness value (10).

Glide calculations were run on a local installation of the Schrödinger suite (2021-2 release) (Fig. 6E). The protein structures were first prepared following the protein preparation workflow implemented in Maestro, which involved addition of hydrogen atoms, modeling of missing side chains and loops with Prime (Jacobson et al., 2004), optimization of the hydrogen bond network and assignment of protonation states of ionizable residues with Propka (pH 7) (Olsson, Sondergaard, Rostkowski, & Jensen, 2011). Protein structures were then energy minimized using restraints on heavy atoms. Water molecules were not removed when present. Indeed, differently from the first two docking programs, Glide can take into account water molecules in the docking calculations and the ones in the bosutinib binding site have been shown to be important for its binding (Levinson & Boxer, 2014). Ligands were prepared using the LigPrep tool for the addition of hydrogen atoms and energy minimization. Different protonation states were generated using Epik (Shelley et al., 2007) (pH 7.0) for ionizable ligands. The position and size of the grid box for the docking was defined by using

PyMOL (Autodock/Vina Plugin (Seeliger & de Groot, 2010)). Protein structures were first aligned by best-fit superimposition of equivalent atoms. A single box was then defined to include the binding sites of all the ligands. The coordinates of the center of this box were used in the Glide grid generation stage for all the target structures. The outer box size was set to $22 \times 22 \times 22\,\text{Å}^3$ and the inner box to $10 \times 10 \times 10\,\text{Å}^3$. Docking calculations were performed in XP precision mode and the XP GlideScore was used to estimate binding affinities. Where applicable, the Epik state penalties were included in the ranking of the binding poses. For all the calculations, the default implementation of OPLS-AA in the Schrödinger suite was used (Lu et al., 2021). Protein-ligand interactions were analyzed with the Pose Viewer tool of Maestro.

Glide was the most accurate at predicting the experimental binding mode of bosutinib (Fig. 6B–E), with an overall root mean square deviation (RMSD) between the predicted and experimental structure of 1.81 Å (Table 1). This is further reduced to 0.22 Å if the flexible group containing the piperazine ring is not considered, indicating an almost perfect match between predicted and experimental structure for the aromatic rings (Fig. 6B and E). The other two docking tools instead misplaced either one (Fig. 6C) or both (Fig. 6D) aromatic systems.

Because of its excellent performance with bosutinib, Glide predictions were extended to other known ligands of HER3 and of the related EGFR receptor (Table 1). Two structures for HER3 bound to AMP-PNP have been reported, as well an EGFR-HER3 co-crystal, which show similar kinase domain structures and AMP-PNP binding mode (Jura, Shan, et al., 2009; Littlefield et al., 2014; Shi et al., 2010). HER3 adopts an inactive conformation with the αC helix rotated outwards compared with active kinase domain structures, with AMP-PNP bound in a similar mode to other inactive kinase structures (Jura, Shan, et al., 2009). Moreover, we considered published co-crystal structures of erlotinib and lapatinib bound to EGFR (Stamos et al., 2002; Wood et al., 2004) to test Glide on a different HER family member. A good performance was obtained in all cases except for PDB ID 3LMG (Shi et al., 2010), where the main source of deviation from the experimental structure was due to the position of the ribose ring. For lapatinib and erlotinib, RMSD values of ~1 Å or below were observed when the flexible groups attached to the quinazoline moiety were excluded, indicating again an excellent agreement with the experimental structure for the aromatic core of the ligands.

5.2 AC3573 docking to HER3 and EGFR family kinases domain structures

The AC3573 inhibitor was docked to all the experimental HER3 and EGFR structures considered in this work (Table 1) using our optimized Glide protocol (PDB ID 3KEX to PDB ID 4RIW in Table 2). The Glide docking protocol and parameters described above were used with the following exceptions. For each target structure, two docking runs were first performed, one with standard XP precision sampling parameters and a second one with expanded sampling, where all the initially generated poses were considered in the rough scoring phase (EXPANDED_SAMPLING = True) and the number of poses in the energy minimization phase was increased from 800 to 1000. The top-ranking binding pose from both runs was then retained as the best one. At this stage, two groups of binding orientations were observed across the different target structures. In the "main" binding orientation, the indole end of the molecule was found to bind deep in the binding site, while in the "flipped" orientation it was located closer to the site entrance. The main orientation was observed more frequently than the flipped one and tended to be associated with stronger binding. In a second stage, each binding pose was transferred from its original target structure to each of the others and the binding affinity was recalculated after energy-minimizing the pose in the new host structure. For some of the target structures, this led to binding poses with better affinities, and these were then used as new best binding poses. After this

Table 2 Predicted binding affinity of AC3573 toward HER3 and EGFR target structures.

Target structure	Binding affinity[a]
1M17 (EGFR)	−10.29
1XKK (EGFR)	−9.01
3KEX (HER3)	−8.96
3LMG (HER3)	−9.40
6OP9 (HER3)	−10.14
4RIW (HER3)	−9.35
HER3 from active HER2-HER3 model	−8.31
HER3 from inactive HER2-HER3 EGFR-like model	−10.01
HER3 from inactive HER2-HER3 HER3-like model	−9.83

[a]XP GlideScore (kcal/mol).

cross-docking procedure, no flipped orientations were observed, confirming that the main binding pose is the most favored from the energetic point of view independently from the conformation of the target.

The optimal AC3573 docking position shows the aminopyrimidine and sulfonamide groups form hydrogen-bond interactions with the backbone of the hinge region residues L790 and P791 (Fig. 7A–E). Importantly, the predicted interactions between AC3573 and this hinge region are supported by the hydrogen-deuterium exchange mass spectrometry where these residues are protected from exchange (Colomba et al., 2020). In all cases, overlaying the position of AC3573 with AMP-PNP shows that the central aminopyrimidine group of AC3573 occupies the adenine binding position in the HER3:AMP-PNP complexes (Fig. 7F–H). In most of the structures, the indole NH group of AC3573 also forms a hydrogen bond with the critical aspartate residue in the DFG motif. The only exception is PDB ID 6OP9 (Campbell et al., 2022), where the aspartate is too distal to form direct interactions with the molecule. In all the structures, hydrophobic contacts are formed with L715, V723, Y789, L790, G793 and weakly P791. Finally, in all optimized binding modes, AC3573 makes contact with C740. This raises the possibility that AC3573 derivatives could be developed to bind irreversibly to HER3. Importantly, cysteine in this position is unique in the EGFR family, and this could be exploited to improve both specificity and durability of inhibitory compound effects. This residue has been previously exploited to develop irreversible nucleotide pocket binding compounds by the Gray laboratory (Xie et al., 2014).

Some variation is seen in the rotation of the indole group across the different HER3 structures. In the HER3 head-to-head proposed inhibitory dimer (PDB ID 3KEX) (Jura, Shan, et al., 2009) and the HER3-EGFR active dimer conformation (PDB ID 4RIW), the indole is aligned with the ribose ring position in the HER3:AMP-PNP complexes (Fig. 7A/F and C/H). In PDB ID 3LMG and PDB ID 6OP9, the indole group is rotated to bind deeper into the ATP pocket (Fig. 7B/G and D) and it occupies a position similar to the methoxyanilino group in the HER3-bosutinib complex. In these two structures, AC3573 forms additional hydrophobic contacts with the apolar portions of K742 and T787. The similarity of the binding affinity values (Table 2) and of the HER3 structures in this region suggests that both indole orientations are possible. Subtle differences in the accessibility of the deep cavity adjacent to the ATP binding site might slightly favor one orientation over the other. Calculations involving Molecular Dynamics (MD) simulations and more accurate estimates of the binding affinity will be needed to further investigate this point.

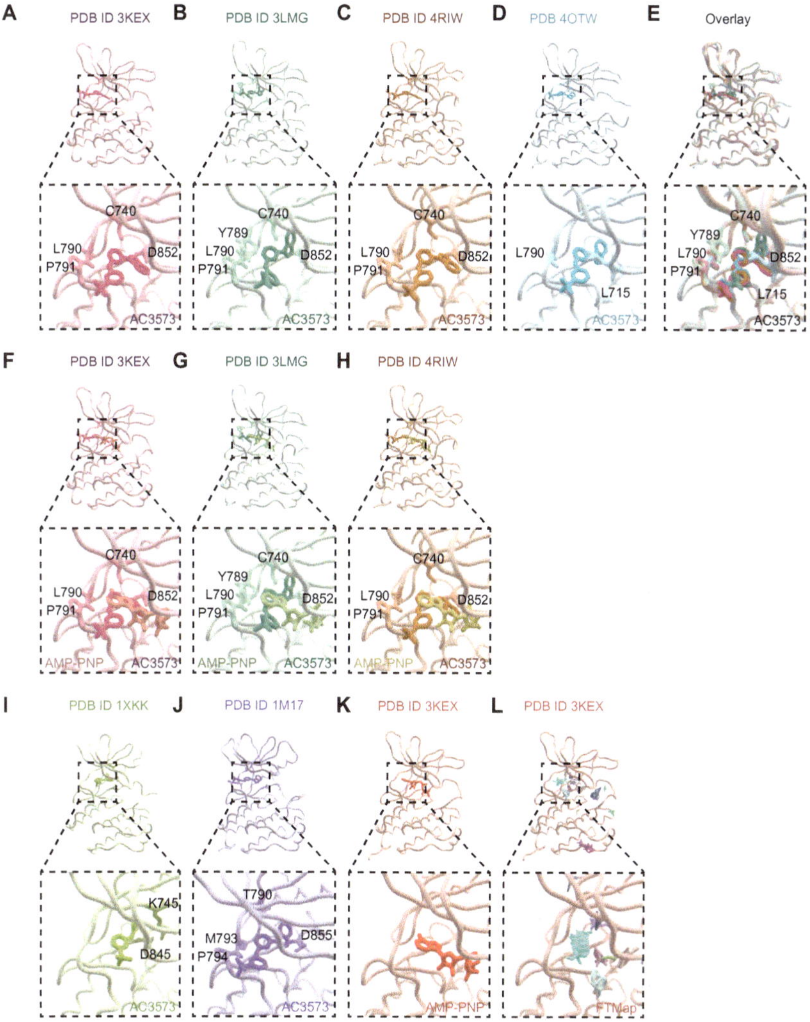

Fig. 7 *In silico* docking of AC3573 in HER3 kinase domain structures. (A) Docking of AC3573 into the HER3 kinase domain structure from PDB ID 3KEX with notable interacting residues highlighted. (B) AC3573 docked into the HER3 kinase domain structure from PDB ID 3LMG (Shi et al., 2010). (C) Docking of AC3573 into the HER3 kinase domain from the EGFR:HER3 co-crystal structure PDB ID 4RIW (Littlefield et al., 2014). (D) AC3573 docked into the bosutinib-bound crystal structure of HER3, PDB ID 6OP9 (Campbell et al., 2022). (E) Overlay of all binding orientations of (A–D). (F) Comparison of crystal structure binding of AMP-PNP to docked AC3573 in PDB ID 3KEX (Jura, Shan, et al., 2009). (G) Comparison of AC3573 docking orientation to the binding of AMP-PNP in PDB ID 3LMG (Shi et al., 2010). (H) Comparison of AC3573 docking to the AMP-PNP binding orientation in PDB ID 4RIW (Littlefield et al., 2014). (I) Docking of AC3573 into the active conformation, erlotinib-bound structure of EGFR, PDB ID 1M17 (Stamos et al., 2002). (J) Docking of AC3573 into the inactive conformation, lapatinib-bound structure of EGFR, PDB ID 1XKK (Wood et al., 2004). (K) Crystallographic binding orientation of AMP-PNP in the HER3 kinase domain pseudoactive site from PDB ID 3KEX (Jura, Shan, et al., 2009). (L) FTMap analysis of potential small molecule binding pockets in the HER3 kinase domain from PDB ID 3KEX. All docking performed using Glide.

Docking of AC3573 to EGFR structures produced binding affinity values comparable to those obtained for HER3 (Table 2), suggesting that this compound could have some level of activity also toward this HER family member. Interestingly, in both EGFR structures the binding mode of AC3573 (Fig. 7I and J) seems to reproduce at least partially that of the larger inhibitors lapatinib (PDB ID 1XKK) (Wood et al., 2004) and erlotinib (PDB ID 1M17) (Stamos et al., 2002). In PDB ID 1XKK, AC3573 binds as deep as lapatinib, with the indole ring occupying the same position as the fluorophenyl group in lapatinib. In PDB ID 1M17, the indole and aminopyrimidine groups are perfectly aligned with the ethynylphenyl and quinazoline groups of erlotinib, respectively.

In order to further characterize the potential binding space inside the HER3 kinase domain, we performed an FTMap analysis (Brenke et al., 2009; de Souza Neto et al., 2020; Kozakov et al., 2015). FTMap uses a fast Fourier transform approach to find potential binding sites for 16 small organic molecules. It then clusters these binding sites based on where multiple of these small organic molecules could be docked, creating so-called consensus sites. Comparing the consensus sites from the FTMap analysis to the ATP-analogue binding of the crystal structure used to perform FTMap analysis (Fig. 7K, L) shows the HER3 nucleotide binding pocket to be the main region of the kinase domain where potential consensus sites would occur. The two exceptions are consensus sites found in the active dimer interaction interface at the bottom of the C-lobe, as well as around the highly flexible HER3 activation loop. Whether these potential small-molecule binding sites can be exploited to interfere with HER2: HER3 dimerization or function remains to be explored.

While most of the consensus sites found in the nucleotide binding pocket occupy similar space to the binding site of AMP-PNP in the original HER3 kinase domain crystal structure, there are two clusters of note in the context of our AC3573 docking described above. First, the main consensus site (Fig. 7L, teal) appears to follow the HER3 hinge region in a way not dissimilar to the docking orientations of the aminopyrimidine and sulfonamide moieties of AC3573 in Fig. 7A–E. Second, an additional consensus site (Fig. 7L, lilac) found behind this main clustering region approximates the rotated state of the indole group seen in the docking of AC3573 in PDB ID 3LMG (Shi et al., 2010) and PDB ID 6OP9 (Campbell et al., 2022) (Fig. 7B, D).

The FTMap analysis independently indicates the potential of the hinge region to provide a binding opportunity for HER3-directed small molecules, aligning with our Glide-based docking analysis as well as our previously reported hydrogen-deuterium exchange mass spectrometry results (Colomba et al., 2020).

Finally, a recent structure-function analysis of the HER2:HER3 complex (Campbell et al., 2022) has revealed conformational coupling of the pseudoactive site with a distinct hydrophobic AP-2 pocket in the N-lobe, analogous to the AP-2 pocket described for EGFR (Jura, Endres, et al., 2009). Unbiased fragment binding MD simulations imply that binding of small molecules to this pocket could have the potential to allosterically disrupt the HER3 pseudoactive site and inhibit function.

Together, these studies highlight the existence of alternative small molecule binding sites with the potential to disrupt HER3 function.

5.3 AC3573 docking to activating and inhibitory HER3 dimer models

Finally, we docked AC3573 into the model HER3 structures we generated in our previous study examining the allosteric activation of HER2:HER3 heterodimers by ATP competitive inhibitors of both HER2 and HER3 (Claus et al., 2018); only the HER3 monomer was included in the docking calculations. In all the models, AC3573 adopts the same binding mode (Fig. 8B–D), similar to the mode observed for AC3573 docking to all X-ray structures in Fig. 7. Moreover, in all cases the indole orientation is the same as that found for the Shi et al. structure (PDB ID 3LMG), with the indole group sitting deep in the binding site. A larger number of unfavorable clashes were found between AC3573 and the HER3 model from the active dimer compared to the inactive ones (in particular for residues C740, K742 and D852), which explains the weaker binding affinity (Table 2) and indicates a less favorable match between AC3573 and the shape of the binding site in this structure. Further investigations will be needed to confirm if this difference in the HER3 models is correlated to the different HER2:HER3 interface. In particular, molecular dynamics (MD) simulations performed on the complete dimer in the presence of the inhibitor would allow induced fit effects to be taken into account, together with a possible allosteric coupling between the HER2:HER3 interface and the inhibitor binding site.

All models again predict extensive interaction with hinge residues. Scoring of interactions suggest stronger interaction with the inactive structure. The extensive hinge interactions and high score with inactive structures make it tempting to speculate that AC3573 may act by stabilizing

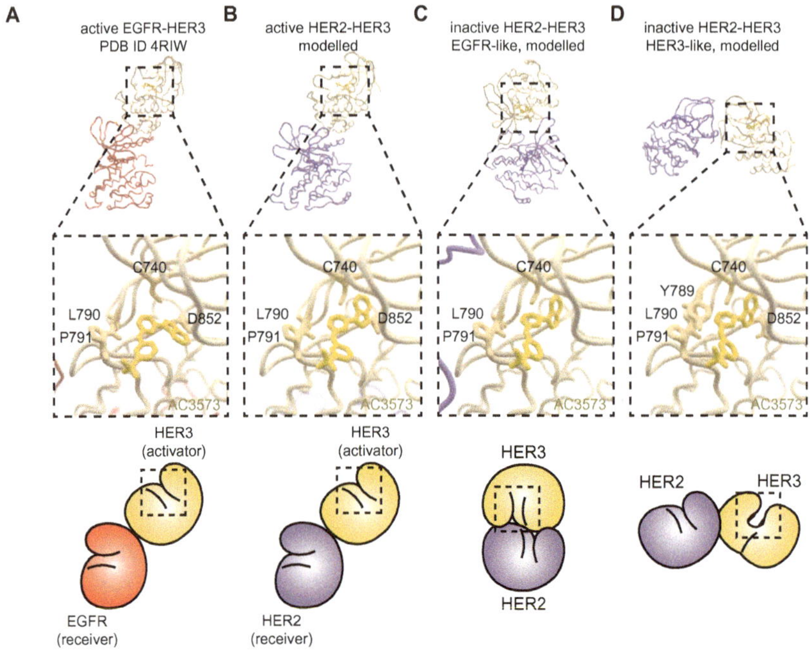

Fig. 8 *In silico* docking of AC3573 in modeled HER2:HER3 dimerization interfaces. (A) Docking of AC3573 into the co-crystal structure of the active, asymmetric EGFR: HER3 kinase heterodimer from PDB ID 4RIW (Littlefield et al., 2014). (B) AC3573 docked into the modeled asymmetric HER2:HER3 active dimer orientation from Claus et al. (2018). (C) Docking of AC3573 into the symmetric EGFR homodimer-like orientation of the HER2:HER3 heterodimer as modeled in Claus et al. (2018). (D) Docking of AC3573 into the symmetric HER3 homodimer-like orientation of the HER2:HER3 heterodimer as modeled in Claus et al. (2018). All docking performed using Glide.

pre-existing inactive HER3 dimers or oligomers; interaction with the hinge might be expected to limit flexibility. This may in turn reduce the ability of growth factor binding to induce the conformational rearrangements necessary to drive active dimers with HER2.

6. Summary and perspectives

Here we have described methods for identifying and validating small molecule inhibitors of HER3. While TSA is a powerful method for identification of HER3 binders, identification of compounds which act to disrupt HER3 mediated signaling is challenging. The majority of nucleotide pocket binders appear to stabilize HER3 in a signaling competent conformation (Claus et al., 2018; Colomba et al., 2020; Macdonald-Obermann et al., 2013). As HER3 has evolved through gene-duplication of its active and likely homo-dimer partner ancestor, it may be expected to share significant topological similarities between its pseudoactive site and the active site of its cognate binding partner, HER2. Nevertheless, as demonstrated for the Janus Kinase (JAK) family, pseudokinases can develop inhibitory allosteric roles where the pseudokinase domain can suppress activation of its cognate kinase domain binding partner (Babon, Lucet, Murphy, Nicola, & Varghese, 2014; Brooks et al., 2014; Mace & Murphy, 2021; Varghese et al., 2014). Trapping such inhibitory conformers with small molecules may offer a novel means for targeting such pseudokinase:kinase pairs.

To practically address the *inhibitor as activator* problem, a parallel two-phase screening approach may be best to identify allosteric HER3 inhibitors: (i) HER3 binding (e.g., TSA or NanoBRET) and (ii) HER3: HER2 dimer disruption (e.g., FRET or Split-luciferase). While rare, our screening approach was capable of identifying the proof-of-concept inhibitor AC3573 (Colomba et al., 2020). This compound binds to HER3, but not HER2, and effectively disrupts HER2:HER3 dimers and inhibits downstream signaling. It should be noted that this two-phase screening approach may not identify inhibitors which stabilize HER2:HER3 inhibitory dimer conformers, which may theoretically exist. Additional cellular functional screens would be necessary to discover such compounds. Many biophysical, biochemical, imaging and modeling approaches can be taken to characterize the mode and impact of AC3573 binding. Existing data indicate that this compound binds to the nucleotide pocket but does not induce higher order oligomers or receptor degradation (Colomba et al., 2020). In this chapter, we also used *in silico* docking to examine models of AC3573 binding to the HER3 kinase domain. By scoring binding of AC3573 to HER2:HER3 dimer models we generated in a previous study (Claus et al., 2018), we provide preliminary evidence that AC3573 may preferentially bind to HER3 in inactive dimer conformations. Stabilization of pre-existing inhibitory HER3 dimers or oligomers may thus represent the inhibitory mode of action for AC3573. Exploring binding modes further with molecular dynamics and X-ray crystallography, alongside functional studies with HER3 kinase domain mutants will be key to accurately define the A3573 inhibitory mechanism. We conclude that development of clinically useful small molecule inhibitors of HER3 remains a tantalizing prospect, but detailed studies are required to understand mechanism of action to guide rational approaches to drug development.

Acknowledgments

A.C. is supported by a HEFCE award to Barts Cancer Institute and Cancer Research UK Centre Grants to Barts Cancer Institute (C355/A25137) and the City of London Centre (C7893/A26233).

References

Alimandi, M., Romano, A., Curia, M. C., Muraro, R., Fedi, P., Aaronson, S. A., et al. (1995). Cooperative signaling of ErbB3 and ErbB2 in neoplastic transformation and human mammary carcinomas. *Oncogene*, *10*(9), 1813–1821. Retrieved from https://www.ncbi.nlm.nih.gov/pubmed/7538656.

Amin, D. N., Sergina, N., Ahuja, D., McMahon, M., Blair, J. A., Wang, D., et al. (2010). Resiliency and vulnerability in the HER2-HER3 tumorigenic driver. *Science Translational Medicine*, *2*(16), 16ra17. https://doi.org/10.1126/scitranslmed.3000389.

Babon, J. J., Lucet, I. S., Murphy, J. M., Nicola, N. A., & Varghese, L. N. (2014). The molecular regulation of Janus kinase (JAK) activation. *The Biochemical Journal*, *462*(1), 1–13. https://doi.org/10.1042/BJ20140712.

Baell, J. B., & Holloway, G. A. (2010). New substructure filters for removal of pan assay interference compounds (PAINS) from screening libraries and for their exclusion in bioassays. *Journal of Medicinal Chemistry*, *53*(7), 2719–2740. https://doi.org/10.1021/jm901137j.

Baell, J. B., & Nissink, J. W. M. (2018). Seven year itch: Pan-assay interference compounds (PAINS) in 2017-utility and limitations. *ACS Chemical Biology*, *13*(1), 36–44. https://doi.org/10.1021/acschembio.7b00903.

Barber, P. R., Ameer-Beg, S. M., Gilbey, J., Carlin, L. M., Keppler, M., Ng, T. C., et al. (2009). Multiphoton time-domain fluorescence lifetime imaging microscopy: Practical application to protein-protein interactions using global analysis. *Journal of the Royal Society Interface*, *6*, S93–S105. https://doi.org/10.1098/rsif.2008.0451.focus.

Barber, P. R., Weitsman, G., Lawler, K., Barrett, J. E., Rowley, M., Rodriguez-Justo, M., et al. (2020). HER2-HER3 heterodimer quantification by FRET-FLIM and patient subclass analysis of the COIN colorectal trial. *Journal of the National Cancer Institute*, *112*(9), 944–954. https://doi.org/10.1093/jnci/djz231.

Bodey, B., Bodey, B., Jr., Groger, A. M., Luck, J. V., Siegel, S. E., Taylor, C. R., et al. (1997). Clinical and prognostic significance of the expression of the c-erbB-2 and c-erbB-3 oncoproteins in primary and metastatic malignant melanomas and breast carcinomas. *Anticancer Research*, *17*(2B), 1319–1330. Retrieved from https://www.ncbi.nlm.nih.gov/pubmed/9137492.

Bonner, J. A., Harari, P. M., Giralt, J., Cohen, R. B., Jones, C. U., Sur, R. K., et al. (2010). Radiotherapy plus cetuximab for locoregionally advanced head and neck cancer: 5-Year survival data from a phase 3 randomised trial, and relation between cetuximab-induced rash and survival. *The Lancet Oncology*, *11*(1), 21–28. https://doi.org/10.1016/S1470-2045(09)70311-0.

Boudeau, J., Miranda-Saavedra, D., Barton, G. J., & Alessi, D. R. (2006). Emerging roles of pseudokinases. *Trends in Cell Biology*, *16*(9), 443–452. https://doi.org/10.1016/j.tcb.2006.07.003.

Brenke, R., Kozakov, D., Chuang, G. Y., Beglov, D., Hall, D., Landon, M. R., et al. (2009). Fragment-based identification of druggable 'hot spots' of proteins using Fourier domain correlation techniques. *Bioinformatics*, *25*(5), 621–627. https://doi.org/10.1093/bioinformatics/btp036.

Brooks, A. J., Dai, W., O'Mara, M. L., Abankwa, D., Chhabra, Y., Pelekanos, R. A., et al. (2014). Mechanism of activation of protein kinase JAK2 by the growth hormone receptor. *Science*, *344*(6185), 1249783. https://doi.org/10.1126/science.1249783.

Byrne, D. P., Foulkes, D. M., & Eyers, P. A. (2017). Pseudokinases: Update on their functions and evaluation as new drug targets. *Future Medicinal Chemistry*, *9*(2), 245–265. https://doi.org/10.4155/fmc-2016-0207.

Cameron, A. J. (2011). Occupational hazards: Allosteric regulation of protein kinases through the nucleotide-binding pocket. *Biochemical Society Transactions*, *39*(2), 472–476. https://doi.org/10.1042/BST0390472.

Cameron, A. J., Escribano, C., Saurin, A. T., Kostelecky, B., & Parker, P. J. (2009). PKC maturation is promoted by nucleotide pocket occupation independently of intrinsic kinase activity. *Nature Structural & Molecular Biology*, *16*(6), 624–630. https://doi.org/10.1038/nsmb.1606.

Campbell, M. R., Ruiz-Saenz, A., Peterson, E., Agnew, C., Ayaz, P., Garfinkle, S., et al. (2022). Targetable HER3 functions driving tumorigenic signaling in HER2-amplified cancers. *Cell Reports*, *38*(5), 110291. https://doi.org/10.1016/j.celrep.2021.110291.

Clarke, D. T., & Martin-Fernandez, M. L. (2019). A brief history of single-particle tracking of the epidermal growth factor receptor. *Methods Protoc*, *2*(1), 12. https://doi.org/10.3390/mps2010012.

Claus, J., Cameron, A. J., & Parker, P. J. (2013). Pseudokinase drug intervention: A potentially poisoned chalice. *Biochemical Society Transactions*, *41*(4), 1083–1088. https://doi.org/10.1042/BST20130078.

Claus, J., Patel, G., Autore, F., Colomba, A., Weitsman, G., Soliman, T. N., et al. (2018). Inhibitor-induced HER2-HER3 heterodimerisation promotes proliferation through a novel dimer interface. *eLife*, 7, e32271. https://doi.org/10.7554/eLife.32271.

Claus, J., Patel, G., Ng, T., & Parker, P. J. (2014). A role for the pseudokinase HER3 in the acquired resistance against EGFR- and HER2-directed targeted therapy. *Biochemical Society Transactions*, *42*(4), 831–836. https://doi.org/10.1042/BST20140043.

Clayton, A. H., Tavarnesi, M. L., & Johns, T. G. (2007). Unligated epidermal growth factor receptor forms higher order oligomers within microclusters on A431 cells that are sensitive to tyrosine kinase inhibitor binding. *Biochemistry*, *46*(15), 4589–4597. https://doi.org/10.1021/bi700002b.

Clayton, A. H., Walker, F., Orchard, S. G., Henderson, C., Fuchs, D., Rothacker, J., et al. (2005). Ligand-induced dimer-tetramer transition during the activation of the cell surface epidermal growth factor receptor-a multidimensional microscopy analysis. *The Journal of Biological Chemistry*, *280*(34), 30392–30399. https://doi.org/10.1074/jbc.M504770200.

Colomba, A., Fitzek, M., George, R., Weitsman, G., Roberts, S., Zanetti-Domingues, L., et al. (2020). A small molecule inhibitor of HER3: A proof-of-concept study. *The Biochemical Journal*, *477*(17), 3329–3347. https://doi.org/10.1042/BCJ20200496.

Cortes, J., Saura, C., Bellet, M., Munoz-Couselo, E., Ramirez-Merino, N., Calvo, V., et al. (2011). HER2 and hormone receptor-positive breast cancer- -blocking the right target. *Nature Reviews. Clinical Oncology*, *8*(5), 307–311. https://doi.org/10.1038/nrclinonc.2010.185.

Cui, J., Zhu, Q., Zhang, H., Cianfrocco, M. A., Leschziner, A. E., Dixon, J. E., et al. (2017). Structure of Fam20A reveals a pseudokinase featuring a unique disulfide pattern and inverted ATP-binding. *eLife*, *6*, e23990. https://doi.org/10.7554/eLife.23990.

Dahlin, J. L., & Walters, M. A. (2016). How to triage PAINS-full research. *Assay and Drug Development Technologies*, *14*(3), 168–174. https://doi.org/10.1089/adt.2015.674.

Dar, A. C., & Shokat, K. M. (2011). The evolution of protein kinase inhibitors from antagonists to agonists of cellular signaling. *Annual Review of Biochemistry*, *80*, 769–795. https://doi.org/10.1146/annurev-biochem-090308-173656.

de Souza Neto, L. R., Moreira-Filho, J. T., Neves, B. J., Maidana, R., Guimaraes, A. C. R., Furnham, N., et al. (2020). In silico strategies to support fragment-to-Lead optimization in drug discovery. *Frontiers in Chemistry*, *8*, 93. https://doi.org/10.3389/fchem.2020.00093.

Diwanji, D., Trenker, R., Thaker, T. M., Wang, F., Agard, D. A., Verba, K. A., et al. (2021). Structures of the HER2-HER3-NRG1beta complex reveal a dynamic dimer interface. *Nature, 600*(7888), 339–343. https://doi.org/10.1038/s41586-021-04084-z.

Dou, X. Q., Wang, H., Zhang, J., Wang, F., Xu, G. L., Xu, C. C., et al. (2018). Aptamer-drug conjugate: Targeted delivery of doxorubicin in a HER3 aptamer-functionalized liposomal delivery system reduces cardiotoxicity. *International Journal of Nanomedicine, 13*, 763–776. https://doi.org/10.2147/IJN.S149887.

Engelman, J. A., Zejnullahu, K., Mitsudomi, T., Song, Y., Hyland, C., Park, J. O., et al. (2007). MET amplification leads to gefitinib resistance in lung cancer by activating ERBB3 signaling. *Science, 316*(5827), 1039–1043. https://doi.org/10.1126/science.1141478.

Fedorov, O., Niesen, F. H., & Knapp, S. (2012). Kinase inhibitor selectivity profiling using differential scanning fluorimetry. *Methods in Molecular Biology, 795*, 109–118. https://doi.org/10.1007/978-1-61779-337-0_7.

Feng, Y., De Franceschi, G., Kahraman, A., Soste, M., Melnik, A., Boersema, P. J., et al. (2014). Global analysis of protein structural changes in complex proteomes. *Nature Biotechnology, 32*(10), 1036–1044. https://doi.org/10.1038/nbt.2999.

Friesner, R. A., Banks, J. L., Murphy, R. B., Halgren, T. A., Klicic, J. J., Mainz, D. T., et al. (2004). Glide: A new approach for rapid, accurate docking and scoring. 1. Method and assessment of docking accuracy. *Journal of Medicinal Chemistry, 47*(7), 1739–1749. https://doi.org/10.1021/jm0306430.

Friman, T. (2020). Mass spectrometry-based cellular thermal shift assay (CETSA(R)) for target deconvolution in phenotypic drug discovery. *Bioorganic & Medicinal Chemistry, 28*(1). https://doi.org/10.1016/j.bmc.2019.115174, 115174.

Gala, K., & Chandarlapaty, S. (2014). Molecular pathways: HER3 targeted therapy. *Clinical Cancer Research, 20*(6), 1410–1416. https://doi.org/10.1158/1078-0432.CCR-13-1549.

Gandullo-Sanchez, L., Capone, E., Ocana, A., Iacobelli, S., Sala, G., & Pandiella, A. (2020). HER3 targeting with an antibody-drug conjugate bypasses resistance to anti-HER2 therapies. *EMBO Molecular Medicine, 12*(5), e11498. https://doi.org/10.15252/emmm.201911498.

Gschwind, A., Fischer, O. M., & Ullrich, A. (2004). The discovery of receptor tyrosine kinases: Targets for cancer therapy. *Nature Reviews. Cancer, 4*(5), 361–370. https://doi.org/10.1038/nrc1360.

Haikala, H. M., & Janne, P. A. (2021). Thirty years of HER3: From basic biology to therapeutic interventions. *Clinical Cancer Research, 27*(13), 3528–3539. https://doi.org/10.1158/1078-0432.CCR-20-4465.

Hanker, A. B., Brown, B. P., Meiler, J., Marin, A., Jayanthan, H. S., Ye, D., et al. (2021). Co-occurring gain-of-function mutations in HER2 and HER3 modulate HER2/HER3 activation, oncogenesis, and HER2 inhibitor sensitivity. *Cancer Cell, 39*(8), 1099–1114.e1098. https://doi.org/10.1016/j.ccell.2021.06.001.

Hatzivassiliou, G., Song, K., Yen, I., Brandhuber, B. J., Anderson, D. J., Alvarado, R., et al. (2010). RAF inhibitors prime wild-type RAF to activate the MAPK pathway and enhance growth. *Nature, 464*(7287), 431–435. https://doi.org/10.1038/nature08833.

Holbro, T., Beerli, R. R., Maurer, F., Koziczak, M., Barbas, C. F., 3rd, & Hynes, N. E. (2003). The ErbB2/ErbB3 heterodimer functions as an oncogenic unit: ErbB2 requires ErbB3 to drive breast tumor cell proliferation. *Proceedings of the National Academy of Sciences of the United States of America, 100*(15), 8933–8938. https://doi.org/10.1073/pnas.1537685100.

Huang, Y., Bharill, S., Karandur, D., Peterson, S. M., Marita, M., Shi, X., et al. (2016). Molecular basis for multimerization in the activation of the epidermal growth factor receptor. *eLife, 5*, e14107. https://doi.org/10.7554/eLife.14107.

Huynh, K., & Partch, C. L. (2015). Analysis of protein stability and ligand interactions by thermal shift assay. *Curr Protoc Protein Sci, 79*. https://doi.org/10.1002/0471140864.ps2809s79. 28 29 21–28 29 14.

Jacob, W., James, I., Hasmann, M., & Weisser, M. (2018). Clinical development of HER3-targeting monoclonal antibodies: Perils and progress. *Cancer Treatment Reviews, 68*, 111–123. https://doi.org/10.1016/j.ctrv.2018.06.011.

Jacobson, M. P., Pincus, D. L., Rapp, C. S., Day, T. J., Honig, B., Shaw, D. E., et al. (2004). A hierarchical approach to all-atom protein loop prediction. *Proteins, 55*(2), 351–367. https://doi.org/10.1002/prot.10613.

Jaiswal, B. S., Kljavin, N. M., Stawiski, E. W., Chan, E., Parikh, C., Durinck, S., et al. (2013). Oncogenic ERBB3 mutations in human cancers. *Cancer Cell, 23*(5), 603–617. https://doi.org/10.1016/j.ccr.2013.04.012.

Jasial, S., Hu, Y., & Bajorath, J. (2017). How frequently are pan-assay interference compounds active? Large-scale analysis of screening data reveals diverse activity profiles, low global hit frequency, and many consistently inactive compounds. *Journal of Medicinal Chemistry, 60*(9), 3879–3886. https://doi.org/10.1021/acs.jmedchem.7b00154.

Jerabek-Willemsen, M., Wienken, C. J., Braun, D., Baaske, P., & Duhr, S. (2011). Molecular interaction studies using microscale thermophoresis. *Assay and Drug Development Technologies, 9*(4), 342–353. https://doi.org/10.1089/adt.2011.0380.

Jonker, D. J., O'Callaghan, C. J., Karapetis, C. S., Zalcberg, J. R., Tu, D., Au, H. J., et al. (2007). Cetuximab for the treatment of colorectal cancer. *The New England Journal of Medicine, 357*(20), 2040–2048. https://doi.org/10.1056/NEJMoa071834.

Jura, N., Endres, N. F., Engel, K., Deindl, S., Das, R., Lamers, M. H., et al. (2009). Mechanism for activation of the EGF receptor catalytic domain by the juxtamembrane segment. *Cell, 137*(7), 1293–1307. https://doi.org/10.1016/j.cell.2009.04.025.

Jura, N., Shan, Y., Cao, X., Shaw, D. E., & Kuriyan, J. (2009). Structural analysis of the catalytically inactive kinase domain of the human EGF receptor 3. *Proceedings of the National Academy of Sciences of the United States of America, 106*(51), 21608–21613. https://doi.org/10.1073/pnas.0912101106.

Kozakov, D., Grove, L. E., Hall, D. R., Bohnuud, T., Mottarella, S. E., Luo, L., et al. (2015). The FTMap family of web servers for determining and characterizing ligand-binding hot spots of proteins. *Nature Protocols, 10*(5), 733–755. https://doi.org/10.1038/nprot.2015.043.

Kumai, T., Ohkuri, T., Nagato, T., Matsuda, Y., Oikawa, K., Aoki, N., et al. (2015). Targeting HER-3 to elicit antitumor helper T cells against head and neck squamous cell carcinoma. *Scientific Reports, 5*, 16280. https://doi.org/10.1038/srep16280.

Kung, J. E., & Jura, N. (2019). Prospects for pharmacological targeting of pseudokinases. *Nature Reviews. Drug Discovery, 18*(7), 501–526. https://doi.org/10.1038/s41573-019-0018-3.

Larcombe-Young, D., Papa, S., & Maher, J. (2020). PanErbB-targeted CAR T-cell immunotherapy of head and neck cancer. *Expert Opinion on Biological Therapy, 20*(9), 965–970. https://doi.org/10.1080/14712598.2020.1786531.

Lemmon, M. A., & Schlessinger, J. (2010). Cell signaling by receptor tyrosine kinases. *Cell, 141*(7), 1117–1134. https://doi.org/10.1016/j.cell.2010.06.011.

Levinson, N. M., & Boxer, S. G. (2014). A conserved water-mediated hydrogen bond network defines bosutinib's kinase selectivity. *Nature Chemical Biology, 10*(2), 127–132. https://doi.org/10.1038/nchembio.1404.

Li, Y., Macdonald-Obermann, J., Westfall, C., Piwnica-Worms, D., & Pike, L. J. (2012). Quantitation of the effect of ErbB2 on epidermal growth factor receptor binding and dimerization. *The Journal of Biological Chemistry, 287*(37), 31116–31125. https://doi.org/10.1074/jbc.M112.373647.

Littlefield, P., Liu, L., Mysore, V., Shan, Y., Shaw, D. E., & Jura, N. (2014). Structural analysis of the EGFR/HER3 heterodimer reveals the molecular basis for activating HER3 mutations. *Science Signaling*, 7(354), ra114. https://doi.org/10.1126/scisignal.2005786.

Lochhead, P. A., Tucker, J. A., Tatum, N. J., Wang, J., Oxley, D., Kidger, A. M., et al. (2020). Paradoxical activation of the protein kinase-transcription factor ERK5 by ERK5 kinase inhibitors. *Nature Communications*, *11*(1), 1383. https://doi.org/10.1038/s41467-020-15031-3.

Lovekin, C., Ellis, I. O., Locker, A., Robertson, J. F., Bell, J., Nicholson, R., et al. (1991). c-erbB-2 oncoprotein expression in primary and advanced breast cancer. *British Journal of Cancer*, *63*(3), 439–443. https://doi.org/10.1038/bjc.1991.101.

Lu, C., Wu, C., Ghoreishi, D., Chen, W., Wang, L., Damm, W., et al. (2021). OPLS4: Improving force field accuracy on challenging regimes of chemical space. *Journal of Chemical Theory and Computation*, *17*(7), 4291–4300. https://doi.org/10.1021/acs.jctc.1c00302.

Macdonald-Obermann, J. L., Adak, S., Landgraf, R., Piwnica-Worms, D., & Pike, L. J. (2013). Dynamic analysis of the epidermal growth factor (EGF) receptor-ErbB2-ErbB3 protein network by luciferase fragment complementation imaging. *The Journal of Biological Chemistry*, *288*(42), 30773–30784. https://doi.org/10.1074/jbc.M113.489534.

Macdonald-Obermann, J. L., Piwnica-Worms, D., & Pike, L. J. (2012). Mechanics of EGF receptor/ErbB2 kinase activation revealed by luciferase fragment complementation imaging. *Proceedings of the National Academy of Sciences of the United States of America*, *109*(1), 137–142. https://doi.org/10.1073/pnas.1111316109.

Mace, P. D., & Murphy, J. M. (2021). There's more to death than life: Noncatalytic functions in kinase and pseudokinase signaling. *The Journal of Biological Chemistry*, *296*, 100705. https://doi.org/10.1016/j.jbc.2021.100705.

Martin-Fernandez, M., Clarke, D. T., Tobin, M. J., Jones, S. V., & Jones, G. R. (2002). Preformed oligomeric epidermal growth factor receptors undergo an ectodomain structure change during signaling. *Biophysical Journal*, *82*(5), 2415–2427. https://doi.org/10.1016/S0006-3495(02)75585-9.

Mendes, D., Alves, C., Afonso, N., Cardoso, F., Passos-Coelho, J. L., Costa, L., et al. (2015). The benefit of HER2-targeted therapies on overall survival of patients with metastatic HER2-positive breast cancer- -a systematic review. *Breast Cancer Research*, *17*, 140. https://doi.org/10.1186/s13058-015-0648-2.

Mishra, R., Patel, H., Alanazi, S., Yuan, L., & Garrett, J. T. (2018). HER3 signaling and targeted therapy in cancer. *Oncology Reviews*, *12*(1), 355. https://doi.org/10.4081/oncol.2018.355.

Muller, W. J., Sinn, E., Pattengale, P. K., Wallace, R., & Leder, P. (1988). Single-step induction of mammary adenocarcinoma in transgenic mice bearing the activated c-neu oncogene. *Cell*, *54*(1), 105–115. https://doi.org/10.1016/0092-8674(88)90184-5.

Murphy, J. M., Lucet, I. S., Hildebrand, J. M., Tanzer, M. C., Young, S. N., Sharma, P., et al. (2014). Insights into the evolution of divergent nucleotide-binding mechanisms among pseudokinases revealed by crystal structures of human and mouse MLKL. *The Biochemical Journal*, *457*(3), 369–377. https://doi.org/10.1042/BJ20131270.

Murphy, J. M., Zhang, Q., Young, S. N., Reese, M. L., Bailey, F. P., Eyers, P. A., et al. (2014). A robust methodology to subclassify pseudokinases based on their nucleotide-binding properties. *The Biochemical Journal*, *457*(2), 323–334. https://doi.org/10.1042/BJ20131174.

Niesen, F. H., Berglund, H., & Vedadi, M. (2007). The use of differential scanning fluorimetry to detect ligand interactions that promote protein stability. *Nature Protocols*, *2*(9), 2212–2221. https://doi.org/10.1038/nprot.2007.321.

Oh, D. Y., & Bang, Y. J. (2020). HER2-targeted therapies - a role beyond breast cancer. *Nature Reviews. Clinical Oncology*, *17*(1), 33–48. https://doi.org/10.1038/s41571-019-0268-3.

Okuzumi, T., Fiedler, D., Zhang, C., Gray, D. C., Aizenstein, B., Hoffman, R., et al. (2009). Inhibitor hijacking of Akt activation. *Nature Chemical Biology*, *5*(7), 484–493. https://doi.org/10.1038/nchembio.183.

Olsson, M. H., Sondergaard, C. R., Rostkowski, M., & Jensen, J. H. (2011). PROPKA3: Consistent treatment of internal and surface residues in empirical pKa predictions. *Journal of Chemical Theory and Computation*, 7(2), 525–537. https://doi.org/10.1021/ct100578z.

Owens, M. A., Horten, B. C., & Da Silva, M. M. (2004). HER2 amplification ratios by fluorescence in situ hybridization and correlation with immunohistochemistry in a cohort of 6556 breast cancer tissues. *Clinical Breast Cancer*, *5*(1), 63–69. https://doi.org/10.3816/cbc.2004.n.011.

Papa, F. R., Zhang, C., Shokat, K., & Walter, P. (2003). Bypassing a kinase activity with an ATP-competitive drug. *Science*, *302*(5650), 1533–1537. https://doi.org/10.1126/science.1090031.

Parsons, M., & Ng, T. (2002). Intracellular coupling of adhesion receptors: Molecular proximity measurements. *Methods in Cell Biology*, *69*, 261–278. https://doi.org/10.1016/s0091-679x(02)69017-x.

Peter, M., Ameer-Beg, S. M., Hughes, M. K., Keppler, M. D., Prag, S., Marsh, M., et al. (2005). Multiphoton-FLIM quantification of the EGFP-mRFP1 FRET pair for localization of membrane receptor-kinase interactions. *Biophysical Journal*, *88*(2), 1224–1237. https://doi.org/10.1529/biophysj.104.050153.

Poulikakos, P. I., Zhang, C., Bollag, G., Shokat, K. M., & Rosen, N. (2010). RAF inhibitors transactivate RAF dimers and ERK signalling in cells with wild-type BRAF. *Nature*, *464*(7287), 427–430. https://doi.org/10.1038/nature08902.

Qiu, C., Tarrant, M. K., Choi, S. H., Sathyamurthy, A., Bose, R., Banjade, S., et al. (2008). Mechanism of activation and inhibition of the HER4/ErbB4 kinase. *Structure*, *16*(3), 460–467. https://doi.org/10.1016/j.str.2007.12.016.

Roberts, S. K., Hirsch, M., McStea, A., Zanetti-Domingues, L. C., Clarke, D. T., Claus, J., et al. (2018). Cluster analysis of endogenous HER2 and HER3 receptors in SKBR3 cells. *Bio-Protocol*, *8*(23), e3096. https://doi.org/10.21769/BioProtoc.3096.

Rolfe, D. J., McLachlan, C. I., Hirsch, M., Needham, S. R., Tynan, C. J., Webb, S. E., et al. (2011). Automated multidimensional single molecule fluorescence microscopy feature detection and tracking. *European Biophysics Journal*, *40*(10), 1167–1186. https://doi.org/10.1007/s00249-011-0747-7.

Roskoski, R., Jr. (2019). Small molecule inhibitors targeting the EGFR/ErbB family of protein-tyrosine kinases in human cancers. *Pharmacological Research*, *139*, 395–411. https://doi.org/10.1016/j.phrs.2018.11.014.

Saraon, P., Pathmanathan, S., Snider, J., Lyakisheva, A., Wong, V., & Stagljar, I. (2021). Receptor tyrosine kinases and cancer: Oncogenic mechanisms and therapeutic approaches. *Oncogene*, *40*(24), 4079–4093. https://doi.org/10.1038/s41388-021-01841-2.

Scaltriti, M., & Baselga, J. (2006). The epidermal growth factor receptor pathway: A model for targeted therapy. *Clinical Cancer Research*, *12*(18), 5268–5272. https://doi.org/10.1158/1078-0432.CCR-05-1554.

Scheeff, E. D., Eswaran, J., Bunkoczi, G., Knapp, S., & Manning, G. (2009). Structure of the pseudokinase VRK3 reveals a degraded catalytic site, a highly conserved kinase fold, and a putative regulatory binding site. *Structure*, *17*(1), 128–138. https://doi.org/10.1016/j.str.2008.10.018.

Seeliger, D., & de Groot, B. L. (2010). Ligand docking and binding site analysis with PyMOL and Autodock/Vina. *Journal of Computer-Aided Molecular Design*, *24*(5), 417–422. https://doi.org/10.1007/s10822-010-9352-6.

Shelley, J. C., Cholleti, A., Frye, L. L., Greenwood, J. R., Timlin, M. R., & Uchimaya, M. (2007). Epik: A software program for pK(a) prediction and protonation state generation for drug-like molecules. *Journal of Computer-Aided Molecular Design*, *21*(12), 681–691. https://doi.org/10.1007/s10822-007-9133-z.

Shi, F., Telesco, S. E., Liu, Y., Radhakrishnan, R., & Lemmon, M. A. (2010). ErbB3/HER3 intracellular domain is competent to bind ATP and catalyze autophosphorylation. *Proceedings of the National Academy of Sciences of the United States of America, 107*(17), 7692–7697. https://doi.org/10.1073/pnas.1002753107.

Spriestersbach, A, Kubicek, J, Schäfer, F, Block, H, & Maertens, B. (2015). Purification of his-tagged proteins. *Methods in Enzymology, 559*, 1–15. https://doi.org/10.1016/bs.mie.2014.11.003.

Stamos, J., Sliwkowski, M. X., & Eigenbrot, C. (2002). Structure of the epidermal growth factor receptor kinase domain alone and in complex with a 4-anilinoquinazoline inhibitor. *The Journal of Biological Chemistry, 277*(48), 46265–46272. https://doi.org/10.1074/jbc.M207135200.

Todd, M. J., & Gomez, J. (2001). Enzyme kinetics determined using calorimetry: A general assay for enzyme activity? *Analytical Biochemistry, 296*(2), 179–187. https://doi.org/10.1006/abio.2001.5218.

Trott, O., & Olson, A. J. (2010). AutoDock Vina: Improving the speed and accuracy of docking with a new scoring function, efficient optimization, and multithreading. *Journal of Computational Chemistry, 31*(2), 455–461. https://doi.org/10.1002/jcc.21334.

van Lengerich, B., Agnew, C., Puchner, E. M., Huang, B., & Jura, N. (2017). EGF and NRG induce phosphorylation of HER3/ERBB3 by EGFR using distinct oligomeric mechanisms. *Proceedings of the National Academy of Sciences of the United States of America, 114*(14), E2836–E2845. https://doi.org/10.1073/pnas.1617994114.

Varghese, L. N., Ungureanu, D., Liau, N. P., Young, S. N., Laktyushin, A., Hammaren, H., et al. (2014). Mechanistic insights into activation and SOCS3-mediated inhibition of myeloproliferative neoplasm-associated JAK2 mutants from biochemical and structural analyses. *The Biochemical Journal, 458*(2), 395–405. https://doi.org/10.1042/BJ20131516.

Wallasch, C., Weiss, F. U., Niederfellner, G., Jallal, B., Issing, W., & Ullrich, A. (1995). Heregulin-dependent regulation of HER2/neu oncogenic signaling by heterodimerization with HER3. *The EMBO Journal, 14*(17), 4267–4275. Retrieved from https://www.ncbi.nlm.nih.gov/pubmed/7556068.

Webb, S. E., Needham, S. R., Roberts, S. K., & Martin-Fernandez, M. L. (2006). Multidimensional single-molecule imaging in live cells using total-internal-reflection fluorescence microscopy. *Optics Letters, 31*(14), 2157–2159. https://doi.org/10.1364/ol.31.002157.

Willemsen-Seegers, N., Uitdehaag, J. C. M., Prinsen, M. B. W., de Vetter, J. R. F., de Man, J., Sawa, M., et al. (2017). Compound selectivity and target residence time of kinase inhibitors studied with surface plasmon resonance. *Journal of Molecular Biology, 429*(4), 574–586. https://doi.org/10.1016/j.jmb.2016.12.019.

Wood, E. R., Truesdale, A. T., McDonald, O. B., Yuan, D., Hassell, A., Dickerson, S. H., et al. (2004). A unique structure for epidermal growth factor receptor bound to GW572016 (Lapatinib): Relationships among protein conformation, inhibitor off-rate, and receptor activity in tumor cells. *Cancer Research, 64*(18), 6652–6659. https://doi.org/10.1158/0008-5472.CAN-04-1168.

Xie, T., Lim, S. M., Westover, K. D., Dodge, M. E., Ercan, D., Ficarro, S. B., et al. (2014). Pharmacological targeting of the pseudokinase Her3. *Nature Chemical Biology, 10*(12), 1006–1012. https://doi.org/10.1038/nchembio.1658.

Yang, K. S., Ilagan, M. X., Piwnica-Worms, D., & Pike, L. J. (2009). Luciferase fragment complementation imaging of conformational changes in the epidermal growth factor receptor. *The Journal of Biological Chemistry, 284*(12), 7474–7482. https://doi.org/10.1074/jbc.M808041200.

Yarden, Y. (2001). The EGFR family and its ligands in human cancer. Signalling mechanisms and therapeutic opportunities. *European Journal of Cancer, 37*(Suppl. 4), S3–S8. https://doi.org/10.1016/s0959-8049(01)00230-1.

Yaziji, H., Goldstein, L. C., Barry, T. S., Werling, R., Hwang, H., Ellis, G. K., et al. (2004). HER-2 testing in breast cancer using parallel tissue-based methods. *JAMA*, *291*(16), 1972–1977. https://doi.org/10.1001/jama.291.16.1972.

Yonesaka, K., Takegawa, N., Watanabe, S., Haratani, K., Kawakami, H., Sakai, K., et al. (2019). An HER3-targeting antibody-drug conjugate incorporating a DNA topoisomerase I inhibitor U3-1402 conquers EGFR tyrosine kinase inhibitor-resistant NSCLC. *Oncogene*, *38*(9), 1398–1409. https://doi.org/10.1038/s41388-018-0517-4.

Zanetti-Domingues, L. C., Korovesis, D., Needham, S. R., Tynan, C. J., Sagawa, S., Roberts, S. K., et al. (2018). The architecture of EGFR's basal complexes reveals auto-inhibition mechanisms in dimers and oligomers. *Nature Communications*, *9*(1), 4325. https://doi.org/10.1038/s41467-018-06632-0.

Zeqiraj, E., Filippi, B. M., Goldie, S., Navratilova, I., Boudeau, J., Deak, M., et al. (2009). ATP and MO25alpha regulate the conformational state of the STRADalpha pseudokinase and activation of the LKB1 tumour suppressor. *PLoS Biology*, 7(6), e1000126. https://doi.org/10.1371/journal.pbio.1000126.

Zhang, X., Gureasko, J., Shen, K., Cole, P. A., & Kuriyan, J. (2006). An allosteric mechanism for activation of the kinase domain of epidermal growth factor receptor. *Cell*, *125*(6), 1137–1149. https://doi.org/10.1016/j.cell.2006.05.013.

Zhang, Q., Park, E., Kani, K., & Landgraf, R. (2012). Functional isolation of activated and unilaterally phosphorylated heterodimers of ERBB2 and ERBB3 as scaffolds in ligand-dependent signaling. *Proceedings of the National Academy of Sciences of the United States of America*, *109*(33), 13237–13242. https://doi.org/10.1073/pnas.1200105109.

CHAPTER SEVENTEEN

Considerations for studying phosphorylation of the mitotic checkpoint pseudokinase BUBR1

Luciano Gama Braga[a,b,c,d], Chantal Garand[b,c], and Sabine Elowe[a,b,c,d,*]
[a]Biologie Cellulaire et Moléculaire, Faculté de Médicine, Université Laval, Québec, QC, Canada
[b]Centre de Recherche du Centre Hospitalier Universitaire (CHU) de Québec-Université Laval, Axe de Reproduction, Santé de la Mère et de l'Enfant, Québec, QC, Canada
[c]PROTEO-Regroupement Québécois de Recherche sur la Fonction, l'Ingénierie et les Applications des Protéines, Québec, QC, Canada
[d]Département de Pédiatire, Faculté de Médicine, Université Laval et le Centre de Recherche sur le Cancer de l'Université Laval, Québec, QC, Canada
*Corresponding author: e-mail address: sabine.elowe@crchudequebec.ulaval.ca

Contents

Methods in Enzymology, Volume 667
ISSN 0076-6879
https://doi.org/10.1016/bs.mie.2022.03.045

Abstract

Budding uninhibited by benzimidazole 1-related protein 1 (BUBR1) is a mitotic checkpoint (better known as the spindle assembly checkpoint) protein that forms part of an inhibitory complex required to delay mitosis when errors occur in the attachment between chromosomes and the mitotic spindle. If these errors remain uncorrected, it could result in unequal distribution of genetic material to each of the nascent daughter cells, leading to potentially disastrous consequences at both the cellular and organismal level. In some higher eukaryotes including vertebrates, BUBR1 has a C-terminal kinase fold that is largely thought to be inactive, whereas in many species this domain has been lost through evolution and the truncated protein is known as mitotic arrest deficient 3 (MAD3). Here we present advice and practical considerations for the design of experiments, their analysis and interpretation to study the functions of the vertebrate BUBR1 during mitosis with emphasis on analysis implicating the pseudokinase domain.

1. Introduction

Equal distribution of the duplicated genome during symmetric cell division is primordial for the health of offspring cells in unicellular eukaryotes and the entire organism in metazoans. This process relies on the proper attachment of the duplicated chromosome (known as a sister chromatid pair) to microtubules emanating from opposite poles of the mitotic spindle and their subsequent alignment to the spindle equator. The site of microtubule attachment is the kinetochore, a large protein network that assembles onto chromosomes upon mitotic entry. In addition to its structural role, the enrichment of kinases, phosphatases, and other modifying enzymes to its various substructures supports its function as a signaling platform during mitosis (Hara & Fukagawa, 2018; Musacchio & Desai, 2017; Navarro & Cheeseman, 2021).

To safeguard the fidelity of kinetochore-microtubule attachments, the cell has evolved a surveillance pathway known as the spindle assembly checkpoint (SAC). SAC signaling is initiated at kinetochores in response to lack of attachment or aberrant attachment between kinetochores and spindle microtubules. The result of SAC signaling is a delay in anaphase, allowing the cell time to correct attachment defects (Fig. 1). The importance of this pathway to cell and organismal health is underscored by its almost universal conservation across eukaryotes, and the reader is referred to many excellent reviews on this subject (Musacchio, 2011; Lara-Gonzalez, Pines, & Desai, 2021; Musacchio, 2015). The target of the SAC is a large E3 ubiquitin

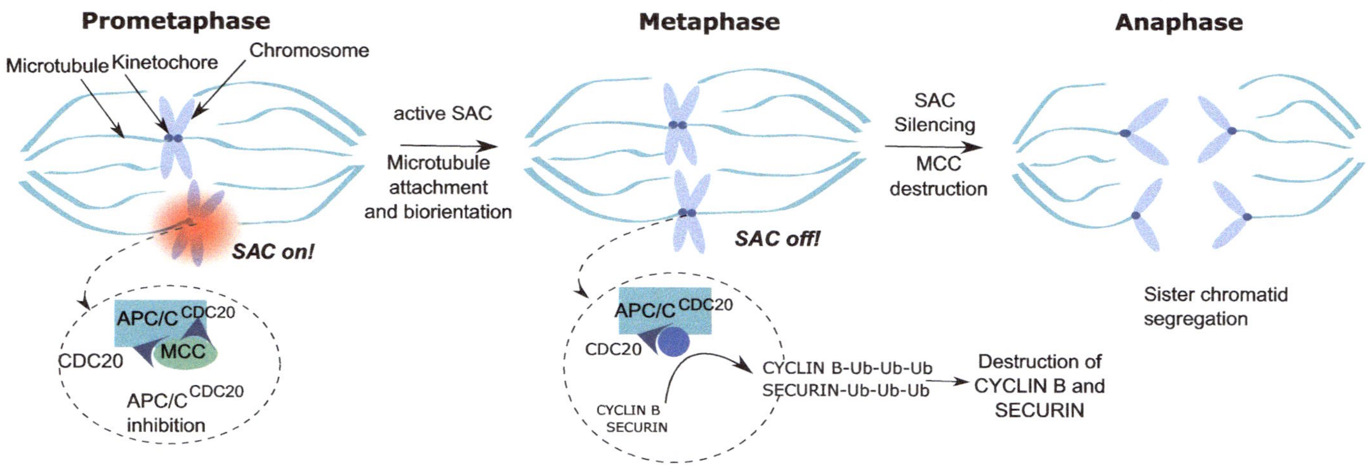

Fig. 1 Schematic of the spindle assembly checkpoint (SAC)-dependent regulation of mitotic progression. During early mitosis, unattached or incorrectly attached kinetochores (Red) catalyze the formation of the mitotic checkpoint complex (MCC) which is then releases and diffuses throughout the cell cytoplasm (red cloud) to inhibit APC/C^{CDC20}, a ubiquitin ligase which is responsible for sister chromatid separation and mitotic exit. The two major targets of the APC/C^{CDC20} in this context are Cyclin B, the obligatory coactivator of the master mitotic kinase CDK1 and securin which holds the caspase-like protein separase in check. Destruction of securin releases the inhibition on separase which is required to cleave the cohesion between sister chromatids to allow for their physical separation and segregation into the nascent daughter cells. In this manner, the SAC ensures that the transition to anaphase only occurs after all kinetochores are attached effectively to spindle microtubules.

ligase, the Anaphase Promoting Complex/Cyclosome (APC/C) when bound to its co-activator subunit Cell Division Cycle 20 (CDC20) to generate APC/C^{CDC20}. This enzyme is responsible for the destruction of Cyclin B, the obligatory subunit of the master mitotic kinase Cyclin-dependent kinase 1 (CDK1), and securin, a protein indirectly responsible for keeping sister chromatids tethered to each other (Barford, 2020; Watson, Brown, Peters, Stark, & Schulman, 2019). Destruction of these two proteins is a hallmark of anaphase onset (Fig. 1). Ultimately, attachment errors at the kinetochore catalyze the formation of an APC/C^{CDC20} inhibitor, the mitotic checkpoint complex (MCC), which is then released into the cytoplasm where it is free to interact with and inhibit cytoplasmic APC/C^{CDC20}, resulting in rapid mitotic arrest.

The core components of the MCC are well-conserved and include the Budding Uninhibited by Benzimidazole 1-Related protein 1 (BUBR1) in higher eukaryotes. Curiously, this protein which has a C-terminal kinase fold in vertebrates and certain insects, is functionally replaced by a truncated version which lacks the kinase but retains most of the N-terminal motifs that are implicated directly in either MCC formation or APC/C^{CDC20} inhibition (Fig. 2). The role of the C-terminal kinase fold is therefore unclear.

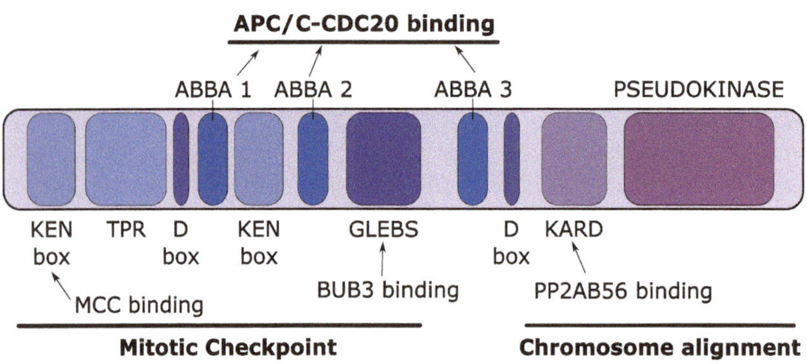

Fig. 2 The domain and motif architecture of human BUBR1. BUBR1 orthologs are modular proteins and most of the SAC functionality is included in the N-terminal half of the protein which is very highly conserved amongst eukaryotes. The SAC motifs and domains include the KEN (Lys-Asp-Asn) box, the TPR (tetratricopeptide repeat) domain, the D (*D*estruction box), the ABBA (cyclin *A*, *B*UBR1, *B*UB1, and *A*cm1) motif and the GLEBS (*GLE B*inding *S*equence). Collectively these motifs collaborate to form the MCC or to directly associate with and inhibit the APC/C^{CDC20}. In contrast, the C-terminal part of hBUBR1 includes the KARD (*K*inetochore *A*lignment *R*egulatory *D*omain) and the pseudokinase which collectively cooperate to maintain the balance of kinase-phosphatase signaling at the outer kinetochore.

In vertebrates, some early studies suggested that BUBR1 is a genuine kinase that responds to the status of microtubule attachments and is activated/disactivated in parallel with the SAC (Guo, Kim, Ahmad, Zhang, & Mao, 2012; Huang et al., 2019; Mao, Abrieu, & Cleveland, 2003; Mao, Desai, & Cleveland, 2005). More recent evidence however has led to the reclassification of BUBR1 as a rather unusual pseudokinase; indeed while the catalytic triad residues in vertebrate BUBR1 are easily identifiable, several other motifs important for catalysis are missing or have degenerated including the glycine-rich loop, and the catalytic loop (Breit et al., 2015; Gama Braga et al., 2020; Suijkerbuijk et al., 2012). In this chapter we describe experimental methods and considerations for studying the pseudokinase domain of human BUBR1 and pseudokinase mutations in mitotic cells.

2. Experimental and theoretical considerations for studying BUBR1 during mitosis

2.1 Choice of BUBR1 ortholog to study

The idea that human BUBR1 in particular is a pseudokinase remains contentious, as recent work from different labs continue to debate this point (Breit et al., 2015; Gama Braga et al., 2020; Guo et al., 2012; Huang et al., 2019; Suijkerbuijk, van Dam, et al., 2012). Our laboratory and others have not been able to associate phosphotransfer activity with or ATP binding to hBUBR1 and for the purposes of this work here we consider it catalytically inactive(Breit et al., 2015; Gama Braga et al., 2020; Murphy et al., 2014; Suijkerbuijk, van Dam, et al., 2012). However, it is important to note that proving constitutive absence of kinase activity is very difficult and one cannot rule out that kinase activity could, under specific conditions, be attributed to hBUBR1. Nevertheless, and considering this controversy in question, a number of important considerations must be taken into account when studying this domain in BUBR1 in eukaryotes. Most importantly is the choice of BUBR1 orthologue to study, which must be carefully considered. As noted above, while many in the field consider vertebrate BUBR1 proteins to be pseudokinases, it is clear now that in certain *Drosophila* and *Anopheles* species, this is not the case and the BUBR1 orthologue in these organisms is very likely a *bona fide* kinase (Rahmani, Gagou, Lefebvre, Emre, & Karess, 2009; Tang et al., 2020). Therefore, here we will focus on tips and practical advice on how to study vertebrate BUBR1 pseudokinases in mitosis with emphasis on hBUBR1.

2.2 Possible alternative functions of the pseudokinase domain

In lieu of catalysis, several alternative functions to explain the retention of the BUBR1 kinase domain in vertebrate lineages have been proposed and suggests that it represents an evolutionary advantage (Gama Braga et al., 2020; Suijkerbuijk et al., 2010; Suijkerbuijk, van Dam, et al., 2012). Mosaic-variegated aneuploidy (MVA, OMIM: 257300) syndrome is a rare childhood disorder caused by biallelic mutations in a small number of genes, most notably BUBR1 (Matsuura et al., 2006; Sieben et al., 2020; Simmons et al., 2019; Suijkerbuijk et al., 2010). MVA is characterized by increased chromosome missegregation, and a broad spectrum of clinical features including various cancers, congenital defects, and progeroid pathologies. MVA-causing BUBR1 mutations are found in and around the pseudokinase domain and ultimately function through destabilization of the entire BUBR1 protein, significantly reducing the cellular levels of full-length BUBR1 (Gama Braga et al., 2020; Suijkerbuijk et al., 2010). During mitosis, mutation of the BUBR1 kinase domain or its removal entirely results in the loss of an important pool of protein phosphatase 2A (PP2A) at the outer-kinetochore, which ultimately leads to an imbalance in phosphosignaling at the kinetochore resulting in aberrant mitotic divisions (Hein, Hertz, Garvanska, Kruse, & Nilsson, 2017; Suijkerbuijk, Vleugel, Teixeira, & Kops, 2012; Xu, Raetz, Kitagawa, Virshup, & Lee, 2013). Moreover, and like many other pseudokinases (for recent reviews see (Mace & Murphy, 2021; Murphy, Mace, & Eyers, 2017)), hBUBR1 can function as scaffold, and associates with the large mitotic motor protein centromere protein E (CENP-E) through a C-terminal helix of the kinase fold (Ciossani et al., 2018; Legal et al., 2020). CENP-E plays a critical role in fine-tuning chromosome movements as it drives the alignment of chromosomes that are lagging close to spindle poles (Craske & Welburn, 2020). Given the number of non-catalytic activities attributed to the hBUBR1 pseudokinase domain, there is a clear need to understand the molecular details of its functions and regulation, in particular during mitosis.

2.3 Expression of exogenous BUBR1 plasmids in cultured cells

The stability of the BUBR1 protein is dependent on several core residues, as well as amino acids at both N- and C-terminal extremities of the pseudokinase domain. As noted above, hBUBR1 proteolytic instability as a result of mutations at these residues is a key player in MVA associated pathologies. Mutations in the C-terminal portion of BUBR1 have been shown to reduce protein stability, while truncation of the entire domain stabilized the remaining protein (Gama Braga et al., 2020; Sieben et al., 2020; Suijkerbuijk et al., 2010). Because the functions of BUBR1 in mitosis are exquisitely linked to its expression levels, protein levels of various BUBR1 pseudokinase domain mutants must be equalized to avoid errors in functional assessment of pseudokinase domain mutants. To circumvent this issue, the expression levels of BUBR1 can be regulated to achieve relatively equalized levels of BUBR1 mutants, which are set as close as possible to endogenous levels. This can be accomplished, in our experience, through the Flp-In™ T-Rex™ System as we have recently reported (Gama Braga et al., 2020), or a similar system where exogenous protein expression can be titrated through induction.

The development of stable cell lines in the Flp-In™ T-Rex™ System is an excellent solution for strict regulation of exogenous BUBR1 expression, and to avoid heterogenous expression. In order to develop stable cell lines expressing different forms of BUBR1, we use the T-REx™-HeLa cell line, which stably expresses the tetracycline repressor protein, as well as a single FRT integration site at a transcriptionally active genomic locus, therefore providing tighter control over protein expression. The reader is referred to the manufacturer's manual for a more detailed explanation of the Flp-In™ T-Rex™ System (https://www.thermofisher.com/order/catalog/product/K650001).

3. Generation of stable BUBR1 cell lines using the Flp-In™ T-Rex™ system

To generate isogenic, stable cell lines expressing the BUBR1 constructs of interest, we routinely use the Flp-In™ T-Rex system (ThermoFisher Scientific, catalog number K650001 for the core Kit). Several cell lines engineered for use with this system have already been generated and are available commercially through ThermoFisher Scientific. For BUBR1 studies, we use the HeLa T-Rex system, which was a kind gift Dr. Arshad Desai, (Ludwig Cancer Institute, La Jolla, CA). Briefly, to generate stable lines, cells are transfected with a Flp-In™ expression vector and a Flp recombinase (Flippase) expression plasmid and are incubated for approximately 2 weeks under selection with hygromycin and blasticidin according to the manufacturer's instructions. Serial dilution is then used to isolate a single clone for analysis, which consists of a Western blot assay with the whole cell lysate to evaluate if the exogenous BUBR1 is expressed. As we use a fluorescent tag (GFP), Fluorescence-activated cell sorting (FACS) may also be a suitable alternative at this step to identify stably transfected cell populations. Alternatively, if single clones are not desired, the individual colonies can be pooled. A gradient of tetracycline or doxycycline (we generally use tetracycline and doxycycline interchangeably) ranging from 25 ng/mL to 1 μg/mL is then used to compare the expression of the various mutants with BUBR1 wild-type (WT) by Western blotting. Finally, BUBR1 levels are evaluated through immunofluorescence to verify uniform expression of BUBR1 in mitotic cells. As a general rule, unstable forms of BUBR1, such as the K795R and D911A mutants, usually require substantially higher levels of tetracycline to achieve similar expression to BUBR1-WT. In this section, we describe the generation and validation of the BUBR1 stables cell lines. Detailed protocols for Western blotting and Immunofluorescence techniques required to validate these lines are described in Sections 4 and 6, respectively.

3.1 Key resources table

Reagent or resource	Source	Identifier
Chemicals, peptides, and recombinant proteins		
Tetracycline hydrochloride	Sigma	T7660-5G
Zeocin	Thermofisher	R25001
Blacticidine	Invivogene	Ant-bl-1
Hygromycin B	Wisent	Cat# 450-141-XL
jetPRIME®	Polyplus	101000027
jetPRIME® Buffer	Polyplus	201000003
Experimental models: cell lines		
T-Rex doxycycline-inducible Flp-In system cell line	Thermofisher	K650001
Oligonucleotides		
dsiBUBR1 5'-GUCUCACAGAUUG CUGCCUCAGAGC-3'	Integrated DNA technologies (IDT)	Custom design
Recombinant DNA		
pcDNA5/TO/FRT	Thermofisher	V601020
pOG44 Flp-recombinase expression vector	Thermofisher	V600520

3.2 Step-by-step method details

3.2.1 Creation of stable cell lines

The following steps describe the creation of stable cell lines for BUBR1 protein expression. The timeline for this procedure is shown in Fig. 3A. Prior to the generation of the cell lines however, the BUBR1 sequence must be cloned into a Flp-In™ expression vector, along with a suitable tag using the preferred method. Our group uses BUBR1 N-terminally tagged with triple MYC fused to GFP, however N-terminal tagging with other epitopes has been shown to perform well too in cell-culture experiments (Lischetti, Zhang, Sedgwick, Bolanos-Garcia, & Nilsson, 2014; Malureanu et al., 2009; Suijkerbuijk, van Dam, et al., 2012). Note that along with the various pseudokinase domain mutants, cell lines expressing BUBR1-WT as a control must be generated.

1. Parental HeLa T-Rex cells are routinely grown in Dulbecco's modified Eagle medium (DMEM,) containing 10% (vol/vol) of fetal bovine serum (FBS) and Pen/Strep (100 mg/mL) with additional 100 μg/mL zeocin and 10 μg/mL blasticidin at 37 °C and 5% CO_2. Zeocin is to maintain the flippase recognition target (FRT) integration site whereas the blasticidin is for maintaining the Tet repressor. When the gene of interest is introduced in the FRT site, the zeocin cassette is replace by the hygromycin creating a cell line resistant to hygromycin but sensitive to zeocin (Fig. 3B). The HeLa T-Rex cells are adherent and are grown regularly on cell-culture treated dishes such as the Bio-Lite series from ThermoFisher, although other suitable options have also been successfully tested (Falcon standard Tissue culture dishes, VWR Cell culture dishes).
2. On the day prior to transfection, seed 2×10^6 cells (counted using a hemocytometer or a cell counter such as TC20 automated cell counter from Bio-Rad laboratories) into a single 10 cm tissue culture dish in 10 mL of the media under the conditions listed above, and allow to adhere and spread for 24 h.
3. Transfect cells with the desired, tagged BUBR1 mutants cloned into the Flp-In™ expression vector containing the BUBR1 sequence and the Flp recombinase vector (called pOG44 in the Flp-IN T-Rex kit) 24 h after seeding according to the ratio recommended by the manufacturer. Change media 24 h post infection to exclude zeocin but retain blasticidin.
4. The next day, perform a serial dilution to isolate single clones for analysis, if desired. Typically, we prepare plates of transfected cells diluted 20-,

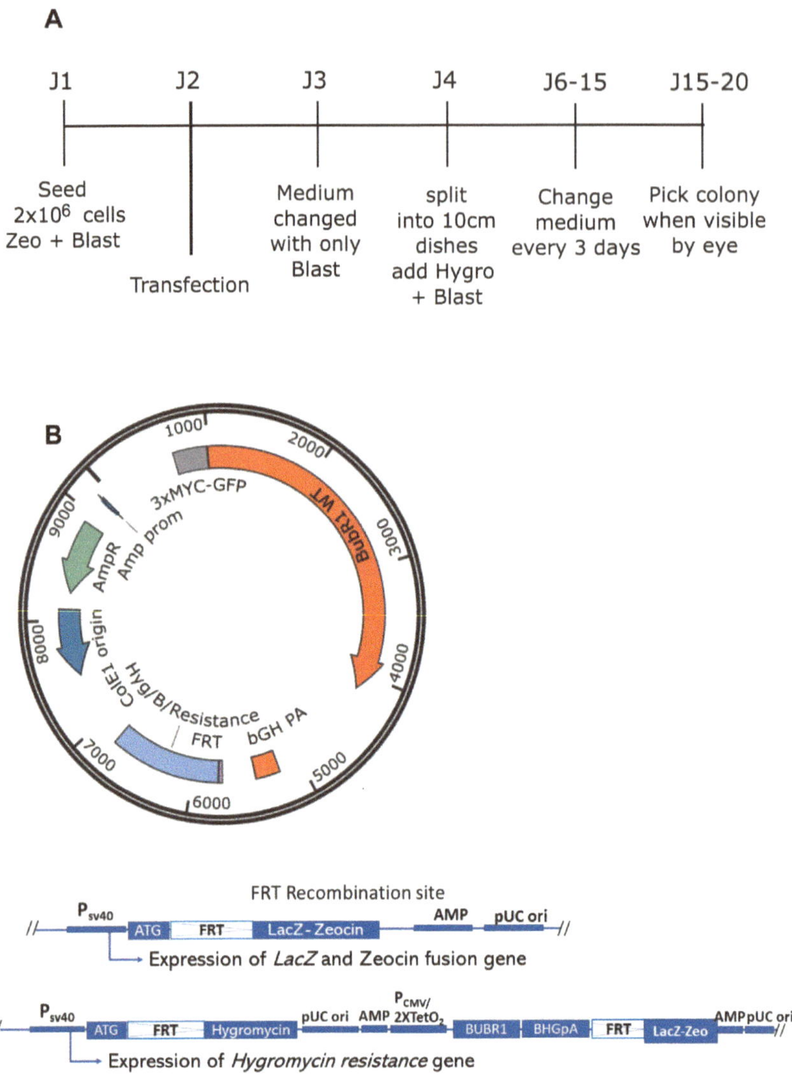

Fig. 3 Strategy for expression of BUBR1 constructs using the Flp-In T-Rex system. (A) Timeline for the generation of single-cell colonies expressing BUBR1 constructs. (B) Vector design for the donor plasmid expressing BUBR1 constructs and the recombination strategy. *Figures modified from the manufacturer's instructions (https://www.thermofisher.com/document-connect/document-connect.html?url=https%3A%2F%2Fassets.thermofisher.com%2FTFS-Assets%2FLSG%2Fmanuals%2Fflpintrex_man.pdf).*

10- and 5-fold in 10 cm dishes. Ultimately, a minimum of 10 clones for each construct are picked and expanded for testing expression. Dilutions are performed in media containing 10 μg/mL blasticidin and 200 μg/mL hygromycin to commence selection. The exact concentration of hygromycin for the selection process is dependent on the parental cell line and must be optimized prior to the start of the experiment.

5. Select cells with hygromycin and blasticidin as indicated above, changing the media every 3 days for approximately 2 weeks or until visible colonies are detected.
6. Colonies can be picked using standard methods. We routinely pick colonies in the tissue culture hood using an inverted microscope to locate them. A detailed protocol for picking colonies has been recently described (Cerbini, Luo, Rao, & Zou, 2015). Colonies are propagated in complete media containing the same concentration of hygromycin and blasticidin used for selection.
7. Expand the isolated colonies in the presence of selection until cells can be harvested for testing BUBR1 expression. Typically, we keep at minimum 1×10^5 cells for each clone for maintenance and a minimum of 5×10^5 cells are harvested for testing expression by Western blotting after 24h induction with 1 μg/mL of tetracycline. A detailed description of Western blotting for BUBR1 is indicated in Section 6.
8. Once identified, colonies are maintained in the presence of 10 μg/mL blasticidin and 200 μg/mL hygromycin.

3.2.2 Equalizing the level of exogenous BUBR1 expression

Considering that various mutants of BUBR1 within the kinase fold render the BUBR1 molecule unstable, it is prudent to optimize and calibrate the concentrations of tertacycline required to achieve relatively equal levels of expression of the various BUBR1 mutants. We noticed that unstable forms of BUBR1, such as K795R and D911A, usually require around a 100-fold higher tetracycline concentration compared to wild-type BUBR1.

1. A gradient of tetracycline (usually from 25 ng/mL to 1 μg/mL) is used to induce expression of BUBR1 in HeLa T-Rex cells for 24h. We typically perform this experiment using approximately 3×10^5 cells per condition grown as indicated in above (Section 3.2.1) and plated in 3.5 cm dishes or an individual well of six-well plate.
2. Perform Westerns blots (see Section 6 for detailed protocol) using approximately 50–80 μg whole cell lysates to evaluate expression of BUBR1. To compare the expression with wild-type BUBR1 with the various mutants, a Western blot for the tag is performed, whereas for comparing to the endogenous protein, a Western blot against BUBR1 itself is used. Note that we regularly use N-terminal GFP and triple MYC tags which are large enough to distinguish between endogenous and exogenous protein.

3. If expression of the transgene is detected by Western blotting, BUBR1 levels are then further validated by immunofluorescence of mitotic cells to validate that expression of BUBR1 at the level of individual cells and individual kinetochores. The detailed immunofluorescence protocol is described below in Section 4.
4. Quantitative analysis of BUBR1 levels at kinetochores can be determined by the approach described in Section 5.
5. Note that once cell lines are generated, we recommend testing an experimental hypothesis in more than one clone to ensure that the results are not due to a clonal effect.

3.2.3 Preparing BUBR1 cell lines for an experimental protocol

In a typical experiment, cells are treated with siRNA targeting endogenous BUBR1 approximately 24h after seeding which allows sufficient time for attachment and spreading. The depletion of endogenous BUBR1 is necessary to avoid interference and competition with the exogenously expressed protein. Expression of the BUBR1 transgene is induced 6h after siRNA treatment which is generally efficiently achieved by 48h. In this manner, as endogenous BUBR1 levels decrease in cells, the exogenous protein is expressed. This is critical because BUBR1 is an essential protein, loss of which is incompatible with cellular and organismal survival (Wang et al., 2004). At this point, cells can be harvested or treated as required by the experimental protocol. We describe here a typical experiment of cells grown in a 10 cm dish for reference, but it can be scaled down or up as required.

1. Plate 1–1.5 x 10^6 cells in a 10 cm dish containing 10 mL of DMEM, 10% (vol/vol) FBS and Pen/Strep (100 mg/mL, Hyclone) as well as 10 μg/mL blasticidin and 200 μg/mL hygromycin at 37 °C and 5% CO_2. Allow 24h for cells to attach and spread.
2. Deplete endogenous BUBR1 with dsiBUBR1 oligos. In our experience, 30 nM dsiBUBR1 (sequence provided above) is sufficient for depletion of endogenous BUBR1 in several cell lines. For siRNA-mediated depletion, we generally use jetPRIME® according to the manufacturer's instructions.
3. After 6h of dsiBUBR1 treatment, add tetracycline at the optimized concentration (see Section 3.2.2) to induce BUBR1 transgene expression.
4. After 24h of dsiBUBR1 treatment, remove the medium to eliminate debris and add 10 mL of fresh media with tetracycline.
5. 48h after dsiRNA treatment, cells should be harvested for Western blotting or fixed for immunofluorescence (as detailed below) according to the design of the experiment.

4. Use of phosphospecific antibodies to infer function

We recently found that mutation or complete removal of the BUBR1 pseudokinase domain results in attenuation of the BUBR1 phosphorylation in mitosis and decreased binding to the PP2A-B56 phosphatase complex, resulting in slower mitotic exit and defects in alignment of chromosomes at metaphase as a result of kinase-phosphatase imbalance (Gama Braga et al., 2020). How the pseudokinase domain regulates phosphorylation remains unclear. However, direct measurement of phosphorylation sites in BUBR1 can be performed, and phosphorylation of S670 (a cyclin dependent kinase 1 target site) and S676 (a polo-like kinase 1 target site), in particular, can be measured using published phosphospecific antibodies (Elowe et al., 2010; Elowe, Hümmer, Uldschmid, Li, & Nigg, 2007). These sites lie within the PP2A-B56 binding domain of BUBR1 and when phosphorylated can increase PP2A-B56 affinity up to 11-fold (Kruse et al., 2013). Because BUBR1 phosphorylation and localization at kinetochores occurs only in mitosis, it is important to ensure a robust enrichment for mitotic cells. Furthermore, as the kinetochore is a subcellular structure that is not easily purified biochemically, indirect immunofluorescence is used routinely to study BUBR1 localization and phosphorylation. This approach offers the additional advantage of being able to observe and quantify signals, not only at the level of a single cell, but also at the level of a single kinetochore. We use this section here to detail the steps required for performing immunofluorescence experiments in the HeLa T-Rex system described above.

For measuring BUBR1 phosphorylation levels accurately, pS676 and pS670 must be normalized against the bulk BUBR1 signal. Importantly, in the context of BUBR1 pseudokinase mutants that are exogenously expressed, the normalization is performed against this pool which can be detected using the co-expressed tag. Alternatively, normalization can be performed using another kinetochore protein whose levels are known to be unaltered by the experimental conditions. In this context, ACA (anti-centromeric antigen) or CENP-C are often used for normalization. After image acquisition at the microscope, the measurement of the resulting images can be performed with image analysis software of choice. Image J is a free image analysis software commonly used for immunofluorescence analysis, offers many highly specialized analysis macros and plugins (Schneider, Rasband, & Eliceiri, 2012). The steps to measure BUBR1 phosphoantibody signals at kinetochores using Image J are described below.

5. Before you begin

The solutions required are:

- *Synchronization buffer*: 200 mM thymidine
- *Culture medium*: Dulbecco's modified Eagle medium (DMEM, Hyclone) containing 10% (vol/vol) FBS, blasticidin (10 μg/mL) + hygromycin (2 mg/mL), and Pen/Strep (100 mg/mL)
- *PTEMF fixation and extraction buffer*: 0.2% Triton X-100, 20 mM PIPES pH 6.9, 1 mM $MgCl_2$, 10 mM EGTA and 4% formaldehyde
- *Blocking solution*: 1x PBS with BSA (3% m/v)
- *Washing solution*: 1x PBS withTween-20 (0.05% v/v)
- ddH_2O
- Clear nail polish to seal coverslips
- Kimwipes or Delicate Task Wipes

5.1 Key resources table

Reagent or resource	Source	Identifier
Antibodies		
pS670, Rabbit polyclonal	-Elowe et al. (2010). Use at 1 μg/mL for immunofluorescence and Western blotting	–
pS676, Rabbit polyclonal	-Elowe et al. (2010). Use at 1 μg/mL for immunofluorescence and Western blotting	–
ACA, human serum	-ImmunoVision. Use at 1:1000	-HCT-0100
CENP-C, guinea pig polyclonal	-MBL International. Use at 1 μg/mL for immunofluorescence	-PD030
Hoechst 33342	Cell signaling technology. Use at 1 μg/mL for immunofluorescence	4082
Anti-BUBR1, mouse monoclonal antibody, clone 68-3-9	Invitrogen. Use at 1 μg/mL for immunofluorescence and Western blotting	41-2800

—cont'd

Reagent or resource	Source	Identifier
Anti-MYC, mouse monoclonal antibody, clone 9E10	Fisher Scientific. Use at 1 μg/mL for immunofluorescence and Western blotting	MA19801MG
Alexa Fluor® 488-AffiniPure Donkey Anti-Mouse IgG	Jackson ImmunoResearch. Use at 1 μg/mL for immunofluorescence and Western blotting	Cat# 715-545-150; RRID: AB_2340846
Alexa Fluor® 647-AffiniPure Donkey Anti-Guinea Pig IgG	Jackson ImmunoResearch. Use at 1 μg/mL for immunofluorescence and Western blotting	Cat# 706-605-148; RRID: AB_2340476
Alexa Fluor® 647-AffiniPure Donkey Anti-Human IgG	Jackson ImmunoResearch. Use at 1 μg/mL for immunofluorescence and Western blotting	Cat# 709-605-149; RRID: AB_2340578
Alexa Fluor® 594-AffiniPure Donkey Anti-Rabbit IgG	Jackson ImmunoResearch. Use at 1 μg/mL for immunofluorescence and Western blotting	Cat# 711-585-152; RRID: AB_2340621
Experimental models: Cell lines		
Human cell lines expressing BUBR1	Generated as per described above	–
Software and algorithms		
ImageJ or fiji	https://imagej.net/software/fiji/	https://doi.org/10.1038/nmeth.2019
Other		
Mounting media: Fluoromount-G	-Thermofisher	-00-4958-02

5.2 Materials and equipment

- Microscope suitable for immunofluorescence imaging. We routinely use a spinning disc confocal microscopy set-up. This includes an Olympus IX80 inverted confocal microscope with a WaveFX-Borealin-SC Yokagawa spinning disc from Quorum Technologies and an Orca

Flash 4.0 camera from Hamamatsu. We perform our image acquisitions with the Metamorph software (Molecular Devices).

- *Microscopy Slides*: Fisherbrand Superfrost Microscope slides, cat# 12-550-12.
- *Coverslips*: Fisherbrand™ Coverglass for Growth™ Cover Glasses (CVR GLAS CIR 15MM GRWTH PK/1OZ) Fisherbrand™ ER15CIR1DFS (cat# 12-545-83), non-coated.

5.3 Step-by-step method details

The steps performed for the immunofluorescence assay using BUBR1 phosphospecific antibodies are described below. Note that all washing and solution exchange steps must be performed gently to reduce cell detachment from coverslips.

1. Place glass coverslips inside a six-well cell-culture plate (BioLite, or other options, as indicated above). Rinse with 1x PBS and remove to attach coverslips to the bottom of the well by capillarity. We generally use 15 mm round coverslips, of which three can be placed in a 3.5 cm dish or well of a six-well plate. In this manner, multiple combinations of antibodies for indirect immunofluorescence can be performed for each individual experiment.
2. Seed cells (1×10^5 to 2×10^5 per 3.5 cm dish/well) and incubate until they are attached to the coverslip. Cells are seeded in 2 mL of DMEM containing 10% (vol/vol) FBS, blasticidin (10 μg/mL), hygromycin (2 mg/mL), and Pen/Strep (100 mg/mL) at 37 °C and 5% CO_2, as indicated above. Attachment and spreading is typically achieved within 24 h.
3. After 24 h, transfect cells with siRNA and induce BUBR1 transgene expression as described in Section 3.2.3 above.
4. To enrich for cells in mitosis, incubation with 2 mM thymidine for 18–24 h (typically overnight) is performed to synchronize cells in the G1 phase of the cell cycle. The release is performed by removing the media and gently adding around 2–3 mL of PBS to the side of the dish to avoid disturbing the cells. Repeat the washing three to five times. After washing, add 2 mL of fresh culture medium to each well to allow the cells to continue cycling. The mitotic peak is typically observed after 8–10 h of release in fresh media and is easily detected in phase contrast under a light microscope as mitotic cells are rounded (Fig. 4). If cells and the experimental timeline can tolerate it, a double thymidine arrest can be performed which involves two consecutive

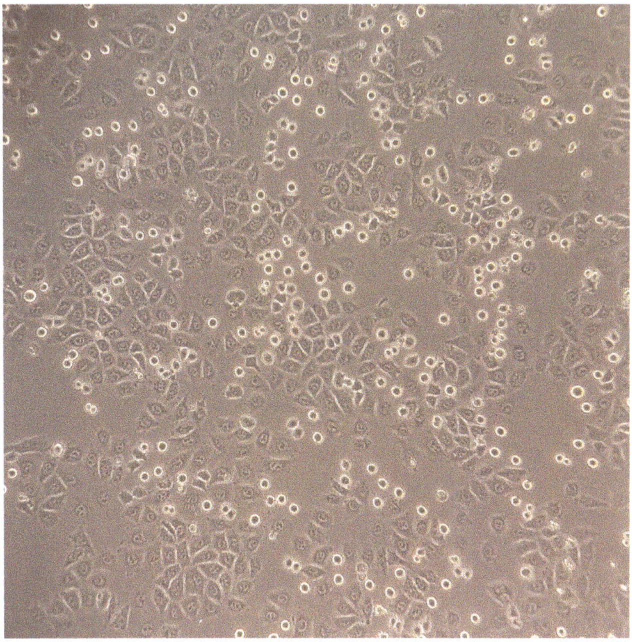

Fig. 4 Rounded appearance of mitotic cells as observed by phase contrast under a light microscope. Cells were treated overnight with CDK1 inhibitor R0-3306 at 4 μM to enrich for cells in mitosis.

thymidine arrest-release cycles, where the second round of thymidine is added approximately 5–8 h after release from the first. In many cell lines, this can significantly increase the efficiency of synchronization.

5. Fix cells with PTEMF (as specified at the beginning of Section 5) by removing culture media and replacing it with enough PTEMF to cover the coverslips completely, typically 0.5 mL for a 3.5 cm dish. After a 10 min incubation at room temperature, remove the PTEMF solution and rinse cells gently 3x with 2–3 mL of 1x PBS. The plate with coverslips (in fresh PBS) can now be stored at 4 °C. For long-term storage, plates are sealed with parafilm to avoid dehydration. Alternatively proceed immediately with the immunofluorescence protocol in step 6.
6. Incubate coverslips with the blocking solution (defined at the beginning of Section 5) for 30 min at room temperature.
7. Rinse coverslips with 3x with 1x PBS and incubate with a primary antibody mix diluted in the blocking media for at least 2 h at room temperature or overnight at 4 °C in a humidified chamber. We find that overnight incubation generally works better for detecting kinetochore

proteins. In either case, for 15 mm coverslips, a minimal volume of 80 μL of the primary antibody mix is required. Working dilutions for the most commonly used antibodies in our studies of BUBR1 are listed in the key resources table above. In addition to the BUBR1-specific antibodies (i.e., antibodies against BUBR1 itself, its tags or phosphorylation sites) we consistently co-stain with another kinetochore protein (we have listed here CREST and CENP-C) against which BUBR1 levels can be normalized (see Section 5 below).

8. After incubation with the primary antibody mix, rinse coverslips 3x with the washing solution (as specified in the start of Section 5) and incubate with their respective conjugated secondary antibodies (see "key resources table" above) together with a DNA dye (DAPI or HOECHST) diluted (1 μg/mL) in blocking media for 1–2 h at room temperature, in the dark to avoid loss of fluorescence.
9. Rinse slides 3x with washing solution followed by a final rinse with ddH_2O.
10. Mount the coverslips on a microscope slide: Gently remove as much of the ddH_2O as possible by capilarity using a kimwipe placed on the edge coverslip, being careful not to disrupt the cells. Place the coverslip (cell side down) on a drop of mounting media that has been freshly pipetted onto the microscopy slide. Gently remove any excess mounting media using Kimwipes. Nail polish should then be applied to the coverslip extremities to seal the coverslip. This prevents the mounting solution from drying and helps fix the coverslip in place. Store slides at −20 °C in the dark until ready for imaging.
11. Image acquisition is performed with an objective capable of resolving kinetochores. In our experience, a 60x objective or, optimally, a 100x objective offer sufficient resolution in a spinning disc confocal set-up to observe individual kinetochores. A sequence of images collected at different planes perpendicular to the optical axis is necessary to ensure that all kinetochores are present in the final image. Therefore, we recommend taking enough images in the Z-axis to cover the entire depth of the cell. Typically, we acquire 12–15 stacks at intervals of 0.2 μm covering the depth of a HeLa cell.

6. Quantification and statistical analysis

6.1 Quantifying kinetochore signals using ImageJ

To accurately compare phosphorylation signals between the wild-type and mutated BUBR1, acquisition as well as analysis parameters of different cells and experimental conditions must be kept constant. Ideally, kinetochore signals from a minimum of 30 to 50 cells per experiment should be acquired to allow for statistical analysis. The next steps outlined below consist of the measurement using ImageJ of the signal intensities from each channel.

1. Image analysis starts by converting stack of images (i.e. all the focal planes) for each channel into a signal-averaged single image. We routinely use the maximum projection tool in ImageJ (Image\Stacks \Z project\Max project). At the end of this step, each image should have a single projection for each of the channels that were imaged. An image showing the merged Z-projection of stacks in two channels (representing anti-BUBR1 and anti-CENP-C staining) in a typical immunofluorescence experiment is shown in Fig. 5.
2. In the second step, the objective is to find the region of the image which delimits the kinetochores for one single cell. For this, select the cell of interest first then eliminate the signal outside using the function "Edit\Clear outside." We typically use the chromatin signal from the HOECHST/DAPI channel to select the appropriate area that is then copied to the images from other channels, and the same operation ("Edit\Clear outside" to eliminate undesirable signals) is performed. The resulting image now contains the signal that delimits the selected kinetochore (i.e. the tag on the BUBR1 transgene, CREST or CENP-C).

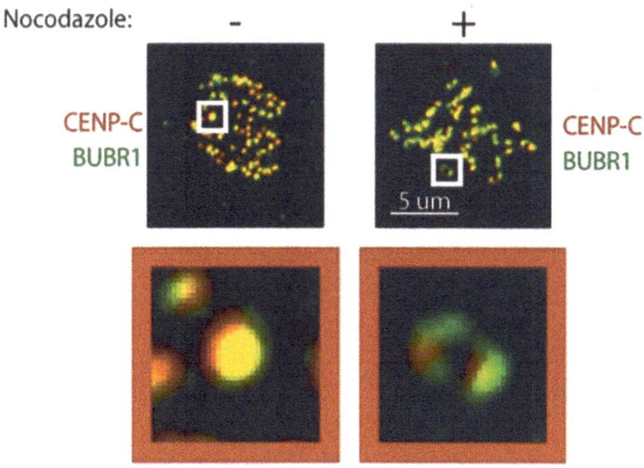

Fig. 5 A representative image of BUBR1 at kinetochores as detected by immunofluorescence in untreated cells or cells treated with Nocodazole. The image illustrates a Z-axis projection of multiples planes taken at 0.2 μM intervals, and the merge of two independent channel. The kinetochore is marked by CENP-C (red) whereas BUBR1 is shown in green. Note the movement of BUBR1 to the outer kinetochore in the presence of nocodazole.

3. Next, select the kinetochores with the function "Image\Adjust \Threshold", which will create a binary image or 'mask' of the selected area. The mask can now be adjusted to threshold the image and to create a demarcation corresponding to the kinetochores in the image. Next, create a selection containing the kinetochores with the "Edit \Selection\Create selection" function. It should be noted that selections in ImageJ are area demarcations in images. Therefore, the creation of a selection containing the kinetochores in one image corresponds to the spatial kinetochore positioning in the images from other channels.
4. Propagate the selected kinetochore region to the other channels (for example to the BUBR1 and phospho-BUBR1 channels) with the "Edit\Selection\Restore Selection" function. Since this selection corresponds to the kinetochore region, the signal inside this region in the BUBR1 channel image reflects the precise BUBR1 signal originating exclusively from the kinetochores. The intensity of the average signal inside this region is measured with the "Analyze\Measure" function in ImageJ. The process is repeated for the other channels. The values measured are recorded for posterior analysis and can be directly exported from ImageJ into the spreadsheet of choice.
5. Finally, to calculate the levels of BUBR1 (endogenous or MYC-GFP-tagged) or phospho-BUBR1 at kinetochores, signal for these must be normalized for each cell individually. This step is crucial to account for potentially different BUBR1 levels at the kinetochores. To normalize phospho-BUBR1, normalize the values obtained for each cell from the image corresponding to the phospho-antibodies by the value from the same cells corresponding to the BUBR1 transgene (usually, a tag). In the same manner, to determine the heterogeneity and inherent intercellular variability in total BUBR1 levels, the values corresponding to exogenous BUBR1 (MYC-GFP signals) are normalized relative to a marker kinetochore protein (CREST, CENP-C etc.). Note that this process (step 1-5) represents the analysis of BUBR1 phosphorylation or levels at kinetochore of a single cell and must be repeated for each individual cell.
6. The measurements from all cells in one treatment, either phospho-BUBR1/BUBR1 or BUBR1/kinetochore signal, can be compared to the measurements of all cells from another signal by the appropriate statistical method.

7. Use of autophosphorylation and the characteristic mitotic upshift in SDS-PAGE as a readout

In addition to detecting BUBR1 phosphorylation by immunofluorescence as indicated above, bulk BUBR1 phosphorylation, which occurs exclusively in mitosis, can be measured by Western blotting. BUBR1 typically migrates as a doublet during mitosis, with the slower migrating band corresponding to hyperphosphorylated BUBR1 (Elowe et al., 2007; Matsuura et al., 2006). This upper band is lost or severely decreased in various pseudokinase domain mutants indicating that the pseudokinase fold regulates mitotic phosphorylation (Gama Braga et al., 2020). An important consideration here is the percentage of gel used to resolve the two bands; we routinely use 7.5% SDS-page gels for this assay. Another aspect to take into consideration is the viability of cells under mitotic arrest and the choice of method for cell cycle synchronization. Although nocodazole, the microtubule poison that is used to synchronize cells in mitosis is extremely effective, prolonged exposure can cause cell death, and thus timely sample recovery and processing of the cultured cells are essential to maintain BUBR1 phosphorylation intact for analysis. HeLa cells have a doubling time close to 24 h, and most microtubule poisons result in efficient enrichments after treatment for 12 to 16 h. In most cases, our experience shows that 12 h incubation with nocodazole at 300 nM is ideal, since it does not exacerbate cellular stress, but allows for enrichment in mitosis. Pre-synchronization with thymidine (as described above) enhances subsequent mitotic synchronization with nocodazole. In certain circumstances, for examples when the mitotic spindle needs to remain intact for the experimental protocol, alternative methods to nocodazole will be needed for synchronization (e.g. thymidine arrest and release or treatments with taxane drugs that stabilize microtubules, which also results in mitotic arrest). Finally, the quantity of starting material needed is inherently larger for biochemical experiments than for immunofluorescence and this must be also considered when planning all aspects of this experiment.

8. Before you begin

The solutions used in this protocol are:

- *RIPA lysis buffer*: Tris–HCl 150 mM pH 7.5, NaCl 150 mM, NaF 10 mM, NP-40 1%, Na-deoxycholate 0.1%, B-glycerophosphate

20 mM, sodium vanadate 0.1 mM, sodium pyrophosphate 10 mM, Leupeptin 1 mg/mL, 1 mg/mL Aprotinin, and 1 mM AEBSF

- *1x PBS*: 137 mM NaCl, 2.7 mM KCl, 10 mM Na_2HPO_4, 1.8 mM KH_2PO, pH7.4
- *1xTBS-T*: 50 mM Tris, 150 mM NaCl pH7.5, 0.02% Tween-20
- SDS-PAGE sample buffers, running buffers and transfer buffers of choice. We use the following:
  - *SDS-PAGE loading buffer 1X (Laemmli buffer or sample buffer)*: 60 mM Tris–HCl pH 6.8, 2% SDS, 10% glycerol, 5% b-mercaptoethanol, 0.01% bromophenol blue
  - *SDS-PAGE lower gel buffer*: 1.5 M Tris–HCl pH 8.8, 0.4% SDS
  - *SDS-PAGE upper gel buffer*: 0.5 M Tris–HCl pH 6.8, 0.4% SDS
  - *SDS-PAGE running buffer*: 25 mM Tris, 192 mM Glycine 0.1% SDS.
  - *Transfer Buffer (Bjerrum Schafer-Nielsen Buffer)*: 48 mM Tris, 39 mM glycine, 20% methanol, 0.1% SDS.
- *Kits/buffers for Chemiluminescence detection*: Western blotting signals from overexpressed BUBR1 can generally be readily detected using the following enhanced chemiluminescence buffers: 100 mM Tris–HCl pH 8.5, 0.225 mM p-coumaric acid (MP Biochemical, 102576), 1.25 mM Luminol (Sigma, A8511-5G/Fisher Scientific, S25395A). 3 μL of 30% hydrogen peroxide (Fisher, H325-500) is freshly added before the membrane is incubated in this buffer for 5 min. The chemiluminescence signal can now be visualized. Alternatively, if the signals are difficult to detect, we use SuperSignal West Femto Chemiluminescent Substrate (Fisher Scientific, PI34095).

8.1 Key resources table

See also Key resources table 5.1 for antibodies

Reagent or resource	Source	Identifier
Chemicals, peptides, and recombinant proteins		
Nocodazole	Sigma	M1404
Acryl/Bis solution(37.5:1) 40% (w/v)	Bio Basic	cat# A0008
BLUelf Prestained Protein ladder	Froggabio	PM008
Immobulon-P PVDF transfer membrane, 0.45 μM pore size	Millipore	IPVH00010

—cont'd

Reagent or resource	Source	Identifier
Experimental models: cell lines		
Human cell lines expressing BUBR1 (exogenous or endogenous)	–	–
Secondary antibodies for western blotting		
Peroxidase AffiniPure Goat Anti-Mouse IgG	Jackson ImmunoResearch (1:10000)	Cat# 115-035-003; RRID: AB_10015289
Peroxidase AffiniPure Goat Anti-Rabbit IgG	Jackson ImmunoResearch (1:10000)	Cat# 111-035-144; RRID:AB_2307391

8.2 Materials and equipment

- Tube rotator, preferentially in a cold room at 4 °C
- Protein quantification kit compatible with RIPA lysis buffer, such as the pierce BCA kit
- SDS-page running, transfer and chemiluminescence visualization equipment of your choice. We use the Bio-Rad mini-PROTEAN tetraSystem for SDS-PAGE, the Trans-Blot Turbo semi-dry transfer system V.1.02, and the Bio-Rad Chemi-Doc MP imaging system
- Falcon tubes
- A Büchner or vacuum flask connected to vacuum

8.3 Step-by-step method details

1. Plate cells expressing the various BUBR1 constructs for a minimum of 24 h or until cells have adhered and spread. For a 10 cm tissue dish 1.5–2 x 10^6 cells are seeded in 10 mL of DMEM containing 10% (vol/vol) FBS, Pen/Strep (100 mg/mL), 200 μg/mL hygromycin and 10 μg/mL blasticidin at 37 °C and 5% CO_2. Treat cells with nocodazole (300 nM–3.3 μM) for 12–16 h. This treatment is often done overnight for convenience.
2. Verify that cells have arrested efficiently in mitosis. This can be visualized by the presence of rounded cells by phase contrast in a light microscope (Fig. 4). Ideally, a mitotic index (percentage of mitotic cells) of 50-60% is desirable and indicates an efficient mitotic arrest. A mitotic index of more than 70% arrested cells may cause difficulties in cell

detachment and decrease the anticipated yield. This will vary with cell line and concentration of nocodazole used and may require additional optimization.

3. Wash cells with 1x PBS and collect the mitotic fractions by shake-off by gently tapping the side of the plate to dislodge the loosely attached mitotic cells but not the well-attached interphasic cells. Collect cells in 50 mL falcon tubes. More precisely, collect the medium into the conical tube leaving approximately 10% on the plate. Then, gently hit/tap the dish to dislodge the loosely attached mitotic cell which can now be collected with the remaining medium. A solution of PBS (5 mL) is added to the plate and the plate is tapped again. Repeat the process until most mitotic cells have been collected and mostly interphase cells are observed under the microscope.
4. Centrifuge cells in table-top centrifuge suitable for cell culture at low speed (500 x *g* for 5 min) and collect the pellet. Wash the pellet with 1 x PBS up to three times to remove the culture medium. To remove the supernatant after every wash, carefully use a Pasteur pipette equipped with a 200 μL micropipette tip on the narrow end and a tube connected to a Büchner or vacuum flask on the large end. The pressure should be low enough to avoid turbulence and the resuspension of the pellet. After the final wash, remove as much supernatant as possible. The cell pellet can now be stored at −80 °C if necessary.
5. After removing the final PBS wash, add RIPA buffer and solubilize the pellet. As a rule of thumb, for every 5 μL of pellet we use 30 μL of RIPA buffer. We routinely use 100–120 μL for a 10 cm plate.
6. Pipette up and down to disrupt the cells and begin the lysis. Incubate cells in a tube rotator at 4 °C for 30 to 45 min to lyse cells. We usually place the rotator in a cold room to keep the temperature low.
7. Centrifuge samples at 14,000 x *g*, 4 °C for 10 min to pellet the membranes fraction and cellular debris. Recover the supernatant.
8. Quantify the total protein concentration in the samples. We routinely use the Pierce BCA protein Assay Kit. A 20 μL aliquot of the lysate is used to generate quadruplicate samples for protein measurement at two independent dilutions according to the manufacturer's instructions.
9. Prepare equivalent protein amounts of the different cell extracts in SDS-PAGE sample buffer and run samples on 7.5% SDS-PAGE gels. Equalized amount of each lysate/condition should be mixed with 5X SDS-PAGE sample buffer (to give a final concentration of approximately 1X). Heat samples at 90 °C for 5 min and load in the wells of

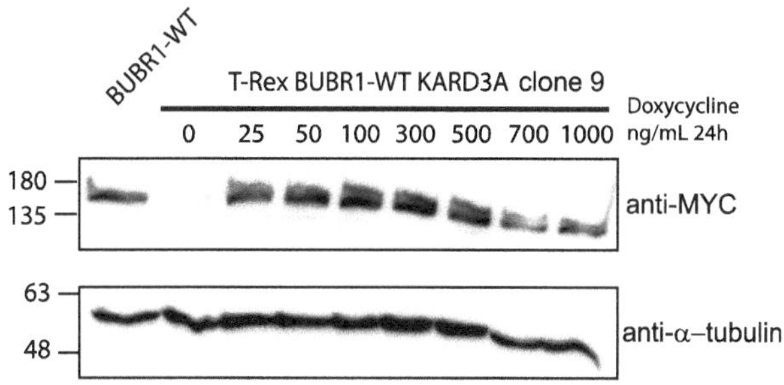

Fig. 6 A representative Western blot illustrating the expression of BUBR1 from mitotic samples at various concentrations of tetracycline. Note the double band of BUBR1 illustrating the mitotic-specific hyperphosphorylation.

the SDS-PAGE gel. The gel is run in the running buffer indicated above at 120 V until the blue dye front passes the stacking gel, after which the voltage is increased to 150 V.

10. Transfer the proteins to a membrane according to standard protocols and perform the immunoblot assay to visualize BUBR1. We use PVDF membranes and the Bio-Rad Trans-Blot Turbo semi-dry system and the Bjerrum Schafer-Nielsen Buffer to transfer. For a single mini-gel, transfer conditions are 25 V, 1 amp for 30 mins.
11. Transferred membranes are then blotted for BUBR1 using standard protocols. Briefly, membranes are blotted for 30 min in 5% (W/V) milk powder in 1xTBS-T and incubated overnight at 4 °C in primary antibody (dilution indicated in the "key resources table" in Section 4). Subsequently, membranes are washed in 3 x 10 min in 1xTBS-T at room temperature before adding the secondary antibody solution (1:10000 dilution of the anti-mouse/rabbit HRP diluted in 1xTBS-T) for 2 h at room temperature. The membrane is then washed again for 3x 10 min in 1xTBS-T at room temperature before developing the Western using a detection system. An example Western Blot for BUBR1 is shown in Fig. 6.

9. Summary

Studies of hBUBR1 have been challenging in the field of mitosis, in part due to the controversy regarding its catalytic/non-catalytic nature, which is compounded by the destabilizing effect of mutating the kinase fold on the entire peptide and the lack of an atomic structure of this domain. Interestingly, mutating kinase residues important for the phosphotransfer reaction have also been noted to destabilize another pseudokinase, the integrin-linked kinase arguing that pseudokinases may also play an important structural role (Fukuda, Knight, Piszczek, Kothary, & Qin, 2011). This chapter describes some of the challenges associated with studying pseudokinases like BUBR1 and provides some suggestions and considerations for overcoming the destabilizing effect of mutating the kinase fold in this protein. We also describe assays that can be used indirectly to measure a non-catalytic function of this domain, namely its effect on hBUBR1 phosphorylation in mitosis using two independent approaches.

References

Barford, D. (2020). Structural interconversions of the anaphase-promoting complex/cyclosome (APC/C) regulate cell cycle transitions. *Current Opinion in Structural Biology*, *61*, 86–97. https://doi.org/10.1016/j.sbi.2019.11.010.

Breit, C., Bange, T., Petrovic, A., Weir, J. R., Müller, F., Vogt, D., et al. (2015). Role of intrinsic and extrinsic factors in the regulation of the mitotic checkpoint kinase Bub1. *PLoS One*, *10*(12), 1–24. https://doi.org/10.1371/journal.pone.0144673.

Cerbini, T., Luo, Y., Rao, M. S., & Zou, J. (2015). Transfection, selection, and colony-picking of human induced pluripotent stem cells TALEN-targeted with a GFP gene into the AAVS1 safe harbor. *Journal of Visualized Experiments*, *96*, 52504. https://doi.org/10.3791/52504.

Ciossani, G., Overlack, K., Petrovic, A., Huis In 't Veld, P. J., Koerner, C., Wohlgemuth, S., et al. (2018). The kinetochore proteins CENP-E and CENP-F directly and specifically interact with distinct BUB mitotic checkpoint Ser/Thr kinases. *Journal of Biological Chemistry*, *293*(26), 10084–10101. https://doi.org/10.1074/jbc.RA118.003154.

Craske, B., & Welburn, J. P. I. (2020). Leaving no-one behind: How CENP-E facilitates chromosome alignment. *Essays in Biochemistry*. https://doi.org/10.1042/EBC20190073.

Elowe, S., Dulla, K., Uldschmid, A., Li, X., Dou, Z., & Nigg, E. A. (2010). Uncoupling of the spindle-checkpoint and chromosome-congression functions of BubR1. *Journal of Cell Science*, *123*(1), 84–94. https://doi.org/10.1242/jcs.056507.

Elowe, S., Hümmer, S., Uldschmid, A., Li, X., & Nigg, E. A. (2007). Tension-sensitive Plk1 phosphorylation on BubR1 regulates the stability of kinetochore–microtubule interactions. *Genes & Development*, *21*(17), 2205–2219. https://doi.org/10.1101/gad.436007.

Fukuda, K., Knight, J. D. R., Piszczek, G., Kothary, R., & Qin, J. (2011). Biochemical, proteomic, structural, and thermodynamic characterizations of integrin-linked kinase (ILK): Cross-validation of the pseudokinase. *The Journal of Biological Chemistry*, *286*(24), 21886–21895. https://doi.org/10.1074/jbc.M111.240093.

Gama Braga, L., Cisneros, A. F., Mathieu, M. M., Clerc, M., Garcia, P., Lottin, B., et al. (2020). BUBR1 pseudokinase domain promotes kinetochore PP2A-B56 recruitment, spindle checkpoint silencing, and chromosome alignment. *Cell Reports*, *33*(7), 108397. https://doi.org/10.1016/j.celrep.2020.108397.

Guo, Y., Kim, C., Ahmad, S., Zhang, J., & Mao, Y. (2012). CENP-E-dependent BubR1 autophosphorylation enhances chromosome alignment and the mitotic checkpoint. *The Journal of Cell Biology*, *198*(2), 205–217. https://doi.org/10.1083/jcb.201202152.

Hara, M., & Fukagawa, T. (2018). Kinetochore assembly and disassembly during mitotic entry and exit. *Current Opinion in Cell Biology*, *52*, 73–81. https://doi.org/10.1016/j.ceb.2018.02.005.

Hein, J. B., Hertz, E. P. T., Garvanska, D. H., Kruse, T., & Nilsson, J. (2017). Distinct kinetics of serine and threonine dephosphorylation are essential for mitosis. *Nature Cell Biology*, *ncb3634*. https://doi.org/10.1038/ncb3634.

Huang, Y., Lin, L., Liu, X., Ye, S., Yao, P. Y., Wang, W., et al. (2019). BubR1 phosphorylates CENP-E as a switch enabling the transition from lateral association to end-on capture of spindle microtubules. *Cell Research*, *29*(7), 562. https://doi.org/10.1038/s41422-019-0178-z.

Kruse, T., Zhang, G., Larsen, M. S. Y., Lischetti, T., Streicher, W., Kragh Nielsen, T., et al. (2013). Direct binding between BubR1 and B56-PP2A phosphatase complexes regulate mitotic progression. *Journal of Cell Science*, *126*(Pt 5), 1086–1092. https://doi.org/10.1242/jcs.122481.

Lara-Gonzalez, P., Pines, J., & Desai, A. (2021). Spindle assembly checkpoint activation and silencing at kinetochores. *Seminars in Cell & Developmental Biology*, *117*, 86–98. https://doi.org/10.1016/j.semcdb.2021.06.009.

Legal, T., Hayward, D., Gluszek-Kustusz, A., Blackburn, E. A., Spanos, C., Rappsilber, J., et al. (2020). The C-terminal helix of BubR1 is essential for CENP-E-dependent chromosome alignment. *Journal of Cell Science*. https://doi.org/10.1242/jcs.246025.

Lischetti, T., Zhang, G., Sedgwick, G. G., Bolanos-Garcia, V. M., & Nilsson, J. (2014). The internal Cdc20 binding site in BubR1 facilitates both spindle assembly checkpoint signalling and silencing. *Nature Communications*, *5*, 5563. https://doi.org/10.1038/ncomms6563.

Mace, P. D., & Murphy, J. M. (2021). There's more to death than life: Noncatalytic functions in kinase and pseudokinase signaling. *Journal of Biological Chemistry*, *296*, 100705. https://doi.org/10.1016/j.jbc.2021.100705.

Malureanu, L. A., Jeganathan, K. B., Hamada, M., Wasilewski, L., Davenport, J., & Deursen, J. M.v. (2009). BubR1 N terminus acts as a soluble inhibitor of cyclin B degradation by APC/CCdc20 in interphase. *Developmental Cell*, *16*(1), 118–131. https://doi.org/10.1016/j.devcel.2008.11.004.

Mao, Y., Abrieu, A., & Cleveland, D. W. (2003). Activating and silencing the mitotic checkpoint through CENP-E-dependent activation/inactivation of BubR1. *Cell*, *114*(1), 87–98. https://doi.org/10.1016/S0092-8674(03)00475-6.

Mao, Y., Desai, A., & Cleveland, D. W. (2005). Microtubule capture by CENP-E silences BubR1-dependent mitotic checkpoint signaling. *The Journal of Cell Biology*, *170*(6), 873–880. https://doi.org/10.1083/jcb.200505040.

Matsuura, S., Matsumoto, Y., Morishima, K., Izumi, H., Matsumoto, H., Ito, E., et al. (2006). Monoallelic BUB1B mutations and defective mitotic-spindle checkpoint in seven families with premature chromatid separation (PCS) syndrome. *American Journal of Medical Genetics Part A*, *140A*(4), 358–367. https://doi.org/10.1002/ajmg.a.31069.

Murphy, J. M., Mace, P. D., & Eyers, P. A. (2017). Live and let die: Insights into pseudoenzyme mechanisms from structure. *Current Opinion in Structural Biology*, *47*, 95–104. https://doi.org/10.1016/j.sbi.2017.07.004.

Murphy, J. M., Zhang, Q., Young, S. N., Reese, M. L., Bailey, F. P., Eyers, P. A., et al. (2014). A robust methodology to subclassify pseudokinases based on their nucleotide-binding properties. *Biochemical Journal*, *457*(2), 323–334. https://doi.org/10.1042/BJ20131174.

Musacchio, A. (2011). Spindle assembly checkpoint: The third decade. *Philosophical Transactions of the Royal Society B: Biological Sciences*, *366*(1584), 3595–3604. https://doi.org/10.1098/rstb.2011.0072.

Musacchio, A. (2015). The molecular biology of spindle assembly checkpoint signaling dynamics. *Current Biology*, *25*(20), R1002–R1018. https://doi.org/10.1016/j.cub.2015.08.051.

Musacchio, A., & Desai, A. (2017). A molecular view of kinetochore assembly and function. *Biology*, *6*(1). https://doi.org/10.3390/biology6010005.

Navarro, A. P., & Cheeseman, I. M. (2021). Kinetochore assembly throughout the cell cycle. *Seminars in Cell & Developmental Biology*, *117*, 62–74. https://doi.org/10.1016/j.semcdb.2021.03.008.

Rahmani, Z., Gagou, M. E., Lefebvre, C., Emre, D., & Karess, R. E. (2009). Separating the spindle, checkpoint, and timer functions of BubR1. *The Journal of Cell Biology*, *187*(5), 597–605. https://doi.org/10.1083/jcb.200905026.

Schneider, C. A., Rasband, W. S., & Eliceiri, K. W. (2012). NIH image to ImageJ: 25 years of image analysis. *Nature Methods*, *9*(7), 671–675. https://doi.org/10.1038/nmeth.2089.

Sieben, C. J., Jeganathan, K. B., Nelson, G. G., Sturmlechner, I., Zhang, C., van Deursen, W. H., et al. (2020). BubR1 allelic effects drive phenotypic heterogeneity in mosaic-variegated aneuploidy progeria syndrome. *The Journal of Clinical Investigation*, *130*(1), 171–188. https://doi.org/10.1172/JCI126863.

Simmons, A. J., Park, R., Sterling, N. A., Jang, M.-H., van Deursen, J. M. A., Yen, T. J., et al. (2019). Nearly complete deletion of BubR1 causes microcephaly through shortened mitosis and massive cell death. *Human Molecular Genetics*, *28*(11), 1822–1836. https://doi.org/10.1093/hmg/ddz022.

Suijkerbuijk, S. J. E., Osch, M. H. J.v., Bos, F. L., Hanks, S., Rahman, N., & Kops, G. J. P. L. (2010). Molecular causes for BUBR1 dysfunction in the human cancer predisposition syndrome mosaic variegated aneuploidy. *Cancer Research*, *70*(12), 4891–4900. https://doi.org/10.1158/0008-5472.CAN-09-4319.

Suijkerbuijk, S. J. E., van Dam, T. J. P., Karagöz, G. E., von Castelmur, E., Hubner, N. C., Duarte, A. M. S., et al. (2012). The vertebrate mitotic checkpoint protein BUBR1 is an unusual pseudokinase. *Developmental Cell*, *22*(6), 1321–1329. https://doi.org/10.1016/j.devcel.2012.03.009.

Suijkerbuijk, S. J. E., Vleugel, M., Teixeira, A., & Kops, G. J. P. L. (2012). Integration of kinase and phosphatase activities by BUBR1 ensures formation of stable kinetochore-microtubule attachments. *Developmental Cell*, *23*(4), 745–755. https://doi.org/10.1016/j.devcel.2012.09.005.

Tang, R., Jiang, Z., Chen, F., Yu, W., Fan, K., Tan, J., et al. (2020). The kinase activity of Drosophila BubR1 is required for insulin signaling-dependent stem cell maintenance. *Cell Reports*, *31*(12), 107794. https://doi.org/10.1016/j.celrep.2020.107794.

Wang, Q., Liu, T., Fang, Y., Xie, S., Huang, X., Mahmood, R., et al. (2004). BUBR1 deficiency results in abnormal megakaryopoiesis. *Blood*, *103*(4), 1278–1285. https://doi.org/10.1182/blood-2003-06-2158.

Watson, E. R., Brown, N. G., Peters, J.-M., Stark, H., & Schulman, B. A. (2019). Posing the APC/C E3 ubiquitin ligase to orchestrate cell division. *Trends in Cell Biology*, *29*(2), 117–134. https://doi.org/10.1016/j.tcb.2018.09.007.

Xu, P., Raetz, E. A., Kitagawa, M., Virshup, D. M., & Lee, S. H. (2013). BUBR1 recruits PP2A via the B56 family of targeting subunits to promote chromosome congression. *Biology Open*, *2*(5), 479–486. https://doi.org/10.1242/bio.20134051.

CHAPTER EIGHTEEN

The pseudokinase domain in receptor guanylyl cyclases

Avipsa Bose and Sandhya S. Visweswariah*
Department of Molecular Reproduction, Development and Genetics, Indian Institute of Science, Bangalore, India
*Corresponding author: e-mail address: sandhya@iisc.ac.in

Contents

Abstract

Cyclic GMP is produced by enzymes called guanylyl cyclases, of which the membrane-associated forms contain an intracellular pseudokinase domain that allosterically regulates the C-terminal guanylyl cyclase domain. Ligand binding to the extracellular domain of these single transmembrane-spanning domain receptors elicits an increase in cGMP levels in the cell. The pseudokinase domain (or kinase-homology domain) in

Methods in Enzymology, Volume 667
ISSN 0076-6879
https://doi.org/10.1016/bs.mie.2022.03.046

these receptors appears to be critical for ligand-mediated activation. While the pseudokinase domain does not possess kinase activity, biochemical evidence indicates that the domain can bind ATP and thereby allosterically regulate the catalytic activity of these receptors. The pseudokinase domain also appears to be the site of interaction of regulatory proteins, as seen in the retinal guanylyl cyclases that are involved in visual signal transduction. In the absence of structural information on the pseudokinase-guanylyl cyclase domain organization of any member of this family of receptors, biochemical evidence has provided clues to the physical interaction of the pseudokinase and guanylyl cyclase domain. An α-helical linker region between the pseudokinase domain and the guanylyl cyclase domain regulates the basal activity of these receptors in the absence of a stimulatory ligand and is important for stabilizing the structure of the pseudokinase domain that can bind ATP. Here, we present an overview of salient features of ATP-mediated regulation of receptor guanylyl cyclases and describe biochemical approaches that allow a clearer understanding of the intricate interplay between the pseudokinase domain and catalytic domain in these proteins.

Abbreviations

ANP atrial natriuretic peptide
ATP adenosine triphosphate
BNP B-type natriuretic peptide
cGMP guanosine 3′,5′-cyclic monophosphate
CNP C-type natriuretic peptide
GTP guanosine triphosphate

1. Introduction

Guanosine 3′,5′-cyclic monophosphate (cGMP) was first discovered in rat urine in 1963 (Ashman, Lipton, Melicow, & Price, 1963). Levels of this second messenger are maintained inside the cell by conversion of GTP to cGMP by guanylyl cyclases and hydrolysis of cGMP to GMP by cyclic nucleotide phosphodiesterases. Mammals express four soluble guanylyl cyclase (sGC) subunits (α_1, α_2, β_1, and β_2) and seven membrane-spanning or receptor guanylyl cyclases (rGC), namely, GC-A, GC-B, GC-C, GC-D, GC-E, GC-F, and GC-G (Potter, 2011a). In primates, GC-D and GC-G exist as pseudogenes, with multiple insertions and deletions accumulated during the course of evolution, that would generate truncated and/or non-functional proteins (Manning, Whyte, Martinez, Hunter, & Sudarsanam, 2002). The functions of these receptors are determined by the tissue and cell types in which they are located. They have essential roles to play in phototransduction, intravascular blood volume, vascular tone, bone development, and gut physiology (Potter, 2011a). rGCs can be activated by diverse peptides, but the extracellular ligands for rGCs involved in visual signal transduction are unknown.

Receptor GCs are believed to function as homodimers because of the requirement for dimerization of the guanylyl cyclase domain for catalytic activity (Liu, Ruoho, Rao, & Hurley, 1997; Thompson & Garbers, 1995). These receptors have a complex multidomain organization that includes an N-terminal, extracellular domain required for ligand binding, a single transmembrane domain, a juxtamembrane domain, followed by the pseudokinase or kinase-homology domain (KHD) responsible for allosteric regulation of the receptor (Potter, 2011b). Indeed, the first recognition of a "pseudokinase-like domain" in a protein was noted by David L. Garbers in a study related to the cloning of a plasma membrane-associated guanylyl cyclase from the sea urchin (Singh et al., 1988). The KHD is connected to the catalytic guanylyl cyclase domain through a linker region (Fig. 1). Phylogenetic analysis of receptor GCs suggests coevolution of the KHD and its associated cyclase domain, and conservation of the sequence and length of a predicted helical region between these two domains (Biswas, Shenoy, Dutta, & Visweswariah, 2009).

A broader phylogenetic analysis of the KHDs of rGCs with protein kinases, in general, show that they branch close to tyrosine kinases (Fig. 2) (Kwon et al., 2019; Letunic & Bork, 2021). Shown in Fig. 3 is an alignment of the KHDs of rGCs along with Src tyrosine kinase, where critical blocks of residues required for kinase activity are shown. The KHD in these receptors are considered pseudokinases since they lack the aspartate residue present in the catalytic loop of classical protein kinases that binds the Mg^{2+} ion required for catalysis and transfer of the γ-phosphate of ATP to the kinase substrate (Knape & Herberg, 2017).

Several studies have described the critical role of the KHD in regulating the activity of these receptors (Bhandari et al., 2001; Potter, 2011a). Consequently, multiple human disorders are associated with mutations in the KHD (Table 1). Therefore, understanding the regulatory mechanisms of the KHD will aid in the design of molecules that either activate or inhibit the receptor as required. It is important to note that structural information is only available for the ECD of GC-A (Ogawa, Qiu, Ogata, & Misono, 2004; Padayatti, Pattanaik, Ma, & van den Akker, 2004). Therefore, the pseudokinase domain of these receptors and residues involved in binding ATP have so far been characterized only through mutational and biochemical analyses.

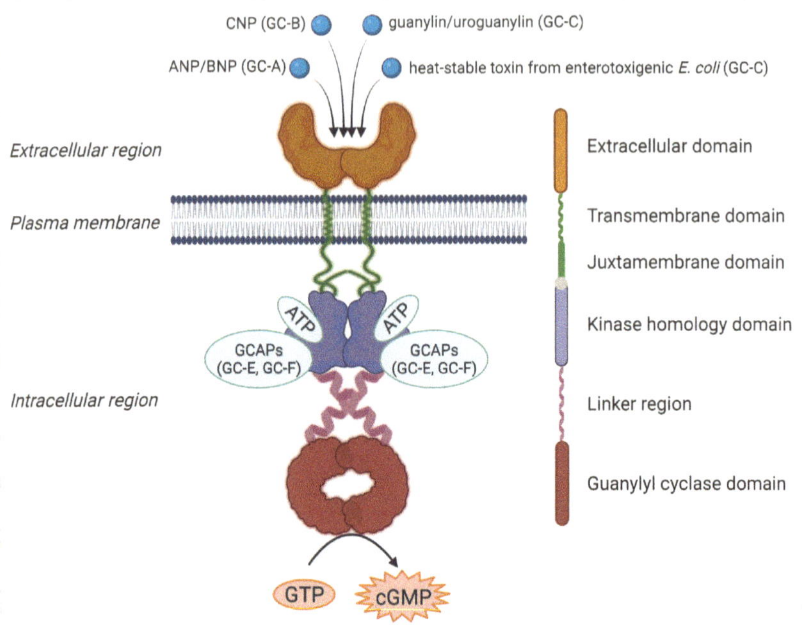

Fig. 1 Schematic representation of overall domain organization in receptor guanylyl cyclases. Receptor guanylyl cyclases function as homodimers where each monomer possesses six domains, namely, an N-terminal extracellular domain (ECD) for ligand binding, a single membrane spanning transmembrane domain followed by adjacent juxtamembrane domain, a pseudokinase or kinase homology domain (KHD) which allosterically regulates the receptor, and an α-helical linker region connecting the KHD with the catalytic guanylyl cyclase domain. GC-C has an additional C-terminal domain of ~60 amino acids that helps in anchoring to the cytoskeleton (not shown). Numbers in the bracket indicate the approximate length of each domain (GC-C has been used as a reference). Diverse extracellular ligands can bind to the extracellular domain, and the ligands thus far identified are shown. The pseudokinase domain also serve as a binding site for guanylyl cyclase activating proteins (GCAPs) which interact with Ca^{2+} to regulate guanylyl cyclase activity. ATP binds to the KHD of all rGCs.

2. Allosteric regulation of rGCs by the pseudokinase domain

2.1 GC-A and GC-B

Guanylyl cyclase A (GC-A), also called natriuretic peptide receptor A (NPR-A or NPR1), is activated by atrial natriuretic peptide (ANP) and B-type natriuretic peptide (BNP) produced by atria and ventricles, respectively (Potter, 2011b; Suga et al., 1992). GC-A is expressed in the lung,

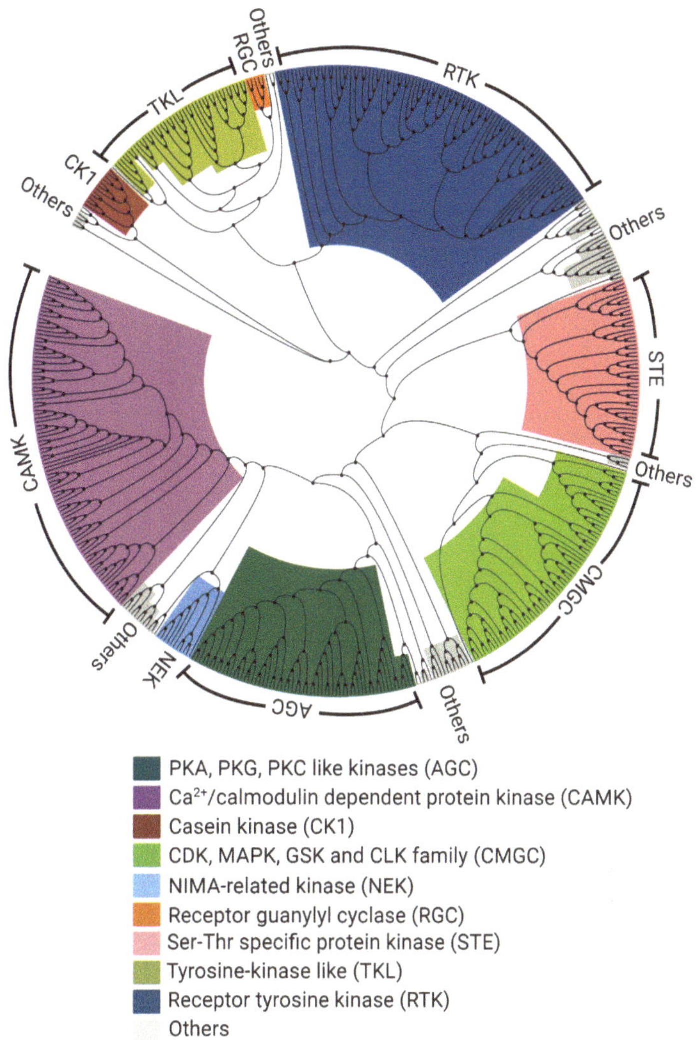

Fig. 2 Phylogenetic analysis of human kinases. Phylogenetic analysis of 497 kinases from the human genome classifies them into 9 groups, namely, PKA, PKG, PKC like kinases (AGC), Ca^{2+}/calmodulin dependent protein kinase (CAMK), casein kinase 1 (CK1), CDK, MAPK, GSK and CDK family (CMGC), NIMA-related kinase (NEK), receptor guanylyl cyclase (RGC), Ser-Thr specific protein kinase (STE), tyrosine-kinase like (TKL), and receptor tyrosine kinase (RTK). The receptor guanylyl cyclases cluster as a separate class close to receptor tyrosine kinases. The tree has been prepared using iTOL (https://itol.embl.de) (Letunic & Bork, 2021).

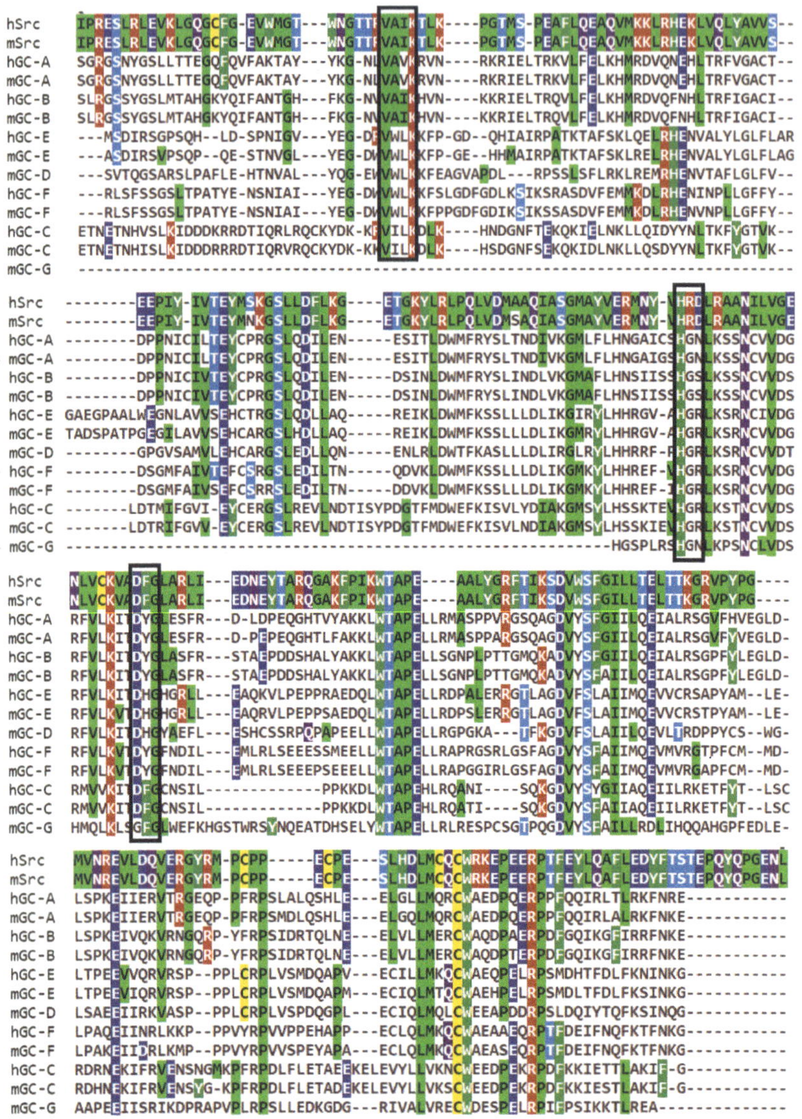

Fig. 3 Multiple sequence alignment of kinase homology domains (KHDs) from human and mouse receptor guanylyl cyclases (rGCs) with Src tyrosine kinase. Comparison of rGCs, from human and mouse, with Src tyrosine kinase. Mutations in the 3 motifs responsible for kinase activity (highlighted in black box) are shown. The Lys residue in the VAIK motif and Asp residue in the DFG motif, critical for interacting with ATP and metal ion is conserved across all rGCs. However, the Asp residue in the HRD motif responsible for phosphotransfer activity of a kinase is mutated to Asn (in GC-A), Ser (in GC-B) or Arg (in GC-C, GC-E, and GC-F).

Table 1 Mutations in the pseudokinase domain of receptor guanylyl cyclases associated with human disease.

Mutation	**Protein change**	**Condition(s)**	**dbSNP ID**
GC-B mutations			
c.1589C>A	A530D, A533D	Acromesomelic dysplasia, Maroteaux type | Epiphyseal chondrodysplasia, miura type	
c.1636A>T	N546Y, N549Y		
c.1670G>A	R557H, R560H		
c.1673T>C	I558T		rs751324720
c.1684C>T	R562W		rs566096931
c.1714A>T	R572[a]	Not provided	rs1554673485
c.1739C>G	T580S		rs1131692010
c.1801C>T	R601C		rs1563988849
c.1802G>A	R604H, R601H	Acromesomelic dysplasia, Maroteaux type | Epiphyseal chondrodysplasia, miura type	
c.1802G>T	R601L		
c.1813C>T	Q605[a]		rs969576919
c.1839C>G	I616M, I613M		
c.1844T>G	L615W, L618W		
c.1922C>T	S641L, S644L		
c.1948T>G	C650G, C653G		
c.1963C>T	R655C		rs587777596
c.1968dup	V657fs		rs1554673888
c.2065C>T	P689S, P692S		
c.2105T>A	M702K		rs200129431
c.2143C>T	Q715[a]		
c.2150T>C	I717T, I720T		
c.2162_2172del	S721fs	Trident hand | Limb undergrowth | Growth delay | Craniosynostosis syndrome	rs1057518817
c.2246G>A	R749Q, R752Q	Short stature with nonspecific skeletal abnormalities	

Continued

Table 1 Mutations in the pseudokinase domain of receptor guanylyl cyclases associated with human disease.—cont'd

<table>
<tr><th>Mutation</th><th>Protein change</th><th>Condition(s)</th><th>dbSNP ID</th></tr>
<tr><td>c.2252G>A</td><td>S751N</td><td rowspan="3">Acromesomelic dysplasia, Maroteaux type</td><td></td></tr>
<tr><td>c.2260C>T</td><td>R754W</td><td>rs371968545</td></tr>
<tr><td>c.2261G>A</td><td>R754Q</td><td>rs763488261</td></tr>
<tr><td>c.2266C>T</td><td>Q756[a], Q759[a]</td><td>Intellectual disability|Epiphyseal chondrodysplasia, miura type</td><td></td></tr>
<tr><td>c.2299C>T</td><td>R767[a], R770[a]</td><td>Short stature with nonspecific skeletal abnormalities</td><td></td></tr>
<tr><td>c.2302T>C</td><td>C768R</td><td rowspan="8">Acromesomelic dysplasia, Maroteaux type</td><td>rs1057519333</td></tr>
<tr><td>c.2321C>T</td><td>A774V</td><td>rs369154896</td></tr>
<tr><td>c.2327G>A</td><td>R776Q</td><td>rs780293535</td></tr>
<tr><td>c.2351G>A</td><td>G784D</td><td></td></tr>
<tr><td>c.2359C>T</td><td>R787W</td><td>rs114147262</td></tr>
<tr><td>c.2360G>A</td><td>R787Q, R790Q</td><td></td></tr>
<tr><td>c.2362C>T</td><td>R788C, R791C</td><td></td></tr>
<tr><td>c.2363G>A</td><td>R788H, R791H</td><td></td></tr>
<tr><td colspan="4">GC-C mutations</td></tr>
<tr><td>c.1540A>G</td><td>I514V</td><td rowspan="2">Diarrhea 6|Meconium ileus</td><td></td></tr>
<tr><td>c.1544T>C</td><td>L515P</td><td></td></tr>
<tr><td>c.1616T>C</td><td>I539T</td><td>Not provided</td><td></td></tr>
<tr><td>c.1798G>A</td><td>G600R</td><td rowspan="3">Intestinal obstruction in the newborn due to guanylate cyclase 2C deficiency</td><td></td></tr>
<tr><td>c.1997A>G</td><td>Y666C</td><td></td></tr>
<tr><td>c.2008G>A, c.2728T>C</td><td>A670T, C928R</td><td>rs587784572, rs587784573</td></tr>
<tr><td>c.2067T>G</td><td>N689K</td><td rowspan="2">Not specified</td><td>rs138497004</td></tr>
<tr><td>c.2086G>T</td><td>E696[a]</td><td>rs1057520701</td></tr>
<tr><td>c.2155G>C</td><td>E719Q</td><td>Meconium ileus</td><td>rs895214647</td></tr>
<tr><td colspan="4">GC-E mutations</td></tr>
<tr><td>c.1608G>A</td><td>M536I</td><td>Cone-rod dystrophy 6|Leber congenital amaurosis 1</td><td></td></tr>
</table>

Table 1 Mutations in the pseudokinase domain of receptor guanylyl cyclases associated with human disease.—cont'd

Mutation	Protein change	Condition(s)	dbSNP ID
c.1615A>G	I539V	Not provided	rs61749753
c.1618C>T	R540C	Not provided	rs61749754
c.1619G>T	R540L	Leber congenital amaurosis 1|Cone-rod dystrophy 6	
c.1624G>A	G542S		
c.1633C>T	Q545[a]		rs1290420698
c.1660G>A	V554I		
c.1672G>A	D558N		rs188568530
c.1694T>C	F565S		rs61749755
c.1707G>T	Q569H		
c.1713A>G	I571M		rs886043672
c.1717A>G	I573V		rs61749756
c.1720C>T	R574C		rs137853897
c.1721G>A	R574H		rs560270873
c.1724C>T	P575L		rs28743021
c.1758G>T	E586D		
c.1763G>A	R588Q		
c.1766A>G	H589R		
c.1771A>G	N591D		
c.1773del	N591fs		rs794727952
c.1774G>A	V592M		
c.1804C>T	R602W		
c.1806_1830del	A604fs		rs63749078
c.1817G>A	G606D		
c.1825G>A	A609T		
c.1828C>A	L610I		
c.1841A>G	N614S		
c.1852G>A	V618I		

Continued

Table 1 Mutations in the pseudokinase domain of receptor guanylyl cyclases associated with human disease.—cont'd

Mutation	Protein change	Condition(s)	dbSNP ID
c.1868C>T	T623M		
c.1870C>T	R624W		
c.1871G>A	R624Q		
c.1877C>G	S626C		
c.1887C>A	D629E		
c.1901G>C	R634T		
c.1904A>G	E635G		
c.1924T>G	F642V		
c.1937T>C	L646P		
c.1943T>C	L648P		
c.1957G>A	G653R	Not provided	rs61750160
c.1964G>A	R655K	Retinal dystrophy	
c.1972C>T	H658Y	Cone-rod dystrophy 6	rs1598149154
c.1974C>A	H658Q		
c.1978C>T	R660[a]		rs61750161
c.1979G>A	R660Q		rs61750162
c.1984G>A	V662M		
c.1991A>C	H664P		
c.1992T>G	H664Q		rs1598149187
c.1996C>T	R666W		
c.1997G>A	R666Q		
c.2008C>T	R670W		
c.2015G>A	C672Y		
c.2056G>A	G686S		
c.2059C>T	H687Y		
c.2078_2085del	A693fs	Not provided	rs61750163
c.2080C>A	Q694K		rs61750164

Table 1 Mutations in the pseudokinase domain of receptor guanylyl cyclases associated with human disease.—cont'd

Mutation	Protein change	Condition(s)	dbSNP ID
c.2093C>G	P698R	Leber congenital amaurosis 1|Cone-rod dystrophy 6	
c.2101C>T	P701S		rs34598902
c.2108C>T	A703V		
c.2122T>C	W708R		rs1064797217
c.2129C>T	A710V		rs781725943
c.2132C>T	P711L		
c.2164C>T	R722W		rs34331388
c.2165G>A	R722Q		
c.2171C>T	T724M		
c.2179G>A	G727S		
c.2182G>A	D728N		
c.2204T>C	I735T		
c.2224C>T	R742C		
c.2234del	P745fs		rs1598149659
c.2237A>G	Y746C	Not provided	rs61750166
c.2248G>T	E750[a]		rs61750167
c.2260G>A	E754K	Leber congenital amaurosis 1|Cone-rod dystrophy 6	
c.2281C>T	R761W		
c.2290C>T	P764S		
c.2302C>T	R768W		rs61750168
c.2303G>A	R768Q		rs750889782
c.2312T>A	V771E		
c.2317A>C	M773L		rs61750169
c.2318T>C	M773T		
c.2325G>T	Q775H		
c.2335G>T	E779[a]		rs541299023
c.2338T>C	C780R		

Continued

Table 1 Mutations in the pseudokinase domain of receptor guanylyl cyclases associated with human disease.—cont'd

Mutation	Protein change	Condition(s)	dbSNP ID
c.2345T>A	L782H		rs8069344
c.2359T>G	C787G		
c.2377del	E793fs		rs1555635668
c.2383C>T	R795W		
c.2384G>A	R795Q		rs61750171
c.2384G>T	R795L	Not provided	rs61750171
c.2393T>G	M798R	Leber congenital amaurosis 1	
c.2395_2398dup	H800fs		
c.2431G>A	G811S		rs1196292120

[a]The introduction of a stop codon after the indicated amino acid.
Data taken from ClinVar https://www.ncbi.nlm.nih.gov/clinvar/.

kidney, brain, liver, adrenal, vasculature, adipose, and endothelial tissues and at lower levels in the heart (Bryan et al., 2006; Dickey et al., 2007; Potter, Abbey-Hosch, & Dickey, 2006). Guanylyl cyclase B (GC-B) shows 78% identity to GC-A in the intracellular region and is also called the natriuretic peptide receptor B (NPR-B or NPR2) (Schulz, Green, Yuen, & Garbers, 1990). GC-B is stimulated by C-type natriuretic peptide (CNP) and enhances long bone growth and oocyte maturation (Koller et al., 1991; Yasoda et al., 1998; Zhang, Su, Sugiura, Xia, & Eppig, 2010). GC-B is highly expressed in the brain, heart, lung, bone, ovary, fibroblast, and vascular smooth muscle cells (Chrisman, Schulz, Potter, & Garbers, 1993; Nagase, Katafuchi, Hirose, & Fujita, 1997; Schulz et al., 1989).

GC-A exists as a homodimer, and ligand binding does not induce further oligomerization (Chinkers & Wilson, 1992). Early on, it was hypothesized that the conformational changes seen in GC-A and GC-B upon ligand binding propagate through the KHD to the catalytic domain. Phosphorylation of the KHD at its N-terminus is essential for ligand-mediated signal transduction from the ECD to the catalytic domain in both GC-A and GC-B, although the identities of the kinases that phosphorylate these receptors remain unknown. The phosphorylated sites have been chemically determined in rat and human GC-A and GC-B. They are Ser^{487}, Ser^{497}, Thr^{500}, Ser^{502}, Ser^{506}, Ser^{510}, and Thr^{513} (in rat GC-A) and Ser^{513}, Thr^{516}, Ser^{518}, Ser^{523},

Ser^{526}, and Thr^{529} (in rat GC-B) (Potter, 1998; Potter & Garbers, 1992; Potter & Hunter, 1998a, 1998b; Schroter et al., 2010; Yoder, Stone, Griffin, & Potter, 2010). Prolonged exposure to natriuretic peptides in intact cells, or treatment of membrane fractions with protein phosphatase 2A, reduces the phosphate content of these receptors. This is associated with a loss of natriuretic peptide-dependent guanylyl cyclase activity (Potter & Garbers, 1992). GC-A and GC-B show increased guanylyl cyclase activity (as monitored by cGMP production by radioimmunoassay (RIA) or enzyme-linked immunosorbent assays (ELISAs)) in the presence of non-ionic detergents, by unclear mechanisms. Phosphorylation/dephosphorylation does not affect detergent-mediated cGMP production (Joubert, Labrecque, & De Lean, 2001; Potter, 1998).

Deleting the pseudokinase domain in GC-A led to a constitutively active receptor that was not stimulated further upon ligand addition. On the contrary, partial deletion of the KHD results in an inactive receptor (Chinkers & Garbers, 1989; Koller, de Sauvage, Lowe, & Goeddel, 1992).

To differentiate the allosteric effects of ATP from its phosphorylation effects, the non-hydrolyzable analog AMP-PNP and dephosphorylation resistant analog ATP-γ-S were used in assays with crude preparations of membranes prepared from cells or tissues (Chang et al., 1990; Gazzano, Wu, & Waldman, 1991; Kurose, Inagami, & Ui, 1987). ATP and AMP-PNP augmented ligand-independent or dependent GC-A activity in vitro when Mg^{2+} ions (required for the two-metal ion catalysis shown by nucleotidyl cyclases; see below) were provided in assays along with GTP. However, guanylyl cyclase activity was inhibited when Mn^{2+} ions were provided along with GTP, or assays were performed in the presence of non-ionic detergents (Chang et al., 1990; Gazzano et al., 1991; Kurose et al., 1987).

Detailed enzymatic analyses demonstrated that ATP shifts the kinetics from positive cooperativity in the absence of ATP to classical Michaelis-Menten kinetics in its presence. This results in a 10-fold reduction of the Michaelis' constant for GTP, allowing the receptor to function at physiological concentration of GTP in the presence of natriuretic peptides (Robinson & Potter, 2012). The crystal structure of the ECD of GC-A revealed that ligand binding results in a 24° counter-clockwise rotation in the ECD of one monomer with respect to the other (Ogawa et al., 2004). This conformational change is thought to propagate along the receptor through the KHD and ultimately relay the signal to the catalytic domain to form a functional catalytic domain dimer. Comparison with Src tyrosine kinase shows that the ATP-binding Lys residue in the VAIK motif and Asp residue in the DFG motif of GC-A and GC-B is conserved (Fig. 3). However, the Asp residue in the HRD motif required for phosphotransfer activity is replaced by Asn in GC-A and GC-B (Fig. 3). Alanine substitutions in the VAIK and DYG kinase motifs of GC-A and GC-B did not alter the phosphate content of the receptors or ligand-mediated activation, but prevented the ATP-dependent reduction of Michaelis' constant for GTP (Edmund, Walseth, Levinson, & Potter, 2019). To understand the mechanism of allosteric regulation by ATP, mutations were made to rigidify the regulatory and catalytic spine of the KHD (A533W in GC-A, M571F, and I583W in GC-B). This mimics the stabilization of the spine on ATP binding to protein kinases (Taylor, Meharena, & Kornev, 2019). These mutations increased guanylyl cyclase activity and reduced the Michaelis' constant for GTP, even in the absence of ATP (Edmund et al., 2019).

To determine the effect of ATP on cooperativity, substrate-velocity experiments were performed after deletion of the KHD. The absence of the KHD led to a shift from positive cooperativity to linear kinetics but failed to reduce the Michaelis' constant in the presence of ATP. Therefore, ATP binding to the catalytic domain of GC-A and GC-B is sufficient for its effect on cooperativity, but the reduction of the Michaelis' constant requires ATP binding to the KHD, coupled with ligand binding to the ECD (Edmund et al., 2019).

2.2 GC-C

Guanylyl cyclase C (GC-C; gene *GUCY2C*) has a similar domain architecture to GC-A and GC-B but possesses a C-terminal domain of 60 amino acids that anchors the receptor to the cellular cytoskeleton (Basu, Arshad, & Visweswariah, 2010). GC-C is predominantly expressed on the apical surface of intestinal epithelial cells and plays an important role in regulating intestinal fluid-ion homeostasis (Arshad & Visweswariah, 2012; De Jonge, 1975a, 1975b; Swenson, Mann, Jump, Witte, & Giannella, 1996). It is stimulated by the endogenous peptide hormones, guanylin and uroguanylin, and exogenous heat-stable enterotoxin (STa) from enterotoxigenic *E. coli*. STa acts as the super-agonist of GC-C and therefore results in diarrhea in infected individuals (Bose, Banerjee, & Visweswariah, 2020; Schulz et al., 1990). GC-C is also expressed in the kidney, airway epithelium, perinatal liver, stomach, brain, adrenal glands, and reproductive tissues (Forte, Krause, & Freeman, 1989; Jaleel, London, Eber, Forte, & Visweswariah, 2002). GC-C has been reported to exist both as a dimer and trimer (Vaandrager, van der Wiel, & de Jonge, 1993;

Vaandrager, van der Wiel, Hom, Luthjens, & de Jonge, 1994). Similar to GC-A and GC-B, ligand binding to the ECD results in elaborate conformational changes along the length of the receptor, which relays the signal to the catalytic domain through the KHD. Unlike GC-A and GC-B, phosphorylation of GC-C is not required for its activity (Vaandrager et al., 1993). Activating protein kinase C using phorbol esters increases the phosphate content and activity of GC-C and results in a down-regulation of GC-C transcripts in colon carcinoma cell lines (Crane & Shanks, 1996; Roy et al., 2001; Vaandrager et al., 1993). In addition, phosphorylation by c-Src tyrosine kinase at the Y820 residue in the catalytic domain inhibits the catalytic activity of GC-C (Basu, Bhandari, Natarajan, & Visweswariah, 2009).

Although the KHD of GC-C lacks a glycine-rich loop typically seen in conventional kinases, a KHD-linker construct, expressed and purified from insect cells, bound ATP-agarose (Bhandari et al., 2001; Jaleel, Saha, Shenoy, & Visweswariah, 2006). Binding was abolished upon removal of the linker region (Jaleel et al., 2006). Thus, residues in the linker region outside the KHD are essential to fold the KHD domain into a structure capable of binding ATP. Homology modeling coupled with mutational and biochemical analyses revealed that K516 is the residue responsible for ATP binding and ligand-mediated regulation in GC-C (Bhandari et al., 2001). Failure of a monoclonal antibody to bind to an epitope C-terminal to K516 in the presence of ATP indicated that conformational changes occur in the KHD-linker region upon ATP binding (Jaleel et al., 2006). Ligand-addition to membrane fractions containing GC-C resulted in increased cGMP production in the presence of ATP, with Mg^{2+} as the metal cofactor in assays (Bhandari et al., 2001; Parkinson, Carrithers, & Waldman, 1994). However, ATP inhibits non-ionic detergent-mediated receptor activation (Bhandari, Suguna, & Visweswariah, 1999). ATP inhibits guanylyl cyclase activity of GC-C in the presence of Mn^{2+} as a cofactor, and this inhibition was also observed in the K516A mutant receptor (Bhandari et al., 1999). This indicates that ATP can also bind to an allosteric site in the catalytic domain and regulate guanylyl cyclase activity.

Tyrphostins, receptor tyrosine kinase inhibitors that bind to the kinase domain of protein kinases, inhibit the catalytic activity of GC-C, suggesting that they may bind to the KHD. However, we showed that these molecules interact with the guanylyl cyclase domain (Jaleel, Shenoy, & Visweswariah, 2004). There is a very high similarity between adenylyl and guanylyl cyclases, with a few substrate-specifying residues determining whether ATP or GTP is used as a substrate (Beuve, 1999). Therefore, tyrphostins may bind to a region in the guanylyl cyclase domain that is the structural homolog of the catechol estrogen binding site in adenylyl cyclases (Steegborn et al., 2005).

2.3 GC-E and GC-F

Guanylyl cyclase E (GC-E) and guanylyl cyclase F (GC-F), also referred to as retinal GCs (RetGC1 and RetGC2, respectively), play a crucial role in phototransduction. GC-E expression has been detected in the retina and pineal gland, while GC-F expression is limited only to the retina (Lowe et al., 1995; Shyjan, de Sauvage, Gillett, Goeddel, & Lowe, 1992; Yang, Foster, Garbers, & Fulle, 1995). Although the domain architecture of GC-E and GC-F are similar to GC-A, GC-B and GC-C, extracellular ligands of GC-E and GC-F have not been identified yet. Instead, 20–24 kDa, soluble, intracellular calcium-binding proteins called guanylyl cyclase activator proteins (GCAPs) regulate the activity of GC-E and GC-F (Dizhoor et al., 1995; Dizhoor, Lowe, Olshevskaya, Laura, & Hurley, 1994; Gorczyca, Gray-Keller, Detwiler, & Palczewski, 1994; Palczewski et al., 1994). GCAPs are associated with the KHD of the cyclases constitutively, and their interaction is determined by the structure of the KHD. Cyclic GMP synthesis is induced by conformational changes in the GCAP/RetGC complex brought upon by Ca^{2+} binding. These conformational changes are such that the GCAP/RetGC complex produces cGMP only under low calcium concentrations. In contrast, an increase in intracellular calcium levels leads to inhibition of cyclase activity by the GCAPs (Laura & Hurley, 1998). Although GC-E has multiple serine and threonine residues in the kinase homology domain, phosphorylation is not essential for activation of the receptor (Bereta et al., 2010).

ATP enhanced GC-E activity both in the presence and absence of GCAPs (Aparicio & Applebury, 1996; Gorczyca, Van Hooser, & Palczewski, 1994). GC-E activity was increased two-fold in bovine rod outer segments in the presence of ATP, by increasing the V_{max} of the enzyme. However, GCAP did not further potentiate activity in the presence of ATP (Gorczyca, Van Hooser, & Palczewski, 1994). Retinal GC is rapidly inactivated at physiological temperatures, and this inactivation could be prevented in the presence of GCAP2, ATP and AMP-PNP (Tucker, Laura, & Hurley, 1997; Yamazaki et al., 2003). This suggested that ATP binding to the pseudokinase domain could occur when the rGCs are bound to GCAPs, and the enhanced activation seen in the presence of ATP is a consequence of stabilization of GC-E.

AMP-PNP and ATP-γ-S were more potent than ATP in mediating the increased activation/stabilization (Aparicio & Applebury, 1996). Interestingly, this study showed that an intrinsic protein kinase activity was seen with the purified receptor, though this result has not been confirmed by other investigators.

2.4 Non-primate rGCs: GC-D and GC-G

Guanylyl cyclase D (GC-D) is expressed in the olfactory bulb of rodents (Fulle et al., 1995; Juilfs et al., 1997; Meyer, Angele, Kremmer, Kaupp, & Muller, 2000). Guanylyl cyclase G (GC-G) is expressed in the lung, intestine, skeletal muscle, testes, and at lower levels in the kidney of the rat and mouse (Kuhn et al., 2004; Schulz, Wedel, Matthews, & Garbers, 1998). Carbon dioxide and bicarbonate ions can stimulate both GC-D and GC-G, and uroguanylin has been reported to activate GC-D (Arakawa, Kelliher, Zufall, & Munger, 2013; Chao et al., 2010; Guo, Zhang, & Huang, 2009; Leinders-Zufall et al., 2007; Sun et al., 2009). This is surprising since the sequence similarity in the ECD between GC-D and GC-C is low. GC-D plays an important role in olfaction and enhancing the food preference response in mice in diverse social contexts (Munger et al., 2010; Zimmerman, Nagy, & Munger, 2020).

Loss of GC-G leads to renal ischemia in mice (Lin et al., 2008). Both GC-D and GC-G have accumulated multiple mutations which introduce premature stop codons and truncations (Manning et al., 2002). The allosteric regulation mediated by the KHD of these receptors is yet to be understood.

2.5 Unusual guanylyl cyclases associated with a kinase domain

Many receptor-like kinases in plants contain a guanylyl cyclase catalytic center embedded within the C-terminal region of a kinase domain (Freihat, Muleya, Manallack, Wheeler, & Irving, 2014). Interleukin 1 receptor-associated kinase 3 (IRAK3) also contains a "guanylyl cyclase centre" and cGMP production appears to mediate downstream signaling of IRAK3 (Freihat et al., 2019). However, a very recent structure of the IRAK3 pseudokinase domain showed that the ~14 residues comprising the guanylyl cyclase core formed a part of the C-terminal lobe of the pseudokinase domain (Lange, Nelen, Cohen, & Kulathu, 2021). Therefore, this unusual report of cGMP production by this protein could result purely from interactions with soluble guanylyl cyclases in the cell, and not be a direct consequence of the guanylyl cyclase "core" present within the pseudokinase domain.

2.6 Equipment

The following laboratory equipment are needed to perform the experimental procedures described.

- Table-top centrifuge for 1.5 and 2 mL microcentrifuge tubes (Centrifuge 5415 D, Eppendorf)
- Cooling table-top centrifuge for 1.5 and 2 mL microcentrifuge tubes set to 4 °C (MicroCL 17R, Thermo Scientific)
- Centrifuge capable of reaching 3200 × *g* and reducing temperature to 4 °C for centrifuging RIA vials and 50 mL conical tubes (Eppendorf centrifuge 5810R, rotor A-4-62)
- Centrifuge capable of reaching 100,000 × *g* and reducing temperature to 4 °C (L8-60M Ultracentrifuge, Beckman-Coulter)
- Sonicator (IKA T 10 basic ULTRA-TURRAX with S 10 N–5 G dispersing tool, Branson Digital Sonifier 450 with a 1/8″ tapered microtip)
- Dounce homogenizer
- Spectrophotometer (Tecan)
- MonoQ anion exchanger column (GE Healthcare Systems)
- Akta FPLC system (GE Healthcare Systems)
- Superdex 200 10/30 column (GE Healthcare Systems)
- Dry bath (Benchmark)
- Water bath (Julabo)
- Thermomixer (Eppendorf)
- End over rocker (Bionova)
- Horizontal rocker (Neolab)
- Chemiluminescence imaging system (Biorad Chemidoc XRS+)
- Magnetic stirrer (SpinIt)
- Dialysis bag (Merck)
- 10 K MWCO spin filter (Pall Corporation)
- Mini-Protean tetra vertical electrophoresis cell (Biorad)
- Mini-Trans blot electrophoretic transfer cell (Biorad)
- Mini-Protean short plates and 1 mm spacer plates (Biorad)
- Hemocytometer
- 21-gauge needle
- Light microscope
- Humidified incubator at 37 °C and 5% CO_2
- Shaker incubator capable of attaining 37 °C and 200 rpm
- Conical flask
- 15 and 50 mL conical tubes

- 1.5 mL microcentrifuge tubes
- Flasks and dishes treated for tissue culture
- Sterile pipettes for tissue culture
- BSL2 laminar air flow hood for tissue culture and radioactive experiments
- Glass chamber for thin layer chromatography
- Table-top vortex
- Plastic vials for radioimmunoassay
- Gamma counter (Perkin Elmer)

2.7 Reagents

- HEPES buffer (pH 7.5)
- MOPS buffer (pH 7.1)
- Tris-Cl buffer (pH 7.5)
- Sodium phosphate buffer (pH 8.0)
- Potassium phosphate buffer (pH 6.8)
- Sodium acetate buffer (pH 4.75)
- Phosphate buffer saline (pH 7.5)
- Sodium chloride (Merck)
- Magnesium chloride (Merck)
- Manganese chloride (Merck)
- Calcium chloride (Merck)
- Potassium chloride (Merck)
- Imidazole (Merck)
- Dipyridamole (Merck, cat. no.: D9766)
- Glycerol (Merck)
- Ethylenediamine tetraacetic acid (EDTA) (Merck)
- Ethylene glycol-bis(β-aminoethyl ether)-N,N,N′,N′-tetraacetic acid (EGTA) (Merck)
- Dithiothreitol (Merck)
- 2-mercaptoethanol (MP Biomedicals)
- Soyabean trypsin inhibitor (Merck)
- Leupeptin (Merck)
- Aprotinin (Merck)
- Phenylmethylsulfonyl fluoride (PMSF) (Merck)
- Pepstatin A (Merck)
- Microcystin (Merck, cat. no.: M2912)
- Creatine phosphate (Merck, cat. no.: CRPHO-RO)

- Creatine phosphokinase (Merck, cat. no.: C3755)
- Adenosine 5′ triphosphate (ATP) (Merck, cat. no.: A2383)
- Guanosine 5′ triphosphate (GTP) (Merck, cat. no.: G8877)
- AMP-PNP (Merck, cat. no.:10102547001)
- Atrial natriuretic peptide (ANP) and C-type natriuretic peptide (CNP) (Merck, cat. no.: A1663, N8768)
- Stable toxin peptide (Bachem, cat. no.: YMC31034–163-2 CB915)
- Sodium orthovanadate (Merck)
- Sodium pyrophosphate (Merck)
- Sodium fluoride (Merck)
- Acrylamide and bisacrylamide (Merck)
- Pre-stained protein ladder (Biorad, cat. no.: 161-0374)
- Immun-Blot PVDF membrane (Biorad, cat. no.: 1620177)
- Whatman no. 1 and no. 3 filter paper (GE Healthcare Systems)
- Substrate solution for horse radish peroxidase during western blot (Immobilon, Millipore)
- Isobutyl methyl xanthine (IBMX) (Merck, cat. no.: I5879)
- DNaseI (Thermo Fisher Scientific)
- RNase (Thermo Fisher Scientific)
- NP-40 (Merck)
- Triton X-100 (Merck)
- Lubrol-PX (Merck)
- 1-propanol (Merck)
- Myristic acid (Merck, cat. no.: M3128)
- Butanol (Merck)
- Glacial acetic acid (Merck)
- Isopropyl β-d-1-thiogalactopyranoside (IPTG) (G Biosciences)
- Bovine serum albumin (Merck)
- Bradford reagent [prepare 5× stock solution by dissolving 100 mg of Coomassie Brilliant Blue G (Merck) in 50 mL of 95% ethanol followed by addition of 100 mL concentrated orthophosphoric acid (Merck). Make up volume till 200 ml with distilled water and filter through Whatman no. 1 filter paper (GE Healthcare Systems)].
- Ni-NTA beads (GE Healthcare Systems, Cat. No.: 17-5628-01)
- ATP-agarose beads (Innova Biosciences 502-0002; Merck A2767)
- Luria Bertani broth bacteriological grade (Pronadisa)
- Agar powder bacteriological grade (Applied Life Sciences)
- Dulbecco's modified Eagle medium (DMEM) (Merck) with 10% fetal bovine serum (FBS) for culturing HEK293 cells

- TC-100 medium (Merck) with 10% FBS for culturing Sf21 insect cells
- Dulbecco's phosphate-buffered saline (DPBS) (Merck)
- *E. coli* DH10Bac cells for propagation of bacmid DNA (Invitrogen). Use either chemically competent or electrocompetent cells for transformation
- Bac-to-Bac baculovirus expression system (Invitrogen)
- *E. coli* BL21 DE3 pLysS for protein expression
- Monosuccinyl cGMP tyrosine methyl ester (Sc-cGMP-TME) (Biolog, cat. no.: M 023-01s)
- $Na^{125}I$ (half-life: 60 days) (ARC, USA, cat. no.: ARI 102)
- Chloramine T (Merck, cat. no.: 402869)
- Sodium metabisulfite (Merck, cat. no.: S1516)
- Activated charcoal (Merck)

3. Protocol

3.1 Preparation of membranes for studying GC-A, GC-B, and GC-C activity (Edmund et al., 2019; Mishra et al., 2021)

1. Isolate required cells from the target tissue following standard protocols. The yield of receptor available for analysis will depend on the tissue and the cell type. For adherent cells in culture (e.g., HEK293 cells transfected with plasmids harboring cDNAs encoding the receptor of interest and cultured on DMEM containing 10% fetal calf serum), wash the adherent monolayer of cells in a confluent 10 cm tissue culture dish with ice-cold PBS (10 mL) to remove serum and media components. For suspended cells (i.e., Sf21 cells grown in glass bottles or Erlenmeyer flasks), resuspend the cells ($\sim 10^7$) by pipetting in the culture medium, centrifuge the cells at $300 \times g$ at 4° C in a 15 mL plastic conical tube. Carefully remove the supernatant and resuspend the pellet in chilled PBS (10 mL), and centrifuge again at $300 \times g$. Remove the supernatant and keep the pellet.
2. Resuspend $\sim 10^7$ cells in the pellet in 1 mL homogenization buffer containing 50 mM HEPES pH 7.5, 100 mM NaCl, 5 mM EDTA, 1 mM dithiothreitol, 5 μg/mL soybean trypsin inhibitor (SBTI), 5 μg/mL leupeptin, 5 μg/mL aprotinin, 2 mM PMSF, 10 mM sodium orthovanadate, 1 mM sodium pyrophosphate, 20 mM sodium fluoride, and 500 μM isobutyl methyl xanthine (IBMX).
3. Lyse cells by sonication (IKA T 10 basic ULTRA-TURRAX with S 10 N–5 G dispersing tool) at 60 cycles, 60% amplitude for 10 pulses (each pulse 5 s).

4. Centrifuge the suspension at 12,000 × *g*, 4 °C for 60 min. Aspirate the supernatant and retain for samples that contain intracellular domains of rGCs (i.e., the KHD-linker of the entire intracellular domain). Resuspend the pellet containing the membrane fraction in a phosphatase inhibitor buffer with 50 mM HEPES pH 7.5, 20% glycerol, 5 µg/mL SBTI, 5 µg/mL leupeptin, 5 µg/mL aprotinin, 2 mM PMSF, 10 mM sodium orthovanadate, 1 mM sodium pyrophosphate and 20 mM sodium fluoride.
5. Estimate protein concentration in the crude membrane fraction using the Bradford reagent (see Section 3.2) and use for ATP binding assay or in vitro guanylyl cyclase assay. Advantages of estimating protein by this method are rapidity, reproducibility, minimal interferences, and high stability of protein-dye complex.

3.2 Preparation of membrane fractions from rod and cone outer segments for studying GC-E and GC-F activity (Yamazaki et al., 2003)

1. Isolate outer segment from 25 bovine retinas in dim to red light (to avoid photobleaching of rhodopsin) and resuspend them in 5 ml of Buffer A (5 mM HEPES pH 7.5, 1 mM DTT, 5 mM $MgCl_2$, 100 µM $CaCl_2$, 0.1 mM PMSF, 5 µM leupeptin, and 5 µM pepstatin A).
2. Homogenize the suspension at 4 °C by passing it through a needle with a 21-gauge diameter. Repeat 7 times.
3. Centrifuge at 100,000 × *g* for 15 min at 4 °C to separate the membrane fraction.
4. Wash the pellet containing the membrane fraction in 5 mL of buffer B (5 mM HEPES pH 7.5, 1 mM DTT, 5 mM $MgCl_2$, 2 mM EGTA, 0.1 mM PMSF, 5 µM leupeptin, and 5 µM pepstatin A). Repeat 7 times.
5. Resuspend the membrane fraction in buffer C (10 mM HEPES pH 7.5, 1 mM DTT, 2 mM $MgCl_2$, 0.1 mM PMSF, 5 µM leupeptin, and 5 µM pepstatin A) at a concentration of 5 mg/mL.
6. Estimate the concentration of rhodopsin in the membrane fraction (see Section 3.6.3) and use for ATP binding assay or in vitro guanylyl cyclase assay.

3.3 Preparation of membrane fractions from mammalian cells for studying GC-E and GC-F activity (Dizhoor et al., 1994)

1. Wash monolayer of adherent cells expressing GC-E or GC-F with ice-cold PBS.

2. Scrape the cells into ice cold 10 mM MOPS pH 7.1 containing 1 mM EDTA, 1 mM DTT, 0.5 μg/mL leupeptin, 0.5 μg/mL pepstatin and homogenize with 12 strokes in a Dounce homogenizer.
3. Remove debris by centrifugation at 200 × *g* for 15 min at 4 °C.
4. Separate the membrane fraction by centrifuging at 100,000 × *g* for 30 min at 4 °C.
5. Resuspend the pellet in 10 mM MOPS pH 7.1 and 1 mM DTT.
6. Estimate protein concentration by the Bradford assay and use for ATP binding assay or in vitro guanylyl cyclase assay.

3.4 Purification of proteins

3.4.1 Purification of the intracellular domain of GC-A with KHD (GC-A_{ID}) or without KHD (GC-A_{CC+GC}) (Pattanaik, Fromondi, Ng, He, & van den Akker, 2009)

1. Recombinant Bacmid DNA using DH10Bac cells and the Bac-to-Bac Expression system from Invitrogen was used to allow expression of a protein comprising amino acids 465 to 1029 of the entire intracellular domain of GC-A with an N-terminal a 6 × His-tag. High-Five cells were used for protein expression and harvested 72 h following infection with recombinant baculovirus (5–10 multiplicity of infection; cell density ~10^6 cells/mL). An additional construct that did not contain the KHD but only the coiled-coil domain and the guanylyl cyclase domain, from residues 776–1029, was cloned and expressed in insect cells in a similar manner.
2. Lyse cells by sonicating in 50 mM sodium phosphate pH 8.0 containing 40 mM imidazole, 300 mM NaCl, 1 mM DTT, 10% glycerol, 5 μg/mL DNAseI, 10 μg/mL RNAse, and protease inhibitor cocktail.
3. Add NP-40 to a final concentration of 1%.
4. Incubate the lysates at 4 °C for 1 h in a conical tube placed on a rocking platform.
5. Remove cell debris by centrifugation at 30,000 × *g* for 30 min at 4 °C.
6. Add washed Ni-NTA beads (1 mL per 10 mL of packed cells) to the supernatant and incubate for 2 h at 4 °C with rocking.
7. Centrifuge at 200 g at 4 °C to pellet the beads and pack them into a 5 mL column.
8. Pass lysis buffer containing 1% NP-40 (3 × bed volume) through the column.
9. Wash the beads 15 times with the lysis buffer without NP-40 (10 × bed volume).

10. Elute the protein in 3× bed volume of buffer containing 50 mM sodium phosphate pH 8.0, 250 mM imidazole, 300 mM NaCl, 1 mM DTT, 10% glycerol and the protease inhibitor cocktail.
11. Immediately after elution, add EDTA (pH 7.0) to a final concentration of 10 mM.
12. Exchange the buffer by dialysis against 10 volumes of buffer containing 20 mM Tris-Cl pH 8.0, 1 mM DTT, and 10% glycerol, with 3 changes of buffer.
13. The sample is then applied to a MonoQ anion exchanger column (GE Healthcare Systems) and sample eluted with a 1-1 M NaCl gradient with a flow rate of 1 mL/per minute on an AKTA FPLC system (GE Healthcare Systems).
14. Fractions containing the protein were then subjected to size exclusion chromatography on a Superdex 200 10/30 column (GE Healthcare Systems) in a buffer containing 20 mM Tris-Cl, pH 8.0, 1 mM DTT, 150 mM NaCl and 10% glycerol with a flow rate of 0.2 mL/min. Estimate protein using the Bradford method with BSA as a standard and use for ATP binding assay or in vitro guanylyl cyclase assay.

3.4.2 Purification of the entire intracellular domain of GC-C (GC-C_{ID}) or KHD-linker region of GC-C (GC-C_{KHD+CC}) (*Jaleel et al., 2006*)

1. The entire intracellular domain of GC-C (residues 455–1073) and the KHD and coiled-coil region (residues 455–810) were cloned and expressed in Sf21 insect cells using the Bac-to-Bac expression system from Invitrogen (see Section 3.4.1).
2. For expression, cells are grown in 100 mL of TC-100 medium containing 10% fetal calf serum, with a starting cell density of 5×10^5/mL. Cells are grown with gentle shaking in a 500 mL conical flask at 100 rpm for 72 h at 25 °C.
3. The cell suspension is centrifuged at 1000 × *g* and the pellet resuspended in lysis buffer (3 mL) containing 50 mM HEPES pH 7.5, 100 mM NaCl, 5 mM 2-mercaptoethanol, 5 µg/mL aprotinin, 5 µg/mL leupeptin, 2 mM PMSF, 10% glycerol.
4. Lyse cells by sonication at 4 °C (Branson Digital Sonifier 450 with a 1/8″ tapered microtip) at 50% power till the cells lyse (usually 5 min).
5. Centrifuge at 12,000 × *g* for 60 min at 4 °C and harvest the supernatant which contains the soluble target protein (GC-C_{ID} or GC-C_{KHD+CC}). Estimate total cytosolic protein using Bradford reagent with BSA as a standard and use for ATP binding assay or in vitro guanylyl cyclase assay.

3.4.3 Purification of guanylyl cyclase-activating protein 1 (GCAP-1) *(Krylov, Niemi, Dizhoor, & Hurley, 1999)*

1. Transform pET expression plasmid encoding GCAP-1 into *E. coli* BL21 DE3pLysE harboring pBB131 encoding yeast N-myristoyl transferase (NMT) and kanamycin resistance (Krylov et al., 1999). For transformation, mix 1 μg plasmid DNA with 50 μL of *E. coli* BL21 DE3pLysE cells made chemically competent by calcium chloride, incubate on ice for 30 min and give a heat shock for 2 min at 37 °C followed by immediate incubation on ice for 5 min. Add 1 mL of Luria broth without antibiotics and allow recovery for 1 h at 37 °C. Plate the cells on LB agar plate containing kanamycin (50 μg/mL) and incubate at 37 °C for 12–16 h.
2. Use a single colony to grow a culture (500 mL) in Luria Broth with shaking in a 2 L conical flask at 37 °C and 200 rpm. Grow to an OD_{600} of 1.
3. Supplement bacterial media with free myristic acid (50 μg/mL) 30 min before induction of expression with 1 mM IPTG.
4. After expression for 2–5 h at 37 °C and 200 rpm, lyse cells by sonication (Branson Digital Sonifier 450 with a 1/8″ tapered microtip) on ice in 5 mL lysis buffer containing 40 mM Tris-Cl pH 7.5, 0.5 mM EDTA, 1 mM 2-mercaptoethanol, 20 μg/mL leupeptin, 100 μM PMSF.
5. Centrifuge the lysate at 30,000 × *g* for 30 min at 4 °C.
6. Wash the pellet twice with 10 mL lysis buffer by resuspension and centrifugation at 30,000 × *g* for 30 min at 4 °C.
7. Dissolve the pellet in 5 mL of 6 M urea and dialyze 3 times against 1000 volumes of lysis buffer without protease inhibitors.
8. Estimate protein concentration by the Bradford assay and use for guanylyl cyclase assays of GC-E and GC-F.

3.4.4 Purification of guanylyl cyclase activating protein 2 (GCAP-2) *(Olshevskaya, Hughes, Hurley, & Dizhoor, 1997)*

1. Transform pET expression plasmid into *E. coli* BL21 DE3pLysS harboring pBB131 encoding yeast N-myristoyl transferase (NMT) and kanamycin resistance as described above.
2. Use a single colony to grow a culture (500 mL) in Luria Broth with shaking in a 2 L conical flask at 200 rpm. Grow to an OD_{600} of 1.
3. Supplement bacterial media with free myristic acid (50 μg/mL) 30 min before induction of expression with 1 mM IPTG.
4. After expression for 3.5 h at 200 rpm and 37 °C, harvest bacterial cells by centrifugation at 8000 × *g* for 20 min at 4 °C.

5. Lyse the cells by sonication (Branson Digital Sonifier 450 with a 1/8″ tapered microtip) in 3 mL lysis buffer containing 40 mM Tris-Cl pH 7.5, 0.5 mM EDTA, 1 mM 2-mercaptoethanol, 20 μg/mL leupeptin, 100 μM PMSF.
6. Centrifuge the suspension at 20,000 × *g* for 10 min and wash the pellet thrice with 10 mL of buffer A containing 20 mM Tris-Cl pH 7.5, 1 mM EDTA, 7 mM 2-mercaptoethanol, 100 μM PMSF, 20 μg/mL leupeptin.
7. Extract GCAP-2 from the pellet by homogenization in buffer A supplemented with 100 mM 2-mercaptoethanol, 1 mM EDTA, 6 M urea for 30 min at 4 °C (5 mL).
8. Dialyze twice against 300–1000 volumes of buffer A overnight at 4 °C.
9. Remove precipitate by centrifugation at 30,000 × *g* for 10 min.
10. Purify GCAP-2 by gel filtration on a Superdex 200 10/30 on an Akta FPLC system equilibrated with 10 mM Tris-Cl pH 7.5 and 10 mM 2-mercaptoethanol, flow rate 0.2 ml/min.
11. Estimate protein concentration by the Bradford method and use for guanylyl cyclase assays of GC-E and GC-F.

3.5 ATP-agarose binding assay (Edmund et al., 2019; Jaleel et al., 2006)

1. Crude whole cell lysates, membrane fractions containing full-length rGCs or cytosolic fractions containing the soluble intracellular domains of rGCs are prepared from mammalian or insect cells expressing the desired protein as described above. The amount of material taken would depend on the expression levels of the protein of interest in particular cell lines or extracts. This would have to be empirically determined for each receptor, based on the method of detection used to monitor presence of the protein. If western blot is used, that would depend on the affinity of the antibody used for blotting and would vary for each protein.
2. For membrane bound receptors, the membrane would first need to be suspended in 50 mM HEPES pH 7.5, 150 mM NaCl, 10 mM $MgCl_2$, 1 mM DTT, 10 μg/mL leupeptin, 10 μg/mL aprotinin, containing 1% Triton X-100 at a concentration of 5 mg protein/mL. The suspension is then placed in a 15 mL conical tube and placed on a rocker for 1 h at 4 °C and gently mixed. This would solubilize membrane-bound proteins. Centrifuge the sample at 16,000 × *g* at 4 °C and retain the supernatant.

3. Dilute the supernatant or cytosolic fraction (1:10, v/v) in ATP-agarose interaction buffer. For GC-A, the buffer used is 50 mM HEPES pH 7.4, 150 mM NaCl, 10 mM $MgCl_2$. For GC-C, buffer used is 50 mM HEPES pH 7.5, 150 mM NaCl, 10 mM $MgCl_2$, 1 mM DTT, 10 μg/ml leupeptin, 10 μg/mL aprotinin, 0.1% Triton X-100.
4. Remove endogenous ATP by passing the protein mix from step 2 or 3 through a 10K MWCO spin filter. The protein of interest will be retained, and ATP will be removed in the flow-through. Bring down the volume by 90% and repeat this step twice in ATP-agarose interaction buffer.
5. Wash the agarose beads conjugated to ATP in the ATP-agarose interaction buffer thrice. Beads are available with a number of suppliers (e.g., Innova Biosciences 502-0002; Merck A2767) but ensure that the conjugation is through the γ-phosphate.
6. Add the required volume of washed ATP-agarose beads to 500 μL of the protein mix from step 3 such that the final concentration of ATP on the beads is 250 μM (for GC-A) or 200 μM (for GC-C). The conjugation of ATP to agarose beads is usually at 1–5 μmoles/mL, so for 500 μL of sample volume, 100–125 μL beads would be sufficient.
7. Incubate the reaction mixture in a 1.5 mL tube for 1 h at 4 °C with gentle mixing in an Eppendorf Thermomixer.
8. Wash the ATP-agarose beads thrice with chilled ATP-agarose interaction buffer (1 mL each wash for GC-A) or twice with ATP-agarose interaction buffer with Triton X-100 and once without Triton X-100 (1 mL each wash for GC-C).
9. Boil the beads in 200 μL of reducing Laemmli buffer for 5 min prior to loading on the gel. SDS-PAGE is carried out in a 10% gel with 0.7–1.2% crosslinker, depending on the size of the protein of interest and the separation required. Gels are run at 200 V till the dye reaches the end of the gel. Markers used are from Biorad.
10. Transfer proteins to a PVDF membrane (Biorad, cat. no.: 1620177) in buffer containing 25 mM Tris base, 192 mM glycine, 20% methanol, pH 8.3 at 200 V and 160 mA for 120 min using Mini-Trans blot electrophoretic transfer cell (Biorad).
11. Rinse the PVDF membrane in 10 mM Tris-Cl pH 7.2, 100 mM NaCl, 0.1% Tween-20 (TBST) and block for 1 h at room temperature in 5% blocking solution (GE Healthcare systems) on a gentle rocking platform.

12. Incubate the membrane with respective primary antibodies for 12–14 h at 4 °C on a gentle rocking platform. Dilute the primary antibody in TBST containing 5% BSA at the necessary dilution recommended by the manufacturer or standardized upon trial.
13. Wash the membrane thrice with TBST by incubating for 10 min each on a fast-rocking platform at room temperature.
14. Incubate membrane with secondary antibody diluted (at a concentration recommended by manufacturer or standardized upon trial) in TBST containing 5% BSA for 1 h at room temperature on a gentle rocking platform. The type of secondary antibody will depend on the host species in which the primary antibody was raised.
15. Wash the membrane thrice with TBST by incubating for 10 min each on a fast-rocking platform at room temperature.
16. Visualize immunoreactive bands by chemiluminescence using Immobilon reagent (Millipore) on a Chemidoc XRS+ system (Biorad).

3.6 Guanylyl cyclase assay

The catalytic guanylyl cyclase domain of rGCs belongs to the family of class III nucleotide cyclases (Biswas et al., 2009; Dove, 2017). Class III nucleotide cyclases undergo a head-to-tail dimerization to attain catalytic activity (Shenoy & Visweswariah, 2004; Zhang, Liu, Ruoho, & Hurley, 1997), with the active site formed by the two subunits at the dimer interface. Class III nucleotide cyclases convert ATP or GTP to their equivalent 3′,5′-cyclic nucleotide following a two-metal supported mechanism (Liu et al., 1997). Receptor guanylyl cyclases homodimerize to form two active sites and bind two substrate molecules per dimer (Dove, 2017). Three residues form the functional catalytic center. An Asp residue from one protomer assists in proton abstraction and an Asn/Arg pair from another protomer stabilizes the transition state. One metal ion (B) stays bound to the β and γ-phosphates of ATP, while the other free metal ion (A) is called the catalytic ion. Ion A binds to the 3′-OH of ribose of GTP and reduces its pK_a, thus generating a nucleophile that attacks the α-phosphoanhydride bond in GTP and cleaves it through a direct displacement reaction. The transition state stabilizing Asn stabilizes the 3′ oxyanion generated by proton removal, while the Arg stabilizes a pentavalent transition state of the α-phosphate while it is subject to a nucleophilic attack by the 3′ oxyanion leading to the formation of cGMP (Shenoy & Visweswariah, 2004; Zhang et al., 1997).

Guanylyl cyclase activity is assayed by measuring the amount of cGMP derived from GTP in the presence of $MgCl_2$ or $MnCl_2$, a phosphodiesterase inhibitor (isobutyl methyl xanthine; IBMX), and a GTP regenerating system. Free metal ions or allosteric regulators are added to the reaction mix as per requirement.

3.6.1 GC-A and GC-B *(Edmund et al., 2019)*

1. Dilute the protein (usually 10–50 µg total protein) in 20 µL of phosphatase inhibitor buffer (50 mM HEPES pH 7.5, 20% glycerol, 5 µg/mL soyabean trypsin inhibitor (SBTI), 5 µg/mL leupeptin, 5 µg/mL aprotinin, 2 mM PMSF, 10 mM sodium orthovanadate, 1 mM sodium pyrophosphate, 20 mM sodium fluoride).
2. Add 60 µL of reaction cocktail prewarmed in a water bath to 25 °C to obtain final reaction conditions of 25 mM HEPES pH 7.5, 50 mM NaCl, 500 µM IBMX, 0.5 µM microcystin, 1 mM EDTA, 0.1% BSA, 0.1 mM GTP and a nucleotide triphosphate regeneration system containing 5 mM creatine phosphate and 0.1 µg/mL creatine phosphokinase.
3. To study effect of activators, add 20 µL of a 5 × activation mix to obtain 5 mM $MgCl_2$ or $MnCl_2$, 1 mM ATP, 1 µM ANP or CNP, 1% v/v Triton X-100 in the final reaction mix as required.
4. Incubate the reaction mix at 37 °C for 10 min in a water bath.
5. Stop the reaction with 400 µL of ice-cold 50 mM sodium acetate buffer (pH 4.75) containing 5 mM EDTA. Heat samples for 5 min in a boiling hot water bath in tightly capped Eppendorf tubes followed by centrifugation at 10,000 × g for 5 min at 4 °C. Collect the supernatant.
6. Estimate cGMP levels by radioimmunoassay (see Section 3.7) in the supernatant fractions.

3.6.2 GC-C *(Saha, Biswas, Kondapalli, Isloor, & Visweswariah, 2009)*

1. Dilute membrane or cytosolic protein with buffer (50 mM HEPES pH 7.5, 20% glycerol, 5 µg/mL soybean trypsin inhibitor (SBTI), 5 µg/mL leupeptin, 5 µg/mL aprotinin, 2 mM PMSF, 10 mM sodium orthovanadate, 1 mM sodium pyrophosphate, 20 mM sodium fluoride) containing $MgCl_2$ or $MnCl_2$ (final concentration 10 mM) in a total volume of 20 µL. Usually, 10–50 µg of protein is sufficient for the assay.
2. To this add 20 µL of reaction cocktail prewarmed in a water bath to 25 °C, to final reaction conditions of 60 mM Tris-HCl pH 7.5, 500 µM IBMX, and a nucleotide triphosphate regeneration system containing 7.5 mM creatine phosphate and 10 µg of creatine phosphokinase.

3. Add GTP, $MgCl_2$ or $MnCl_2$ stock solutions in a final volume of 10 μL to achieve a final concentration of 1 mM of each reactant.
4. To study the effect of activators, add 1 mM Mg-ATP, 10^{-7} M or 10^{-6} M STa (for human and mouse respectively), or 0.1% Lubrol-PX to the protein mix (step 1).
5. Incubate the reaction mix at 37 °C for 10 min.
6. Stop the reaction by adding 400 μL of ice-cold 50 mM sodium acetate buffer (pH 4.75). Heat samples for 5 min in a heating block set to 95 °C in tightly capped Eppendorf tubes and centrifuge at 10,000 × g at 4 °C for 5 min. Collect the supernatant for measurement of cGMP by radioimmunoassay (see Section 3.7).

3.6.3 GC-E and GC-F *(Tucker et al., 1997)*

1. Dilute the membrane fraction to a rhodopsin concentration of 0.48 μg/μL (estimate total protein; 80% of total protein in washed retinal outer segments is rhodopsin (Papermaster & Dreyer, 1974)) in 2× GC buffer (200 mM KCl, 100 mM MOPS, 14 mM 2-mercaptoethanol, 20 mM $MgCl_2$ or $MnCl_2$, 16 mM NaCl, 2 mM EGTA). For kinetic analysis at least 10 concentrations of substrate (GTP) would be required. Therefore, ~150 μL total volume would be needed.
2. Combine 12.5 μL of the diluted mix from the previous step (6 μg rhodopsin) with 10 μM dipyridamole and additional factors (detailed in step 3) to a total volume of 20 μL.
3. Measure basal activity in the presence of 0.9 mM $CaCl_2$. For stimulated activity, preincubate membrane fractions in 2 × GC buffer in 30 °C in the presence of 0.5 mM $CaCl_2$ coupled with 0.5 mM ATP, 0.75 mM AMP-PNP, 5.1 μM (0.12 mg/mL) GCAP-1 or 0.84 μM (0.02 mg/mL) GCAP-2 (purified as described in Sections 3.4.3 and 3.4.4) or 1% Triton X-100 for the required time durations in a total volume of 45 μL.
4. Initiate the reaction by adding 5 μL of 5 mM GTP to attain a final concentration of 1 mM GTP in the reaction.
5. Incubate the reaction mix for 10 min at 30 °C.
6. Stop the reaction by adding 450 μL of ice-cold 50 mM sodium acetate buffer (pH 4.75).
7. Estimate cGMP concentration by radioimmunoassay as described below.

3.7 Radio-iodination of cGMP and radioimmunoassay (RIA)

All precautions required for use of radioactivity must be followed. Refer to Fig. 4 for schematic illustration.

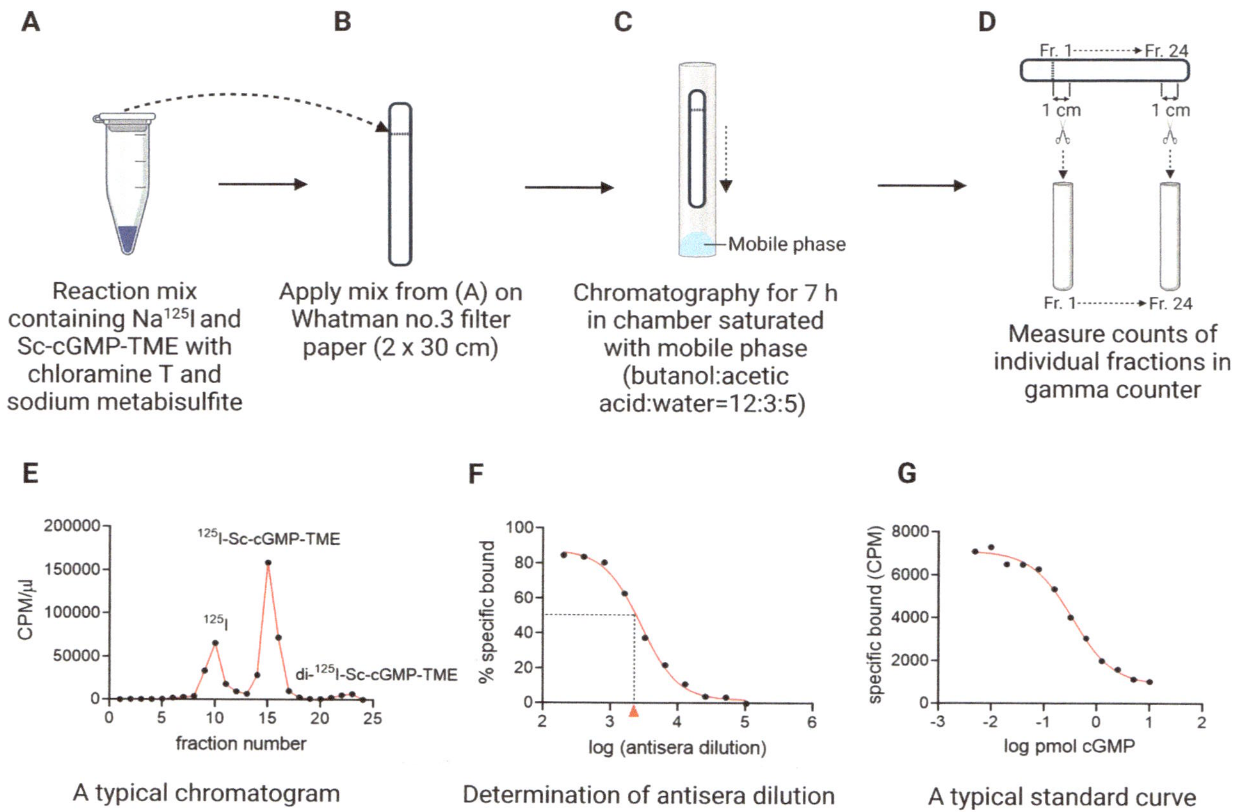

Fig. 4 Schematic representation of radioiodination of cGMP and radioimmunoassay. (A–D) Workflow for radioiodination of cGMP with I^{125}. (E) A typical chromatogram post cGMP radioiodination depicting free iodine, mono and di iodinated Sc-cGMP-TME distributed across the fractions. (F) Curve depicting the determination of the dilution of cGMP antisera to be used in radioimmunoassay. Red arrowhead indicates antisera dilution corresponding to 50% specific binding. This dilution of antisera is to be used for 25,000 CPM of I^{125} Sc-cGMP-TME. (G) A typical standard curve of radioimmunoassay with cGMP concentration ranging from 5 fmol to 10 pmol. Concentration of cGMP in samples can be determined by interpolating against the standard curve.

1. For radio-iodination, mix 400 ng of monosuccinyl cGMP tyrosine methyl ester (Sc-cGMP-TME) with 600 μCi of NaI^{125} 0.5 M potassium phosphate buffer pH 6.8.
2. Initiate the reaction by adding 5 μL of oxidizing agent chloramine T (1 mg/mL in potassium phosphate buffer pH 6.8) to a final reaction volume of 50 μL.
3. Incubate the reaction mix on ice for 60 s and stop it by adding 50 μL of sodium metabisulfite (5 mg/mL) in potassium phosphate buffer (pH 6.8).
4. Subject the reaction mixture to descending chromatography in a solvent system comprising 1-butanol: glacial acetic acid: water (12:3:5) for 7 h in an overnight saturated glass chamber using a Whatman 3 filter paper (~2*30 cm) as the solid phase.
5. Dry the chromatogram and cut it horizontally into thin strips of 1 cm width to obtain individual fractions post chromatographic separation. Unlabeled, mono-iodinated and di-iodinated cGMP-TME migrate to different extents on the paper chromatogram and will therefore be present in separate fractions.
6. Incubate the strips in 1 mL of 50 mM sodium acetate buffer pH 4.75 for 1 h at 4 °C followed by vortexing at maximum speed at room temperature for 30 s for mechanical elution.
7. After allowing the paper strip to settle, estimate the radioactivity in each fraction by aliquoting 10 μL of each fraction in an RIA vial (see Section 3.2) which is placed in a γ-counter. Tubes are counted for 30 s.
8. A typical chromatogram (Fig. 4) will result in 3 peaks: di-iodinated cGMP migrating the furthest distance followed by mono-iodinated cGMP and free I^{125}.
9. Use the fraction containing mono-iodinated Sc-cGMP-TME for radioimmunoassay after adding an equal volume of 1-propanol to fractions containing the labeled cGMP. Store at −20° C.
10. Set up the radioimmunoassay in a total volume of 300 μL of 50 mM sodium acetate buffer pH 4.75 containing 5 mg/mL BSA in RIA vials (see Section 3.2).
11. Provide a total input of ~25,000 CPM of ^{125}I-cGMP-TME in each tube. Prepare dilutions of the available antiserum and add to the radiolabeled cGMP. Incubate the tubes at 4 °C for 16 h. Add 1 ml of ice-cold activated charcoal suspension (2 mg/L) in 50 mM potassium phosphate buffer pH 6.3 containing 1 mg/mL BSA. Free radiolabeled cGMP will bind to the charcoal.

12. Stand the tubes for 2 min and centrifuge at 3000 × *g* for 20 min at 4 °C. Decant off the supernatant and place the tubes containing the charcoal pellet in the γ-counter. Count for 15 s. Determine the dilution of antibody required to achieve ~50% binding of the input radioactivity. This dilution will be used for the radioimmunoassay to estimate cGMP in samples.
13. For the assay to estimate cGMP in samples, assay volumes are 300 μL and the amount of antiserum is that required for ~50% binding if ~25,000 CPM is added.
14. Prepare a set of 12 tubes for a standard curve having a range of 5 fmol to 10 pmol unlabeled cGMP per tube. To make the standards, dilute the cGMP in 100 μL of 50 mM sodium acetate (pH 4.75). Add 10–100 μL of samples to be tested to a separate set of tubes and make up volume till 100 μL with 50 mM sodium acetate (pH 4.75). Prepare ^{125}I-cGMP-TME solution in 50 mM sodium acetate (pH 4.75) containing 5 mg/mL BSA such that there is ~25,000 CPM/100 μL and add 100 μL per tube. Dilute antiserum to the required amount (determined in step 13) in 50 mM sodium acetate (pH 4.75) containing 5 mg/mL BSA and add 100 μL of that to each tube. Set up two tubes which contain buffer but to which no antibody is added. These two tubes serve to estimate non-specific binding.
15. Incubate the reaction at 4 °C for 12–16 h.
16. Separate the free cGMP from cGMP bound to the antibody by adding 1 mL of ice-cold activated charcoal suspension (2 mg/L) in 50 mM potassium phosphate buffer pH 6.3 containing 1 mg/ml BSA.
17. Centrifuge the RIA vials containing the reaction mix at 3200 × *g* for 20 min at 4 °C.
18. Decant the supernatant and estimate the radioactivity of the free I^{125}-cGMP-TME adsorbed on the charcoal using a gamma counter.
19. Set up a standard curve after subtracting non-specific binding from total counts bound (counts bound = input radioactivity − counts bound to charcoal).
20. Estimate the concentration of cGMP in samples from the standard curve

4. Summary

We present methods to monitor the allosteric regulation of rGCs by the pseudokinase domain. There is no structural information to date on the intracellular domain of this class of receptors. In its absence, one can speculate on interdomain interactions and regions critical for signal transmission using mutagenesis (Saha et al., 2009). Specific amino acids critical for the relay of information following ligand binding are identifiable, based on mutations in these residues and association with human disorders (Table 1). It is anticipated that the pseudokinase domain in these receptors could be targeted by small-molecule activators or inhibitors to regulate receptor activity. Several inhibitors of protein tyrosine kinases in therapeutic use indicate that the kinase domain is amenable to specific targeting (Wu, Nielsen, & Clausen, 2016). High-throughput screening approaches, perhaps using drugs already in use, using sensors to monitor cGMP production could be utilized for such studies in the future, but this would necessitate distinguishing molecules that interact with the guanylyl cyclase domain from those that bind solely to the pseudokinase domain. Alternatively, engineered constructs harboring the pseudokinase domain coupled to fluorescence sensors will allow direct monitoring of conformational changes that occur on binding of either ATP or small molecules.

Acknowledgments

A.B. acknowledges support from the Indian Institute of Science and the Commonwealth Scholarship Foundation, UK, with award number INCN-2020-140. S.S.V. is a JC Bose National Fellow (SB/S2/JCB-18/2013) and a Margdarshi Fellow supported by the Wellcome Trust DBT India Alliance (IA/M/16/502606).

References

Aparicio, J. G., & Applebury, M. L. (1996). The photoreceptor guanylate cyclase is an autophosphorylating protein kinase. *The Journal of Biological Chemistry*, *271*(43), 27083–27089. https://doi.org/10.1074/jbc.271.43.27083.

Arakawa, H., Kelliher, K. R., Zufall, F., & Munger, S. D. (2013). The receptor guanylyl cyclase type D (GC-D) ligand uroguanylin promotes the acquisition of food preferences in mice. *Chemical Senses*, *38*(5), 391–397. https://doi.org/10.1093/chemse/bjt015.

Arshad, N., & Visweswariah, S. S. (2012). The multiple and enigmatic roles of guanylyl cyclase C in intestinal homeostasis. *FEBS Letters*, *586*(18), 2835–2840. https://doi.org/10.1016/j.febslet.2012.07.028.

Ashman, D. F., Lipton, R., Melicow, M. M., & Price, T. D. (1963). Isolation of adenosine 3′, 5′-monophosphate and guanosine 3′, 5′-monophosphate from rat urine. *Biochemical and Biophysical Research Communications*, *11*, 330–334. https://doi.org/10.1016/0006-291x(63)90566-7.

Basu, N., Arshad, N., & Visweswariah, S. S. (2010). Receptor guanylyl cyclase C (GC-C): Regulation and signal transduction. *Molecular and Cellular Biochemistry*, *334*(1–2), 67–80. https://doi.org/10.1007/s11010-009-0324-x.

Basu, N., Bhandari, R., Natarajan, V. T., & Visweswariah, S. S. (2009). Cross talk between receptor guanylyl cyclase C and c-src tyrosine kinase regulates colon cancer cell cytostasis. *Molecular and Cellular Biology*, *29*(19), 5277–5289. https://doi.org/10.1128/MCB.00001-09.

Bereta, G., Wang, B., Kiser, P. D., Baehr, W., Jang, G. F., & Palczewski, K. (2010). A functional kinase homology domain is essential for the activity of photoreceptor guanylate cyclase 1. *The Journal of Biological Chemistry*, *285*(3), 1899–1908. https://doi.org/10.1074/jbc.M109.061713.

Beuve, A. (1999). Conversion of a guanylyl cyclase to an adenylyl cyclase. *Methods*, *19*(4), 545–550. https://doi.org/10.1006/meth.1999.0896.

Bhandari, R., Srinivasan, N., Mahaboobi, M., Ghanekar, Y., Suguna, K., & Visweswariah, S. S. (2001). Functional inactivation of the human guanylyl cyclase C receptor: Modeling and mutation of the protein kinase-like domain. *Biochemistry*, *40*(31), 9196–9206. https://doi.org/10.1021/bi002595g.

Bhandari, R., Suguna, K., & Visweswariah, S. S. (1999). Guanylyl cyclase C receptor: Regulation of catalytic activity by ATP. *Bioscience Reports*, *19*(3), 179–188. https://doi.org/10.1023/a:1020273619211.

Biswas, K. H., Shenoy, A. R., Dutta, A., & Visweswariah, S. S. (2009). The evolution of guanylyl cyclases as multidomain proteins: Conserved features of kinase-cyclase domain fusions. *Journal of Molecular Evolution*, *68*(6), 587–602. https://doi.org/10.1007/s00239-009-9242-5.

Bose, A., Banerjee, S., & Visweswariah, S. S. (2020). Mutational landscape of receptor guanylyl cyclase C: Functional analysis and disease-related mutations. *IUBMB Life*, *72*(6), 1145–1159. https://doi.org/10.1002/iub.2283.

Bryan, P. M., Smirnov, D., Smolenski, A., Feil, S., Feil, R., Hofmann, F., et al. (2006). A sensitive method for determining the phosphorylation status of natriuretic peptide receptors: cGK-Ialpha does not regulate NPR-A. *Biochemistry*, *45*(4), 1295–1303. https://doi.org/10.1021/bi051253d.

Chang, C. H., Kohse, K. P., Chang, B., Hirata, M., Jiang, B., Douglas, J. E., et al. (1990). Characterization of ATP-stimulated guanylate cyclase activation in rat lung membranes. *Biochimica et Biophysica Acta*, *1052*(1), 159–165. https://doi.org/10.1016/0167-4889(90)90071-k.

Chao, Y. C., Cheng, C. J., Hsieh, H. T., Lin, C. C., Chen, C. C., & Yang, R. B. (2010). Guanylate cyclase-G, expressed in the Grueneberg ganglion olfactory subsystem, is activated by bicarbonate. *The Biochemical Journal*, *432*(2), 267–273. https://doi.org/10.1042/BJ20100617.

Chinkers, M., & Garbers, D. L. (1989). The protein kinase domain of the ANP receptor is required for signaling. *Science*, *245*(4924), 1392–1394. https://doi.org/10.1126/science.2571188.

Chinkers, M., & Wilson, E. M. (1992). Ligand-independent oligomerization of natriuretic peptide receptors. Identification of heteromeric receptors and a dominant negative mutant. *The Journal of Biological Chemistry*, *267*(26), 18589–18597.

Chrisman, T. D., Schulz, S., Potter, L. R., & Garbers, D. L. (1993). Seminal plasma factors that cause large elevations in cellular cyclic GMP are C-type natriuretic peptides. *The Journal of Biological Chemistry*, *268*(5), 3698–3703.

Crane, J. K., & Shanks, K. L. (1996). Phosphorylation and activation of the intestinal guanylyl cyclase receptor for Escherichia coli heat-stable toxin by protein kinase C. *Molecular and Cellular Biochemistry*, *165*(2), 111–120. https://doi.org/10.1007/BF00229472.

De Jonge, H. R. (1975a). The localization of guanylate cyclase in rat small intestinal epithelium. *FEBS Letters*, *53*(2), 237–242. https://doi.org/10.1016/0014-5793(75)80028-7.

de Jonge, H. R. (1975b). Properties of guanylate cyclase and levels of cyclic GMP in rat small intestinal villous and crypt cells. *FEBS Letters*, *55*(1), 143–152. https://doi.org/10.1016/0014-5793(75)80980-x.

Dickey, D. M., Flora, D. R., Bryan, P. M., Xu, X., Chen, Y., & Potter, L. R. (2007). Differential regulation of membrane guanylyl cyclases in congestive heart failure: Natriuretic peptide receptor (NPR)-B, Not NPR-A, is the predominant natriuretic peptide receptor in the failing heart. *Endocrinology*, *148*(7), 3518–3522. https://doi.org/10.1210/en.2007-0081.

Dizhoor, A. M., Lowe, D. G., Olshevskaya, E. V., Laura, R. P., & Hurley, J. B. (1994). The human photoreceptor membrane guanylyl cyclase, RetGC, is present in outer segments and is regulated by calcium and a soluble activator. *Neuron*, *12*(6), 1345–1352. https://doi.org/10.1016/0896-6273(94)90449-9.

Dizhoor, A. M., Olshevskaya, E. V., Henzel, W. J., Wong, S. C., Stults, J. T., Ankoudinova, I., et al. (1995). Cloning, sequencing, and expression of a 24-kDa Ca(2+)-binding protein activating photoreceptor guanylyl cyclase. *The Journal of Biological Chemistry*, *270*(42), 25200–25206. https://doi.org/10.1074/jbc.270.42.25200.

Dove, S. (2017). Mammalian nucleotidyl cyclases and their nucleotide binding sites. *Handbook of Experimental Pharmacology*, *238*, 49–66. https://doi.org/10.1007/164_2015_34.

Edmund, A. B., Walseth, T. F., Levinson, N. M., & Potter, L. R. (2019). The pseudokinase domains of guanylyl cyclase-A and -B allosterically increase the affinity of their catalytic domains for substrate. *Science Signaling*, *12*(566). https://doi.org/10.1126/scisignal.aau5378.

Forte, L. R., Krause, W. J., & Freeman, R. H. (1989). *Escherichia coli* enterotoxin receptors: Localization in opossum kidney, intestine, and testis. *The American Journal of Physiology*, *257*(5 Pt. 2), F874–F881. https://doi.org/10.1152/ajprenal.1989.257.5.F874.

Freihat, L., Muleya, V., Manallack, D. T., Wheeler, J. I., & Irving, H. R. (2014). Comparison of moonlighting guanylate cyclases: Roles in signal direction? *Biochemical Society Transactions*, *42*(6), 1773–1779. https://doi.org/10.1042/BST20140223.

Freihat, L. A., Wheeler, J. I., Wong, A., Turek, I., Manallack, D. T., & Irving, H. R. (2019). IRAK3 modulates downstream innate immune signalling through its guanylate cyclase activity. *Scientific Reports*, *9*(1), 15468. https://doi.org/10.1038/s41598-019-51913-3.

Fulle, H. J., Vassar, R., Foster, D. C., Yang, R. B., Axel, R., & Garbers, D. L. (1995). A receptor guanylyl cyclase expressed specifically in olfactory sensory neurons. *Proceedings of the National Academy of Sciences of the United States of America*, *92*(8), 3571–3575. https://doi.org/10.1073/pnas.92.8.3571.

Gazzano, H., Wu, H. I., & Waldman, S. A. (1991). Adenine nucleotide regulation of particulate guanylate cyclase from rat lung. *Biochimica et Biophysica Acta*, *1077*(1), 99–106. https://doi.org/10.1016/0167-4838(91)90531-4.

Gorczyca, W. A., Gray-Keller, M. P., Detwiler, P. B., & Palczewski, K. (1994). Purification and physiological evaluation of a guanylate cyclase activating protein from retinal rods. *Proceedings of the National Academy of Sciences of the United States of America*, *91*(9), 4014–4018. https://doi.org/10.1073/pnas.91.9.4014.

Gorczyca, W. A., Van Hooser, J. P., & Palczewski, K. (1994). Nucleotide inhibitors and activators of retinal guanylyl cyclase. *Biochemistry*, *33*(11), 3217–3222. https://doi.org/10.1021/bi00177a011.

Guo, D., Zhang, J. J., & Huang, X. Y. (2009). Stimulation of guanylyl cyclase-D by bicarbonate. *Biochemistry*, *48*(20), 4417–4422. https://doi.org/10.1021/bi900441v.

Jaleel, M., London, R. M., Eber, S. L., Forte, L. R., & Visweswariah, S. S. (2002). Expression of the receptor guanylyl cyclase C and its ligands in reproductive tissues of the rat: A potential role for a novel signaling pathway in the epididymis. *Biology of Reproduction*, *67*(6), 1975–1980. https://doi.org/10.1095/biolreprod.102.006445.

Jaleel, M., Saha, S., Shenoy, A. R., & Visweswariah, S. S. (2006). The kinase homology domain of receptor guanylyl cyclase C: ATP binding and identification of an adenine nucleotide sensitive site. *Biochemistry*, *45*(6), 1888–1898. https://doi.org/10.1021/bi052089x.

Jaleel, M., Shenoy, A. R., & Visweswariah, S. S. (2004). Tyrphostins are inhibitors of guanylyl and adenylyl cyclases. *Biochemistry*, *43*(25), 8247–8255. https://doi.org/10.1021/bi036234n.

Joubert, S., Labrecque, J., & De Lean, A. (2001). Reduced activity of the NPR-A kinase triggers dephosphorylation and homologous desensitization of the receptor. *Biochemistry*, *40*(37), 11096–11105. https://doi.org/10.1021/bi010580s.

Juilfs, D. M., Fulle, H. J., Zhao, A. Z., Houslay, M. D., Garbers, D. L., & Beavo, J. A. (1997). A subset of olfactory neurons that selectively express cGMP-stimulated phosphodiesterase (PDE2) and guanylyl cyclase-D define a unique olfactory signal transduction pathway. *Proceedings of the National Academy of Sciences of the United States of America*, *94*(7), 3388–3395. https://doi.org/10.1073/pnas.94.7.3388.

Knape, M. J., & Herberg, F. W. (2017). Metal coordination in kinases and pseudokinases. *Biochemical Society Transactions*, *45*(3), 653–663. https://doi.org/10.1042/BST20160327.

Koller, K. J., de Sauvage, F. J., Lowe, D. G., & Goeddel, D. V. (1992). Conservation of the kinase like regulatory domain is essential for activation of the natriuretic peptide receptor guanylyl cyclases. *Molecular and Cellular Biology*, *12*(6), 2581–2590. https://doi.org/10.1128/mcb.12.6.2581-2590.1992.

Koller, K. J., Lowe, D. G., Bennett, G. L., Minamino, N., Kangawa, K., Matsuo, H., et al. (1991). Selective activation of the B natriuretic peptide receptor by C-type natriuretic peptide (CNP). *Science*, *252*(5002), 120–123. https://doi.org/10.1126/science.1672777.

Krylov, D. M., Niemi, G. A., Dizhoor, A. M., & Hurley, J. B. (1999). Mapping sites in guanylyl cyclase activating protein-1 required for regulation of photoreceptor membrane guanylyl cyclases. *The Journal of Biological Chemistry*, *274*(16), 10833–10839. https://doi.org/10.1074/jbc.274.16.10833.

Kuhn, M., Ng, C. K., Su, Y. H., Kilic, A., Mitko, D., Bien-Ly, N., et al. (2004). Identification of an orphan guanylate cyclase receptor selectively expressed in mouse testis. *The Biochemical Journal*, *379*(Pt. 2), 385–393. https://doi.org/10.1042/BJ20031624.

Kurose, H., Inagami, T., & Ui, M. (1987). Participation of adenosine 5′-triphosphate in the activation of membrane-bound guanylate cyclase by the atrial natriuretic factor. *FEBS Letters*, *219*(2), 375–379. https://doi.org/10.1016/0014-5793(87)80256-9.

Kwon, A., Scott, S., Taujale, R., Yeung, W., Kochut, K. J., Eyers, P. A., et al. (2019). Tracing the origin and evolution of pseudokinases across the tree of life. *Science Signaling*, *12*(578). https://doi.org/10.1126/scisignal.aav3810.

Lange, S. M., Nelen, M. I., Cohen, P., & Kulathu, Y. (2021). Dimeric structure of the pseudokinase IRAK3 suggests an allosteric mechanism for negative regulation. *Structure*, *29*(3), 238–251 e234. https://doi.org/10.1016/j.str.2020.11.004.

Laura, R. P., & Hurley, J. B. (1998). The kinase homology domain of retinal guanylyl cyclases 1 and 2 specifies the affinity and cooperativity of interaction with guanylyl cyclase activating protein-2. *Biochemistry*, *37*(32), 11264–11271. https://doi.org/10.1021/bi9809674.

Leinders-Zufall, T., Cockerham, R. E., Michalakis, S., Biel, M., Garbers, D. L., Reed, R. R., et al. (2007). Contribution of the receptor guanylyl cyclase GC-D to chemosensory function in the olfactory epithelium. *Proceedings of the National Academy of Sciences of the United States of America*, *104*(36), 14507–14512. https://doi.org/10.1073/pnas.0704965104.

Letunic, I., & Bork, P. (2021). Interactive Tree Of Life (iTOL) v5: An online tool for phylogenetic tree display and annotation. *Nucleic Acids Research*, *49*(W1), W293–W296. https://doi.org/10.1093/nar/gkab301.

Lin, H., Cheng, C. F., Hou, H. H., Lian, W. S., Chao, Y. C., Ciou, Y. Y., et al. (2008). Disruption of guanylyl cyclase-G protects against acute renal injury. *Journal of the American Society of Nephrology*, *19*(2), 339–348. https://doi.org/10.1681/ASN.2007050550.

Liu, Y., Ruoho, A. E., Rao, V. D., & Hurley, J. H. (1997). Catalytic mechanism of the adenylyl and guanylyl cyclases: Modeling and mutational analysis. *Proceedings of the National Academy of Sciences of the United States of America*, *94*(25), 13414–13419. https://doi.org/10.1073/pnas.94.25.13414.

Lowe, D. G., Dizhoor, A. M., Liu, K., Gu, Q., Spencer, M., Laura, R., et al. (1995). Cloning and expression of a second photoreceptor-specific membrane retina guanylyl cyclase (RetGC), RetGC-2. *Proceedings of the National Academy of Sciences of the United States of America, 92*(12), 5535–5539. https://doi.org/10.1073/pnas.92.12.5535.

Manning, G., Whyte, D. B., Martinez, R., Hunter, T., & Sudarsanam, S. (2002). The protein kinase complement of the human genome. *Science, 298*(5600), 1912–1934. https://doi.org/10.1126/science.1075762.

Meyer, M. R., Angele, A., Kremmer, E., Kaupp, U. B., & Muller, F. (2000). A cGMP-signaling pathway in a subset of olfactory sensory neurons. *Proceedings of the National Academy of Sciences of the United States of America, 97*(19), 10595–10600. https://doi.org/10.1073/pnas.97.19.10595.

Mishra, V., Bose, A., Kiran, S., Banerjee, S., Shah, I. A., Chaukimath, P., et al. (2021). Gut-associated cGMP mediates colitis and dysbiosis in a mouse model of an activating mutation in GUCY2C. *The Journal of Experimental Medicine, 218*(11). https://doi.org/10.1084/jem.20210479.

Munger, S. D., Leinders-Zufall, T., McDougall, L. M., Cockerham, R. E., Schmid, A., Wandernoth, P., et al. (2010). An olfactory subsystem that detects carbon disulfide and mediates food-related social learning. *Current Biology, 20*(16), 1438–1444. https://doi.org/10.1016/j.cub.2010.06.021.

Nagase, M., Katafuchi, T., Hirose, S., & Fujita, T. (1997). Tissue distribution and localization of natriuretic peptide receptor subtypes in stroke-prone spontaneously hypertensive rats. *Journal of Hypertension, 15*(11), 1235–1243. https://doi.org/10.1097/00004872-199715110-00007.

Ogawa, H., Qiu, Y., Ogata, C. M., & Misono, K. S. (2004). Crystal structure of hormone-bound atrial natriuretic peptide receptor extracellular domain: Rotation mechanism for transmembrane signal transduction. *The Journal of Biological Chemistry, 279*(27), 28625–28631. https://doi.org/10.1074/jbc.M313222200.

Olshevskaya, E. V., Hughes, R. E., Hurley, J. B., & Dizhoor, A. M. (1997). Calcium binding, but not a calcium-myristoyl switch, controls the ability of guanylyl cyclase-activating protein GCAP-2 to regulate photoreceptor guanylyl cyclase. *The Journal of Biological Chemistry, 272*(22), 14327–14333. https://doi.org/10.1074/jbc.272.22.14327.

Padayatti, P. S., Pattanaik, P., Ma, X., & van den Akker, F. (2004). Structural insights into the regulation and the activation mechanism of mammalian guanylyl cyclases. *Pharmacology & Therapeutics, 104*(2), 83–99. https://doi.org/10.1016/j.pharmthera.2004.08.003.

Palczewski, K., Subbaraya, I., Gorczyca, W. A., Helekar, B. S., Ruiz, C. C., Ohguro, H., et al. (1994). Molecular cloning and characterization of retinal photoreceptor guanylyl cyclase-activating protein. *Neuron, 13*(2), 395–404. https://doi.org/10.1016/0896-6273(94)90355-7.

Papermaster, D. S., & Dreyer, W. J. (1974). Rhodopsin content in the outer segment membranes of bovine and frog retinal rods. *Biochemistry, 13*(11), 2438–2444. https://doi.org/10.1021/bi00708a031.

Parkinson, S. J., Carrithers, S. L., & Waldman, S. A. (1994). Opposing adenine nucleotide-dependent pathways regulate guanylyl cyclase C in rat intestine. *The Journal of Biological Chemistry, 269*(36), 22683–22690.

Pattanaik, P., Fromondi, L., Ng, K. P., He, J., & van den Akker, F. (2009). Expression, purification, and characterization of the intra-cellular domain of the ANP receptor. *Biochimie, 91*(7), 888–893. https://doi.org/10.1016/j.biochi.2009.04.011.

Potter, L. R. (1998). Phosphorylation-dependent regulation of the guanylyl cyclase-linked natriuretic peptide receptor B: Dephosphorylation is a mechanism of desensitization. *Biochemistry, 37*(8), 2422–2429. https://doi.org/10.1021/bi972303k.

Potter, L. R. (2011a). Guanylyl cyclase structure, function and regulation. *Cellular Signalling, 23*(12), 1921–1926. https://doi.org/10.1016/j.cellsig.2011.09.001.

Potter, L. R. (2011b). Regulation and therapeutic targeting of peptide-activated receptor guanylyl cyclases. *Pharmacology & Therapeutics, 130*(1), 71–82. https://doi.org/10.1016/j.pharmthera.2010.12.005.

Potter, L. R., Abbey-Hosch, S., & Dickey, D. M. (2006). Natriuretic peptides, their receptors, and cyclic guanosine monophosphate-dependent signaling functions. *Endocrine Reviews*, *27*(1), 47–72. https://doi.org/10.1210/er.2005-0014.

Potter, L. R., & Garbers, D. L. (1992). Dephosphorylation of the guanylyl cyclase-A receptor causes desensitization. *The Journal of Biological Chemistry*, *267*(21), 14531–14534.

Potter, L. R., & Hunter, T. (1998a). Identification and characterization of the major phosphorylation sites of the B-type natriuretic peptide receptor. *The Journal of Biological Chemistry*, *273*(25), 15533–15539. https://doi.org/10.1074/jbc.273.25.15533.

Potter, L. R., & Hunter, T. (1998b). Phosphorylation of the kinase homology domain is essential for activation of the A-type natriuretic peptide receptor. *Molecular and Cellular Biology*, *18*(4), 2164–2172. https://doi.org/10.1128/MCB.18.4.2164.

Robinson, J. W., & Potter, L. R. (2012). Guanylyl cyclases A and B are asymmetric dimers that are allosterically activated by ATP binding to the catalytic domain. *Science Signaling*, *5*(240), ra65. https://doi.org/10.1126/scisignal.2003253.

Roy, N., Guruprasad, M. R., Kondaiah, P., Mann, E. A., Giannella, R. A., & Visweswariah, S. S. (2001). Protein kinase C regulates transcription of the human guanylate cyclase C gene. *European Journal of Biochemistry*, *268*(7), 2160–2171. https://doi.org/10.1046/j.1432-1327.2001.02101.x.

Saha, S., Biswas, K. H., Kondapalli, C., Isloor, N., & Visweswariah, S. S. (2009). The linker region in receptor guanylyl cyclases is a key regulatory module: Mutational analysis of guanylyl cyclase C. *The Journal of Biological Chemistry*, *284*(40), 27135–27145. https://doi.org/10.1074/jbc.M109.020032.

Schroter, J., Zahedi, R. P., Hartmann, M., Gassner, B., Gazinski, A., Waschke, J., et al. (2010). Homologous desensitization of guanylyl cyclase A, the receptor for atrial natriuretic peptide, is associated with a complex phosphorylation pattern. *The FEBS Journal*, *277*(11), 2440–2453. https://doi.org/10.1111/j.1742-4658.2010.07658.x.

Schulz, S., Green, C. K., Yuen, P. S., & Garbers, D. L. (1990). Guanylyl cyclase is a heat-stable enterotoxin receptor. *Cell*, *63*(5), 941–948. https://doi.org/10.1016/0092-8674(90)90497-3.

Schulz, S., Singh, S., Bellet, R. A., Singh, G., Tubb, D. J., Chin, H., et al. (1989). The primary structure of a plasma membrane guanylate cyclase demonstrates diversity within this new receptor family. *Cell*, *58*(6), 1155–1162. https://doi.org/10.1016/0092-8674(89)90513-8.

Schulz, S., Wedel, B. J., Matthews, A., & Garbers, D. L. (1998). The cloning and expression of a new guanylyl cyclase orphan receptor. *The Journal of Biological Chemistry*, *273*(2), 1032–1037. https://doi.org/10.1074/jbc.273.2.1032.

Shenoy, A. R., & Visweswariah, S. S. (2004). Class III nucleotide cyclases in bacteria and archaebacteria: Lineage-specific expansion of adenylyl cyclases and a dearth of guanylyl cyclases. *FEBS Letters*, *561*(1–3), 11–21. https://doi.org/10.1016/s0014-5793(04)00128-0.

Shyjan, A. W., de Sauvage, F. J., Gillett, N. A., Goeddel, D. V., & Lowe, D. G. (1992). Molecular cloning of a retina-specific membrane guanylyl cyclase. *Neuron*, *9*(4), 727–737. https://doi.org/10.1016/0896-6273(92)90035-c.

Singh, S., Lowe, D. G., Thorpe, D. S., Rodriguez, H., Kuang, W. J., Dangott, L. J., et al. (1988). Membrane guanylate cyclase is a cell-surface receptor with homology to protein kinases. *Nature*, *334*(6184), 708–712. https://doi.org/10.1038/334708a0.

Steegborn, C., Litvin, T. N., Hess, K. C., Capper, A. B., Taussig, R., Buck, J., et al. (2005). A novel mechanism for adenylyl cyclase inhibition from the crystal structure of its complex with catechol estrogen. *The Journal of Biological Chemistry*, *280*(36), 31754–31759. https://doi.org/10.1074/jbc.M507144200.

Suga, S., Nakao, K., Hosoda, K., Mukoyama, M., Ogawa, Y., Shirakami, G., et al. (1992). Receptor selectivity of natriuretic peptide family, atrial natriuretic peptide, brain natriuretic peptide, and C-type natriuretic peptide. *Endocrinology*, *130*(1), 229–239. https://doi.org/10.1210/endo.130.1.1309330.

Sun, L., Wang, H., Hu, J., Han, J., Matsunami, H., & Luo, M. (2009). Guanylyl cyclase-D in the olfactory CO_2 neurons is activated by bicarbonate. *Proceedings of the National Academy of Sciences of the United States of America, 106*(6), 2041–2046. https://doi.org/10.1073/pnas.0812220106.

Swenson, E. S., Mann, E. A., Jump, M. L., Witte, D. P., & Giannella, R. A. (1996). The guanylin/STa receptor is expressed in crypts and apical epithelium throughout the mouse intestine. *Biochemical and Biophysical Research Communications, 225*(3), 1009–1014. https://doi.org/10.1006/bbrc.1996.1286.

Taylor, S. S., Meharena, H. S., & Kornev, A. P. (2019). Evolution of a dynamic molecular switch. *IUBMB Life, 71*(6), 672–684. https://doi.org/10.1002/iub.2059.

Thompson, D. K., & Garbers, D. L. (1995). Dominant negative mutations of the guanylyl cyclase-A receptor. Extracellular domain deletion and catalytic domain point mutations. *The Journal of Biological Chemistry, 270*(1), 425–430. https://doi.org/10.1074/jbc.270.1.425.

Tucker, C. L., Laura, R. P., & Hurley, J. B. (1997). Domain-specific stabilization of photoreceptor membrane guanylyl cyclase by adenine nucleotides and guanylyl cyclase activating proteins (GCAPs). *Biochemistry, 36*(39), 11995–12000. https://doi.org/10.1021/bi971212k.

Vaandrager, A. B., van der Wiel, E., & de Jonge, H. R. (1993). Heat-stable enterotoxin activation of immunopurified guanylyl cyclase C. Modulation by adenine nucleotides. *The Journal of Biological Chemistry, 268*(26), 19598–19603.

Vaandrager, A. B., van der Wiel, E., Hom, M. L., Luthjens, L. H., & de Jonge, H. R. (1994). Heat-stable enterotoxin receptor/guanylyl cyclase C is an oligomer consisting of functionally distinct subunits, which are non-covalently linked in the intestine. *The Journal of Biological Chemistry, 269*(23), 16409–16415.

Wu, P., Nielsen, T. E., & Clausen, M. H. (2016). Small-molecule kinase inhibitors: An analysis of FDA-approved drugs. *Drug Discovery Today, 21*(1), 5–10. https://doi.org/10.1016/j.drudis.2015.07.008.

Yamazaki, A., Yu, H., Yamazaki, M., Honkawa, H., Matsuura, I., Usukura, J., et al. (2003). A critical role for ATP in the stimulation of retinal guanylyl cyclase by guanylyl cyclase-activating proteins. *The Journal of Biological Chemistry, 278*(35), 33150–33160. https://doi.org/10.1074/jbc.M303678200.

Yang, R. B., Foster, D. C., Garbers, D. L., & Fulle, H. J. (1995). Two membrane forms of guanylyl cyclase found in the eye. *Proceedings of the National Academy of Sciences of the United States of America, 92*(2), 602–606. https://doi.org/10.1073/pnas.92.2.602.

Yasoda, A., Ogawa, Y., Suda, M., Tamura, N., Mori, K., Sakuma, Y., et al. (1998). Natriuretic peptide regulation of endochondral ossification. Evidence for possible roles of the C-type natriuretic peptide/guanylyl cyclase-B pathway. *The Journal of Biological Chemistry, 273*(19), 11695–11700. https://doi.org/10.1074/jbc.273.19.11695.

Yoder, A. R., Stone, M. D., Griffin, T. J., & Potter, L. R. (2010). Mass spectrometric identification of phosphorylation sites in guanylyl cyclase A and B. *Biochemistry, 49*(47), 10137–10145. https://doi.org/10.1021/bi101700e.

Zhang, G., Liu, Y., Ruoho, A. E., & Hurley, J. H. (1997). Structure of the adenylyl cyclase catalytic core. *Nature, 386*(6622), 247–253. https://doi.org/10.1038/386247a0.

Zhang, M., Su, Y. Q., Sugiura, K., Xia, G., & Eppig, J. J. (2010). Granulosa cell ligand NPPC and its receptor NPR2 maintain meiotic arrest in mouse oocytes. *Science, 330*(6002), 366–369. https://doi.org/10.1126/science.1193573.

Zimmerman, A. D., Nagy, C. R., & Munger, S. D. (2020). Sensory neurons expressing the atypical olfactory receptor guanylyl cyclase D are required for the acquisition of odor preferences by mice in diverse social contexts. *Physiology & Behavior, 227*, 113150. https://doi.org/10.1016/j.physbeh.2020.113150.

CHAPTER NINETEEN

Methods for discovering catalytic activities for pseudokinases

Miles H. Black[a,†], Marcin Gradowski[b,†], Krzysztof Pawłowski[a,b,*], and Vincent S. Tagliabracci[a,c,d,e,*]

[a]Department of Molecular Biology, University of Texas Southwestern Medical Center, Dallas, TX, United States
[b]Department of Biochemistry and Microbiology, Institute of Biology, Warsaw University of Life Sciences, Warsaw, Poland
[c]Harold C. Simmons Comprehensive Cancer Center, University of Texas Southwestern Medical Center, Dallas, TX, United States
[d]Hamon Center for Regenerative Science and Medicine, University of Texas Southwestern Medical Center, Dallas, TX, United States
[e]Howard Hughes Medical Institute, University of Texas Southwestern Medical Center, Dallas, TX, United States
*Corresponding authors: e-mail address: krzysztof.pawlowski@utsouthwestern.edu; vincent.tagliabracci@utsouthwestern.edu

Contents

† These authors contributed equally.

Methods in Enzymology, Volume 667
ISSN 0076-6879
https://doi.org/10.1016/bs.mie.2022.03.047

Abstract

Pseudoenzymes resemble active enzymes, but lack key catalytic residues believed to be required for activity. Many pseudoenzymes appear to be inactive in conventional enzyme assays. However, an alternative explanation for their apparent lack of activity is that pseudoenzymes are being assayed for the wrong reaction. We have discovered several new protein kinase-like families which have revealed how different binding orientations of adenosine triphosphate (ATP) and active site residue migration can generate a novel reaction from a common kinase scaffold. These results have exposed the catalytic versatility of the protein kinase fold and suggest that atypical kinases and pseudokinases should be analyzed for alternative transferase activities. In this chapter, we discuss a general approach for bioinformatically identifying divergent or atypical members of an enzyme superfamily, then present an experimental approach to characterize their catalytic activity.

1. Introduction

The lack of expected catalytic residues in homologs of known enzymes typically results in compromised or absent catalytic activity. For example, many such pseudo-enzymes have been shown to act as scaffolds or allosteric regulators (Murphy, Farhan, & Eyers, 2017; Ribeiro et al., 2019). However, the atypical kinase and pseudokinase research has taught us that distant, atypical members of the protein kinase-like clan/superfamily (or any enzyme superfamily) can exhibit unexpected or long sought-after enzymatic activities, e.g., phosphorylation of secreted proteins (Tagliabracci et al., 2012), phosphorylation of hydroxy-lysine (Veiga da Cunha, Hadi, Balligand, Stroobant, & Van Schaftingen, 2012) or AMPylation (Sreelatha et al., 2018). Many of these atypical kinase-like proteins were initially predicted to be inactive as judged by the lack of the classical set of typical active site motifs, but later found to perform enzymatic activities by using "redesigned" or migrated active sites. We argue that by broadly searching for novel, most distant members of the protein kinase-like superfamily one may expect to find more such interesting enzymatic activities.

Here, we outline the steps that can be taken, starting from bioinformatic prediction of a putative kinase-like enzyme. Once a candidate kinase family has been identified, we offer an experimental framework toward biochemical demonstration of an alternative catalytic activity. While there is no single approach that will uncover the activity of every candidate, we suggest that intact protein mass spectrometry is a powerful method to detect unpredictable post-translational modifications (PTMs). The example of the *Legionella pneumophila* pseudokinase effector SidJ, which catalyzes glutamylation instead of phosphorylation (Bhogaraju et al., 2019; Black et al., 2019; Gan et al., 2019; Osinski et al., 2021; Sulpizio et al., 2019), is presented to illustrate the steps one might take toward discovering catalytic activities of pseudokinases.

2. Bioinformatics-based search for novel atypical kinases/pseudokinases. Sequence-based approach

Search for novel distant members of an established structural clan/superfamily consists of three general steps:

- "High throughput" screening the target protein set (e.g., proteome) for distant, significant, and sub-significant similarities to the query (here, kinase-like families)
- Filtering out the "already known" clan/superfamily members
- Validating the similarity of the remaining "potentially novel" clan/superfamily members

Below, we sketch this approach by presenting two examples, whereas the screening step is performed either applying remote sequence similarity search, or three-dimensional structure similarity search using 3D structure models of the target protein set.

The first approach involves sequence similarity search, comparing a protein set of interest (here, the *L. pneumophila* proteome) against a library of known protein domains (here, PDB, Pfam and scop databases).

2.1 Downloading a "target" sequence dataset

The screen for novel kinase-like proteins must start from a defined set of sequences, e.g., proteome of an organism of interest. The UniProt database (UniProt, 2021) provides users sets of protein sequences with semi-automatic functional annotation. Also, it provides full proteomes of organisms.

2.1.1 Websites/webtools

UniProtKB, www.uniprot.org

alternatively, NCBI Protein, www.ncbi.nlm.nih.gov/protein

2.1.2 Procedure

(1) From the Uniprot main page, navigate to Proteomes (www.uniprot.org/proteomes). Find the proteome of interest (here, we use the proteome of the strain "*L. pneumophila* subsp. pneumophila strain Philadelphia 1").

(2) Go to the proteome record (e.g., www.uniprot.org/proteomes/UP000000609) and select the option "Download one protein sequence per gene (FASTA)" to fetch a FASTA file containing sequences of all the proteins of the selected organism.

2.1.3 Notes

Alternatively, one can download the sequences from the NCBI Protein database, Refseq section (www.ncbi.nlm.nih.gov/refseq/) or any preferred sequence database.

2.2 Reducing redundancy in the target sequence dataset

Redundancy in a sequence dataset is usually not desirable because it may lead to unnecessary time-consuming processing of highly similar sequences. One can remove redundancy by dividing the sequence set into clusters with high similarity within, and lower similarity between them, e.g., at 40% sequence identity threshold using CD-HIT (then, one representative sequence per cluster is selected for subsequent analysis).

2.2.1 Websites/tools

CD-HIT Suite webserver, http://weizhong-lab.ucsd.edu/cdhit-web-server/cgi-bin/index.cgi.

2.2.2 Procedure

(1) Go to the CD-HIT webserver and select the "cd-hit" tab.

(2) Upload the protein dataset (FASTA file from step 2.1) to CD-HIT, choose "40%" identity threshold and hit "Submit."

(3) Download ".fas.1" file from the CD-HIT results. It contains representative sequences of the clusters.

2.2.3 Notes

(1) Alternatively, one can use MMseqs2 (https://toolkit.tuebingen.mpg.de/tools/mmseqs2).
(2) For the *L. pneumophila* proteome dataset, a CD-HIT run (40% threshold) reduces the dataset only by 10%. It is more effective when one clusters several related proteomes, e.g. when clustering two *Legionella* proteomes of different species, the number of proteins may decrease by 30%.
(3) One can also do clustering using the h-cd-hit option, which performs hierarchical, multiple clustering runs. First, proteins are clustered at high sequence identity threshold (e.g., 90%), and representatives of clusters of similar sequences are used in the next clustering step at lower threshold (e.g., 60%). Subsequently, cd-hit can perform yet another third clustering step at a lower threshold (e.g., 40%). It is effective when one is clustering a large number of sequences. Then, multistep clustering can be faster than one step clustering.
(4) Alternatively, one can use "one step" clustering at 90% or 50% threshold within the UniProt database when selecting the set of proteins for download, by choosing the clustering threshold shown to the left of the table containing the protein entries.

2.3 Splitting sequences of the dataset (optional)

All clustered sequences can be split into fragments to increase the chances of detecting remote sequence similarity in longer, multidomain proteins. For example, we "cut" proteins into 300 amino acid (aa) long fragments, with 100 aa overlap. Thus, we analyze fragments 1–300, 201–500, 401–700, etc.

2.3.1 Websites/webtools

Sequence Manipulation Suite—split FASTA, http://www.bioinformatics.org/sms2/split_fasta.html.

2.3.2 Procedure

(1) Paste the contents of the FASTA file created in step 2.2. Set length of fragments to 300 amino acids and 100 amino acid overlap and click "submit" button.
(2) Save results to a file.

2.3.3 Notes

"Split FASTA" has an input limit of 500,000,000 characters (e.g., more than 800,000 sequences of 600 amino acids each). If one needs to split a larger number of sequences, one can use the program "emboss splitter" (https://www.bioinformatics.nl/cgi-bin/emboss/splitter).

2.4 Remote homology detection

The well-known BLAST searches typically detect homologies at >30% sequence identity, but are in most cases unable to detect more distant similarity, e.g., at about 10–15% identity (Geourjon, Combet, Blanchet, & Deleage, 2001). PSI-BLAST overcomes these limitations to some extent but is not best suited for large-scale screening. Several even more sensitive distant homology search tools can be used to this purpose. These methods are based on the construction of amino acid frequency profiles, e.g., FFAS (Xu, Jaroszewski, Li, & Godzik, 2014) or Hidden Markov Models, e.g., HHblits (Steinegger et al., 2019). The query profile/model is then compared to a library of profiles/models of known protein families. In this way, by filtering and meticulously reviewing the results, we can find new enzyme families.

The Fold and Function Assignment System (FFAS) method is amino acid frequency profile-based. Specific profile generation is an important step that distinguishes the FFAS program from other profile-based methods. FFAS first uses 5 iterations of PSI-BLAST with a threshold E-value of 0.005 to generate set of homologs for the profile. To facilitate distant homology detection, the more distant from other sequences in the set a sequence is, the more weight it has in the profile (Rychlewski, Jaroszewski, Li, & Godzik, 2000; Xu et al., 2014).

2.4.1 Websites/tools

FFAS server, https://ffas.godziklab.org.

2.4.2 Procedure

(1) The split sequences from the target protein set will be used to search for distant homology to known kinase domains.

(2) One can use FFAS Web server online or the standalone version. The server version has input limit: 100 sequences. Thus, we recommend installing standalone FFAS for the Linux system (https://ffas.godziklab.org/ffas-cgi/cgi/download.pl).

(3) Run the FFAS search, against selected databases of FFAS profiles. The recommended databases are SCOP (https://scop.berkeley.edu/) (Fox, Brenner, & Chandonia, 2014), Pfam (https://pfam.xfam.org/), PDB (https://www.rcsb.org/).

2.4.3 Notes

(1) For FFAS search, one might consider adding other well-annotated protein domain databases, e.g., ecod (http://prodata.swmed.edu/ecod/) (Cheng et al., 2014). The results of the standalone version are flat text files that can be reviewed using simple custom scripts or even manually. For example, one can use notepad++ (Windows) or gedit (linux) to search keywords like "kinase" in results.

(2) The searches are more sensitive if profiles are built using the most recent version of the NR and ENV_NR sequence databases from NCBI (https://www.ncbi.nlm.nih.gov/; "non-redundant protein sequences" and "metagenomic proteins"). This applies to query profiles, but also to profile libraries (PDB, Pfam, SCOP).

(3) Although the FFAS authors recommend hits with *Z*-scores lower than −9.5 as likely true, for searches of remote homologs one should analyze also less significant, "borderline" hits. We typically used Z-scores up to the value of −6.

(4) A good alternative to FFAS is HHblits/HHpred (HH-suite) (Zimmermann et al., 2018). Although FFAS and HHpred usually agree, there is anecdotal evidence that in some cases only one of these methods provides correct prediction. Also, in our hands the FFAS method is more simple to use for large datasets and requires less space.

2.5 Filtering out the "known" kinases

Obviously, the FFAS search results would in principle include all the known kinases in the target set. They can be filtered out using several non-exclusive approaches:

- Using automated annotations available in the UniProt database (e.g., assignments to kinase-like Pfam domains, PF00069, etc.)
- Running an automated annotation program, e.g., Batch CD-Search
- Ignoring very high confidence FFAS hits
- Performing filtering by annotation (i.e., ignoring all hits annotated as kinases)
- Performing a manual or semi-manual literature search

2.6 Authentication of distant homology results

Since the search for novel remote kinase-like proteins leads to borderline significance hits, they need to be validated to discard false positives and gain more confidence in the remaining predictions. This usually requires a tailored case-by-case approach, below we present a non-exhaustive list of few options.

- Use of other methods to predict distant homology such as HHpred and Phyre also considering the predicted secondary structure (Kelley, Mezulis, Yates, Wass, & Sternberg, 2015; Steinegger et al., 2019).
- Use of Artificial Intelligence (AI)-based methods for 3D structure prediction such as AlphaFold and RoseTTAFold to build structure models and compare them with known kinase structures (Baek et al., 2021; Jumper et al., 2021).
- Analysis of sequence conservation, in particular the elements of the active site, e.g., by collecting homologs and building sequence logos (Crooks, Hon, Chandonia, & Brenner, 2004).
- Consideration of agreement between top hits. If several of top FFAS predictions are kinases, even when they are of low statistical significance, they are more likely to be true.

Here, we use an example of SidJ from *L. pneumophila* with poor similarity to known protein kinase domains. The FFAS search showed that few of the top hits were kinase-like, but the significance of these similarities was low. In addition, the conserved protein kinase active site motifs (HRD and DFG) appeared to be missing from the alignments (Fig. 1A).

2.6.1 Websites/tools

(1) HHpred server, https://toolkit.tuebingen.mpg.de/tools/hhpred
(2) PSI-BLAST, https://blast.ncbi.nlm.nih.gov/Blast.cgi
(3) MAFFT, https://mafft.cbrc.jp/alignment/server/index.html
(4) Weblogo, http://weblogo.threeplusone.com

2.6.2 Procedure—Use of complementary method to predict distant homology

(1) In HHpred, choose pdb70, ECOD, CDD and pfam databases, paste the "suspect" kinase sequence of SidJ, and submit.
(2) Inspect the results, search for protein kinase-like hits—see Fig. 1B. Here, these hits are not statistically significant. However, some features

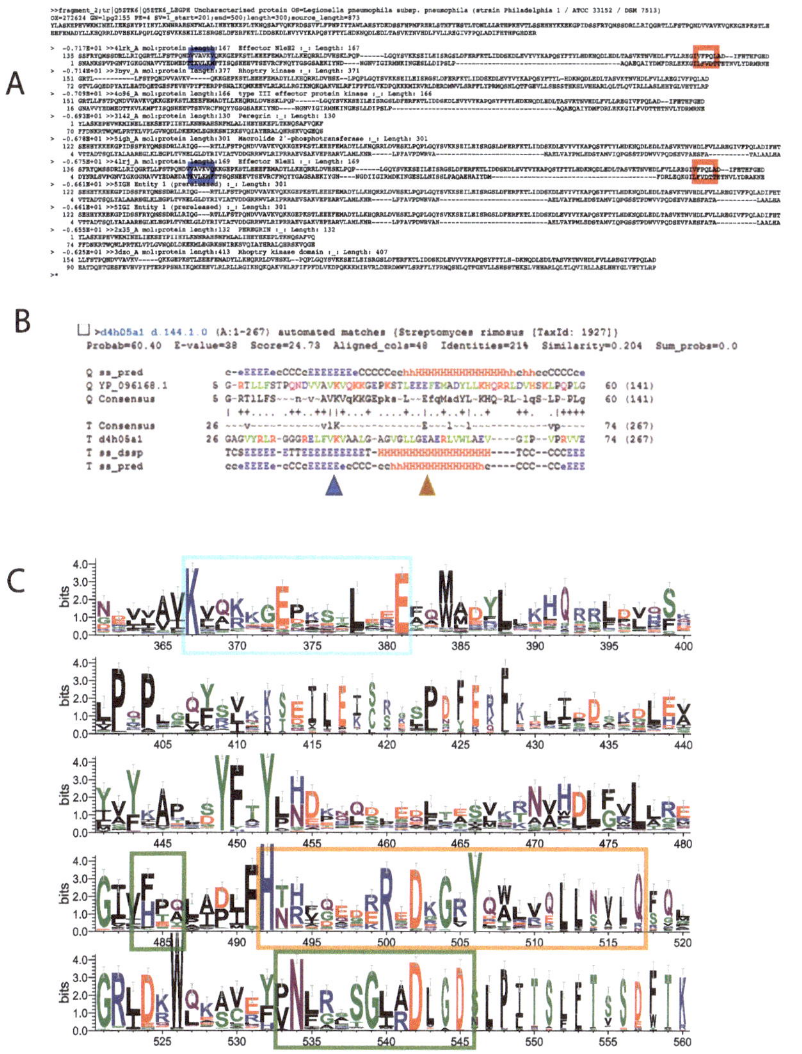

Fig. 1 Sequence search for novel kinases. (A) Archival FFAS results. *L. pneumophila* proteome compared against the PDB. Results for the SidJ protein shown, with noticeable similarity to protein kinases. Blue box—equivalent of the PKA catalytic residue, K72. Red box—equivalent of the PKA catalytic residue, D166. (B) Archival results, HHpred analysis of SidJ. Similarity to N-lobe of protein kinases. Blue triangle—equivalent of PKA K72. Brown triangle—equivalent of PKA E91. (C) Sequence logo of SidJ homologs. Blue box: K72-E91 ion pair region. Green boxes: canonical kinase active site motifs. Orange box: migrated kinase active site motifs.

of kinases (the counterparts of the protein kinase A (PKA) residues, K72 and E91) and secondary structures for a fragment of the kinase domain appear to be conserved.

2.6.3 Procedure—Analyzing sequence conservation by sequence logo

(1) Use PSI-BLAST, HHblits or JackHMMer to collect sequence homologs.
(2) Remove sequence redundancy, e.g., at 80% sequence identity threshold, using CD-HIT.
(3) Align the representative homologs using a multiple alignment program, e.g., MAFFT.
(4) Create sequence logo in the Weblogo server.

For SidJ, the sequence logo (see Fig. 1C) shows remarkable conservation of the K72-E91 ion pair important for kinase catalysis (PKA numbering) and of the several conserved motifs downstream, some of them reminiscent of the classic DFG metal ion binding motif. The catalytic HRD motif appears to be absent. However, sequence analysis did not allow unequivocal assignment of catalytic residues, since two sets of motifs appeared plausible. This ambiguity turned out to reflect reality, as SidJ was found to possess two active sites, the "canonical" one and a "migrated" one—see Fig. 1C.

2.6.4 Notes

(1) These are just examples of steps one can take to gain more confidence in a weakly predicted similarity to protein kinases. Each such prediction may require special customized steps.
(2) The logos are more informative if one uses a custom script to remove gaps in the query sequence from the alignment (see Supplement for a sample script in Supplementary Material 1 in the online version at https://doi.org/10.1016/bs.mie.2022.03.047).
(3) In the example of SidJ protein shown in Fig. 1, it is important to note that this *L. pneumophila* protein has been already characterized experimentally as a deubiquitinase (Qiu et al., 2017). Thus, although experimental validation is the ultimate arbiter of truth, it is essential to keep in mind that interpretation of experimental data can be wrong or misleading. Here, although published experimental data were claimed to prove protease function for SidJ, our later bioinformatics prediction of a kinase-like protein and kinase-like active site was in the end proved correct by extensive structural, biochemical and *Legionella* infection model experiments. Thus, experimental evidence contradicting bioinformatic predictions need not always mean the prediction is wrong.

3. Bioinformatics-based search for novel atypical kinases/pseudokinases. Approach using predicted three-dimensional protein structures

This approach involves three-dimensional structure similarity searches, comparing a representative set of protein domains of interest (here, protein kinase-like domains) against a set of 3D structure models built for a protein set of interest (here: AlphaFold models of the *Leishmannia infantum* proteome) (Varadi et al., 2022).

3.1 Downloading the database of protein structure models (search domain) and the set of representative kinase-like structures (query set)

3.1.1 Websites/webtools

(1) AlphaFold (AF) database—protein structure models, http://alphafold.ebi.ac.uk/download#proteomes-section

(2) ecod40 dataset of representative protein domain structures, http://prodata.swmed.edu/ecod

3.1.2 Procedure

(1) Download the AlphaFold structure models of the proteome of interest (here, *L. infantum*: https://ftp.ebi.ac.uk/pub/databases/alphafold/UP000008153_5671_LEIIN.tar)

(2) Download the ecod F40 dataset of representative protein domain structures http://prodata.swmed.edu/ecod/distributions/ecod.latest.F40.pdb.tar.gz

(3) Using the ecod classification available at http://prodata.swmed.edu/ecod/distributions/ecod.latest.F40.domains.txt

Note: out of the ecod F40 representative structures, select those of the "protein kinase" (T) group

3.1.3 Notes

(1) Alternatively, one could use other representative sets of kinase-like structures (query set), e.g., from SCOP, CATH, Pfam. However, in our experience, the ecod database (Schaeffer, Liao, Cheng, & Grishin, 2017) is most up-to-date, and also has the structural domains very well curated from the point of structural bioinformatics.

(2) Nevertheless, one may want to complement the ecod40 set with a manually prepared set of additional domains, including AlphaFold models of reliably predicted kinases. Here, we added the structures of Lpg2603, MavQ, CotH, SelO, and SidJ kinase-like proteins (pdb codes 6vve (Sreelatha et al., 2020), 7m7a (Hsieh et al., 2021), 5jda (Nguyen et al., 2016), 6eac (Sreelatha et al., 2018), 6oqq (Black et al., 2019), respectively) as well as AlphaFold models of human FAM69A and DIA1 predicted kinases (Dudkiewicz, Lenart, & Pawłowski, 2013).

(3) In principle, one might use only a single kinase-like query. However, in our experience the use of multiple, diverse queries increases the likelihood of detecting distant structural neighbors, in the spirit of cascading sequence searches (Li, Pio, Pawlowski, & Godzik, 2000).

(4) Also in this spirit, one might use the novel kinase-like proteins found by the structure search as subsequent queries in an iterative procedure.

(5) Instead of selecting ecod representative structures at the topology (T) level, one might use different levels of the ecod structural hierarchy, e.g., the higher levels of possible homology (X) or homology (H).

(6) As an alternative to AlphaFold, one could use a database of custom-made 3D models built using other tools, e.g., RoseTTAFold (Baek et al., 2021). However, as of writing this paper, AlphaFold provides the only comprehensive structure model database for entire proteomes, including human. Also, the AlphaFold database is expected to grow significantly in the coming months.

3.2 Comparing structures of query representative kinase structures with the models

TMalign is a fast and sensitive protein structure comparison program providing reliable measures of similarity (Zhang & Skolnick, 2005).

3.2.1 Websites/webtools

TMalign structure comparison program, https://zhanggroup.org/TM-align

3.2.2 Procedure

(1) Download and install the standalone TMalign program https://zhanggroup.org/TM-align

Note: compiling the code under Linux on the user's machine is preferable to achieve best speed, although an executable for the Linux system is also provided.

(2) Run TMalign, preferably using a script, performing pairwise comparisons of all query structures (representative kinases) vs all the models of the proteome of interest (here—*L. infantum*—see Supplement for a sample script in Supplementary Material 2 in the online version at https://doi.org/10.1016/bs.mie.2022.03.047).
Use TM-score normalization by the query (kinase) structure.

(3) For every comparison, save similarity-related parameters: TM-score, alignment length, sequence identity and RMSD (C_α root mean square deviation).

3.2.3 Notes

Alternatively, one could use other structure comparison methods, e.g., Dali or FATCAT servers (however, in our experience, TMalign was preferable for its speed).

3.3 Analyzing the results of structure comparison

Similarly, to the novel kinase search using sequence analysis, the results need be filtered to remove known kinases—see Section 2.5. Then, formerly unpredicted, possibly novel kinases, if any, need be spotted among the majority of non-kinase structures. Although TMalign does not provide a clear-cut statistical similarity measure such as E-value in BLAST or HHpred, the TMscore is a good proxy. Because a novel atypical kinase may exhibit TMalign-detectable structural similarity only to few known kinases, it is prudent not to rely on measures such as "average TMscore for comparison of protein X vs kinases." Instead, we propose to search for such proteins that exhibit even just one or only few relatively high TMscores in comparison to kinases. To this end, we found visual inspection of graphical representation of the data very convenient.

3.3.1 Websites/webtools

For graphical presentation, any data-plotting program, e.g., a spreadsheet application, will work. Here, we used the Spotfire program (from TIBCO), which handles large datasets and allows flexible use of annotations and filtering.

3.3.2 Procedure

(1) Plot TMscore vs alignment length for all TMalign results.

(2) Filter plot by alignment length (e.g., requiring at least 100 residues). Filter out "known" kinases, e.g. using annotations of Pfam domains provided by Uniprot—(see Fig. 2A).
(3) Seek "promising" kinase-like candidates as outliers in the plot (see Fig. 2B). If needed, use additional plots to inspect sequence identity and RMSD.

3.3.3 Notes

Alternatively, one could entertain a statistical approach to TMscore analysis. For example, one could evaluate, using a *t*-test, the difference between average TMscore for comparing structure of a query protein with known kinase structures to the average TMscore between the query and all non-kinase proteins. However, in our experience, visual inspection was sufficient.

3.4 Validating the structure similarity

The candidate novel kinases require the validity and extent of the detected structural similarity to be checked. As with novel kinase searches by sequence, this step is most variable and typically will require a customized approach. Here, we present a non-exhaustive list of possible steps. In our experience, using an alternative structure comparison tool and visual inspection of structural superposition is most important.

3.4.1 Websites/webtools

(1) Structure comparison: Dali server, http://ekhidna2.biocenter.helsinki.fi/dali/
(2) Structure comparison: FATCAT server, https://fatcat.godziklab.org
(3) Structure viewing: Pymol: https://pymol.org/2/
(4) Structure viewing: Chimera, https://www.cgl.ucsf.edu/chimera/

3.4.2 Procedure

(1) Superpose the candidate kinase-like protein and the closest known kinase indicated by TMalign search, using Dali or FATCAT pairwise comparison options, or the cealign command in Pymol.

Inspect the statistical parameters describing the similarity (FATCFAT *P*-value, Dali *Z*-score).

(2) Visualize the superposed structures in a structure viewer, e.g., Pymol or Chimera.

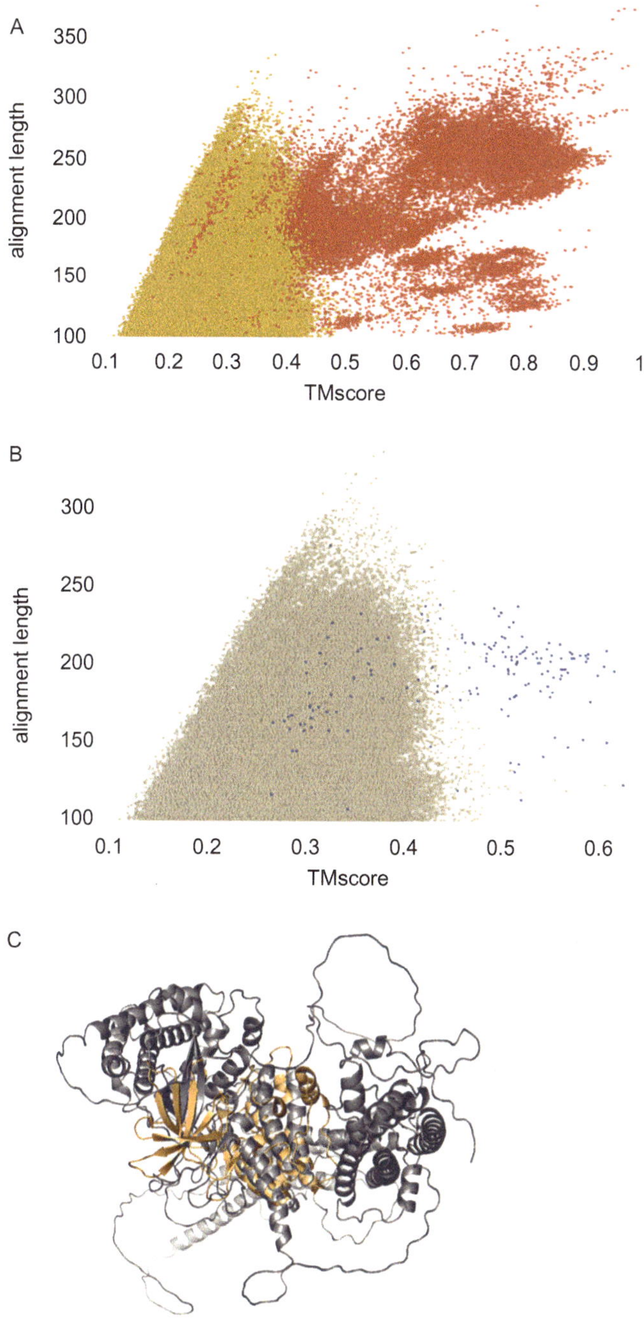

Fig. 2 Structure search for novel kinases. (A) TMalign results, *Leishmania infantum* proteome vs kinases. Known kinases marked in red. (B) As in (A), but known kinases filtered out. Data points corresponding to *Leishmania* E9AHR4 shown in blue. (C) Structural superposition of an AlphaFold model of *Leishmania* E9AHR4 (gray) vs PDB 2zv2 (CaMKK2 kinase).

(3) Additionally, build a sequence logo for the candidate kinase (see Section 2.6.3).
(4) Inspect the conservation of the crucial elements of the kinase active site.

3.4.3 Notes

(1) In addition, one might use tools such as the CONSURF server (Ashkenazy et al., 2016) to map sequence conservation directly onto structure, and to evaluate evolutionary conservation of a putative kinase-like active site.
(2) In addition, one might use tools docking tools such as the HADDOCK server (van Zundert et al., 2016), to investigate the likelihood of nucleotide binding in the putative kinase-like nucleotide binding pocket.
(3) One might additionally include the information on residue-by-residue "model confidence" provided by AlphaFold to assess reliability of the putative kinase-like structure.

In the example shown in Fig. 2C, the uncharacterized *L. infantum* protein E9AHR4 can be seen to align well to a known kinase structure although sequence identity is as low as 5%. Also, sequence conservation of the putative active site motifs supports the prediction of an active kinase-like enzyme.

4. Cloning a candidate novel kinase for bacterial expression

To maximize one's chances of detecting a novel activity, one should strive to obtain soluble, homogenous, and active candidate protein. The investigator should anticipate a requirement for cofactors, binding partners, or activating modifications that are not apparent from the primary sequence. When it can be predicted, the substrate of the candidate enzyme should also be cloned and co-expressed. Ideally, a cell-based or phenotypic assay will also exist to determine if the activity is meaningful in a biological context. The following protocol outlines the approach taken to discover the glutamylation activity of the *Legionella* effector SidJ.

4.1 Designing stable truncations that encompass the kinase domain

We follow design principles from the field of protein crystallography to produce constructs for recombinant protein expression. While these have been reviewed in crystallographic texts (Deller, Kong, & Rupp, 2016). some of the most important considerations are the species of origin and the truncations tested.

Select a homolog from the candidate kinase family, considering the full diversity of species available. Ideal candidates are single domain proteins that comprise the core novel kinase fold. Transmembrane features, large, disordered regions, and poorly conserved insertions that interrupt the kinase domain are undesirable. These features can be detected by analysis of sequence conservation, secondary structure predictions (by Phyre or similar programs), and AI-based methods for 3D structure prediction (RoseTTAFold or AlphaFold) as outlined in Section 2.6. Design truncations to remove these features.

4.1.1 Tools and equipment

(1) Software for codon optimization and primer design: Any sequence analysis program such as SnapGene software (from Insightful Science, available at snapgene.com) or MacVector (from Oxford Molecular Group).

(2) Polymerase Chain Reaction (PCR) template: Genomic DNA or cDNA clones may be used if available. Chemically synthesized custom DNA templates such as gBlocks from Integrated DNA Technologies can provide sequences from uncommon organisms and offer the advantage of codon optimization for the expression host.

(3) Thermocycler, heat blocks, or adjustable-temperature water baths

4.1.2 Buffers and reagents

The reagents will depend on the cloning method that the investigator chooses to pursue. New England BioLabs sells reagents and buffers for various cloning methods including traditional restriction enzyme ligation, TA ligation, recombination, Gibson Assembly® (Gibson et al., 2009), and others.

Vector backbone: we clone into a modified pET28a bacterial expression vector (ppSUMO) containing an N-terminal 6×-His tag followed by the yeast SUMO (smt3) coding sequence. The 6×-His-SUMO tag serves as a solubility tag to promote protein folding, and as an affinity tag that can be purified by nickel affinity (described later). The 6×-His-SUMO tag is easily removed by the SUMO Protease (ULP, UniProt Q02724).

4.1.3 *Protocol*

Follow lab cloning protocol adapted to your lab. Green, Sambrook, and Sambrook (2012) provides an authoritative manual of molecular cloning techniques.

4.1.4 *Notes*

(1) Consider cloning homologs from thermophilic organisms or substrates of bacterial secretion systems, which tend to adopt stable protein folds. Any strain or species may be selected.

(2) It is desirable to test multiple truncations of the kinase domain, as it can be difficult to predict how truncation will affect protein expression and stability. One reasonable approach is to make 10 different truncations every 5–15 amino acid residues around the approximate boundaries of the kinase domain. Truncations beginning at secondary structural features predicted by Phyre or Alpha fold are preferable to truncations that interrupt such features. We performed serial N- and C- terminal truncations of SidJ to identify a stable fragment consisting of amino acid residues 59–851.

(3) Recombinant full length proteins could be subjected to limited proteolysis with a dilution series of enzymes, such as trypsin or chymotrypsin, to remove unstructured regions of the protein and identify a folded core domain. A stable core domain, as observed by SDS-PAGE, can be characterized by mass spectrometry or N-terminal peptide sequencing to identify its composition for recloning and expression as a recombinant protein.

4.2 Introducing and testing mutations of predicted catalytic residues

Predicted catalytic mutants serve as important experimental controls. We select at least one of the conserved residues corresponding to kinase catalytic motifs (see Section 2.6.3) to mutate to alanine. Alanine substitutions are usually well-tolerated and do not compromise the solubility of the protein.

4.2.1 *Tools*

(1) For designing primers for targeted mutagenesis, we use the Agilent QuikChange Primer design tool: https://www.genomics.agilent.com

(2) Thermocycler

(3) 42 °C water bath

(4) Incubator set to 37 °C with rotating tube rack

(5) Incubator set to 37 °C with orbital shaker

4.2.2 Buffers and reagents

(1) PfuTurbo DNA polymerase and 10 × reaction buffer (Thermo Fisher Scientific)
(2) Dimethyl Sulfoxide (DMSO)
(3) *Dpn*I Endonuclease (New England BioLabs)
(4) Chemically competent DH5α *E. coli* (Thermo Fisher Scientific)
(5) SOC recovery media: Dextrose, 3.603 g/L KCl, 0.186 g/L $MgSO_4$, 4.8 g/L Tryptone, 20 g/L, Yeast extract, 5 g/L
(6) LB agar: 10 g/L Peptone, 5 g/L Yeast Extract, 5 g/L NaCl, 12 g/L Agar
(7) LB Broth: 10 g/L Peptone, 5 g/L Yeast Extract, 5 g/L NaCl
(8) Kanamycin sulfate (Millipore Sigma). Dilute in H_2O and add to LB at 50 μg/mL final
(9) GeneJET plasmid Miniprep DNA kit or other similar kit (Thermo Fisher Scientific, K0502)

4.2.3 Protocol

(1) Perform QuikChange site-directed mutagenesis with PfuTurbo DNA polymerase (Agilent) according to the manufacturer's protocol
(2) Digest reaction products with DpnI endonuclease overnight
(3) Add 5 μL PCR into ~50 μL chemically competent DH5α *E. coli* on ice, then heat shock at 42 °C for 45 s
(4) Briefly return the *E. coli* to ice, then add 500 μL SOC. Incubate at 37 °C with rotating tube rack for at least 30 min
(5) Plate the *E. coli* on LB agar plates and incubate overnight at 37 °C
(6) Pick 2 single colonies and inoculate 5 mL LB broth with kanamycin selection
(7) Extract plasmid DNA with the miniprep kit
(8) Confirm mutations by Sanger sequencing

4.2.4 Notes

(1) Catalytic residues may be hard to predict, especially when there are few sequence homologs available to build a weblogo. In these cases, make several mutants.
(2) In some cases, published reports may exist that identify important residues. For instance, the predicted metal-binding "DFG" motif of SidJ (Fig. 1) corresponds to a DxxD motif previously shown to be required for SidJ's role in *Legionella* virulence (Jeong, Sexton, & Vogel, 2015; Qiu et al., 2017). Both D542 and the predicted ion-pair lysine (K367) were mutated to alanine as controls.

(3) Addition of DMSO to 4% to PCR conditions may improve the success rate for difficult mutations.

(4) If available, a phenotypic assay is performed to determine if kinase catalytic residues are required for the enzyme's activity in cells. We frequently choose the yeast *Saccharomyces cerevisiae* as a proxy to study novel enzymes due to the ease of genetic manipulations in this organism. *S. cerevisiae* has been particularly useful for the study of bacterial effectors, which often target cellular processes that are conserved between yeast and other eukaryotes and therefore may cause a growth inhibition phenotype which can be readily detected. For example, expression of the SidE proteins in *S. cerevisiae* results in a growth inhibition phenotype that can be suppressed by co-expression of SidJ (Havey & Roy, 2015; Jeong et al., 2015). Using this system, we were able to demonstrate that the kinase catalytic residues of SidJ are required for its role in suppression of SidE-mediated toxicity, lending assurance to the hypothesis that SidJ harbors catalytic activity. A yeast phenotypic assay also allows the investigator to test whether truncations of the candidate enzyme, as designed for protein expression in Section 4.1, retain activity.

4.3 Identifying interacting proteins

While some enzymes possess intrinsic activity, others may only adopt a catalytic conformation if induced by a post-translational modification, a protein-protein interaction, or by binding to a small molecule cofactor (i.e., cAMP-dependent protein kinase (Johnson, Akamine, Radzio-Andzelm, Madhusudan, & Taylor, 2001) and calmodulin-dependent protein kinases (Haridas et al., 1998)). These interactions may be impossible to predict from the primary amino acid sequence of the candidate, so they should be actively sought using a combination of affinity-based and mass spectrometry approaches. These protocols are beyond the scope of this methods paper, but we briefly highlight key points to consider. If a cellular phenotype has been identified, the cofactor may be sought in the same system. For example, although we did not detect kinase activity of SidJ in vitro, SidJ was active against SidE in *S. cerevisiae*. We therefore sought to identify interacting proteins or potential substrates by affinity purification and mass spectrometry. This led to the identification of calmodulin (CaM) as a SidJ-interacting protein. In some cases, sequence motifs may help to predict interacting partners. For instance, the Calmodulin Target Database (http://calcium.uhnres.utoronto.ca/ctdb/ctdb/sequence.html) can analyze a protein sequence for possible CaM-interaction motifs. In the case of SidJ, a well-known CaM-binding IQ motif (Putkey, Kleerekoper, Gaertner, & Waxham, 2003) conserved in SidJ suggested CaM as an interactor.

Confirm that any predicted protein-protein interactions are direct by performing binding studies with purified components. Isothermal titration calorimetry (ITC) directly analyzes association kinetics to determine the binding constants and stoichiometry of an interaction (Pierce, Raman, & Nall, 1999). ITC is also a powerful assay to test mutations designed to disrupt the interaction. Other biophysical techniques including Microscale Thermophoresis (MST) or Surface Plasmon Resonance (SPR) may be appropriate. MST can be performed with much smaller quantities of protein but requires that the analyte molecule is conjugated to a fluorophore, which may disrupt the interaction. SPR also can be performed with low quantities of analyte but requires derivatization of the analyte so it can be immobilized to a sensor surface.

5. Preparation of an active kinase complex from bacteria

The next step toward biochemical and structural studies is production of recombinant, catalytically competent holoenzyme. We present an approach to co-express SidJ and CaM, which could be adapted to other enzyme-cofactor pairs.

5.1 Design of a co-expression system

To express the active SidJ-CaM complex, SidJ residues 59–851 and yeast CaM were cloned into the pETDuet1 vector (Novagen), which contains two multiple cloning sites preceded by a T7 promoter/*lac* operator. SidJ was cloned with N-terminal 6 × His and SUMO tags, while CaM was untagged (pETDuet1 SUMO-SidJ/CaM, Fig. 3A). This enabled the simultaneous expression of both proteins from a single construct and the subsequent enrichment of the SidJ-CaM complex by virtue of their high-affinity interaction (Fig. 3B). Ni^{2+} NTA is a convenient and inexpensive purification method, but any affinity-based system could be utilized.

5.1.1 Tools

(1) Incubators with orbital shakers, set to 37 °C

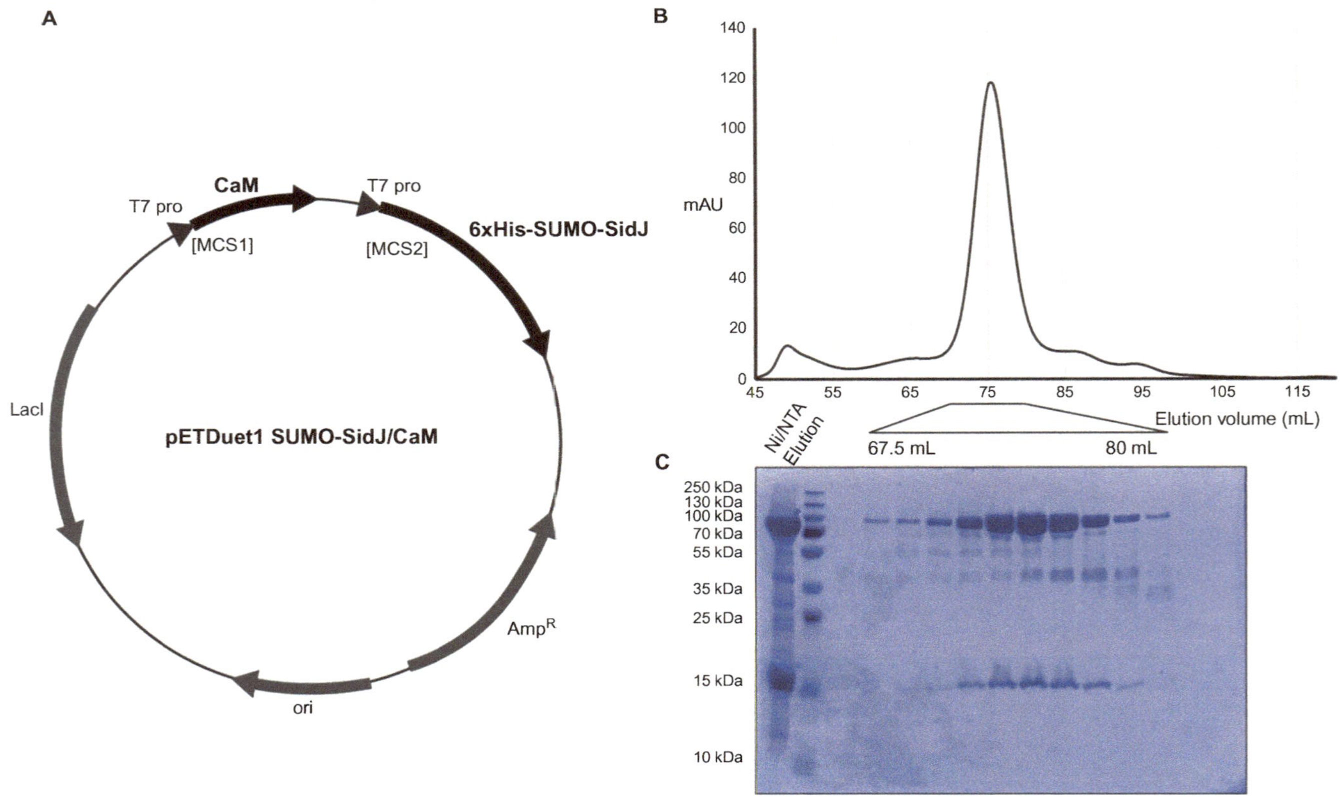

Fig. 3 Purification of the active SidJ-CaM complex. (A) Plasmid map of pETDuet1 SUMO-SidJ/CaM used to co-express SidJ and CaM. Black arrows indicate the coding sequences of CaM and SidJ cloned into the first and second multiple cloning sites ([MCS]), respectively, under control of a T7 promoter (T7 pro.) The map also indicates the ampicillin resistance cassette (AmpR), pUC high-copy number origin if replication (ori) and the coding sequence of the lac repressor (LacI). (B) Superdex 200 size exclusion chromatography trace of the SidJ-CaM complex purified from *E. coli* as described in Section 6.1.2. (C) Fractions that eluted from 67.5 to 80 mL were separated by reduced 6% SDS-PAGE and visualized with Coomassie blue, demonstrating co-elution of SidJ (the ~100 kDa species) and CaM (the ~15 kDa species).

(2) Spectrophotometer and cuvettes for measuring bacterial OD_{600} and protein concentration determination by Bradford assay
(3) High-speed floor-standing centrifuge that accommodates a fixed-angle rotor, such as the Sorvall LYNX 4000
(4) Sonicator cell disruptor. We use a VCX-750 ultrasonic processor coupled to a CV334 converter and 3mm stepped micro tip probe (Biomedical Solutions, Inc.)
(5) 4°C work environment for protein purification steps
(6) Chromatography system equipped for fast protein liquid chromatography (FPLC). We use the AKTA Pure FPLC chromatography system coupled to a Superdex 200 column (GE Healthcare) with flow rate set to 1mL/min
(7) Heat block at 98°C

5.1.2 Buffers and reagents

(1) Rosetta™ DE3 *E. coli* (Novagen Cat#70954) (This host strain is a BL21 derivative that supplies tRNAs for codons that are rare in *E. coli*), and expression construct (i.e., pETDuet1 SUMO-SidJ/CaM)
(2) LB media (see Section 4.2.2)
(3) Kanamycin sulfate (Millipore Sigma K1377) and Ampicillin sodium salt (Millipore Sigma A9518). The final concentrations are Kan, 50μg/mL final, and Amp, 100μg/mL
(4) Isopropyl β-D-1 thiogalactopyranoside (IPTG). Prepare a stock at 400mM in H_2O and sterile filter the solution
(5) Lysis buffer: 50mM Tris-HCl pH 8, 300mM NaCl, 1mM Dithiothreitol (DTT) (Millipore Sigma D0632), 1mM phenylmethylsulfonyl fluoride (PMSF). Prepare fresh and keep ice cold. PMSF is prepared as a 100mM stock in isopropanol on the day of the prep
(6) Hispure™ Ni-Nitriloacetic acid (NTA) resin (Thermo Scientific 88,221). The resin is stored in 20% EtOH, so it is recommended to wash the resin in H_2O and wash buffer before adding it to a sample
(7) Gravity flow columns, such as Bio-Rad cat#7321010
(8) Wash Buffer: 50mM Tris-HCl pH 8, 300mM NaCl, 1mM Dithiothreitol DTT, 25mM Imidazole (Millipore Sigma). Imidazole stock is prepared at 3M in H_2O and the pH is adjusted to 8.0 with HCl. All buffers with DTT should be prepared on the day of the prep
(9) Elution Buffer: 50mM Tris-HCl pH 8, 300mM NaCl, 1mM DTT, 300mM Imidazole

(10) Gel filtration buffer: 50 mM Tris-HCl pH 8, 300 mM NaCl, 1 mM DTT
(11) ULP protease: full-length ULP can be cloned from yeast and expressed as a 6×-His fusion in *E. coli*. We aliquot ULP at 5 mg/mL in 20 μL volumes, and store the enzyme at −80 °C
(12) 50,000 Da Molecular Weight Cut-Off (MWCO) centrifugal filters (such as Amicon® Ultra-15 filter units, Millipore Sigma, UFC905008)
(13) Bradford assay reagent (Thermo Fisher Scientific)
(14) Sodium Dodecyl Sulfate Polyacrylamide Gel electrophoresis (SDS-PAGE) gels. For routine applications, we use hand cast discontinuous gels with a 4% acrylamide stacking gel (pH 6.8) and a 12% resolving gel (pH 8.8). This gives good resolution of both SidJ and CaM. Refer to a Bio-Rad bulletin for detailed instructions for preparing SDS-PAGE gels: https://www.bio-rad.com/webroot/web/pdf/lsr/literature/Bulletin_6040.pdf
(15) SDS-PAGE gel tank and power supply (Bio-Rad)
(16) SDS-PAGE running buffer: 25 mM Tris, 192 mM glycine, 0.1% SDS
(17) 5× SDS-PAGE reducing Laemmli sample buffer: 12.5 mM Tris PO_4 pH 6.8, 10% (*w/v*) glycerol, 1.25% (w/v) SDS, 0.02% (w/v) bromophenol blue and 1% β-ME
(18) Coomassie Blue staining reagents
(19) Molecular weight marker such as the PageRuler™ Plus Prestained Protein ladder (ThermoFisher)

5.1.3 Procedures

(1) Transform pETDuet1 SUMO-SidJ/CaM into chemically competent Rosetta DE3 cells (see Section 4.2.3).
(2) The next day, take a streak of bacteria and inoculate several 5 mL cultures of LB. Grow for ~2 h at 37C with orbital shaking at 250 rpm. When the cultures are visually turbid, inoculate 1 L flasks of LB media with 10 mL starter culture. Grow bacteria for ~5 h to an OD_{600} of 0.6–1.1 before cooling to room temperature and inducing protein expression with 0.4 mM IPTG.
(3) Allow protein expression for 16–18 h at room temperature, usually overnight.
(4) Collect cells by centrifugation at 4000 × *g* for 15 min in Nalgene™ 1 L sealed centrifuge bottles using a F10-4 × 1000 LEX rotor or equivalent. Pour off the LB media.

(5) Resuspend the bacterial pellet in ~1/100 volume of lysis buffer on ice. Aliquot into 50 mL conical tubes for sonication.
(6) Sonicate ~40 mL volumes with a microtip set to deliver five 30 s pulses at 40% amplitude, each pulse followed by a 1 min recovery on ice.
(7) Clear the lysate in high-speed centrifuge tubes (such as Nalgene™ Oak Ridge High-Speed PPCO Centrifuge Tubes) by centrifugation at 30,000–35,000 × *g* for 30 min, at 4 °C. Use a F21-8x50y rotor or equivalent. Decant and save the supernatant, which contains soluble protein.
(8) Incubate the cleared lysate with Ni-NTA beads for a minimum of 1 h at 4 °C, nutating, in a 50 mL polyethylene conical tube. The binding capacity is 60 mg protein/mL rein, so 0.5 to 1 mL of packed resin is usually sufficient.
(9) Load the lysate-resin mixture onto a gravity-flow column. Allow the lysate to flow through and the beads to settle.
(10) Wash the beads with 20 mL of wash buffer in 1 mL increments. Do not allow the resin to dry.
(11) Elute the protein by adding 10–15 mL elution buffer all at once. Collect the elution, and measure the protein concentration using a commercial assay, such as the Bradford assay, to estimate yield. Take a small aliquot of elution and boil in 1 × SDS-PAGE buffer.
(12) Add 6 ×-His tagged ULP SUMO protease (~2 μg/mL) to cleave the 6xHis-SUMO tag from SUMO-SidJ. Incubate static overnight at 4 °C.
(13) The next morning, take another sample for SDS-PAGE to determine efficiency of ULP treatment. Then, concentrate the protein in a 50,000 Da-cutoff centrifugal filter to remove SUMO and ULP. Use a centrifuge cooled to 4 °C and spin at 3000 × *g*. Aim for a final volume that can be accommodated by your FPLC system, usually 1–2 mL. As a final step before gel filtration, spin the protein at 10,000 × *g* at 4 °C to remove any aggregates.
(14) Further purify the SidJ-CaM complex by gel filtration chromatography using a Superdex 200 gel filtration column. Resolve samples from each fraction by SDS-PAGE to assess purity and separation. A representative separation is shown in Fig. 3B. The complex elutes between 65 and 80 mL on this column system.
(15) Combine the peak fractions and concentrate the SidJ-CaM complex to 5–10 mg/mL. Dilute the protein into 50 mM Tris-HCl pH 8, 300 mM NaCl, 1 mM DTT with 5% glycerol. Aliquot and snap-frozen in liquid nitrogen for storage at −80 °C.

5.1.4 Notes

(1) This protocol is one version of many that have been published for expression and purification of affinity-tagged proteins from *E. coli* (Spriestersbach, Kubicek, Schafer, Block, & Maertens, 2015). Follow a protocol adapted to your laboratory. Keep in mind that any protocol will likely require modifications and customization to optimize expression.

5.2 Initial assays to detect AMPylation and phosphorylation

A standard kinase assay using γ^{32}P-ATP (see Tagliabracci, Wen, & Xiao, 2016 for an example) is an appropriate initial experiment to detect if a predicted kinase catalyzes phosphotransfer. It is important to include an active canonical kinase as a positive control, and to test predicted catalytic mutants of the novel kinase fold. These controls help distinguish genuine activity from non-enzymatic binding of reagents to substrate proteins or any activity that is co-purified with the candidate. Generic kinase substrates, such as casein or myelin basic protein, may be tested.

Some atypical protein kinase-like folds instead transfer adenosine monophosphate (AMP) to their substrates. SelO is an example of such a kinase-like fold (Sreelatha et al., 2018). To detect AMPylation activity, we perform assays with α^{32}P-ATP as a cosubstrate. Reactions with α^{32}P-ATP are otherwise performed identically to kinase assays. Substrates, pH, and metals may be varied to find optimal conditions for phosphorylation or AMPylation. YdiU, the *E. coli* homolog of SelO, is an ideal positive control for AMPylation assays. YdiU expressed as a N-terminal 6xHis-SUMO fusion AMPylates the generic substrate myelin basic protein (Sreelatha et al., 2018).

6. Assigning novel catalytic activities

Our general approach for assigning a catalytic activity to a novel kinase fold is to co-express or incubate the candidate enzyme with its substrate, then analyze the substrate for mass shifts using intact protein mass spectrometry. Co-expression provides metabolites and cofactors from the milieu of the host cell which may participate in the reaction. To the same effect, cell lysates may be added to reactions with purified components. Whenever possible, include control conditions with predicted catalytic mutants of the candidate kinase. If the substrate of a candidate enzyme cannot be predicted, intact mass measurement of the candidate itself may reveal its activity. We have observed that many enzymes modify themselves when overexpressed and when substrates are provided in excess. Once an activity can be predicted from intact mass analysis, it should be reconstituted with purified components.

6.1 Design of a co-expression system with predicted substrates

SidJ is a genetic suppressor of toxicity caused by the SidE family effectors (Havey & Roy, 2015; Jeong et al., 2015). Therefore, we hypothesized that these are direct substrates of SidJ, and sought to co-express SidJ, CaM, and the SidE family member SdeA in *E. coli*. We designed a system to express SidJ and CaM from pETDuet1 (Amp^R), while SdeA is expressed from the modified pET28a vector ppSUMO (Kan^R) (Fig. 4A). *E. coli* maintained under double selection with kanamycin and ampicillin will retain both plasmids. SdeA was expressed as an N-terminal 6xHis-SUMO fusion to enable Ni-NTA affinity purification, while SidJ and Calmodulin were untagged as not to be retained by the resin (unless bound to SdeA). As controls, the same experiment was performed with the D542A predicted catalytic mutant of SidJ, empty pETDuet1 plasmid, and pETDuet1 CaM without SidJ. Following co-expression, SdeA was purified as described in Section 6.1.3 and analyzed by SDS-PAGE. An electrophoretic mobility shift of SdeA appeared only when WT SidJ and CaM were co-expressed (Fig. 4B).

6.1.1 Tools

Strains and reagents for expressing recombinant protein in *E. coli*, as detailed in Section 6.1.1.

6.1.2 Procedures

(1) Transform Rosetta DE3 cells with plasmids pETDuet1-SidJ/CaM and ppSUMO-SdeA. As a control condition, transform cells with plasmids pETDuet1-SidJ D542A/CaM and ppSUMO-SdeA. Plate the transformations on LB agar plates with double selection using kanamycin (50 μg/mL) and ampicillin (100 μg/mL). Maintain double selection throughout subsequent steps to ensure *E.coli* retain both plasmids.

(2) Follow protocol in Section 6.1.3 to purify SdeA.

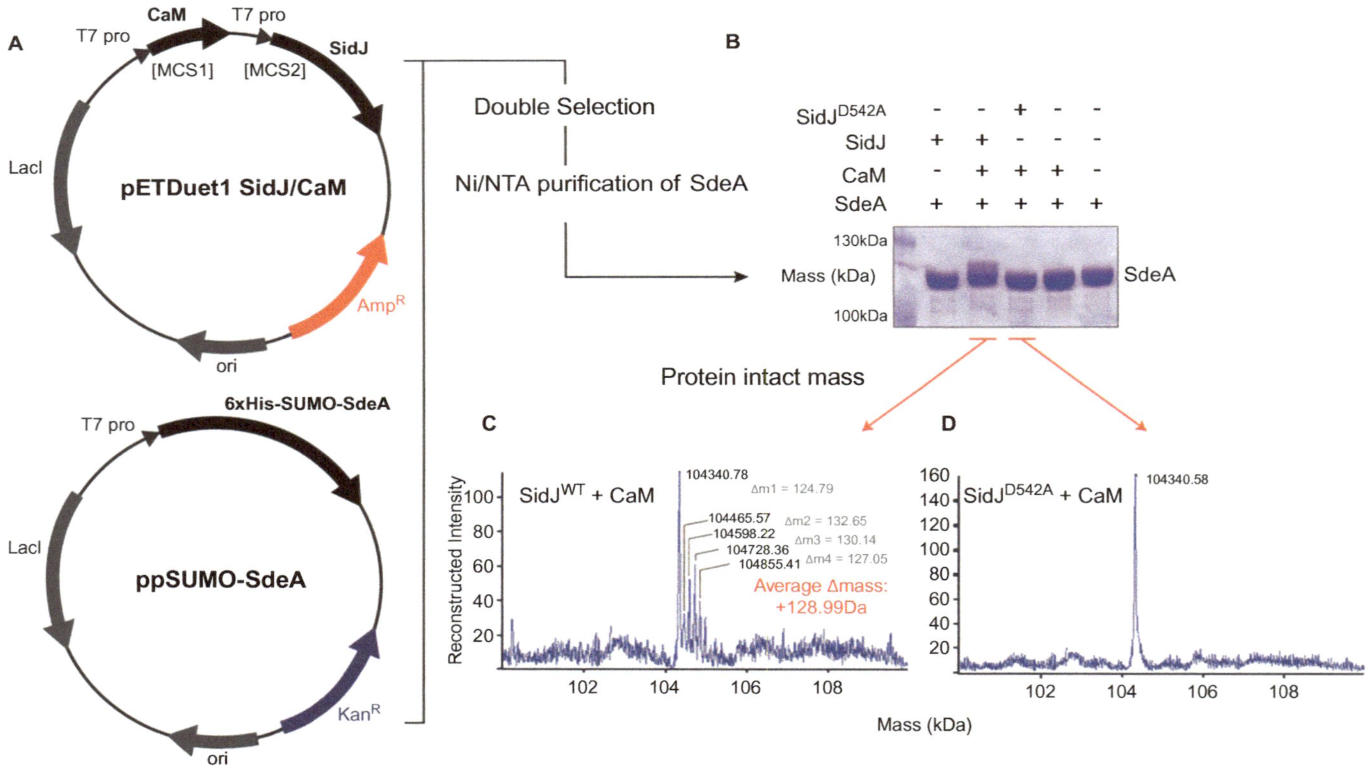

Fig. 4 Discovery of glutamylation of SdeA after co-expression with SidJ and CaM. (A) Plasmid maps of pETDuet1 SidJ/CaM and pSUMO-SdeA, annotated as in Fig. 3. (B) Reducing 6% SDS-PAGE analysis and Coomassie staining of SdeA isolated from *E. coli* following co-expression with CaM, SidJ, or the indicated mutants. Note the electrophoretic mobility shift of SdeA co-expressed with SidJ and CaM. (C) Intact mass LC/MS spectra of SdeA from lane 2 of (B). The theoretical MW of SdeA is 104,340 Da. The labeled peaks demonstrate up to five mass shifts of +129 Da. D. LC/MS spectra of SdeA from lane 3 of (B), in which the D542A mutant of SidJ was co-expressed. *Panels (B)–(D): Adapted from fig. 2 in Black, M. H., Osinski, A., Gradowski, M., Servage, K. A., Pawlowski, K., Tomchick, D. R., et al. (2019). Bacterial pseudokinase catalyzes protein polyglutamylation to inhibit the SidE-family ubiquitin ligases.* Science, *364(6442), 787–792. doi:https://doi.org/10.1126/science.aaw7446.*

(3) Resolve FPLC fractions by SDS-PAGE to assess purity and separation. Visualize protein bands using Coomassie blue staining. Because SdeA is over 100 kDa, run a gel with a 6% polyacrylamide resolving section to achieve better resolution. Small electrophoretic mobility shifts indicate a PTM (see Note 2).

(4) Pool the peak fractions from each co-expression experiment and concentrate. Dilute each sample to the same protein concentration, using the Bradford assay or similar assay. Boil 2–5 μg of protein in Laemmli sample buffer for 10 min. Run different experimental conditions side-by-side on denaturing SDS-PAGE to appreciate subtle shifts that may have been overlooked (Fig. 4B). Load between 2 and 5 μg for optimal visualization and pick a gel acrylamide percentage to achieve the best possible resolution of the band.

6.1.3 Notes

(1) Confirm that untagged proteins have been expressed by immunoblotting of cell lysates after induction. If secondary antibodies against the untagged protein(s) are not available, it may be helpful to express them as fusions with tags that are not enriched by Ni/NTA (such as the FLAG or Myc tags).

(2) Analyze both the FPLC fractions and pooled protein by SDS-PAGE. PTMs can often be appreciated as electrophoretic mobility shifts. Keep in mind that some PTMs may not withstand the sample preparations given here (reducing agents, boiling, and denaturing gel conditions). Native PAGE is a less harsh method of visualizing proteins that may preserve labile PTMs. However, Native PAGE should be interpreted cautiously when comparing an enzyme to its catalytic mutants because even single amino acid mutations can change the native mass-to-charge ratio of a protein, and therefore its migration pattern.

6.2 Detecting post-translational modifications by mass spectrometry

Liquid Chromatography-tandem Mass Spectrometry (LC-MS/MS) is a powerful tool to detect post-translational modifications of proteins, so long as the PTM can be predicted a priori. Unexpected PTMs change the precursor peptide mass such that it may not be identified in mass spectrometry databases. This precludes the use of LC-MS/MS to screen for unknown PTMs. One possible solution is to use software tools with a wide precursor mass tolerance (Solntsev, Shortreed, Frey, & Smith, 2018). Intact mass analysis, in contrast, is a sensitive technique that removes bias entirely from the process of detecting mass shifts. We routinely subject novel candidate enzymes and their predicted substrates to intact mass analysis for this purpose.

Intact mass analysis of SdeA co-expressed with SidJ/CaM revealed a series of mass increases in increments of ~128.99 Da (Fig. 4C). These mass shifts were not present when SidJ harbored a mutation in the catalytic D542 (Fig. 4D). Referencing to an online database generated the hypothesis that SidJ catalyzes glutamylation.

Importantly, even if the substrate of a candidate cannot be deduced, intact mass analysis of the candidate enzyme itself may reveal auto-modification with an unexpected PTM. As an example, SidJ co-expressed with CaM demonstrates mass shifts of +129 Da, consistent with auto-glutamylation and auto-polyglutamylation (Fig. 5).

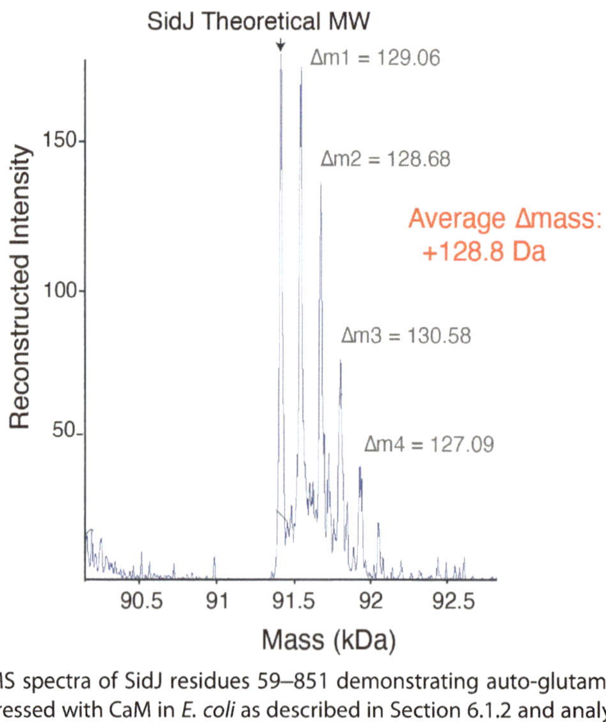

Fig. 5 LC/MS spectra of SidJ residues 59–851 demonstrating auto-glutamylation. SidJ was co-expressed with CaM in *E. coli* as described in Section 6.1.2 and analyzed by protein intact mass spectrometry.

6.2.1 Tools

(1) Mass spectrometer equipped for protein intact mass analysis. Institutional proteomics or biochemistry core facilities typically offer this service. Our experiments are performed on a Sciex X500B Q-ToF mass spectrometer coupled to an Agilent 1290 Infinity II HPLC with a POROS R1 reverse-phase column (2.1×30 mm, 20 μm particle size, 4000 Å pore size)
(2) Software to deconvolute mass spectra, such as BioPharmaView v. 2.1 software (Sciex)
(3) Software to determine the theoretical molecular weight of the target, such as Expasy (https://web.expasy.org/compute_pi)
(4) Database of post-translational modifications and their corresponding mass tags (abrf.org/deltamass).

6.2.2 Buffers and reagents

(1) HPLC Buffer A: 0.1% formic acid in H_2O
(2) HPLC Buffer B: 0.1% formic acid in acetonitrile

6.2.3 Procedures

(1) Dilute 50–100 μg protein in a MS-compatible buffer (Note 1) to between 5 and 10 mg/mL.
(2) Inject sample into the reverse-phase column.
(3) Run Buffer A at 300 μL/min to elute salts. (Note 2).
(4) Increase Buffer B stepwise until the protein elutes from the column.
(5) Ionize the protein sample gas using the following settings: Gas 1, 15 psi, ion source gas 2, 30 psi, curtain gas 35, CAD gas 7, temperature 200C, spray voltage 5200 V, declustering potential 80 V, collision energy 15 V.
(6) Acquire data from 1400 to 3600 Da with a 1 s accumulation time and 80 time bins summed.
(7) Deconvolute the acquired mass spectra to obtain the molecular weights using BioPharmaView v. 2.1 software (Sciex).
(8) Compare the acquired molecular weight to the theoretical one and make note of any discrepancies that appear (Note 3). Protein that is incompletely modified may appear as multiple peaks on the mass spectrum.
(9) Reference a database of mass tags (such as abrf.org/deltamass) to generate a hypothesis about the identity of the mass shift.

6.2.4 *Notes*

(1) All buffer components cause signal suppression to some degree. The strongest suppressors include detergents (SDS, Tween-20, and Triton X-100), and metallic salts including magnesium chloride. Half-maximum suppression concentrations (Donnelly et al., 2019) can be referenced to develop a compatible buffer. Sample cleanup by ultrafiltration through a MWCO membrane or a desalting spin column (such as Bio-Rad Micro Bio-spin™ 6 gel columns, cat#7326221) may be necessary. Refer to manufacturer documentation for instructions on specific filter unit protocols.

(2) Be aware that some PTMs (such as those with phosphoramidate bonds) may undergo acid-catalyzed hydrolysis under the liquid chromatography conditions.

(3) Fully sequence the coding sequence of all expression constructs. Overlooked amino acid substitutions can lead to confusion when interpreting intact mass data. Similarly, scars left by restriction enzyme sites may introduce amino acids to the termini of proteins. Alpha-N-6-phosphogluconoylation of polyhistidine peptides in 6xHis tags can induce unexpected mass shifts of 178 and 258 Da (Geoghegan et al., 1999). Oxidation of methionine or cysteine may introduce a +16 Da or +32 Da shifts. It is important to have catalytic mutants or control conditions to determine if a modification is catalyzed by the candidate enzyme.

7. Conclusion

The advent of high-accuracy structure prediction algorithms allows researchers to quickly generate hypotheses about protein function and enzymatic activity. However, the three-dimensional fold of a protein does not always predict its function. This chapter offers a general approach to obtain experimental evidence of enzymatic activity. Discovering novel activities of pseudokinases is a daunting prospect, but with inclusion of predicted catalytic mutants as controls, a thorough search for cofactors, and careful attention to experimental results, one can maximize the chances of uncovering an enzymatic activity. While we have focused on members of the kinase superfamily, similar methodology could be applied to diverse groups of enzymes.

Acknowledgments

Research in the Tagliabracci lab is supported by a Welch Foundation Grant (I-1911), a W. M. Keck Foundation Medical Research Grant, the Howard Hughes Medical Institute (HHMI) and NIH grants R01GM135189 and DP2 GM137419. M.G. is supported by a Polish National Science Centre grant 2019/35/N/NZ2/02844 and K.P. is supported by Polish National Science Centre grant 2019/33/B/NZ2/01409.

References

Ashkenazy, H., Abadi, S., Martz, E., Chay, O., Mayrose, I., Pupko, T., et al. (2016). ConSurf 2016: An improved methodology to estimate and visualize evolutionary conservation in macromolecules. *Nucleic Acids Research*, *44*(W1), W344–W350. https://doi.org/10.1093/nar/gkw408.

Baek, M., DiMaio, F., Anishchenko, I., Dauparas, J., Ovchinnikov, S., Lee, G. R., et al. (2021). Accurate prediction of protein structures and interactions using a three-track neural network. *Science*, *373*(6557), 871–876. https://doi.org/10.1126/science.abj8754.

Bhogaraju, S., Bonn, F., Mukherjee, R., Adams, M., Pfleiderer, M. M., Galej, W. P., et al. (2019). Inhibition of bacterial ubiquitin ligases by SidJ-calmodulin catalysed glutamylation. *Nature*, *572*(7769), 382–386. https://doi.org/10.1038/s41586-019-1440-8.

Black, M. H., Osinski, A., Gradowski, M., Servage, K. A., Pawlowski, K., Tomchick, D. R., et al. (2019). Bacterial pseudokinase catalyzes protein polyglutamylation to inhibit the SidE-family ubiquitin ligases. *Science*, *364*(6442), 787–792. https://doi.org/10.1126/science.aaw7446.

Cheng, H., Schaeffer, R. D., Liao, Y., Kinch, L. N., Pei, J., Shi, S., et al. (2014). ECOD: An evolutionary classification of protein domains. *PLoS Computational Biology*, *10*(12), e1003926. https://doi.org/10.1371/journal.pcbi.1003926.

Crooks, G. E., Hon, G., Chandonia, J. M., & Brenner, S. E. (2004). WebLogo: A sequence logo generator. *Genome Research*, *14*(6), 1188–1190.

Deller, M. C., Kong, L., & Rupp, B. (2016). Protein stability: A crystallographer's perspective. *Acta Crystallographica Section F: Structural Biology and Crystallization Communications*, *72*(Pt. 2), 72–95. https://doi.org/10.1107/S2053230X15024619.

Donnelly, D. P., Rawlins, C. M., DeHart, C. J., Fornelli, L., Schachner, L. F., Lin, Z., et al. (2019). Best practices and benchmarks for intact protein analysis for top-down mass spectrometry. *Nature Methods*, *16*(7), 587–594. https://doi.org/10.1038/s41592-019-0457-0.

Dudkiewicz, M., Lenart, A., & Pawłowski, K. (2013). A novel predicted calcium-regulated kinase family implicated in neurological disorders. *PLoS One*, *8*(6), e66427.

Fox, N. K., Brenner, S. E., & Chandonia, J. M. (2014). SCOPe: Structural Classification of Proteins—Extended, integrating SCOP and ASTRAL data and classification of new structures. *Nucleic Acids Research*, *42*(Database issue), D304–D309. https://doi.org/10.1093/nar/gkt1240.

Gan, N., Zhen, X., Liu, Y., Xu, X., He, C., Qiu, J., et al. (2019). Regulation of phosphoribosyl ubiquitination by a calmodulin-dependent glutamylase. *Nature*, *572*(7769), 387–391. https://doi.org/10.1038/s41586-019-1439-1.

Geoghegan, K. F., Dixon, H. B., Rosner, P. J., Hoth, L. R., Lanzetti, A. J., Borzilleri, K. A., et al. (1999). Spontaneous alpha-N-6-phosphogluconoylation of a "His tag" in Escherichia coli: The cause of extra mass of 258 or 178 Da in fusion proteins. *Analytical Biochemistry*, *267*(1), 169–184. https://doi.org/10.1006/abio.1998.2990.

Geourjon, C., Combet, C., Blanchet, C., & Deleage, G. (2001). Identification of related proteins with weak sequence identity using secondary structure information. *Protein Science, 10*(4), 788–797. https://doi.org/10.1110/ps.30001.

Gibson, D. G., Young, L., Chuang, R. Y., Venter, J. C., Hutchison, C. A., 3rd, & Smith, H. O. (2009). Enzymatic assembly of DNA molecules up to several hundred kilobases. *Nature Methods, 6*(5), 343–345. https://doi.org/10.1038/nmeth.1318.

Green, M. R., Sambrook, J., & Sambrook, J. (2012). *Molecular cloning: A laboratory manual* (4th ed.). Cold Spring Harbor, N.Y.: Cold Spring Harbor Laboratory Press.

Haridas, V., Ni, J., Meager, A., Su, J., Yu, G. L., Zhai, Y., et al. (1998). TRANK, a novel cytokine that activates NF-kappa B and c-Jun N-terminal kinase. *Journal of Immunology, 161*(1), 1–6.

Havey, J. C., & Roy, C. R. (2015). Toxicity and SidJ-mediated suppression of toxicity require distinct regions in the SidE family of legionella pneumophila effectors. *Infection and Immunity, 83*(9), 3506–3514. https://doi.org/10.1128/iai.00497-15.

Hsieh, T. S., Lopez, V. A., Black, M. H., Osinski, A., Pawłowski, K., Tomchick, D. R., et al. (2021). Dynamic remodeling of host membranes by self-organizing bacterial effectors. *Science, 372*(6545), 935–941. https://doi.org/10.1126/science.aay8118.

Jeong, K. C., Sexton, J. A., & Vogel, J. P. (2015). Spatiotemporal regulation of a *Legionella pneumophila* T4SS substrate by the metaeffector SidJ. *PLoS Pathogens, 11*(3), e1004695. https://doi.org/10.1371/journal.ppat.1004695.

Johnson, D. A., Akamine, P., Radzio-Andzelm, E., Madhusudan, M., & Taylor, S. S. (2001). Dynamics of cAMP-dependent protein kinase. *Chemical Reviews, 101*(8), 2243–2270. https://doi.org/10.1021/cr000226k.

Jumper, J., Evans, R., Pritzel, A., Green, T., Figurnov, M., Ronneberger, O., et al. (2021). Highly accurate protein structure prediction with AlphaFold. *Nature, 596*(7873), 583–589. https://doi.org/10.1038/s41586-021-03819-2.

Kelley, L. A., Mezulis, S., Yates, C. M., Wass, M. N., & Sternberg, M. J. (2015). The Phyre2 web portal for protein modeling, prediction and analysis. *Nature Protocols, 10*(6), 845–858. https://doi.org/10.1038/nprot.2015.053.

Li, W., Pio, F., Pawlowski, K., & Godzik, A. (2000). Saturated BLAST: An automated multiple intermediate sequence search used to detect distant homology. *Bioinformatics, 16*(12), 1105–1110.

Murphy, J. M., Farhan, H., & Eyers, P. A. (2017). Bio-zombie: The rise of pseudoenzymes in biology. *Biochemical Society Transactions, 45*(2), 537–544. https://doi.org/10.1042/bst20160400.

Nguyen, K. B., Sreelatha, A., Durrant, E. S., Lopez-Garrido, J., Muszewska, A., Dudkiewicz, M., et al. (2016). Phosphorylation of spore coat proteins by a family of atypical protein kinases. *Proceedings of the National Academy of Sciences of the United States of America, 113*(25), E3482–E3491. https://doi.org/10.1073/pnas.1605917113.

Osinski, A., Black, M. H., Pawlowski, K., Chen, Z., Li, Y., & Tagliabracci, V. S. (2021). Structural and mechanistic basis for protein glutamylation by the kinase fold. *Molecular Cell, 81*(21), 4527–4539 e4528. https://doi.org/10.1016/j.molcel.2021.08.007.

Pierce, M. M., Raman, C. S., & Nall, B. T. (1999). Isothermal titration calorimetry of protein-protein interactions. *Methods, 19*(2), 213–221. https://doi.org/10.1006/meth.1999.0852.

Putkey, J. A., Kleerekoper, Q., Gaertner, T. R., & Waxham, M. N. (2003). A new role for IQ motif proteins in regulating calmodulin function. *The Journal of Biological Chemistry, 278*(50), 49667–49670. https://doi.org/10.1074/jbc.C300372200.

Qiu, J., Yu, K., Fei, X., Liu, Y., Nakayasu, E. S., Piehowski, P. D., et al. (2017). A unique deubiquitinase that deconjugates phosphoribosyl-linked protein ubiquitination. *Cell Research, 27*(7), 865–881. https://doi.org/10.1038/cr.2017.66.

Ribeiro, A. J. M., Das, S., Dawson, N., Zaru, R., Orchard, S., Thornton, J. M., et al. (2019). Emerging concepts in pseudoenzyme classification, evolution, and signaling. *Science Signaling*, *12*(594). https://doi.org/10.1126/scisignal.aat9797.

Rychlewski, L., Jaroszewski, L., Li, W., & Godzik, A. (2000). Comparison of sequence profiles. Strategies for structural predictions using sequence information. *Protein Science*, *9*(2), 232–241.

Schaeffer, R. D., Liao, Y., Cheng, H., & Grishin, N. V. (2017). ECOD: New developments in the evolutionary classification of domains. *Nucleic Acids Research*, *45*(D1), D296–D302. https://doi.org/10.1093/nar/gkw1137.

Solntsev, S. K., Shortreed, M. R., Frey, B. L., & Smith, L. M. (2018). Enhanced global post-translational modification discovery with MetaMorpheus. *Journal of Proteome Research*, *17*(5), 1844–1851. https://doi.org/10.1021/acs.jproteome.7b00873.

Spriestersbach, A., Kubicek, J., Schafer, F., Block, H., & Maertens, B. (2015). Purification of his-tagged proteins. *Methods in Enzymology*, *559*, 1–15. https://doi.org/10.1016/bs.mie.2014.11.003.

Sreelatha, A., Nolan, C., Park, B. C., Pawłowski, K., Tomchick, D. R., & Tagliabracci, V. S. (2020). A Legionella effector kinase is activated by host inositol hexakisphosphate. *The Journal of Biological Chemistry*, *295*, 6214–6224. https://doi.org/10.1101/2020.02.18.954925.

Sreelatha, A., Yee, S. S., Lopez, V. A., Park, B. C., Kinch, L. N., Pilch, S., et al. (2018). Protein AMPylation by an evolutionarily conserved pseudokinase. *Cell*, *175*(3), 809–821 e819. https://doi.org/10.1016/j.cell.2018.08.046.

Steinegger, M., Meier, M., Mirdita, M., Vohringer, H., Haunsberger, S. J., & Soding, J. (2019). HH-suite3 for fast remote homology detection and deep protein annotation. *BMC Bioinformatics*, *20*(1), 473. https://doi.org/10.1186/s12859-019-3019-7.

Sulpizio, A., Minelli, M. E., Wan, M., Burrowes, P. D., Wu, X., Sanford, E. J., et al. (2019). Protein polyglutamylation catalyzed by the bacterial calmodulin-dependent pseudokinase SidJ. *eLife*, *8*. https://doi.org/10.7554/eLife.51162.

Tagliabracci, V. S., Engel, J. L., Wen, J., Wiley, S. E., Worby, C. A., Kinch, L. N., et al. (2012). Secreted kinase phosphorylates extracellular proteins that regulate biomineralization. *Science*, *336*, 1150–1153.

Tagliabracci, V. S., Wen, J., & Xiao, J. (2016). Methods to purify and assay secretory pathway kinases. *Methods in Molecular Biology*, *1496*, 197–215. https://doi.org/10.1007/978-1-4939-6463-5_16.

The UniProt Consortium. (2021). UniProt: The universal protein knowledgebase in 2021. *Nucleic Acids Research*, *49*(D1), D480–D489. https://doi.org/10.1093/nar/gkaa1100.

van Zundert, G. C., Rodrigues, J. P., Trellet, M., Schmitz, C., Kastritis, P. L., Karaca, E., et al. (2016). The HADDOCK2.2 web server: User-friendly integrative modeling of biomolecular complexes. *Journal of Molecular Biology*, *428*(4), 720–725. https://doi.org/10.1016/j.jmb.2015.09.014.

Varadi, M., Anyango, S., Deshpande, M., Nair, S., Natassia, C., Yordanova, G., et al. (2022). AlphaFold Protein Structure Database: Massively expanding the structural coverage of protein-sequence space with high-accuracy models. *Nucleic Acids Research*, *50*(D1), D439–D444. https://doi.org/10.1093/nar/gkab1061.

Veiga da Cunha, M., Hadi, F., Balligand, T., Stroobant, V., & Van Schaftingen, E. (2012). Molecular identification of hydroxylysine kinase and of ammoniophospholyases acting on 5-phosphohydroxy-L-lysine and phosphoethanolamine. *The Journal of Biological Chemistry*, *287*(10), 7246–7255. https://doi.org/10.1074/jbc.M111.323485.

Xu, D., Jaroszewski, L., Li, Z., & Godzik, A. (2014). FFAS-3D: Improving fold recognition by including optimized structural features and template re-ranking. *Bioinformatics*, *30*(5), 660–667. https://doi.org/10.1093/bioinformatics/btt578.

Zhang, Y., & Skolnick, J. (2005). TM-align: A protein structure alignment algorithm based on the TM-score. *Nucleic Acids Research*, *33*(7), 2302–2309. https://doi.org/10.1093/nar/gki524.

Zimmermann, L., Stephens, A., Nam, S. Z., Rau, D., Kübler, J., Lozajic, M., et al. (2018). A completely reimplemented MPI bioinformatics toolkit with a new HHpred server at its core. *Journal of Molecular Biology*, *430*(15), 2237–2243. https://doi.org/10.1016/j.jmb.2017.12.007.

CHAPTER TWENTY

Efficient expression, purification, and visualization by cryo-EM of unliganded near full-length HER3

Devan Diwanji[a,b,†], Raphael Trenker[a,†], Natalia Jura[a,c,d,*], and Kliment A. Verba[c,e,*]

[a]Cardiovascular Research Institute, University of California San Francisco, San Francisco, CA, United States
[b]Medical Scientist Training Program, University of California San Francisco, San Francisco, CA, United States
[c]Quantitative Biosciences Institute (QBI), University of California San Francisco, San Francisco, CA, United States
[d]Department of Cellular and Molecular Pharmacology, University of California San Francisco, San Francisco, CA, United States
[e]Department of Pharmaceutical Chemistry, University of California San Francisco, San Francisco, CA, United States
*Corresponding authors: e-mail address: natalia.jura@ucsf.edu; kliment.verba@ucsf.edu

Contents

† Authors contributed equally to the work.

Methods in Enzymology, Volume 667
ISSN 0076-6879
https://doi.org/10.1016/bs.mie.2022.03.048

Abstract

Biochemical analyses of membrane receptor kinases have been limited by challenges in obtaining sufficient homogeneous receptor samples for downstream structural and biophysical characterization. Here, we report a suite of methods for the efficient expression, purification, and visualization by cryo-electron microscopy (cryo-EM) of near full-length Human Epidermal Growth Factor Receptor 3 (HER3), a receptor tyrosine pseudokinase, in the unliganded state. Through transient mammalian cell expression, a two-step purification with detergent exchange into lauryl maltose neopentyl glycol (LMNG), and freezing devoid of background detergent micelle, we obtained ~6 Å reconstructions of the ~60 kDa fully-glycosylated unliganded extracellular domain of HER3 from just 30 mL of suspension culture. The reconstructions reveal previously unappreciated extracellular domain dynamics and glycosylation sites.

1. Introduction: The human epidermal growth factor receptor 3 (HER3)

Receptor kinases play critical roles in cell signaling and disease (Du & Lovly, 2018; Lemmon & Schlessinger, 2010). Structural biology and biophysical investigations of receptor kinases in the full-length context are invaluable in probing the fundamental mechanisms by which an extracellular signal may be transmitted across the membrane to modulate the activity of the intracellular kinases (Diwanji, Thaker, & Jura, 2019). Unfortunately, to date, there are only a handful of structural studies with receptor kinases in their full-length or near full-length forms, all at low resolution (Chen, Unger, & He, 2015; Gutmann, Kim, Grzybek, Walz, & Coskun, 2018; Huang et al., 2020; Li, Choi, Yu, & Bai, 2019; Mi et al., 2008, 2011; Opatowsky et al., 2014; Uchikawa, Choi, Shang, Yu, & Bai, 2019). Major barriers to these structural pursuits are governed by the intrinsic biology of receptor kinases that limit their expression, homogeneous purification, and visualization at high resolution. Low expression levels, for example, may be attributed to cellular downregulation or cytotoxicity (Goh & Sorkin, 2013). Isolated full-length receptors may exist in numerous oligomeric and/or conformational states or associate with other cellular components yielding an unacceptable level of compositional heterogeneity for structural biology (Diwanji et al., 2019). Even if a homogeneous detergent-solubilized full-length receptor sample is obtained (often from liters of mammalian culture), the sample must be sufficiently concentrated for visualization by single-particle cryo-electron microscopy (cryo-EM) or X-ray crystallography. It is evident that numerous technical bottlenecks have hampered full-length receptor kinase structural biology. Here, we describe the efficient expression, purification, and visualization of near full-length HER3 that may serve as a blueprint for the biophysical investigations of other receptor kinases in their full-length forms.

Receptor tyrosine kinases (RTKs) are single spanning membrane proteins with extracellular domains that bind ligands and intracellular kinases whose catalytic activity regulates downstream signaling. The HER receptor subfamily consists of EGFR, HER2, HER3, and HER4 (Lemmon & Schlessinger, 2010). HER3 is a unique HER receptor as a pseudokinase with divergent residues in the kinase domain that abrogate its catalytic activity (Guy, Platko, Cantley, Cerione, & Carraway, 1994; Jura, Shan, Cao, Shaw, & Kuriyan, 2009). Despite evolving into a pseudokinase, HER3 has retained its ability to bind to extracellular ligands such as neuregulin-1β (NRG1β) which triggers HER3 heterodimerization with other catalytically active HER receptors, notably HER2 (Hsieh & Moasser, 2007; Moasser, 2007). HER3 allosterically activates HER2 through the asymmetric kinase domain dimer (Zhang, Gureasko, Shen, Cole, & Kuriyan, 2006). In the absence of ligand, the HER3 extracellular domain exists in an inhibited conformation that does not permit its association with co-receptors (Cho & Leahy, 2002). The extracellular domains of human HER receptors contain four subdomains: domains I-IV. Domains II and IV form an intramolecular tether in the autoinhibited, "tethered" state. Ligand binding relieves the autoinhibition to promote an extended conformation with dimerization interfaces accessible by co-receptors (Kovacs, Zorn, Huang, Barros, & Kuriyan, 2015). In previously solved liganded HER receptor structures, the β-hairpin dimerization arm located in domain II coordinates the dimeric interaction (Ferguson et al., 2003; Kovacs et al., 2015; Lu et al., 2010). This critical dimerization arm remains inaccessible in the unliganded state. Interestingly, unlike all previously solved structures of liganded HER receptors, the HER3 dimerization arm does not engage HER2 owing to the singly-liganded nature of the HER2/HER3 heterodimer (Diwanji et al., 2021).

Heterodimerization enables transphosphorylation of tyrosine residues on the HER3 C-terminal tail (Choi et al., 2020). The HER3 C-terminal tail is remarkable, not only for being the longest tail of the HER family, but also for containing six PI3K consensus binding motifs (YxxM). It is through these phosphorylated YxxM motifs that the HER3 C-terminal tail acts as a "super-scaffold" to potently recruit phosphoinositide-3 kinase (PI3K) to regulate cell survival and proliferation (Hellyer, Cheng, & Koland, 1998).

The PI3K signaling pathway constitutes one of the most frequently deregulated pathways in cancer, rendering HER3 and PI3K as important therapeutic targets (Kandoth et al., 2013). Structural investigations of HER3 in the inactivated, unliganded state may reveal important regulatory mechanisms that keep HER3-PI3K signaling in check.

Despite our growing knowledge of HER receptor mechanism, our understanding of signal transduction across the membrane is limited without structures containing intracellular and extracellular domains. Numerous questions exist about the unliganded state, especially the organization of the receptor relative to the membrane. For example, how are the unliganded tethered extracellular domains oriented relative to the membrane? Does the membrane mediate potential autoinhibitory interactions with the intracellular juxtamembrane and/or kinase domain? To date, the unliganded state of HER receptors has been characterized by X-ray crystallography with partially deglycosylated extracellular domain samples to facilitate crystallization (Bouyain, Longo, Li, Ferguson, & Leahy, 2005; Cho & Leahy, 2002). Deglycosylation itself may alter structure. Several studies with HER receptors have shown that the glycosylation state modulates binding to growth factors, dimerization, and recognition by therapeutic agents (Azimzadeh Irani, Kannan, & Verma, 2017). For example, mutating the HER3 extracellular N-linked glycosylation site, N437, has been shown to promote heterodimerization with HER2 in the absence of ligand (Takahashi et al., 2013; Yokoe et al., 2007). Molecular dynamic simulations provide evidence that EGFR N-linked glycosylation plays a critical role in the binding of growth factors (Azimzadeh Irani et al., 2017). HER2 glycosylation decreases the affinity of the monoclonal antibody Herceptin for HER2 (Peiris et al., 2017). Clearly, structural studies with glycosylated full-length HER receptors can provide invaluable insight into the fundamental mechanism of signal transduction.

1.1 Key resources tables

1.1.1 Chemicals, resins, and peptides

Reagent or resource	Source
TRIS (Trizma Base)	Millipore Sigma, #T6066
NaCl	Millipore Sigma, #SX0420
Ethylenediaminetetraacetic acid (EDTA)	Millipore Sigma, #324503

—cont'd

Reagent or resource	Source
Sodium orthovanadate ($NaVO_3$)	Millipore Sigma, #S6508
NaF	Millipore Sigma, # SX0550
cOmplete EDTA-free Protease Inhibitor Cocktail Mini Tablets	Roche, sold by Millipore Sigma, #11836170001
DNase I, grade II, from bovine pancreas	Roche, sold by Millipore Sigma, #10104159001
n-Dodecyl-β-D-Maltopyranoside (DDM), for gel filtration	Anatrace, #D310
n-Dodecyl-β-D-Maltopyranoside (DDM), for membrane extraction	Inalco, #1758-1350 Inalco brand was used to save costs during extraction
2,2-didecylpropane-1,3-bis-β-D-maltopyranoside (LMNG)	Anatrace, #NG310
DYKDDDDK synthetic peptide (Flag Peptide)	Sino Biological, #PP101274
Bosutinib (10 mM in DMSO)	SelleckChem, #SKI-606
Anti-Flag resin (Anti-DYKDDDDK G1 Affinity Resin)	GenScript (#L00432-10)
Flag-HER3 (Δ1022-1342)-TS construct	In-house preparation
Uranyl formate	Electron Microscopy Sciences, #22450
GDN	Anatrace, #GDN101
MNA-C12	Anatrace, #MNA-C12

1.1.2 Cell culture

Reagent or resource	Source
Expi293	ThermoFisher Scientific
Expifectamine Kit	ThermoFisher Scientific
Expi293 Media	ThermoFisher Scientific

1.1.3 Software

Reagent or resource	Source
PRISM	GraphPad
Rosetta	Downloaded from dimaiolab.ipd.uw.edu
CryoSPARC	Structura Biotechnology
Relion	Downloaded from www2.mrc-lmb.cam.ac.uk
MotionCor2	Downloaded from emcore.ucsf.edu
ChimeraX	Downloaded from rbvi.ucsf.edu

1.2 Materials and equipment

125 mL flask non-baffled, vented cap flask, (VWR, #89095-258)
Incubator at 37 °C and 8% CO_2 with shaker (Heracell VIOS 160i, Thermo Scientific)
Class II Biosafety Cabinet
Disposable Serological Pipets 5, 10, 25 mL (VWR, #89130)
Conical tubes 15, 50 mL (VWR, #21008)
Liquid nitrogen
Table-top microcentrifuge (5424 R, Eppendorf)
Table-top centrifuge (for 15/50 mL conical tubes, Allegra X-15R, Beckman Coulter)
-80 °C freezer
Nutating mixer (Fisher scientific)
Micro Bio-Spin columns, 0.8 mL (Bio-Rad, #7326204)
Amicon Ultra-0.5 Centrifugal Filter Units, MWCO 100 kDa (Millipore Sigma, #UFC510096)
Superose6 10/300 Increase (GE Healthcare, now Cytiva, #29-0915-96)
Akta purification system (Akta Pure, Cytiva)
FEI-Tecnai T12
4 k CCD camera (Gatan)
GPU-equipped workstation (at least 11GB of GPU memory, at least 128GB of system memory, at least 4 TB of fast SSD storage and 20 TB of spinning disc storage)
Glow discharger (PELCO easyGlow, Ted Pella)
Formvar/carbon 400 mesh copper grids (Ted Pella, #01754-F)

Quantifoil R1.2/1.3 mesh Au holey-carbon grids Electron Microscopy Sciences, #Q310AR1.3
Vitrobot Mark IV (FEI)
Talos-Arctica 200 keV (ThermoFisher)
K3 Direct electron detector camera (Gatan)

2. Construct design and expression of near full-length HER3

2.1 Near full-length construct design

We sought to optimize expression and purification of unliganded HER3 to enable reconstitution efforts of the HER2/HER3 heterodimer *in vitro*. We introduced an N-terminal 1 × FLAG tag and a C-terminal twin-Strep (TS) tag to the full-length construct of human HER3 in a pFastBac vector (Thermo Fisher Scientific, #10360014) in which a baculovirus promoter was replaced by a CMV promoter for mammalian cell protein expression (Sung et al., 2014). Alternatively, standard pCDNA3 or pCDNA4 vectors may be used for this protocol. As C-terminal tails of HER family receptors are predicted to be unstructured, we chose to express HER3 with most of its C-terminal tail truncated (Δ1022-1342) as the tail could destabilize the receptor or hamper expression. Our near full-length HER3 construct therefore contains the extracellular domain (ECD), the transmembrane domain (TMD), the intracellular kinase domain (KD), and a brief segment of the C-terminal tail retained for kinase stability (Fig. 1A and B).

2.2 Efficient expression in mammalian suspension cultures

Previous structural studies with full-length or near full-length RTKs used adherent or suspension mammalian cell culture for expression (Chen et al., 2015; Gutmann et al., 2018; Huang et al., 2020; Li et al., 2019; Mi et al., 2008, 2011; Uchikawa et al., 2019). We initially expressed our HER3 constructs in adherent format on 15-cm dishes and found that cell material from at least 20 15-cm dishes would be required to be spectroscopically detectable during gel filtration chromatography after purification. Transfection with different lipophilic reagents or titrating plasmid concentrations did not improve expression efficiency. Furthermore, large-scale suspension expression in human embryo kidney (HEK) 293S with BacMAM required culture volumes of more than 6 L to net disappointingly low sample yields. Such expression strategies are not cost effective nor time efficient for optimization.

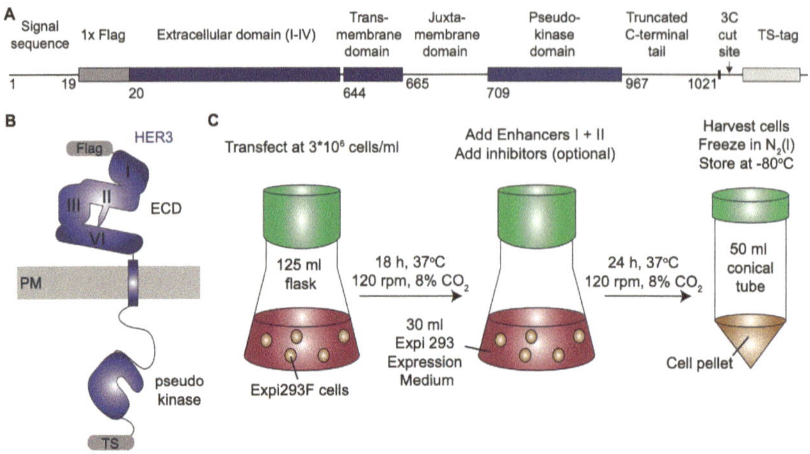

Fig. 1 Construct Design and Mammalian Cell Expression of Near Full-Length HER3. (A) Construct map of N-terminal Flag tagged HER3 (Δ1022-1342) with a C-terminal 3C protease site and twin-strep tag. (B) Cartoon of construct tagging scheme of near Full-Length HER3 comprising the extracellular, transmembrane, intracellular juxta-membrane, and pseudokinase domains. A 1 × Flag tag precedes the N-terminus and the C-terminus is followed by a 3C protease cleavable twin-Strep tag. (C) Cartoon of expression method.

We therefore turned to the Expi293 HEK transient expression system (Life Technologies). In this system, cells cultured to densities of 2–3 × 10^6 are transfected with a lipophilic transfection reagent and enhanced with two proprietary enhancers 18 h post-transfection (Fig. 1C). Our initial expression trials with near full-length HER3 revealed a remarkable boost in expression with maximal expression at 24 h post-enhancer. We estimated that 30 mL of culture in the Expi293 system achieved a 100-fold increase in expression compared with 20 15-cm dishes (data not shown). The boost in expression combined with the ease of suspension culture and short incubation times allowed us to parallelize optimization at a small scale which would have simply not been feasible with previous expression strategies.

3. Purification and initial visualization of homogeneous near full-length HER3 sample

3.1 Streamlined purification of near full-length HER3

Purification of unliganded near full-length HER3 begins with detergent extraction from flash-frozen cell pellets from 30 mL of suspension culture. Our lysis buffer contains ethylenediaminetetraacetic acid (EDTA) to chelate residual Mg^{2+} and $NaVO_3$ to inhibit phosphatases. We reasoned that kinase inactivation may improve sample stability by reducing trans-phosphorylation. We found that extraction in the presence of 1% w/v (n-Dodecyl-β-D-maltoside) DDM worked best for yields and homogeneity, consistent with earlier studies that purified near full-length EGFR (Mi et al., 2008, 2011). Following solubilization in detergent micelles, excess cell debris were removed by centrifugation and the clarified lysate was subject to overnight Flag pulldown with Flag resin. The receptor-bound beads were washed in serial-batch mode in a buffer containing 0.5 mM DDM (2–3 × critical micelle concentration, CMC) and eluted in 5 column volumes (CVs) of buffer with Flag peptide. Curiously, two other species were stoichiometrically co-enriched with near full-length HER3 after the one-step purification (see Fig. 2A). Eluate was then concentrated in 0.5 mL 100 kDa cutoff spin concentrators (Amicon) and injected over a Superose6 10/300 Increase column (GE Healthcare) pre-equilibrated in 0.5 mM DDM buffer. A representative gel filtration chromatogram is shown in Fig. 2B, blue curve). A step-by-step purification protocol is provided below.

3.1.1 Purification protocol of near full-length HER3 in DDM

Buffers

Lysis Buffer: 50 mM TrisCl, 150 mM NaCl, 1 mM EDTA, pH 7.5 at 4 °C, 1 mM $NaVO_3$, 1 mM NaF, Roche Protease Inhibitor Tablet Cocktail-1 pellet dissolved in 50 mL of buffer, DNase, and 1% w/v DDM (Inalco).

Flag Equilibration Buffer: 50 mM TrisCl, 150 mM NaCl, pH 7.5 at 4 °C, 0.5 mM DDM (Anatrace).

Flag Elution Buffer: 50 mM TrisCl, 150 mM NaCl, pH 7.5 at 4 °C, 0.5 mM DDM (Anatrace), 250 mg/mL of Flag peptide (SinoBiological)

SEC Buffer: same as Flag Equilibration Buffer

1. *Cell lysis.* Pellets stored in −80°C were thawed on ice for ~30 min. To soften the pellet, 10 mL of chilled lysis buffer was incubated with the pellet for an additional 30 min. The pellet was gently resuspended in Lysis Buffer by gently pipetting with a 10 mL pipette until homogeneous and placed on a rotator for 1–2 h. Complete lysis was judged by the homogeneity of the resulting solution and the presence of DNA aggregate.
2. *Clarification.* Cells were spun down in 50 mL conical tubes at 4000 × *g* for 20 min at 4°C in an Allegra X-15R (Beckman Coulter) refrigerated centrifuge to separate cell debris from detergent solubilized lysate.

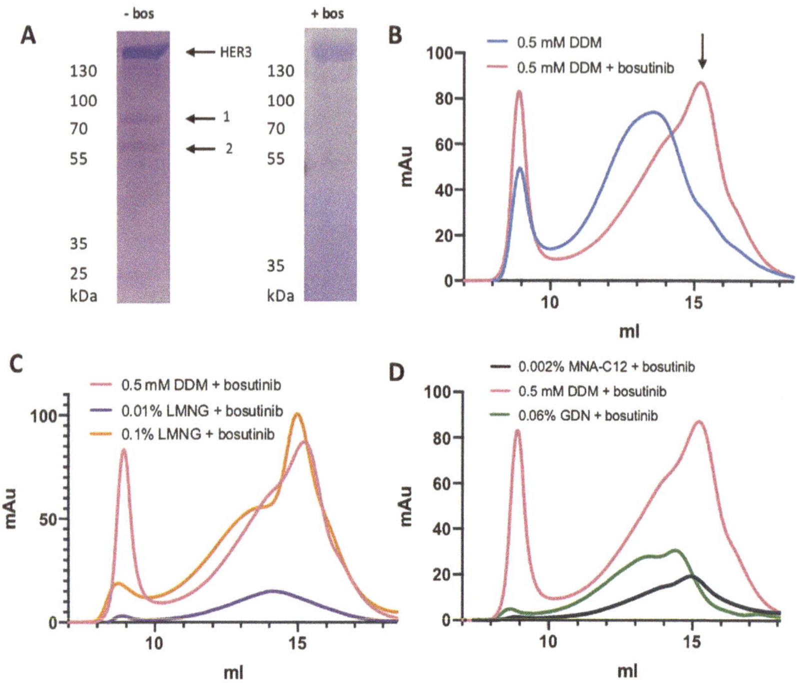

Fig. 2 Purification of Near Full-Length HER3. (A) Left, Flag eluate sample of near full-length HER3 in the absence of bosutinib (−bos) immunoprecipitated with two other bands (1 and 2). Right, the addition of bosutinib (+bos) during expression reduces the other two species. (B) Superimposed gel filtration chromatograms of near full-length HER3 on a Superose6 Increase 10/300 GL column in the presence and absence of bosutinib. Both samples were run in the presence of 0.5 mM DDM. (C) Superimposed gel filtration chromatograms of near full-length HER3 on a Superose6 Increase 10/300 GL column after exchange to 0.01% LMNG or 0.1% over Flag wash. Both LMNG samples were run on gel filtration in a buffer without detergent. The 0.5 mM DDM is the same as in (B). (D) Superimposed gel filtration chromatograms of near full-length HER3 on a Superose6 Increase 10/300 GL column after exchange to 0.002% MNA-C12 or 0.06% GDN over Flag wash.

3. *Equilibration of Flag resin.* 2 mL of 50% resin slurry of Genscript G1 Flag resin was mixed with 8 mL of Flag Equilibration Buffer in a conical tube. The resin was gently pelleted at 500 × g in an Allegra X-15R (Beckman Coulter) refrigerated centrifuge, for 2 min, 4 °C and the residual Flag Equilibration Buffer was aspirated. The wash was repeated for a total of 3 serial washes in Flag Equilibration Buffer.
4. *Incubation with Flag resin.* The clarified lysate was incubated with Flag Resin overnight at 4 °C on a gently rocking platform.

5. *Flag resin wash.* The clarified lysate was separated from receptor-bound Flag Resin through gentle centrifugation at 500 × *g* in an Allegra X-15R (Beckman Coulter) for 2 min, 4 °C in conical tubes. The Flag Resin was then mixed in 10 column volumes (CVs) of Flag Equilibration Buffer and gently pelleted by centrifugation at 500 × *g* in an Allegra X-15R. The washes were serially repeated for a total of 4 times.
6. *Flag elution.* To elute the receptor from the resin, 3 CVs of Flag Elution Buffer was incubated with receptor-bound Flag Resin and incubated for 1 h at 4 °C on a rotating platform. The resin-eluate mixture was placed over a 1 mL gravity flow column to segregate resin from eluate. The residual Flag Resin was regenerated with serial gentle washes in 0.1 M glacial acetic acid and 3 × in water in the Allegra X-15R centrifuge (akin to serial washes in Step 5) for subsequent use.
7. *Concentration.* The eluted receptor was serially concentrated in 5 min increments with a 0.5 mL 100 kDa cutoff spin concentrators (Amicon) pre-equilibrated with Flag Equilibration Buffer at 3000 × *g*, 4 °C to a final volume of 0.5 mL.
8. *Gel filtration:* A Superose6 10/300 Increase (GE Healthcare) column was equilibrated with SEC buffer for two CVs using an Akta Pure chromatography instrument. The concentrated eluate was injected over the column at a flow rate of 0.5 mL/min at 4 °C and 0.5 mL fractions were collected for further analysis

The broader than expected peak on gel filtration is likely indicative of both receptor oligomeric and compositional heterogeneity. The two lower molecular weight bands are present in all fractions of the peak (data not shown). Mass spectrometry analysis identified band 1 in 2A as the chaperone heat shock protein 70 (Hsp70). Band 2 was not clearly identified by mass spectrometry. The tight association with chaperones may be an artifact of overexpression or could represent a physiologic association of the HER3 intracellular domains with chaperones to regulate activity. Nevertheless, the sample heterogeneity poses a significant obstacle to structure determination.

We hypothesized that kinase domain stabilization with small molecule inhibitors may improve homogeneity. Bosutinib, a Src inhibitor, has been shown to bind and stabilize the HER3 kinase (Davis et al., 2011; Levinson & Boxer, 2014). A unique advantage of our expression system is that we may express our receptors in the presence of high affinity kinase inhibitors that would be cost inefficient at higher culture volumes. We therefore expressed near full-length HER3 with bosutinib (10 μM final concentration) added contemporaneously with the enhancers to the culture medium.

We observed increased homogeneity after only one affinity purification step and a more monodisperse peak by gel filtration (Fig. 2B, pink curve). Our extrapolated final yield after purification was ~10 mg/L rendering only 30 mL of mammalian suspension culture sufficient for cryo-electron microscopy (Cryo-EM).

3.2 Characterization by negative stain electron microscopy

Diluted sample from the peak fraction of HER3 expressed with bosutinib was applied on carbon coated copper grids, stained with 0.75% uranyl formate, and imaged on an FEI-Tecnai T12 with a 4k CCD camera (Gatan). Negative stain micrographs were processed in Relion and resulting class averages revealed a "tri-lobular" shape (Fig. 3A). Only two of the lobes could be explained by a lateral projection of the previously published crystal structure of the HER3 unliganded extracellular domain (Cho & Leahy, 2002) (Fig. 3B). The third lobe may correspond to the micelle or a combination of the micelle and kinase.

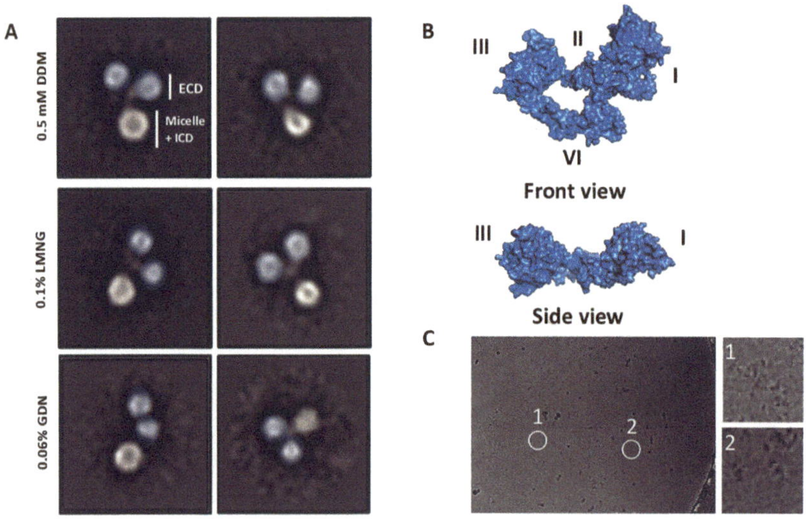

Fig. 3 Negative Stain and Cryo-EM Freezing of Near Full-Length HER3. (A) Top, representative NS-EM class averages of near full-length HER3 in the presence of 0.5 mM DDM. Middle, representative NS-EM class averages of near full-length HER3 in the presence of 0.1% LMNG. Bottom, representative NS-EM class averages of near full-length HER3 in the presence of 0.06% GDN. Density corresponding to putative extracellular domains I and III are colored in blue. (B) Surface cartoon of the HER3 extracellular domain crystal structure (PDB ID: 1M6B) viewed from the front and side. (C) Representative cryo-EM micrograph with HER3 particles solubilized in 0.01% LMNG more clearly visualized in the insets.

3.3 Initial attempts at visualization by single particle cryo-electron microscopy

Near full-length HER3 expressed with bosutinib was concentrated to 7.7 mg/mL (66 μM) and 3 μL was applied onto negatively glow discharged Quantifoil R1.2/1.3 mesh Au holey-carbon grids under standard blotting conditions using a Vitrobot Mark IV (FEI) and plunge frozen in liquid ethane (negative glow discharge 30 s, blot force 4, 100% humidity, 4 °C, 6 s blot time). We found aggregated particles at 7.7 mg/mL and no particles in the holes at lower concentrations. We hypothesized that the background DDM micelle may preferentially coat the air-water interface to prevent HER3 particles from entering thin ice and thus necessitating high receptor concentrations that promote sample aggregation.

4. The detergent exchange strategy

4.1 Detergent exchange into LMNG and visualization by cryo-electron microscopy

We rationalized that either exchanging the detergent or eliminating the background detergent in the buffer would reduce the need to achieve high sample concentrations. We therefore selected a limited set of detergents to exchange the sample into based on chemical properties and previous success in high resolution membrane protein structural studies. We evaluated sample homogeneity through gel filtration and negative stain EM.

The ultra-low CMC detergent, LMNG, is structurally similar to DDM yet does not require background detergent to maintain the micelle (Stetsenko & Guskov, 2017). Glyco diosgenin (GDN) is structurally similar to digitonin and previous works on EGFR and HER2/HER3 heterodimer suggests that digitonin facilitates kinase stability (Choi et al., 2020; Mi et al., 2008). Like LMNG, the amphiphile 1,2,5,6-tetra-beta-D-glucopyranoside-3,4-O-Di-dodecyl-D-mannitol (MNA-C12) has an ultra-low CMC and has been successfully employed in structure determination (Benton et al., 2018).

We exchanged HER3 with bosutinib into 4 different conditions (0.01% w/v LMNG, 0.1% w/v LMNG, 0.06% w/v GDN, and 0.002% w/v MNA-C12) in a stepwise manner over the Flag wash step. Two concentrations of LMNG were chosen given that both were successfully applied in single particle cryo-EM. A detailed protocol is provided below:

4.1.1 Detergent exchange purification protocol

Additional buffers:

- *LMNG Wash Buffers:* 50 mM TrisCl, 150 mM NaCl, pH 7.5 at 4 °C, 0.1% w/v (or 0.01% w/v) LMNG (Anatrace)
- *MNA-C12 Wash Buffer:* 50 mM TrisCl, 150 mM NaCl, pH 7.5 at 4 °C, 0.002% w/v MNA-C12 (Anatrace)
- *GDN Wash Buffer Buffer:* 50 mM TrisCl, 150 mM NaCl, pH 7.5 at 4 °C, 0.06% w/v GDN (Anatrace)
- *LMNG Flag Elution Buffer:* 50 mM TrisCl, 150 mM NaCl, pH 7.5 at 4 °C, 0.1% w/v (or 0.01% w/v) LMNG (Anatrace), 250 mg/mL of Flag peptide (SinoBiological)
- *MNA-C12 Flag Elution Buffer:* 50 mM TrisCl, 150 mM NaCl, pH 7.5 at 4 °C, 0.002% w/v MNA-C12 (Anatrace), 250 mg/mL of Flag peptide (SinoBiological)
- *GDN Flag Elution Buffer:* 50 mM TrisCl, 150 mM NaCl, pH 7.5 at 4 °C, 0.06% w/v GDN (Anatrace), 250 mg/mL of Flag peptide (SinoBiological)
- *TBS:* 50 mM TrisCl, 150 mM NaCl, pH 7.5 at 4 °C

Steps 1–4 the same as Purification Protocol for Near-Full Length HER3.

5. *Flag resin wash and detergent exchange.* The clarified lysate was separated from receptor-bound Flag Resin through gentle centrifugation at 500 × *g* in an Allegra X-15R (Beckman Coulter) for 2 min, 4°C in conical tubes. The Flag Resin was then mixed with 10 CVs of Flag Equilibration Buffer containing DDM and gently pelleted by centrifugation (500 × *g*, Allegra X-15R, 5 min, 4°C). 10 mL of a 50/50 v/v mixture of DDM containing Flag Equilibration Buffer and LMNG Wash Buffer was made. The same 50/50 v/v mixture was made for MNA-C12 and GDN samples. The receptor-bound resin was then incubated with this mixture for 5 min and the resin was gently pelleted by centrifugation as before. The supernatant was discarded and another 10 CVs of 100% LMNG/MNA-C12/GDN Wash Buffer were applied followed by centrifugation (centrifuge settings same as first and second washes) to separate the supernatant from the resin. This was repeated once more in a 4th wash step.
6. *Flag elution.* To elute the receptor from the resin, 3 CVs of LMNG/MNA-C12/GDN Flag Elution Buffer was incubated with receptor-bound Flag Resin and incubated for 1 h at 4 °C on a rotating platform. The resin-eluate mixture was placed over a 1 mL gravity flow column to segregate resin from eluate. The residual Flag Resin was regenerated for subsequent use.

7. *Concentration.* The eluted receptor was serially concentrated with a 0.5 mL 100 kDa cutoff spin concentrators (Amicon) pre-equilibrated with LMNG/MNA-C12/GDN Flag Wash Buffer at 3000 × *g*, 4 °C to a final volume of 0.5 mL. In serial concentration, 0.5 mL is first applied to the filter device and centrifuged. Buffer passes through to the cellulose membrane to the filtrate collection tube while the concentrated protein resides in the filter device. Additional eluted receptor is applied to the filter device to 0.5 mL and subjected to an additional round centrifugation. This process is repeated until the entire eluted receptor volume in the filter device is 0.5 mL.
8. *Gel filtration*: A Superose6 10/300 Increase (GE Healthcare) column was equilibrated using an Akta Pure chromatography instrument with TBS for two CVs for the LMNG samples. The MNA-C12 and GDN receptor samples were equilibrated in respective Wash Buffers. The concentrated eluate was injected over the column via the Akta Pure at a flow rate of 0.5 mL/min at 4 °C and 0.5 mL fractions were collected for further analysis.

Interestingly, we found a striking difference in receptor stability between 0.1% LMNG and 0.01% LMNG conditions. Although both samples were run on gel filtration without any background detergent, the majority of the 0.01% LMNG sample precipitated during concentration and the peak elution volume resembles a species larger than that of DDM and 0.1% LMNG samples (Fig. 2C). The detergent in the 0.01% LMNG sample is perhaps not high enough to completely exchange with DDM, resulting in exposed unstable hydrophobic transmembrane domains. The 0.1% LMNG sample recapitulated the DDM sample in both the gel filtration profile and NS-EM (Fig. 3A, middle class averages). Although 0.06% GDN resulted in similar tri-lobular NS-EM class averages from sample eluting near 15 mL (Fig. 2D and Fig. 3A, bottom class averages), the presence of background detergent rendered this sample less promising for cryo-EM. Like 0.01% LMNG, 0.002% MNA-C12 did not stabilize the sample.

Fractions corresponding to the gel filtration elution peak of the 0.1% LMNG sample were pooled and concentrated to 3.5 mg/mL (30 μM) and applied as with the DDM samples onto Quantifoil grids. A range of concentrations were frozen with a lower bound of 0.11 mg/mL. Remarkably, we found adequate particle density without aggregation at 0.11 mg/mL (0.9 μM) (Fig. 3C).

5. Reconstructions of the fully-glycosylated HER3 extracellular domain

5.1 Data collection and processing

The grid was imaged on a 200-keV ThermoFisher Talos Arctica electron microscope with a K3 direct electron detector (Gatan) operating in super-resolution mode at a physical pixel size of 1.82 Å/pix with a dose rate of ~10 e^- per pixel per second. Images were recorded with a 15 s exposure over 115 frames with a dose rate of 0.4 $e^-/Å^2$/frame. A total of 1411 micrographs were collected.

1411 micrographs were corrected for motion and radiation damage with MotionCor2 and the resulting sums were imported in CryoSPARC2. The dataset underwent micrograph curation with a final stack of 1119 micrographs. Micrograph Contrast Transfer Function (CTF) parameters were estimated with the patch CTF job in CryoSPARC2. Particles were picked initially using a Gaussian blob as a reference to minimize bias and the resulting picks were extracted with 2 × Fourier cropping and subjected to iterative rounds of *ab initio* and heterogeneous refinements. The particles subclassified into two reconstructions representing two conformations (178,000 and 160,000 particles each) of the HER3 extracellular domain in solution (see Section 5.4). Collection and processing parameters are provided in Table 1.

5.2 Inconsistency between NS-EM and cryo-EM

Our extracellular domain cryo-EM reconstructions are consistent with the previously published model of the unliganded HER3 extracellular domain; domains I-IV of the extracellular module are resolved in the tethered, autoinhibited conformation. However, NS-EM class averages indicating a tri-lobular architecture are not recapitulated in 2D class averages or 3D volumes from cryo-EM (Fig. 4A). Two of the three lobes may be explained by bulky domains I and III of the HER3 extracellular domain. The third lobe, potentially micelle or kinase and micelle, is completely missing in cryo-EM. While it is plausible that unliganded HER3 may be highly dynamic, especially at the juxtamembrane segments, we would expect to see such dynamics in NS-EM processing. The fact that there is a discordance between NS-EM and cryo-EM with the same sample indicates that another phenomenon may account for the discrepancy, such as denaturation at the air-water interface. Recent studies with full-length Insulin Receptor (InsR),

Table 1 Collection Statistics.

	HER3 with Bosutinib
Data collection	
Microscope	Talos-Arctica
Voltage (keV)	200
Nominal Mag	22,000 ×
Exposure navigation	Stage position/beam and image shift
Cumulative dose ($e^-/Å^2$)	46
Requested defocus range (um)	0.5–2.5
Detector	Gatan K3
Detector Operation Mode	CDS
Pixel size (physical pixel, Å)	1.82
Dose rate (e^-/physical pixel/s)	10
Total exposure time (s)	15.0
Exposure per frame (s)	0.13
Micrographs collected	1411

Insulin-like Growth Factor Receptor, and EGFR homodimers have also failed to reveal micelle or submicelle density (Huang et al., 2020; Li et al., 2019; Uchikawa et al., 2019). This phenomenon could explain why to date we do not have a high-resolution reconstruction of a full-length receptor kinase. Our recent work on the HER2/HER3/NRG1β heterocomplex with graphene oxide (GO) supported grids suggests that submicelle density may be recovered through particle recentering (see chapter "An effective strategy for ligand-mediated pulldown of the HER2/HER3/NRG1β heterocomplex and cryo-EM structure determination at low sample concentrations" by Trenker et al., in this issue). Further investigation in this area is warranted.

5.3 Identification of glycans

Docking the crystal structure into the ~6 Å (gold-standard Fourier Shell Correlation (FSC) at 0.143 as estimated in CryoSPARC) reconstruction revealed additional density corresponding to glycosylation sites (Fig. 4B). The FSC is a statistics-based resolution measurement based on correlations

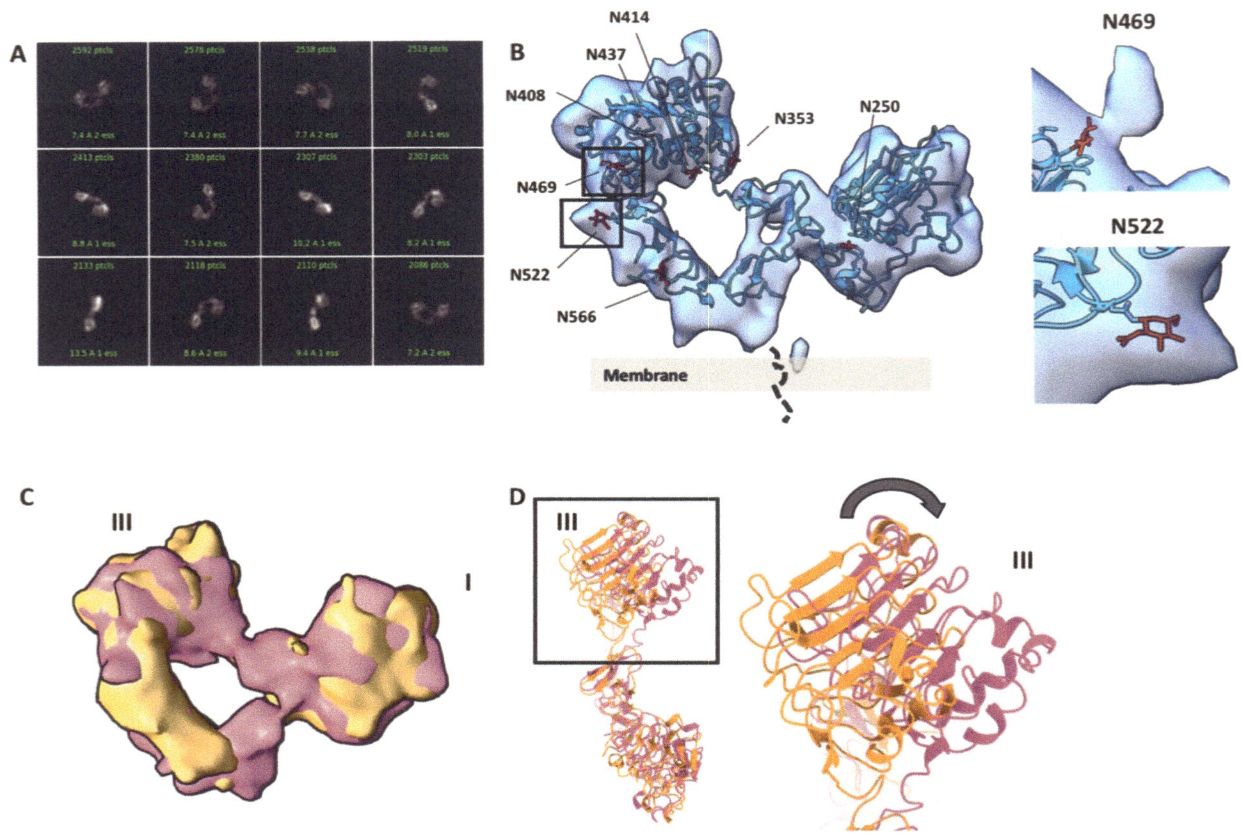

Fig. 4 Reconstructions of the HER3 extracellular domain reveal glycosylation sites and dynamic properties. (A) Representative cryo-EM 2D class averages of near full-length HER3 exchanged into 0.1% LMNG. (B) Rosetta fit model of the HER3 extracellular domain (PDB ID 1M6B) into the resulting cryo-EM reconstruction with N-linked glycans labeled. Insets highlight densities corresponding to N469 and N522. (C) Left, two resulting volumes after subclassification are superimposed. (D) Rosetta fit models for each volume overlayed to show movement of domain III relative to domain I in the inset.

between two independent half reconstructions *vs* resolution, with the 0.143 value commonly accepted as the "resolution" of a map (Rosenthal & Henderson, 2003). We could identify density for the previously noted glycosylation sites in the crystal structure (N250, N353, N408, N414, N437, N469, N522, and N566). Even at these moderate resolutions, sufficient density was found extending beyond the first GlcNAc present in the crystal structure at several sites such as at N469 (Fig. 4B).

5.4 Extracellular dynamics

Subsequent 3D classification revealed two conformational states of the HER3 ECD with roughly equal particle distribution. The two states differ by a subtle rigid-body rotation of domain I relative to III resulting in slight movements of the flexible domains II and IV (Fig. 4C). We used Rosetta flexible fitting protocol in torsion space to fit the crystal structure of unliganded HER3 extracellular domain (PDB ID: 1M6B) into each cryo-EM class and superimposed the resulting models (Wang et al., 2016). The movement of domain III relative to domain I is readily visualized through the models (Fig. 4D). The presence of multiple receptor conformations in solution is intriguing and may potentially resemble receptor states with different ligand affinities. These dynamics may provide an additional and previously unappreciated layer of receptor regulation.

6. Conclusion

Biophysical pursuits on full-length receptor kinases have been hampered by challenges in expression, purification, and visualization. Here, we demonstrate the markedly efficient expression of near full-length HER3 in Expi293 mammalian suspension cell culture such that only 30 mL are sufficient for cryo-EM. The efficiency enables small-scale parallelizability to accelerate optimization in sample homogeneity. Although this expression technique may not be generalizable to all receptor kinases, it provides a valuable starting point from which further modifications can be made. Similarly, our detergent exchange strategy from DDM to LMNG enables freezing at lower concentrations for cryo-EM which may be beneficial for samples prone to aggregation or receptors with inherently low yields. The ability to isolate near full-length HER3 enables future biophysical applications including the potential to reconstitute HER3 in lipid nanodiscs to probe receptor-membrane interactions or evaluate therapeutics in the near full-length context.

References

Azimzadeh Irani, M., Kannan, S., & Verma, C. (2017). Role of N-glycosylation in EGFR ectodomain ligand binding. *Proteins, 85*(8), 1529–1549. https://doi.org/10.1002/prot.25314.

Benton, D. J., Nans, A., Calder, L. J., Turner, J., Neu, U., Lin, Y. P., et al. (2018). Influenza hemagglutinin membrane anchor. *Proceedings of the National Academy of Sciences of the United States of America, 115*(40), 10112–10117. https://doi.org/10.1073/pnas.1810927115.

Bouyain, S., Longo, P. A., Li, S., Ferguson, K. M., & Leahy, D. J. (2005). The extracellular region of ErbB4 adopts a tethered conformation in the absence of ligand. *Proceedings of the National Academy of Sciences of the United States of America, 102*(42), 15024–15029. https://doi.org/10.1073/pnas.0507591102.

Chen, P. H., Unger, V., & He, X. (2015). Structure of full-length human PDGFRbeta bound to its activating ligand PDGF-B as determined by negative-stain electron microscopy. *Journal of Molecular Biology, 427*(24), 3921–3934. https://doi.org/10.1016/j.jmb.2015.10.003.

Cho, H.-S., & Leahy, D. J. (2002). Structure of the extracellular region of HER3 reveals an interdomain tether. *Science, 297*, 1330–1333.

Choi, B., Cha, M., Eun, G. S., Lee, D. H., Lee, S., Ehsan, M., et al. (2020). Single-molecule functional anatomy of endogenous HER2-HER3 heterodimers. *eLife, 9*. https://doi.org/10.7554/eLife.53934.

Davis, M. I., Hunt, J. P., Herrgard, S., Ciceri, P., Wodicka, L. M., Pallares, G., et al. (2011). Comprehensive analysis of kinase inhibitor selectivity. *Nature Biotechnology, 29*(11), 1046–1051. https://doi.org/10.1038/nbt.1990.

Diwanji, D., Thaker, T., & Jura, N. (2019). More than the sum of the parts: Toward full-length receptor tyrosine kinase structures. *IUBMB Life, 71*(6), 706–720. https://doi.org/10.1002/iub.2060.

Diwanji, D., Trenker, R., Thaker, T. M., Wang, F., Agard, D. A., Verba, K. A., et al. (2021). Structures of the active HER2/HER3 receptor complex reveal dynamics at the dimerization interface induced by binding of a single ligand. *bioRxiv*. https://doi.org/10.1101/2021.05.03.442258.

Du, Z., & Lovly, C. M. (2018). Mechanisms of receptor tyrosine kinase activation in cancer. *Molecular Cancer, 17*(1), 58. https://doi.org/10.1186/s12943-018-0782-4.

Ferguson, K., Berger, M., Mendrola, J., Cho, H.-S., Leahy, D. J., & Lemmon, M. A. (2003). EGF activates its receptor by removing interactions with autoinhibit ectodomain dimerization. *Molecular Cell, 11*, 507–517.

Goh, L. K., & Sorkin, A. (2013). Endocytosis of receptor tyrosine kinases. *Cold Spring Harbor Perspectives in Biology, 5*(5), a017459. https://doi.org/10.1101/cshperspect.a017459.

Gutmann, T., Kim, K. H., Grzybek, M., Walz, T., & Coskun, U. (2018). Visualization of ligand-induced transmembrane signaling in the full-length human insulin receptor. *The Journal of Cell Biology, 217*(5), 1643–1649. https://doi.org/10.1083/jcb.201711047.

Guy, P. M., Platko, J. V., Cantley, L., Cerione, R. A., & Carraway, K. L. (1994). Inset cell-expressed p180erbB3 possesses an impaired tyrosine kinase activity. *Proceedings of the National Academy of Sciences of the United States of America, 91*, 8132–8136.

Hellyer, N., Cheng, K., & Koland, J. G. (1998). ErbB3 (HER3) interaction with the p85 regulatory subunit of phosphoinositide 3-kinase. *Biochemical Journal, 333*, 757–763.

Hsieh, A. C., & Moasser, M. M. (2007). Targeting HER proteins in cancer therapy and the role of the non-target HER3. *British Journal of Cancer, 97*(4), 453–457. https://doi.org/10.1038/sj.bjc.6603910.

Huang, Y., Ognjenović, J., Karandur, D., Merk, A., Subramaniam, S., & Kuriyan, J. (2020). A structural mechanism for the generation of biased agonism in the epidermal growth factor receptor. *bioRxiv*. https://doi.org/10.1101/2020.12.08.417006.

Jura, N., Shan, Y., Cao, X., Shaw, D. E., & Kuriyan, J. (2009). Structural analysis of the catalytically inactive kinase domain of the human EGF receptor 3. *Proceedings of the National Academy of Sciences of the United States of America*, *106*(51), 21608–21613. https://doi.org/10.1073/pnas.0912101106.

Kandoth, C., McLellan, M. D., Vandin, F., Ye, K., Niu, B., Lu, C., et al. (2013). Mutational landscape and significance across 12 major cancer types. *Nature*, *502*(7471), 333–339. https://doi.org/10.1038/nature12634.

Kovacs, E., Zorn, J. A., Huang, Y., Barros, T., & Kuriyan, J. (2015). A structural perspective on the regulation of the epidermal growth factor receptor. *Annual Review of Biochemistry*, *84*, 739–764. https://doi.org/10.1146/annurev-biochem-060614-034402.

Lemmon, M. A., & Schlessinger, J. (2010). Cell signaling by receptor tyrosine kinases. *Cell*, *141*(7), 1117–1134. https://doi.org/10.1016/j.cell.2010.06.011.

Levinson, N. M., & Boxer, S. G. (2014). A conserved water-mediated hydrogen bond network defines Bosutinib's kinase selectivity. *Nature Chemical Biology*, *10*(2), 127–132. https://doi.org/10.1038/nchembio.1404.

Li, J., Choi, E., Yu, H., & Bai, X. C. (2019). Structural basis of the activation of type 1 insulin-like growth factor receptor. *Nature Communications*, *10*(1), 4567. https://doi.org/10.1038/s41467-019-12564-0.

Lu, C., Mi, L. Z., Grey, M. J., Zhu, J., Graef, E., Yokoyama, S., et al. (2010). Structural evidence for loose linkage between ligand binding and kinase activation in the epidermal growth factor receptor. *Molecular and Cellular Biology*, *30*(22), 5432–5443. https://doi.org/10.1128/MCB.00742-10.

Mi, L. Z., Grey, M. J., Nishida, N., Walz, T., Lu, C., & Springer, T. A. (2008). Functional and structural stability of the epidermal growth factor receptor in detergent micelles and phospholipid nanodiscs. *Biochemistry*, *47*, 10314–10323.

Mi, L. Z., Lu, C., Li, Z., Nishida, N., Walz, T., & Springer, T. A. (2011). Simultaneous visualization of the extracellular and cytoplasmic domains of the epidermal growth factor receptor. *Nature Structural & Molecular Biology*, *18*(9), 984–989. https://doi.org/10.1038/nsmb.2092.

Moasser, M. M. (2007). The oncogene HER2: Its signaling and transforming functions and its role in human cancer pathogenesis. *Oncogene*, *26*(45), 6469–6487. https://doi.org/10.1038/sj.onc.1210477.

Opatowsky, Y., Lax, I., Tome, F., Bleichert, F., Unger, V. M., & Schlessinger, J. (2014). Structure, domain organization, and different conformational states of stem cell factor-induced intact KIT dimers. *Proceedings of the National Academy of Sciences of the United States of America*, *111*(5), 1772–1777. https://doi.org/10.1073/pnas.1323254111.

Peiris, D., Spector, A. F., Lomax-Browne, H., Azimi, T., Ramesh, B., Loizidou, M., et al. (2017). Cellular glycosylation affects Herceptin binding and sensitivity of breast cancer cells to doxorubicin and growth factors. *Scientific Reports*, *7*, 43006. https://doi.org/10.1038/srep43006.

Rosenthal, P. B., & Henderson, R. (2003). Optimal determination of particle orientation, absolute hand, and contrast loss in single-particle electron cryomicroscopy. *Journal of Molecular Biology*, *333*(4), 721–745. https://doi.org/10.1016/j.jmb.2003.07.013.

Stetsenko, A., & Guskov, A. (2017). An overview of the top ten detergents used for membrane protein crystallization. *Crystals*, 7(7). https://doi.org/10.3390/cryst7070197.

Sung, L. Y., Chen, C. L., Lin, S. Y., Li, K. C., Yeh, C. L., Chen, G. Y., et al. (2014). Efficient gene delivery into cell lines and stem cells using baculovirus. *Nature Protocols*, *9*(8), 1882–1899. https://doi.org/10.1038/nprot.2014.130.

Takahashi, M., Hasegawa, Y., Ikeda, Y., Wada, Y., Tajiri, M., Ariki, S., et al. (2013). Suppression of heregulin beta signaling by the single N-glycan deletion mutant of soluble ErbB3 protein. *The Journal of Biological Chemistry*, *288*(46), 32910–32921. https://doi.org/10.1074/jbc.M113.491902.

Uchikawa, E., Choi, E., Shang, G., Yu, H., & Bai, X. C. (2019). Activation mechanism of the insulin receptor revealed by cryo-EM structure of the fully liganded receptor-ligand complex. *eLife*, *8*. https://doi.org/10.7554/eLife.48630.

Wang, R. Y., Song, Y., Barad, B. A., Cheng, Y., Fraser, J. S., & DiMaio, F. (2016). Automated structure refinement of macromolecular assemblies from cryo-EM maps using Rosetta. *eLife*, *5*. https://doi.org/10.7554/eLife.17219.

Yokoe, S., Takahashi, M., Asahi, M., Lee, S. H., Li, W., Osumi, D., et al. (2007). The Asn418-linked N-glycan of ErbB3 plays a crucial role in preventing spontaneous heterodimerization and tumor promotion. *Cancer Research*, *67*(5), 1935–1942. https://doi.org/10.1158/0008-5472.CAN-06-3023.

Zhang, X., Gureasko, J., Shen, K., Cole, P. A., & Kuriyan, J. (2006). An allosteric mechanism for activation of the kinase domain of epidermal growth factor receptor. *Cell*, *125*(6), 1137–1149. https://doi.org/10.1016/j.cell.2006.05.013.

CHAPTER TWENTY-ONE

An effective strategy for ligand-mediated pulldown of the HER2/HER3/NRG1β heterocomplex and cryo-EM structure determination at low sample concentrations

Raphael Trenker[a,†], Devan Diwanji[a,b,†], Kliment A. Verba[c,d,*], and Natalia Jura[a,e,*]

[a]Cardiovascular Research Institute, University of California San Francisco, San Francisco, CA, United States
[b]Medical Scientist Training Program, University of California San Francisco, San Francisco, CA, United States
[c]Quantitative Biosciences Institute (QBI), University of California San Francisco, San Francisco, CA, United States
[d]Department of Pharmaceutical Chemistry, University of California San Francisco, San Francisco, CA, United States
[e]Department of Cellular and Molecular Pharmacology, University of California San Francisco, San Francisco, CA, United States
*Corresponding authors: e-mail address: kliment.verba@ucsf.edu; natalia.jura@ucsf.edu

Contents

† Authors contributed equally to the work.

Methods in Enzymology, Volume 667
ISSN 0076-6879
https://doi.org/10.1016/bs.mie.2022.03.049

Abstract

Obtaining high-resolution structures of Receptor Tyrosine Kinases that visualize extracellular, transmembrane and intracellular kinase regions simultaneously is an eagerly pursued but still unmet challenge of structural biology. The Human Epidermal Growth Factor Receptor 3 (HER3) that has a catalytically inactive kinase domain (pseudokinase) forms a potent signaling complex upon binding of growth factor neuregulin 1β (NRG1β) and upon dimerization with a close homolog, the HER2 receptor. The HER2/HER3/NRG1β complex is often referred to as an oncogenic driver in breast cancer and is an attractive target for anti-cancer therapies. After overcoming significant hurdles in isolating sufficient amounts of the HER2/HER3/NRG1β complex for structural studies by cryo-electron microscopy (cryo-EM), we recently obtained the first high-resolution structures of the extracellular portion of this complex. Here we describe a step-by-step protocol for obtaining a stable and homogenous HER2/HER3/NRG1β complex for structural studies and our recommendation for collecting and processing cryo-EM data for this sample. We also show improved EM density for the transmembrane and kinase domains of the receptors, which continue to evade structural determination at high resolution. The discussed strategies are tunable and applicable to other membrane receptor complexes.

1. Introduction

Receptor Tyrosine Kinases (RTKs) are key mediators of transmembrane signaling and are one of the most often deregulated receptor classes in disease (Arteaga & Engelman, 2014; Du & Lovly, 2018; Lemmon & Schlessinger, 2010). RTKs couple binding of the extracellular ligands to their ectodomains with activation of the intracellular tyrosine kinase domains, resulting in phosphorylation of the flexible C-terminal receptor tails that then recruit downstream signaling molecules (Lemmon & Schlessinger, 2010). Activation occurs via receptor dimerization, or sometimes higher-order oligomerization, which is initiated by binding of the extracellular ligands (Garrett et al., 2002; Himanen & Nikolov, 2003; Lemmon & Schlessinger, 2010; Liu, Chen, Focia, & He, 2007; Trenker & Jura, 2020; Ullrich & Schlessinger, 1990; van Lengerich, Agnew, Puchner, Huang, & Jura, 2017; Wiesmann et al., 1997). Disease mutations have been shown to map across RTK domains and perturb normal receptor regulation resulting in oncogenic signaling (Du & Lovly, 2018). Understanding the detailed mechanisms by which ligand binding is transduced into kinase activation is of importance for understanding the pathology of disease mutations, and for enabling the development of therapeutics that target this important family of receptors.

Most of what is known about the structural changes that the receptors undergo during ligand-induced activation is based on structures of isolated extracellular, transmembrane and intracellular domains (Kovacs, Zorn, Huang, Barros, & Kuriyan, 2015). While such studies had brought fundamental insights into modes of ligand/receptor binding and architecture of the individual receptor domains, the information about allosteric coupling between receptor domains across the plasma membrane is inevitably lost in such investigations. Several recent structural studies by single particle negative-stain (NS) and cryo-electron microscopy (cryo-EM) have focused on full-length RTK preparations—signaling the beginning of the exciting new era in structural studies of RTKs (Chen, Unger, & He, 2015; Diwanji, Thaker, & Jura, 2019; Gutmann, Kim, Grzybek, Walz, & Coskun, 2018; Huang et al., 2021; Li, Choi, Yu, & Bai, 2019; Mi et al., 2008, 2011; Opatowsky et al., 2014; Uchikawa, Choi, Shang, Yu, & Bai, 2019). Despite the efforts, all reported studies fell short of reconstructing high-resolution structures of receptor complexes that visualize all functional domains simultaneously. Thus, such structures remain an unsolved problem in RTK structural biology and are being eagerly pursued.

Human epidermal growth factor receptors (HERs) are a subfamily of RTKs that encompass four closely related receptors, EGFR, HER2, HER3 and HER4 (Lemmon & Schlessinger, 2010). Activation of HERs proceeds through homo- or heterodimerization induced by ligand binding (Alroy & Yarden, 1997; Freed et al., 2017; Rush, Peterson, & Ceresa, 2018; van Lengerich et al., 2017). Depending on a type of bound ligand and receptors in a dimer, the active HER dimer will elicit unique downstream signaling responses (Freed et al., 2017; Huang et al., 2021; Sweeney & Carraway 3rd., 2000; Trenker & Jura, 2020). HER3 is a unique family member because it has a catalytically-impaired kinase domain—a pseudokinase (Guy, Platko, Cantley, Cerione, & Carraway 3rd., 1994; Jura, Shan, Cao, Shaw, & Kuriyan, 2009; Sierke, Cheng, Kim, & Koland, 1997). Despite having negligible catalytic activity, HER3 can form signaling-competent complexes through heterodimerization with other HER receptors. This is induced by binding of the HER3 extracellular domain (ECD) to protein ligands, which belong to the neuregulin (NRG) family of growth factors. In the heterodimeric complexes, the HER3 pseudokinase domain serves as an allosteric activator of the kinase domain of the catalytically-active HER dimerization partner (Littlefield et al., 2014; Zhang, Gureasko, Shen, Cole, & Kuriyan, 2006). HER3 preferentially dimerizes with HER2, which also depends on heterodimerization with other HER receptors for signaling

(Sliwkowski et al., 1994; Wallasch et al., 1995). This is because HER2 is an orphan receptor for which no known activating ligands have been identified. Thus, in the active HER2/HER3/NRG1β signaling complex HER3 is the ligand-sensing and HER2 is the kinase-active component.

The functionally asymmetric HER2/HER3 heterodimer is one of the most potent signaling HER receptor complexes. Its potency is in large part determined by the ability of phosphorylated HER3 to recruit and activate multiple copies of PI3K (Wallasch et al., 1995). HER2 does not bind PI3K directly and, on its own, HER2 does not efficiently activate PI3K. Thus, heterodimerization with HER3 is an essential component of HER2-dependent transformation and is considered to be a driving force of cancer progression in several tissues, most predominantly in the breast (Moasser, 2007). Despite its importance, no structures of the HER2/HER3 complexes have been solved until our recent study in which we report cryo-EM reconstructions of the extracellular domain module of the NRG1β-bound HER3 with the wild type HER2 or an oncogenic cancer mutant HER2-S310F (Diwanji et al., 2021). These structures have revealed unique features of the functionally asymmetric HER2/HER3 complex, including an unusual mode of engagement of the dimerization arms, which are key structural elements contributing to the dimerization of the HER receptor extracellular domains. We found that HER2 is unable to engage a dimerization arm from HER3, making dimerization reliant on HER3's ability to engage the HER2 arm. We hypothesized that this is why HER2 cannot form homodimers itself, which had been a curious puzzle in the field. By solving the structure of HER3 in complex with an oncogenic HER2 mutant (S310F) that mutation restores HER3 dimerization arm engagement, which provided a mechanistic explanation for its activating effect on HER2/HER3 signaling (Diwanji et al., 2021).

The structural characterization of the HER2/HER3/NRG1β complex was enabled by the development of the experimental protocols for its purification and reconstitution. Despite the fact that we have only resolved the extracellular domain module of the HER2/HER3 heterodimer, our studies had taught us that the inclusion of the intracellular domains in our constructs was necessary for purification of the stable dimers. While we have not yet been able to obtain a high-resolution reconstruction of the intracellular domain module of the HER2/HER3 dimer, we made progress in data collection and analysis that focus on resolution of these domains. Here we describe a detailed protocol for assembling the stable HER2/HER3/NRG1β and HER2-S310F/HER3/NRG1β complexes and for collecting and processing cryo-EM data with improved EM density for transmembrane and kinase domains (Diwanji et al., 2021). These strategies are tunable and applicable to other membrane receptor complexes.

2. Expression and purification of NRG1β

All ligands known to bind to HER receptors feature an EGF-like domain, which is required and sufficient for receptor binding. To reconstitute the HER2/HER3/NRG1β complex, we first established a protocol for expression and purification of the EGF-like domain of the HER3-binding growth factor ligand, NRG1β. The EGF-like domain is held together by three conserved disulfide bonds, which pose a major challenge for recombinant expression of the correctly folded HER receptor ligands. Based on a previous protocol, we express the NRG1β EGF-like domain (residues 177–236) as a fusion protein to thioredoxin (TrxA) and C-terminal FLAG (DYKDDDDK)- and 6× polyhistidine (His)-tag in *E. coli* Origami B (DE3) pLysS cells (Diwanji et al., 2021; Liu et al., 2012) (Fig. 1A and B). This *E. coli* strain features several mutations that lower the reducing capability of the *E. coli* cytoplasm which aids the formation of disulfide bonds. The TrxA may also assist in disulfide bond formation and may support solubility of the fusion protein (LaVallie et al., 1993). The TrxA-NRG1β-FLAG-His expression construct features a 3C protease cleavage site between TrxA and NRG1β (Fig. 1A). The His-tag in this construct is utilized for affinity purification of the protein from *E. coli* origami cells (see protocol below, Section 2). As discussed in Section 3, the additional FLAG-tag can be used to prepare a NRG1β-coated resin by incubating the purified NRG1β with anti-FLAG resin. This resin then allows for pulldown of the HER3 receptor from HER3-expressing cell lysates (protocol in Section 3).

2.1 Key resources and equipment

Key resources and consumables

- *E. coli* origami B (DE3) pLysS (made chemically competent in house and are stored at −80 °C at an OD600 of ~12.5. Can be purchased from Millipore Sigma, #70839)
- pET32a expression plasmid with Ampicillin resistance (LaVallie et al., 1993). Empty plasmid available from Millipore Sigma, # 69015). Construct was cloned in house and encodes a Thioredoxin-NRG1β (177–236)-FLAG-His fusion protein with 3C cleavage site between thioredoxin and NRG1β (Fig. 1).

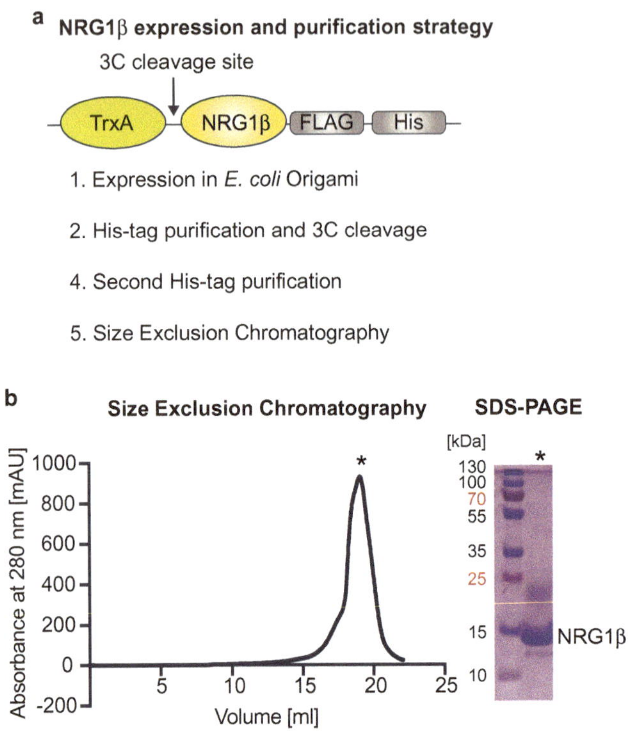

Fig. 1 Expression and purification of NRG1β. (A) Overview of the expression and purification pipeline. Thioredoxin-NRG1β (TrxA-NRG1β-FLAG-His) fusion protein is expressed in *E. coli* origami B (DE3) pLysS cells and purified via a His-tag. The TrxA protein is cleaved off with the 3C protease and NRG1β is purified using another His-tag purification followed by Size Exclusion Chromatography (SEC). (B) The SEC chromatogram of NRG1β loaded onto a Superdex 200 increase column shows a homogenous peak of NRG1β elution at 18–19 mL in high purity as shown on the Coomassie-stained SDS-PAGE gel with 10 μL of sample loaded.

- Tryptone (Thermo Fisher Scientific, # BP9726–2)
- Yeast Extract (Thermo Fisher Scientific, # BP97275)
- Glycerol (Millipore Sigma, # GX0185)
- Potassium phosphate, Monobasic, (KH_2PO_4 (Millipore Sigma, # PX1565)
- Potassium phosphate, Dibasic (K_2HPO_4, Millipore Sigma, # PX1570))
- LB Broth with agar (Millipore Sigma, # L3147)
- LB-Agar plates (made in house using 40 g of LB Agar per 1 L ddH_2O)
- Chloramphenicol (AllStar Scientific, #480-045)
- Kanamycin Sulfate (VWR, #97061-598)

- Tetracycline (Tetracycline Hydrochloride, Gold Biotechnology, # T-101-25)
- Ampicillin (Calbiochem, sold by Millipore Sigma, #171254)
- Two-Sided Disposable Plastic Cuvettes, Semi-Micro 1.3–3 mL (VWR, 97000-586)
- TRIS (Trizma base Millipore Sigma, #T6066)
- Sodium chloride (NaCl, Millipore Sigma, #SX0420)
- Imidazole (Millipore Sigma, #5720)
- Isopropyl β-D-1-thiogalactopyranoside (IPTG, Gold Biotechnology, #I2481C100)
- Phenylmethylsulfonyl fluoride (PMSF, VWR, #0754)
- DNase I, grade II, from bovine pancreas (Roche, sold by Millipore Sigma, #10104159001)
- HisPur Ni-NTA Resin (Thermo Fisher Scientific, PI88222)
- 3C protease (prepared in house with non-cleavable His-tag, stored in 1 mg aliquots at −80 °C at ~1.5 mg/mL, adjusted from (Lee, Lolicato, Arrigoni, & Minor Jr., 2021))
- Liquid nitrogen
- Amicon Ultra-15 Centrifugal Filter Units, 3000 and 10,000 MWCO, (Millipore Sigma, #UFC900324 and #UFC901024, respectively)
- Conical tubes, 15 and 50 mL, polypropylene (VWR, #21008)
- Microcentrifuge tubes, 1.7 mL (VWR, #87003-294)
- Econo-Pac(R) Chromatography Columns, 20 mL bed volume (Bio-Rad, #7321011)
- Syringes and Millex-HV Syringe Filter Unit, 0.45 μm, Millipore Sigma, (#SLHVM33RS)
- Liquid nitrogen

Buffers and media

- TBS (50 mM TRIS pH 7.4, 150 mM NaCl)
- 5 M imidazole stock pH 8.0
- Terrific Broth (1 L: 12 g Tryptone, 24 g Yeast Extract, 4 mL glycerol, dissolve in 900 mL ddH_2O and autoclave. Before use, add 100 mL Phosphate buffer at 0.17 M KH_2PO_4 and 0.72 M K_2HPO_4)
- L Broth (1 L: 10 g Tryptone, 10 g NaCl, 5 g Yeast Extract, adjust to pH 7)
- IPTG stock solution, 1 M (2.38 g IPTG for 10 mL, dissolve in ddH_2O)

Equipment

- 37 °C static incubator (VWR)
- 37 °C shaking incubator (New Brunswick Innova 4300, Marshall Scientific, #NB-4300)

- Sonicator for cell lysis (S-4000 Ultrasonic Liquid Processor, Misonix, probe diameter 1 cm)
- Non-baffled 2 L sterile glass flasks
- Non-baffled 500 mL sterile glass flasks
- Glass beaker (50 mL volume per bacterial pellet)
- Glass bottle (50 mL per 1 L bacterial pellet)
- Centrifuge to pellet 1 L bacterial cultures (Avanti J-26 XP, Beckman Coulter, rotor JLA-8.1000)
- Polypropylene Bottle Assembly, 1000 mL (Beckman Coulter, # C31597)
- Ultracentrifuge (Optima L-90K, Beckman Coulter, rotor Type 45 Ti)
- Polycarbonate Centrifuge Tubes, 70 mL with cap (Beckman Coulter, #355620)
- Table-top centrifuge (for 15/50 mL conical tubes, Allegra X-15R, Beckman Coulter)
- Magnetic stirrer plate
- Nutating mixer (Fisher Scientific)
- Spectrophotometer optical density measurements of bacterial cultures (Nanodrop 2000c, Thermo Scientific)
- Superdex 200 Increase 10/300 column (GE HEalthcare, now Cytiva #28990944)
- Akta purification system (Akta Pure, Cytiva)

2.2 Protocol

Note: *E. coli* Origami cells grow slowly thus allowing for longer expansion times as outlined below is recommended. The protocol below is optimized for 6 L cultures and typically yields in ~15–20 mg of the pure NRG1β EGF-like domain.

2.2.1 Transformation

1. Transform 1 μg of DNA into 50 μL chemically competent *E. coli* Origami B (DE3) pLysS cells and plate onto LB-Agar plates containing Ampicillin (100 μg/mL), Kanamycin (30 μg/mL), Chloramphenicol (35 μg/mL) and Tetracycline (10 μg/mL) using any standard transformation protocol with a recovery time of 1 h.
2. Incubate at 37 °C in a static incubator for 24–40 h. Cells grow very slowly and we advise transforming in the morning 1 day or in the evening 2 days before starter cultures are grown.

2.2.2 Expression

3. In the morning, streak across 5–10 colonies on the LB-Agar plate and inoculate a 10 mL starter culture in LB containing Ampicillin (100 μg/mL), Kanamycin (30 μg/mL), Chloramphenicol (35 μg/mL) and Tetracycline (10 μg/mL) and grow throughout the day at 37 °C and 200 rpm.
4. Set up overnight culture. Transfer 10 mL starter culture into 100 mL LB supplemented with antibiotics (see step 3) and grow overnight at 37 °C and 200 rpm in a non-baffled 500 mL flask.
5. The next day, add 15 mL of the overnight culture to 1 L Terrific Broth + antibiotics and grow to an OD600 of 1 at 37 °C and 200 rpm in a non-baffled 2 L flask. This usually takes 3–4 h. OD600 measurements are performed using a Nanodrop in Semi-Micro Plastic Cuvettes.
6. Induce with 1 mM IPTG (1 mL of a 1 M fresh stock solution per 1 L culture) and reduce temperature of the incubator to 25 °C.
7. The next morning, pellet cells by centrifugation using 1000 mL Polypropylene Bottles at 4000 rpm (3397 × *g*, Avanti J-26 XP, Beckman Coulter JLA-8.1000 rotor) for 20 min at 25 °C and flash freeze the pellet in liquid nitrogen. Either store pellet at −80 °C or continue with the purification.

2.2.3 Purification

8. Thaw bacterial pellet(s) in a cold-water bath and resuspend in 40 mL TBS supplemented with 1 mM PMSF and 1 mg/mL DNase I per 1 L pellet.
9. Sonicate in a glass beaker on ice for at least 10 min total at power 40, 4 s on, 4 s off, or until thoroughly lysed.
10. Transfer lysate into required numbers of 70 mL Polycarbonate Centrifuge Tubes and pellet cell debris by ultracentrifugation at 40,000 rpm (1,865,511 × *g*, Optima L-90K, Beckman Coulter, rotor Type 45 Ti) for 1 h at 4 °C. If required, add TBS to fill tubes to the top.
11. Filter supernatant through a 0.45 μm syringe filter into a glass bottle.
12. Equilibrate ~1 mL Ni-NTA resin per 1 L expression culture in TBS, add to the lysate and incubate overnight at 4 °C while stirring the beads on a magnetic stirrer plate.
13. Transfer the lysate + beads into Econo-Pac Chromatography Columns. The beads usually appear brown for well-expressing proteins.
14. Wash beads with at least 5 column volumes of TBS + 20 mM imidazole, followed by 5 column volumes of TBS + 50 mM imidazole.

15. Elute in 3–5 × column volumes of TBS + 300 mM imidazole, or until beads return to their initial blue color, into a fresh 50 mL conical tube.
16. To reduce the imidazole concentration to 30 mM or lower, concentrate the eluate to a volume of 4 mL or less using an Amicon-15 Centrifugal Filter Unit with a molecular weight cut-off (MWCO) of 10 kDa and spinning at 4000 × *g* for 20 min intervals in a table-top centrifuge at 4 °C. Transfer the sample to a 50 mL conical tube and add TBS to a total volume of 40 mL.
17. Add 3C protease to cleave off thioredoxin from the NRG1β-FLAG-His ligand, mix by inverting the tube 2–3 × and incubate overnight at 4 °C without agitation. We use 1 mg of protease for 6 L of expression culture.
18. The following day, add ~1 mL equilibrated Ni-NTA resin per 1 L expression culture to the sample and incubate overnight at 4 °C while rocking the tube on a nutating mixer.
19. Transfer beads to an Econo-Pac Chromatography Column and wash with 5 column volumes of TBS.
20. Elute in 3–5 column volumes TBS + 300 mM imidazole into a fresh 50 mL conical tube.
21. Concentrate the eluate to a volume of 0.5 mL per liter of expression culture using an Amicon-15 Centrifugal Filter Unit with a MWCO of 3 kDa and spinning at 4000 × *g* for 20 min intervals in a table-top centrifuge at 4 °C. The eluate usually appears yellow/brown in color.
22. Transfer the sample into a microcentrifuge tube, flash freeze in liquid nitrogen and store at −80 °C or proceed directly to Size Exclusion Chromatography (SEC). For larger preparations of more than 1 L, freeze 500 μL aliquots for each liter expression culture. If precipitation occurs, remove precipitate by centrifugation at ~21,000 × *g* for 3 min and 4 °C in a table-top microcentrifuge before freezing and/or SEC.
23. Equilibrate a Superdex 200 increase 10/300 GE column with TBS using an Akta Pure instrument, and load 500 μL of sample on the equilibrated column. Collect fractions of high-purity NRG1β around an elution volume of 18 mL (Fig. 1A and B).
24. **Alternatively**, the second Ni-NTA step can be replaced by ion exchange chromatography to remove the protease and thioredoxin A as described in a similar protocol (Liu et al., 2012). We have also successfully performed on-column digests after step 14 prior to elution. Transfer the protein-bound Ni-NTA beads into a 50 mL conical tube and adjust to a total volume of 40 mL with TBS. Add 3C protease

and incubate overnight at 4 °C while rocking the tube on a nutating mixer. After the on-column digest, transfer the resin into an Econo-Pac Chromatography Column, wash in 5 bed volumes of TBS + 50 mM imidazole and elute in TBS + 300 mM imidazole. Afterward proceed with step 21.

25. Freeze protein as 1 mg aliquots in microcentrifuge tubes in liquid nitrogen and store at −80 °C. The extinction coefficient of the NRG1β-FLAG-His ligand is 6335 with a molecular weight of 9.73 kDa (Abs 0.1% = 0.651 g/L, determined via the Expasy ProtParam online tool). Absorbance readings are performed using a nanodrop.

Expected yield: We typically obtain ∼15–20 mg of the pure NRG1β EGF-like domain for 6 L of culture.

3. Purification, crosslinking and preparation of the HER2/HER3/NRG1β sample for single particle electron microscopy imaging

Previous efforts aimed at generating a stable HER2/HER3/NRG1β complex using isolated extracellular domains of the receptors were unsuccessful (Ferguson, Darling, Mohan, Macatee, & Lemmon, 2000). We set to assemble this signaling complex between nearly full-length receptors, which only had their flexible C-terminal tails truncated to facilitate expression, purification and structural analysis. HER3 (1-1022) was expressed with a C-terminal Twin-Strep (TS) tag, HER2 (1-1029) featured C-terminal TS and Maltose-Binding-Protein (MBP) tags (Fig. 2A). Both receptors were expressed individually in Expi293F cells using ExpiFectamine-based transfection. Receptor pellets were lysed and the complex was purified from mixed cell lysates using tandem affinity purification to pull specifically on HER3 and then HER2. In the first step, HER3 was specifically pulled down via NRG1β-coated beads. In a second step HER2 was specifically pulled down via the MBP-tag followed by SEC. Below we describe a detailed procedure for purifying the HER2/HER3/NRG1β complex, discuss optimization of the constructs and the additional steps we took to obtain an optimal sample for structure determination by cryo-EM.

3.1 Key resources and equipment

Key resources and consumables

- Plasmid DNA for expression in mammalian cells encoding for HER2(1-1029)-MBP-TS and HER3(1-1022)-TS. Here pFastBac1

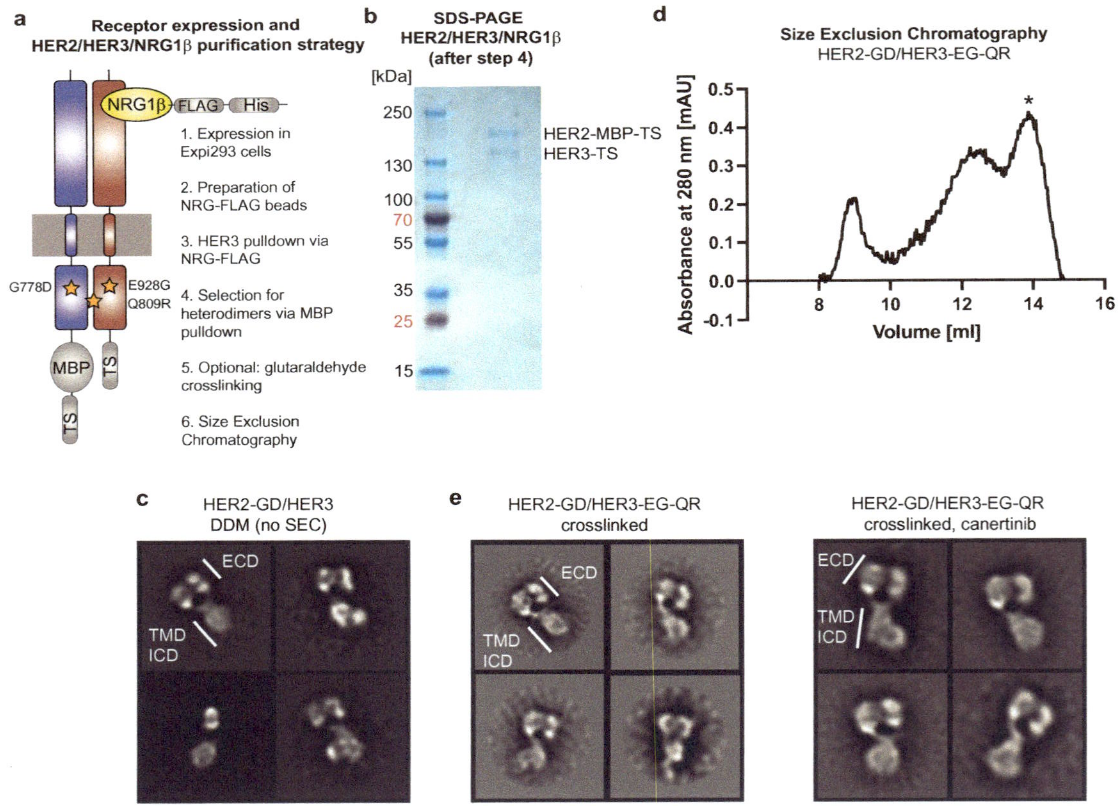

Fig. 2 See figure legend on opposite page.

plasmids (Thermo Fisher Scientific, #10360014) were use in which the baculovirus promoter was replaced by a CMV promoter that is active in mammalian cells (Sung et al., 2014). We also commonly use pcDNA3.1 and pcDNA4/TO for this protocol (Thermo Fisher Scientific, # V79020 and # V102020).

- ExpiFectamine 293 Transfection Kit (Thermo Fisher Scientific, #A14524)
- Expi293 Expression Medium (Thermo Fisher Scientific, #A1435101)
- Expi 293F human cells (Thermo Fisher Scientific, #A14527)
- Opti-MEM (Thermo Fisher Scientific, #31985070)
- Canertinib, 10 mM in DMSO (MedChem Express, #HY-10367)
- NRG1β-FLAG (in house, see Section 2)
- n-Dodecyl-β-D-Maltopyranoside (DDM, Anatrace, #D310)
- n-Dodecyl-β-D-Maltopyranoside (DDM, Inalco, #1758–1350, used only for mammalian cell lysis)
- TRIS (Triszma base, Millipore Sigma, #T6066)
- HEPES (Thermo Fisher Scientific, # FLBP3101)
- Sodium chloride (NaCl, Millipore Sigma, #SX0420)
- Ethylenediaminetetraacetic acid (EDTA, Millipore Sigma, #324503)
- Sodium orthovanadate (Millipore Sigma, #S6508)
- Sodium fluoride (NaF, Millipore Sigma, # SX0550)
- cOmplete EDTA-free Protease Inhibitor Cocktail Mini Tablets (Roche, sold by Millipore Sigma, #11836170001)
- DNase I, grade II, from bovine pancreas (Roche, sold by Millipore Sigma, #10104159001)

Fig. 2 Expression and purification of the HER2/HER3/NRG1β complexes. (A) Cartoon of the HER2/HER3/NRG1β complex with overview of the expression/purification pipeline. The optimized HER2 construct has the G778D (GD) mutation, and HER3 has the E928G (EG) and Q809R (QR) mutations. Receptors are expressed in Expi293 suspension cells and purified via tandem affinity chromatography using NRG1β-coated FLAG beads and amylose resin, followed by SEC. (B) Coomassie-stained SDS-PAGE gel with 10 μL of the ~4 mL elution volume loaded shows a stoichiometric ration of HER2 and HER3 receptors. NRG1β is less visible due to small molecular weight. (C) NS-EM 2D class averages of the HER2-GD/HER3/NRG1β complex after tandem affinity purification with the expected heart-shaped ECD dimers in some classes. Images were collected on a FEI Tecnai T12 microscope at 52,000× magnification. (D) Representative SEC chromatogram shows a profile of the crosslinked HER2-GD/HER3-EG-QR/NRG1β complex with the heterodimer peak eluting at 14 mL (marked with *). (E) NS-EM 2D class averages of the cross-linked HER2-GD/HER3-EG-QR/NRG1β complexes with and without canertinib. Images were collected on a FEI Tecnai T12 microscope at 52,000× magnification.

- Anti-FLAG resin (Anti-DYKDDDDK G1 Affinity Resin, GenScript, #L00432-10)
- FLAG peptide (DYKDDDDK synthetic peptide, Sino Biological, #PP101274)
- Glutaraldehyde Aqueous 8%, EM Grade (Electron Microscopy Sciences, #16019)
- Amylose resin (NEB, E8021L)
- Maltose monohydrate (Millipore Sigma, # M5895)
- Erlenmeyer Flasks, 125 mL, sterile (VWR, #89095-258)
- Conical tubes, 15 and 50 mL, polypropylene (VWR, #21008)
- Microcentrifuge tubes, 1.7 mL (VWR, #87003-294)
- Micro Bio-Spin columns, 0.8 mL (Bio-Rad, #7326204)
- Amicon Ultra-0.5 Centrifugal Filter Units, MWCO 100 kDa (Millipore Sigma, #UFC510096)
- Disposable Serological Pipets 5, 10, and 25 mL (VWR, #89130)
- Liquid nitrogen and appropriate vessel for freezing tubes of 1.5–50 mL volume
- Uranyl formate (Electron Microscopy Sciences, #22450)
- Formvar/carbon 400 mesh copper grids (Ted Pella, #01754-F)
- Graphene-oxide (GO)-coated Quantifoil R1.2/1.3300 mesh Au holey-carbon grids (in house, provided by Feng Wang; Palovcak et al., 2018).
- Liquid ethane, (Matheson, #UN1035).
- Whatman qualitative filter paper, grade 1 for preparation of negative-stain EM grids (Millipore Sigma, #WHA1001090)
- Whatman qualitative filter paper, grade 597 for cryoEM grid freezing using a Vitrobot Mark IV (FEI) (Millipore Sigma, #WHA10311807)

Buffers

- 1 M TRIS pH 7.4
- TBS (50 mM TRIS pH 7.4 and 150 mM NaCl)
- Lysis Buffer (50 mM TRIS pH 7.4, 150 mM NaCl, 1% DDM (Inalco), 1 mM EDTA, 1 mM sodium orthovanadate, 1 mM NaF, one tablet of cOmplete EDTA-free Protease Inhibitor Cocktail Mini Tablets per 50 mL lysis buffer, and 1 mg/mL DNase)
- TBS-DDM (50 mM TRIS pH 7.4, 150 mM NaCl and 0.5 mM DDM (Anatrace))
- HBS-DDM (50 mM HEPES pH 7.4, 150 mM NaCl and 0.5 mM DDM (Anatrace))
- FLAG elution buffer (TBS-DDM + 0.25 mg/mL FLAG peptide)

Software
- MotionCor2 (Zheng et al., 2017)
- CryoSPARC2 (Punjani, Rubinstein, Fleet, & Brubaker, 2017)
- RELION3 (Scheres, 2012)
- UCSF PyEM v0.5 ("Asarnow, Palovcak, & Cheng, 2019; asarnow/pyem: UCSF pyem v0.5 (v0.5). Zenodo. https://doi.org/10.5281/zenodo.3576630")
- UCSF Chimera (Pettersen et al., 2004)

Equipment
- Incubator at 37 °C and 8% CO_2 with shaker (Heracell VIOS 160i, Thermo Scientific)
- Tissue culture hood
- Table-top microcentrifuge (5424 R, Eppendorf)
- Table-top centrifuge (for 15/50 mL conical tubes, Allegra X-15R, Beckman Coulter)
- Nutating mixer (Fisher scientific)
- Superose 6 10/300 Increase column (GE Healthcare, now Cytiva, #29–0915-96)
- Akta purification system (Akta Pure, Cytiva)
- FEI-Tecnai T12
- 4k CCD camera (Gatan)
- Glow discharger (PELCO easyGlow, Ted Pella)
- GPU-equipped workstation (at least 11GB of GPU memory, at least 128GB of system memory, at least 4TB of fast SSD storage and 20TB of spinning disc storage)
- Vitrobot Mark IV (FEI)
- 300-keV Titan Krios (Thermo Fisher Scientific) with K3 direct electron detector (Gatan) and BioQuantum energy filter (Gatan)
- Tweezers, Style N5AC, Electron Microscopy Sciences (#0203-N5AC-PO)
- −80 °C freezer

3.2 Protocol

3.2.1 Transfection of Expi293 cultures

Expi293F cells were cultured in Expi293 Expression Medium and were transfected with the expression constructs using the ExpiFectamine 293 Transfection Kit. Maintain 30 mL suspension cell cultures in 125 mL flasks in Expi Expression Medium at 37 °C and 8% CO_2 shaking at 120 rpm at a cell density between 0.5 and 4×10^6 cells/mL. Cells double approximately once every 24 h. Separate cultures for HER2 and HER3 expression were prepared. Following optimization, we found that two 30 mL expression flasks per receptor yielded enough sample for heterodimer reconstitution to allow preparation of at least 12 cryo-EM grids. The purification protocol below was optimized for a total of four 30 mL cell pellets.

1. One day prior to transfection, seed 30 mL of Expi293F suspension cultures into 125 mL flasks at a density of 2×10^6 cells/mL to reach a density of $\sim 4 \times 10^6$ cells/mL at the time of transfection. Depending on the number of flasks needed, plan to have sufficient cultures at $\sim 4 \times 10^6$ cells/mL on this day to seed transfection cultures. For example, we typically seed multiple 30 mL cultures at 0.5×10^6 cells/mL on Fridays and seed transfection cultures on Mondays by diluting approximately 1:1 with fresh medium to be transfected on Tuesdays.
2. In the afternoon, prepare transfection mix according to the manufacturer's protocol and add to cells. We transfect the suggested 30 μg of plasmid DNA for each 30 mL culture.
3. The next morning (16–18 h after transfection), add 150 μL Transfection Enhancer 1 and 1.5 mL Transfection Enhancer 2 to each 30 mL flask of transfected cells according to the manufacturer's protocol. At this stage, any other desired supplement can be added to the expression cultures. As described below, for expression of the HER2 constructs, we obtained best results by growing cells in 10 μM canertinib. Canertinib is a covalent inhibitor of HER2 and was only added at this stage of the purification.
4. Grow cells for another 24 h.
5. Harvest cell pellets via centrifugation of the cultures at $1500 \times g$ for 5 min.
6. Flash freeze cell pellets in liquid nitrogen and store at −80 °C.

3.2.2 Purification and sample preparation for negative stain and cryo-EM

7. Thaw cells on ice and resuspend each 30 mL cell pellet in 10 mL of the Lysis Buffer and combine lysates of HER2 and HER3 in a 50 mL conical tube.
8. Allow cells to lyse for 2–3 h by gently rocking the lysate at 4 °C on a nutating mixer.
9. In the meantime, prepare NRG1β-coated anti-FLAG resin. Transfer 0.5 mL anti-FLAG resin for four 30 mL pellets to a 15 mL conical tube. Equilibrate the resin in TBS and add 1 mg NRG1β (EGF-like domain)

for 0.5 mL anti-FLAG resin. Incubate by gently rocking at 4 °C for 1 h and then wash the beads with 3 × 10 mL TBS using cycles of centrifugation (500 × *g* for 2 min at 4 °C) and resuspension in TBS. Remove excess TBS. Beads can be kept on ice until receptor lysates are ready.

10. Remove cell debris via centrifugation at 4000 × *g* for 20 min at 4 °C and transfer the clarified lysate (supernatant) into a fresh 50 mL conical tube.

11. Add the NRG1β-bound anti-FLAG resin and incubate by gently rocking at 4 °C overnight.

12. The next morning, wash resin by centrifugation at 500 × *g* for 2 min at 4 °C and resuspension in 10 mL TBS-DDM in 3 serial cycles. Remove as much of the buffer from the beads as possible.

13. Add 4 bed volumes (2 mL for 500 μL resin) of FLAG elution buffer to the resin and incubate by gently rocking for 1 h at 4 °C.

14. Tape a Bio-Spin column at the top of a fresh 15 mL conical tube and elute anti-FLAG resin suspension stepwise through the column by gravity flow. Run additional two column volumes of FLAG elution buffer over the column and collect the flow-through in the same tube.

15. Equilibrate 500 μL of Amylose resin with TBS-DDM and load the eluate obtained in the previous step on this resin.

16. Incubate for 2 h gently rocking at 4 °C.

17. Wash resin by centrifugation at 500 × *g* for 2 min at 4 °C and resuspension in TBS-DDM in 2 serial cycles. Add 4 column volumes of TBS-DDM + 20 mM maltose and incubate overnight gently rocking at 4 °C.

18. Elute through a Bio-Spin column, as described in step 14, into a fresh 15 mL conical tube. Run additional two column volumes of TBS-DDM + 20 mM maltose over the column and collect the flow-through into the same tube.

19. Concentrate the eluate using Amicon Ultra-0.5 centrifugal concentration units with 100 kDa cutoff filters to 400 μL via repeated spins at 4000 × *g* for 4 min and 4 °C.

20. Load the concentrated sample on the Superose 6 10/300 Increase column equilibrated in TBS-DDM using an Akta Pure chromatography instrument.

21. Collect peak fractions corresponding to the elution volume of ~14 mL. In our experience, the sample may be flash frozen in liquid nitrogen and stored at −80 °C at this point.

22. To prepare the grids for negative stain EM (NS-EM), we added 3 μL of sample equivalent in concentration to a 4 mAU SEC peak, which we empirically determined to be an ideal protein concentration for NS-EM imaging. For example, for the SEC profile shown in Fig. 2D, we took the 2 mL peak fraction of peak 2 (peak height of ~0.4 mAU) and concentrated the sample to 200 μL to prepare negative stain grids. Samples were applied to negatively glow-discharged carbon coated copper grids, stained with 0.75% uranyl-formate, and imaged on an FEI-Tecnai T12 with an 4k CCD camera (Gatan). Note that it is important to empirically determine the optimal protein concentration for NS-EM imaging based on an individual protein and chromatography system. This empirical concentration can then be used to estimate the concentration for cryo-EM imaging on Graphene Oxide (GO)-coated holey carbon grids.
23. The same sample was then further concentrated to 20–40 μL to prepare GO grids for cryo-EM. The resulting sample concentration for cryo-EM is 5–10 fold higher than that prepared for the NS-EM, and is equivalent to a peak of 20–40 mAU range. See section 4 for further rationale and details. Sample was applied to graphene-oxide coated Quantifoil R1.2/1.3300 mesh Au holey-carbon grids prepared as previously described (Palovcak et al., 2018), blotted using a Vitrobot Mark IV (FEI) and plunge frozen in liquid ethane (no glow discharge, 30 s wait time, room temperature, 100% humidity, 4–8 s blot time and 0 blot force using Whatman qualitative filter paper, grade 597).

3.2.3 *Optional receptor crosslinking*

We were able to increase the homogeneity and quality of NS-EM class averages further by mildly crosslinking the HER2/HER3/NRG1β sample with a non-specific, amine-reactive crosslinker glutaraldehyde (Fig. 2E). Further rationale to apply crosslinking was to increase the chance of observing density for intracellular domains. As discussed below, crosslinking resulted in a slight overall improvement of the extracellular domain resolution, but only marginally improved the density of the kinase domains after initial processing.

Since glutaraldehyde is an amine-reactive crosslinker, it is important not to perform the crosslinking reaction in buffers that contain primary amines such as TBS. Therefore, prior to crosslinking, we buffer-exchanged the sample over the second affinity step of the purification into the HBS-DDM buffer.

1. Follow steps 1–16 of the HER2/HER3/NRG1β purification protocol above (Section 3.2.1 + Section 3.2.2).
2. Wash resin by centrifugation at 500 × *g* for 2 min at 4 °C and resuspension in HBS-DDM in 3 serial cycles.
3. Add 4 column volumes of HBS-DDM + 20 mM maltose and incubate overnight at 4 °C on a rotator.
4. Transfer into a Bio-Spin column and elute into a fresh 15 mL conical tube by gravity flow. Run additional two column volumes of HBS-DDM + 20 mM maltose over the column and collect the flow-through in the same tube.
5. Concentrate the eluate using Amicon Ultra-0.5 centrifugal concentration units with 100 kDa cutoff filters to 400 μL. Spin at 4000 × *g* at 4 °C and repeated spins of 4 min.
6. Crosslink the sample on ice by adding 40 μL of aqueous 2% glutaraldehyde stock to the 400 μL of receptor from the previous step. Quench the reaction after 40 min by adding 40 μL of 1 M TRIS pH 7.4 buffer.
7. Load the sample on the Superose 6 10/300 Increase column equilibrated in TBS-DDM.
8. Collect peak fractions. In our experience, the HER2/HER3/NRG1β preparations may be flash frozen in liquid nitrogen and stored at -80 °C.
9. For preparation of EM samples, see steps 23 and 24 of the sample transfection and purification protocol (Section 3.2.1 + Section 3.2.2).

3.3 Analysis and sample optimization

We screened various constructs of HER2 and HER3 to maximize the yields of NRG1β-induced heterodimers. Using the wild type sequence of both receptors, we were able to purify sufficient amounts of the heterodimer after tandem affinity purification for characterization by NS-EM (Fig. 2C). 2D NS-EM class averages revealed heart-shaped particles, resembling those previously seen for the EGF-bound extracellular EGFR dimer, providing first structural characterization of the HER2/HER3 heterodimers, and showing, for the first time, that HER3 adopts an extended conformation upon ligand binding (discussed in detail in Diwanji et al., 2021). However, these initial samples appeared insufficiently homogeneous for further imaging by cryo-EM. We could never obtain high enough sample yields, even from a 120 mL Expi293 cell expression culture. We reasoned that the limiting factor is the amount of HER2 that was pulled down with HER3 in the first step of the purification, suggesting that the WT heterodimer complex might be inherently weak. This observation is consistent with previous findings that the extracellular domains of HER2 and HER3 do not heterodimerize in solution (Ferguson et al., 2000) and suggests that heterodimerization may be reliant on considerable contributions from the transmembrane and/or intracellular receptor domains.

To further increase the stability of the heterocomplex, yields of the purification and the chance of visualizing the kinase domains, we rationally introduced stabilizing mutations into the HER2 and HER3 kinase domains. The best yields were observed with HER2 constructs featuring the G778D (GD) mutation which confers independence from the cytosolic chaperone Hsp90 (Xu et al., 2005). We introduced two mutations in the HER3 kinase domain: E928G (EG) and Q809R (QR). These cancer-associated mutations reside in the interface of the asymmetric kinase dimer and increase kinase/kinase and kinase/JM interactions between the activator (HER3) and the receiver (HER2) kinase, respectively (Fig. 2A) (Hanker et al., 2021; Littlefield et al., 2014). In particular, there is good evidence that the oncogenic HER3 E928G mutation drives the formation of active asymmetric kinase dimers by increasing the affinity of the two kinase domains and stabilizing the active conformation of HER2 (Hanker et al., 2021; Littlefield et al., 2014). The introduction of stabilizing kinase mutations yielded sufficient amounts of heterocomplex from 4×30 mL expression cultures to be visualized by SEC. Fig. 2D shows a representative SEC profile of the crosslinked HER2-GD/HER3-EG-QR/NRG1β sample on a Superose 6 10/300 Increase column with the NRG1β-bound heterodimer eluting as the main species around 14 mL (Fig. 2D, marked with (*)). NS-EM 2D class averages of this peak showed the expected heart-shaped structure of the HER2/HER3 extracellular domain dimer with improved overall homogeneity and structural detail (Fig. 2E). In the presence of these mutations, we observed that addition of the HER2 inhibitor, canertinib, further increases the yield and quality of the HER2/HER3/NRG1β sample with improved resolution of transmembrane domains and intracellular domains. As described above, to obtain canertinib-bound HER2, a final concentration of 10 μM canertinib was added to the HER2 expression culture contemporaneously with the enhancers. Thus, introduction of stabilizing mutations in the receptor intracellular domains, crosslinking, and the use of the HER2 kinase inhibitor improved overall homogeneity of the samples as judged by the NS-EM 2D class average analysis (Fig. 2E).

4. Sample preparation for freezing and imaging by cryo-EM

4.1 Grid selection for cryo-EM analysis of the HER2/HER3/NRG1β complex

In addition to the heart-shaped extracellular domain density observed in the 2D class averages obtained by NS-EM analysis of the purified HER2/HER3/NRG1β complex, we observed a density that we interpreted as a micelle and the kinase domains (Fig. 2C + E; ICD). However, we were unable to resolve this density at high resolution by cryo-EM. Other recent studies that have reported on cryo-EM analysis of full-length RTKs similarly failed to resolve the intracellular regions of the receptors. Likewise, in our accompanying paper describing the structural analysis of the unliganded HER3 by cryo-EM (see chapter "Efficient expression, purification, and visualization by cryo-EM of unliganded near full-length HER3" by Diwanji et al.), we only resolved the extracellular domain by cryo-EM, despite observing the extra density in the NS-EM 2D average analysis.

There are likely several reasons for the lack of high resolution in the intracellular region, which is currently a universal problem in the cryo-EM analysis of RTKs (Huang et al., 2021; Li et al., 2019; Uchikawa et al., 2019). One possibility is that the juxtamembrane/transmembrane domain junction between the extracellular and intracellular domains of RTKs is too flexible to stabilize these domains relative to each other, with the computational analysis biasing resolution of the extracellular domains due to their typically larger molecular weight and distinct structural features. This scenario, however, does not explain why the density corresponding to the intracellular domains can be observed by NS-EM, for HER receptors for example (Fig. 2E). We hypothesized that the intracellular domains might be more prone to unfolding at the air-water interface when samples are frozen on uncoated holey carbon grids, and that we could improve their resolution by using different grids for freezing. We turned to graphene oxide (GO)-coated grids because the GO coating creates a hydrophilic support layer on which proteins settle during freezing, similar to the conditions present during negative staining (Palovcak et al., 2018; Pantelic, Meyer, Kaiser, Baumeister, & Plitzko, 2010). The GO layer also eliminates one of two air-water interfaces that are present when proteins are frozen in thin ice on uncoated holey carbon grids in the absence of a support layer.

We hypothesized that the similarities between NS and GO grids may recover additional density and enable reconstructions of full-length receptor complexes by cryo-EM.

Another major enabling property of GO grids is sample imaging at low protein concentration. This opens up the studies on protein complexes that have traditionally been difficult, or even impossible, to obtain at high enough amounts for structure determination. We found that a working concentration of HER receptor samples for the NS-EM is an equivalent to a 4 mAU elution peak on SEC and that the optimal density of particles on GO grids can be obtained at only ~5–10 fold higher than that concentration (20–40 mAU; Fig. 3A + B). This corresponds to a concentration of the HER2/HER3/NRG1β complex equivalent to ~0.06–0.12 μM (0.017–0.034 mg/mL). Our typical SEC profile from the HER2/HER3/NRG1β complex purification from 120 mL of Expi293 cells yields a dimer peak with ~0.5–1 mAU at 280 nm over a 2 mL elution volume, which is equivalent to 20–40 mAU when concentrated to 50 μL (Fig. 2D). Typically, such amounts would be insufficient for imaging on uncoated holey carbon grids, which is usually done at concentration ranges at least one order of magnitude higher (Huang et al., 2021; Passmore & Russo, 2016; Uchikawa et al., 2019). Of note is that in order to obtain ideal and reproducible concentrations of HER receptor samples for imaging by cryo-EM on GO grids, we follow concentration estimates based on SEC profiles and subsequent analysis of the NS-EM micrographs, rather than absorbance readings on a spectrophotometer, such as nanodrop, which we found quite unreliable at such low protein concentrations.

4.2 Freezing samples on GO grids

GO grids do not need to be glow-discharged prior to sample application. DDM-solubilized receptors are plunge-frozen into liquid ethane using a Vitrobot Mark IV (FEI) with a wait time of 30 s. The waiting period is required to reduce the presence of round, micelle-like particles in recorded micrographs. This has been empirically observed but the exact process is unclear. For data collection, we found that if GO is not easily visible on medium magnification montages (210–360 × magnification, Fig. 3C), we often cannot find particles in these areas and obtain suboptimal datasets with slightly above 50% particle coverage even when grids with close to 100% GO coverage are used (Palovcak et al., 2018). We only make this observation for samples that contain detergent (DDM), suggesting that DDM

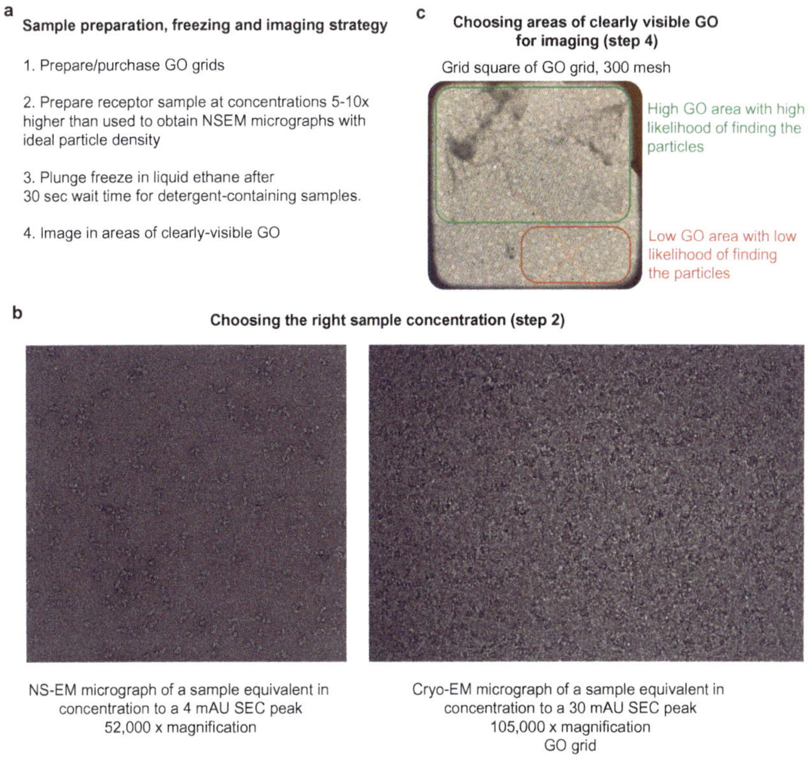

Fig. 3 Cryo-EM imaging of the HER2/HER3/NRG1β complex on graphene oxide (GO) grids. (A) Summary of ideal grid preparation procedure for low-concentration, detergent-solubilized HER2/HER3 heterocomplex. (B) Representative NS-EM and cryo-EM GO micrographs of the heterocomplex. The NS-EM micrograph was recorded at a sample concentration of 4 mAU of the SEC elution peak after concentrating the 0.4 mAU elution peak 10 fold. For GO grid preparation, we found that the optimal sample concentration corresponds to a 5–10 fold concentration of a sample that yielded good NS-EM grids. In the presented example, this corresponds to 30 mAU of the SEC elution peak. (C) Medium magnification montage (360× magnification) of GO grid square showing clear areas of GO and areas without GO.

loosens and removes GO flakes during grid preparation for cryo-EM. We therefore advise imaging the detergent-solubilized membrane proteins in areas with clearly visible GO to increase the yield of micrographs with good particle coverage (Fig. 2B). If GO grids can be made in house, we suggest exploring the use of GO grids with thicker GO coating (grid preparation and how to vary GO thickness is described in detail in Palovcak et al., 2018), which has increased our particle yields for various detergent-containing samples.

4.3 Structure determination of the HER2/HER3/NRG1β extracellular domain dimer and the full-length dimer complex

Application of GO grids allowed us to collect several datasets for the HER2/HER3/NRG1β complex sample on a 300-keV Titan Krios (Thermo Fisher) with a K3 direct electron detector (Gatan) and a BioQuantum energy filter (Gatan). Image stacks were motion-corrected using MotionCor2 and imported into CryoSPARC2 for most processing steps (Diwanji et al., 2021; Punjani et al., 2017). Micrographs were automatically curated to select for Contrast Transfer Function (CTF)-fit resolution better than 4–5 Å and to remove micrographs with thick ice, and again manually curated to remove all micrographs without particles due to a lack of GO in the respective area (in our experience this reflected 40% of data; the efficiency can be improved by being more selective during data acquisition). For particle picking by both Gaussian blob and 2D class average-derived template picking, we split the whole micrograph stack into a low-defocus and a high-defocus stack to ensure that during particle inspections the very high and very low contrast particles are not excluded (Diwanji et al., 2021). For the non-crosslinked sample bound to the inhibitor canertinib, we obtained a resolution of 3.7 Å for the extracellular domain dimer module (Fig. 4B left). Since NS-EM 2D class averages suggested improved sample homogeneity after glutaraldehyde crosslinking, we collected a dataset with 2035 micrographs using the crosslinked sample (Diwanji et al., 2021). Data were processed with CryoSPARC2 to reach a final resolution within the extracellular domain dimer of 2.9 Å, which is a significant improvement over the non-crosslinked sample (Fig. 4B middle). Our final reconstruction of the crosslinked mutant HER2-S310F/HER3/NRG1β complex bound to canertinib reached a final resolution of 3.1 Å (Diwanji et al., 2021) (Fig. 4B right). Features of the structures obtained from these datasets and the in-depth details of cryo-EM processing are discussed in detail in Diwanji et al. (2021).

All high-resolution maps that we obtained of the extracellular domain module of the wild type and mutant HER2/HER3/NRG1β complexes (Diwanji et al., 2021), in addition to the density corresponding to the extracellular domains, featured density towards the expected transmembrane and intracellular domain regions (Fig. 4B). These regions by and large were not seen in our reconstructions of unliganded HER3 obtained from a dataset collected on uncoated holey carbon grids (published in this edition) (see chapter "Efficient expression, purification, and visualization

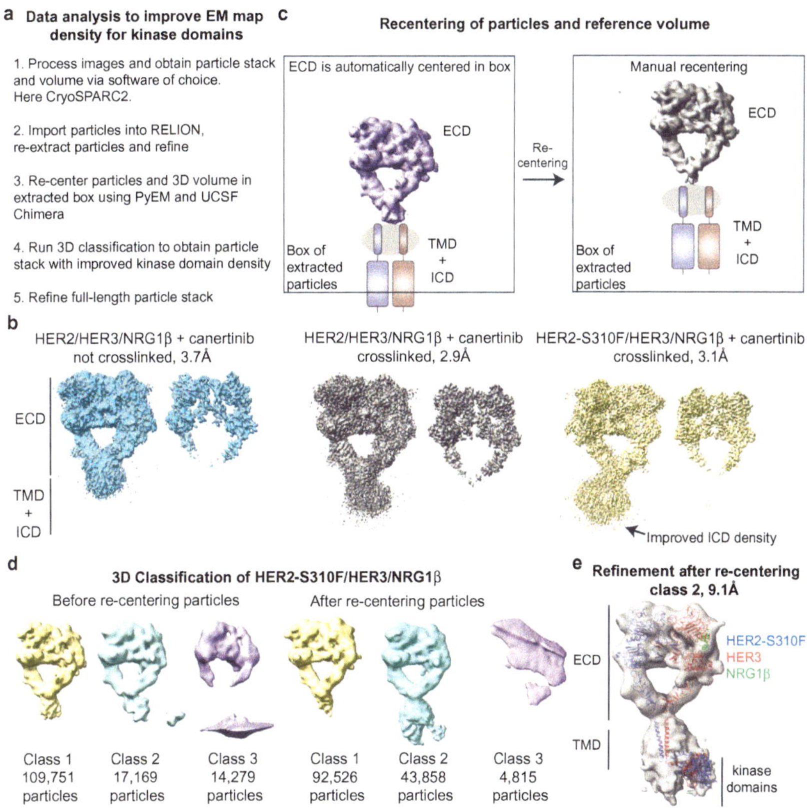

Fig. 4 Processing scheme for obtaining improved maps of full-length complexes. (A) Overview of the data and the processing pipeline for the HER2/HER3/NRG1β and HER2-S310F/HER3/NRG1β complexes. (B) Low- and high-contour EM maps of the complexes show best transmembrane and intracellular domain density for the cross-linked HER2-S310F/HER3/NRG1β complex. (C) Illustration of the re-centering process using UCSF chimera and PyEM. (D) 3D classification in RELION3 of the HER2-S310F/HER3/NRG1β complex before and after re-centering coordinates. (E) 9.1 Å resolution full-length reconstruction of the HER2-S310F/HER3/NRG1β complex. Models used for fitting into cryo-EM map included: PDB ID: 7MN6 for the HER2-S310F/HER3/NRG1β extracellular domain dimer (Diwanji et al., 2021); PDB ID: 2N2A for the transmembrane domains and juxtamembrane-A (JM-A) segments (Bragin et al., 2016); and the model of the asymmetric HER2/HER3 kinase dimer generated based on PDB ID: 3PP0 (Aertgeerts et al., 2011) and PDB ID: 4RIW (Littlefield et al., 2014).

by cryo-EM of fully-glycosylated unliganded HER3" by *Diwanji et al.)* Notably, the density for the transmembrane and intracellular domain regions were more pronounced in the mutant HER2-S310F/HER3/NRG1β than in the wild type HER2/HER3/NRG1β complex, indicating that the disease mutation that stabilized the interactions between the extracellular domains has also stabilizing effects on the intracellular domain interactions (Fig. 4B).

We hypothesized that the drop and eventual loss in resolution towards the kinase domains may be due to a loose connection between the extracellular, transmembrane and intracellular domains. In this case, during image alignment CryoSPARC2 focuses on aligning bulky, featureful extracellular domains thereby losing signal for adjacent transmembrane domains and kinase domains. To uncover a subpopulation of particles with matched orientation of the three subdomains, we exported a large particle stack of the most stable HER2-S310F/HER3/NRG1β complex from CryoSPARC2 into RELION3 using PyEM ("Asarnow et al., 2019; asarnow/pyem: UCSF pyem v0.5 (v0.5). Zenodo. https://doi.org/10.5281/zenodo.3576630"; Punjani et al., 2017; Scheres, 2012), re-extracted particles and performed 3D classification into 3 classes (Fig. 4D left panel). Remarkably, this 3D classification revealed one class with density for only the extracellular domains at relatively high-resolution and another class with overall lower resolution but stronger density for putative transmembrane and kinase domains (Fig. 4D left panel, classes 1 and 2 with RELION estimated resolution of unrefined 3D classes of 7.3 and 12.0 Å, respectively). This observation might suggest that the extracellular and intracellular domains are loosely connected and the more one focuses on the extracellular domains in particle alignment the more definition is lost on the intracellular domains. It is, however, also possible that partial unfolding of transmembrane and intracellular domains at the air-water interface during freezing causes this reduction in cryo-EM density and that this phenomenon is slightly reduced on GO grids.

We consistently observe that CryoSPARC2 focuses on the bulky and featureful extracellular domains during particle picking and image alignment, centering the extracellular domains in the extraction box while placing putative intracellular domains toward the edge of the box (Fig. 4C). This can result in gradual loss of cryo-EM density for the intracellular domains. Using UCSF Chimera (Pettersen et al., 2004) and PyEM ("Asarnow et al., 2019; asarnow/pyem: UCSF pyem v0.5 (v0.5). Zenodo. https://doi.org/10.5281/zenodo.3576630"), we shifted coordinates of particles within the extracted box to center the micelle region instead (Fig. 4C). Particles with adjusted coordinates were then re-extracted in RELION3 and again subjected to 3D classification. Remarkably, this re-centering process increased the density for the transmembrane and intracellular domain regions of the

HER2-S310F/HER3/NRG1β complex. This particle stack and volume were refined and we were able to obtain a 9.1 Å resolution reconstruction of the full-length complex (Fig. 4E).

Overall, while high-resolution maps of full-length RTKs remain challenging to obtain, we find that the combination of GO grids, and careful 3D classification with re-centering can significantly recover previously unresolved kinase domains at very low protein concentration.

Acknowledgments

This work was supported by the UCSF Program for Breakthrough Biomedical Research to K.V. and N.J., DFG German Research Foundation GZ: TR 1668/1-1 to R.T., and by the National Institutes of Health: NIH/NIGMS R35-GM139636 to N.J., pilot project grant within the U54 CA209891 to N.J., and NIH/NCI 1F30CA247147 to D.D.

References

Aertgeerts, K., Skene, R., Yano, J., Sang, B. C., Zou, H., Snell, G., et al. (2011). Structural analysis of the mechanism of inhibition and allosteric activation of the kinase domain of HER2 protein. *The Journal of Biological Chemistry*, *286*(21), 18756–18765.

Alroy, I., & Yarden, Y. (1997). The ErbB signaling network in embryogenesis and oncogenesis: Signal diversification through combinatorial ligand-receptor interactions. *FEBS Letters*, *410*(1), 83–86.

Arteaga, C. L., & Engelman, J. A. (2014). ERBB receptors: From oncogene discovery to basic science to mechanism-based cancer therapeutics. *Cancer Cell*, *25*(3), 282–303.

Asarnow, D., Palovcak, E., & Cheng, Y. (2019). asarnow/pyem: UCSF pyem v0.5 (v0.5). Zenodo. https://doi.org/10.5281/zenodo.3576630.

Bragin, P. E., Mineev, K. S., Bocharova, O. V., Volynsky, P. E., Bocharov, E. V., & Arseniev, A. S. (2016). HER2 transmembrane domain dimerization coupled with self-Association of Membrane-Embedded Cytoplasmic Juxtamembrane Regions. *Journal of Molecular Biology*, *428*(1), 52–61.

Chen, P. H., Unger, V., & He, X. (2015). Structure of full-length human PDGFRbeta bound to its activating ligand PDGF-B as determined by negative-stain electron microscopy. *Journal of Molecular Biology*, *427*(24), 3921–3934.

Diwanji, D., Thaker, T., & Jura, N. (2019). More than the sum of the parts: Toward full-length receptor tyrosine kinase structures. *IUBMB Life*, *71*(6), 706–720.

Diwanji, D., Trenker, R., Thaker, T. M., Wang, F., Agard, D. A., Verba, K. A., et al. (2021). Structures of the HER2-HER3-NRG1beta complex reveal a dynamic dimer interface. *Nature*, *600*(7888), 339–343.

Du, Z., & Lovly, C. M. (2018). Mechanisms of receptor tyrosine kinase activation in cancer. *Molecular Cancer*, *17*(1), 58.

Ferguson, K. M., Darling, P. J., Mohan, M. J., Macatee, T. L., & Lemmon, M. A. (2000). Extracellular domains drive homo- but not hetero-dimerization of erbB receptors. *The EMBO Journal*, *19*(17), 4632–4643.

Freed, D. M., Bessman, N. J., Kiyatkin, A., Salazar-Cavazos, E., Byrne, P. O., Moore, J. O., et al. (2017). EGFR ligands differentially stabilize receptor dimers to specify signaling kinetics. *Cell*, *171*(3), 683–695 e618.

Garrett, T. P., McKern, N. M., Lou, M., Elleman, T. C., Adams, T. E., Lovrecz, G. O., et al. (2002). Crystal structure of a truncated epidermal growth factor receptor extracellular domain bound to transforming growth factor alpha. *Cell*, *110*(6), 763–773.

Gutmann, T., Kim, K. H., Grzybek, M., Walz, T., & Coskun, U. (2018). Visualization of ligand-induced transmembrane signaling in the full-length human insulin receptor. *The Journal of Cell Biology, 217*(5), 1643–1649.

Guy, P. M., Platko, J. V., Cantley, L. C., Cerione, R. A., & Carraway, K. L., 3rd. (1994). Insect cell-expressed p180erbB3 possesses an impaired tyrosine kinase activity. *Proceedings of the National Academy of Sciences of the United States of America, 91*(17), 8132–8136.

Hanker, A. B., Brown, B. P., Meiler, J., Marin, A., Jayanthan, H. S., Ye, D., et al. (2021). Co-occurring gain-of-function mutations in HER2 and HER3 modulate HER2/HER3 activation, oncogenesis, and HER2 inhibitor sensitivity. *Cancer Cell, 39*(8), 1099–1114 e1098.

Himanen, J. P., & Nikolov, D. B. (2003). Eph signaling: A structural view. *Trends in Neurosciences, 26*(1), 46–51.

Huang, Y., Ognjenovic, J., Karandur, D., Miller, K., Merk, A., Subramaniam, S., et al. (2021). A molecular mechanism for the generation of ligand-dependent differential outputs by the epidermal growth factor receptor. *eLife, 10.*

Jura, N., Shan, Y., Cao, X., Shaw, D. E., & Kuriyan, J. (2009). Structural analysis of the catalytically inactive kinase domain of the human EGF receptor 3. *Proceedings of the National Academy of Sciences of the United States of America, 106*(51), 21608–21613.

Kovacs, E., Zorn, J. A., Huang, Y., Barros, T., & Kuriyan, J. (2015). A structural perspective on the regulation of the epidermal growth factor receptor. *Annual Review of Biochemistry, 84*, 739–764.

LaVallie, E. R., DiBlasio, E. A., Kovacic, S., Grant, K. L., Schendel, P. F., & McCoy, J. M. (1993). A thioredoxin gene fusion expression system that circumvents inclusion body formation in the E. coli cytoplasm. *Biotechnology (N Y), 11*(2), 187–193.

Lee, H., Lolicato, M., Arrigoni, C., & Minor, D. L., Jr. (2021). Production of K2P2.1 (TREK-1) for structural studies. *Methods in Enzymology, 653*, 151–188.

Lemmon, M. A., & Schlessinger, J. (2010). Cell signaling by receptor tyrosine kinases. *Cell, 141*(7), 1117–1134.

Li, J., Choi, E., Yu, H., & Bai, X. C. (2019). Structural basis of the activation of type 1 insulin-like growth factor receptor. *Nature Communications, 10*(1), 4567.

Littlefield, P., Liu, L., Mysore, V., Shan, Y., Shaw, D. E., & Jura, N. (2014). Structural analysis of the EGFR/HER3 heterodimer reveals the molecular basis for activating HER3 mutations. *Science Signaling*, 7(354), ra114.

Liu, H., Chen, X., Focia, P. J., & He, X. (2007). Structural basis for stem cell factor-KIT signaling and activation of class III receptor tyrosine kinases. *The EMBO Journal, 26*(3), 891–901.

Liu, P., Cleveland, T. E. T., Bouyain, S., Byrne, P. O., Longo, P. A., & Leahy, D. J. (2012). A single ligand is sufficient to activate EGFR dimers. *Proceedings of the National Academy of Sciences of the United States of America, 109*(27), 10861–10866.

Mi, L. Z., Grey, M. J., Nishida, N., Walz, T., Lu, C., & Springer, T. A. (2008). Functional and structural stability of the epidermal growth factor receptor in detergent micelles and phospholipid nanodiscs. *Biochemistry, 47*(39), 10314–10323.

Mi, L. Z., Lu, C., Li, Z., Nishida, N., Walz, T., & Springer, T. A. (2011). Simultaneous visualization of the extracellular and cytoplasmic domains of the epidermal growth factor receptor. *Nature Structural & Molecular Biology, 18*(9), 984–989.

Moasser, M. M. (2007). The oncogene HER2: Its signaling and transforming functions and its role in human cancer pathogenesis. *Oncogene, 26*(45), 6469–6487.

Opatowsky, Y., Lax, I., Tome, F., Bleichert, F., Unger, V. M., & Schlessinger, J. (2014). Structure, domain organization, and different conformational states of stem cell factor-induced intact KIT dimers. *Proceedings of the National Academy of Sciences of the United States of America, 111*(5), 1772–1777.

Palovcak, E., Wang, F., Zheng, S. Q., Yu, Z., Li, S., Betegon, M., et al. (2018). A simple and robust procedure for preparing graphene-oxide cryo-EM grids. *Journal of Structural Biology*, *204*(1), 80–84.
Pantelic, R. S., Meyer, J. C., Kaiser, U., Baumeister, W., & Plitzko, J. M. (2010). Graphene oxide: A substrate for optimizing preparations of frozen-hydrated samples. *Journal of Structural Biology*, *170*(1), 152–156.
Passmore, L. A., & Russo, C. J. (2016). Specimen preparation for high-resolution Cryo-EM. *Methods in Enzymology*, *579*, 51–86.
Pettersen, E. F., Goddard, T. D., Huang, C. C., Couch, G. S., Greenblatt, D. M., Meng, E. C., et al. (2004). UCSF chimera—A visualization system for exploratory research and analysis. *Journal of Computational Chemistry*, *25*(13), 1605–1612.
Punjani, A., Rubinstein, J. L., Fleet, D. J., & Brubaker, M. A. (2017). cryoSPARC: Algorithms for rapid unsupervised cryo-EM structure determination. *Nature Methods*, *14*(3), 290–296.
Rush, J. S., Peterson, J. L., & Ceresa, B. P. (2018). Betacellulin (BTC) biases the EGFR to dimerize with ErbB3. *Molecular Pharmacology*, *94*(6), 1382–1390.
Scheres, S. H. (2012). RELION: Implementation of a Bayesian approach to cryo-EM structure determination. *Journal of Structural Biology*, *180*(3), 519–530.
Sierke, S. L., Cheng, K., Kim, H. H., & Koland, J. G. (1997). Biochemical characterization of the protein tyrosine kinase homology domain of the ErbB3 (HER3) receptor protein. *The Biochemical Journal*, *322*(Pt 3), 757–763.
Sliwkowski, M. X., Schaefer, G., Akita, R. W., Lofgren, J. A., Fitzpatrick, V. D., Nuijens, A., et al. (1994). Coexpression of erbB2 and erbB3 proteins reconstitutes a high affinity receptor for heregulin. *The Journal of Biological Chemistry*, *269*(20), 14661–14665.
Sung, L. Y., Chen, C. L., Lin, S. Y., Li, K. C., Yeh, C. L., Chen, G. Y., et al. (2014). Efficient gene delivery into cell lines and stem cells using baculovirus. *Nature Protocols*, *9*(8), 1882–1899.
Sweeney, C., & Carraway, K. L., 3rd. (2000). Ligand discrimination by ErbB receptors: Differential signaling through differential phosphorylation site usage. *Oncogene*, *19*(49), 5568–5573.
Trenker, R., & Jura, N. (2020). Receptor tyrosine kinase activation: From the ligand perspective. *Current Opinion in Cell Biology*, *63*, 174–185.
Uchikawa, E., Choi, E., Shang, G., Yu, H., & Bai, X. C. (2019). Activation mechanism of the insulin receptor revealed by cryo-EM structure of the fully liganded receptor-ligand complex. *eLife*, *8*.
Ullrich, A., & Schlessinger, J. (1990). Signal transduction by receptors with tyrosine kinase activity. *Cell*, *61*(2), 203–212.
van Lengerich, B., Agnew, C., Puchner, E. M., Huang, B., & Jura, N. (2017). EGF and NRG induce phosphorylation of HER3/ERBB3 by EGFR using distinct oligomeric mechanisms. *Proceedings of the National Academy of Sciences of the United States of America*, *114*(14), E2836–E2845.
Wallasch, C., Weiss, F. U., Niederfellner, G., Jallal, B., Issing, W., & Ullrich, A. (1995). Heregulin-dependent regulation of HER2/neu oncogenic signaling by heterodimerization with HER3. *The EMBO Journal*, *14*(17), 4267–4275.
Wiesmann, C., Fuh, G., Christinger, H. W., Eigenbrot, C., Wells, J. A., & de Vos, A. M. (1997). Crystal structure at 1.7 A resolution of VEGF in complex with domain 2 of the Flt-1 receptor. *Cell*, *91*(5), 695–704.
Xu, W., Yuan, X., Xiang, Z., Mimnaugh, E., Marcu, M., & Neckers, L. (2005). Surface charge and hydrophobicity determine ErbB2 binding to the Hsp90 chaperone complex. *Nature Structural & Molecular Biology*, *12*(2), 120–126.

Zhang, X., Gureasko, J., Shen, K., Cole, P. A., & Kuriyan, J. (2006). An allosteric mechanism for activation of the kinase domain of epidermal growth factor receptor. *Cell*, *125*(6), 1137–1149.

Zheng, S. Q., Palovcak, E., Armache, J. P., Verba, K. A., Cheng, Y., & Agard, D. A. (2017). MotionCor2: Anisotropic correction of beam-induced motion for improved cryo-electron microscopy. *Nature Methods*, *14*(4), 331–332.

CHAPTER TWENTY-TWO

Enabling pseudokinases as potential drug targets

Franziska Preuss[a,b], Deep Chatterjee[a,b], Verena Dederer[a,b], Stefan Knapp[a,b,*], and Sebastian Mathea[a,b]

[a]Structural Genomics Consortium, Buchmann Institute for Molecular Life Sciences, Frankfurt am Main, Germany
[b]Institute for Pharmaceutical Chemistry, Johann Wolfgang Goethe-University, Frankfurt am Main, Germany
*Corresponding author: e-mail address: knapp@pharmchem.uni-frankfurt.de

Contents

Abstract

Pseudokinases play significant roles in disease development. Similar to active kinases, their cellular functions can be targeted pharmacologically. But notably, instead of inhibiting an enzymatic activity, drug-like molecules act by stabilizing distinct pseudokinase conformations, by interfering with protein interactions, or by inducing proteasomal degradation. Herein, we describe our approach of enabling particular pseudokinases as potential drug targets. The method starts with obtaining recombinant

Methods in Enzymology, Volume 667
ISSN 0076-6879
https://doi.org/10.1016/bs.mie.2022.03.050

proteins for assay development and for biochemical evaluation. The next step is to probe the pseudoactive site as a binding pocket for small molecules, providing initial insight into binding modes and even candidate chemotypes. Finally, structural features of pseudokinase:inhibitor complexes are explored. Taken together, we provide detailed method descriptions for essential inhibitor development technologies.

1. Introduction

The development of small-molecule binders for pseudokinases is a worthwhile endeavor both for understanding the physiological role of a given pseudokinase (Mace & Murphy, 2021) and for validating pseudokinases as potential therapeutic targets (Kung & Jura, 2019). In the cell, specific binders may act as inhibitors, as molecular glues or as PROTAC handles. But developing such molecules is a difficult and time-consuming process that usually requires the availability of the purified target protein for biophysical studies, activity assays and structural studies. With the protocol described herein, we have had a high success rate for obtaining recombinant pseudokinases (Preuss et al., 2020; Sheetz et al., 2020). The protocol takes advantage of publicly available computational resources but also includes a trial-and-error component in order to achieve robustness and broad applicability.

2. Construct design, cloning and test purification

Designing suitable expression constructs is the first step in producing recombinant proteins. Here, we focus primarily on generating cytosolic pseudokinase domains. The description can be adapted to other proteins including secreted or integral membrane proteins. However, these might need more tailored expression constructs. Once cloning has been completed, a small-scale test purification is performed to determine construct stability before the expression is scaled-up (Fig. 1).

2.1 Construct design

The available information on predicted secondary structure elements from the PSIPRED server (Jones, 1999) and predicted tertiary structure from the AlphaFold server (Jumper et al., 2021) is collected. With this information, N- and C-terminal boundaries of the expression constructs are chosen. The boundaries are preferably located in flexible loop regions because disruption of secondary structure elements often leads to destabilization of recombinant

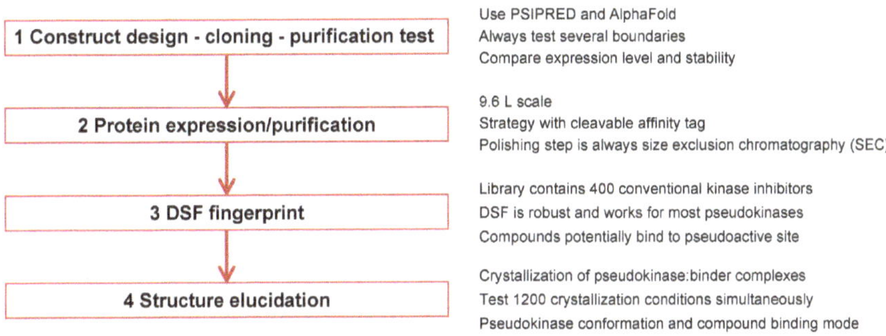

Fig. 1 Our workflow to enable pseudokinases as potential drug targets. It ranges from making recombinant protein fragments to solving pseudokinase:inhibitor structures.

proteins. In our experience, predicting expression level and stability of a construct is difficult. Therefore, we apply a trial-and-error strategy and recommend exploring at least three to four boundaries at the N- as well as the C-terminus (Fig. 2A).

2.2 Ligation-independent cloning (LIC)

A quick and efficient method to obtain various expression constructs in a medium-throughput manner is provided by ligation-independent cloning (LIC) (Strain-Damerell, Mahajan, Fernandez-Cid, Gileadi, & Burgess-Brown, 2021). LIC cloning makes use of the $3' \rightarrow 5'$ exonuclease activity of T4 DNA polymerase to create complementary overhangs between the linearized vector and the PCR-amplified inserts. After annealing of the complementary fragments, DNA nicks are repaired by transformation into *E. coli*. This procedure bypasses limitations of conventional cloning using restriction enzymes such as internal restriction sites in the gene of interest.

2.3 Step-by-step protocol

2.3.1 Cloning into the expression vector pFB-6HZB (Strain-Damerell et al., 2021)

1. Amplify the gene of interest using primers which contain a 15 bp sequence complementary to the expression vector. These will later become the overhangs for annealing of insert and vector. We perform PCR amplification of multiple different targets in 96-well plates applying a touch-down protocol. Herculase II fusion DNA polymerase (Agilent #600679) allows for amplification of inserts between 200 and 15,000 bp simultaneously.

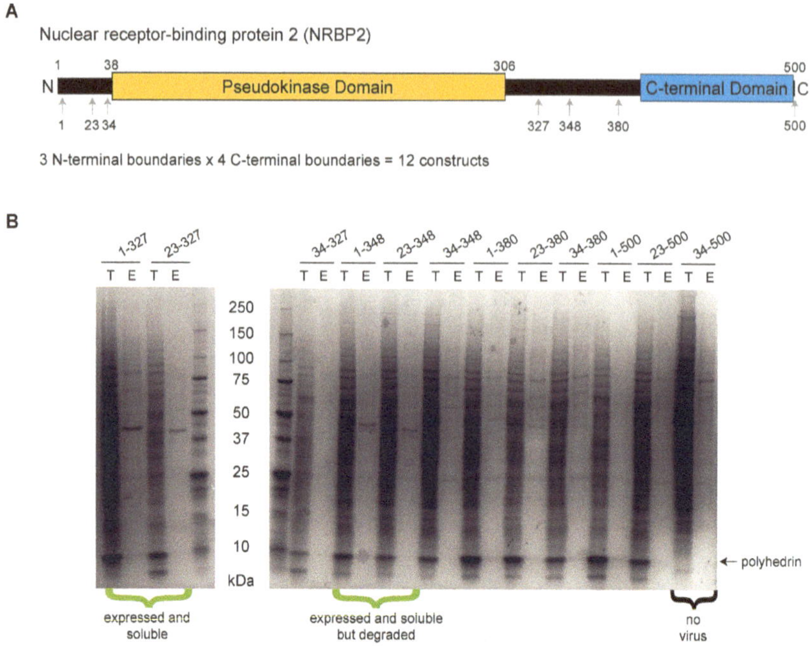

Fig. 2 Expression of the recombinant pseudokinase NRBP2. (A) Domain architecture of the full-length NRBP2. Arrows indicate 3 N- and 4 C-terminal boundaries to clone 12 different constructs for test purification. (B) After a one-step test purification, 4 of the 12 constructs show a Coomassie signal in the elution fraction 'E'. While the constructs 1–327 and 23–327 result in one single band, 1–348 and 23–348 reveal a double band indicating that these constructs were degraded at their C-termini. For construct 34–500 no virus was produced as can be seen by the lack of the polyhedrin band.

Reaction mixture composition:

Component	**Volume (μL)**
HERCII buffer (5 ×)	5
dNTP mix (10 mM)	1.5
HERCII DNA polymerase	0.25
Primer mix (5 μM each)	1.5
Template (2.5 ng/μL)	2.5
Water	14.25

Thermal cycling procedure:
95 °C, 10 min.
(95 °C, 30 s; 68 °C, 30 s; 68 °C, 2 min*) ×5.
(95 °C, 30 s; 60 °C, 30 s; 68 °C, 2 min*) ×5.
(95 °C, 30 s; 55 °C, 30 s; 68 °C, 2 min*) ×5.
(95 °C, 30 s; 50 °C, 30 s; 68 °C, 2 min*) ×20.
68 °C, 10 min.
*The time for amplification depends on the length of the longest PCR product according to the rule of thumb: length (bp)/1000 plus 1 min.

2. Treat the PCR reaction mixture with *Dpn*I (NEB, #R0176S) and incubate (1 h, 37 °C) to remove the template DNA.

Reaction mixture composition:

Component	Volume (μL)
PCR mixture	25
CutSmart (10 ×)	2.5
*Dpn*I (1 U/μL)	1

3. Purify the PCR products using MultiScreen 96-well PCR clean-up plates (Merck Millipore, #MSNU03010). Load the reaction mixtures to the wells of the filter plate and apply vacuum from the bottom to remove the liquid. Wash with 50 μL water. Finally, add 30 μL water, resuspend the PCR products by gentle shaking and transfer the purified PCR products to a new plate. Incubate with T4 DNA polymerase (NEB #M0203L) in presence of dCTP to create complementary sequence overhangs (first at 22 °C for 30 min, then at 75 °C for 30 min).

Reaction mixture composition:

Component	Volume (μL)
Purified PCR product	5
Water	2.15
NEB Buffer 2.1 (10 ×)	1
dCTP (25 mM)	1
BSA (10 mg/mL)	0.1
DTT (100 mM)	0.5
T4 DNA polymerase	0.25

4. Digest the vector pFB-6HZB by adding *Bse*RI (NEB # R0581L), incubate (2 h, 37 °C) and purify the linearized vector using a DNA clean-up kit (Invitrogen, #K310001) according to the manufacturer's instructions.

Reaction mixture composition:

Component	**Volume (μL)**
pFB-6HZB (57 ng/μL)	88
CutSmart Buffer (10 ×)	10
*Bse*RI	2

5. Incubate the linearized vector pFB-6HZB with T4 DNA polymerase (NEB #M0203L) in presence of dGTP to create complementary sequence overhangs (first at 22 °C for 30 min, then at 75 °C for 30 min).

Reaction mixture composition:

Component	**Volume (μL)**
Linearized pFB-6HZB vector	50
Water	21.5
NEB Buffer 2.1 (10 ×)	10
dGTP (25 mM)	10
BSA (10 mg/mL)	1
DTT (100 mM)	5
T4 DNA polymerase	2.5

6. Anneal vector and insert at room temperature. In our experience a vector-to-insert volume ratio of 1:4 in 10 μL total volume yields the highest number of colonies. However, if this fails, we recommend varying the ratio from 1:4, 3:2, 2:3 to 4:1 while decreasing the sample volume to 5 μL.
7. Transform the annealed plasmid into *E. coli* for nick repair, selection and propagation. Add 5 μL of annealing mixture to 40 μL of competent Mach1 *E. coli* cells on ice. After incubation on ice for 20 min, perform a heat shock (42 °C, 45 s). Incubate for another 5 min on ice, then add 80 μL of LB medium and incubate (37 °C, 1 h). Plate the mixture on

LB agar plates (25 g/L LB medium, 15 g/L agar, 100 μg/mL ampicillin, 5% sucrose) and incubate (overnight, 37 °C). The addition of sucrose allows for selection of clones that contain successfully generated plasmids over undigested or re-annealed vector that still contains the SacB gene.

8. Inoculate 1 mL LB medium (with 100 μg/mL ampicillin) with a single colony each in a 96-well block and incubate under shaking (overnight, 37 °C). Perform plasmid miniprep (Merck Millipore #LSKP09624) according to the manufacturer's instructions. Validate successful cloning by PCRs with a primer pair that anneals upstream and downstream of the insert in the vector backbone as described above and analyze the lengths of the PCR products by agarose gel electrophoresis.

2.3.2 Transposition and bacmid isolation

Next, the expression cassette is transposed from the donor plasmid into an engineered bacmid suitable for insect cell transfection and virus generation (Bac-to-bac system) (Luckow, Lee, Barry, & Olins, 1993).

9. Transform validated donor plasmids encoding the pseudokinase of interest into chemically competent DH10Bac *E. coli* cells as described above. After adding the LB medium, let cells rest (5 h, 37 °C) before plating on Spezial LB agar plates (25 g/L LB medium, 15 g/L agar, 50 μg/mL kanamycin, 7 μg/mL gentamycin, 10 μg/mL tetracycline, 100 μg/mL Bluo-gal, and 40 μg/mL IPTG). The resulting colonies will appear either blue or white. Blue colonies still contain the parental empty bacmid, while white colonies indicate complete transposition. Incubation for 40 h at 37 °C allows for more convenient identification of white colonies.

10. Pick the whitest colony, streak on a new Spezial LB agar plate and incubate (24 h, 37 °C) to verify that there is no parental empty bacmid left.

11. Inoculate 5 mL LB medium (containing 50 μg/mL kanamycin to keep selection pressure) from a white colony and incubate under shaking (overnight, 37 °C). Isolate bacmid DNA from 4 mL of this culture. Common miniprep kits are not suitable for bacmid preparation due to the length of the bacmids (~132 kb). However, we use components of a commercially available plasmid purification kit (Merck #PLN70-1KT) to perform the alkaline cell lysis. After centrifugation of the lysate (10 min, 20,000 × *g*, 4 °C) mix the supernatant with 0.1 volume 3 M sodium acetate pH 5.2 and precipitate the genomic DNA by adding 1 volume cold isopropanol and incubating on ice for 30 min. The bacmid DNA is pelleted by centrifugation (15 min, 20,000 × *g*, 4 °C).

12. The bacmid DNA is then washed with 1 volume 80% ethanol, dried and dissolved in 50 μL TE buffer (10 mM Tris pH 8.0, 1 mM EDTA). The concentration is assessed by measuring the absorbance at 260 nm using NanoDrop2000 (Thermo Scientific). It is usually between 2 and 4 μg/μL. Store bacmids at 4 °C until further use.

2.3.3 Virus production and test purification

For recombinant baculovirus production, the bacmids are transfected into adherent Sf9 cells (derived from *Spodoptera frugiperda*, ATCC #CRL-1711). The P0 baculovirus stock can then be used for infection of TriEx Sf9 suspension cells (Novagen #71023) to amplify the virus and to identify solubly expressed protein constructs (Fig. 2B). On average, TriEx cells express higher protein levels than Sf9 cells (own observations, no presentable statistics available), but they are difficult to be transfected. Therefore, we commonly use adherent Sf9 cells for the initial transfection and TriEx suspension cells for virus amplification and protein expression.

13. Detach exponentially growing adherent Sf9 cells by vigorous rinsing with culture medium (Insect XPRESS, Lonza #BELN12-730Q), count cells using a Neubauer chamber and dilute with fresh medium to 0.2×10^6 cells/mL. Seed cells to a 24-well plate (1 mL per well) and allow to attach to the bottom of the plate (no shaking, 16 h, 27 °C).
14. For transfection, mix 49 μL medium with 1 μL bacmid stock (corresponds to 3 μg DNA). Also, mix 46 μL medium with 4 μL Cellfectin II transfection reagent (Thermo Fisher #10362100). Finally, combine both mixes and incubate (15 min, RT) to allow for the formation of DNA-lipid particles.
15. Add the transfection mix dropwise to the respective well of the 24-well plate and incubate the plate (no shaking, 7 days, 27 °C). During this time, transfected cells will produce recombinant viruses that infect neighboring cells. The supernatant is referred to as the P0 virus stock.
16. For P1 virus stock production, dilute exponentially growing TriEx cells to 2×10^6 cells/mL and add 3 mL each to the wells of a 24-well block. Infect by adding 100 μL of the respective P0 virus stock and incubate (shaking as fast as possible without spilling the medium, 4 days, 27 °C).
17. Spin the block to separate the more dense cells from the virus suspension (20 min, $1000 \times g$, 4 °C). Transfer the suspension to a new 24-well block and store at 4 °C. It is referred to as the P1 virus stock. The virus titer might drop with time, but it will be high enough to generate fresh P2 stocks for years.

18. The P1 cell pellets from the block are used for an initial test purification. Resuspend the pellets in 2 mL lysis buffer by pipetting up and down (50 mM HEPES pH 7.4, 500 mM NaCl, 0.5 mM TCEP, 5% glycerol, 20 mM imidazole) and lyse cells by sonication on ice using a 24-tip horn (35% amplitude, 5 s pulse / 10 s pause, 3 min total pulse time). Transfer the lysates into 2-mL tubes and centrifuge to remove cell debris (30 min, 13,000 rpm, 4 °C).
19. In the meantime, prepare small gravity-flow columns (such as Pierce #29920) by adding 2 mL lysis buffer and 20 μL Ni-Sepharose beads (Cytiva #17531803). Equilibrate with 2 mL lysis buffer, load the cleared lysate by simple pouring it from the 2-mL tube to the equilibrated column and wash with 2 mL lysis buffer. Finally, elute the potentially bound pseudokinase by adding 80 μL elution buffer (lysis buffer with 300 mM imidazole).
20. Analyze the lysate (before centrifugation) and the elution by SDS-PAGE. For sample preparation, mix 5 μL loading dye and 10 μL of the respective sample. Load 10 μL per well to commercial gels (Bio-Rad #3450125) and perform electrophoresis according to the manufacturer's instructions. An example of one such Coomassie-stained gel is shown in Fig. 2B. A band in the elution lane indicates the soluble expression of the target protein.

3. Protein expression and purification

The production of proteins with the baculoviral expression system involves three steps (Mathea, Salah, & Knapp, 2019). The first step is the amplification of the recombinant virus stock. Once an amplified virus stock is generated, the large-scale expression of the target protein is performed. For that, exponentially growing insect cells in suspension are infected using the amplified virus stock and incubated with shaking. The recombinant virus takes over the cellular protein production machinery and overexpresses the desired protein. The final step is protein purification, which includes standard methods such as breaking of cells, separation of insoluble and soluble components of the lysate and finally isolation of the target protein using different chromatography techniques. To validate the quality of the protein we use four complementary methods. Using this approach, high-quality proteins are generated at amounts sufficient for the development of biophysical assays and structure elucidation.

3.1 Advantages of the baculoviral expression system over other expression systems

Recombinant proteins can be generated using several expression systems which differ in yield, success, rate and costs (Table 1). When targeting a so far unexplored pseudokinase, we first test whether it can be solubly expressed in *E. coli*. However, the success rate of this approach is low for most pseudokinase targets. We then switch to the baculoviral expression system. Advantages of this system are low cost, versatility and assisted protein folding due to its close resemblance with mammalian expression system. Disadvantages are that the glycosylation pattern and the subcellular localization differ from the mammalian system. However, this is not a major concern when expressing cytosolic pseudokinase domains. The limiting factors

Table 1 Suitability of expression systems for pseudokinase domain production.

Expression system	Advantages	Disadvantages
E. coli	Fast—low cost—high yield	Many human pseudokinases not solubly expressed
Baculovirus in insect cells	Modular workflow—moderate yield—medium cost	Intracellular protein processing might differ from human cells
Stably transfected insect cells	Ideal for secreted proteins Easily scalable	Requires clonal selection
BacMam in HEK cells	Based on Baculovirus system Expression of protein complexes	High biosafety level High cost
Transiently transfected HEK cells	Versatile—fast	Requires huge amounts of DNA Very low yield
Stably transfected HEK cells	Ideal for secreted proteins Easily scalable	Requires clonal selection Low yield
Adenovirus in human cells	Modular workflow Applicable in broad host cell range	High biosafety level High cost

for our follow-up experiments are always the yield and purity of the protein, where insect cells perform substantially better compared to mammalian cells.

3.2 Step-by-step protocol

3.2.1 Virus amplification

1. Start with exponentially growing TriEx suspension cells and determine cell density using a Neubauer counting chamber. Dilute with Insect-XPRESS medium (Lonza #BELN12-730Q) to a density of 2×10^6 cells/mL. Inoculate 250 mL of the cell suspension in a 3 L-Erlenmeyer glass flask with 50 µL P1 virus stock of the selected targets and incubate under shaking (96 h, 90 rpm, 27 °C).
2. Harvest by centrifugation (20 min, 1000 × *g*, 4 °C). The supernatant is the P2 virus suspension. Look out for the three signs of a successful virus preparation: the supernatant is a bit turbid, the cell pellet is smaller compared to the not infected control and the color of the cell pellet is more grayish. The virus titer is not routinely determined because both the expression yield and the reproducibility are usually satisfying. Virus suspensions are stored in 50-mL tubes at 4 °C for further use.

3.2.2 Large-scale protein expression

3. The TriEx suspension cells are scaled up, counted using the Neubauer chamber and diluted to a density of 2×10^6 cells/mL. 12 × 800 mL cell suspension in 3-L glass flasks are inoculated with the P2 virus suspension (12.5 mL virus suspension / 800 mL cell suspension) and incubated under shaking (66 h, 90 rpm, 27 °C) to allow for protein expression.
4. Cells are harvested by centrifugation in 1-L buckets (20 min, 1000 × *g*, 4 °C) and the pellets are stored at −20 °C.

3.2.3 Protein purification

5. Resuspend the harvested pellet in lysis buffer (50 mM HEPES pH 7.4, 500 mM NaCl, 20 mM imidazole, 0.5 mM TCEP, 5% glycerol) by vigorous vortexing. The pellets from 9.6 L culture will result in 300 mL cell suspension. All purification steps are performed at 4 °C.
6. Lyse cells by sonicating on ice using a 13-mm probe (35% amplitude, 5 s pulse / 10 s pause, 3 min total pulse time).
7. Clear lysate by centrifugation (60 min, 100,000 × *g*, 4 °C).
8. Pass supernatant through previously equilibrated Ni-Sepharose beads (Cytiva #17531803, 10 mL beads per 250 mL supernatant) in two gravity flow columns (Bio-Rad #7372512).

9. Wash beads with 30 bead volumes lysis buffer. Speed up the flow by applying some pressure with a syringe.
10. Elute the protein using 8 × 10 mL (40 mL per column) lysis buffer with imidazole (50 mM HEPES pH 7.4, 500 mM NaCl, 300 mM imidazole, 0.5 mM TCEP, 5% glycerol).
11. Dilute the eluted protein with half the volume of buffer containing no NaCl (20 mM HEPES pH 7.4, 0.5 mM TCEP, 5% glycerol). The resulting volume is 125 mL. The aim is reducing the ionic strength to prepare the solution for the ion exchange column.
12. Load the diluted protein onto a 5-mL SP-Sepharose column (Cytiva #17505701) connected to an ÄKTA Basic system.
13. Wash with buffer containing 250 mM NaCl (20 mM HEPES pH 7.4, 250 mM NaCl, 0.5 mM TCEP, 5% glycerol) until the absorbance of the flowthrough is below 5 mAU and elute the pseudokinase by applying a NaCl gradient (100 mL, 250–2.5 mM, 3 mL/min).
14. The elution fractions are collected (fraction size 4 mL) and analyzed by SDS PAGE. The protein-containing fractions are combined and treated with in-house prepared TEV protease overnight (molar ratio TEV to substrate about 1:50) to cleave the purification tag off the pseudokinase.
15. To remove contaminating proteins, potentially uncleaved target protein and TEV protease, a rebinding strategy is applied. Two 5-mL columns are serially combined to achieve optimal purification: First an SP-Sepharose column, then a Ni-Sepharose column. Dilute the protein solution with an equal amount of buffer without NaCl to reduce the ionic strength (20 mM HEPES pH 7.4, 0.5 mM TCEP, 5% glycerol). Then load onto the combined columns with an ÄKTA Basic system at a flowrate of 3 mL/min. Wash with equilibration buffer (20 mM HEPES pH 7.4, 250 mM NaCl, 0.5 mM TCEP, 5% glycerol) until the absorbance of the flowthrough is below 2 mAU (baseline). Disconnect the SP-Sepharose column. Elute unspecifically bound target protein from the Ni-Sepharose column with elution buffer (20 mM HEPES pH 7.4, 250 mM NaCl, 24 mM imidazole, 0.5 mM TCEP, 5% glycerol). During the entire rebinding procedure, collect the flowthrough in 4-mL fractions (also when washing and eluting). Analyze the fractions by SDS-PAGE and combine the fractions that contain the cleaved pseudokinase protein.
16. Concentrate the protein to a volume of 5.2 mL using Amicon Ultra 15 mL Centrifugal Filter Units 10,000 MWCO (Millipore #UFC901024).

Subject this concentrated solution to gel filtration using an ÄKTA Pure system combined with a HiLoad 16/600 Superdex 200 pg gel filtration column (20 mM HEPES pH 7.4, 150 mM NaCl, 0.5 mM TCEP, 5% glycerol) with collecting the flowthrough in 3-mL fractions.

17. Analyze the fractions by SDS-PAGE (percentage of the gels depending on the MW of the target protein) and combine the protein-containing fractions.
18. Finally, concentrate the protein to 2–3 mg/mL as calculated from the absorbance at 280 nm (NanoDrop 2000, Thermo Scientific), flash-freeze in liquid nitrogen and store at −80 °C until further use.

In case the target pseudokinase turns out to be prone to aggregation or degradation, its stability during purification can be improved by increasing the salt concentration or by adding binders or protease inhibitors.

3.3 Protein quality control

We have standardized the quality control (QC) of our recombinant proteins. The applied methods are simple and quick, so they can easily be performed after every protein preparation. The parameters purity, identity, post-translational modifications, degradation, oligomerization state, concentration and stability are analyzed allowing convenient comparison of individual preparations. The QC procedure includes four complementary methods: UV spectroscopy, SDS-PAGE, SEC and ESI mass spectrometry. As an example, the QC documentation of a preparation of the EPHB6 intracellular region is shown in Fig. 3.

4. DSF fingerprint of pseudokinases

Differential scanning fluorimetry (DSF) is a quick and robust *in vitro* screening method to analyze the impact of small molecules on the thermostability of a given protein. In principle, the protein of interest is mixed with a fluorescent dye and an excess of the small molecule in a 96-well plate. The sample is then gradually heated up in a qPCR machine, and denaturation of the protein is monitored *via* the fluorescence of the dye. An interaction between the small molecule and the protein leads to a shift in the melting temperature (T_M) of the protein. For pseudokinases, DSF is an invaluable tool to determine the accessibility of the pseudoactive site for small molecules and to identify potential starting points for drug discovery. Furthermore, DSF has been extensively used to test whether or not

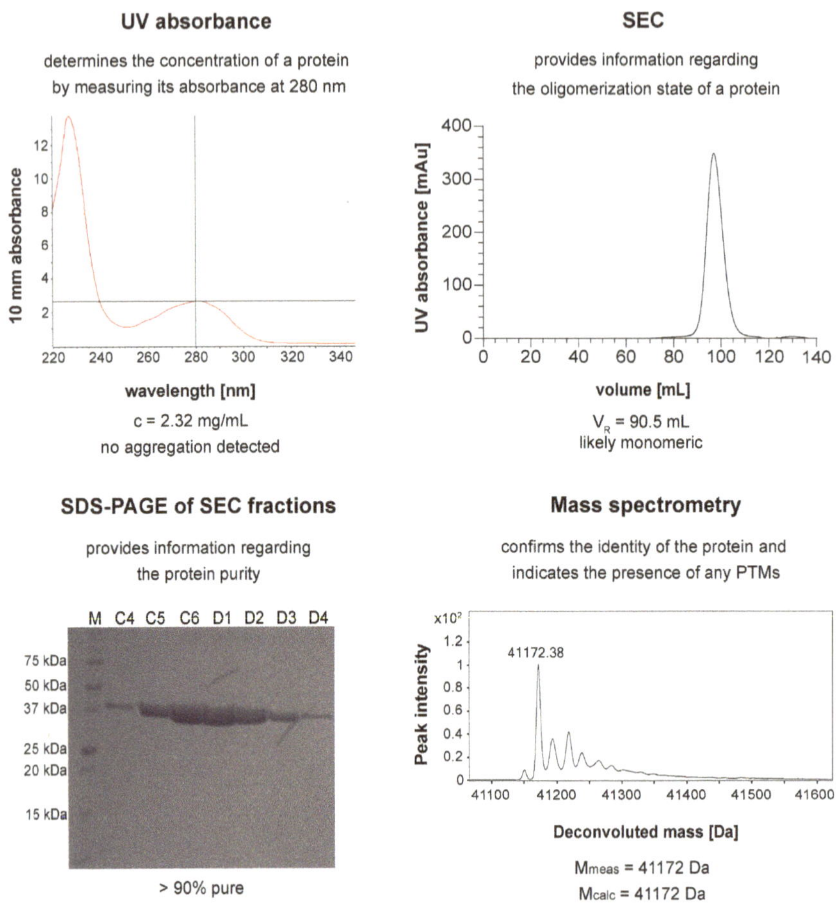

Fig. 3 Quality control of a preparation of the EPHB6 intracellular region showing the UV spectrum, SDS-PAGE, SEC profile and ESI-MS spectrum.

pseudokinases bind to nucleotides (Murphy et al., 2014). Here we describe a method for the screening of small-molecule libraries against recombinant pseudokinases. This is an adaption from a previously published protocol (Niesen, Berglund, & Vedadi, 2007).

4.1 Temperature stabilization is not the same as affinity

Besides the low cost and time requirements, advantages of DSF screening are its robustness and the high reproducibility of the results. Spurious and misleading binding events are precluded by neither the target protein nor the potential ligand bearing any tags. In addition to affinity, parameters such as buffer conditions, binding mode of the compound and stability of the apo protein determine to what extent a pseudokinase is stabilized by a given small molecule. Thus, shifts in stability (ΔT_M) may only be compared within a set of similar compounds such as an SAR series measured under the same experimental conditions. Fluorescent compounds require careful evaluation and should be avoided if possible, as their fluorescence may interfere with the measurement or quench the fluorescent dye. We generally only consider thermal shifts of 2 °C or more as hits to reduce the rate of false positive hits. The increase of thermal stability and the affinity of a compound generally correlate, with higher ΔT_M corresponding to tighter binding of the compound, however, both terms are not equivalent. For determination of the binding affinity or further characterization of the binding event, orthogonal assays such as isothermal titration calorimetry (ITC) and surface-plasmon resonance (SPR) should be applied.

4.2 Compound libraries

For *in vitro* screening of compounds, several commercial libraries are available comprising hundreds or thousands of compounds each. Due to the similarity of pseudokinases with active kinases, we first screen against a set of kinase inhibitors. Here we chose the L1200 library (SelleckChem), that contains a diverse set of around 400 kinase inhibitors that mainly bind to the active site of their target kinases. Though inhibitors targeting Aurora A, B-RAF, EGFR or MEK kinases are overrepresented due to the bias in kinase inhibitor development toward a small set of kinases (Fedorov, Müller, & Knapp, 2010), this library offers us an insight into the accessibility of the active site and provides an inhibitor fingerprint of the pseudokinase. Additionally, identified hits provide tools for co-crystallization experiments, because they may stabilize the pseudokinase in a more compact conformation. If the hit rate is low, we use other libraries such as the FDA-approved drugs library (L1300 library, SelleckChem) or the pharmacologically active compounds library (LOPAC library, SigmaAldrich) for a broader screening. Fragment screening approaches enable the identification of starting points for chemical synthesis. Fragments are small molecules with a MW between 100 and 250 Da and are expected to bind the target protein with only moderate affinity. Therefore, for fragment screening a much higher compound concentration in the assay is recommended. The experiment can be performed similar to how it is described below but at a much higher compound concentration (500 μM fragment in comparison to 10 μM small molecule).

4.3 Step-by-step protocol

The method described here is performed in 96-well plates. This allows for convenient manual pipetting using multichannel pipettes and a quick evaluation of hundreds of compounds. For the screening of larger compound libraries, however, we recommend the use of 384-well plates in order to save time (dispensing via robotics) and resources (using less of the solutions and consumables). In both cases, every screening plate should have at least 4 wells compound-free, where only DMSO is added as reference. Usually, there are no edge effects, so all wells of the plates can be used.

1. Program the Mx3005P real-time PCR instrument (Stratagene) with excitation and emission filters set to 465 and 590 nm, respectively, as follows (Mathea et al., 2021):
   Temperature range: 25–96 °C
   Number of cycles: 71 (1 K/cycle)
   Temperature increase rate: 3 K/min
2. Thaw the 96-well plates containing the compound library (pre-diluted in DMSO to 500 μM and stored at −20 °C) and spin down for 1 min at 1000 rpm.
3. Thaw the pseudokinase of interest quickly in your hand to minimize microencapsulation shear stress.
4. In a 2-mL Eppendorf tube prepare 2 mL of a 2 μM protein solution in ice-cold DSF buffer (20 mM HEPES pH 7.4, 150 mM NaCl, 0.5 mM TCEP, 5% glycerol).
5. Add 2 μL of SYPRO orange fluorescent dye (Invitrogen #S6650).
6. Mix the solution by gently inverting the Eppendorf tube three to five times, then put it on ice.
7. Distribute the protein solution in 20 μL aliquots into the wells of a white 96-well plate. Make sure the protein solution is at the bottom of the wells.
8. Using a multichannel pipette, add 0.4 μL of the 500 μM compound DMSO stock into the corresponding wells of the sample plate to obtain a final compound concentration of 10 μM. Mix by pipetting up and down 10 times.
9. When all compounds and the DMSO references are added, seal the plate using a transparent qPCR seal and spin down briefly (1 min, 1000 rpm, 4 °C) to collect the mixture at the bottom of each well.
10. Insert the plate into the qPCR machine and start the measurement.

4.4 Data evaluation

An exemplary data evaluation is outlined in Fig. 4, showing screening results of the pseudokinase ROR1 against two inhibitor libraries containing kinase inhibitors and FDA-approved drugs (L1200 and L1300, SelleckChem) (Sheetz et al., 2020). We use the Stratagene MxPro software to determine the individual melting temperatures, then Microsoft Excel for the calculation of the thermal shifts and GraphPad Prism for the visualization of the results. The raw result of a DSF measurement is the melting curve of the protein plotted as fluorescence against the temperature. For the reliable determination of the melting temperature (T_M), we plot the negative first derivative of the melting curve, with the minimum corresponding to T_M (Fig. 4A). The thermal shift for each compound is then calculated as followed: $\Delta T_M = T_M$(compound)—T_M(DMSO). Once all thermal shifts have been calculated, the results can be displayed as shown in Fig. 4B. A threshold of $\Delta T_M = \pm 2K$ has been set to reduce the number of false-positive hits.

5. Crystallization and structure elucidation

Obtaining a 3D model of a pseudokinase:inhibitor complex by X-ray crystallography is a crucial step in rational drug design. Though the binding of small molecules can be generally assessed with biophysical assays, only an experimental structure model proves the existence of a defined 1+1

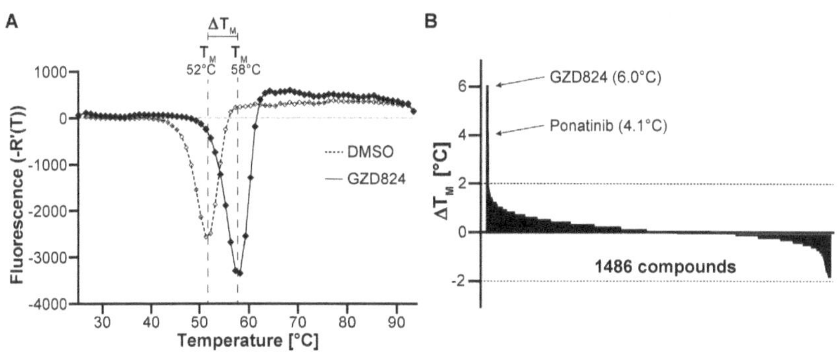

Fig. 4 Example of data evaluation for DSF screening of pseudokinase ROR1. (A) Negative first derivatives of the melting curves of the DMSO control (dashed line) and the hit compound GZD824 (solid line). (B) All T_M shifts measured for the compound libraries L1200 and L1300 sorted by value from highest to lowest.

complex formed by the pseudokinase and the small-molecule binder. Furthermore, a structure model confirms the existence and accessibility of an inhibitor binding pocket. It is possible to chemically modify the bound molecule to develop more specific binders. However, the chemotypes in the L1200 library are well explored and often promiscuous. As a side effect, the established crystallization system might be suitable for crystallographic fragment screening. This approach is more promising for the identification of chemical starting points than repurposing existing kinase inhibitor scaffolds (Douangamath et al., 2020).

5.1 Four coarse screens to crystallize them all

The success of crystallization trials depends largely on the composition of the tested precipitant solutions. To our knowledge, there is no reliable algorithm to predict suitable crystallization conditions, so we usually apply a trial-and-error approach. Certain conditions with a high probability of crystal formation have been identified in the past decades. These conditions have been combined in commercially available coarse screens. We have a high success rate using the JCSG, HIN, HCS and LFS screens (Molecular Dimensions). Testing 4 × 96 conditions in every crystallization trial is a compromise between covering as much as possible "precipitant space" and not wasting too many resources. It is important to note that within these screens, there is some redundancy, but also a lot of diversity regarding precipitants and additives.

5.2 Step-by-step protocol

1. Start with 1 mL of the frozen pseudokinase stock solution (2–3 mg/mL) and thaw at room temperature.
2. Add 10 μL of the respective compound solution (50 mM in DMSO), mix by gentle agitation, and incubate for 5 min on ice. Centrifuge to remove any insoluble compound (10 min, 13,000 rpm, 4 °C).
3. Concentrate the pseudokinase:inhibitor complex to 10–15 mg/mL using Amicon Ultra 4 mL Centrifugal Filter Units (Millipore #UFC8030) in a benchtop centrifuge (at either 4 or 20 °C, depending on the protein, 4000 rpm, in 5 min steps).
4. Prepare four 96-well 3-drop crystallization plates (Swissci) by adding 20 μL precipitant solution to each reservoir well. The four coarse screens described above contain 384 different precipitant solutions.

5. Use a mosquito Xtal3 liquid handler (SPT Labtech) to prepare crystallization drops. Mix the pseudokinase:inhibitor solution and the precipitant solution in the ratios 1 : 2, 1 : 1 and 2 : 1 (total volume 200 nL, the mosquito needs to be programmed accordingly) and seal the plates with transparent pressure seals. In the end, there will be 1152 crystallization drops.
6. Incubate the plates at 20 °C (repeat and try 4 °C in the case that no crystals are obtained). Inspect the drops under the inverted light microscope after 1 week and 4 weeks, respectively, for crystal growth.
7. In case the obtained crystals are too small or layered or of otherwise low quality, the crystallization experiment should be repeated with an optimized screen with a finely tuned matrix of precipitant concentration and pH around the original condition. Other options to improve crystal quality include testing different concentrations of the pseudokinase: inhibitor complex and applying seeding methods.
8. With suitable crystals, the 3D structure of the pseudokinase:inhibitor complex can be determined according to well-established procedures, starting with collecting diffraction data at a synchrotron (Wojdyla et al., 2018).

6. Concluding remarks

An intrinsic obstacle when working on pseudokinases is their lack of catalytic activity. As a consequence, no activity assays can be applied for inhibitor evaluation. It is therefore of particular importance to understand the physiological role of the pseudokinase, in order to develop functional cell assays for inhibitor evaluation. Starting points are often obtained from screening fragment libraries or from repurposing existing kinase binders (Degorce et al., 2020). Developing these starting points usually includes several design-make-test cycles and the close collaboration of medicinal chemists, biochemists and structural biologists. With the recombinant protein in place, the DSF assay established and ideally some structural information on pseudokinase:inhibitor complexes, the scene is set for the development of specific drug-like molecules.

Acknowledgments

The authors are grateful for support by the SGC, a registered charity (no. 1097737) that receives funds from; AbbVie, BayerAG, Boehringer Ingelheim, the Canada Foundation for Innovation, Eshelman Institute for Innovation, Genentech, Genome Canada through

Ontario Genomics Institute [OGI-196], EU/EFPIA/OICR/McGill/KTH/Diamond, Innovative Medicines Initiative 2 Joint Undertaking [EUbOPEN grant 875510], Janssen, Merck KGaA, Merck & Co, Pfizer and Takeda. S.K. is also grateful for support by the German translational cancer network (DKTK) as well as the Frankfurt Cancer Institute (FCI).

References

Degorce, S. L., Tavana, O., Banks, E., Crafter, C., Gingipalli, L., Kouvchinov, D., et al. (2020). Discovery of proteolysis-targeting chimera molecules that selectively degrade the IRAK3 pseudokinase. *Journal of Medicinal Chemistry*, *63*(18), 10460–10473. https://doi.org/10.1021/acs.jmedchem.0c01125.

Douangamath, A., Fearon, D., Gehrtz, P., Krojer, T., Lukacik, P., Owen, C. D., et al. (2020). Crystallographic and electrophilic fragment screening of the SARS-CoV-2 main protease. *Nature Communications*, *11*(1), 5047. https://doi.org/10.1038/s41467-020-18709-w.

Fedorov, O., Müller, S., & Knapp, S. (2010). The (un)targeted cancer kinome. *Nature Chemical Biology*, *6*(3), 166–169. https://doi.org/10.1038/nchembio.297.

Jones, D. T. (1999). Protein secondary structure prediction based on position-specific scoring matrices. *Journal of Molecular Biology*, *292*(2), 195–202. https://doi.org/10.1006/jmbi.1999.3091.

Jumper, J., Evans, R., Pritzel, A., Green, T., Figurnov, M., Ronneberger, O., et al. (2021). Highly accurate protein structure prediction with AlphaFold. *Nature*, *596*(7873), 583–589. https://doi.org/10.1038/s41586-021-03819-2.

Kung, J. E., & Jura, N. (2019). Prospects for pharmacological targeting of pseudokinases. *Nature Reviews. Drug Discovery*, *18*(7), 501–526. https://doi.org/10.1038/s41573-019-0018-3.

Luckow, V. A., Lee, S. C., Barry, G. F., & Olins, P. O. (1993). Efficient generation of infectious recombinant baculoviruses by site-specific transposon-mediated insertion of foreign genes into a baculovirus genome propagated in Escherichia coli. *Journal of Virology*, *67*(8), 4566–4579. https://doi.org/10.1128/JVI.67.8.4566-4579.1993.

Mace, P. D., & Murphy, J. M. (2021). There's more to death than life: Noncatalytic functions in kinase and pseudokinase signaling. *The Journal of Biological Chemistry*, *296*, 100705. https://doi.org/10.1016/j.jbc.2021.100705.

Mathea, S., Salah, E., & Knapp, S. (2019). High-throughput purification of protein kinases from Escherichia coli and insect cells. *Methods in Molecular Biology (Clifton, N.J.)*, *2025*, 191–202. https://doi.org/10.1007/978-1-4939-9624-7_8.

Mathea, S., Salah, E., Tallant, C., Chatterjee, D., Berger, B.-T., Konietzny, R., et al. (2021). Conformational plasticity of the ULK3 kinase domain. *The Biochemical Journal*, *478*(14), 2811–2823. https://doi.org/10.1042/BCJ20210257.

Murphy, J. M., Zhang, Q., Young, S. N., Reese, M. L., Bailey, F. P., Eyers, P. A., et al. (2014). A robust methodology to subclassify pseudokinases based on their nucleotide-binding properties. *The Biochemical Journal*, *457*(2), 323–334. https://doi.org/10.1042/BJ20131174.

Niesen, F. H., Berglund, H., & Vedadi, M. (2007). The use of differential scanning fluorimetry to detect ligand interactions that promote protein stability. *Nature Protocols*, *2*(9), 2212–2221. https://doi.org/10.1038/nprot.2007.321.

Preuss, F., Chatterjee, D., Mathea, S., Shrestha, S., St-Germain, J., Saha, M., et al. (2020). Nucleotide binding, evolutionary insights, and interaction Partners of the Pseudokinase Unc-51-like kinase 4. *Structure (London, England : 1993)*, *28*(11), 1184–1196.e6. https://doi.org/10.1016/j.str.2020.07.016.

Sheetz, J. B., Mathea, S., Karvonen, H., Malhotra, K., Chatterjee, D., Niininen, W., et al. (2020). Structural insights into Pseudokinase domains of receptor tyrosine kinases. *Molecular Cell*, *79*(3), 390–405.e7. https://doi.org/10.1016/j.molcel.2020.06.018.

Strain-Damerell, C., Mahajan, P., Fernandez-Cid, A., Gileadi, O., & Burgess-Brown, N. A. (2021). Screening and production of recombinant human proteins: Ligation-independent cloning. *Methods in Molecular Biology (Clifton, N.J.)*, *2199*, 23–43. https://doi.org/10.1007/978-1-0716-0892-0_3.

Wojdyla, J. A., Kaminski, J. W., Panepucci, E., Ebner, S., Wang, X., Gabadinho, J., et al. (2018). Da+ data acquisition and analysis software at the Swiss light source macromolecular crystallography beamlines. *Journal of Synchrotron Radiation*, *25*(Pt. 1), 293–303. https://doi.org/10.1107/S1600577517014503.

CHAPTER TWENTY-THREE

Identification and characterization of TYK2 pseudokinase domain stabilizers that allosterically inhibit TYK2 signaling

Gregory A. Locke*, Jodi Muckelbauer, John S. Tokarski, Christopher M. Barbieri, Stefan Belić, Bradley Falk, Jeffrey Tredup, and Ying-Kai Wang

Leads Discovery and Optimization, Mechanistic Pharmacology, Bristol Myers Squib, Lawrenceville, NJ, United States

*Corresponding author: e-mail address: gregory.locke@bms.com

Contents

Abstract

Kinase inhibition continues to be a major focus of pharmaceutical research and discovery due to the central role of these proteins in the regulation of cellular processes. One family of kinases of pharmacological interest, due to its role in activation of immunostimulatory

Methods in Enzymology, Volume 667
ISSN 0076-6879
https://doi.org/10.1016/bs.mie.2022.03.051

pathways, is the Janus kinase family. Small molecule inhibitors targeting the individual kinase proteins within this family have long been sought-after therapies. High sequence and structural similarity of the family members makes selective inhibitors difficult to identify but critical because of their inter-related multiple cellular regulatory pathways. Herein, we describe the identification of inhibitors of the important Janus kinase, TYK2, a regulator of type I interferon response. In addition, the biochemical and structural confirmation of the direct interaction of these small molecules with the TYK2 pseudokinase domain is described and a potential mechanism of allosteric regulation of TYK2 activity through stabilization of the pseudokinase domain is proposed.

1. Introduction

BMS has a long history of pursuing kinases as drug targets resulting in several candidates progressing to the clinic such as the BTK inhibitor BMS-986142, the approval of the tyrosine kinase Inhibitor (BCR/AbL) Dasatinib in 2010, and the TYK2 inhibitor deucravacitinib undergoing phase 3 clinical trials (Chaudhry et al., 2021; Gavai et al., 2009; Liu et al., 2013; Lombardo et al., 2004; Misra et al., 2015; Schroeder et al., 2009; Spergel et al., 2019; Wan et al., 2015; Watterson et al., 2019; Wrobleski et al., 2019).

The Janus family of kinases, which includes JAK1, JAK2, JAK3, and TYK2 have garnered significant interest within the pharmaceutical community (Borzilleri, Hart, Moslin, Tokarski, & Wrobleski, 2021; Ghoreschi, Laurence, & O'Shea, 2009; Virtanen, Haikarainen, Raivola, & Silvennoinen, 2019). Each of the JAK family members function as regulators of cytokine signaling in immuno-response pathways (Yamaoka et al., 2004). TYK2, in particular, has been linked to downstream receptor signaling of IL-23, IL-12, and the type I interferon (IFN) IFNα and IFNβ (Dhar & Dyckman, 2017). Upon activation, TYK2, along with either JAK1 or JAK2, participates in the phosphorylation of intracellular receptor domains, which then recruit specific STAT proteins, which are in turn phosphorylated. This results in their dissociation from the receptor and the formation of STAT dimers that subsequently translocate into the nucleus and trigger transcriptional processes, which lead to inflammatory responses (Fig. 1). Indeed, studies of the TYK2-deficient mice have demonstrated that TYK2 is essential for IL-12, IL-23 and type I IFN signaling (Dhar & Dyckman, 2017). In autoimmune diseases such as psoriasis and inflammatory bowel disease (IBD), IL-23 is known to be responsible for promoting T_H17 cell proliferation which has been associated with the pathogenesis

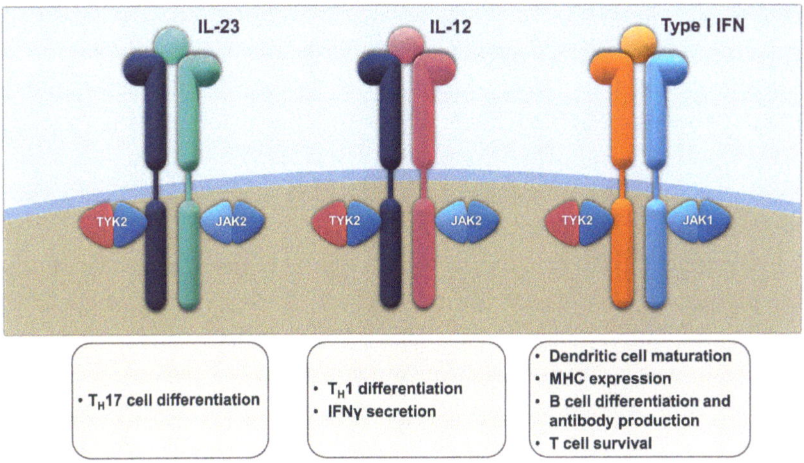

Fig. 1 Illustration of the JAK/STAT signaling pathway and effects for IL-23, IL-12 and Type I IFN signals.

of these diseases (Tang, Chen, Qian, & Huang, 2012). IL-12 on the other hand, is responsible for differentiation of T-cells to T_H1 and induction of IFNγ which is involved in stimulating innate and T-cell mediated immune responses (Liu et al., 2005). T_H1 overstimulation can lead to both tissue damage and chronic inflammation found in several auto-immune diseases (Dhar & Dyckman, 2017).

JAK-targeting inhibitors are important pharmacological treatments for autoimmune diseases such as psoriasis and inflammatory bowel disease as well as acute lymphoblastic leukemia and solid tumors (Flex et al., 2008; Jeong et al., 2008; Liu et al., 2021). Many current JAK inhibitors are non-selective and show strong activity across all Janus kinase family members (Borzilleri et al., 2021; Dhar & Dyckman, 2017). This broad inhibitory profile might provide additional efficacy due to the inter-related regulatory pathways that these kinases are known to modulate, but toxicity is an important concern. Like other disease-modifying antirheumatic therapies, treatment with non-selective JAK inhibitors is associated with an increased incidence of infection (Borzilleri et al., 2021). However, certain viral infections may be more prevalent in non-selective JAK-inhibitor treated individuals. (Curtis, Xie, Yun, Bernatsky, & Winthrop, 2016; Virtanen et al., 2019; Yamaoka et al., 2004). JAK 1/2/3 inhibitors are approved or are in development for the treatment of immune diseases (Borzilleri et al., 2021; Virtanen et al., 2019). However, these JAK inhibitors are associated with various non-immune-related adverse effects, including hematologic changes, hyperlipidemia, and renal and hepatic abnormalities, which may be due to the lack of selectivity of these inhibitors for specific family members (Chimalakonda et al., 2021; Shaw et al., 2003; Vincenti et al., 2012; Virtanen et al., 2019). TYK2 selective inhibition is sought, due to its role in regulation of IL-12, IL-23, and type I and III IFN receptor expression (Shaw et al., 2003). The therapeutic modulation of the type I IFN pathway has proven clinical relevance in the treatment of systemic lupus erythematosus (SLE) by targeting IFNα or its receptor using the biologic drugs sifalimumab and anifrolumab, respectively (Furie et al., 2017; Khamashta et al., 2016). Likewise, inhibition of IL-12 and IL-23 has demonstrated efficacy in plaque psoriasis, psoriatic arthritis, and inflammatory bowel disease (Kashani & Schwartz, 2019; Tang et al., 2012; Teng et al., 2015). Although the therapeutic need for selectivity within the Janus kinase family is clear, achieving selectivity has been difficult due to the high homology of their kinase domain catalytic ATP binding pockets. One approach to enable selective kinase targeting is to identify allosteric binding sites or domains that regulate kinase activity that can be exploited for inhibitor binding (Lu, Smaill, & Ding, 2020).

The Janus family of kinases are multidomain complex proteins consisting of four domains: FERM (JH7, JH6, JH5), SH2 (JH4, JH3), pseudokinase (JH2) and kinase (JH1) (Fig. 2).

The FERM domain is responsible for mediating interactions with cytokine receptors as well as the plasma membrane and binds the kinase domain, where it positively regulates catalytic activity (Yamaoka et al., 2004). It is tightly associated with the SH2 domain to form a single receptor-binding construct (Ferrao & Lupardus, 2017). At the carboxy terminus lies the catalytically active kinase domain along with a pseudokinase domain sandwiched between the SH2 and kinase domain. The presence of a catalytically inactive pseudokinase domain within the same polypeptide as an active kinase is a rare protein architecture (Yamaoka et al., 2004).

The pseudokinase domains are believed to play important regulatory roles in the kinase activity of JAK as demonstrated by various mutational studies (Gakovic, Ragimbeau, Francois, Constantinescu, & Pellegrini, 2008; Saharinen & Silvennoinen, 2002; Saharinen, Takaluoma, & Silvennoinen, 2000; Saharinen, Vihinen, & Silvennoinen, 2003; Yeh,

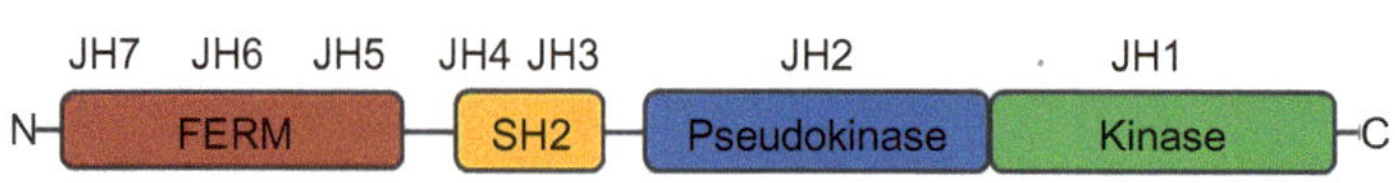

Fig. 2 Janus kinase domain structure.

Dondi, Uze, & Pellegrini, 2000). Also, the constitutive activation of JAK2 that is a major cause of the hematological disorder polycythemia vera, has been shown to be driven by the V617F pseudokinase domain mutation (Kralovics et al., 2005; Percy & McMullin, 2005; Staerk, Kallin, Demoulin, Vainchenker, & Constantinescu, 2005; Zhao, Vainchenker, Krantz, Casadevall, & Constantinescu, 2005; Zhao, Xing, et al., 2005). While small molecules that target the kinase domain of TYK2 can block the IL-12, IL-23, and type I IFN pathways offering the ability to treat a broad spectrum of inflammatory and autoimmune diseases, the challenge has been finding small molecules that are selective for TYK2 over the other Janus family kinases or other kinases in the human kinome (Wrobleski et al., 2019).

BMS was able to overcome this selectivity challenge by discovering potent, selective compounds that bind to and stabilize the pseudokinase domain of TYK2, which consequently prevent receptor-mediated activation of the TYK2-JH1 catalytic domain (Burke et al., 2019; Liu et al., 2019, 2021; Moslin et al., 2017, 2019; Tokarski et al., 2015; Wrobleski et al., 2019).

In the following sections, the strategies and methodologies used to enable the identification and characterization of these TYK2-JH2 pseudokinase small molecule allosteric inhibitors will be described.

2. Identification of TYK2-JH2 pseudokinase domain binders

Below, we describe the approach we used that involved combining kinome-wide inhibition profiles with cellular activity against IL-23-stimulated transcriptional responses to identify TYK2-JH2 and novel inhibitors involved in the IL-23 signal transduction pathway.

2.1 Compound selection

A collection of 21,851 compounds with broad kinase inhibitory profiles (obtained against a panel of 386 unique human kinase and pseudokinase domains at Ambit Biosciences) that were synthesized over the course of various kinase-targeted drug discovery efforts at Bristol Myers Squibb were assembled for this analysis.

Because kinase inhibitors with broad inhibition across the kinome would confound the interpretation of the analysis of kinases involved in IL-23 signaling, compounds were removed from this collection if they were promiscuous kinase binders (bound to more than 20 of 386 kinases in the Ambit kinase panel) or cytotoxic to cells (CDK2, CDK5 and CDK7). In addition, because JAK2 and TYK2 were known to be involved in IL-23R signal transduction, compounds from the collection were excluded if they were known inhibitors of the enzymatic activity of these JAK family kinases. This resulted in a library of 7908 compounds.

2.2 IL-23-stimulated transcriptional response

The library of 7908 compounds was then used to identify inhibitors of IL-23R signal transduction in cells. Hits showing >40% inhibition in the initial screen were confirmed by retesting in triplicate. Confirmed hits were then evaluated in concentration response format to calculate IC50 values. Detailed protocols for this assay can be found in Tokarski et al., 2015.

2.3 Association of cell activity with kinases

Using the results of the IL-23-stimulated transcription assay screened against the 7908-compound library, kinases in the 386-kinase panel were evaluated for enrichment. Kinase enrichment was determined using the "cell active rate" (CAR) defined as the percentage of compounds inhibiting a particular kinase by more than 90% that also inhibited the IL-23-stimulated assay. This allowed for the identification of kinases regulating cellular end points. Identifying inhibitors of a particular kinase with high CAR across the known inhibitors of that kinase provides a more confident approach to identifying kinases truly involved in the pathway. Three kinases had CAR greater than 10%: GSK3α (10.5%), IKKβ (12.2%) and TYK2-JH2 (15.6%) (Fig. 3). GSK3 kinases have been implicated in IL-23R signal transduction so this result was not surprising. IKKβ was not expected but subsequent examination of these inhibitors revealed that these compounds also bound TYK2-JH2 so the apparent CAR for IKKβ and TYK2-JH2 resulted from a common set of compounds that bound both proteins.

These studies demonstrated a novel association of compounds that inhibit the IL-23-stimulated assay with the JH2 pseudokinase domain of TYK2; the cellular activity of these compounds results from their action on the TYK2-JH2 pseudokinase domain.

2.4 Representative compounds identified

Two potent, selective chemotypes were identified from this approach (Fig. 4): Imidazothiazolopyridines (ITPs) (Tokarski et al., 2015) and

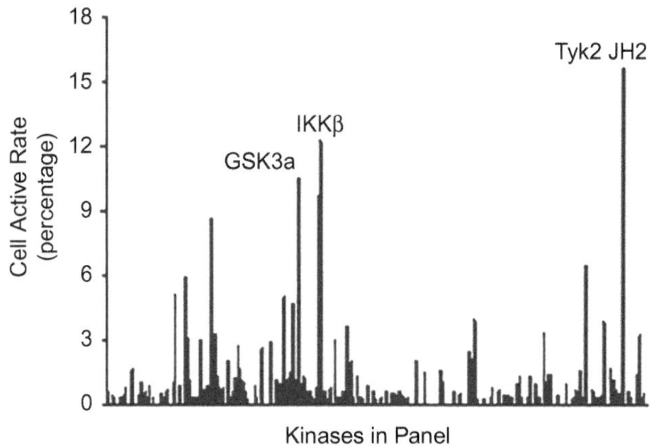

Fig. 3 Identification of Enriched Kinases. The graph shows the 3 kinases (GSK3α, IKKβ and TYK2-JH2) with Cell Active Rates greater than 10%.

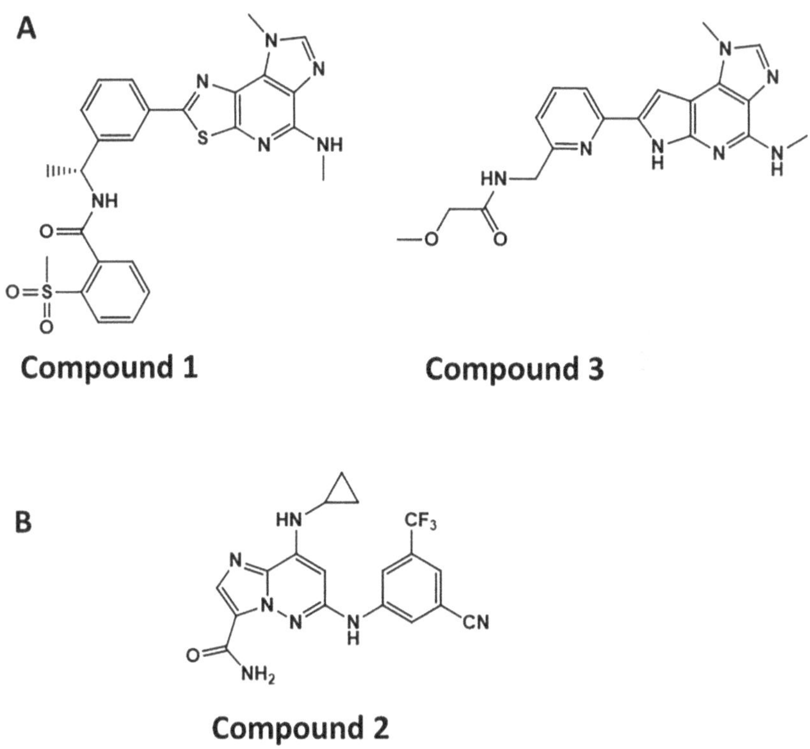

Fig. 4 Structures of representative compounds identified from the approach. (A) Imidazothiazolopyridines (ITPs); Compound 1 and compound 3. (B) Imidazopyridazines (IZPs); compound 2 (Moslin et al., 2017; Tokarski et al., 2015).

imidazopyridazines (IZPs) (Moslin et al., 2017). The ITP compound, Compound 1, had good potency against TYK2-JH2 (IC_{50} = 27 nM) (in a scintillation proximity assay (SPA) binding assay) and IKKβ (IC_{50} = 14 nM), was selective against JAK family kinase domains (IC_{50} ≥ 2 μL), and was active in the IL-23 (1 μM) and IFNα (0.97 μM) assays. The IZP compound, Compound 2, had moderate affinity (SPA) for TYK2-JH2 (IC_{50} = 480 nM) and in the IL23/IFNα assays (8.2/12.6 μM) but had good selectivity against the JAK family kinase domains (IC_{50} ≥ 50 μM) and IKKβ (IC_{50} ≥ 2 μM).

3. Characterization of TYK2-JH2 pseudokinase domain binders

3.1 Cloning, expression, and purification of TYK2-JH2 pseudokinase domain

To characterize the inhibitors that target the TYK2-JH2 pseudokinase domain, recombinant protein was produced for biophysical and structural studies. The human TYK2-JH2 pseudokinase domain (residues 575–869) was cloned for expression in a baculovirus expression system. For in-vitro assay development, the protein was purified using the standard chromatography techniques of Nickel affinity protein separation and size exclusion. Subsequently, the N-terminal hexahistidine affinity tag was removed for biophysical and crystallography studies and the protein was buffer exchanged for storage and analysis (Tokarski et al., 2015).

3.1.1 Equipment

PCR instrument

WAVE-Bioreactor system 20/50 (GE Healthcare).

Branson-550 tip sonifier.

Refrigerated centrifuges and appropriate rotors (see protocol).

AKTA Explorer-100 or FPLC system with UV detection.

1-mL and 5-mL HisTrap FF crude columns (GE Healthcare, now Cytiva)

Amicon Ultra-15 centrifuge filtration units (Millipore)

HiLoad Superdex 75 (16/60), (GE Healthcare)

HiPrep 26/10 desalting column (GE Healthcare)

UV/visible spectrophotometer

3.1.2 Materials

pFastBac1 vector (Invitrogen)

Bac-to-Bac baculovirus expression system (Invitrogen 10360014)

ESF921 insect medium (Expression Systems 96-001-10)

EDTA-free Protease inhibitor tablets (Roche Applied Science 118361 70001)
Benzonase (Sigma E8263-5KU)
His-tobacco vein mottling virus protease (His-TVMV protease)
TCEP (Sigma-Aldrich C4706-10G)
1 M sodium HEPES, pH 7.5—ThermoFisher, #15630080.
5 M Sodium Chloride (Sigma-Aldrich, S6546-1L)
Imidazole (Sigma-Aldrich 56,750-500G)
Glycerol-(Sigma-Aldrich, #G9012)
Triton X-100 (Sigma-Aldrich T9284-100ML)
1 M Manganese Chloride Solution (Sigma-Aldrich M1787-100ML)
0.5 M Sodium Phosphate Buffer, pH 7.20 (Teknova S5072)
DTT-Sigma-Aldrich, #D9779

3.1.3 Buffers

1. Buffer A: 50 mM sodium HEPES, 500 mM NaCl, 25 mM imidazole, 5% (*v*/v) glycerol, 0.1% (v/v) Triton X-100, and 0.5-mM TCEP at pH 7.7
2. Buffer B: 50 mM sodium HEPES, 500 mM NaCl, 350 mM Imidazole, 5% (v/v) glycerol, and 0.5 mM TCEP at pH 7.7
3. Buffer C: 50 mM sodium HEPES, 500 mM NaCl, 1 mM $MnCl_2$, 5% (v/v) glycerol, 0.5 mM TCEP at pH 7.7
4. Buffer D: 50 mM Na_2HPO_4, 500 mM NaCl, 5% (v/v) glycerol, and 5 mM DTT at pH 7.7
5. Buffer E: 10 mM Na_2HPO_4, 500 mM NaCl, 5% (v/v) glycerol, and 5 mM DTT at pH 7.7

3.1.4 Protocol

1. TYK2 pseudokinase domain sequence (residues 575–869; UniProtKB —P29597) was used as a template for PCR. PCR primers were designed to incorporate an *Nde* I restriction site at 5′ and a *Xho* I restriction site at 3′
2. Clone the PCR product into a modified pFastBac1 vector with an N-terminal His-TVMV (Tobacco Vein Mottling Virus protease recognition site) tag to generate His-TVMV-hTYK2 pseudokinase domain (residues 575–869)-pFB
3. Verify the DNA sequence of the PCR product
4. Generate baculovirus using the Bac-to-Bac baculovirus expression system (Invitrogen) according to the manufacturer's protocol
5. Amplify the virus in 100 mL of Sf9 cells in ESF921 insect medium seeded at 2×10^6 cells/mL in a 500 mL shake flask, then infect cells with

initial virus stock at a 1:100 virus/cell ratio and culture the cells for 66 h at 27 °C with shaking at 130 rpm

6. Carry out expression in Sf9 cells grown to a density of 2×10^6 cells/mL in ESF921 medium infect with 1:10 virus/cell ratio at 20-L scale in a 50-L Cellbag using the WAVE-Bioreactor system 20/50. Express for 65 h at 27 °C

   Harvest infected cells by centrifugation in a 2 L bottle in the RC 12BP centrifuge at 200 rpm (H-12,000 rotor, 1000 × *g* for 20 min, at 4 °C rotor, 1000 × *g*) for 20 min, at 4 °C.
7. Dissolve cell pellet in Buffer A (Section 3.1.3), supplemented with protease inhibitor (1 tablet per 50 mL Buffer A) tablets and Benzonase (5 μL Benzonase per 50 mL Buffer A).
8. Lyse cells in a 600 mL plastic beaker (Corning) by tip sonication (Branson Ultrasonics flat tip, 40 pulses, 100% power, cycling at 1.5″ on and 0.7″ off, with 5 min rests on ice repeated three times), and clarified in a 500 mL Nalgene polypropylene centrifuge bottle by centrifugation in the RC 6+ centrifuge at 9500 rpm (17,700 × *g*) in a F10S-6x500Y rotor for 30 min, at 4 °C
9. Purification steps are executed using an AKTA Explorer-100 system at 4 °C. The supernatant is applied to a 5-mL HisTrap FF crude nickel-affinity column, and washed until baseline with Buffer A, and then bump-off eluted with Buffer B (see Section 3.1.3) at 2.50 mL/min, 4 °C, and collected 3 mL fractions
10. Fractions containing the TYK2 SH2 domain are then concentrated to 10 mL with Amicon Ultra-15 centrifugal ultra-filtration units
11. Following concentration, the protein is further purified using size exclusion chromatography on a Hiload Superdex 75 column run in Buffer C (see Section 3.1.3) at 1.25 mL/min, 4 °C, and collected 3 mL fractions
12. For both the nickel-affinity and sizing columns, the His-tagged TYK2-JH2 containing fractions were identified by 4–20% Tris-Glycine SDS PAGE (Invitrogen) analysis and visualized by SimplyBlue Stain (Invitrogen).
13. Removal of N-terminal hexahistidine sequence accomplished by overnight cleavage at 4 °C, using His-TVMV protease at a 10:1 mass ratio of TYK2 to TVMV protease

3.1.5 Protein amino acid sequence prior to his-tag removal

MGSSHHHHHHSSG**ETVRFQ/GH**MNLSQLSFHRVDQKEITQLSHLGQGTRTNVYEGRLRVEGSGDP
EEGKMDDEDPLVPGRDRGQELRVVLKVLDPSHHDIALAFYETASLMSQVSHTHLAFVHGVCVRGP

ENIMVTEYVEHGPLDVWLRRERGHVPMAWKMVVAQQLASALSYLENKNLVHGNVCGRNILLARLG
LAEGTSPFIKLSDPGVGLGALSREERVERIPWLAPECLPGGANSLSTAMDKWGFGATLLEICFDG
EAPLQSRSPSEKEHFYQRQHRLPEPSCPQLATLTSQCLTYEPTQRPSFRTILRDLTRL
With the cleavage site located at: **ETVRFQ/GH**

3.1.6 Protein amino acid sequence after his-tag removal

GHMNLSQLSFHRVDQKEITQLSHLGQGTRTNVYEGRLRVEGSGDPEEGKMDDEDPLVPGRDRGQE
LRVVLKVLDPSHHDIALAFYETASLMSQVSHTHLAFVHGVCVRGPENIMVTEYVEHGPLDVWLRR
ERGHVPMAWKMVVAQQLASALSYLENKNLVHGNVCGRNILLARLGLAEGTSPFIKLSDPGVGLGA
LSREERVERIPWLAPECLPGGANSLSTAMDKWGFGATLLEICFDGEAPLQSRSPSEKEHFYQRQH
RLPEPSCPQLATLTSQCLTYEPTQRPSFRTILRDLTRL

1. To reduce the likelihood of non-specific binding of His-tag cleaved TYK2-JH2 to the 1-mL HisTrap FF crude nickel-affinity column, the overnight cleavage reaction mixture was adjusted to 25 mM imidazole from addition of a 1 M imidazole solution pH 7.7
2. The His-TVMV and cleaved hexahistidine tag are captured by chromatography of the cleavage reaction using a 1-mL HisTrap FF crude nickel affinity column attached to the AKTA Explorer-100 while the processed TYK2 pseudokinase domain (residues 575–869) protein passes through the column at 1 mL/min, 4 °C, and collected the flow through in a 250 mL polypropylene Nalgene beaker
3. The protein is then buffer-exchanged using a HiPrep 26/10 desalting column attached to the AKTA Explorer-100 at 5.00 mL/min, 4 °C, and collected 3 mL fractions into either Buffer D (see Section 3.1.3) for long term storage at −80 °C following flash freezing or into Buffer E (see Section 3.1.3) for ligand-binding and crystallization trials

3.2 Label-free methods to characterize direct interactions with TYK2-JH2

To probe the ability of small molecules identified to bind directly to the TYK2-JH2 domain compound mediated protein thermal-denaturation shift analyses (TSA) and isothermal titration calorimetry (ITC) were employed. These approaches, orthogonal to the kinase profiling panel (Ambit Biosciences) are solution-based and label free, and thus provide strong evidence of a direct interaction between the molecules identified from the screening campaigns and the TYK2-JH2 domain. Furthermore, both methods provide data describing the thermodynamics and stoichiometry underlying these interactions (Figs. 5 and 6).

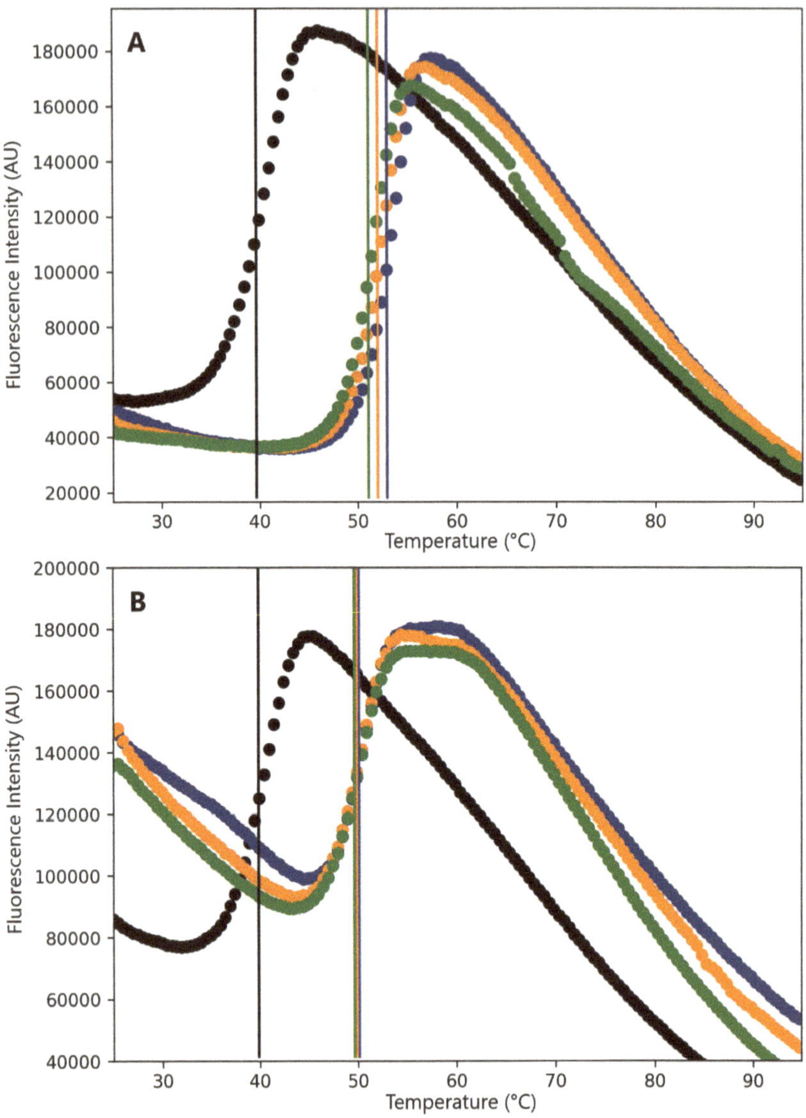

Fig. 5 Thermal shift assay to monitor enhanced stability of His-TYK2-JH2 by Compounds 2 and 3. (A) Compound 3 demonstrates concentration-dependent stabilization of His-TYK2-JH2 by 11.4, 12.4, and 13.3 °C for 50, 100, or 200 μM compound (green, orange, and blue circles, respectively). His-TYK2-JH2 with DMSO control has an average T_M of 39.66 ± 0.17 °C (black). Black circles represent the apo-protein and vertical lines indicate T_M for their respective curves (B) Compound 2 demonstrates concentration dependent stabilization of His-TYK2-JH2 by 9.8, 10.0, and 10.5 °C for 50, 100, or 200 μM compound (green, orange, and blue circles, respectively). His-TYK2-JH2 with DMSO control has an average T_M of 39.66 ± 0.17 °C (black). Black circles represent the apo-protein and vertical lines indicate T_M for their respective curves.

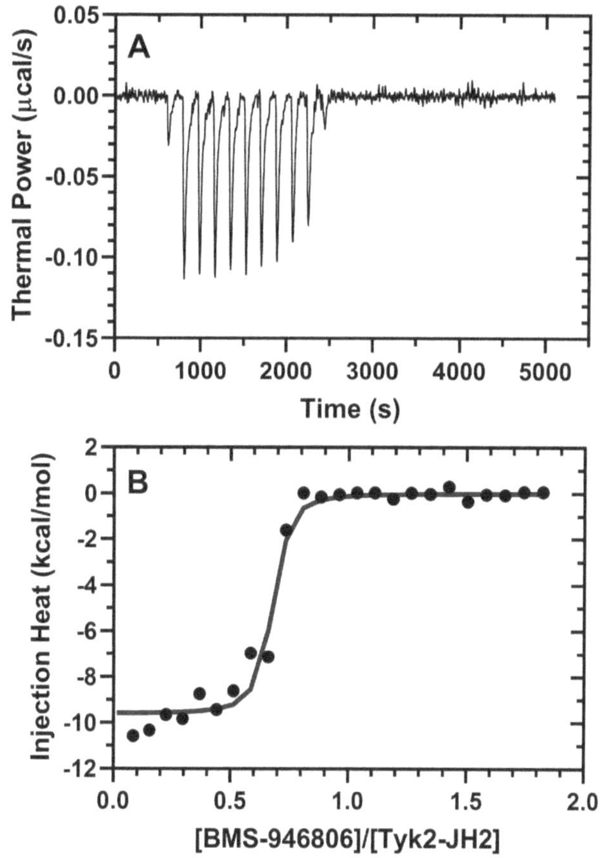

C

Compound	Stoichiometry, N	Affinity, K_D (nM)	Enthalpy, ΔH (kcal/mol)
Compound 1	0.7	9.2	-9.2
Compound 3	0.6	8.4	-15.4

Fig. 6 Isothermal titration calorimetric measurement of interaction of Compound 1 and TYK2-JH2. (A) Time-dependent thermal power evolved during titration of Compound 1 into a solution of TYK2-JH2. The first spike is associated with the injection of 2 μL of 50 μM Compound 1 into approximately 1.8 mL of 5 μM TYK2-JH2, subsequent spikes correspond to injections of 10 μL 50 μM Compound 1. (B) Heat generated per injection of Compound 1 into the solution containing TYK2-JH2 determined from the numerical integration of data in panel (A). Black circles correspond to binding heat generated following each injection of Compound 1 into the solution containing TYK2-JH2. Black line is the non-linear fit of the data using an equation to determine binding affinity, stoichiometry, and enthalpy. Representative data selected from 3 replicated experiments are shown. (C) Thermodynamic data for Compound 1 and Compound 3 derived from experiments identical to those described in panels (A and B).

TSA requires monitoring the change in the optical signature of a protein as it is heated from a temperature where the protein is predominantly in its native folded conformation to a temperature where the protein is predominantly in an unfolded denatured state. The presence of small molecules or other binding partners that interact more strongly with the folded native form of the protein increase the temperature required to denature the protein (Brandts & Lin, 1990). The use of extrinsic fluorescent dyes that weakly interact with denatured protein and show an increase fluorescence quantum yield upon binding, largely eliminates optical interference from buffer components and small molecules and enables the miniaturization and increased throughput to screen multiple compounds for their capacity to increase protein thermal stability (Kranz & Schalk-Hihi, 2011; Matulis, Kranz, Salemme, & Todd, 2005; Pantoliano et al., 2001; Zhang & Monsma, 2010). Fig. 5 depicts the thermal denaturation of the TYK2-JH2 domain alone and in the presence of increasing concentrations of Compound 3 (a close analog of Compound 1) and Compound 2. These data indicate a strong and specific interaction between this protein construct and these compounds, as suggested by the >10 °C increase in TYK2-JH2 melting temperature at stoichiometric and greater compound concentrations (Pantoliano et al., 2001).

The TSA results were further confirmed through ITC studies, where the heat evolved from titration of compound into protein is monitored. These studies, shown in Fig. 6, described a high affinity and enthalpically favorable stoichiometric complex formed between Compound 1 and TYK2-JH2.

These studies provided strong evidence that the ITP and IZP chemotypes interact directly with TYK2-JH2 and that these interactions are likely through multiple strong, non-covalent interactions (Ladbury, Klebe, & Freire, 2010; Tokarski et al., 2015).

3.3 Protocol for TYK2-JH2 thermal shift assay

These assays were performed using a QuantStudio 7 Flex Real-Time PCR System (Thermo Fisher Scientific, Inc.) and QuantStudio Real-Time PCR Software v1.3. Buffer solutions were prepared using water filtered by a BarnStead GenPure Pro water purification system (Thermo Scientific), final solutions were filtered using a 0.22 μm pore, polyethersulfone (PES) sterile filter (Corning). The assay plate used was an Applied Biosystems MicroAmp EnduraPlate Optical 384-well clear reaction plate (Thermo Fisher Scientific, Inc.). SYPRO Orange protein dye (Thermo Fisher Scientific, Inc. Catalog number S-6651) was obtained at a concentration of 5000 × in DMSO.

1. *Compound preparation*: Compounds were dissolved in DMSO to create 10 mM stock solutions. They were then transferred by hand into a 384-well ECHO source plate. DMSO was added into a separate well to be used for dilution
2. *Protein solution preparation*: 2.2 mL of master mix was prepared per 384 well plate by diluting His-TYK2-JH2 (575–869) to 0.2 mg/mL in 25 mM Tris, pH 8.0, 150 mM NaCl, 4 mM $MgCl_2$, 2 mM TCEP, 2% glycerol, and 0.005% Tween-20, and a final dye concentration of 10× (5000× stock diluted 1:500).
3. *Compound titration*: Compounds were transferred from the source plate into a clear 384-well PCR plate using an ECHO 655 Liquid Handler (Beckman). 100 nL of each stock solution was added to a well in the sample plate. 50 and 25 nL of the compounds were added to subsequent wells and backfilled with DMSO to a total volume of 100 nL, creating a 3-point two-fold dilution series from a top concentration of 200 μM with 2% DMSO in the final samples. Three wells (A1, A2, and A3) were filled with 100 nL DMSO only (2%) to be used as a reference
4. *Sample-plate preparation*: 5 μL of the protein solution was added to each compound/DMSO-populated well using a multi-channel pipette. The sample plate was sealed with a translucent seal to prevent evaporation or contamination, centrifuged in an Eppendorf Centrifuge 5430 for 60 s at 100 × *g* at room temperature, then immediately loaded into the PCR instrument
5. *Melt curve method*: The samples underwent a thermal melt using a 30 s step-and-hold melt ramp at 0.5 °C increments from 25 to 95 °C. Optical filters selected: emission 520 nm excitation 470 nm, emission 586 nm excitation 470 nm, and emission 623 nm excitation 580 nm. Fluorescence intensity was measured as a function of temperature

3.3.1 Data analysis

TSA data were analyzed in according to the standard Boltzmann function in the form of $F(T) = f_{min} + \frac{f_{max} - f_{min}}{1 + e^{\frac{Tm-T}{a}}}$ where f_{max} is the maximum fluorescence, f_{min} is the minimum fluorescence, and a was set to 1 for initial conditions (Niesen, Berglund, & Vedadi, 2007). Curve fitting was performed using a program written in Python 3.6. The curve fitting routing was done using the curve_fit function within scipy.optimize to do nonlinear least squares fitting. To handle non-ideal thermal melt curves a baseline correction was applied to yield more realistic and robust values of f_{min} and f_{max} and for initial guesses of T_M a search window could be applied to only find peaks within an expected range relative to the T_M of the reference. TSA data can be similarly analyzed using standard curve fitting software such as GraphPad Prism using the Boltzmann function.

3.4 Protocol for TYK2-JH2 isothermal titration calorimetry

3.4.1 *Equipment*

VP-ITC™ (Microcal, Inc.)

15 mL Centrifugal filter 30,000 NMWCO (Millipore, Inc.)

Syringe sterile filter with 0.2 μm pore size

Temperature controlled swinging bucket centrifuge such as Beckman-Coulter Allegra X-12R

Spectrophotometer such as NanoDrop 2000 (Thermo Scientific, Inc.)

3.4.2 *Chemicals*

His-TVMV-human TYK2-JH2 (575–869) Internally produced at BMS 0.5 M Sodium Phosphate Buffer, pH 7.20 (Teknova S5072).

10 N Sodium Hydroxide (Sigma-Aldrich SX0607N-6).

5 M Sodium Chloride Solution (Sigma-Aldrich S6546-1 L).

Glycerol (Sigma-Aldrich, #G9012).

Dithiothreitol (Sigma-Aldrich, #D9779) DMSO (Sigma-Aldrich, #D8418).

10 mM Compound 1 disolved in DMSO.

3.4.3 *Protocol*

1. Prepare 1.0 L buffer solution for ITC experiments consisting of 10 mM sodium phosphate, 500 mM NaCl, 1% glycerol (v/v), 0.5 mM DTT, and 2% DMSO (v/v) adjust pH to pH 7.7 using 10 N Sodium Hydroxide
2. Degas buffer. This can be done under lab vacuum or through use of Microcal degasser. To facilitate degassing of large volumes a sonicator can be employed.
3. Buffer exchange TYK2-JH2 protein sample to into ITC buffer using 15 mL Centrifugal filter 30,000 NMWCO in swinging bucket centrifuge using suggested centrifugal force by manufacturer. To ensure adequate buffer exchange, sample is concentrated down to approximately 1 mL and redissolved in ITC experimental buffer 10 times then syringe-filtered.

4. Determine final protein concentration using NanoDrop 2000 spectrophotometer that has been blanked using fresh ITC buffer. To ensure protein is not aggregated differential light scattering can be employed.
5. Prepare 1 mL of 50 μM Compound 1 in ITC experimental buffer having 3% DMSO. Add 5 μL of 10 mM Compound 1 DMSO solution and 5 μL DMSO to 990 μL ITC experimental buffer.
6. Prepare 2.25 mL of TYK2-JH2 solution in ITC experimental buffer having 3% DMSO, by adding enough TYK2-JH2 for a 5 μM final concentration from the buffer exchanged stock, along with 22.5 mL DMSO to ITC buffer to yield 2.25 mL.
7. Prepare enough ITC experimental buffer, without compound or protein, by adding an additional 1% DMSO to degassed ITC buffer.
8. Ensure VP-ITC is properly cleaned and maintained by performing water into water test titrations and ITC experimental buffer into ITC experimental buffer titrations. Absolute thermal power changes should not exceed 0.05 μCal/s per injection.
9. Load ITC experimental buffer into VP-ITC reaction cell using loading syringe according to manufacturer's instructions, being careful to eliminate air from the cell.
10. Load injection syringe with 50 μM Compound 1 solution.
11. Set ITC experimental parameters to 20 total injections, cell temperature to 25 °C, reference power to 10 μCal/s, initial delay to 600 s, syringe concentration to 0.050 mM, cell concentration to 0.005 mM and stirring speed to 307 rpm, and gain set to high.
12. Set Injection parameters to enable 1 low volume injection, followed by 19 10 μL injections of 20 s duration, with 180 s between injections, and a filter period of 2.
13. After sufficient time to allow ITC to equilibrate, initiate experiment to determine heat of dilution of Compound 1 into buffer.
14. Following compound dilution experiment, rinse reaction cell with ITC experimental buffer and reload injection syringe with 50 μM Compound 1 solution.
15. Load solution of 5 μM TYK2-JH2 into reaction cell. Keep experimental and injection parameters set as previous and repeat experiment.
16. Load experiment into Microcal Origin analysis sofware, adjust baseline of Thermal power vs. time plot if needed, and integrate to determine heat evolved per injection.

17. Plot injection heat versus [Compound 1]/[TYK2-JH2] and analyze data to determine binding enthalpy (ΔH), stoichiometry (N), and affinity (KA) using a model for 1:1 binding.

3.4.4 Results

As seen in Fig. 5, His-TYK2-JH2 (575–869) showed a single melting transition with an average T_M of 39.66 °C. In the presence of 200 μM Compound 2, TYK2-JH2 showed stabilization with a thermal shift of 10.5 °C. In the presence of 200 μM Compound 3 (a close analog of Compound 1), TYK2-JH2 showed stabilization with a robust thermal shift of 13.3 °C, indicative of a strong interaction selective for the folded native form of the JH2 domain. These findings are corroborated by the ITC study, indicating enthalpically driven stoichiometric binding of Compound 3 and Compound 1 of $\leq$10 nM (Fig. 6).

3.5 Structural studies of TYK2-JH2 with Compound 3

As the binding of ITPs and IZPs to TYK2-JH2 was confirmed by ITC and TSA, it was important to next obtain the X-ray crystal structure of these binders with TYK2-JH2 to fully understand the binding modes of the compounds to the pseudokinase domain of TYK2. We hypothesized that the compounds were binding into the deep cleft between the N- and C-terminal domains (ATP binding pocket for kinase domain proteins) but we wanted structural confirmation and specific details of ligand-protein interactions. Using human TYK2-JH2 (575–869) protein expressed and purified as described above, we pursued crystallization studies.

Although several structures of TYK2-JH2 in complex with ITPs and IZPs were obtained early in the effort, only details of crystallization and structure determination of TYK2-JH2 with Compound 3 (a close analog of the ITP Compound 1) will be detailed and presented in the following sections.

3.5.1 Protocol

3.5.1.1 Crystallization

Given that binding of the compounds showed significant protein stabilization by TSA, we chose to complex the ligand with protein in solution prior to crystallization to enhance protein stability and homogeneity, improving the chances of successful crystallization.

100 μL of His-tagged cleaved TYK2-JH2 (575–869) at 1.2 mg/mL in Buffer E (Section 3.1) was complexed with a 3-fold molar excess of

Compound 3, incubated overnight at 4 °C, and then concentrated to 2.3 mg/mL just prior to crystallization. Initial crystal screens were carried out using a grid of 25% PEG 5000 (methyl ether) and 1.75 M ammonium sulfate versus pH (5.0–7.5) at 0.5 pH intervals with tris-maleate buffer. The initial screens indicated promising crystalline precipitant in the PEG crystallization drops in a pH range of 6.0–6.5. Follow-up crystallization screens were then carried out using a grid of 30% PEG 5000 (methyl ether), 100 mM sodium cacodylate buffer at pH 6.0 and 6.5 with the following salt additives at 0.2 M: sodium acetate, ammonium acetate, lithium sulfate, ammonium sulfate, magnesium chloride. The best crystal leads were obtained with 30% PEG 5000 (methyl ether), 100 mM sodium cacodylate buffer at pH 6.5 with 0.2 M ammonium sulfate or lithium sulfate. The crystals that resulted in the structure of TYK2-JH2 (575-869) in complex with Compound 3 were obtained at room temperature using the hanging drop vapor diffusion method in a drop consisting of 1 μL of protein solution and 1 μL of reservoir solution containing 30% PEG 5000 (methyl ether), 200 mM ammonium sulfate, and 100 mM sodium cacodylate buffer, pH 6.5. Crystals appeared in 1 week and took 2–3 weeks to grow to full size (0.1 × 0.1 × 0.15 mm). Crystals were flash-cooled in liquid nitrogen with 25% glycerol and 75% reservoir solution as cryoprotectant.

3.5.1.2 Data collection

A 1.80 Å resolution data set was collected at the Canadian Light Source beamline 08ID-1 and data was processed using HKL2000 (Otwinowski & Minor, 1997). The crystals had symmetry consistent with the P1 space group with cell dimensions $a = 41.5$ Å, $b = 63.6$ Å, $c = 68.0$ Å ($\alpha = 74.3°$, $\beta = 75.5°$, $\gamma = 88.2°$) and 2 molecules of TYK2-JH2 pseudokinase domain in the asymmetric unit. The structure was determined by molecular replacement using Phaser (McCoy et al., 2007) with Abl kinase (2GQG; Tokarski et al., 2006) as a starting model. To avoid any model bias, the 2GQG starting model residues were stubbed and the P-loop and activation-loop removed. The N- and C-terminal domains of the 2GQG starting model needed to be split apart during the molecular replacement process to obtain a molecular replacement solution.

3.5.1.3 Structure determination

Multiple cycles of model building using Coot (Emsley, Lohkamp, Scott, & Cowtan, 2010) and refinement using refmac (Murshudov et al., 2011) and autoBUSTER (Bricogne et al., 2011) were carried out to build side chains and missing loops. At this point, the electron density maps were of sufficient quality to unambiguously identify electron density for Compound 3. Final statistics for the structure of TYK2-JH2 in complex with Compound 3 are shown in Table 1. The crystal structure of TYK2-JH2 in complex with Compound 3 has been deposited to the protein data bank with a PDB ID of 4WOV.

3.5.2 Analysis of the structure

At the time we determined the first structure of TYK2-JH2, there were no published structures reported and so it was exciting and impactful to obtain and analyze this novel pseudokinase structure. We observed that the structure of the TYK2-JH2 encompasses the main structural features of a protein kinase. Namely, the protein fold is separated into two domains; the N-terminal domain composed primarily of β sheets and the prominent helix α-C and the larger C-terminal domain that is predominantly helical. The two domains are connected through a single polypeptide strand (hinge), which in catalytic kinases is normally involved in binding to the adenine ring of ATP. The site analogous to the ATP binding site in catalytic kinases forms a deep cleft between the two domains and Compound 3 is observed to bind in this cleft. See high level view of Compound 3 binding to TYK2-JH2 illustrated in Fig. 7. Subsequent to our first TYK2-JH2 structure, other structures and papers of TYK2-JH2 in complex with small molecule ligand have been reported (unpublished data; 3ZON (Lupardus et al., 2014); 4OLI (Min et al., 2015); 5C01 and 5C03, (Greenhough et al., 2021); 7AX4)). All of the TYK2-JH2 structures reported adopt very similar conformations and show the ligands binding into the site analogous to the ATP binding site.

Consistent with TYK2-JH2 being defined as a pseudokinase, the crystal structure shows several features that, when compared with catalytic kinases, indicate that this domain is unable to support catalytic activity (Fig. 8):

(1) The typically conserved β3-strand lysine (Lys-642) is present but does not form a salt bridge with a conserved glutamate from helix-αC (important for maintaining an active conformation in catalytic kinases). Instead, Thr-658 occupies the position of the typical conserved glutamate. Lys-64 in the pseudokinase domain structure is instead near hydrogen bonding distance to Asp-759, the aspartate that in catalytic kinases belongs to the conserved DFG motif, which marks the N-terminal portion of the activation loop and normally interacts with the ATP phosphates through Mg^{2+}. The other JAK JH2 members also lack the helix-αC Glu but instead contain an Ala at the corresponding position

Table 1 Data collection and structure refinement statistics for compound 3.

Compound	**Compound 3**
PDB Code	4WOV
Crystal Space Group	P1
Unit Cell Axes	a = 4.15 Å, b = 63.6 Å, c = 68.0 Å
Unit Cell Angles	$\alpha = 74.3°$, $\beta = 75.5°$, $\gamma = 88.2°$
Molecules/AU	2
X-ray Source	CLS Beamline 08ID-1
Temperature (K)	100
Detector	Rayonix MX-300
Wavelength (Å)	0.979
Data Reduction Program	HKL
Resolution Range (Å)	50.00–1.80
Unique Reflections	57,908
Redundancy	3.4
Completeness (%)	97.7
R-Merge (%)	6.7
$I/\sigma I$	16.6
Structure Refinement	
Resolution Range (Å)	36.00–1.80
Number of Reflections Used	57,869
Rwork (%)	19.2
Rfree (%)	21.4
# Protein atoms used	3849
# Heterogen atoms used	81
# Solvent atoms used	257
Mean B-value ($Å^2$)	31.59
R.M.S.D. from ideal	
bond length (Å)	0.010
bond angle (°)	0.97
omega torsion angles (°)	3.31
Other torsion angles (°)	14.45

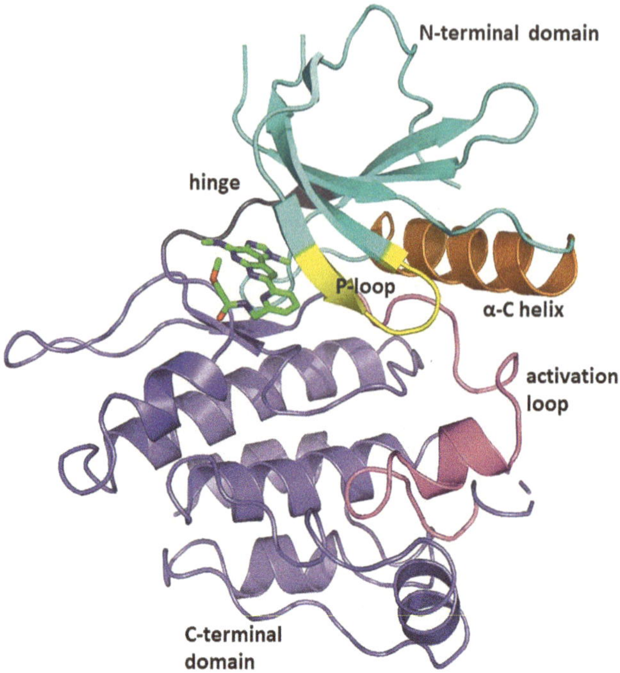

Fig. 7 X-ray crystal structure of TYK2-JH2 in complex with Compound 3. A ribbon diagram representing protein with Compound 3 shown in stick (green carbons). The N-terminal domain is colored in cyan, the P-loop is colored yellow, the α-C helix is colored orange, the hinge is colored gray, the activation loop is colored magenta, and the C-terminal domain is colored in blue.

(2) The DFG motif of kinases is replaced by DPG, whereby proline appears to restrict the position of Asp-759 to a region normally occupied by the helix-αC glutamate. The altered position of Lys-642 and Asp-759 would not optimally interact with Mg^{2+}-ATP. The DPG motif is conserved in the JAK JH2 family

(3) The P-loop that normally serves as a gate to ATP binding in catalytic protein kinases usually contains a conserved glycine-rich sequence motif (G*X*G*X*G), where, usually tyrosine or phenylalanine, caps the site of phosphate transfer, but in TYK2-JH2 the sequence is G*X*G*X*RT, which might be expected to reduce the flexibility of the P-loop due to the loss of the glycine and perhaps limit the binding and release of ATP/ADP. Other JAK JH2 family members also lack the third Gly; JAK1 JH2 shares the Arg at that position with TYK2 but JAK2 and JAK3 JH2 instead possess a Phe

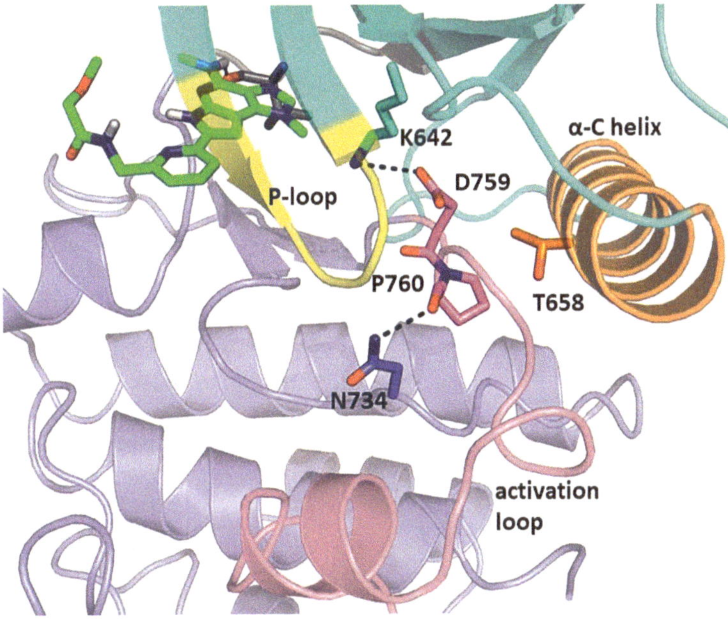

Fig. 8 Structural features that describe TYK2-JH2 as a pseudokinase. P760 appears to restrict the position of D759 (aspartate that normally binds Mg^{2+}) to a region normally occupied by the helix αC glutamate (important for catalytic lysine salt bridge that stabilizes active conformation) which is replaced by T658. The canonical protein kinase catalytic aspartate is replaced by N734. The activation loop which normally serves as a binding platform for the substrate is truncated compared with a typical kinase and adopts a partially helical conformation.

(4) The activation loop is truncated compared with a typical catalytic domain activation loop (20 residues versus typically closer to 30 residues) and adopts a partially helical conformation such that it might hinder substrate docking. This feature is shared with other JAK JH2 family members

(5) The canonical protein kinase catalytic loop motif HRD (D is the catalytic aspartate that contributes to the nucleophilicity of the substrate hydroxyl) is replaced by HGN (Asn-734) in TYK2-JH2 as well as in other JAK JH2 family members. Asn-734 is observed to be involved in a hydrogen bond with the backbone carbonyl of Pro-760 of the DPG motif

Further illustrating the uniqueness of TYK2-JH2, a sequence comparison of the Compound 3 binding site residues with the 491 sequences of the kinome

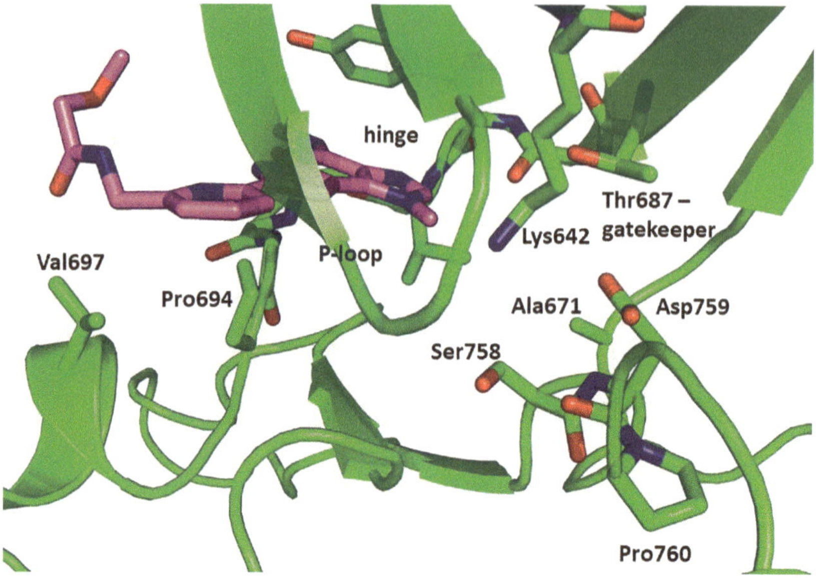

Fig. 9 Uniqueness of TYK2-JH2 relative to other kinases in the ligand binding site. Residues besides catalytic Lys642 and Asp759 are shown that are in low abundance especially in combination compared to the kinome.

including other pseudokinases reveals uncommon residues near the Compound 3 binding site. These differences include (Fig. 9):

(1) Pro-760 of the DPG motif, which is not present within any other kinase other than the JAK JH2 family members

(2) Thr-687, the gatekeeper residue, is present in only 19% of the kinome

(3) Ala-671, the residue below the gatekeeper, is present in only 2% in kinome

(4) Ser-758, the residue before DPG, is present in only 15% of the kinome

(5) Pro-694, in the extended hinge region, is present in only 2% of the kinome

(6) Val-697, in the extended hinge region, is present in less than 1% of the kinome

These residues relative to other kinases in the kinome provide a unique ligand binding environment and indicate that the pocket of TYK2-JH2 is well suited to design highly selective therapeutics relative to typical kinase domains. Additionally, the reduced homology between the pseudokinase domains within the Janus family provides for a greater opportunity to design family specific inhibitors (Fig. 10).

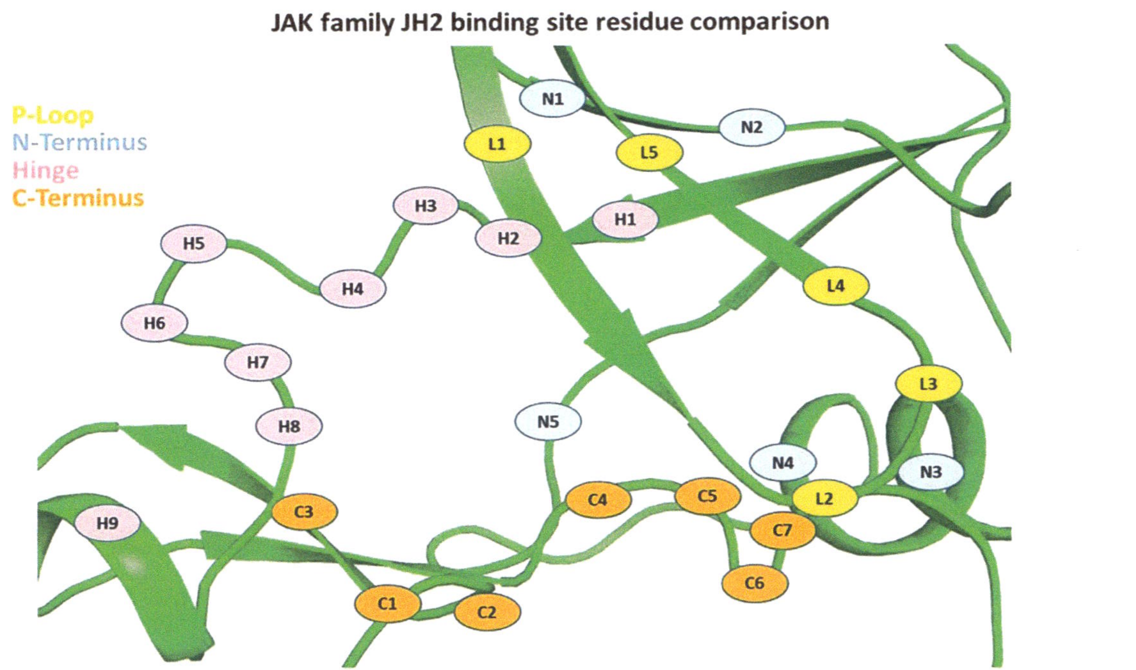

kinase	L1	L2	L3	L4	L5	N1	N2	N3	N4	N5	H1	H2	H3	H4	H5	H6	H7	H8	H9	C1	C2	C3	C4	C5	C6	C7
JAK1	L	T	R	T	I	I	K	A	M	V	E	E	F	V	E	G	G	P	L	K	N	L	S	D	P	G
JAK2	L	T	F	T	I	L	K	A	M	V	Q	E	F	V	K	F	G	S	T	K	N	L	S	D	P	G
JAK3	L	S	F	T	I	L	K	A	M	V	Q	E	F	V	H	L	G	A	M	R	K	L	S	D	P	G
TYK2	L	T	R	T	V	V	K	T	M	A	T	E	Y	V	E	H	G	P	V	R	N	L	S	D	P	G

Fig. 10 Residues across the JAK family JH2 domain that line the ligand binding site. Positions of the residues are color-coded by region.

We have been able to develop a structure platform for TYK2-JH2 and, over the lifetime of our TYK2-JH2 program, obtained numerous structures in support of the program. In addition to establishing a co-crystallization system where protein is complexed with ligand in solution prior to crystallization (described above), we established a crystal soaking system where co-crystals of TYK2-JH2 in complex with Compound 2 are soaked with a new ligand of interest which results in the displacement of Compound 2 and the binding of the new ligand of interest. We often found that we could not obtain large, diffraction quality co-crystals of TYK2-JH2 in complex with Compound 2, therefore, we used crystal seeding about 90% of the time to obtain the crystals needed for structure determination. We used a streak seeding protocol that involves touching the parent crystal with a cat whisker and then streaking the cat whisker through a target crystallization drop. Conditions of the target drop are the same as the parent drop (as described in the crystallization section above). The growth of the crystals is improved if drops are left open for one to 2 min after seeding—this promotes further evaporation and protein concentration.

This method of obtaining co-crystals of TYK2-JH2 with Compound 2 and using a displacement soaking system to obtain structures of interest has been the workhorse system for this project. Crystal structures that used this displacement soaking system are described in the following papers (Liu et al., 2021, 2019; Moslin et al., 2017, 2019; Tokarski et al., 2015; Wrobleski et al., 2019).

4. Probe displacement assays for binding energetics and lead identification

To further investigate the structure activity relationship (SAR) of compounds discovered and to identify other novel compounds that bind to TYK2-JH2 pseudokinase domain, efficient high-throughput assays were needed. To this end, an isotopically labeled form of the TYK2-JH2 specific compound, Compound 1 was generated, enabling probe displacement assays specific for TYK2-JH2 to be developed. Scintillation proximity analysis (SPA) assays using this radioligand provided biochemical SAR for the ITP and IZP chemotypes.

Although potent and highly selective for TYK2 over other kinases, liabilities for these chemotypes were observed. Molecules derived from the ITP series demonstrated potency against IKKβ, another potentially immunomodulatory target, which could confound an understanding of efficacy and increase the risk of compound mediated toxicity (Moslin et al., 2017). The IZP series of compounds eliminated IKKβ activity. Unfortunately, off-target inhibition of PDE4 was detected along with difficult to optimize pharmacokinetic properties and limited metabolic stability (Moslin et al., 2019). The inability to further progress these compounds precipitated the modification of the SPA assay to a high-throughput screening method used to identify novel chemical scaffolds. It was adapted to 1536-well microtiter plate format allowing 120,000 compounds from BMS proprietary collections for broad chemical diversity and selective kinase inhibition to be assessed for displacement of Compound 1 from TYK2-JH2. This led to the identification of a series of compounds containing a nicotinamide core, including Compound 4 (Fig. 11).

The SPA assay was used for further optimization of the nicotinamide compounds, but specific activity of the tritiated probe and detection limitations of the SPA imaging beads made accurate determinations of compound potency difficult. Advantageously, additional fluorescently-tagged promiscuous probes were developed to ascertain compound binding potency against a broad range of pharmacologically relevant kinases. This assay enables quantitative detection of binding over a wide range of potencies using homogeneous time-resolved fluorescence (HTRF) measurement. In these assays, time resolved Forster resonance energy transfer is detected following excitation of a lanthanide cryptate fluorophore that is appended to antibodies selective for specific protein expression tags, such as hexahistine and GST (Bazin, Preaudat, Trinquet, & Mathis, 2001; Chaudhry et al., 2021; Degorce et al., 2009; Mathis, 1995; Selvin, 2002), as described below in Section 4.2.

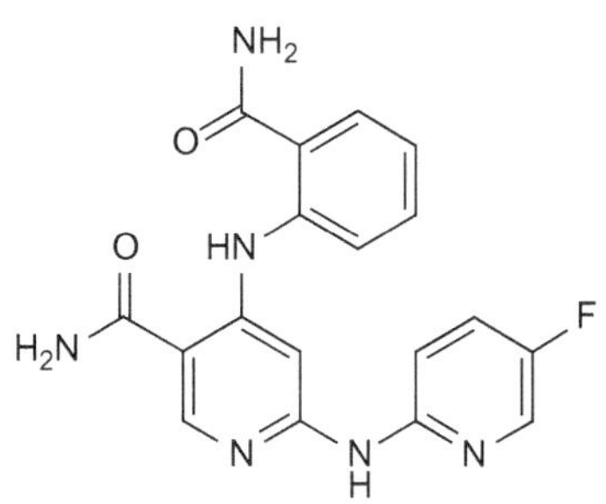

Fig. 11 Structure of nicotinamide Compound 4.

4.1 Scintillation proximity analysis to monitor binding to TYK2-JH2 domain

4.1.1 Equipment

Plate-based radiometric imager, e.g., TopCount™ PerkinElmer (384-well assay) LEADSeeker™ GE Lifesciences (1536-well assay).

His-Tag PVT imaging beads or His-Tagged YOX imaging beads (such as Perkin Elmer RPNQ-0266 or RPNQ-0276).

Microtiter assay plates, Solid white such as Perkin-Elmer 384-well OptiPlate™, 1536-well Corning 3725.

Labcyte Echo® qualified source plate.

4.1.2 Chemicals

DMSO-Sigma-Aldrich, #D8418

1 M Sodium HEPES, pH 7.5-ThermoFisher, #15630080

Tween 20-Sigma-Aldrich, #P1379

Dithiothreitol-Sigma-Aldrich, #D9779

Bovine Serum Albumin Solution, 30% Sigma-Aldrich, #A7030

His-TVMV-human TYK2-JH2 (575–869)

Phosphate buffered saline-Sigma-Aldrich, #806552

Glycerol-Sigma-Aldrich, #G9012

4.1.3 Protocols

1. Assay Buffer contains 50 mM HEPES sodium salt, 0.1% (m/v) BSA, 0.1% (*v*/v) Tween 20 and 1% (v/v) glycerol from fresh reagent aqueous or bulk liquid stocks. During common usage 100 mL is prepared using 5 mL of a 1 M stock solution of pH 7.5 sodium HEPES, 33 μL of 30% BSA, 10 μL Tween 20, 1 mL Glycerol and enough water to bring to the total 100 mL volume
2. Tritiated Probe Solution is prepared containing 50 nM ^{3}H-Compound 1 (4 mCi/L) in Assay Buffer by 250-fold dilution of the probe stock solution, 12.4 μM ^{3}H-Compound 1 in 80% ethanol (v/v) solution with a specific activity of 80.6 Ci/mmol (155 Ci/g Compound 1).
3. SPA Beads are pre-bound to TYK2-JH2 by diluting stock solution of copper-PVT beads to 400 μg/mL and His-tagged TYK2-JH2 to 8.35 nM in Assay Buffer and incubating for at least 1 h with gentle orbital shaking. For no-binding controls SPA Beads are treated identically except that no His-tagged TYK2-JH2 is added

4. Compounds for testing are prepared as 10 mM DMSO stock solutions and may be subsequently serial diluted in 100% DMSO in polypropylene storage plates. Compound solutions are either transferred directly to 384-well OptiPlate assay plates, or to Labcyte Echo™ qualified plates for subsequent microvolume acoustic transfer to assay plates. A constant volume of compound solution in DMSO is transferred to each assay plate. An equivalent volume of DMSO is added to each well not containing a compound
5. 20 μL of Tritiated Probe solution is added to all wells followed by 30 μL of TYK2-JH2 bound bead solution or 30 μL of no-binding control bead solution. For 384-well plate assays 1 column (16 wells) is used for no-binding controls and an additional column is used for full-binding controls where no test compound is added
6. Following a 30-min incubation at room temperature with gentle shaking plates are read using the TopCount radiometric imager
7. Raw counts from each well are normalized relative to the average values from the no-binding and full-binding controls, blank and total values, respectively. % Inhibition of probe binding for each well was determined using the standard equation, where $\%Inhibition = \frac{total - wellcount}{total - blank} \times 100$
8. When compound dilution series are tested IC_{50} values corresponding to the displacement of 50% of the tritiated probe from TYK2-JH2 for each compound are calculated analyzing plots of % inhibition versus compound concentration, and fitting to the standard 4-parameter binding isotherm where

   $\%Inhibition = Bottom + \frac{[Compound]^N \times (Top - Bottom)}{Halfmax^N + [Compound]^N}$. In this relationship Top and bottom are the % Inhibition at infinitely large and infinitely small compound concentrations, respectively, while Halfmax is the compound concentration yielding % Inhibition corresponding to ½ Top-Bottom, and N is the Hill coefficient. IC_{50} values are calculated according to the relationship:

$$\mathrm{IC}_{50} = \mathrm{Halfmax}\left(\sqrt[N]{\frac{50 - \mathrm{Bottom}}{\mathrm{Top} - 50}}\right)$$

9. Adapting the assay to high-density 1536-well microtiter plate format to enable high-throughput screening requires altering the His-tag binding SPA beads to YOX SPA imaging beads, increasing bead concentration to 4 mg/mL, and increasing TYK2-JH2 concentration to 25 nM. An appropriate radiometric imager for high-density microtiter plate detections such as a LEADSeeker™ is also needed

4.2 HTRF compound screening

To address the need to characterize and provide specificity and selectivity data for an ever-growing kinase portfolio and an ever-expanding compound library, BMS embarked on a project to develop an internal method for a high-throughput platform that would enable the discovery team to overcome these hurdles. To achieve this goal, the BMS computer aided drug design (CADD) group and discovery chemistry group collaborated to design fluorescein labeled ATP-competitive probes built from known scaffolds of promiscuous kinase inhibitors (Chaudhry et al., 2021). These probes were then used to build HTRF based assays with either His-tagged or GST-tagged kinases, which measures activity of testing compounds by displacement of the fluorescein labeled probe (Fig. 11).

With this new tool available to us and using internally sourced technological and IT abilities we were able to build an efficient HTS platform that allowed us to efficiently screen dozens of kinase targets against hundreds of new compound submittals on a weekly basis with data turnaround times of 2 weeks or less. Eventually these abilities allowed us to expand our screening efforts to encompass more than 250 kinase SAR assays in an 11-point, 1536 plate-based format. This new tool allowed us to screen dozens of kinase targets and was used to screen TYK2-JH2 (Fig. 12).

To have the ability to carry out high throughput screens and identify binders to TYK2-JH2, we had to characterize probe molecule binding affinities and off-rates in order to identify a preferred fluorescent probe molecule to use for the high throughput HTRF competition assay.

A typical method for measuring affinity of a fluorescent probe to an enzyme is to titrate the probe against a fixed concentration of enzyme. In the case of TYK2-JH2, the following section describes the materials and methodology used in determining probe equilibrium and kinetic dissociation constants.

4.2.1 Equipment

Combi-Multi Drop liquid Dispenser (Thermo-Fisher).

Centrifuge (plates).

Echo 550 Liquid Handler.

Remp Plate.

Echo Source Plate.

Corning 3573 Assay Plate.

Envision Multi-mode Plate Reader.

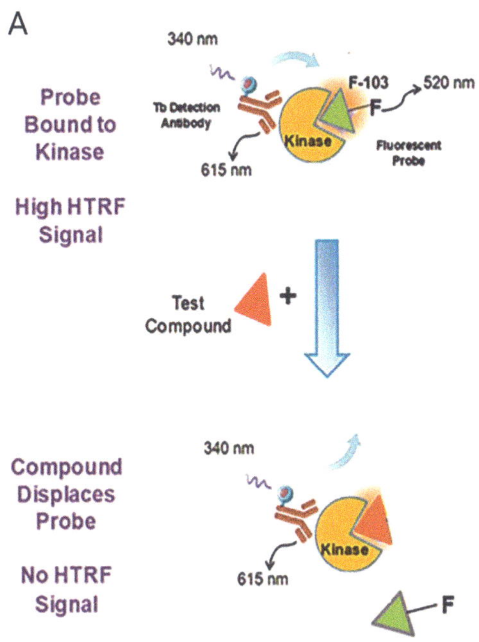

Fig. 12 (A) Schematic diagram detailing the design of the BMS HTRF kinase screening assay. (B) Kinome Profiling Tree. A dot is placed over a kinase that is included in our internal panel showing good coverage of the kinome. The colors of the dots indicate what chemotype the HTRF probe belongs to. Purple star indicates kinome branch for JAK-JH2 pseudokinase domains.

(Continued)

4.2.2 Chemicals

DMSO-Sigma-Aldrich, #D8418.

1 M sodium HEPES, pH 7.5-ThermoFisher, #15630080.

1 M $MgCl_2$-Invitrogen, #AM9530G.

Brij-35-Millipore, #203724.

DTT-Sigma-Aldrich, #D9779.

BSA-Sigma-Aldrich, #A7030

AMP-PNP –Sigma-Aldrich, #A2647

His-TVMV-human TYK2-JH2 (575–869)-Internally produced at BMS

Anti-hexahistidine antibody custom labeled with terbium cryptate (CisBio, corp.). Note that commercially available Tb-labeled anti-hexahistidine antibodies can be substituted, but titration of this antibody and the fluorescence probe should be performed to achieve optimal signal and accuracy.

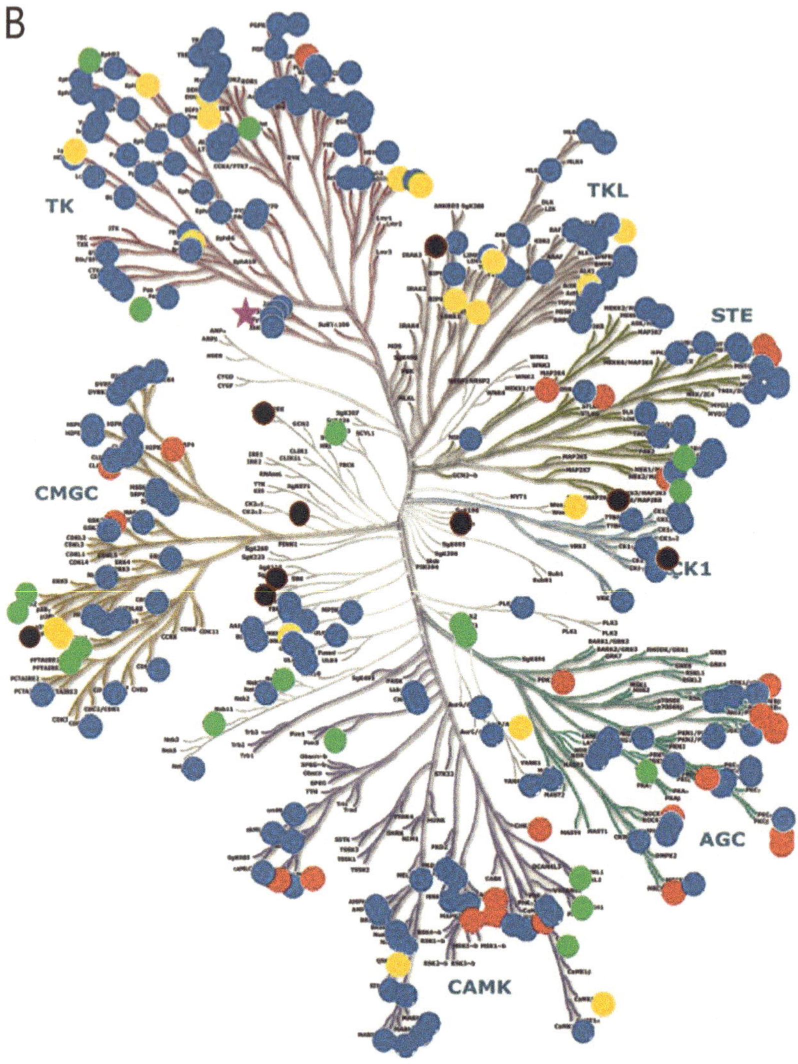

Fig. 12—Cont'd

4.2.3 Protocol

1. 1 mg of each Fluorescent Promiscuous probe was solubilized in 100% DMSO to a final concentration of 10 mM in 200 μL DMSO
2. An intermediate dilution of each probe was made at a concentration of 400 μM, and an 8-point serial dilution was performed down to 3.125 μM in a REMP plate. These dilutions were then transferred to an Echo source plate

3. 25 nL of each probe was transferred acoustically using an Echo 550 liquid handler to two different sections of a 384 well assay plate. The first section being used for background measurements (buffer and antibody only), the second section being used for the mixture containing Enzyme, antibody, and buffer (Fig. 13)
4. A 0.4-nM solution of anti-His Tb-labeled antibody was prepared in 50 mM sodium HEPES 10 mM $MgCl_2$, 1 mM DTT, 0.05 mg/mL BSA, 0.015% (*v*/v) Brij-35 at pH 7.5 (Buffer F) and a separate solution of 2 nM TYK2-JH2 (double the final concentration) was also prepared in Buffer F
5. 10 μL of the anti-His Tb-labeled antibody solution was added to the entire assay plate containing 25 nL of the HTRF promiscuous probe dilution, followed by 10 μL buffer only to section A to enable measurement of background, and subsequently 10 mL of the TYK2-JH2 solution was added to Section B of the assay plate. The plates were then placed into a centrifuge and centrifuged for 1 min at 1000 RPM to ensure mixing
6. The assay plate was incubated for 1 h at room temperature with a foil seal to protect it from light
7. Following the incubation, the plate was read on the Envision Multi-Mode reader in HTRF mode (520/495).
8. After reading, the 520/495 ratio for the antibody only background was subtracted from the 520/495 ratio of the TYK2-JH2/His-Antibody reactions and that difference was plotted against probe concentrations and fit to a One-Site – Specific Binding equation using GraphPad Prism

The probe screening effort yielded a tight-fitting Probe (Probe "A") and Fitting to the One-Site Specific Binding equation showed that this probe has a K_D of 26 nM (Fig. 13).

With the K_d having been determined, the next step in finding a probe suitable for screening efforts is to determine the kinetic dissociation rate (k_{off}) of the probe to ensure that equilibrium binding can be achieved within ~1 h.

In order to determine the k_{off} value for the promiscuous probe, the probe was incubated with 20 μL of 1 nM TYK2-JH2 at the Probe K_d value for a period of 1 h and was then competed off with a non-hydrolyzable ATP analog (AMP-PNP).

4.2.4 Protocol

1. From a 10 mM Probe Stock in DMSO, a 40 μM intermediate dilution of the probe was made and transferred to an Echo source plate

		Background Measurement (Antibody/Buffer only)								TYK2-JH2/His-Antibody/Buffer							
		1	2	3	4	5	6	7	8	9	10	11	12	13	14	15	16
Probe 1	A	3.90625 nM	7.8125 nM	15.625 nM	31.25 nM	62.5 nM	125 nM	250 nM	500 nM	3.90625 nM	7.8125 nM	15.625 nM	31.25 nM	62.5 nM	125 nM	250 nM	500 nM
Probe 2	B	3.90625 nM	7.8125 nM	15.625 nM	31.25 nM	62.5 nM	125 nM	250 nM	500 nM	3.90625 nM	7.8125 nM	15.625 nM	31.25 nM	62.5 nM	125 nM	250 nM	500 nM
Probe 3	C	3.90625 nM	7.8125 nM	15.625 nM	31.25 nM	62.5 nM	125 nM	250 nM	500 nM	3.90625 nM	7.8125 nM	15.625 nM	31.25 nM	62.5 nM	125 nM	250 nM	500 nM
Probe 4	D	3.90625 nM	7.8125 nM	15.625 nM	31.25 nM	62.5 nM	125 nM	250 nM	500 nM	3.90625 nM	7.8125 nM	15.625 nM	31.25 nM	62.5 nM	125 nM	250 nM	500 nM
Probe 5	E	3.90625 nM	7.8125 nM	15.625 nM	31.25 nM	62.5 nM	125 nM	250 nM	500 nM	3.90625 nM	7.8125 nM	15.625 nM	31.25 nM	62.5 nM	125 nM	250 nM	500 nM
Probe 6	F	3.90625 nM	7.8125 nM	15.625 nM	31.25 nM	62.5 nM	125 nM	250 nM	500 nM	3.90625 nM	7.8125 nM	15.625 nM	31.25 nM	62.5 nM	125 nM	250 nM	500 nM
Probe 7	G	3.90625 nM	7.8125 nM	15.625 nM	31.25 nM	62.5 nM	125 nM	250 nM	500 nM	3.90625 nM	7.8125 nM	15.625 nM	31.25 nM	62.5 nM	125 nM	250 nM	500 nM
Probe 8	H	3.90625 nM	7.8125 nM	15.625 nM	31.25 nM	62.5 nM	125 nM	250 nM	500 nM	3.90625 nM	7.8125 nM	15.625 nM	31.25 nM	62.5 nM	125 nM	250 nM	500 nM
Probe 9	I	3.90625 nM	7.8125 nM	15.625 nM	31.25 nM	62.5 nM	125 nM	250 nM	500 nM	3.90625 nM	7.8125 nM	15.625 nM	31.25 nM	62.5 nM	125 nM	250 nM	500 nM
Probe 10	J	3.90625 nM	7.8125 nM	15.625 nM	31.25 nM	62.5 nM	125 nM	250 nM	500 nM	3.90625 nM	7.8125 nM	15.625 nM	31.25 nM	62.5 nM	125 nM	250 nM	500 nM
Probe 11	K	3.90625 nM	7.8125 nM	15.625 nM	31.25 nM	62.5 nM	125 nM	250 nM	500 nM	3.90625 nM	7.8125 nM	15.625 nM	31.25 nM	62.5 nM	125 nM	250 nM	500 nM
Probe 12	L	3.90625 nM	7.8125 nM	15.625 nM	31.25 nM	62.5 nM	125 nM	250 nM	500 nM	3.90625 nM	7.8125 nM	15.625 nM	31.25 nM	62.5 nM	125 nM	250 nM	500 nM
Probe 13	M	3.90625 nM	7.8125 nM	15.625 nM	31.25 nM	62.5 nM	125 nM	250 nM	500 nM	3.90625 nM	7.8125 nM	15.625 nM	31.25 nM	62.5 nM	125 nM	250 nM	500 nM
Probe 14	N	3.90625 nM	7.8125 nM	15.625 nM	31.25 nM	62.5 nM	125 nM	250 nM	500 nM	3.90625 nM	7.8125 nM	15.625 nM	31.25 nM	62.5 nM	125 nM	250 nM	500 nM
Probe 15	O	3.90625 nM	7.8125 nM	15.625 nM	31.25 nM	62.5 nM	125 nM	250 nM	500 nM	3.90625 nM	7.8125 nM	15.625 nM	31.25 nM	62.5 nM	125 nM	250 nM	500 nM
Probe 16	P	3.90625 nM	7.8125 nM	15.625 nM	31.25 nM	62.5 nM	125 nM	250 nM	500 nM	3.90625 nM	7.8125 nM	15.625 nM	31.25 nM	62.5 nM	125 nM	250 nM	500 nM

Fig. 13 Plate layout for probe K_d determination.

2. 15.0 nL of the 40 μM probe stock was transferred via echo acoustic transfer to wells A1, A2, and A3 of a Corning 3573 assay plate. The volume of transfer was dependent on the K_d determined earlier, so that the final concentration of probe in the assay plate matches as closely as possible the K_d determined earlier
3. A solution of 1 nM TYK2-JH2, 0.2 nM His-Tb antibody and a solution of 0.2 nM His-Tb antibody were prepared in Buffer A
4. 20 μL of TYK2-JH2 was then added to wells A1 and A2, while 20 μL of the 0.2 nM His-Tb antibody solution only was added to well A3 for a background determination. This was than incubated at room temperature for a period of 1 h (covered).
5. After the 1-h incubation period, 5 μL of Buffer A was added to wells A1 and A3 (for totals and background measurement respectively), and 5 μL of 50 mM AMP-PNP was added to well A2 to compete off the probe from the E/probe/AB complex in that well
6. The plate was then read on an Envision multi-mode reader in HTRF kinetic mode, with one read every 30 s occurring for a period of 75 min

4.2.5 Analysis of kinetic data

At the end of the run, the kinetic data (520/495 ratio) was parsed, and plotted as fluorescence intensity versus time and fit to a first order exponential decay using GraphPad Prism. The fitted data show probe displacement occurring within the one-hour time frame of a typical HTS screen with a k_{off} of $0.0011\,s^{-1}$ (Fig. 14).

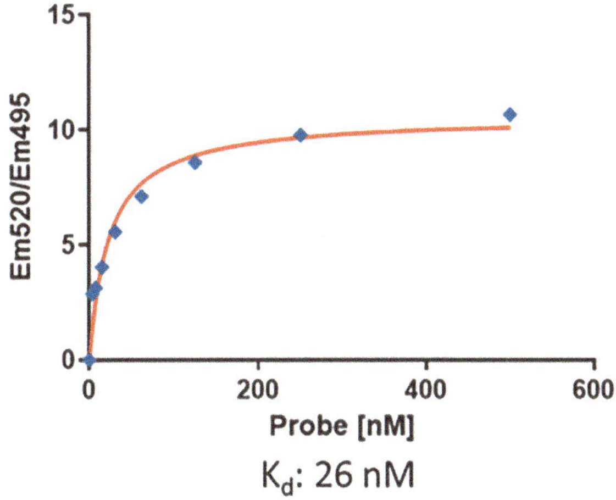

Fig. 14 K_d determination of Promiscuous Probe "A."

In this figure, the displacement of the probe by the ATP analog AMP-PNP is indicated by the descending line comprised of circles and is showing complete displacement within the first 20 min of the reaction. The square symbols at the top of the plot indicate the total signal obtained with enzyme and probe only, while the triangles at the bottom of the plot indicate the background in the absence of enzyme (His-AB and probe only) (Fig. 15).

4.3 SAR screening process

4.3.1 Equipment

Combi-Multi Drop liquid Dispenser (Thermo-Fisher)
Echo 550 Liquid Handler
Remp Plate
Echo Source Plate
Corning 3724 1536 well Assay Plate
Envision Multi-mode Plate Reader

4.3.2 Chemicals

DMSO-Sigma-Aldrich, #D8418
1 M sodium HEPES, pH 7.5—ThermoFisher, #15630080
1 M $MgCl_2$-Invitrogen, #AM9530G
Brij-35-Millipore, #203724
DTT-Sigma-Aldrich, #D9779
BSA-Sigma-Aldrich, #A7030

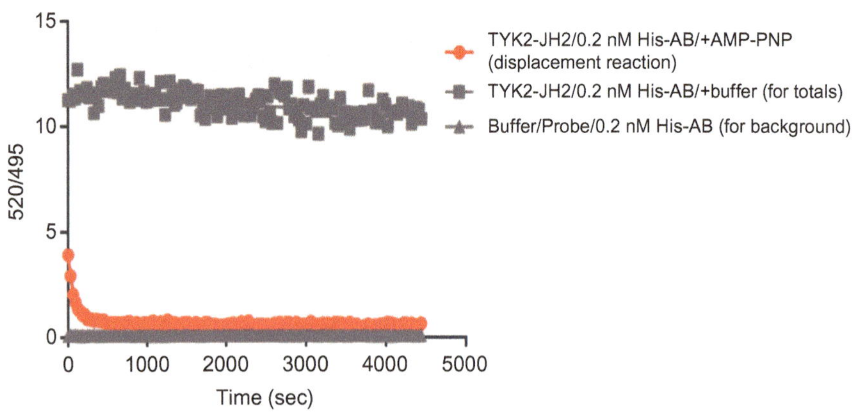

Fig. 15 Displacement of promiscuous probe by the non-hydrolyzable ATP analog AMP-PNP.

His-Terbium Antibody (custom/proprietary prep) or Anti-GST/Terbium antibody

His-TVMV-human TYK2-JH2 (575–869)

4.3.3 Protocol

1. Compounds were serially diluted in an 11 point, 1:3 dose response were provided in a 1536 Echo Diamond plate
2. Using an internally built robotics and IT platform, assay ready plates were created by using an Echo 555 system to transfer 20 nL of compound solution (or DMSO for blanks and totals) to a Corning (3724) 1536 assay plate
3. 2 μL of a solution containing 1 nM Kinase/0.2 nM antibody and probe at probe K_d was then added to columns 3–48 of the assay plate, with antibody and probe only added to columns 1 and 2 for background measurements
4. The plates were than covered, and allowed to incubate at room temperature for a period of 1 h
5. After the 1-h incubation period, the plate was then read on an Envision multi-mode reader in HTRF mode (520/495). IC_{50} determinations were calculated from the 520/495 ratio using the no enzyme reactions for 100% inhibition and DMSO-only reactions for 0% inhibition. Dose – response curves were generated to determine the concentration required for inhibiting 50% of the HTRF signal (IC_{50}) (Fig. 16) (Moslin et al., 2019).

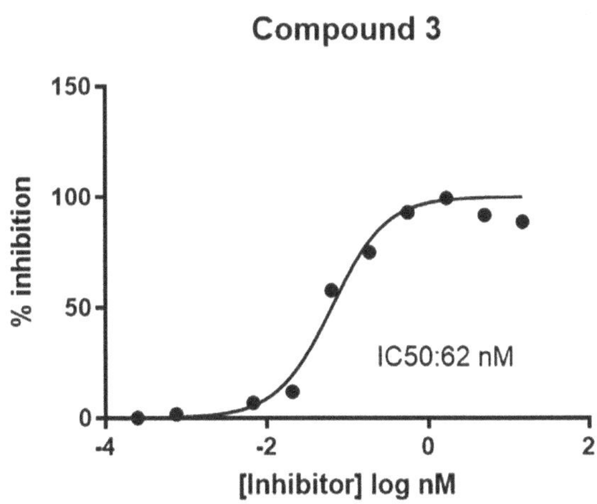

Fig. 16 IC_{50} determination of Compound 3 showing the 11 point dose response curve used in the IC_{50} determination described above with compound 3 having an IC_{50} of 62 nM.

5. Summary

In this chapter we have described methods used to identify, characterize, and optimize small molecule binders to the TYK2-JH2 pseudokinase domain. We reported how BMS used an approach that married kinome-wide inhibition profiles with cellular activity against IL-23-stimulated transcriptional response and identified TYK2-JH2 kinase and novel inhibitors (ITPs and IZPs) that are involved in the IL-23 signal transduction pathway.

We characterized the binding of ITPs and IZPs to TYK2-JH2 by using ITC and TSA and determined, at the time, the first 3-dimensional structure of TYK2-JH2 pseudokinase. We observed in three dimensions the factors that defined TYK2-JH2 as a pseudokinase, detailed the Compound 3 binding site and explained the unique nature of the ligand binding pocket indicating why the JH2 domain of TYK2 is well suited to the design of highly selective small molecule ligands. We also described the optimization of screening/SAR assays which lead to the discovery of the nicotinamide chemotype, which was ultimately optimized to the clinical candidate BMS-986165 (deucravacitinib). Specific screening challenges related to compound throughput and data analysis were discussed, as were the novel approaches we took to develop an HTS screen centered around the use of fluorescent probes based on promiscuous kinase inhibitor scaffolds.

By designing small molecule inhibitors that can only bind the pseudokinase domain, we have been able to selectively inhibit the catalytic activity of TYK2 while limiting activity against the other Janus family members. The mechanism by which binding to the pseudokinase domain inhibits TYK2 is not clearly known but given that the JH2 and JH1 domains form a significant interface that appears to be auto inhibitory and that mutations at this interface cause activation (Lupardus et al., 2014), we postulate that the binding of these allosteric inhibitors to the TYK2 JH2 domain stabilizes an inhibitory conformation, thereby preventing activation. As a result, targeting the JH2 domain provides a method for developing more highly selective TYK2 inhibitors such as BMS-986165 which has been advanced into clinical studies as a potential treatment for patients suffering from a spectrum of IFN- and IL-12/IL-23 driven inflammatory and autoimmune diseases (Wrobleski et al., 2019).

Acknowledgments

The co-authors of this chapter would like to acknowledge Lynda Cook, Donna Pedicord, Kevin O'Malley, Suzanne Edavettal, ChiehYing Chang, Dianlin Xie, Sophie Wu, Mark Witmer, Adriana Zupa-Fernandez, Kristen Pike, Lihong Cheng, Yang Hong, Lisa Elkin, Ramesh Padmanabha, Yuval Blat, William Pitts, James Burke, and David Weinstein for providing and analyzing experimental data. We would like to thank Iyoncy Rodrigo, Max Ruzanov, and Mian Gao for providing details for some of the experimental methods and we humbly thank Ryan Moslin, Lihong Cheng, Qihong Zhao, Stephen Johnson, Brian Arey, Robert Borzilleri, John Throup, and Charu Chaudhry for helpful editorial comments and managerial support.

References

Bazin, H., Preaudat, M., Trinquet, E., & Mathis, G. (2001). Homogeneous time resolved fluorescence resonance energy transfer using rare earth cryptates as a tool for probing molecular interactions in biology. *Spectrochimica Acta. Part A, Molecular and Biomolecular Spectroscopy*, *57*(11), 2197–2211. https://doi.org/10.1016/s1386-1425(01)00493-0.

Borzilleri, R. M., Hart, A. C., Moslin, R., Tokarski, J. S., & Wrobleski, S. T. (2021). JAK family inhibitors for autoimmune diseases. In *Burger's medicinal chemistry and drug discovery* (pp. 1–86). Wiley.

Brandts, J. F., & Lin, L. N. (1990). Study of strong to ultratight protein interactions using differential scanning calorimetry. *Biochemistry*, *29*(29), 6927–6940. https://doi.org/10.1021/bi00481a024.

Bricogne, G., Blanc, E., Brandl, M., Flensburg, C., Keller, P., Paciorek, W., et al. (2011). *BUSTER version 2.11.5*. Cambridge, United Kingdom: Global Phasing Ltd.

Burke, J. R., Cheng, L., Gillooly, K. M., Strnad, J., Zupa-Fernandez, A., Catlett, I. M., et al. (2019). Autoimmune pathways in mice and humans are blocked by pharmacological stabilization of the TYK2 pseudokinase domain. *Science Translational Medicine*, *11*(502). https://doi.org/10.1126/scitranslmed.aaw1736.

Chaudhry, C., Tebben, A., Tokarski, J. S., Borzilleri, R., Pitts, W. J., Lippy, J., et al. (2021). An innovative kinome platform to accelerate small-molecule inhibitor discovery and optimization from hits to leads. *Drug Discovery Today*, *26*(5), 1115–1125. https://doi.org/10.1016/j.drudis.2021.01.018.

Chimalakonda, A., Burke, J., Cheng, L., Catlett, I., Tagen, M., Zhao, Q., et al. (2021). Selectivity profile of the tyrosine kinase 2 inhibitor Deucravacitinib compared with Janus kinase 1/2/3 inhibitors. *Dermatology and Therapy (Heidelb)*, *11*(5), 1763–1776. https://doi.org/10.1007/s13555-021-00596-8.

Curtis, J. R., Xie, F., Yun, H., Bernatsky, S., & Winthrop, K. L. (2016). Real-world comparative risks of herpes virus infections in tofacitinib and biologic-treated patients with rheumatoid arthritis. *Annals of the Rheumatic Diseases*, *75*(10), 1843–1847. https://doi.org/10.1136/annrheumdis-2016-209131.

Degorce, F., Card, A., Soh, S., Trinquet, E., Knapik, G. P., & Xie, B. (2009). HTRF: A technology tailored for drug discovery—A review of theoretical aspects and recent applications. *Current Chemistry Genomics*, *3*, 22–32. https://doi.org/10.2174/1875397300903010022.

Dhar, T. G. M., & Dyckman, A. J. (2017). Evolution of small-molecule immunology research—Changes since CMC II. In D. R. A. S. E. W. S. Chackalamannil (Ed.), *Vol. 5. Comprehensive medicinal chemistry III* (3rd ed., pp. 395–419). Amsterdam: Elsevier.

Emsley, P., Lohkamp, B., Scott, W. G., & Cowtan, K. (2010). Features and development of coot. *Acta Crystallographica. Section D, Biological Crystallography*, *66*(Pt 4), 486–501. https://doi.org/10.1107/S0907444910007493.

Ferrao, R., & Lupardus, P. J. (2017). The Janus kinase (JAK) FERM and SH2 domains: Bringing specificity to JAK-receptor interactions. *Frontiers in Endocrinology (Lausanne)*, *8*, 71. https://doi.org/10.3389/fendo.2017.00071.

Flex, E., Petrangeli, V., Stella, L., Chiaretti, S., Hornakova, T., Knoops, L., et al. (2008). Somatically acquired JAK1 mutations in adult acute lymphoblastic leukemia. *The Journal of Experimental Medicine*, *205*(4), 751–758. https://doi.org/10.1084/jem.20072182.

Furie, R., Khamashta, M., Merrill, J. T., Werth, V. P., Kalunian, K., Brohawn, P., et al. (2017). Anifrolumab, an anti-interferon-alpha receptor monoclonal antibody, in moderate-to-severe systemic lupus erythematosus. *Arthritis & Rhematology*, *69*(2), 376–386. https://doi.org/10.1002/art.39962.

Gakovic, M., Ragimbeau, J., Francois, V., Constantinescu, S. N., & Pellegrini, S. (2008). The Stat3-activating Tyk2 V678F mutant does not up-regulate signaling through the type I interferon receptor but confers ligand hypersensitivity to a homodimeric receptor. *The Journal of Biological Chemistry*, *283*(27), 18522–18529. https://doi.org/10.1074/jbc.M801427200.

Gavai, A. V., Fink, B. E., Fairfax, D. J., Martin, G. S., Rossiter, L. M., Holst, C. L., et al. (2009). Discovery and preclinical evaluation of [4-[[1-(3-fluorophenyl)methyl]-1H-indazol-5-ylamino]-5-methylpyrrolo[2,1-f][1,2,4]triazin-6-yl]carbamic acid, (3S)-3-morpholinylmethyl ester (BMS-599626), a selective and orally efficacious inhibitor of human epidermal growth factor receptor 1 and 2 kinases. *Journal of Medicinal Chemistry*, *52*(21), 6527–6530. https://doi.org/10.1021/jm9010065.

Ghoreschi, K., Laurence, A., & O'Shea, J. J. (2009). Janus kinases in immune cell signaling. *Immunological Reviews*, *228*(1), 273–287. https://doi.org/10.1111/j.1600-065X.2008.00754.x.

Greenhough, L. A., Clarke, G., Phillipou, A. N., Mazani, F., Karamshi, B., Rowe, S., et al. (2021). Reducing false positives through the application of fluorescence lifetime technology: A comparative study using TYK2 kinase as a model system. *SLAS Discovery*, *26*(5), 663–675. https://doi.org/10.1177/24725552211002472.

Jeong, E. G., Kim, M. S., Nam, H. K., Min, C. K., Lee, S., Chung, Y. J., et al. (2008). Somatic mutations of JAK1 and JAK3 in acute leukemias and solid cancers. *Clinical Cancer Research*, *14*(12), 3716–3721. https://doi.org/10.1158/1078-0432.CCR-07-4839.

Kashani, A., & Schwartz, D. A. (2019). The expanding role of anti-IL-12 and/or anti-IL-23 antibodies in the treatment of inflammatory bowel disease. *Gastroenterology and Hepatology (N Y)*, *15*(5), 255–265. Retrieved from https://www.ncbi.nlm.nih.gov/pubmed/31360139.

Khamashta, M., Merrill, J. T., Werth, V. P., Furie, R., Kalunian, K., Illei, G. G., et al. (2016). Sifalimumab, an anti-interferon-alpha monoclonal antibody, in moderate to severe systemic lupus erythematosus: A randomised, double-blind, placebo-controlled study. *Annals of the Rheumatic Diseases*, *75*(11), 1909–1916. https://doi.org/10.1136/annrheumdis-2015-208562.

Kralovics, R., Teo, S. S., Buser, A. S., Brutsche, M., Tiedt, R., Tichelli, A., et al. (2005). Altered gene expression in myeloproliferative disorders correlates with activation of signaling by the V617F mutation of Jak2. *Blood*, *106*(10), 3374–3376. https://doi.org/10.1182/blood-2005-05-1889.

Kranz, J. K., & Schalk-Hihi, C. (2011). Protein thermal shifts to identify low molecular weight fragments. *Methods in Enzymology*, *493*, 277–298. https://doi.org/10.1016/B978-0-12-381274-2.00011-X.

Ladbury, J. E., Klebe, G., & Freire, E. (2010). Adding calorimetric data to decision making in lead discovery: A hot tip. *Nature Reviews. Drug Discovery*, *9*(1), 23–27. https://doi.org/10.1038/nrd3054.

Liu, J., Cao, S., Kim, S., Chung, E. Y., Homma, Y., Guan, X., et al. (2005). Interleukin-12: An update on its immunological activities, signaling and regulation of gene expression. *Current Immunology Reviews*, *1*(2), 119–137. https://doi.org/10.2174/1573395054065115.

Liu, C., Lin, J., Everlof, G., Gesenberg, C., Zhang, H., Marathe, P. H., et al. (2013). Synthesis and evaluation of carbamoylmethylene linked prodrugs of BMS-582949, a clinical p38alpha inhibitor. *Bioorganic & Medicinal Chemistry Letters*, *23*(10), 3028–3033. https://doi.org/10.1016/j.bmcl.2013.03.022.

Liu, C., Lin, J., Langevine, C., Smith, D., Li, J., Tokarski, J. S., et al. (2021). Discovery of BMS-986202: A clinical Tyk2 inhibitor that binds to Tyk2 JH2. *Journal of Medicinal Chemistry*, *64*(1), 677–694. https://doi.org/10.1021/acs.jmedchem.0c01698.

Liu, C., Lin, J., Moslin, R., Tokarski, J. S., Muckelbauer, J., Chang, C., et al. (2019). Identification of Imidazo[1,2-b]pyridazine derivatives as potent, selective, and orally active Tyk2 JH2 inhibitors. *ACS Medicinal Chemistry Letters*, *10*(3), 383–388. https://doi.org/10.1021/acsmedchemlett.9b00035.

Lombardo, L. J., Lee, F. Y., Chen, P., Norris, D., Barrish, J. C., Behnia, K., et al. (2004). Discovery of N-(2-chloro-6-methyl- phenyl)-2-(6-(4-(2-hydroxyethyl)- piperazin-1-yl)-2-methylpyrimidin-4- ylamino)thiazole-5-carboxamide (BMS-354825), a dual Src/Abl kinase inhibitor with potent antitumor activity in preclinical assays. *Journal of Medicinal Chemistry*, *47*(27), 6658–6661. https://doi.org/10.1021/jm049486a.

Lu, X., Smaill, J. B., & Ding, K. (2020). New promise and opportunities for allosteric kinase inhibitors. *Angewandte Chemie (International Ed. in English)*, *59*(33), 13764–13776. https://doi.org/10.1002/anie.201914525.

Lupardus, P. J., Ultsch, M., Wallweber, H., Bir Kohli, P., Johnson, A. R., & Eigenbrot, C. (2014). Structure of the pseudokinase-kinase domains from protein kinase TYK2 reveals a mechanism for Janus kinase (JAK) autoinhibition. *Proceedings of the National Academy of Sciences of the United States of America*, *111*(22), 8025–8030. https://doi.org/10.1073/pnas.1401180111.

Mathis, G. (1995). Probing molecular interactions with homogeneous techniques based on rare earth cryptates and fluorescence energy transfer. *Clinical Chemistry*, *41*(9), 1391–1397. Retrieved from https://www.ncbi.nlm.nih.gov/pubmed/7656455.

Matulis, D., Kranz, J. K., Salemme, F. R., & Todd, M. J. (2005). Thermodynamic stability of carbonic anhydrase: Measurements of binding affinity and stoichiometry using Thermo Fluor. *Biochemistry*, *44*(13), 5258–5266. https://doi.org/10.1021/bi048135v.

McCoy, A. J., Grosse-Kunstleve, R. W., Adams, P. D., Winn, M. D., Storoni, L. C., & Read, R. J. (2007). Phaser crystallographic software. *Journal of Applied Crystallography*, *40*(Pt 4), 658–674. https://doi.org/10.1107/S0021889807021206.

Min, X., Ungureanu, D., Maxwell, S., Hammaren, H., Thibault, S., Hillert, E. K., et al. (2015). Structural and functional characterization of the JH2 pseudokinase domain of JAK family tyrosine kinase 2 (TYK2). *The Journal of Biological Chemistry*, *290*(45), 27261–27270. https://doi.org/10.1074/jbc.M115.672048.

Misra, R. N., Xiao, H. Y., Kim, K. S., Lu, S., Han, W. C., Barbosa, S. A., et al. (2015). Correction to N-(cycloalkylamino)acyl-2-aminothiazole inhibitors of cyclin-dependent kinase 2. N-[5-[[[5-(1,1-Dimethylethyl)-2-oxazolyl]methyl]thio]-2-thiazolyl]-4-piperidineca rboxamide (BMS-387032), a highly efficacious and selective antitumor agent. *Journal of Medicinal Chemistry*, *58*(18), 7609. https://doi.org/10.1021/acs.jmedchem.5b01294.

Moslin, R., Gardner, D., Santella, J., Zhang, Y., Duncia, J. V., Liu, C., et al. (2017). Identification of imidazo[1,2-b]pyridazine TYK2 pseudokinase ligands as potent and selective allosteric inhibitors of TYK2 signalling. *Medchemcomm*, *8*(4), 700–712. https://doi.org/10.1039/c6md00560h.

Moslin, R., Zhang, Y., Wrobleski, S. T., Lin, S., Mertzman, M., Spergel, S., et al. (2019). Identification of N-methyl nicotinamide and N-methyl pyridazine-3-carboxamide pseudokinase domain ligands as highly selective allosteric inhibitors of tyrosine kinase 2 (TYK2). *Journal of Medicinal Chemistry*, *62*(20), 8953–8972. https://doi.org/10.1021/acs.jmedchem.9b00443.

Murshudov, G. N., Skubak, P., Lebedev, A. A., Pannu, N. S., Steiner, R. A., Nicholls, R. A., et al. (2011). REFMAC5 for the refinement of macromolecular crystal structures. *Acta Crystallographica. Section D, Biological Crystallography*, *67*(Pt 4), 355–367. https://doi.org/10.1107/S0907444911001314.

Niesen, F. H., Berglund, H., & Vedadi, M. (2007). The use of differential scanning fluorimetry to detect ligand interactions that promote protein stability. *Nature Protocols*, *2*(9), 2212–2221. https://doi.org/10.1038/nprot.2007.321.

Otwinowski, Z., & Minor, W. (1997). Processing of X-ray diffraction data collected in oscillation mode. *Methods in Enzymology*, *276*, 307–326. Retrieved from https://www.ncbi.nlm.nih.gov/pubmed/27754618.

Pantoliano, M. W., Petrella, E. C., Kwasnoski, J. D., Lobanov, V. S., Myslik, J., Graf, E., et al. (2001). High-density miniaturized thermal shift assays as a general strategy for drug discovery. *Journal of Biomolecular Screening*, *6*(6), 429–440. https://doi.org/10.1177/108705710100600609.

Percy, M. J., & McMullin, M. F. (2005). The V617F JAK2 mutation and the myeloproliferative disorders. *Hematological Oncology*, *23*(3–4), 91–93. https://doi.org/10.1002/hon.761.

Saharinen, P., & Silvennoinen, O. (2002). The pseudokinase domain is required for suppression of basal activity of Jak2 and Jak3 tyrosine kinases and for cytokine-inducible activation of signal transduction. *The Journal of Biological Chemistry*, *277*(49), 47954–47963. https://doi.org/10.1074/jbc.M205156200.

Saharinen, P., Takaluoma, K., & Silvennoinen, O. (2000). Regulation of the Jak2 tyrosine kinase by its pseudokinase domain. *Molecular and Cellular Biology*, *20*(10), 3387–3395. https://doi.org/10.1128/MCB.20.10.3387-3395.2000.

Saharinen, P., Vihinen, M., & Silvennoinen, O. (2003). Autoinhibition of Jak2 tyrosine kinase is dependent on specific regions in its pseudokinase domain. *Molecular Biology of the Cell*, *14*(4), 1448–1459. https://doi.org/10.1091/mbc.e02-06-0342.

Schroeder, G. M., An, Y., Cai, Z. W., Chen, X. T., Clark, C., Cornelius, L. A., et al. (2009). Discovery of N-(4-(2-amino-3-chloropyridin-4-yloxy)-3-fluorophenyl)-4-ethoxy-1-(4-fluorophenyl)-2-oxo-1,2-dihydropyridine-3-carboxamide (BMS-777607), a selective and orally efficacious inhibitor of the met kinase superfamily. *Journal of Medicinal Chemistry*, *52*(5), 1251–1254. https://doi.org/10.1021/jm801586s.

Selvin, P. R. (2002). Principles and biophysical applications of lanthanide-based probes. *Annual Review of Biophysics and Biomolecular Structure*, *31*, 275–302. https://doi.org/10.1146/annurev.biophys.31.101101.140927.

Shaw, M. H., Boyartchuk, V., Wong, S., Karaghiosoff, M., Ragimbeau, J., Pellegrini, S., et al. (2003). A natural mutation in the Tyk2 pseudokinase domain underlies altered susceptibility of B10.Q/J mice to infection and autoimmunity. *Proceedings of the National Academy of Sciences of the United States of America*, *100*(20), 11594–11599. https://doi.org/10.1073/pnas.1930781100.

Spergel, S. H., Mertzman, M. E., Kempson, J., Guo, J., Stachura, S., Haque, L., et al. (2019). Discovery of a JAK1/3 inhibitor and use of a prodrug to demonstrate efficacy in a model of rheumatoid arthritis. *ACS Medicinal Chemistry Letters*, *10*(3), 306–311. https://doi.org/10.1021/acsmedchemlett.8b00508.

Staerk, J., Kallin, A., Demoulin, J. B., Vainchenker, W., & Constantinescu, S. N. (2005). JAK1 and Tyk2 activation by the homologous polycythemia vera JAK2 V617F mutation: Cross-talk with IGF1 receptor. *The Journal of Biological Chemistry*, *280*(51), 41893–41899. https://doi.org/10.1074/jbc.C500358200.

Tang, C., Chen, S., Qian, H., & Huang, W. (2012). Interleukin-23: As a drug target for autoimmune inflammatory diseases. *Immunology*, *135*(2), 112–124. https://doi.org/10.1111/j.1365-2567.2011.03522.x.

Teng, M. W., Bowman, E. P., McElwee, J. J., Smyth, M. J., Casanova, J. L., Cooper, A. M., et al. (2015). IL-12 and IL-23 cytokines: From discovery to targeted therapies for immune-mediated inflammatory diseases. *Nature Medicine*, *21*(7), 719–729. https://doi.org/10.1038/nm.3895.

Tokarski, J. S., Newitt, J. A., Chang, C. Y., Cheng, J. D., Wittekind, M., Kiefer, S. E., et al. (2006). The structure of Dasatinib (BMS-354825) bound to activated ABL kinase domain elucidates its inhibitory activity against imatinib-resistant ABL mutants. *Cancer Research*, *66*(11), 5790–5797. https://doi.org/10.1158/0008-5472.CAN-05-4187.

Tokarski, J. S., Zupa-Fernandez, A., Tredup, J. A., Pike, K., Chang, C., Xie, D., et al. (2015). Tyrosine kinase 2-mediated signal transduction in T lymphocytes is blocked by pharmacological stabilization of its pseudokinase domain. *The Journal of Biological Chemistry*, *290*(17), 11061–11074. https://doi.org/10.1074/jbc.M114.619502.

Vincenti, F., Tedesco Silva, H., Busque, S., O'Connell, P., Friedewald, J., Cibrik, D., et al. (2012). Randomized phase 2b trial of tofacitinib (CP-690,550) in de novo kidney transplant patients: Efficacy, renal function and safety at 1 year. *American Journal of Transplantation*, *12*(9), 2446–2456. https://doi.org/10.1111/j.1600-6143.2012.04127.x.

Virtanen, A. T., Haikarainen, T., Raivola, J., & Silvennoinen, O. (2019). Selective JAKinibs: Prospects in inflammatory and autoimmune diseases. *BioDrugs*, *33*(1), 15–32. https://doi.org/10.1007/s40259-019-00333-w.

Wan, H., Schroeder, G. M., Hart, A. C., Inghrim, J., Grebinski, J., Tokarski, J. S., et al. (2015). Discovery of a highly selective JAK2 inhibitor, BMS-911543, for the treatment of myeloproliferative neoplasms. *ACS Medicinal Chemistry Letters*, *6*(8), 850–855. https://doi.org/10.1021/acsmedchemlett.5b00226.

Watterson, S. H., Liu, Q., Beaudoin Bertrand, M., Batt, D. G., Li, L., Pattoli, M. A., et al. (2019). Discovery of Branebrutinib (BMS-986195): A strategy for identifying a highly potent and selective covalent inhibitor providing rapid in vivo inactivation of Bruton's tyrosine kinase (BTK). *Journal of Medicinal Chemistry*, *62*(7), 3228–3250. https://doi.org/10.1021/acs.jmedchem.9b00167.

Wrobleski, S. T., Moslin, R., Lin, S., Zhang, Y., Spergel, S., Kempson, J., et al. (2019). Highly selective inhibition of tyrosine kinase 2 (TYK2) for the treatment of autoimmune diseases: Discovery of the allosteric inhibitor BMS-986165. *Journal of Medicinal Chemistry*, *62*(20), 8973–8995. https://doi.org/10.1021/acs.jmedchem.9b00444.

Yamaoka, K., Saharinen, P., Pesu, M., Holt, V. E., 3rd, Silvennoinen, O., & O'Shea, J. J. (2004). The Janus kinases (Jaks). *Genome Biology*, *5*(12), 253. https://doi.org/10.1186/gb-2004-5-12-253.

Yeh, T. C., Dondi, E., Uze, G., & Pellegrini, S. (2000). A dual role for the kinase-like domain of the tyrosine kinase Tyk2 in interferon-alpha signaling. *Proceedings of the National Academy of Sciences of the United States of America*, *97*(16), 8991–8996. https://doi.org/10.1073/pnas.160130297.

Zhang, R., & Monsma, F. (2010). Fluorescence-based thermal shift assays. *Current Opinion in Drug Discovery & Development*, *13*(4), 389–402. Retrieved from https://www.ncbi.nlm.nih.gov/pubmed/20597025.

Zhao, Z. J., Vainchenker, W., Krantz, S. B., Casadevall, N., & Constantinescu, S. N. (2005). Role of tyrosine kinases and phosphatases in polycythemia vera. *Seminars in Hematology*, *42*(4), 221–229. https://doi.org/10.1053/j.seminhematol.2005.05.019.

Zhao, R., Xing, S., Li, Z., Fu, X., Li, Q., Krantz, S. B., et al. (2005). Identification of an acquired JAK2 mutation in polycythemia vera. *The Journal of Biological Chemistry*, *280*(24), 22788–22792. https://doi.org/10.1074/jbc.C500138200.

CHAPTER TWENTY-FOUR

A suite of *in vitro* and *in vivo* assays for monitoring the activity of the pseudokinase Bud32

Jonah Beenstock[a,*], Samara Mishelle Ona[a,b], and Frank Sicheri[a,b,c,*]

[a]The Lunenfeld-Tanenbaum Research Institute, Mount Sinai Hospital, Toronto, ON, Canada
[b]Department of Molecular Genetics, University of Toronto, Toronto, ON, Canada
[c]Department of Biochemistry, University of Toronto, Toronto, ON, Canada
*Corresponding authors: e-mail address: beenstock@lunenfeld.ca; sicheri@lunenfeld.ca

Contents

Methods in Enzymology, Volume 667
ISSN 0076-6879
https://doi.org/10.1016/bs.mie.2022.03.052

Abstract

Bud32 is a member of the protein kinase superfamily that is invariably conserved in all eukaryotic and archaeal organisms. In both of these kingdoms, Bud32 forms part of the KEOPS (*K*inase, *E*ndopeptidase and *O*ther *P*roteins of *S*mall size) complex together with the three other core subunits Kae1, Cgi121 and Pcc1. KEOPS functions to generate the universal and essential tRNA post-transcriptional modification N^6-theronylcarbamoyl adenosine (t^6A), which is present at position A37 in all tRNAs that bind to codons with an A in the first position (ANN decoding tRNAs) and is essential for the fidelity of translation. Mutations in KEOPS genes in humans underlie the severe genetic disease Galloway-Mowat syndrome, which results in childhood death. KEOPS activity depends on two major functions of Bud32. Firstly, Bud32 facilitates efficient tRNA substrate recruitment to KEOPS and helps in positioning the A37 site of the tRNA in the active site of Kae1, which carries out the t^6A modification reaction. Secondly, the enzymatic activity of Bud32 is required for the ability of KEOPS to modify tRNA. Unlike conventional protein kinases, which employ their enzymatic activity for phosphorylation of protein substrates, Bud32 employs its enzymatic activity to function as an ATPase. Herein, we present a comprehensive suite of assays to monitor the activity of Bud32 in KEOPS *in vitro* and *in vivo*. We present protocols for the purification of the archaeal KEOPS proteins and of a tRNA substrate, as well as protocols for monitoring the ATPase activity of Bud32 and for analyzing its role in tRNA binding. We further present a complementary protocol for monitoring the role Bud32 has in cell growth in yeast.

1. Introduction

The eukaryotic protein kinase superfamily confers the spatial and temporal control over virtually all cellular processes. The importance of kinases for the biology of eukaryotic cells and organisms is perhaps best exemplified by the tight connection that aberrant kinase activity has to the manifestation of diverse diseases in humans, ranging from cancer, to chronic inflammation and to metabolic disorders (reviewed in Dhillon, Hagan, Rath, & Kolch, 2007; Hardie, 2011; Lahiry, Torkamani, Schork, & Hegele, 2010; Reiterer, Eyers, & Farhan, 2014; Zarrin, Bao, Lupardus, & Vucic, 2021). Protein kinases are defined by the presence of the protein kinase domain that consists of two conserved lobes termed the N-terminal and C-terminal lobes (Kornev & Taylor, 2015; Taylor, Knighton, Zheng, Ten Eyck, & Sowadski, 1992). Catalytic and structural elements that localize to both lobes enable the binding of ATP. In particular, the adenine base of ATP is packed against a flexible kinase hinge region that connects both lobes. The phosphate groups are coordinated by the conserved β3 Lys side chain and by a Mg^{2+}/Mn^{2+} ion that is bound to the side chain of an aspartic acid residue in the conserved Asp-Phe-Gly (DFG) motif (Kornev & Taylor, 2015; Taylor et al., 1992). Most kinases fulfill their roles in cells through the ability to transfer the γ-phosphate of the bound ATP molecule to hydroxyl bearing side chains of protein substrates, generating ADP as a biproduct. This transfer event is facilitated by a catalytic Asp residue within a canonical His-Arg-Asp (HRD) motif in a region termed the catalytic loop.

Somewhat surprisingly, a considerable portion of the protein kinase family are classified as pseudokinases and function by other non-conventional mechanisms (Boudeau, Miranda-Saavedra, Barton, & Alessi, 2006; Jacobsen & Murphy, 2017; Kung & Jura, 2016; Reiterer et al., 2014). In humans for example, ~10% of all protein kinases are considered to be pseudokinases (Kung & Jura, 2016). Pseudokinases maintain the canonical bi-lobal domain organization but lack certain key elements within their fold, which leads to the repurposing of the kinase domain for functions other than protein phosphorylation. Many pseudokinases, such as HER3 and STRADα for example, are either completely devoid of ATP hydrolysis activity or have very low activity due to the absence of key catalytic residues and act rather as scaffolds (Jura et al., 2009; Zeqiraj, Filippi, Deak, Alessi, & van Aalten, 2009; Zhang, Gureasko, Shen, Cole, & Kuriyan, 2006). Other pseudokinases maintain the ability to hydrolyze ATP, however, instead of transferring the phosphate to a protein side chain they do so to water and generate ADP + Pi as a by-product. Therefore, these kinases function as ATPase enzymes (Beenstock et al., 2020; Ferreira-Cerca, Kiburu, Thomson, LaRonde, & Hurt, 2014). Pseudokinases that act as ATPases (for simplicity we will refer to these enzymes as kinase-ATPases hereafter) are somewhat harder to distinguish from conventional kinases by sequence analysis and can be erroneously classified as regular kinases. It is therefore unclear what portion of the kinome are kinase-ATPases.

Bud32 is a highly conserved member of the protein kinase superfamily with detectable orthologs in all eukaryotic and archaeal species (Leonard, Aravind, & Koonin, 1998; Thiaville, Iwata-Reuyl, & de Crecy-Lagard, 2014). Bud32 is part of the KEOPS (Kinase, Endopeptidase and Other Proteins of Small size) complex together with the Cgi121, Kae1, and Pcc1 proteins (Downey et al., 2006; Kisseleva-Romanova et al., 2006). In eukaryotes, KEOPS contains a fifth subunit termed Gon7 (Fig. 1)

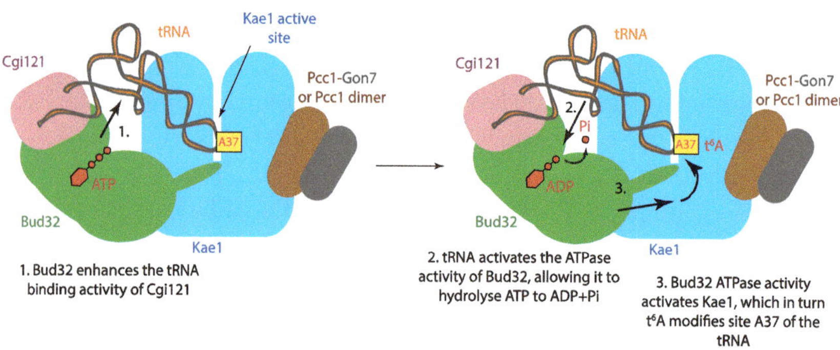

Fig. 1 The roles of the kinase-ATPase Bud32 in the tRNA modifying activity of KEOPS. (left) KEOPS adopts a linear binding architecture. Bud32 is sandwiched between the tRNA recruiting subunit Cgi121 and the tRNA modifying subunit Kae1. Bud32 functions to enhance the tRNA binding activity of Cgi121 and helps in guiding the tRNA A37 modification site to the active site of Kae1. (right) Once the tRNA is engaged by all of KEOPS core subunits, it leads to the activation of Bud32 enzyme activity, which hydrolyzes ATP to ADP + Pi. The catalytic activity of Bud32 in turn serves to activate the tRNA modification activity of Kae1 by an unknown mechanism.

(Downey et al., 2006; Kisseleva-Romanova et al., 2006; Thiaville et al., 2014; Wan et al., 2017). The Kae1 subunit of KEOPS catalyzes the universal tRNA post-transcriptional modification N^6-threonylcarbamoyl adenosine (t^6A) (El Yacoubi et al., 2011; Srinivasan et al., 2011; Wan et al., 2013), while all other subunits including Bud32 play a supporting, yet essential, role for Kae1 activity (Beenstock et al., 2020; Beenstock & Sicheri, 2021). t^6A biosynthesis also depends on the activity of the Sua5 enzyme (El Yacoubi et al., 2009) that generates the reaction intermediate molecule threonylcarbamoyl adenylate (TC-AMP) (Lauhon, 2012), which is used by Kae1 for tRNA modification (Wan et al., 2013) (for a review on the mechanism of tRNA modification by KEOPS, see Beenstock & Sicheri, 2021). t^6A is found specifically at position A37 of ANN-decoding tRNAs (N can be any nucleotide) (Boccaletto et al., 2018), and functions to stabilize the tRNA anticodon loop and to reinforce codon-anticodon binding in the ribosome (Murphy, Ramakrishnan, Malkiewicz, & Agris, 2004; Stuart et al., 2000). As a result, t^6A promotes the fidelity of translation (El Yacoubi et al., 2009, 2011; Thiaville et al., 2016) and is therefore essential for the viability of all organisms (reviewed in Beenstock & Sicheri, 2021). In humans for example, mutations that localize to genes encoding each of the five KEOPS subunits or Sua5 lead to the development of Galloway-Mowat syndrome, which manifests in severe microcephaly and renal dysfunction that result in childhood death (Arrondel et al., 2019; Braun et al., 2017; Edvardson et al., 2017).

Structural analysis revealed that Bud32 has a highly minimalistic kinase domain (Hecker et al., 2008; Mao et al., 2008). Specifically, although Bud32 contains the key canonical catalytic elements characteristic of kinases that are involved in ATP binding and hydrolysis, it lacks elements that are usually involved in substrate binding, such as the C-lobe GHI subdomain as well as an extended activation segment (Mao et al., 2008). Accordingly, no *bona fide* protein substrate has been identified thus far for Bud32 kinase activity *in vivo*. Likewise, although Bud32 exhibits low levels of autophosphorylation activity *in vitro* (Hecker et al., 2008; Mao et al., 2008), this activity does not seem to be relevant for its function *in vivo*. Nonetheless, the catalytic activity of Bud32 is an important feature of its functionality and enzyme-dead mutants of Bud32 cannot support its biochemical and biological function related to t^6A synthesis (Lopreiato et al., 2004; Mao et al., 2008).

This apparent contradiction was resolved by the discovery that Bud32 functions in eukaryotes and archaea as a kinase-ATPase enzyme within KEOPS (Beenstock et al., 2020). The kinase-ATPase activity of Bud32 in KEOPS is intricately regulated by its binding partner Cgi121 and by the binding of tRNA to the assembled KEOPS complex (Fig. 1) (Beenstock et al., 2020; Mao et al., 2008). Corresponding with the essential role the catalytic activity of Bud32 has in supporting cell growth, its kinase-ATPase activity is required for the ability of KEOPS to modify tRNA (Beenstock et al., 2020). Bud32 also promotes the tRNA substrate binding activity of KEOPS, but this function is unrelated to its enzymatic activity. The Cgi121 subunit recruits tRNA to KEOPS, however, maximal tRNA binding activity in eukaryotes and in archaea is only achieved in the presence of Bud32 (Beenstock et al., 2020). Bud32 elevates tRNA binding, most likely by allosterically stabilizing a high affinity binding state in Cgi121 and by forming direct contacts with the tRNA (Fig. 1).

Herein, we provide a comprehensive suite of methods that serve as a versatile platform to address the most central aspects of the activity of archaeal Bud32 proteins *in vitro* as well as a complementary assay in yeast to monitor the role of Bud32 proteins *in vivo*. Firstly, we describe how to purify archaeal Bud32 and its accompanying proteins in KEOPS as well as protocols to purify its tRNA binding partner $tRNA^{Lys}_{UUU}$ (Lys refers to the amino acid that is attached to this tRNA and UUU refers to the sequence of its anticodon triplet). We then provide detailed methods for quantitatively monitoring Bud32 kinase-ATPase activity and tRNA binding activity, using the ADP Glo assay and a fluorescence polarization-based assay, respectively. We also describe a method for measuring the tRNA modification activity of KEOPS *in vitro*, an activity that depends on Bud32, using a high-performance liquid chromatography (HPLC) analysis of nucleoside composition. Finally, we present a method for monitoring Bud32 activity *in vivo* by evaluating the effect Bud32 has on the ability of yeast to grow.

2. Purification of GST- and His-tagged KEOPS proteins and of *in vitro* transcribed tRNA

2.1 Protocols for the purification of archaeal hexahistidine tagged Bud32 and GST-tagged Cgi121, Kae1 and Pcc1 (for a list of key reagents for all protocols, see Table 1)

2.1.1 Protein expression plasmids information

For protein expression (all protein sequences described in this chapter were obtained from the UniProt website), a sequence encoding the *Methanococcus jannaschii* (mj) Bud32 protein (residues 328–535 of the naturally occurring

Table 1 Key resources table.

Reagent or resource	Source	Identifier
Bacterial and Virus Strains		
DH5α *E. coli* competent cells	Thermo Fisher Scientific	18265017
BL21(DE3) *E. coli* competent cells	Thermo Fisher Scientific	EC0114
Chemicals, peptides, and recombinant proteins		
His-tagged TEV protease	NEB	P8112S
T7 RNA polymerase	NEB	M0251L
Quick CIP	NEB	M0525L
Critical commercial assays		
ADP Glo™ assay	Promega	V9101
Plasmid DNA Maxiprep kit	Qiagen	12,162
Experimental models: Cell lines		
W303 *bud32Δ* yeast cells		DDY2067, (Mao et al., 2008)
Oligonucleotides		
647-CCA probe (Alexa-647-5′-CCGCCA-3′)	IDT	
Recombinant DNA		
tRNA expression cassette (sequences shown in Table 3).		
Software and algorithms		
Prism	GraphPad	V9
Chromeleon	Thermo Fisher Scientific	V6.3

Kae1-Bud32 fusion protein mj1130) was ordered as a synthetic clone (Thermo Fisher Scientific, GeneArt Gene Synthesis). The coding sequence was then amplified by PCR using the KOD Hot Start DNA Polymerase (Sigma, Cat# 71086) according to manufacturer's guidelines (for a list of oligonucleotides, see Table 2), while adding 5′ *BamHI* and 3′ *Xho*I sites. The

Table 2 Sequences of oligonucleotides used for subcloning mjBud32, mjKae1, mjCgi121, pfPcc1 and scBud32 into their respective expression plasmids.

Oligonucleotide name	Sequence (from 5′ to 3′)	Destination plasmid
mjBud32 forward	ACGTGGATCCGAAAGGCAAGAAGAGAAAGATTCC	pProEX
mjBud32 reverse	GCCTCGAGTTATTACTCTACATATCTTGC	
mjKae1 forward	CATGGATCCGATGATTTGCTTAGG	pGEX
mjKae1 reverse	GCCTCGAGTTATTATTTAATTTCTTTTATCCAATTAAC	
mjCgi121 forward	CCATGGATCCGATGATAATTAGGGGAATAAG	pGEX
mjCgi121 reverse	CATGCCTCGAGTTACTCTTTCTTTAATTCAATTAAAGC	
pfPcc1 forward	CCATGGATCCGATGAAAGCTAAACGTGTACAGGC	pGEX
pfPcc1 reverse	CATGCCTCGAGTTAAACCTCGATTACGTCGATCG	
scBud32 forward	GCGAGCTCCCCAAACGTCGATTTTACCAGCGACCC	pRS415
scBud32 reverse	GCAAGCTTGAGAATTAATAGATAACAGGGTCAGGTTTGGG	

amplified product was then digested with BamHI and XhoI (NEB, Cat# R3136S and R0146S, respectively) and subcloned into the pProEX HTA backbone (Creative Biogene, Cat# VET1545) in frame with an N-terminal hexahistidine (His) tag followed by the Glu-Asn-Leu-Tyr-Phe-**Gln-Gly** consensus sequence that is recognized by the *Tobacco Etch Virus* (TEV) endoprotease (cleavage site is located between the Gln-Gly residues that are highlighted in bold). Sequences encoding mjKae1 (residues 1–327 of the naturally occurring Kae1-Bud32 fusion protein mj1130), mjCgi121 and *Pyrococcus furiosus* (pf) Pcc1 proteins were similarly ordered as synthetic clones, PCR amplified and subcloned into the pGEX-2T vector *BamHI* and *XhoI* sites (Addgene, Plasmid #1128) in frame with an N-terminal GST-tag followed by a TEV consensus sequence. Expression using the pProEX and pGEX vectors is controlled by IPTG (Isopropyl β-D-1-thiogalactopyranoside) and both vectors contain an ampicillin resistance gene for selection in bacteria.

2.1.2 Equipment, buffers, strains, and reagents

- Chemicals for preparing protein purification buffers and solutions:
- HEPES (Bioshop, Cat# HEP001.5), NaCl (Bioshop, Cat# SOD002-10), imidazole (Thermo Fisher, Cat# 1467-16-9) EDTA (Thermo Fisher, Cat# 6381-92-6,) and dithiothreitol (DTT) (Bioshop, Cat# DTT001-100), tris(2-carboxyethyl) phosphine (TCEP) (Bioshop, Cat# TCE101-50).
- Shaker incubator (Multitron Pro, Infors HT) and static incubator (Boekel Scientific, Cat# 133000).
- High speed centrifuge (Thermo Scientific, Sorvall RC 6+, Cat# 12121680) with F21 8x50y (Thermo Scientific, Cat# 096-084019) and F9 4x1000y (Thermo Scientific, Cat# 096-041032) rotors.
- Benchtop centrifuge (Beckman Coulter Allegra X-15R).
- IPTG (UBPBio, Cat# 1010100).
- Homogenizer (Avestin, EmulsiFlex-C3) or alternatively a Sonicator (Qsonica, Cat# Q500).
- Peristaltic pump (Gilson, MINIPULS 3, Cat# F155009) and tubing (Gilson, Cat# F117970) for loading columns.
- Fast Protein Liquid Chromatography (FPLC) unit (GE, AKTA Purifier 10 FPLC System w/UPC-900).
- 5 mL Ni-NTA column (Cytiva, Hi TrapTM 5 mL chelating HP, Cat# 17040903) and 1 mL Ni-NTA column (Cytiva, Hi TrapTM chelating FF, Cat# 17040801).

- Open gravity columns (Bio-Rad, Econo-Column® Chromatography Columns, Cat# 7372507) and glutathione Sepharose 4B resin (GE, Cat# 17-0756-04).
- Electrophoresis unit (Bio-Rad, Cat# 1658004).
- Synergy NEO plate reader with a Take3 Micro-Volume Plate and Gen5 software (Bio Tek).
- Four PYREX Erlenmeyer Shaker Flasks (Thomas Scientific, Cat# 1234F86), each with 1 L of sterile liquid terrific broth (TB) media (Sigma, Cat# T0918) supplemented with 100 μg/L ampicillin.
- 1 L polypropylene bottle (Beckman, Cat# A98814) and 50 mL polypropylene tubes (Thermo Scientific, Cat# UZ0632747) for high a speed centrifuge.
- Luria-Bertani (LB) agar plates with ampicillin (10 g/L tryptone, 10 g/L NaCl and 5 g/L yeast extract, 20 g/L agar, 100 μg/L ampicillin).
- S200 sizing column (Cytiva, Superdex™ 200 Increase 10/300 GL, Cat# GE28-9909-44).
- *E. coli* BL21(DE3) chemical-competent cells (Thermo Fisher Scientific, Cat# EC0114).
- 50 mL of sterile LB liquid media in a 200 mL PYREX Erlenmeyer shaker flasks.
- Ni-NTA column buffer A (50 mM HEPES pH 7.5, 500 mM NaCl, 10 mM imidazole pH 7.5) and buffer B (50 mM HEPES pH 7.5, 500 mM NaCl, 300 mM imidazole pH 7.5).
- 15% acrylamide SDS-PAGE gels, SDS-PAGE Sample Buffer (6×), Tris-glycine running buffer and Coomassie Brilliant Blue staining solution prepared according to CSH Protocols (http://cshprotocols.cshlp.org/).
- Protein molecular weight marker (Biorad, Cat# 1610374).
- Edvotek™ White Light Box for visualizing SDS-PAGE gels (Fisher Scientific, Cat# S11552).
- Dialysis buffer (50 mM HEPES pH 7.5, 500 mM NaCl and 2 mM TCEP).
- GST column buffer (50 mM HEPES pH 7.5, 500 mM NaCl, 5 mM EDTA, 2 mM DTT).
- Sizing column buffer (20 mM HEPES pH 7.5, 100 mM NaCl, 2 mM DTT).
- Recombinant His-tagged TEV protease (NEB, Cat# P8112S).
- 10 kDa and 3 kDa Amicon Ultra-15 centrifugal filter units (Millipore, Cat# UFC901024 and UFC900324).

- 3.5 kDa membrane tubing kit including closure clips (Spectra/Por, Cat# 132724).
- 200 mM stock of PMSF (phenylmethylsulfonyl fluoride) (Sigma, Cat# p7626) made in 100% ethanol.
- 0.44 μm syringe filters (Cytiva, Whatman GD/X 25 mm, Cat# 6873-2504) and 30 mL sterile syringes (Becton Dickinson, Cat# 302832).

2.1.3 Step-by-step protocol for the expression of Bud32, Kae1, Cgi121 and Pcc1 proteins (4 days). For an overview of the expression and purification process, see (Fig. 2A)

- Transform 100 ng of plasmid DNA into BL21(DE3) competent *E. coli* cells using a standard heat shock protocol, plate transformed cells on LB agar plates supplemented with ampicillin and grow overnight at 37 °C in a static incubator.
- The following day, pick a single colony with a sterile pipette tip into 50 mL liquid LB media supplemented with 100 μg/mL ampicillin and grow overnight in a shaker incubator at 37 °C.
- The following morning, dilute 10 mL of the overnight grown culture into 1 L of fresh TB media supplemented with 100 μg/mL ampicillin.
- Grow fresh cultures at 37 °C in a shaker incubator until the culture reaches an optical density (OD) at 600 nm of approximately 0.8 (typically 2–3 h).
- Reduce shaker temperature to 18 °C. Once cultures have cooled down, induce protein expression by adding IPTG at a final concentration of 0.3 mM. Continue growing cells after IPTG addition overnight at 18 °C.
- The following morning, in 1 L polypropylene bottles, spin down cells using a Sorvall RC 6+ centrifuge with an F9 4x1000y rotor (6000 × *g*, 10 min, 21 °C). Discard supernatant. With a clean spatula, transfer cell pellets into clean beakers or a 50 mL falcon tube and flash freeze pellets by placing them in liquid nitrogen.
- Pellets can be stored at −20 °C for months and even years.

2.1.4 Step-by-step protocol for the purification of Bud32 (2 days)

- All steps are done at room temperature unless specifically stated otherwise.
- Thaw pellet of BL21 cells. Resuspend the pellet in 50 mL Ni-NTA column buffer A (25 mL of buffer per liter of TB used when expressing the protein).

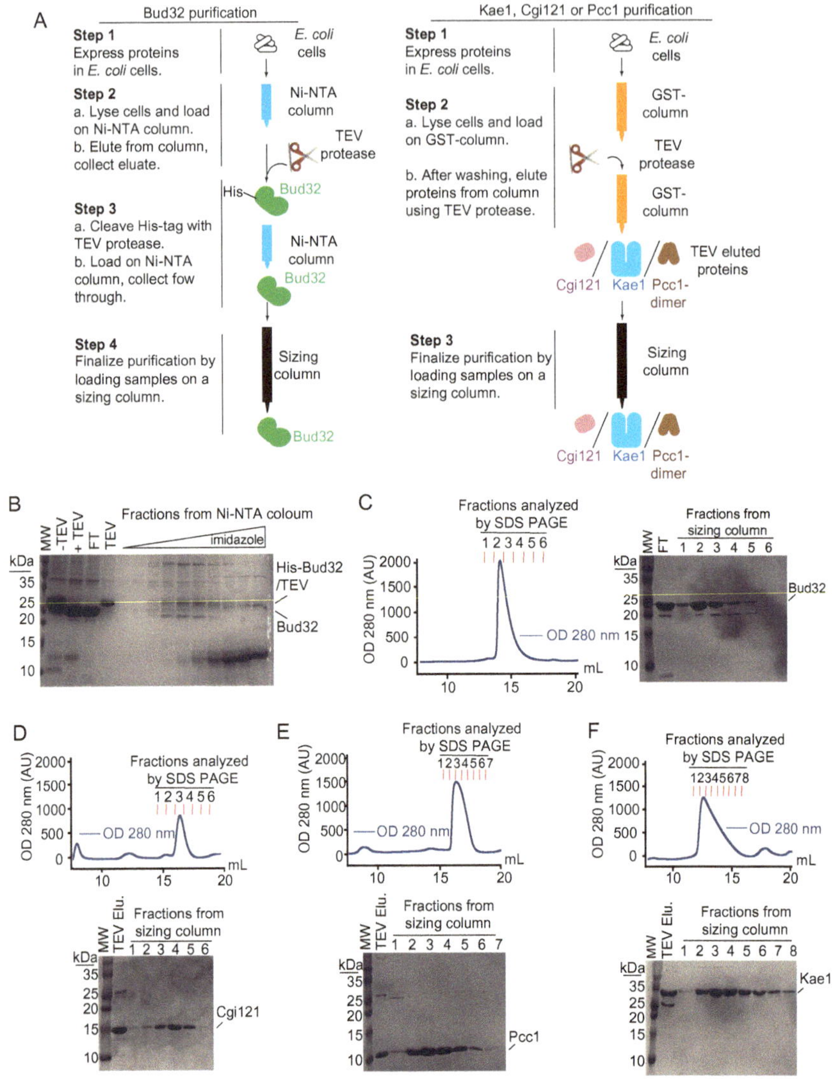

Fig. 2 Purification of *M. jannaschii* Bud32 and its binding partners *M. jannaschii* Cgi121, *M. jannaschii* Kae1 and *P. furiosus* Pcc1. (A) Schematic representation of the major steps in the purification of the KEOPS proteins (left) mjBud32 and (right) mjKae1, mjCgi121 and pfPcc1. (B) A 15% reducing SDS-PAGE was stained with Coomassie, illustrating the His-tag removal from mjBud32 by TEV protease and the subsequent purification of mjBud32 by Ni-NTA subtraction. *Lanes 1–2*: The His-tag is removed from His-mjBud32 by TEV-mediated proteolysis, generating tag-free mjBud32 (denoted by +TEV). *Lane 3*: The mjBud32 protein samples are further purified after TEV cleavage using a Ni-NTA column. Since mjBud32 no longer has a His-tag, it is in the flow-through fraction (denoted by FT). *Lane 4*: recombinant TEV protease. *Lanes 5–14*: Eluted fractions from the Ni-NTA column do not contain the mjBud32 protein. (C–F) Chromatograms and corresponding 15% reducing SDS-PAGE gels stained with Coomassie summarizing the final step of the purification of mjBud32 (C), mjCgi121 (D), pfPcc1 (E) and mjKae1 (F) using an S200 sizing column.

- Once cells are fully resuspended, supplement with 1 mM PMSF and lyse cells by one of the following methods:
  - o Pass cell lysate through a homogenizer 2–3 times, making sure that the homogenizer reaches ~15,000 psi during each round of lysis.
  - o Using a Qsonica sonicator with a macro tip, sonicate cell lysate for 5 rounds. Each round of sonication consists of 20 1 second (s) pulses at an amplitude of 60%. Each sonication pulse is composed of 0.9 s with the sonicator on and 0.1 s with the sonicator off. Make sure to wait 2–3 min between each round of sonication to ensure that the samples do not overheat.
- In 50 mL polypropene tubes, spin down lysed cells for 30 min at 40,000 × *g* at 21 °C using a Sorvall RC 6+ centrifuge with a F21 8x50y rotor to separate supernatant (contains His-mjBud32 proteins) from cell debris. Discard pellet.
- Sonicate supernatant to reduce viscosity using a macro tip. Set sonicator to an amplitude of 40% and perform 3 pulses of 10 s sonication, with 20 s intervals between each pulse.
- Filter supernatant by transferring it through a 0.44 μm syringe filter using a 30 mL syringe.
- Load filtered supernatant on a 5 mL Ni-NTA column that was pre-equilibrated with Ni-NTA column buffer A using a peristaltic pump with a flow rate of ~1 mL per minute.
- Collecting 4 mL fractions, elute bound proteins from 5 mL Ni-NTA column using an FPLC with the following program:
  - o 4 column volumes (CV) 4% Ni-NTA column buffer B and 96% Ni-NTA column buffer A.
  - o 4 CV 8% Ni-NTA column buffer B and 92% Ni-NTA column buffer A.
  - o 4 CV 12% Ni-NTA column buffer B and 88% Ni-NTA column buffer A.
  - o A gradient of 12–100% Ni-NTA column buffer B over 20 CV.
  - o 4 CV 100% Ni-NTA column buffer B.
- Analyze fractions with high absorbance at 280 nm from the gradient elution step using SDS-PAGE. Mix 10 μL from each fraction with 2 μL of 6 × SDS-PAGE Sample Buffer. Load 10 μL on a 15% SDS-PAGE gel, run gel at 200 V for ~45 min or until dye front reaches the end of the gel and stain with Coomassie blue at room temperature with mild agitation for 10 min. Destain with water using mild agitation for 30–60 min and visualize bands by placing the gel on a light box. Pool fractions with

proteins at the molecular weight of ~27 kDa, which corresponds to the molecular weight of His-mjBud32. Save a sample from these pooled fractions for SDS-PAGE (Fig. 2B, lane labeled TEV).

- Supplement the pooled fractions with 2 mM TCEP and 2 mg of recombinant TEV protease and load the mixture into a 3.5 kDa dialysis tubing with clip closures attached to both ends so that the sample will not leak from the tubing. Dialyze with gentle stirring the TEV + His-mjBud32 sample at room temperature overnight against a volume of dialysis buffer that will reduce the imidazole concentration to ~15 mM, following the equation of:

$$\textit{Volume of dialysis buffer} = \frac{15\ \text{mM imidazole} * \textit{volume of His} - \textit{mjBud32 sample}}{\text{estimated concentration of imidazole in His} - \text{mjBud32 sample}}$$

For example, if the volume of the His-mjBud32 sample before dialysis is 50 mL and the estimated concentration of imidazole is 150 mM, dialyze sample against 500 mL of dialysis buffer.

This step will achieve two things:

  - Reduce imidazole concentration to 15 mM.
  - Remove His-tag from mjBud32 proteins. The removal of the His-tag also enables the high confidence identification of mjBud32 proteins on SDS PAGE since the migration of mjBud32 on an SDS PAGE is notably faster than that of His-mjBud32 (Fig. 2B, compare lanes labeled − TEV and + TEV).

- The following morning, using a peristaltic pump load the sample after TEV-digestion on a 1 mL Ni-NTA column that was pre-equilibrated with Ni-NTA column buffer A.
- Collect flow through (contains mjBud32 proteins without the His-tag) and, using an FPLC, elute bound fraction from the 1 mL Ni-NTA column with a gradient of 0–100% Ni-NTA column buffer B over 20 CV. Collect 1 mL fractions. Analyze flow-tduce imidazole concentration hrough fraction and fractions eluted from the column using SDS-PAGE as described above (Fig. 2B).
- Concentrate the mjBud32 sample using a centrifugal filter unit with a 10 kDa cut-off membrane by centrifugation at 4000 × *g* at 21 °C until sample reaches a volume of 1 mL. Load concentrated sample on a 24 mL S200 sizing column pre-equilibrated with sizing column buffer and elute by running 1.5 column volumes of sizing column buffer through the column at a flow rate of 0.5 mL per minute. Collect 0.75 mL fractions.

- Analyze fractions from the sizing column using SDS-PAGE as described above. Bud32 should elute from the column according to its molecular weight of ~24 kDa (Fig. 2C).
- Concentrate Bud32 containing fractions using a centrifugal filter unit with a 10 kDa cut-off membrane by centrifugation at 4000 × *g* at 21 °C. Determine concentration by measuring absorbance at 280 nm using a divisor of 1.063 that is derived from the extinction coefficient estimated using ProtParam (https://web.expasy.org/protparam/). Samples can be concentrated to 20–30 mg/mL. Continue concentrating until this concentration or any desired lower concentration is reached. The typical yield for the purified mjBud32 protein is 5–10 mg of protein per liter of BL21 cells. Aliquot 5–20 μL into 1.5 mL tubes, flash freeze in liquid nitrogen and store at −80 °C. Frozen protein samples are stable for months and even years.

2.1.5 Step-by-step protocol for the purification of mjCgi121, pfPcc1 and mjKae1 proteins (2 days)

- All steps of purification are done at room temperature unless stated otherwise.
- Thaw and resuspend BL21 cell pellets in GST column buffer. Once fully resuspended supplement with 1 mM PMSF.
- Lyse resuspended cells, spin down lysates, and filter cell lysates as described in the protocol for mjBud32 purification.
- Load filtered cell lysates on an open gravity column containing 5 mL glutathione Sepharose resin (denoted GST column) that was pre-equilibrated with GST column buffer.
- Wash GST column with 20 CV GST column buffer.
- Stop column flow. Add 10 mL GST column buffer supplemented with 2 mg of recombinant TEV protease. Incubate overnight without agitation. This step should release mjCgi121, pfPcc1 and mjKae1 proteins from the GST tag, which will remain bound to the column resin.
- The following morning collect the flow through from the GST column, which should contain mjCgi121, pfPcc1 and mjKae1 proteins (Fig. 2D–F, lanes labeled TEV Elu.). Concentrate using a centrifugal filter unit with a 10 kDa cut-off membrane for mjKae1 and mjCgi121 or a 3 kDa cut-off membrane for pfPcc1. Load the concentrated samples separately on a 24 mL S200 sizing column pre-equilibrated with sizing column buffer and elute by running 1.5 column volumes of sizing column buffer at a flow rate of 0.5 mL per minute. Collect 0.75 mL fractions.

- Analyze fractions with high absorbance at 280 nm from sizing column using 15% SDS-PAGE as described for mjBud32. mjKae1 should elute according to its molecular weight of ~36 kDa. mjCgi121 should elute according to its molecular weight of ~16 kDa. pfPcc1 should elute according to the molecular weight of the pfPcc1 dimer of ~18 kDa (Fig. 2D–F).
- Concentrate protein-containing fractions using a centrifugal filter unit with a 10 kDa cut-off membrane for mjKae1 and mjCgi121 or a 3 kDa cut-off membrane for pfPcc1 by centrifugation at 4000 × *g* at 21 °C until a concentration of 15–20 mg/mL, or any desired lower concentration is reached. Determine concentrations by measuring absorbance at OD 280 nm, using divisors calculated from the extinction coefficients: 0.421, 0.949, 1.193 for mjCgi121, mjKae1 and pfPcc1, respectively. Aliquot, flash freeze in liquid nitrogen and store at −80 °C. Frozen protein samples are stable for months and even years.

2.1.6 Expected outcomes

The expected yield of purified mjBud32, mjCgi121 and pfPcc1 is approximately 5–10 mg of protein per liter of BL21 cells. The expected yield of purified mjKae1 is approximately 0.5–1 mg of protein per liter of BL21 cells.

2.1.7 Safety considerations and standards

The components of these protocols do not present any special health and safety hazard. Regular lab safety procedures should be implemented.

2.1.8 Alternative methods/procedures

As an alternative to purifying the archaeal KEOPS proteins, protocols have been established for purifying yeast KEOPS proteins (Collinet et al., 2011; Zhang et al., 2015) and for purifying human KEOPS proteins (Arrondel et al., 2019; Beenstock et al., 2020; Wan et al., 2017).

2.2 Protocols for the *in vitro* transcription and purification of *M. jannashci* $tRNA^{Lys}_{UUU}$

2.2.1 tRNA synthesis vector information and purification

For tRNA synthesis a synthetic tRNA expression cassette (Thermo Fischer Scientific, GeneArt) was ordered containing the DNA sequence corresponding to the mature *M. jannashci* $tRNA^{Lys}_{UUU}$ from the GtRNAdb web site (http://gtrnadb.ucsc.edu/) 3′ to the promoter sequence recognized by the T7 RNA polymerase and 5′ to the sequence of the hepatitis delta virus ribozyme followed by the sequence recognized by the BamHI restriction enzyme (see Fig. 3A for a schematic representation of the tRNA expression

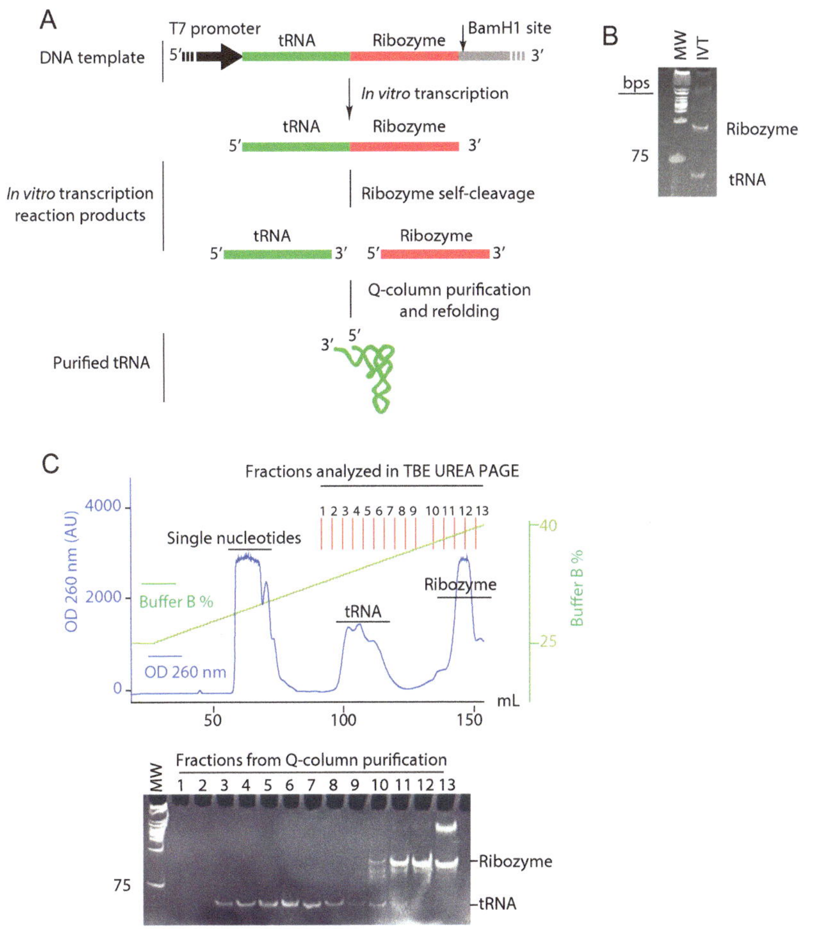

Fig. 3 *In vitro* transcription and purification of *M. jannaschii* tRNA$^{Lys}_{UUU}$. (A) Schematic representation of the major steps of the *in vitro* transcription and purification tRNA$^{Lys}_{UUU}$. (B) 10% TBE-UREA PAGE stained with RedSafe analysis of the products of an *in vitro* transcription reaction generating tRNA$^{Lys}_{UUU}$ and the hepatitis delta virus ribozyme. (C) (top) Chromatogram and the corresponding (bottom) 10% TBE-UREA PAGE stained with RedSafe exemplifying the purification of tRNA$^{Lys}_{UUU}$ from the *in vitro* transcription reaction using a Q column.

cassette and see Table 3 for sequences). In principle, this tRNA expression cassette can be inserted into a variety of different high copy plasmids. We have successfully used the pUC19 (Addgene, Cat# 50005) and pMA (https://genecollections.nci.nih.gov/MGC/Vectors/prot_pMA) plasmids as backbones. Plasmid DNA can be amplified in DH5α *E. coli* cells and purified using DNA maxiprep following the manufacturer's protocols (Qiagen, Cat# 12162).

Table 3 Sequences for generating the DNA template for the *in vitro* transcription of $mjtRNA^{Lys}_{UUU}$.

T7 Promoter	$mjtRNA^{Lys}_{UUU}$	Ribozyme	BamHI
5′-TAATA CGACTCACTATA-3′	5′-GGGCCCGTAGCTCAGTCTGG CAGAGCGCCTGGCTTTTAACCGGTGGTCGA GGGTTCAAATCCCTTCGGGCCCGCCA-3′	5′-GGCCGGCATGGTCCCAGCCTCCTC GCTGGCGCCGGCTGGGCAACATTCCGAGGGGACCGT CCCCTCGGTAATGGCGAATGGGACCCAGGCTTAG TATAGCGAGGTTTAGCTACACTCGTGCTGAGCC-3′	5′-GGATTC-3′

2.2.2 Equipment

- Static 37 °C incubator (Boekel Scientific, Cat# 133000).
- Benchtop centrifuge (Beckman Coulter, Allegra X-15R).
- Peristaltic pump (Gilson, MINIPULS 3, Cat# F155009) and tubing (Gilson, Cat# F117970) for loading columns.
- FPLC unit (GE, AKTA Purifier 10 FPLC System w/UPC-900).
- 5 mL Q column (Cytiva, HiTrap® Q High Performance, Cat# 17115401).
- Electrophoresis system (Bio-Rad, Cat# 1658004).
- Synergy NEO plate reader with a Take3 Micro-Volume Plate and Gen5 software (Bio Tek).
- Gel camera with UV plate.

2.2.3 Buffers and reagents

- Chemicals for making buffers and solutions used for the transcription and purification of the tRNA:

  Tris-HCl (Bioshop, Cat# TRS004.1), Bis-Tris (Sigma, Cat# 6976370), NaCl (Bioshop, Cat# SOD002-10), DTT (Bioshop, Cat# DTT001-100), $MgCl_2$ (Sigma, Cat# 1374248), Triton X-100 (Sigma, Cat# 9002931) EDTA (Thermo Fisher Scientific, Cat# 6381926), diethyl pyrocarbonate (DEPC) (Sigma, Cat# D6758).
- Filtropur V50 Filtration Unit, 500 mL, 0.2 μm (Sarstedt, Cat# 83.3941.001).
- 500 μg of purified plasmid DNA containing tRNA expression cassette.
- T7 RNA polymerase (NEB, Cat# M0251L).
- BamHI-HF (NEB, Cat# R3136S) with rCutSmart™ Buffer (NEB, Cat# B6004S).
- 100 mM stocks of ATP (Sigma, Cat# A3377), GTP (Sigma, Cat# G8877), CTP (Sigma, Cat# C1506) and UTP (Sigma, Cat# U6750) buffered to pH 8.0 in DEPC treated water.
- Resuspension buffer (10 mM Bis-Tris pH 6.5, 10 mM $MgCl_2$).
- 100 mM spermidine (Sigma, Cat# 85578) in DEPC water.
- Thermostable inorganic phosphate (TIPP) (NEB, Cat# M0296L).
- RiboLock (Thermo Fisher Scientific, Cat# EO0381).
- Buffer A (DEPC treated water, filtered at 0.2 μm).
- Buffer B (2 M NaCl made in DEPC treated water, filtered at 0.2 μm).
- 100% Isopropanol (Thermo Fisher, Cat# 67630), 100% ethanol (Sigma, Cat# 64175) and 80% ethanol diluted with DEPC water.
- 10% TBE-urea gels and TBE running buffer (prepared according to Summer, Gramer, & Droge, 2009) and RedSafe nucleic acid staining solution (FrogBio, Cat# 21141).

- Formamide (Sigma, Cat# F9037) and bromophenol blue (Sigma, Cat# B0126) for preparing RNA loading dye (90% formamide, 2 mM EDTA supplemented with bromophenol blue) and molecular weight marker (Thermo Fisher Scientific, GeneRuler 1 kb Plus DNA Ladder, Cat# SM1334).
- Heat blocks set to 37, 50 and 90 °C (VWR, Cat# 75838-318).

2.2.4 Step-by-step protocol for run off in vitro *transcription of* M. jannaschii *$tRNA^{Lys}_{UUU}$ (~5 h)*

- In a 1.5 mL tube, set up a digestion reaction overnight at 37 °C in a heat block using 500 μg of the purified tRNA expression plasmid DNA with 200 units (10 μL) BamHI in a volume of 200 μL.
- The following morning, in a 15 mL tube, set up an *in vitro* transcription reaction by adding the DNA from overnight digestion to a reaction mix (see Table 4 for recipe) with the final concentrations of 100 mM Tris pH 8.0, 4 mM ATP/GTP/CTP/UTP, 10 mM DTT, 1 mM spermidine, 0.1% Triton X-100, 25 mM $MgCl_2$, 4 U/mL T7 RNA polymerase, 10 U/mL TIPP and 200 U/mL RiboLock.

Table 4 *In vitro* transcription reaction recipe.

Reagent	Stock concentration (mM)	Final concentration (mM)	Volume (μL)
Tris pH 8.0	1000	100	1000
ATP	100	4	400
GTP	100	4	400
UTP	100	4	400
CTP	100	4	400
DTT	1000	10	100
Spermadine	100	1	100
Triton X-100	20	0.1	50
$MgCl_2$	1000	25	250
T7 RNA polymerase	50,000 (U/mL)	4 (U/mL)	20
TIPP	2000 (U/mL)	10 (U/mL)	50
Ribolock	40,000 (U/mL)	200 (U/mL)	50
DEPC DDW			6830

- Incubate reaction mix for 4 h at 37 °C in a static incubator. Reaction product can be assessed using TBE-UREA PAGE (Fig. 3B). Mix 4 μL from the *in vitro* transcription reaction with 4 μL of RNA loading buffer. In a separate lane, load 10 μL of the molecular weight marker. Run the TBE-UREA gel at 200 V for 25 min. Transfer the gel to a clean box that is slightly larger than the gel and add to the box enough DEPC DDW to cover the gel. Add 1 μL of the RedSafe nucleic acid staining solution and stain for 10 min at room temperature with mild agitation. Visualize gel with UV camera. Expect products of 76 bps (tRNA), ~300 bps (ribozyme) and of larger molecular weights (tRNA-ribozyme fusions for example). Transfer reactions to −20 °C for storage. Reaction products are stable and can be stored for weeks or even months.

2.2.5 *Protocol for purification of tRNA$^{Lys}_{UUU}$ from* in vitro *transcription reaction (3–5 h)*

- Thaw *in vitro* transcription reaction products and add 40 mL of Resuspension buffer.
- Load mixture on a 5 mL HiTrap Q column using a peristaltic pump.
- Elute from Q column using an FPLC program starting with 4 column volumes of 25% buffer B and 75% buffer A followed by a linear gradient of 25–40% buffer B over 20 column volumes. Excess nucleotides from *in vitro* transcription reaction should elute during the first 4 column volumes of 25% buffer B. tRNA elutes at around 29–32% buffer B and the ribozyme elutes at 33.3–38% buffer B. Analyze fractions with high absorbance at 260 nm using TBE-UREA PAGE as described above and visualize with a UV camera after staining with RedSafe for 10–15 min (Fig. 3C).
- Pool tRNA-containing fractions into a 50 mL falcon tube and precipitate nucleic acid by adding to the tube an equal volume of 100% isopropanol and incubating at −20 °C overnight.
- Using a benchtop centrifuge, spin down and pellet the tRNA sample from the overnight precipitation (10 min at 4000 × *g* at room temperature). Discard supernatant.
- Using a vortex or pipetting, wash pellet with 1 mL 80% ethanol until pellet detaches from tube. The pellet may break up, but it will not be solubilized. Transfer resuspended pellet to a 1.5 mL tube, spin down again (5 min at 14,000 × *g* at room temperature). Aspirate as much of the ethanol as possible and allow pellet to air dry for 5–10 min.
- Resuspend pellet in 100 μL of 10 mM Tris pH 8.

- Refold tRNA:
  - Using a heat block, heat the sample at 90 °C for 2 min.
  - Cool the sample on ice for 3 min.
  - With a heat block, re-heat the sample to 50 °C for 2 min and then add $MgCl_2$ to a final concentration of 2 mM.
  - Slowly cool down sample to room temperature by switching off heat block (should take 1–2 h).
- Determine concentration by measuring absorbance of 1.5 μL at 260/280 nm using the Synergy NEO plate reader with a Take3 Micro-Volume Plate or a nano-drop spectrophotometer.
- Snap-freeze aliquots and store at −80 °C. Frozen samples are stable for months and even years.

2.2.6 Expected outcomes

From a single *in vitro* transcription reaction, expect 2–3 mg of purified tRNA, which is 0.75–1.5 mM in 100 μL.

2.2.7 Safety considerations and standards

The components of this protocol do not present any special health and safety hazard. Regular lab safety procedures should be implemented.

2.2.8 Alternative methods/procedures

As an alternative to these protocols, tRNA can be synthesized *in vitro* using a PCR-based template generation protocol (Korencic, Soll, & Ambrogelly, 2002) and purified using a PAGE-based purification protocol (Sherlin et al., 2001).

3. Monitoring the ATPase activity of Bud32 using the ADP Glo assay

3.1 Equipment, software, buffers, and reagents

- 96 well plates with a V-shaped bottom (Sarstedt, Cat# 79.1980.202).
- Sizing buffer (20 mM HEPES pH 7.5, 100 mM NaCl, 2 mM DTT).
- tRNA buffer (10 mM Tris pH 8.0).
- Chemicals for making the 2 × ADP-Glo reaction buffer (for a recipe, see Table 5) (80 mM Tris pH 7.5, 100 mM NaCl, 20 mM $MgCl_2$, 4 mM $MnCl_2$, 0.2 mg/mL BSA, 0.02% Brij 35, and 2 mM DTT):

  Tris-HCl (Bioshop, Cat# TRS004.1), NaCl (Bioshop, Cat# SOD002-10), $MgCl_2$ (Sigma, Cat# 1374248), $MnCl_2$ (Sigma, Cat# 13446349), BSA (Thermo Fisher Scientific, Cat# 9048468) Brij 35 (Thermo Fisher Scientific, Cat# 9002920), DTT (Bioshop, Cat# DTT001-100).

Table 5 2× ADP GLO reaction buffer recipe.

Reagent	Stock concentration (mM)	Final concentration (mM)	Volume (μL)
Tris pH 7.5	2000	80	20
NaCl	2000	100	25
$MgCl_2$	1000	20	10
$MnCl_2$	1000	4	2
BSA	10 (mg/mL)	0.2	10
Brij 35	10 (%)	0.02	1
DTT	1000	2	1
DEPC DDW			431

- 384 well white plate (VWR, Lumitrac 200, Cat# 10027-746).
- Synergy NEO plate reader with Gen5 software (Bio Tek).
- Heat block for a 96 well plate or a PCR machine (Eppendorf Mastercycler nexus gradient).
- Purified KEOPS proteins and tRNA.
- ADP-Glo™ kit (Promega, Cat# V9101).
- GraphPad Prism v9 software (GraphPad).

3.2 Step-by-step protocol for analyzing the ATPase activity of Bud32 using the ADP Glo assay (2–4 h)

- In a 96 well plate, set up the reaction at a final volume of 20 μL at room temperature by mixing the reaction components in the following order (for a recipe, see Table 6):
  - 10 μL 2× ADP Glo reaction buffer.
  - 2 μL of purified mjBud32 from a 3 μM stock of protein diluted in sizing buffer.
  - 2 μL of a mix of all other desired purified KEOPS proteins from a 3.5 μM stock of protein diluted in sizing buffer.
  - 2 μL of purified $tRNA^{Lys}_{UUU}$ from a 30 μM stock of tRNA diluted in tRNA buffer, or an equivalent volume of tRNA buffer without tRNA as a control.
  - Adjust volume to 20 μL using DEPC treated DDW, considering the volume of ATP to be added in the next step.
  - 2 μL of ATP (from the ADP Glo kit) from a 100 μM stock diluted in DEPC DDW.
  - Mix well by pipetting.

Table 6 ADP GLO reaction mix recipe.

Reagent	**Volume (μL)**
2 × ADP Glo reaction buffer	10
mjBud32 (diluted to 3 μM in sizing buffer)	2
A mixture of other KEOPS proteins (diluted to 3.5 μM in sizing buffer)	2
tRNA (diluted to 30 μM in tRNA buffer), or tRNA buffer without tRNA	2
DEPC DDW	2
ATP (diluted to 100 μM in DEPC DDW)	2

- Cover plate with a sealing foil and start ATPase reactions by transferring the 96-well plate to a heat block preheated to 55 °C. Incubate reactions for 60 min. In this step, mjBud32 proteins will convert ATP to ADP + Pi (see schematic in Fig. 4A).
- Terminate reactions by mixing 10 μL from each reaction with 10 μL of ADP Glo Reagent in a 384-well plate. Mix well by pipetting. Incubate at room temperature for 40 min. This step will deplete all unconsumed ATP, leaving only the ADP generated by Bud32.
- Add 20 μL of the ADP Glo kinase detection reagent. Mix well by pipetting and incubate for 30 min at room temperature. This step will convert all the ADP generated by Bud32 to ATP, which is utilized for a luminescence emitting reaction (see schematic in Fig. 4A).
- Measure luminescence using a plate reader.
- Plot experimental data as an *xy* scatter plot or as columns using GraphPad Prism software or alternatively using Excel.

3.3 Expected outcomes

In the absence of mjCgi121, mjBud32 proteins manifest very low ATPase activity and signals from such reactions should be comparable to background. The presence of mjCgi121 leads to an increase in ATPase activity (~10-fold over background). Addition of tRNA should further potentiate ATPase activity, but only if the mjKae1 protein is present in the reaction (~50-fold over background) (Fig. 4B).

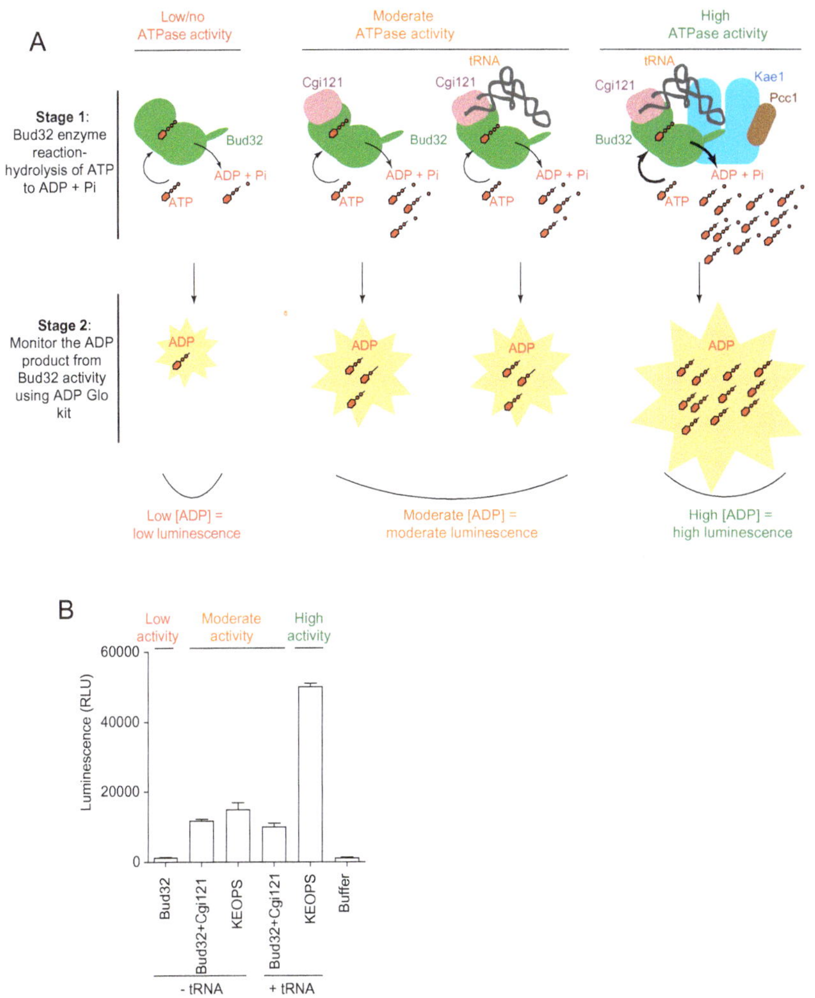

Fig. 4 Monitoring the ATPase activity of Bud32 proteins using the ADP Glo assay. (A) Schematic representation of the main steps in the analysis of the activity of mjBud32-containg complexes and the expected outcomes from each complex. (top) mjBud32 hydrolyzes ATP to ADP+Pi. The highest activity is achieved when the full KEOPS complex is assembled and bound to tRNA. (bottom) The ADP product that results from mjBud32 activity is quantified using the ADP Glo assay. This assay generates luminescence proportional to the amount of ADP produced in each reaction. (B) Column bar graph representing the activities of the indicated mjBud32-containing complexes determined by the ADP Glo assay. Error bars represent standard error of independent data points ($n = 3$).

3.4 Quantification and statistical analysis

Luminescence measurements are done using a BioTek Synergy Neo plate reader. Experimental data points should be obtained minimally from triplicates to enable the generation of error bars. Experimental data are plotted using GraphPad Prism software either as *xy* scatter plots or as bar columns where applicable.

3.5 Advantages

The ADP Glo assay is a versatile, quantitative, and safe assay that enables parallel analysis of multiple samples. Unlike the radioactive-based autophosphorylation assay that has been previously used for measuring the ATPase activity of Bud32 (see below), this assay provides a measurement of Bud32s actual activity in hydrolyzing the γ-phosphate of ATP to water and is therefore more reliable and less prone to misinterpretation.

3.6 Limitations

Since this assay provides a readout for ADP concentrations it is important to be sure that the source of the ADP is indeed from the activity of Bud32 proteins and not from Kae1 (which has some ATPase activity) or a contaminating ATPase that co-purifies with one of the KEOPS proteins. To avoid such pitfalls, the ATPase activity of each component should be measured individually. We recommend including catalytically inactive mutants to selectively control for activity of one enzyme at a time (such as the K52A or D162R mutations in mjBud32 and the K12A and G155E mutations in mjKae1). Likewise, control reactions without any protein should be setup in each experiment to monitor the baseline signal due to the natural breakdown of ATP to ADP + Pi in the reaction buffer.

3.7 Optimization and troubleshooting

Protein, tRNA and ATP concentrations as well as the length of the reaction should be optimized before starting experiments. Optimal assay conditions will be those in which:

- The increase in ATPase activity is linear respective to the reaction time and the enzyme concentration. In other words, make sure that at the chosen concentrations and timepoint the signal doubles when reaction time and enzyme concentrations are doubled.
- Signal is at least 10-fold higher than background.

3.8 Safety considerations and standards

The components of this assay do not present any special health and safety hazard. Regular lab safety procedures should be implemented.

3.9 Alternative methods/procedures

As an alternative indirect measurement of the ATPase activity of Bud32, a radioactive gel-based autophosphorylation assay may be considered and has been used previously (for example, see Hecker et al., 2008; Mao et al., 2008). However, although Kae1 and tRNA do not inhibit the ATPase activity of Bud32 (Beenstock et al., 2020), these components inhibit its autophosphorylation activity (Perrochia, Guetta, Hecker, Forterre, & Basta, 2013; and Sicheri lab unpublished data), most likely due to steric clashes with the accepting sites for autophosphorylation. We therefore do not recommend using this assay.

4. Monitoring the tRNA binding activity of Bud32-containing complexes using a 647-CCA probe displacement fluorescence polarization assay

4.1 Equipment, software, buffers, and reagents

- Chemicals for making fluorescence polarization (FP) buffer (20 mM HEPES pH 7.5, 100 mM NaCl, 2 mM DTT).
  HEPES (Bioshop, Cat# HEP001.5), NaCl (Bioshop, Cat# SOD002-10) and DTT (Bioshop, Cat# DTT001-100).
- Purified KEOPS proteins and tRNA.
- Fluorescent 647-CCA RNA probe:
  5′-CCGCCA-3′ oligonucleotide with an Alexa-647 fluorescent tag fused to its 5′ (Can be made by custom synthesis by IDT Inc.).
- 96-well plates with a V-shaped bottom (Sarstedt, Cat# 79.1980.202).
- 384-well flat bottom black plate (Corning, Cat# 3573).
- BioTek Synergy Neo plate reader (BioTek) with Gen5 software.
- GraphPad Prism 9 software (GraphPad).

4.2 Step-by-step protocol for analyzing the role of Bud32 in tRNA binding using a 647-CCA probe displacement fluorescence polarization assay (2 – 3 h). For a schematic, see Fig. 5A

- In a full row in a 96-well plate, generate a serial dilution of tRNA ranging from 0.05 to 200 μM in FP buffer mixed with the mjBud32-mjCgi121 complex at a set concentration of 0.6 μM (see Fig. 5B, for a schematic representation of the major steps in setting up this serial dilution and assay). Each well should have a final volume of 15 μL.

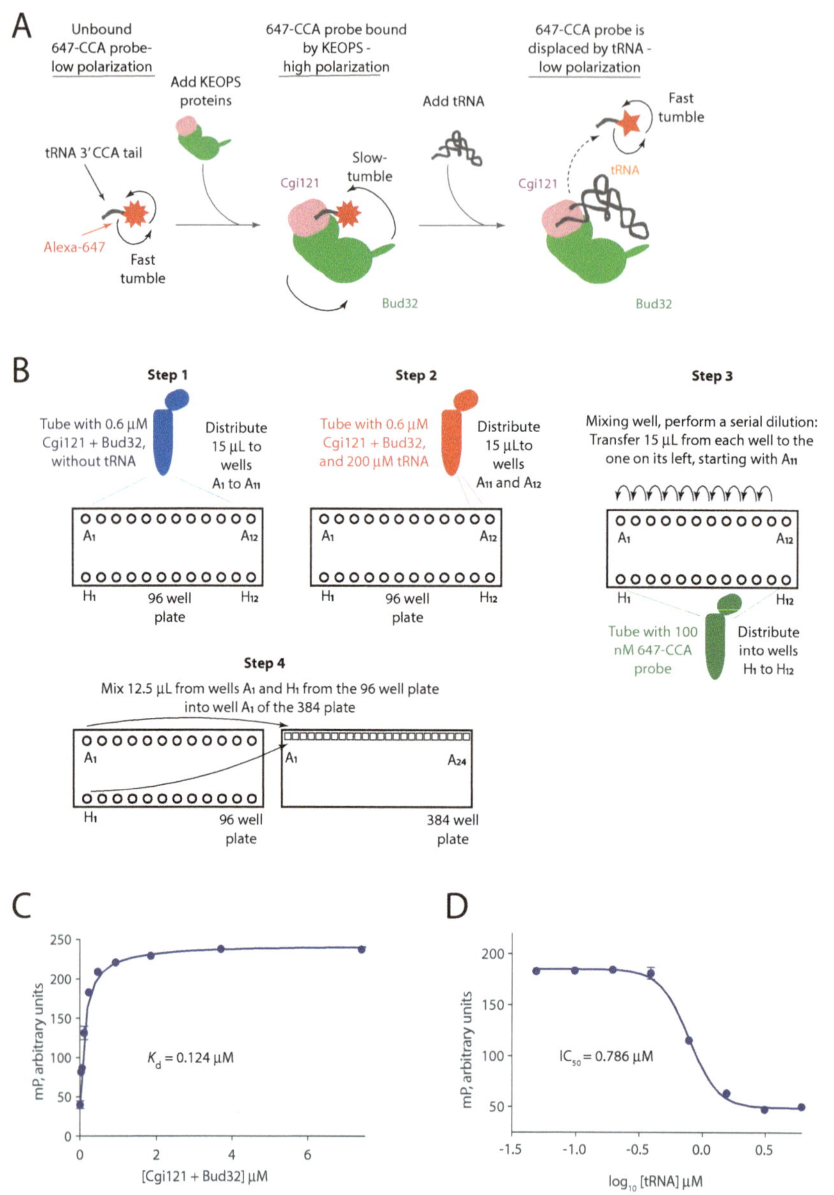

Fig. 5 Monitoring the role of Bud32 proteins in tRNA binding using fluorescence polarization. (A) Schematic representation of the method for measuring the tRNA binding activity of mjBud32-containing complexes using the 647-CCA probe. (left) The 647-CCA probe is composed of the last five nucleotides of *M. jannaschii* $tRNA^{Lys}_{UUU}$ (5′-CGCCA-3′) with an Alexa Fluor 647 group attached to the 5′ end. (middle) The binding of mjCgi121-mjBud32 to the 647-CCA probe leads to an increase in fluorescence polarization (FP) due to a slower tumbling time related to an increase in complex size. (right) tRNA displaces the probe and leads to a decrease in FP. (B) Schematic representation of the major steps in setting up the 647-CCA FP assay. **Step 1**. In a 96 well plate, distribute

- Repeat this mixing in two more rows, generating three rows with tRNA and the mjBud32-mjCgi121 complex.
- In a separate row of the same plate, add in each of the 12 wells 45 μL of the 647-CCA probe diluted to 100 nM in FP buffer.
- With a multi-channel pipette, mix 12.5 μL from the protein/tRNA wells with 12.5 μL from the 647-CCA probe wells into a 384-well plate. This brings the mjBud32-mjCgi121 complex to a final concentration of 0.3 μM (roughly twice the concentration of the K_d, see Fig. 5C), and the concentration of the tRNA to 0.025–100 μM and the concentration of the 647-CCA probe to 50 nM. Perform this mixing in three separate wells to generate triplicate data points.
- As a control, in 2 wells in the 384-well plate mix FP buffer without any protein or tRNA with 50 nM of the 647-CCA probe. This will serve to determine the base line FP signal in which the probe is not bound to protein.
- Cover plate with aluminum foil and incubate at room temperature for 1 h.
- Read FP using a BioTek Synergy Neo plate reader with excitation and absorbance set to 620/680 nm.

---

15 μL of mjCgi121 + mjBud32 at a concentration of 0.6 μM into wells 1–11 in rows A-C. For simplicity, only rows A and H are shown. **Step 2**. Distribute 15 μL of mjCgi121 + mjBud32 at 0.6 μM mixed with 200 μM tRNA into wells 11–12 in rows A–C. **Step 3**. In rows A–C, mix well 11 by pipetting and transfer 15 μL to well 10. Repeat this mixing until reaching well 1. Distribute 45 μL of the 647-CCA probe diluted to 100 nM in FP buffer into wells 1–12 in row H. **Step 4**. With a multichannel pipettor, transfer and mix 12.5 μL from all the wells in rows A and H in the 96-well plate to the wells with the even numbers in row A in a 384-well plate. Repeat this step with 12.5 μL from all the wells in rows B and H in the 96-well plate to the wells with the odd numbers in row A in a 384-well plate. Repeat this step with 12.5 μL from all the wells in rows C and H in the 96-well plate to the wells with the even numbers in row B in a 384-well plate. (C) The binding affinity of the mjCgi121-mjBud32 complex towards the 647-CCA probe determined by FP. FP of the 647-CCA probe was measured and plotted at the indicated concentrations of the mjCgi121-mjBud32 complex. A K_d value was determined by non-linear regression curve fitting to the experimental data. Error bars represent standard error of independent data points ($n = 3$). (D) The binding affinity of the mjCgi121-mjBud32 complex towards tRNA determined by FP. The mjCgi121-mjBud32 complex was mixed with the 647-CCA probe at a concentration of 0.3 μM, roughly $2 \times K_d$. FP was then measured and plotted at the indicated concentrations of tRNA. Note that the higher the tRNA concentrations used, the more the 647-CCA probe is displaced, and the more the FP signal decreases. An IC_{50} value was determined after a $\log_{10}$ transformation of the tRNA concentration values by non-linear regression curve fitting to the experimental data. Error bars represent standard error of independent data points ($n = 3$).

4.3 Expected outcomes

At a concentration of the mjBud32 + mjCgi121 complex that is roughly two times the K_d towards the 647-CCA probe (determined experimentally, see Fig. 5C and Section 4.7) and with low tRNA concentrations, the 647-CCA probe will be bound by mjBud32-mjCgi121 complex, which will lead to an increase in the FP signal. An increase in tRNA concentrations will lead to the release of the probe from the mjBud32-mjCgi121 complex, causing a decrease in the FP signal (Fig. 5D).

4.4 Quantification and statistical analysis

Plot data in an *xy* scatter plot using GraphPad Prism software:

- Perform a $\log_{10}$ transformation of the tRNA concentrations used in each sample. The resulting values from this will be the *x* axis values of the binding curve.
- Analyze the displacement of the 647-CCA probe by monitoring the effect tRNA has on the FP signal. To obtain IC_{50} values, perform a non-linear regression curve fit using the "log(inhibitor) vs. response-variable slope" option.

FP measurements are done using a BioTek Synergy Neo plate reader with Gen5 software. To ensure that the changes in FP are indicative of tRNA displacing the 647-CCA probe, compare signals of samples containing high tRNA concentrations to the values that were obtained in the control reactions without mjBud32-mjCgi121 proteins. The values from these samples should be similar. IC_{50} values can be extrapolated from binding curves. To calculate error bars, each data point must be obtained minimally from a triplicate. To calculate IC_{50} standard deviation values, minimally three independent displacement experiments should be conducted.

4.5 Advantages

The FP 647-CCA probe displacement assay is a quantitative and sensitive assay that enables parallel analysis of multiple samples with relative ease.

4.6 Limitations

The FP 647-CCA probe displacement assay does not allow to generate a K_d value to determine the binding affinity between mjBud32-mjCgi121 and the tRNA. For obtaining further insight into tRNA binding activities, the filter binding assay (see below) may be considered.

4.7 Optimization and troubleshooting

Prior to starting experiments, the concentrations of mjBud32-mjCgi121 proteins used in the displacement assay should be determined by obtaining a K_d value for the binding of KEOPS proteins to the 647-CCA probe. These experiments are executed in a similar way to the 647-CCA probe displacement assays:

- In a 96-well plate, prepare a serial dilution of KEOPS proteins ranging from 0 to 50 μM in FP buffer at a final volume of 15 μL.
- Repeat this dilution in two more rows, generating three rows with protein samples.
- In separate row of the same plate, mix 100 nM of the 647-CCA probe in FP buffer at a final volume of 45 μL.
- In a 384-well plate, mix 12.5 μL from the wells containing the KEOPS proteins with 12.5 μL from the wells containing the 647-CCA probe. Perform this mixing three times in separate wells to generate triplicates.
- Cover plate with aluminum foil and incubate at room temperature for 1 h.
- Read fluorescence polarization using a BioTek Synergy Neo plate reader with excitation and absorbance set to 620/680 nm.
- Plot data in an *xy* scatter plot using GraphPad Prism software. To obtain K_d values, perform a non-linear regression curve fit using the "One site-specific binding" option.

4.8 Safety considerations and standards

The components of this assay do not present any special health and safety hazard. Regular lab safety procedures should be implemented.

4.9 Alternative methods/procedures

The filter binding assay can be considered as an alternative approach for analyzing the tRNA binding activities of mjBud32-containing complexes. This assay measures the co-retention of tRNA and protein complexes on a nitrocellulose membrane. The negatively charged tRNA does not bind by itself to the hydrophobic nitrocellulose membrane.

Main steps (for more details, see Beenstock et al., 2020):

- 5′ label purified tRNA with ^{32}P.
- Co-incubate the mjBud32-mjCgi121 complex and the radioactive tRNA probe.
- Load samples onto a nitrocellulose membrane using a filtration device.
- Expose membranes to X-ray film or to a phosphorimager plate.

The filter binding assay is very informative but has some disadvantages relative to the 647-CCA FP displacement assay. Most notably the filter binding assay is semi-quantitative, and it involves working with radioactive material. More so, the filter binding assay is more labor intensive and takes more time (2–3 days relative to 2–3 h) compared with the FP 647-CCA probe displacement assay.

5. Monitoring the role of Bud32 in the t6A modification of tRNA using an HPLC analysis of nucleoside composition

5.1 Equipment, software, buffers, and reagents

- Chemicals for making the 4 × t^6A modification reaction buffer (for a recipe, see Table 7) (200 mM Tris HCl pH 8, 600 mM NaCl, 10 mM DTT, 2 mM threonine, 2 mM $NaHCO_3$, 8 mM ATP, 1 mM $MnCl_2$, 4 mM $MgCl_2$, 10 mM spermidine, 20 μM mjSua5):

  Tris-HCl (Bioshop, Cat# TRS004.1), NaCl (Bioshop, Cat# SOD 002-10), $MgCl_2$ (Sigma, Cat# 1374248), $MnCl_2$ (Sigma, Cat# 13446349), DTT (Bioshop, Cat# DTT001-100), ATP (Sigma, Cat# A3377), spermidine (Sigma, Cat# 85578), L-threonine (Bioshop, Cat# 72195), $NaHCO_3$ (Sigma, Cat# S5761).

Table 7 4 × t^6A reaction buffer recipe.

Reagent	Stock concentration (mM)	Final concentration (mM)	Volume (μL)
Tris pH 8.0	1000	200	100
NaCl	2000	600	150
DTT	1000	10	5
Thr	100	2	10
Na HCO_3	100	2	10
ATP	100	8	40
$MnCl_2$	250	1	2
$MgCl_2$	1000	4	2
spermadine	100	10	50
mjSua5	3800	20	2.6
DEPC DDW			128

- The mjSua5 protein can be expressed in *E. coli* and purified as a His-tagged protein cloned in the pProEX plasmid using a similar protocol used for the purification of the mjBud32 protein as described above.
- DEPC (Sigma, Cat# D6758)- treated water.
- Purified KEOPS proteins and substrate tRNA.
- Heat blocks set to 37, 55 and 95 °C (VWR, Cat# 75838-318).
- Bench top centrifuge for 1.5 mL tubes (Eppendorf).
- Nuclease P1 (Sigma, Cat# N8630) diluted in P1 storage buffer (20 mM Na acetate pH 5.3, 5 mM $ZnCl_2$, 50 mM NaCl, 50% glycerol) to a concentration of 0.8 U/mL. The P1 nuclease is stored at −20 °C and is stable for at least 6 months.
- $ZnSO_4$ (AnalaR, Cat# 7446200).
- 0.5 M Tris HCl pH 8.3.
- Quick calf intestinal phosphatase (CIP) (NEB, Cat# M0525S).
- Discovery C18 (15 cm × 4.6 mm, 5 μm) reverse-phase column (Supelco Analytical, Cat# SU504955) with a Luna 50 × 4.6 mm 5 μm C18 Guard Column (Phenomenex, Cat# 00B-4252-E0).
- Sizing buffer (20 mM HEPES pH 7.5, 100 mM NaCl, 2 mM DTT made in DEPC treated water).
- Dionex Ultimate 3000 HPLC Unit (Thermo Fisher Scientific, Cat# 9094).
- Chromeleon HPLC software v6.8 (Thermo Fisher Scientific).
- Solvents for preparing HPLC Buffers A, B and C (250 mM ammonium acetate pH 6.5, 40% acetonitrile diluted in HPLC grade water, and 100% methanol, respectively):

  Ammonium acetate (Sigma, Cat# 631618), acetonitrile (Caledon, Cat# 75058), HPLC grade water (Caledon, Cat# 8801740) and methanol (Caledon, CAT# 67561).
- HPLC autosampler 250 μL PP vials (Thermo Fisher Scientific, Cat# 160134) and 11 mm autosampler snap caps (Thermo Fisher Scientific, Cat# 0339143).
- t^6A standard molecule, as a control for the identification of t^6A tRNA modification in the HPLC chromatogram (BioLog, Cat# C022).

5.2 Step-by-step protocol for setting up the tRNA t^6A modification assay and for the analysis of nucleoside composition using an HPLC (2–3 days, depending on the number of samples)

- At room temperature mix in a 1.5 mL tube in the following order (for a recipe, see Table 8):

Table 8 t^6A reaction mix.

Reagent	Volume (μL)
4 × t^6A reaction buffer	5
mjBud32 (diluted to 10 μM in sizing buffer), or sizing buffer without mjBud32	2
A mixture of other KEOPS proteins (diluted to 10 μM in sizing buffer)	2
tRNA (diluted to 500 μM in tRNA buffer)	4
DEPC DDW	7

- o 5 μL 4 × t^6A modification reaction buffer.
- o mjCgi121, mjKae1 and pfPcc1 proteins at a final concentration of 1 μM.
- o mjtRNA$^{Lys}_{UUU}$ at a final concentration of 100 μM.
- o At a final concentration of 1 μM, add either:
  - mjBud32 wild-type.
  - mjBud32 enzyme dead mutant (K52A or D162R).
  - An equivalent volume of sizing buffer as a control for no mjBud32 protein.
- o Adjust volume to 20 μL using DEPC treated DDW.

- Transfer tubes to a heat block set to 55 °C. Incubate for 20 min. In this step KEOPS will t^6A modify the tRNA substrate (for a schematic representation, see Fig. 6A).
- Transfer tubes to a heat block set to 95 °C and heat for 2 min to terminate the reaction.
- Using a benchtop centrifuge, spin down samples at room temperature at 4000 × *g* for 20 s.
- Add to each sample 35 μL of 15 μM $ZnSO_4$.
- Add to each sample 10 μL of P1 nuclease and incubate overnight at 37 °C. In this step the nuclease will digest the tRNA into single nucleotides.
- Using a benchtop centrifuge, spin down samples at room temperature at 4000 × *g* for 20 s.
- After the spin down, add to each sample 10 μL of 0.5 M Tris pH 8.3 and 1 μL of Quick CIP. Mix well by pipetting. Incubate for 2 h at 37 °C in a heat block. In this step, phosphate groups will be removed from the nucleotides, which will convert them to nucleosides.

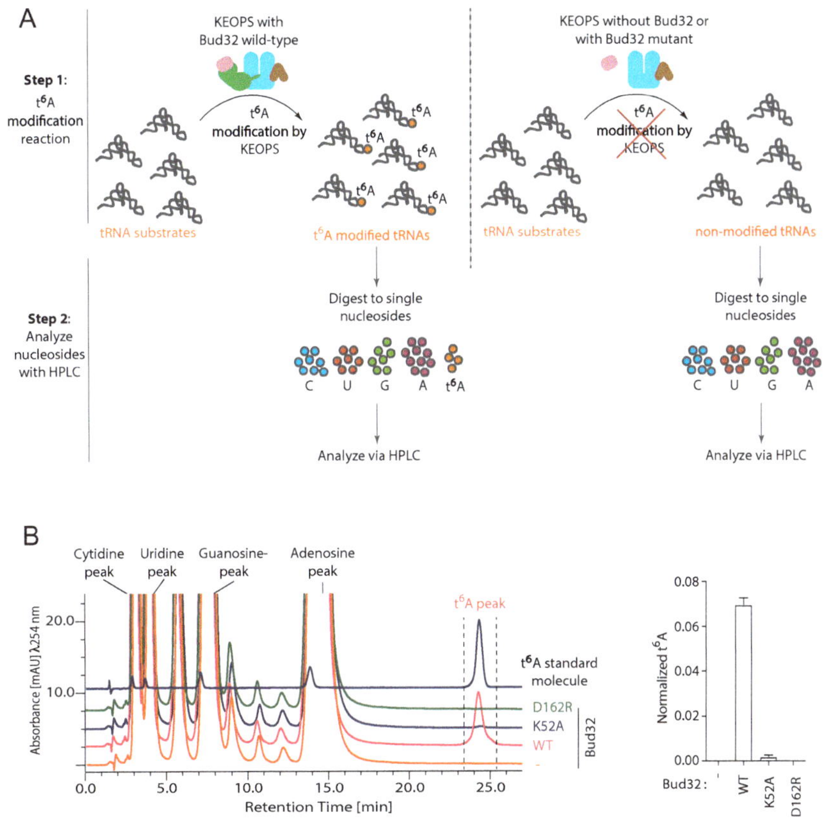

Fig. 6 Analyzing the role of Bud32 in the tRNA modifying activity of KEOPS using an HPLC analysis of nucleoside composition. (A) Schematic representation of the main steps in monitoring the role of Bud32 in the tRNA modification activity of KEOPS. (left) Fully reconstituted KEOPS t^{6}A modifies tRNA substrates in an *in vitro* reaction. (right) KEOPS without Bud32 (or with select Bud32 mutants) does not t^{6}A modify the tRNA substrates. The tRNAs from both different reactions are digested to nucleosides (C- cytidine, U- uridine, G- Guanosine, A-adenosine) and subjected to analysis by HPLC. For simplicity, Sua5 is not shown in the reactions. (B) KEOPS with Bud32 wild-type t^{6}A modifies tRNA while KEOPS without Bud32 or with the Bud32 K52A or D162R enzyme dead mutants does not. (left) chromatograms from the HPLC analysis of nucleoside composition of representative tRNA modification assays with KEOPS with the indicated Bud32 proteins. The chromatogram of a t^{6}A standard molecule is shown to indicate the correct identification of t^{6}A. (right) A column bar graph displaying the normalized t^{6}A contents (t^{6}A/uridine) in each reaction setup. Error bars represent standard error between independent data points ($n = 3$).

- Spin down samples at room temperature at full speed for 20 s.
- Transfer 75 μL from the tubes to HPLC vials. Add 125 μL DDW to each vial to adjust the volume to 200 μL and close vials with a snap cap.
- Analyze the nucleoside content using an HPLC:
  - In separate runs, inject 125 μL from each vial on a C18 reverse phase column equilibrated in 98% HPLC buffer A and 2% HPLC buffer B.
  - Elute from the column with a linear gradient of 98:2 to 87.5:12.5 HPLC buffers A and B respectively over 50 min set at a rate of 1.5 μL per minute.
  - End each HPLC run with 10 min of 98% HPLC buffer A and 2% HPLC buffer B to re-equilibrate the column before the next sample is injected.
  - At the end of the last sample, run with 10 min of 100% HPLC buffer C for storing the column.

5.3 Expected outcomes

The HPLC chromatogram should exhibit four major peaks that elute from the column between 3 and 16 min. These peaks correspond to the four non-modified nucleosides that elute in the following order: cytidine (~3 min retention time), uridine (~4 min retention time), guanosine (~8 min retention time) and adenosine (~14.5 min retention time) (Fig. 6B). The adenosine peak is the largest of all four due to excess ATP that remains from the tRNA modification reaction buffer. The 3–16 min time range includes several smaller peaks that correspond to other unidentified components of the tRNA modification reaction buffer. The t^6A nucleoside elutes from the column after ~24.5 min. The size of this peak is considerably smaller than the size of all other nucleoside peaks due to the low frequency of this nucleoside relative to the other non-modified nucleosides in the tRNA (1 out of 76). Note that retention times may change slightly between different columns and if buffer concentrations are slightly off. The identity of the t^6A peak can be further confirmed by using a commercially available standard t^6A molecule (as shown in Fig. 6B) and by identification by mass (Wan et al., 2013).

KEOPS with mjBud32 wild-type will exhibit t^6A tRNA modification activity while KEOPS without mjBud32 or with an enzyme dead mjBud32 protein will not modify tRNA (Fig. 6B).

5.4 Quantification and statistical analysis

HPLC is an analytical instrument that allows the quantification of KEOPS tRNA modification activity. For quantification of the t^6A content:

- Using Chromeleon software, measure the area beneath the peak that corresponds to t^6A.
- For comparing between different samples, measure the area beneath the peak corresponding to uridine (the second major peak in the chromatogram). Divide the area of the t^6A peak by the area of the uridine peak. Since all samples should have a similar uridine content, this will normalize the t^6A content and will enable sample to sample comparison. Normalization could be done in principle to any of the four nucleoside peaks, we recommend not normalizing to the adenosine peak since it is most prone for fluctuations since it can originate from the tRNA and the reaction buffer. Normalized t^6A can be plotted as a bar graph using Prism software or Excel. For generating error bar statistics, each experimental sample should be done at least in triplicates.

5.5 Advantages

Using an HPLC readout enables a direct and unequivocal analysis of the tRNA modifying activity of KEOPS and to determine the effect of Bud32 activity on tRNA modification.

5.6 Limitations

The main limitation is that analyzing samples using an HPLC takes a long time, since each sample takes 1 h. Therefore, HPLC runs can last over very long times depending on the size of experiments. For large experiments that run for more than 24 h, we recommend performing at least one more reaction with wild-type KEOPS proteins and analyze it last to compare if the results change from the beginning of the HPLC analysis to the end. The sensitivity of t^6A detection using an HPLC is not very high, and the assay consumes a relatively large amount of the tRNA substrate.

5.7 Optimization and troubleshooting

Before running experiments, it is strongly recommended to identify the expected retention times of the four non-modified nucleosides and of t^6A individually. The concentration of KEOPS proteins and tRNA as well as the length of reaction times should also be optimized to ensure that concentrations and time points that are chosen are linear respective to the reaction time and concentrations. In other words, make sure that at the chosen conditions, the signal doubles when reaction time and enzyme concentrations are doubled.

5.8 Safety considerations and standards

The components of this assay do not present any special health and safety hazard. Regular lab safety procedures should be implemented.

5.9 Alternative methods/procedures

A radioactive based assay for monitoring the tRNA modification activity of KEOPS has been established, which measures the incorporation of ^{14}C-labeled threonine into substrate tRNA (Perrochia et al., 2013). This assay can be considered as an alternative for analyzing the t^6A content using an HPLC.

6. Drop assay for monitoring the role of Bud32 in yeast cell growth *in vivo*

6.1 Yeast expression plasmid information

For protein expression in yeast, the coding region of the *Saccharomyces cerevisiae* (sc) Bud32 protein and its ±500 base pairs genomic sequence (according to https://www.yeastgenome.org/locus/S000003494/sequence) in frame with an N-terminal FLAG-tag was ordered as a synthetic clone (Thermo Fisher Scientific, GeneArt). The coding sequence was then amplified by PCR using the KOD Hot Start DNA Polymerase (Sigma, Cat# 71086) according to manufactures guidelines (for a list of oligonucleotides, see Table 2), adding 5′ *Sac*I and 3′ *Hin*dIII sites. The amplified product was then digested with SacI and HindIII (NEB, Cat# R3156S and R3104S respectively) and subcloned into the low copy pRS415 plasmid (Sikorski & Hieter, 1989). This plasmid contains an ampicillin resistance gene for amplification in *E. coli* and a *LEU2* auxotrophic selection marker for yeast. For a scBud32 cover plasmid, the YCplac111-BUD32 high copy plasmid (denoted pBUD32-*URA3*) with a *URA3* auxotrophic selection marker was used (Downey et al., 2006).

6.2 Equipment, yeast strains, buffers, and reagents

- Shaker and static 30 °C incubators.
- 42 °C heat bath.
- 25 mL Erlenmeyer glass flask (Cole-Parmer, Cat# UZ-34502-55).
- Cooling centrifuge (Beckman Coulter Allegra X-15R).
- Sterile 10× Tris-EDTA (10× TE, according to http://cshprotocols.cshlp.org/).
- 5-fluoroorotic acid (5-FOA) (Thermo Fisher Scientific, Cat# R0811).

- Dimethyl sulfoxide (DMSO) (Thermo Fisher Scientific, Cat# 67685).
- W303 *bud32Δ* yeast strain harboring a *ura3* cover plasmid, as described in (Downey et al., 2006) [*MATa, ade2-1, trp1-1, can1-100, leu2-3112 his3-11,15 ura3-1, GAL+, psi+, ssd1-d2, RAD5+, bud32::NATMX* (pBUD32-*URA3*)]. For simplicity, we will term these yeast W303 *bud32Δ* (+pBUD32-*URA3*).
- Yeast transformation reagents:
  - o salmon sperm DNA (ssDNA) (Sigma, Cat# D1626).
  - o 0.1 M Li-acetate (Sigma, Cat# 6108174) made in 1 × TE.
  - o 0.1 M Li-acetate with 40% polyethylene glycol (PEG) 400 (Fluka, Cat# 81170) made in 1 × TE (Li-PEG solution).
- Sterile double distilled water (DDW).
- Sterile 96 well plates.
- Synthetic complete agar plates and liquid media made (for detailed protocols on how to make these plates, see Treco & Lundblad, 2001):
  - o Without uracil (−ura).
  - o Without uracil and leucine (−ura, −leu).
  - o Without leucine and supplemented with 0.8 mg/mL 5-fluoroorotic acid (−leu, 5-FOA).

6.3 Step-by-step protocol for performing a drop assay for monitoring the role of scBud32 *in vivo* in yeast (10–12 days). For a schematic overview, see Fig. 7A

- Transform W303 *bud32Δ* (+pBUD32-*URA3*) yeast with pRS415 scBud32 expression plasmids:
  - o In a 25 mL Erlenmeyer glass flask, grow a W303 *bud32Δ* (+pBUD32-*URA3*) culture in 10 mL −ura liquid media overnight in a 30 °C shaker incubator. The following morning, dilute culture to an OD_{600} nm of 0.2 in fresh 10 mL −ura liquid media in a 25 mL Erlenmeyer glass flask and allow growth for 2–3 h at in a 30 °C shaker incubator. Transfer culture to a sterile 50 mL Falcon tube and spin down (3000 × *g* for 5 min at room temperature). Aspirate supernatant and wash cells by resuspending the pellet in 10 mL 0.1 M Li-acetate. Spin down again (3000 × *g* for 5 min at room temperature) and aspirate supernatant. Resuspend pellet in 1 mL 0.1 M Li-acetate.
  - o While the cells are spinning, heat ssDNA (2 min at 95 °C) and cool for 2–10 min in an ice bucket.
- In separate sterile 1.5 mL tubes add 5 μL of either:
  - ▪ pRS415 empty plasmid.
  - ▪ pRS415 with FLAG-Bud32 insert.

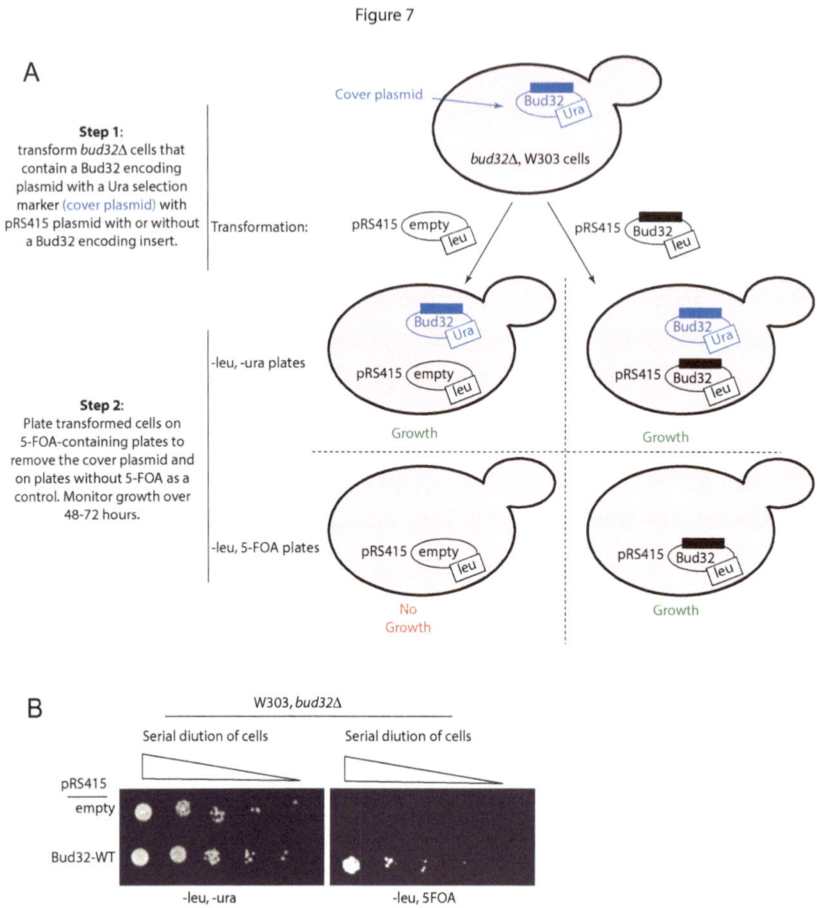

Fig. 7 Analyzing Bud32 activity *in vivo* by monitoring the role Bud32 has in cell growth in *Saccharomyces cerevisiae*. (A) Schematic representation of the main steps in monitoring the role of Bud32 *in vivo* using a drop assay in yeast, in which yeast cultures are subjected to serial dilutions and spotted in drops on plates to monitor their growth. (top) *bud32Δ* W303 cells are transformed to contain two different plasmids. The first is a high copy plasmid with a *URA3* selection marker that expresses Bud32 proteins using a constitutive promotor (denoted as "Cover plasmid"). The second plasmid is a low copy plasmid with a *LEU2* selection marker and the endogenous promotor of Bud32, with or without the region encoding the Bud32 protein (denoted as "pRS415-Bud32 WT" or "pRS415-empty," respectively). (middle) Due to the presence of the Cover plasmid, all cells are expected to grow on -leu, −ura plates. (bottom) Plating the cells on 5-FOA-containing plates leads to the loss of the Cover plasmid, therefore yeast growth depends on the insert within the pRS415 plasmid and will only grow if the plasmid encodes a functional Bud32 protein. (B) The growth of W303 *bud32Δ* cells monitored by a drop assay. W303 *bud32Δ* (+pBUD32-*URA3*) cells transformed with pRS415 empty, or pRS415 Bud32-WT, were subjected to five-fold serial dilutions and plated on −leu, ura plates and on −leu, 5-FOA plates. Growth on both plates was monitored after 48–96 h.

    - sterile DDW as a control (mock transformed).
- To each tube add in the following order:
    - 10 μL ssDNA.
    - 100 μL of the W303 *bud32Δ* cells.
    - 700 μL Li-PEG solution.
- Mix by inverting closed tubes 5–10 times.
- Incubate at 30 °C for 30–60 min.
- Add 86 μL DMSO to each tube. Mix again by inverting closed tubes 4–5 times.
- Heat shock cells using a water bath set to 42 °C for 10 min.
- Spin cells down (3000 × g for 1 min at room temperature), aspirate supernatant and resuspend pellets in 100 μL sterilized DDW.
- Plate cells on −ura, −leu plates and incubate for 3–4 days at 30 °C for colonies to appear.
- Monitor that there is no growth in the mock-transformed plate.
- Patch cells:
    - o Pick 3–5 individual colonies from each transformed plate using a sterile toothpick or an inoculation loop.
    - o Transfer picked colonies to a ∼1 by 1 cm square on a fresh −ura, −leu plate and grow for 24–48 h in a 30 °C incubator or until the squares are filled with cells.
- Grow 4 mL overnight cultures in −ura, −leu liquid media in an incubator shaker at 30 °C from cells picked from ∼1/4 of each square. The following morning, dilute the cultures to an OD_{600} nm of 0.2 in 4 mL of fresh − ura, −leu liquid media. The remaining cells from the over-night liquid cultures can be stored at −20 °C and used for Western blot analysis.
- Grow diluted cell cultures for 2 h. Measure OD_{600} nm and adjust volume so that OD_{600} nm will be 1. For example, if OD_{600} nm is 0.4, spin down 400 μL of cell culture (3000 × g for 5 min at room temperature) discard supernatant and resuspend pellet in 160 μL sterile DDW.
- Transfer 100 μL of the resuspended cells into the first column of a 96-well plate (A_1 well for example). In the adjacent four following wells (A_{2-5} if the A_1 well was used), add 80 μL sterile DDW. For each different culture analyzed, prepare a similar set of wells.
- Generate a five-fold serial dilution of cells by transferring 20 μL from the first column of the plate into the DDW in the second column. Mix well and then transfer cells in a similar way from column 2 to column 3 and so on until column 5.
- With a multi-channel pipette, spot 5 μL from each well onto a − leu, 5-FOA plate. This step will lead to the removal of the pBUD32-URA3

cover plasmid and will enable analysis of the effect of the pRS415 plasmid on growth. As a control, spot 5 μL from each well onto a − leu, −ura plate to show that changes in growth rates observed are dependent on the contents of the pRS415 plasmid.

- Allow drops to be absorbed completely into the plates (1 − 2h) before flipping plates over.
- Transfer plates to a 30 °C static incubator and monitor growth of colonies in each drop over 2–5 days (Fig. 7B).

6.4 Expected outcomes

Within 24–48 h, growth on the −leu, 5-FOA plates will be observed in W303 *bud32Δ* cells expressing a Bud32 wild-type protein. Within 5 days, very little growth will be observed in W303 *bud32Δ* cells with an empty pRS415 plasmid. Normal growth on −ura, −leu plates for all cultures should be observed within 24–48 h (Fig. 7B).

6.5 Quantification and statistical analysis

This method is non-quantitative and therefore quantification and statistical analysis does not apply.

6.6 Advantages

The drop assay is a relatively easy assay to monitor the role Bud32 has *in vivo* since its activity correlates to the ability of yeast cells to grow.

6.7 Limitations

The correlation between yeast growth and the activity of Bud32 is indirect and therefore insight from the drop assay should be supported by biochemical analysis of other aspects of Bud32 function, such as its role in the t^6A modification of tRNAs *in vivo* and the tRNA binding activity of KEOPS (for protocols, see Beenstock et al., 2020).

6.8 Optimization and troubleshooting

The magnitude of serial dilutions used can be optimized experimentally to gain the best insight into the different growth rates.

6.9 Safety considerations and standards

The components of this assay do not present any special health and safety hazard. Regular lab safety procedures should be implemented.

6.10 Alternative methods/procedures

As an alternative and quantitative readout for the growth rates of yeast strains, cell growth can be monitored using a plate reader that measures absorbance at OD_{600} nm (Wan et al., 2013). However, growth in liquid media does not enable counter selection of the cover plasmid.

7. Summary and conclusions

In this methods chapter, we have provided the protocols of a versatile suite of assays for addressing functional questions on the activity of Bud32 in KEOPS. Using these assays, the remaining mysteries of how this enzyme is regulated and how it exerts its regulatory role on the tRNA modifying function of KEOPS may be addressed.

Acknowledgments

The authors wish to thank Mr. Stephen Orlicky with his help in establishing and performing the ADP Glo experiments. Work in the F.S. lab is supported by a grant from the CIHR (FDN 143277).

References

Arrondel, C., Missoury, S., Snoek, R., Patat, J., Menara, G., Collinet, B., et al. (2019). Defects in t(6)A tRNA modification due to GON7 and YRDC mutations lead to Galloway-Mowat syndrome. *Nature Communications*, *10*(1), 3967.

Beenstock, J., Ona, S. M., Porat, J., Orlicky, S., Wan, L. C. K., Ceccarelli, D. F., et al. (2020). A substrate binding model for the KEOPS tRNA modifying complex. *Nature Communications*, *11*(1), 6233.

Beenstock, J., & Sicheri, F. (2021). The structural and functional workings of KEOPS. *Nucleic Acids Research*, *49*(8), 10818–10834.

Boccaletto, P., Machnicka, M. A., Purta, E., Piatkowski, P., Baginski, B., Wirecki, T. K., et al. (2018). MODOMICS: A database of RNA modification pathways. 2017 update. *Nucleic Acids Research*, *46*(D1), D303–D307.

Boudeau, J., Miranda-Saavedra, D., Barton, G. J., & Alessi, D. R. (2006). Emerging roles of pseudokinases. *Trends in Cell Biology*, *16*(9), 443–452.

Braun, D. A., Rao, J., Mollet, G., Schapiro, D., Daugeron, M. C., Tan, W., et al. (2017). Mutations in KEOPS-complex genes cause nephrotic syndrome with primary microcephaly. *Nature Genetics*, *49*(10), 1529–1538.

Collinet, B., Friberg, A., Brooks, M. A., van den Elzen, T., Henriot, V., Dziembowski, A., et al. (2011). Strategies for the structural analysis of multi-protein complexes: Lessons from the 3D-repertoire project. *Journal of Structural Biology*, *175*(2), 147–158.

Dhillon, A. S., Hagan, S., Rath, O., & Kolch, W. (2007). MAP kinase signalling pathways in cancer. *Oncogene*, *26*(22), 3279–3290.

Downey, M., Houlsworth, R., Maringele, L., Rollie, A., Brehme, M., Galicia, S., et al. (2006). A genome-wide screen identifies the evolutionarily conserved KEOPS complex as a telomere regulator. *Cell*, *124*(6), 1155–1168.

Edvardson, S., Prunetti, L., Arraf, A., Haas, D., Bacusmo, J. M., Hu, J. F., et al. (2017). tRNA N6-adenosine threonylcarbamoyltransferase defect due to KAE1/TCS3 (OSGEP) mutation manifest by neurodegeneration and renal tubulopathy. *European Journal of Human Genetics*, *25*(5), 545–551.

El Yacoubi, B., Hatin, I., Deutsch, C., Kahveci, T., Rousset, J. P., Iwata-Reuyl, D., et al. (2011). A role for the universal Kae1/Qri7/YgjD (COG0533) family in tRNA modification. *The EMBO Journal*, *30*(5), 882–893.

El Yacoubi, B., Lyons, B., Cruz, Y., Reddy, R., Nordin, B., Agnelli, F., et al. (2009). The universal YrdC/Sua5 family is required for the formation of threonylcarbamoyladenosine in tRNA. *Nucleic Acids Research*, *37*(9), 2894–2909.

Ferreira-Cerca, S., Kiburu, I., Thomson, E., LaRonde, N., & Hurt, E. (2014). Dominant Rio1 kinase/ATPase catalytic mutant induces trapping of late pre-40S biogenesis factors in 80S-like ribosomes. *Nucleic Acids Research*, *42*(13), 8635–8647.

Hardie, D. G. (2011). AMP-activated protein kinase: A cellular energy sensor with a key role in metabolic disorders and in cancer. *Biochemical Society Transactions*, *39*(1), 1–13.

Hecker, A., Lopreiato, R., Graille, M., Collinet, B., Forterre, P., Libri, D., et al. (2008). Structure of the archaeal Kae1/Bud32 fusion protein MJ1130: A model for the eukaryotic EKC/KEOPS subcomplex. *The EMBO Journal*, *27*(17), 2340–2351.

Jacobsen, A. V., & Murphy, J. M. (2017). The secret life of kinases: Insights into non-catalytic signalling functions from pseudokinases. *Biochemical Society Transactions*, *45*(3), 665–681.

Jura, N., Endres, N. F., Engel, K., Deindl, S., Das, R., Lamers, M. H., et al. (2009). Mechanism for activation of the EGF receptor catalytic domain by the juxtamembrane segment. *Cell*, *137*(7), 1293–1307.

Kisseleva-Romanova, E., Lopreiato, R., Baudin-Baillieu, A., Rousselle, J. C., Ilan, L., Hofmann, K., et al. (2006). Yeast homolog of a cancer-testis antigen defines a new transcription complex. *The EMBO Journal*, *25*(15), 3576–3585.

Korencic, D., Soll, D., & Ambrogelly, A. (2002). A one-step method for in vitro production of tRNA transcripts. *Nucleic Acids Research*, *30*(20), e105.

Kornev, A. P., & Taylor, S. S. (2015). Dynamics-driven Allostery in protein kinases. *Trends in Biochemical Sciences*, *40*(11), 628–647.

Kung, J. E., & Jura, N. (2016). Structural basis for the non-catalytic functions of protein kinases. *Structure*, *24*(1), 7–24.

Lahiry, P., Torkamani, A., Schork, N. J., & Hegele, R. A. (2010). Kinase mutations in human disease: Interpreting genotype-phenotype relationships. *Nature Reviews. Genetics*, *11*(1), 60–74.

Lauhon, C. T. (2012). Mechanism of N6-threonylcarbamoyladenonsine (t(6)A) biosynthesis: Isolation and characterization of the intermediate threonylcarbamoyl-AMP. *Biochemistry*, *51*(44), 8950–8963.

Leonard, C. J., Aravind, L., & Koonin, E. V. (1998). Novel families of putative protein kinases in bacteria and archaea: Evolution of the "eukaryotic" protein kinase superfamily. *Genome Research*, *8*(10), 1038–1047.

Lopreiato, R., Facchin, S., Sartori, G., Arrigoni, G., Casonato, S., Ruzzene, M., et al. (2004). Analysis of the interaction between piD261/Bud32, an evolutionarily conserved protein kinase of Saccharomyces cerevisiae, and the Grx4 glutaredoxin. *The Biochemical Journal*, *377*(Pt 2), 395–405.

Mao, D. Y., Neculai, D., Downey, M., Orlicky, S., Haffani, Y. Z., Ceccarelli, D. F., et al. (2008). Atomic structure of the KEOPS complex: An ancient protein kinase-containing molecular machine. *Molecular Cell*, *32*(2), 259–275.

Murphy, F. V. T., Ramakrishnan, V., Malkiewicz, A., & Agris, P. F. (2004). The role of modifications in codon discrimination by tRNA(Lys)UUU. *Nature Structural & Molecular Biology*, *11*(12), 1186–1191.

Perrochia, L., Crozat, E., Hecker, A., Zhang, W., Bareille, J., Collinet, B., et al. (2013). In vitro biosynthesis of a universal t6A tRNA modification in Archaea and Eukarya. *Nucleic Acids Research*, *41*(3), 1953–1964.

Perrochia, L., Guetta, D., Hecker, A., Forterre, P., & Basta, T. (2013). Functional assignment of KEOPS/EKC complex subunits in the biosynthesis of the universal t6A tRNA modification. *Nucleic Acids Research*, *41*(20), 9484–9499.

Reiterer, V., Eyers, P. A., & Farhan, H. (2014). Day of the dead: Pseudokinases and pseudophosphatases in physiology and disease. *Trends in Cell Biology*, *24*(9), 489–505.

Sherlin, L. D., Bullock, T. L., Nissan, T. A., Perona, J. J., Lariviere, F. J., Uhlenbeck, O. C., et al. (2001). Chemical and enzymatic synthesis of tRNAs for high-throughput crystallization. *RNA*, 7(11), 1671–1678.

Sikorski, R. S., & Hieter, P. (1989). A system of shuttle vectors and yeast host strains designed for efficient manipulation of DNA in Saccharomyces cerevisiae. *Genetics*, *122*(1), 19–27.

Srinivasan, M., Mehta, P., Yu, Y., Prugar, E., Koonin, E. V., Karzai, A. W., et al. (2011). The highly conserved KEOPS/EKC complex is essential for a universal tRNA modification, t6A. *The EMBO Journal*, *30*(5), 873–881.

Stuart, J. W., Gdaniec, Z., Guenther, R., Marszalek, M., Sochacka, E., Malkiewicz, A., et al. (2000). Functional anticodon architecture of human tRNALys3 includes disruption of intraloop hydrogen bonding by the naturally occurring amino acid modification, t6A. *Biochemistry*, *39*(44), 13396–13404.

Summer, H., Gramer, R., & Droge, P. (2009). Denaturing urea polyacrylamide gel electrophoresis (urea PAGE). *Journal of Visualized Experiments*, *32*, 1485.

Taylor, S. S., Knighton, D. R., Zheng, J., Ten Eyck, L. F., & Sowadski, J. M. (1992). Structural framework for the protein kinase family. *Annual Review of Cell Biology*, *8*, 429–462.

Thiaville, P. C., Iwata-Reuyl, D., & de Crecy-Lagard, V. (2014). Diversity of the biosynthesis pathway for threonylcarbamoyladenosine (t(6)A), a universal modification of tRNA. *RNA Biology*, *11*(12), 1529–1539.

Thiaville, P. C., Legendre, R., Rojas-Benitez, D., Baudin-Baillieu, A., Hatin, I., Chalancon, G., et al. (2016). Global translational impacts of the loss of the tRNA modification t(6)A in yeast. *Microbial Cell*, *3*(1), 29–45.

Treco, D. A., & Lundblad, V. (2001). Preparation of yeast media. *Current Protocols in Molecular Biology*, 11. Chapter 13, Unit 13.

Wan, L. C., Maisonneuve, P., Szilard, R. K., Lambert, J. P., Ng, T. F., Manczyk, N., et al. (2017). Proteomic analysis of the human KEOPS complex identifies C14ORF142 as a core subunit homologous to yeast Gon7. *Nucleic Acids Research*, *45*(2), 805–817.

Wan, L. C., Mao, D. Y., Neculai, D., Strecker, J., Chiovitti, D., Kurinov, I., et al. (2013). Reconstitution and characterization of eukaryotic N6-threonylcarbamoylation of tRNA using a minimal enzyme system. *Nucleic Acids Research*, *41*(12), 6332–6346.

Zarrin, A. A., Bao, K., Lupardus, P., & Vucic, D. (2021). Kinase inhibition in autoimmunity and inflammation. *Nature Reviews. Drug Discovery*, *20*(1), 39–63.

Zeqiraj, E., Filippi, B. M., Deak, M., Alessi, D. R., & van Aalten, D. M. (2009). Structure of the LKB1-STRAD-MO25 complex reveals an allosteric mechanism of kinase activation. *Science*, *326*(5960), 1707–1711.

Zhang, W., Collinet, B., Graille, M., Daugeron, M. C., Lazar, N., Libri, D., et al. (2015). Crystal structures of the Gon7/Pcc1 and Bud32/Cgi121 complexes provide a model for the complete yeast KEOPS complex. *Nucleic Acids Research*, *43*(6), 3358–3372.

Zhang, X., Gureasko, J., Shen, K., Cole, P. A., & Kuriyan, J. (2006). An allosteric mechanism for activation of the kinase domain of epidermal growth factor receptor. *Cell*, *125*(6), 1137–1149.

CHAPTER TWENTY-FIVE

Dissecting protein function in vivo: Engineering allelic series in mice using CRISPR-Cas9 technology

Annelise M. Cassidy[a], Emin Kuliyev[b], Destinée B. Thomas[a], Hanying Chen[a], and Stephane Pelletier[a,*]

[a]Department of Medical and Molecular Genetics, Indiana University School of Medicine, Indianapolis, IN, United States
[b]Department of Immunology, St. Jude Children's Research Hospital, Memphis, TN, United States
*Corresponding author: e-mail address: spellet@iu.edu

Contents

Methods in Enzymology, Volume 667
ISSN 0076-6879
https://doi.org/10.1016/bs.mie.2022.03.053

Abstract

Allelic series are extremely valuable genetic tools to study gene function and identify essential structural features of gene products. In mice, allelic series have been engineered using conventional gene targeting in embryonic stem cells or chemical mutagenesis. While these approaches have provided valuable information about the function of genes, they remain cumbersome. Modern approaches such as CRISPR-Cas9 technologies now allow for the precise and cost-effective generation of mouse models with specific mutations, facilitating the development of allelic series. Here, we describe procedures for the generation of three types of mutations used to dissect protein function in vivo using CRISPR-Cas9 technology. This step-by-step protocol describes the generation of missense mutations, large in-frame deletions, and insertions of genetic material using *SCY1-like 1* (*Scyl1*) as a model gene.

Abbreviations

Cas9	CRISPR-associated protein 9
CRISPR	clustered regularly interspaced short palindromic repeats
DSB	double strand break
HDR	homology directed repair
HR	homologous recombination
NHEJ	non-homologous end joining
PAM	protospacer adjacent motif
SCYL1	SCY1-like 1
SCYL2	SCY1-like 2
SCYL3	SCY1-like 3
sgRNA	single guide RNA

1. Introduction

Members of the SCY1-like (SCYL) family of protein kinases are evolutionarily conserved and ubiquitously expressed proteins characterized by an N-terminal protein kinase domain, centrally located HEAT (Huntingtin, elongation factor 3, protein phosphatase 2A, yeast kinase TOR1) repeats and an overall disordered C-terminal segment containing coiled coil domains (Fig. 1A). Three members have been described in humans encoded by three distinct genes: *SCYL1*, *SCYL2* and *SCYL3* (Fig. 1B). SCYL proteins are considered pseudokinases due to alterations in all three motifs essential for catalytic activity in functional kinases (Fig. 1C) (Pelletier, 2016). Biochemical studies have also revealed that SCYL1 is incapable of binding ATP and magnesium ions, further supporting the notion that SCYL proteins are catalytically inactive and likely function through noncatalytic mechanisms (Murphy et al., 2014).

A

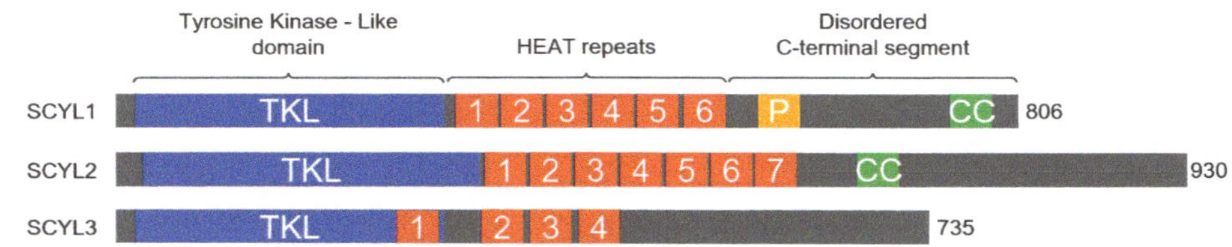

C

Active kinases	SCYL1	SCYL2	SCYL3
VAIK	VAKA	VAVF	FVYK
HRD	HNN	HGN	HNN
DFG	DYM	DFC	GME

B

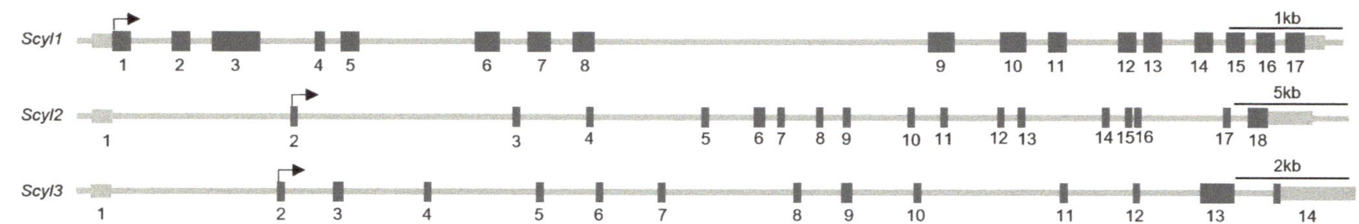

D

	SCYL1 deficiency	SCYL2 deficiency	SCYL3 deficiency
Human	• Liver failure, developmental delay, spine and chest abnormalities, joint laxity, respiratory phenotypes, range of neurological phenotypes including motor dysfunction	• Developmental delay, neurogenic arthrogryposis, microcephaly, brain malformation (absent corpus callosum), optic nerve atrophy, limb fractures, early lethality	• No deficiency reported • NTRK-SCYL3 fusion in colorectal cancer
Mouse	• Developmental delay, spine and chest abnormalities, motor dysfunction leading to complete hindlimb paralysis, motor neuron degeneration	• Developmental delay, early lethality, sensory motor phenotype, loss of hippocampal neurons	• No visible abnormalities • Exacerbates motor phenotype caused by the lack of Scyl1

Fig. 1 Schematic representation of SCYL proteins and genes as well as phenotypes associated with SCYL deficiency in humans and mice. (A) Schematic representation of SCYL1, SCYL2, and SCYL3. TKL, tyrosine kinase like domain; 1–6, HEAT repeats; CC, coiled coil domain; P, proline riche region. SCYL3 contains a myristoylation consensus sequence at its N-terminus (not shown). (B) Schematic representation of *SCYL1*, *SCYL2*, and *SCYL3* genes. Gray lines, introns; light gray boxes, 5′ or 3′ untranslated regions; dark gray boxes, exons. (C) Sequence alteration of key conserved catalytic residues in SCYL1, SCYL2, and SCYL3. (D) Phenotypes associated with *SCYL1*, *SCYL2*, or *SCYL3* inactivation in humans and mice.

In humans, inactivating *SCYL1* mutations have been associated with a rare pediatric multisystem disorder characterized by recurrent episodes of liver failure, global developmental delay, spine and chest deformities, as well as various neurological abnormalities including ataxia and motor dysfunction (Lenz et al., 2018; McNiven et al., 2021; Shohet et al., 2019; Spagnoli, Frattini, Salerno, & Fusco, 2019). Similarly, inactivation of the *SCYL2* gene has been associated with a rare pediatric disorder characterized by severe developmental delay and a range of neurological phenotypes including neurogenic arthrogryposis, microcephaly, and brain malformations (Seidahmed et al., 2020) (Fig. 1D). Although no inactivating mutations have been reported for *SCYL3*, a microdeletion resulting in the fusion of *SCYL3* with *NTRK1* has been identified as a driver mutation in colorectal cancer (Milione et al., 2017). In mice, the inactivation of SCYL genes recapitulates most phenotypic changes seen in humans (Blot, Poirier, & Dreyfus, 1995; Gingras et al., 2015; Pelletier, Gingras, Howell, Vogel, & Ihle, 2012; Schmidt et al., 2007) and further reveals an overlapping role for SCYL1 and SCYL3 in regulating motor neuron function and viability (Kuliyev et al., 2018).

The molecular mechanisms underlying SCYL deficiency syndromes have remained elusive. Several essential cellular functions have been proposed for SCYL proteins including membrane protein trafficking, signal transduction, RNA shuttling, and protein expression and turnover (Pelletier, 2016). However, whether these functions are in fact regulated by SCYL proteins is currently debated (Gingras et al., 2015; Gingras, Kuliyev, & Pelletier, 2017; Kuliyev et al., 2018; Pelletier, 2016; Pelletier et al., 2012), and more importantly whether deregulation of these functions contribute to disease pathogenesis is unknown.

To gain insights into SCYL functions and identify molecular mechanisms underlying SCYL deficiency syndromes, we have engineered allelic series of SCYL mutations in mice. Allelic series are powerful genetic tools to study gene function and identify functionally important protein domains in the context of a living animal and the endogenous gene locus. Until recently, allelic series in mice have been engineered using conventional gene targeting in embryonic stem cells or chemical mutagenesis. While these approaches have provided a wealth of knowledge about the function of genes and gene products, they are time-consuming, relatively imprecise, and very expensive. Modern genome editing technologies, such as CRISPR-Cas9, now allow for time- and cost-effective generation of mouse models with specific mutations and can be used to rapidly engineer allelic series. Engineering mutations in mice using this technology relies on the activation and hijacking of DNA repair pathways triggered by CRISPR-Cas9-mediated insertion of DNA double strand breaks (Pelletier, Gingras, & Green, 2015).

In this chapter, we describe the generation of mouse models with three types of mutations commonly used to dissect protein function. These include large in-frame deletions, which are used to assess the role of protein domain(s); missense mutations, which are used to assess the importance of certain residues at active sites or prevent specific molecular interactions; and insertion of genetic materials, such as epitope tags, which are used to facilitate protein detection or purification. More specifically, using *Scyl1* as a model gene, we present the generation of three mouse models: a large-in frame deletion in which five of the six HEAT repeats have been removed in order to assess their role in SCYL1 function; a missense mutation at position 454 changing a highly conserved lysine for an alanine to study the role of SCYL1 in tRNA trafficking; and the insertion of a FLAG epitope at the N-terminus of *Scyl1* to look for SCYL1 interacting partners in vivo.

2. Before you begin

2.1 Missense mutations

Missense mutations in mice are engineered by the co-injection of zygotes with the RNA-guided endonuclease SpCas9, in the form of an mRNA transcript or recombinant protein; a single guide RNA transcript (sgRNA), formed from the fusion between the CRISPR RNA (crRNA) and the trans-activating CRISPR RNA (tracrRNA); and a homology directed repair (HDR) template, in the form of a single stranded DNA molecule. HDR templates are designed to encode the missense mutation(s) together with additional silent mutation(s) encoding a restriction enzyme recognition sequence as well as additional silent mutations preventing further cleavage by Cas9 upon repair of the DNA break. These mutations are typically flanked by homology arms of 25–70 nucleotides in length (Fig. 2). Silent mutations encoding restriction sites are incorporated to facilitate genotyping.

2.2 Large in-frame deletions

Large in-frame deletions in mice are engineered by the co-injection of zygotes with SpCas9, two sgRNA molecules, and an HDR template overlapping the two target sites (Fig. 3). The homology directed repair template is designed so that upon repair of the DSB by the homology directed repair

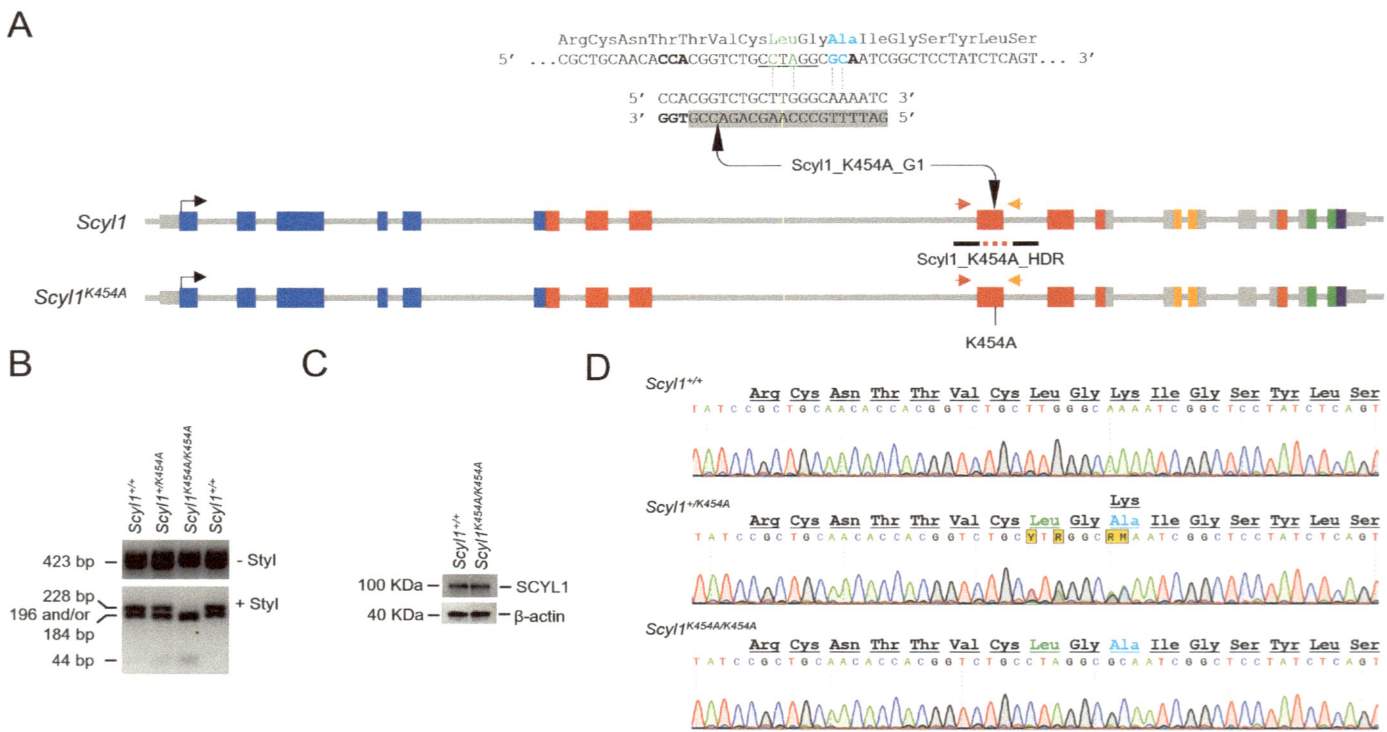

Fig. 2 See figure legend on opposite page.

pathway, the fusion between the two break sites is in frame. 25–70 nucleotide-long homology arms on each side of the fusion point are usually sufficient for proper recombination. Out-of-frame repair will likely result in the formation of a premature stop codon within the target gene, and degradation of the mRNA transcript via the nonsense mediated mRNA decay pathway (Popp & Maquat, 2016). Out of frame deletions are commonly used to inactivate genes (Gingras et al., 2017; Kuliyev et al., 2018).

2.3 Insertion of genetic material

Insertion of genetic material in mice is usually engineered by the co-injection of zygotes with SpCas9, a single sgRNA molecule, and an HDR template encoding the DNA segment to be inserted flanked by homology arms of 25–70 nucleotides in length (Fig. 4). Although not essential, insertion of DNA sequences encoding a restriction site can be used to simplify the genotyping process. DNA sequences encoding the FLAG epitope, as detailed here, can be designed to also encode a *Psi*I restriction site.

3. Allele design

Prior to generating reagents for microinjection, the mutant allele(s) must be designed, preferably using a molecular biology application or software. The design of the allele must include a PCR-based genotyping

Fig. 2 *Scyl1*^{K454A} targeting strategy. (A) Schematic representation of the wild-type and mutant mouse *Scyl1* alleles, the sgRNA and HDR template used to introduce the K454A amino acid substitution. Black arrowheads indicate the location of the Cas9 cleavage site in exon 9. Boxes indicate exons, gray lines indicate introns. Red and yellow arrows represent the genotyping primers. Gray shaded area indicates the guide target sequence which is antisense with respect to the gene. The PAM sequence is indicated in bold. The HDR template is illustrated above the guide sequence and contains missense mutations (blue) and silent mutations (green) generating a StyI restriction site that can be used to distinguish between the wild-type and the mutant alleles. (B) PCR genotyping of the *Scyl1* K454 region in *Scyl1*$^{+/+}$, *Scyl1*$^{+/K454A}$, *Scyl1*$^{K454A/K454A}$ mice. In the absence of restriction digestion with *Sty*I, a band of 423 bp, corresponding to the *Scyl1*$^{+/+}$, *Scyl1*$^{+/K454A}$, *Scyl1*$^{K454A/K454A}$ alleles is obtained. Following restriction enzyme digest, fragments of 228 and 195 correspond to the wild-type allele and fragments of 195, 185, and 44 bp correspond to the mutant allele. (C) SCYL1 expression in the brain of *Scyl1*$^{+/+}$ and Scyl1$^{K454A/K454A}$ mice, as detected by western blot. Whole cell lysates were resolved by SDS-PAGE, transferred to LF-PVDF membranes, and immunoblotted with an anti-SCYL1 antibody as well as an anti-β-actin antibody, as a loading control. (D) Representative Sanger sequencing chromatogram obtained from *Scyl1*$^{+/+}$, Scyl1$^{+/K454A}$, and Scyl1$^{K454A/K454A}$ mice.

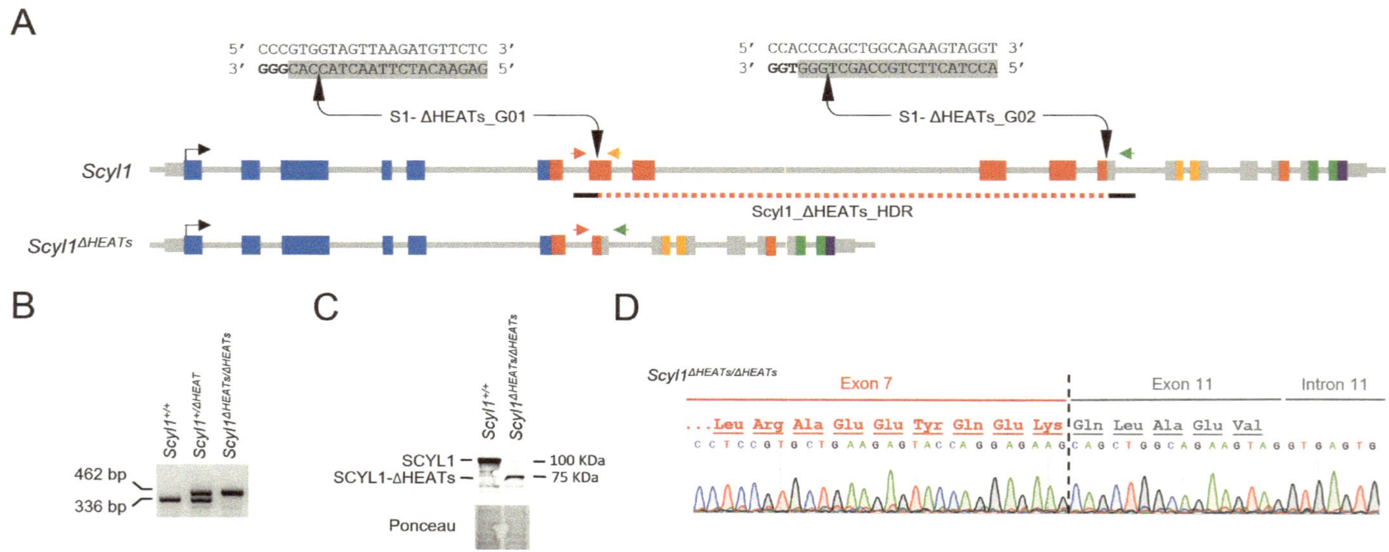

Fig. 3 *Scyl1*$^{\Delta HEATs}$ targeting strategy. (A) Schematic representation of the wild-type and mutant mouse *Scyl1* alleles, the sgRNA and HDR template used to eliminate the HEAT repeats 2–6 of SCYL1. Black arrowheads indicate the location of Cas9 cleavage site in exon 7 and 11. Boxes indicate exons, gray lines indicate introns. Red, yellow, and green arrows represent the genotyping primers. Gray shaded areas indicate the guide target sequences which are antisense of the gene. PAM sequences are indicated in bold. The HDR template contains homology arms flanking the region to be deleted. (B) PCR genotyping of the *Scyl1* alleles in *Scyl1*$^{+/+}$, *Scyl1*$^{+/\Delta HEATs}$, *Scyl1*$^{\Delta HEATs/\Delta HEATs}$ mice. PCR fragment of 336 bp corresponds to the wild-type allele and a fragment of 462 bp corresponds to the mutant allele. (C) SCYL1 expression in the brain of *Scyl1*$^{+/+}$ and *Scyl1*$^{\Delta HEATs/\Delta HEATs}$ mice, as detected by western blot. Whole cell lysates were resolved by SDS-PAGE, transferred to LF-PVDF membranes, and immunoblotted with an anti-SCYL1 antibody. Ponceau S staining was also performed as a loading control. (D) Representative Sanger sequencing chromatogram obtained from a *Scyl1*$^{\Delta HEATs/\Delta HEATs}$ mouse.

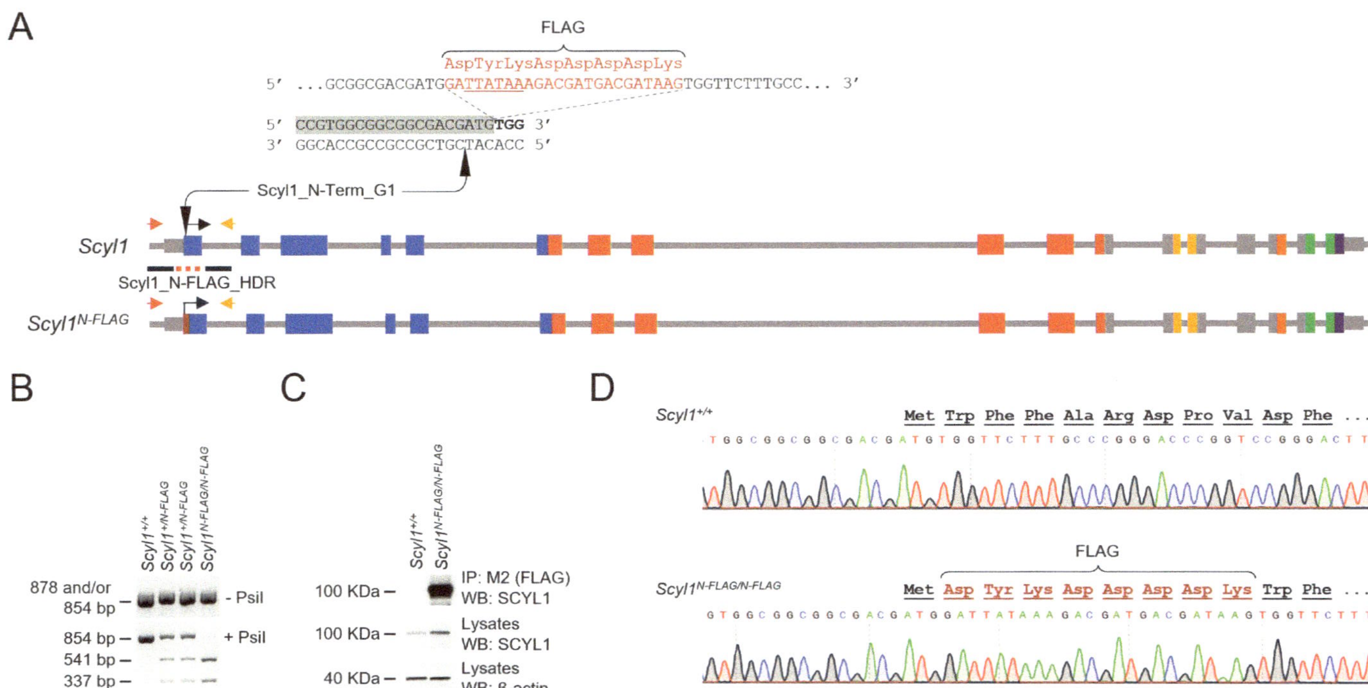

Fig. 4 See figure legend on next page.

strategy to distinguish the mutant from the wild-type allele as well as the selection of the guide sequence(s) and HDR template(s).

3.1 Gene annotation

3.1.1 Timing: 2h

To assist with allele design, we recommend using a molecular biology software. The following steps describe the procedures to design alleles using SnapGene (https://www.snapgene.com/).

1. Download *Scyl1* gene information form NCBI—Gene:
   - **(a)** Go to https://www.ncbi.nlm.nih.gov/gene/ and type "Scyl1" in the query box and click "Search."
   - **(b)** From the list of results, select "Scyl1 from *Mus musculus*, ID: 78891."
   - **(c)** Scroll down to "Genomic regions, transcripts, and products" and click on: "GenBank."
   - **(d)** Click on "Send to" and select "Complete record," "File," "GenBank" format and "Show GI." Click on "Create file." This will download the *Scyl1* GenBank file to your browser
   - **(e)** Open the file in SnapGene
2. Generate the desired mutation(s) in silico:
   - **(a)** For the missense mutation K454A
     - Locate the lysine 454 (K454).
     - Make the following amino acid substitution K454A by changing the nucleotide sequence AAA to GCA

Fig. 4 *Scyl1*$^{N\text{-}FLAG}$ targeting strategy. (A) Schematic representation of the wild-type and mutant mouse *Scyl1* alleles, the sgRNA and HDR template used to insert the FLAG epitope at the N terminus of the *Scyl1* gene. Black arrowheads indicate the location of Cas9 cleavage site in exon 1. Boxes indicate exons, gray lines indicate introns. Red and yellow arrows represent the genotyping primers. Gray shaded area indicates the guide target sequence which is sense of the gene. PAM sequence is indicated in bold. The HDR template contains homology arms flanking the sequence encoding the FLAG epitope. (B) PCR genotyping of the *Scyl1* alleles in *Scyl1*$^{+/+}$, *Scyl1*$^{+/N\text{-}FLAG}$, *Scyl1*$^{N\text{-}FLAG/N\text{-}FLAG}$ mice. In the absence of restriction digestion with *Psi*I, a band of 854bp, corresponding to the *Scyl1*$^{+/+}$, and 874bp corresponding to the mutant allele are obtained. Following restriction enzyme digest, fragments of 854bp corresponding to the wild-type allele and fragments of 541 and 337bp corresponding to the mutant allele are obtained. (C) SCYL1 expression in the liver of *Scyl1*$^{+/+}$ and *Scyl1*$^{N\text{-}FLAG/N\text{-}FLAG}$ mice. Liver lysates from *Scyl1*$^{+/+}$ and *Scyl1*$^{N\text{-}FLAG/N\text{-}FLAG}$ mice were subjected to immunoprecipitation (IP) using the M2-anti-FLAG antibody. Immune complexes and whole cell lysates (Lysates) were resolved by SDS-PAGE, transferred to LF-PVDF membranes, and immunoblotted (WB) with an anti-SCYL1 antibody. (D) Representative Sanger sequencing chromatogram obtained from *Scyl1*$^{+/+}$ and *Scyl1*$^{N\text{-}FLAG/N\text{-}FLAG}$.

- Save the file under a different name

(b) For the large in-frame deletion removing HEAT repeats 2–6

- Locate the amino acid K349 and Q546 and delete the intervening region
- Save the file under a different name

(c) For the insertion of a FLAG epitope at the N-terminus of SCYL1

- Locate the first methionine of SCYL1
- Insert the following DNA sequence: GATTATAAAGACGAT GACGATAAG
- Save the file under a different name

3.2 Guide sequence selection

3.2.1 Timing: 2h

For the most part, the successful manipulation of the mouse genome using CRISPR-Cas9 technology depends on the selection of efficient and selective guide sequences. Guide sequences are selected based on 3 main criteria:

- The proximity of the cleavage site relative to the editing site
- The selectivity of the guide sequence
- The efficiency of a guide sequence to cleave the target site

When possible, select the guide sequence that is closest to the mutation to be engineered, has the least potential off-target sites, and has a high predicted efficiency at promoting SpCas9 activity.

Guide sequences having less than 3 mismatches should be avoided. Off-target cleavage sites can be identified using Cas-Offinder (http://www.rgenome.net/cas-offinder/). The identification of potential off-target cleavage sites is discussed below.

Cas9 activity is affected by nucleotide composition of the sgRNA. In particular, the nature of the nucleotide immediately upstream of the protospacer adjacent motif (PAM), a short sequence located 3′ to the target site (protospacer element) that is essential for Cas9 function, can have negative or positive effects on the endonuclease activity. The presence of a thymine at that position reduces SpCas9 activity in cultured cells, whereas the presence of a guanine residue has the opposite effect (Doench et al., 2016). Similarly, the presence of a cytosine in position -3 (relative to the PAM) increases SpCas9 efficiency whereas a thymine at position -2, -3 or -4 reduces it (Xu et al., 2015).

Table 1 shows the effect of nucleotide composition on the activity of SpCas9 based on Xu et al., 2015. This table should be used only to compare

Table 1 Effect of guide sequence nucleotide composition on SpCas9 activity.

	Nucleotide composition						
Positive effect	T	A	A	G	C	G	G
Negative effect	G			T	T	T/C	T
Position relative to PAM (NGG)	−7	−6	−5	−4	−3	−2	−1

between 2 or more possible guides. For example, the presence of a thymine at position -1 does not completely abolish Cas9 activity, it reduces it.

Guide sequences with approximately 50% GC content are preferred.

To assist with guide selection, we prefer to use the Moreno-Mateos Efficiency score instead of the Doench Rule Set 1, as it considers transcription from the T7 promoter. The Moreno-Mateos score can be found on the UCSC Genome Browser CRISPR Targets track. These scores are based on selectivity and efficiency of the guide at promoting SpCas9 activity. The higher the score, the more specific and efficient the guide is.

To select guide sequences:

3. Open UCSC Genome Browser (https://genome.ucsc.edu/)
4. Click on "BLAT"
5. Copy the target region from SnapGene and paste into the query box
6. Under "Assembly," select "GRCm39/mm39" and click "Submit"
7. From the "BLAT Search Results" list, click the "browser" hyperlink with the highest "SCORE" value
8. On the "CRISPR/Cas9 -NGG Targets, Whole genome" track, identify possible guide sequences. Those located nearest to the target site are preferred. Guide sequences are color coded and appear under the genomic sequence

(a) To load the CRISPR track onto UCSC Genome Browser
   - Scroll down to "Genes and Gene Predictions"
   - Under "CRISPR Targets", select: "Pack"
   - Click on the "Refresh" button

9. Guide sequences selected for the engineering of all three alleles are shown in Figs. 2–4
10. After selecting guide sequences, make annotations using the "Add feature" function in SnapGene
    (a) Copy the guide sequences from UCSC Genome Browser and retrieve that sequence in SnapGene using the Ctrl + F function, in the wild-type gene

   (b) Click on "Add feature" and provide a name for the guide sequence (e.g., Scyl1_K454A_G1).

11. Generate a list of potential off-target sites using Cas-Offinder (http://www.rgenome.net/cas-offinder/)

   (a) Go to the Cas-Offinder website
   (b) Copy and paste your guide sequence in the box under "Query Sequence"
   (c) Under "Mismatch Number", select 3
   (d) Under "PAM Type", Select "SpCas9 from *Streptococcus pyogenes*: 5'-NRG-3' (R=A or G)"
   (e) Under "Target Genome", select "Vertebrate" and then "*M. musculus* (mm10)–Mouse."
   (f) Click "Submit"
   (g) After the list is generated, click on "Download result" under "Status."
   (h) Keep the list of potential off-target sites for future analysis of potential off-target cleavage

12. Order the oligonucleotides listed in Table 2 (specifications described below), which will be used to synthesize the guide RNAs (underlined). We typically order oligonucleotides from Integrated DNA Technology (IDT). DNA oligonucleotides from other sources are also acceptable.

3.3 Homology directed repair template design

3.3.1 Timing: 1 h

For the three types of DNA modifications presented in this chapter, all the HDR templates used are single stranded DNA (ssDNA) molecules of 100–200 nucleotides in length.

For missense mutations, HDR templates contain:

- The nucleotide substitutions encoding the desired amino acid change
- Silent mutations coding for a restriction site that, if possible, is absent from the target region
    - o This restriction site will be used to identify potentially edited founder mice using PCR amplification of the target locus and digestion
- Additional silent mutations disrupting the target sequence or the PAM. These changes will prevent further cleavage of the target site by SpCas9. A minimum of 3 substitutions should be introduced
- Homology arms of 25–70 nucleotides on each side of the engineered substitutions

Table 2 List of oligonucleotides used to synthesize the sgRNAs.

Oligonucleotide Name	Sequence	Synthesis
Scyl1_K454A_G1_F1	TAATACGACTCACTATAGGGATTTTGCCCAAGCAGACCG GTTTTAGAGCTAGAAATAGCA	Custom oligonucleotide 25 nmol scale Lyophilized Standard desalted
S1-ΔHEATs_G01-F1	TAATACGACTCACTATAGGGAGAACATCTTAACTACCAC GTTTTAGAGCTAGAAATAGCA	Custom oligonucleotide 25 nmol scale Lyophilized Standard desalted
S1-ΔHEATs_G02-F1	TAATACGACTCACTATAGGACCTACTTCTGCCAGCTGGG GTTTTAGAGCTAGAAATAGCA	Custom oligonucleotide 25 nmol scale Lyophilized Standard desalted
Scyl1_N-TERM_G1_F	TAATACGACTCACTATAGGCCGTGGCGGCGGCGACGATG GTTTTAGAGCTAGAAATAGCA	Custom oligonucleotide 25 nmol scale Lyophilized Standard desalted
sgRNA-Scaffold-R	AAAAAAGCACCGACTCGGTGCCACTTTTTC AAGTTGATAACGGACTAGCCTTATTTTAAC TTGCTATTTCTAGCTCTAAAAC	Ultramer® technology 4 nmol scale Lyophilized Desalted
T7–19	TAATACGACTCACTATAGG	Custom oligonucleotide 25 nmol scale Lyophilized Standard desalted
sgRNA-R	AAAAGCACCGACTCGGTGCC	Custom oligonucleotide 25 nmol scale Lyophilized Standard desalted

Note: Guide sequences are underlined.

For large in-frame deletions, HDR templates contain:

- Homology arms of 50–70 nucleotides flanking the region to be deleted

For insertion of DNA material, HDR templates contain:

- The nucleotide sequence to be inserted
- Homology arms of 25–70 nucleotides on each side of the insertion

For simplicity, we always select the sense strand, with respect to the gene, for the design of HDR templates. Previous studies in cultured cells have reported differences in targeting efficiencies between sense and antisense HDR templates (Richardson, Ray, DeWitt, Curie, & Corn, 2016). However, for the generation of mouse models, we have found no major effect caused by the orientation of HDR templates (S. Pelletier, unpublished data from more than 200 mouse models engineered using this technology).

For the generation of the *Scyl1*^{K454A} allele, the following HDR template was used:

TTGCAAGGCTACAAGCCAAGGACGACCAGGGTCCTATCCGCTGCAACACCACGG

TCTGCCTAGGCGC**A**AATCGGCTCCTATCTCAGTGCTAGTGTGAGTGTCCTGCAAAG

CTGTGGGAGGATGGCAGCCTCATCTTGGGCTC

This HDR template contains:

- Substitutions encoding the desired amino acid changes (AAA to GC**A**), in blue
- Silent mutations (in green) encoding the *Sty*I site (CCTAGG), underlined, and
- Homology arms of 51 and 70 nucleotides on each side of the nucleotide substitutions

For the generation of the *Scyl1*$^{\Delta HEATs}$ allele, the following HDR template was used:

TCCTTCCTCAGGTGGGAAAATCCCTCCGTGCTGAAGAGTACCAGGAGAAGCAGC

TGGCAGAAGTAGGTGAGTGACCCACACTTGTGTTCTTTCTATGCCT

This HDR template contains:

- Homology arms of 50 nucleotides upstream (red) and downstream (black) of the region to be deleted

For the generation of the *Scyl1*$^{N\text{-}FLAG}$ allele, the following HDR template was used:

AGGACCCGGAGTGGAGGCTCCCGAGCCCGTGGCGGCGGCGACGATGGATTAT

AAAGACGATGACGATAAGTGGTTCTTTGCCCGGGACCCGGTCCGGGACTTCCC

GTTCGAGCTGAGCCCTGAG

This HDR template contains:

- Nucleotide sequence to be inserted (GATTATAAAGACGATGACGATAAG) in red, encoding the FLAG tag and the *Psi*I site, underlined
- Homology arms of 46 and 54 nucleotides on each side of the insertion

13. Once the HDR template is designed, make nucleotide substitutions in SnapGene, as described previously under "Allele Design."

14. Order HDR templates from IDT. For ssDNA fragments shorter than 200 nucleotides, select:

(a) Ultramer® technology

(b) 4 nmol scale

(c) Lyophilized

(d) Desalted

3.4 On- and off-target genotyping primers selection

3.4.1 Timing: 1 h

For each target site, use the "BLAT" function from UCSC Genome Browser to retrieve location of the target sequence. Download the DNA sequence overlapping the target site with an additional ~500 bp on each side of the target site. Design PCR primers to amplify ~500–800 bp amplicons. For on-target PCR genotyping, primer locations should be outside the homology arms of the HDR template. For Off-target sites, PCR primers should be at least 200 nucleotides from the potential cleavage sites. Primers should have an annealing temperature of approximately 60 °C. The following steps describes the procedure to design PCR primers using NCBI Primer-BLAST.

15. Go to the NCBI Primer-BLAST website (https://www.ncbi.nlm.nih.gov/tools/primer-blast/)

16. Paste your target region, in FASTA, in the query box. Set the following parameters as follows:

(a) "PCR product size": set "Min" to 500 and "Max" to 800

- These fragment sizes are convenient for amplification and Sanger sequencing

(b) "Search mode": Select "No user guidance."

(c) Database: Select "Genomes for selected organisms (primary reference assembly only)."

(d) "Organism": Type the name of the species "*M. musculus*" and select "*M. musculus* (taxid: 10090)" from the drop-down menu

(e) Select "Show results in a new window" and "Use new graphic view" boxes

17. Click on "Get Primers" and select an appropriate primer pair

18. Order primers from IDT, with the following specifications:
    (a) Standard synthesis
    (b) 25 nmol scale
    (c) Lyophilized
    (d) Desalted
19. Repeat the procedure for each possible off-target site. Sites with 3 or more mismatches are typically not cleaved by SpCas9 when generating mice via zygote injection of SpCas9 mRNA transcript. Also, off-target sites with 1 or more mismatches or those with the sub-optimal 5′-NAG-3′ PAM sequence are usually not cleaved

3.5 On-target genotyping banding pattern

3.5.1 Timing: 30 min

20. After selecting your primers:
    (a) Add on-target primers to SnapGene files (wild-type and modified alleles).
    (b) Record the PCR genotyping banding patterns obtained from both wild-type and modified alleles. Include banding pattern result if restriction digest is involved in the PCR genotyping strategy

Tables 3–5 show the primer pairs or trios and their respective banding patterns for genotyping the wild-type and all three mutant alleles.

Table 3 Genotyping primers and banding pattern for the *Scyl1*^{K454A} mutation.

Allele	**Scyl1_K454A_F51 + Scyl1_K454A_R52**	**Scyl1_K454A_F51 + Scyl1_K454A_R52 + StyI digest**
Scyl1	423 bp	195 and 228 bp
Scyl1^{K454A}	423 bp	195 and 185 and 44 bp

Table 4 Genotyping primers and banding pattern for the *Scyl1*$^{\Delta HEATs}$ mutation.

Allele	**S1-ΔHEATs-F51 + S1-ΔHEATs-R52 + S1-ΔHEATs-R32**
Scyl1	336 bp
Scyl1$^{\Delta HEATs}$	462 bp

Table 5 Genotyping primers and banding pattern for the *Scyl1*$^{N\text{-}FLAG}$ mutation.

Allele	**S1-N-TERM-F51 + S1-N-TERM-R52**	**S1-N-TERM-F51 + S1-N-TERM-R52 + PsiI digest**
Scyl1	854 bp	854 bp
Scyl1$^{N\text{-}FLAG}$	878 bp	541 and 337 bp

Key resources table.

Reagent or resource	**Source**	**Identifier**
Antibodies		
Anti-SCYL1 7645-AP	Pelletier lab	
Anti-β-actin	Sigma-Aldrich	A5441
Bacterial and virus strains		
N/A		
Biological Samples		
N/A		
Chemicals, peptides, and recombinant proteins		
Proteinase K	Invitrogen	25530031
dNTP Mix, PCR Grade (800 μL)	Qiagen	201901
Tail biopsies lysis buffer (KCl, 500 mM; Tris pH 8.3, 100 mM; Gelatin, 0.1 mg/mL; NP40, 1% (*v*/v); Tween 20, 1% (v/v); Proteinase K, 500 μg/mL)	Pelletier lab	
StyI	New England Biolab	R3500S
PsiI	New England Biolab	R0744S
*Pme*I	New England Biolab	R0560S
Critical commercial assays		
Taq DNA Polymerase (1000 U)	Qiagen	201205
QiaQuick Gel Extraction Kit	Qiagen	28704
QiaQuick PCR Purification Kit	Qiagen	28104
MEGAshortscript™ T7 Transcription Kit	Invitrogen	AM1354

—cont'd

Reagent or resource	Source	Identifier
MEGAclear™ Transcription Clean-Up Kit	Invitrogen	AM1908
mMESSAGE mMACHINE® T7 Ultra Kit	Invitrogen	AM1345
ExoSAP-IT™ PCR Product Cleanup Reagent	Invitrogen	78201
BigDye™ Terminator v3.1 Cycle Sequencing Kit	ThermoFisher Scientific	4337456
BigDye XTerminator™ Purification Kit	ThermoFisher Scientific	4376487
Phusion® High-Fidelity DNA Polymerase	New England Biolabs	M0530L
Deposited data		
N/A		
Experimental models: Cell lines		
N/A		
Experimental models: Organisms/strains		
$Scyl1^{\Delta HEATs}$ mouse model	Pelletier lab	N/A
$Scyl1^{K454A}$ mouse model	Pelletier lab	N/A
$Scyl1^{N\text{-}FLAG}$ mouse model	Pelletier lab	N/A
Oligonucleotides		
sgRNA-Scaffold-R	Integrated DNA Technologies	AAAAAAGCACCGACTC GGTGCCACTTTTTC AAGTTGATAACGGACT AGCCTTATTTTAAC TTGCTATTTCTAGC TCTAAAAC
T7–19	Integrated DNA Technologies	TAATACGACTCACTATAGG
sgRNA-R	Integrated DNA Technologies	AAAAGCACCGACTC GGTGCC

Continued

—cont'd

Reagent or resource	Source	Identifier
S1-ΔHEATs_G01	Integrated DNA Technologies	GAGAACATCTTAAC TACCAC
S1-ΔHEATs_G01-F1	Integrated DNA Technologies	TAATACGACTCACTATA GGGAGAACATCTT AACTACCACGTTTTAGA GCTAGAAATAGCA
S1-ΔHEATs_G02	Integrated DNA Technologies	ACCTACTTCTGCC AGCTGGG
S1-ΔHEATs_G02-F1	Integrated DNA Technologies	TAATACGACTCACTATA GGACCTACTTCTG CCAGCTGGGTTTTA GAGCTAGAAATAGCA
S1-ΔHEATs_HDR	Integrated DNA Technologies	TCCTTCCTCAGGTGGGA AAATCCCTCCGTG CTGAAGAGTACCAGGAG AAGCAGCTGGCAG AAGTAGGTGAGTGACC CACACTTGTGTTCT TTCTATGCCT
S1-ΔHEATs-F51	Integrated DNA Technologies	AGGTCCTGCTGAGG AAATGC
S1-ΔHEATs-R52	Integrated DNA Technologies	TAGGTAGCCAAAG CACAGGC
S1-ΔHEATs-R32	Integrated DNA Technologies	CCAGCTTTCGGATG CTCTCA
Scyl1_K454A_G1	Integrated DNA Technologies	GATTTTGCCCAAGCA GACCG

—cont'd

Reagent or resource	Source	Identifier
Scyl1_K454A_G1_F1	Integrated DNA Technologies	TAATACGACTCACTATAG GGATTTTGCCCA AGCAGACCGGTTTTAGA GCTAGAAATAGCA
Scyl1_K454A_HDR	Integrated DNA Technologies	TTGCAAGGCTACAAG CCAAGGACGACCAGG GTCCTATCCGCTGCA ACACCACGGTCTGCC TAGGCGCAATCGGCT CCTATCTCAGTGCTA GTGTGAGTGTCCTGC AAAGCTGTGGAGGA TGGCAGCCTCATCT TGGGCTC
Scyl1_K454A_F51	Integrated DNA Technologies	AATATGCTAGTT CGGGCGGG
Scyl1_K454A_R52	Integrated DNA Technologies	GCACAAGTTCGA GTGCTGTG
Scyl1_N-TERM_G1	Integrated DNA Technologies	CCGTGGCGGCG GCGACGATG
Scyl1_N-TERM_G1_F	Integrated DNA Technologies	TAATACGACTCACTATA GGCCGTGGCGGCG GCGACGATGGTTTTA GAGCTAGAAATAGCA
Scyl1_N-FLAG_HDR	Integrated DNA Technologies	AGGACCCGGAGTGG AGGCTCCCGAGCCCGT GGCGGCGGCGACG ATGGATTATAAAGACGA TGACGATAAGTGG TTCTTTGCCCGGGACCC GGTCCGGGACTT CCCGTTCGAGCTGAGCCC TGAG

Continued

—cont'd

Reagent or resource	Source	Identifier
S1-N-TERM-F51	Integrated DNA Technologies	TCACGCTCACTTCC TGAACC
S1-N-TERM-R52	Integrated DNA Technologies	TCTTCAGCTCCCG GTTTCAC
Recombinant DNA		
pCDNA3.3TOPO-T7-hCas9	Addgene	161876
Software and algorithms		
NCBI Gene		https://www.ncbi.nlm.nih.gov/gene/
UCSC Genome Browser		https://genome.ucsc.edu/
Cas-OFFinder		http://www.rgenome.net/cas-offinder/
Primer-BLAST		http://www.ncbi.nlm.nih.gov/tools/primer-blast/index.cgi?LINK_LOC=BlastHome
SnapGene		https://www.snapgene.com/
Other		
N/A		

4. Step-by-step method details

4.1 Generation of Cas9 mRNA transcript

4.1.1 Timing: 1 day

The following steps describe how to generate the SpCas9 mRNA transcript. This transcript is produced from a human codon optimized SpCas9 plasmid (Pelletier et al., 2015) and is synthesized in vitro using the mMESSAGE mMACHINE® T7 Ultra Kit (Invitrogen, AM1345). To prevent transcription beyond the spCas9 ORF, the plasmid (pCDNA3.3TOPO-T7-hCas9) is first linearized with *Pme*I, purified, and used directly in the RNA transcription assay. RNA transcripts are subsequently purified using the MEGAclear™ Kit (Invitrogen, AM1908M) and analyzed using the 2100 Bioanalyzer from Agilent Technologies (Fig. 5).

1. Generate the SpCas9 DNA template:
   - (a) Linearize pCDNA3.3TOPO-T7-hCas9 by setting up a restriction digest, as shown in Table 6
   - (b) Incubate the reaction in a heat block at 37 °C for 2–4 h
   - (c) Run 10 μL of the digested and undigested plasmid with 2 μL 6 × loading dye on a 1% TAE-agarose gel supplemented with 0.5 × GelRed® for 60–90 min in 1 × TAE buffer to verify complete digestion of the plasmid
   - (d) Image gel using a UV transilluminator (ChemiDoc™ MP Imaging System, Bio-Rad) and save the image for your records
   - (e) Purify the rest of the reaction using QiaQuick Gel Extraction Kit (Qiagen, 28,704) according to the manufacturer's instructions.
2. Setup the RNA transcription assay using mMESSAGE mMACHINE® T7 Ultra Kit as described by the manufacturer (Invitrogen, AM1345).

Critical: Be sure to perform the polyadenylation step after synthesis of the mRNA.

3. Purify RNA using MEGAclear Kit (Invitrogen, AM1908M), following the manufacturer's protocol
4. Measure RNA concentration with Nanodrop Spectrophotometer (Thermo Fisher Scientific, 13–400-519).
5. Aliquot 8 μg of RNA transcript in 1.5 mL RNase-free microfuge tubes. Label and store at −80 °C
6. Save 2–3 μL for quality control (1 μL is enough per run).
7. Assess the quality of RNA on 2100 *Bioanalyzer*, according to manufacturer's recommendations
   - (a) Prior to running the sample on the 2100 *Bioanalyzer*, heat the sample diluted in loading buffer for 15 s at 65 °C

4.2 Generation of the sgRNA transcript

4.2.1 Timing: 1 day

sgRNA transcripts are produced in vitro from double stranded DNA templates. These templates are generated by the annealing of two ssDNA oligonucleotides, the fill-in of the protruding ends, and the amplification of the full-length template by PCR, as shown in Fig. 6. The following describes the step-by-step procedure for the generation of the sgRNA transcript.

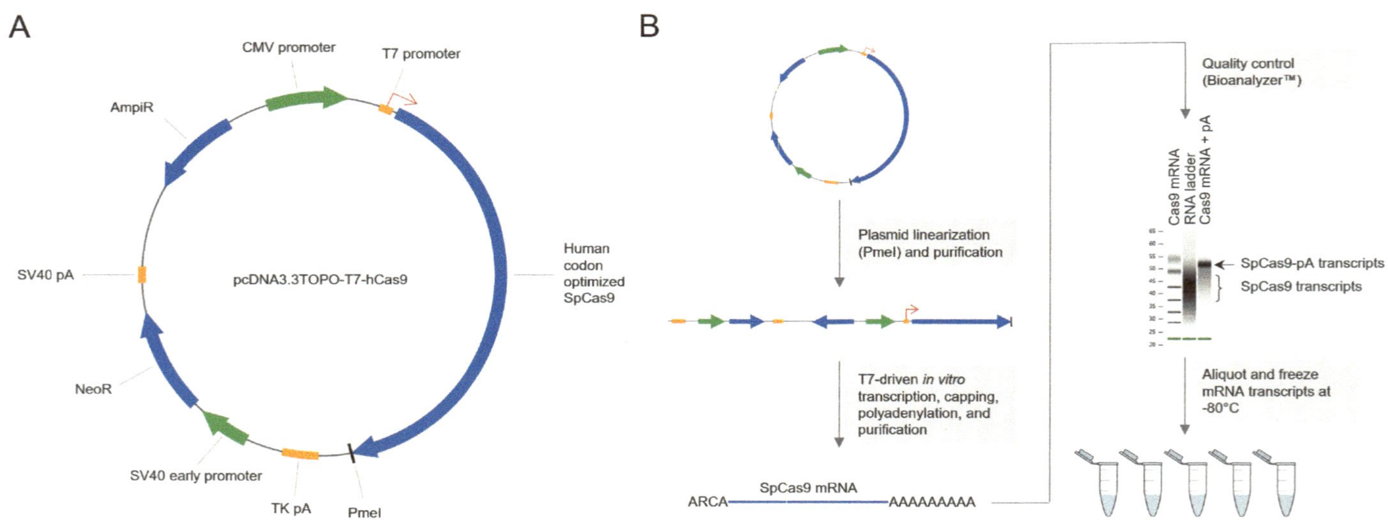

Fig. 5 Workflow of SpCas9 mRNA transcript synthesis. (A) Representation of pcDNA3.3TOPO-T7-hCas9 used for the generation of the human codon optimized Cas9 (hCas9) mRNA transcript. (B) Generation of ARCA-capped polyadenylated (pA) hCas9 transcripts. pcDNA3.3TOPO-T7-hCas9 is linearized using *Pme*I restriction enzyme and purified. The linearized DNA is then used as template for the in vitro transcription assay and purified. mRNA transcripts are then analyzed for quality control using a bioanalyzer (Agilent Technologies, Inc.). Following synthesis and quality control, SpCAs9 mRNA transcript is aliquoted and frozen at −80 °C.

Table 6 Linearization of pCDNA3.3TOPO-T7-hCas9 with PmeI.

Reagents	Amount
pcDNA3.3TOPO-T7-hCas9 plasmid	10 μg
PmeI (R0560S 10,000 units/mL)	2 μL
CutSmart™ Buffer	5 μL
ddH_2O	Up to 50 μL
Total	50 μL

1. Generate the sgRNA DNA template:
   - **(a)** Re-hydrate all oligonucleotides with DNase- and RNase-free ddH_2O to a concentration of 100 μM
   - **(b)** Setup PCR reactions on ice, as shown in Table 7
   - **(c)** Use the cycling conditions described in Table 8
   - **(d)** Load 5 μL of the annealing reaction on a 2% TAE-agarose gel supplemented with 0.5 × GelRed® and run at 120 V for 30–60 min in 1 × TAE buffer
   - **(e)** Image gel using a UV transilluminator (ChemiDoc™ MP Imaging System, Bio-Rad) and keep for your records
   - **(f)** Dilute PCR reaction 1:200 in ddH_2O
   - **(g)** Setup PCR reactions on ice, as detailed in Table 9
   - **(h)** Use the cycling conditions, as indicated in Table 10
   - **(i)** Load 5 μL of PCR products on a 2% TAE-agarose gel supplemented with 0.5 × GelRed® and run at 120 V for 30–60 min in 1 × TAE buffer
   - **(j)** Image gel using a UV transilluminator (ChemiDoc™ MP Imaging System, Bio-Rad) and keep for your reference
   - **(k)** Clean the remainder of the PCR reaction using the QiaQuick PCR Purification Kit (Qiagen, 28,104), using the centrifugation method
   - **(l)** Measure DNA concentration with a Nanodrop Spectrophotometer (Thermo Fisher Scientific, 13–400-519) and store at − 20 °C. On average, 30 ng/μL of DNA is obtained from a single reaction. Use 8 μL of DNA template for the synthesis of sgRNA
2. Synthesize the sgRNA using MEGAshortscript™ T7 kit, as described by the manufacturer (Invitrogen, AM1354).
3. Quantify the RNA (2 μL) using a Nanodrop Spectrophotometer (Thermo Fisher Scientific, 13–400-519). On average, 850 ng/μL of RNA transcript is obtained from a single reaction

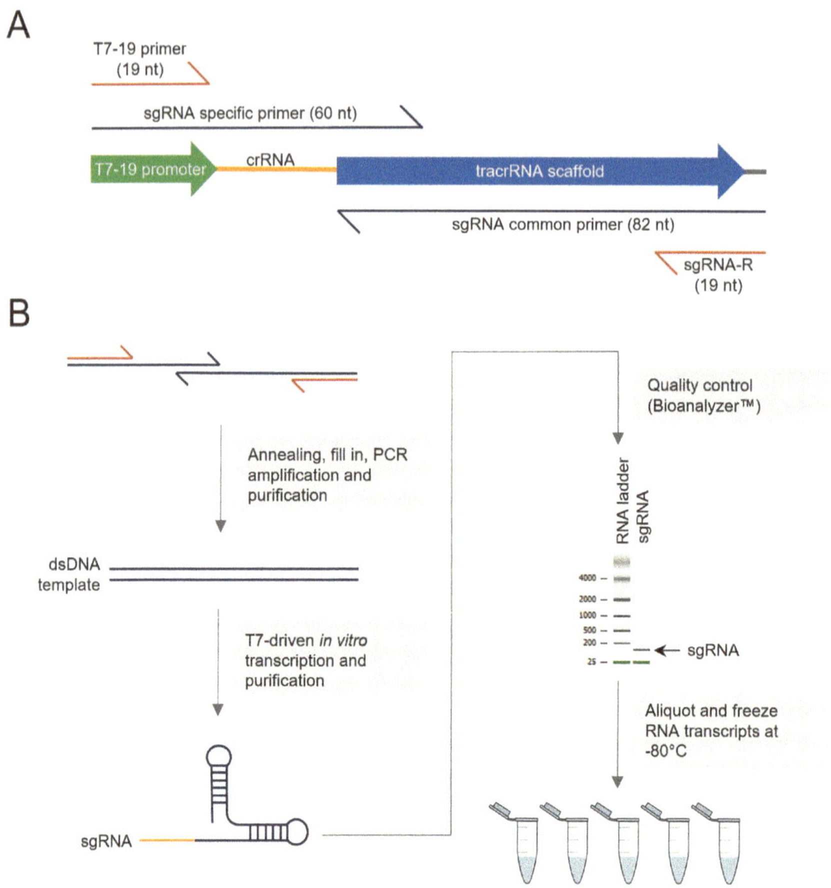

Fig. 6 Schematic workflow of sgRNA transcript synthesis. (A) Schematic representation of the dsDNA template used for sgRNA generation. Green arrow, T7 promoter; yellow line, crRNA specific to the target sequence; blue arrow, tracrRNA scaffold; Orange arrow, T7-19 and sgRNA-R primers; navy blue, sgRNA common and specific primers. sgRNA common and specific primers have overlapping complementary sequence pairing the two oligonucleotides together. (B) Flow chart detailing the steps of sgRNA production. The first step consists of generating a dsDNA template that will subsequently be used in an in vitro transcription assay for short RNA transcripts. The dsDNA template is generated by annealing a target specific oligonucleotide together with the sgRNA common primer. After the annealing reaction, a fill-in reaction produces the dsDNA that will then be amplified using the T7-19 and sgRNA-R primers. The PCR product is then purified and used in an in vitro transcription assay. The transcript is then enriched and analyzed using the 2100 Bioanalyzer from Agilent Technologies, Inc. RNA transcripts are then frozen at −80 °C.

Table 7 Annealing and fill-in reaction for the generation of sgRNA.

	PCR reactions			
	#1	#2	#3	#4
Reagents	Volume (μL)			
5 × HF Phusion buffer	10	10	10	10
dNTPs (10 mM each)	1	1	1	1
sgRNA specific primer (100 μM)	1	1	0	1
sgRNA common primer (100 μM)	1	1	1	0
ddH_2O	36	37	37	37
Phusion (2000 units/mL)	1	0	1	1

Note: Reactions #2, #3, and #4 are used as controls. No PCR product should be obtained from these reactions.

Table 8 Cycling conditions for the annealing and fill-in reaction for the generation of sgRNA.

Step	Number of cycles	Temperature	Time
1	1	98 °C	2 min
2	5	98 °C	10 s
		50 °C	15 s
		72 °C	15 s
3	1	72 °C	10 min
4	1	4 °C	Hold

Table 9 Amplification of the full-length sgRNA template.

Reagents	Volume (μL)
5 × HF Phusion buffer	10
dNTPs, 10 mM each	1
T7–19 primer (100 μM)	0.25
sgRNA-R primer (100 μM)	0.25
ddH_2O	37
Phusion (2000 units/mL)	0.5
DNA (diluted dsDNA from step previous step)	1
Total (μL)	50

Table 10 Cycling conditions for the amplification of the full-length sgRNA template.

Step	Number of cycles	Temperature	Time
1	1	98 °C	2 min
2	25	98 °C	10 s
		55 °C	15 s
		72 °C	15 s
3		72 °C	10 min
4	1	4 °C	Hold

4. Save 2–3 μL for quality control (1 μL is enough per run).
5. Assess the quality of the RNA on 2100 *Bioanalyzer* according to manufacturer's recommendations
   (a) Prior to running the sample on the 2100 *Bioanalyzer*, heat the sample diluted in loading buffer for 15 s at 65 °C
6. Store the sgRNA transcript at −80 °C

Note: In our experience, sgRNAs can undergo several freeze-thaw cycles without loss of activity.

4.3 Alternative strategies

4.3.1 Recombinant Cas9 protein and synthetic sgRNAs

Synthetic sgRNAs can be used in combination with SpCas9 protein for microinjection as an alternative to in vitro transcription for the generation of sgRNAs and Cas9 mRNA. These can be acquired from several commercial sources including IDT, Invitrogen, New England Biolabs, Sigma-Aldrich, and others.

4.3.2 sgRNA quality control

As an alternative strategy to assess the quality of in vitro transcribed sgRNAs, a functional in vitro Cas9 assay can be performed. This is done by combining recombinant Cas9 protein, the in vitro transcribed sgRNA, and the PCR-amplified target locus. This DNA fragment can be obtained by PCR amplification of the target site as described under "PCR optimization." The following steps describe this procedure.

1. For each sgRNA, set up the in vitro Cas9 assay reactions, as indicated in Table 11. The first reaction contains all components of the reactions whereas

Table 11 Setup of the in vitro Cas9 assay.

Components	Reaction #1	Reaction #2	Reaction #3	Reaction #4
	Volume (μL)			
ddH_2O (nuclease-free)	Adjust to 30 μL final	Adjust to 30 μL final	Adjust to 30 μL final	Adjust to 30 μL final
NEB3.1 buffer (NEB, B7203S)	3	3	3	3
Cas9 Nuclease (NEB, M0386S) (1 μM)	1	–	1	1
sgRNA (400 ng/μL)	1	1	–	1
On target DNA (on-target PCR product), 30 nM	3	3	3	–

the other 3 reactions serve as negative controls where either recombinant Cas9 protein, the sgRNA, or the on-target PCR product is omitted

2. Incubate the reactions in a heat block at 37 °C for 30 min
3. Add 1 μL of Proteinase K (500 μg/mL) to each sample and incubate at 55 °C for 10 min, to inactivate Cas9
4. Load 10 μL of each reaction with 2 μL of 6 × loading dye on a 2% TAE-agarose gel supplemented with 0.5 × GelRed® at 120 V for approximately 40 min in 1 × TAE buffer. We typically run the samples alongside a 100 bp ladder to evaluate the relative sizes of the fragments
5. Image gel using a UV transilluminator (ChemiDoc™ MP Imaging System, Bio-Rad) and keep for your records
6. Shorter DNA fragments, corresponding to the cleaved PCR product, should be observed in reaction #1. Full length PCR products should be observed in reaction #2 and reaction #3. No DNA fragments should be observed in reaction #4

4.4 Preparation of microinjection mixes and injections

4.4.1 Timing: 15 min

This step-by-step procedure describes the preparation of microinjection mixes by combining Cas9 mRNA, sgRNA, and HDR template(s).

7. Prepare the Cas9 mRNA transcript, sgRNA, and HDR template as described in Table 12
8. Make 20 μL aliquots and freeze at −80 °C, until ready for microinjection

Table 12 Preparation of microinjection mixes.

Reagents	Amount	Final concentration
Cas9 mRNA transcript	1 aliquot (8 μg)	100 ng/μL
sgRNA	4 μg	50 ng/μL
HDR oligonucleotides (100 μM)	1 μL	1.25 pmol/μL
RNAse-free ddH_20	Up to 80 μL	–

Table 13 Tail lysis buffer composition.

Reagent	Final concentration	Amount
KCl (2 M)	500 mM	12.5 mL
Tris pH 8.3 (1 M)	100 mM	5.0 mL
Gelatin	0.1 mg/mL	5.0 mg
NP40 (100%)	1% (v/v)	0.5 mL
Tween20 (100%)	1% (v/v)	0.5 mL
Proteinase K (50 mg/mL)	500 μg/mL	500 μL
ddH_2O	N/A	31 mL
Total	N/A	50 mL

Note: Gelatin takes several hours to dissolve. Before adding Proteinase K, the solution can be autoclaved to help dissolve gelatin.

4.5 Microinjection of pronuclear stage zygotes

4.5.1 Timing: Several weeks

Microinjection of pronuclear stage zygotes is usually performed by transgenic core facilities. For a detailed procedure, please refer to the following: *Manipulating the Mouse Embryo: A Laboratory Manual*, Fourth Edition (Behringer, 2014).

4.6 PCR optimization

4.6.1 Timing: 2 days

This section describes the procedure to optimize PCR reactions for the characterization of mice obtained from zygote injections. For all genotyping procedures, we use Qiagen Taq Polymerase (Qiagen, 201205). Control genomic DNA can be obtained from a wild-type mouse tail biopsy. Tail biopsies are prepared as follows:

1. Prepare Tail Lysis Buffer, as shown in Table 13

2. In a 1.5 mL microfuge tube, extract genomic DNA by adding 300 μL of Tail Lysis Buffer to the biopsy and incubating in a heat block at 55 °C for 16 h. Following the incubation period, gently flick the tube to mix the DNA. Do not vortex or agitate to prevent shearing of the DNA
3. Following digestion at 55 °C, incubate the genomic DNA at 95 °C in a heat block for 10 min to heat inactivate the Proteinase K

One way of optimizing PCR genotyping conditions is to determine the optimal annealing temperature of the primers by testing identical reactions across a range of annealing temperatures and two concentrations of $MgCl_2$. This procedure should be performed prior to obtaining mice from zygote injection so that when mice are obtained, genotyping can be performed quickly.

4. Prepare PCR mix with low $MgCl_2$ concentration, as indicated in Table 14
5. Prepare PCR mix with high $MgCl_2$ concentration, as detailed in Table 15
6. Distribute 50 μL of master mix per reaction in 8 wells of a 96 well plate to correspond with a different annealing temperature in each well
7. Use the cycling conditions displayed in Table 16
8. Load 10 μL of PCR product on a 2% TAE-agarose gel supplemented with 0.5 × GelRed® and run at 120 V for approximately 40 min in 1 × TAE buffer

Table 14 Low $MgCl_2$ PCR optimization setup.

Reagents	Volume (μL) for 1 Reaction	# of reactions	Volume (μL) for all reactions
10 × CoralLoad PCR Buffer	5	10	50
5 × Q-Solution	10	10	100
$MgCl_2$	0	10	0
dNTP (10 mM each)	1	10	10
Forward primer (100 μM)	0.1	10	1
Reverse primer (100 μM)	0.1	10	1
Taq DNA Polymerase	0.5	10	5
ddH_2O	32.8	10	328
Wild-type genomic DNA	0.5	10	5

Table 15 High $MgCl_2$ PCR optimization setup.

Reagents	Volume (μL) for 1 Reaction	# of reactions	Volume (μL) for all reactions
10 × CoralLoad PCR Buffer	5	10	50
5 × Q-Solution	10	10	100
$MgCl_2$	2	10	20
dNTP (10 mM each)	1	10	10
Forward primer (100 μM)	0.1	10	1
Reverse primer (100 μM)	0.1	10	1
Taq DNA Polymerase	0.5	10	5
ddH_2O	30.8	10	308
Wild-type genomic DNA	0.5	10	5

Table 16 Cycling conditions for PCR optimization.

Step	Number of cycles	Temperature	Time
1	1	94 °C	4 min
2	35	94 °C	30 s
		50–70 °C	30 s
		72 °C	60 s
3	1	72 °C	10 min
4	1	4 °C	Hold

9. Image gel using a UV transilluminator (ChemiDoc™ MP Imaging System, Bio-Rad) and save picture for your records

10. Select the best amplification condition

4.7 On-target genotyping

4.7.1 Timing: 1 week

This procedure describes the characterization, at the genomic level, of mice obtained from zygote injection. To analyze on-target gene editing, the guide target region is PCR amplified and each amplicon is analyzed by restriction enzyme digestion (if applicable), gel electrophoresis, and Sanger sequencing.

11. After receiving biopsies from the transgenic facility, extract genomic DNA by adding 300 μL of Tail Lysis Buffer to each sample in a 1.5 mL microfuge tube and incubating in a heat block at 55 °C for 16 h
12. The next day, incubate the digested biopsies at 95 °C for 10 min to heat inactivate the Proteinase K
13. For each biopsy, setup PCR reactions on ice using previously established PCR conditions. Table 17 describes how to make the PCR mix with optional $MgCl_2$. Also include:
    - **(a)** Positive control (wild-type mouse genomic DNA)
    - **(b)** Negative control (no DNA)
14. Use the cycling conditions, as illustrated in Table 18

Table 17 PCR mix for on-target genotyping with optional $MgCl_2$.

Reagents	Volume (μL) for 1 reaction
10 × CoralLoad PCR Buffer	5
5 × Q-Solution	10
$MgCl_2$ (optional)	2
dNTP (10 mM each)	1
Forward primer (100 μM)	0.1
Reverse primer (100 μM)	0.1
Taq DNA Polymerase	0.5
ddH_2O	30.8
Genomic DNA from Biopsy	0.5

Table 18 Cycling conditions for on-target PCR genotyping.

Step	Number of cycles	Temperature	Time
1	1	94 °C	4 min
2	35	94 °C	30 s
		58 °C (or as determined by PCR optimization)	30 s
		72 °C	60 s
3	1	72 °C	10 min
4	1	4 °C	Hold

15. Following amplification, put aside:
    (a) 10 µL to run on a 2% TAE-agarose gel supplemented with 0.5 × GelRed®
    (b) 10 µL to digest with appropriate restriction enzyme (if applicable) and run on a 2% TAE-agarose gel supplemented with 0.5 × GelRed®
    (c) 25 µL for sample preparation using ExoSAP-IT PCR Product Cleanup Reagent (Affymetrix, 78,200) and DNA sequencing
16. For each PCR reaction, set up a corresponding restriction enzyme digest, as described in Table 19. Incubate in a heat block at 37 °C for 2–18 h
17. Run both digested (30 µL) and undigested (10 µL) PCR products on a 2% TAE-agarose gel supplemented with 0.5 × GelRed® at 120 V for 40–60 min in 1 × TAE buffer
18. Image gel using a UV transilluminator (ChemiDoc™ MP Imaging System, Bio-Rad) and keep for your records
19. Set up ExoSAP-IT PCR Product Cleanup (Affymetrix, 78200) reactions as follows:
    (a) Combine 25 µL of a post-PCR reaction product with 1 µL of ExoSAP-IT reagent (Affymetrix, 78,200) in a well of a 96-well plate or 200 µL microfuge tube
    (b) Incubate in a heat block at 37 °C for 30 min to degrade remaining primers and nucleotides
    (c) Incubate at 80 °C for 15 min to inactivate ExoSAP-IT reagent. The PCR product is now ready for use in DNA sequencing
20. Set up sequencing reactions in a 96-well plate, as demonstrated in Table 20
    (a) Spin down the 96-well plate at 1000 × *g* for 30 s to ensure all reagents are at the bottom of the well

Table 19 Restriction digest of PCR products to distinguish mutant and wild-type alleles.

Reagent	Volume (µL)
Restriction Enzyme Buffer (10 ×)	3
ddH_2O	16.5
Restriction Enzyme	0.5
PCR Product	10

Table 20 Cycle sequencing reaction setup.

Reagent	Volume (μL) per reaction
BigDye™ Terminator v3.1	0.5
5 × Sequencing Buffer	2
Primer (2 μM)	2
Template	5.5

Table 21 Cycling conditions for the cycle sequencing reaction.

	Stage/step				
		Cycling (25 cycles)			
Parameter	**Incubate**	**Denature**	**Anneal**	**Extend**	**Hold**
Ramp rate	–	1 °C/s			
Temperature	96 °C	96 °C	50 °C	60 °C	4 °C
Time (mm:ss)	01:00	00:10	00:05	01:00	Until ready to purify

Table 22 Premix BigDye™ XTerminator and SAM solutions before distributing the mixture to the sequencing reaction plate.

Reagent	Volume (μL) per reaction
BigDye™ XTerminator Solution	10
SAM Solution	45

(b) Transfer the 96-well plate to a thermal cycler with conditions for cycle sequencing, as exemplified in Table 21

(c) When cycle sequencing has finished, centrifuge the 96-well plate at 1000 × *g* for 1 min

(d) Combine the BigDye XTerminator Solution and SAM Solution, as shown in Table 22:

i. Make sure the BigDye Xterminator Solution is homogenous before pipetting

(e) Dispense 55 μL of the premixed XTerminator solution into each well of the sequencing reaction plate

(f) Transfer the total volume (10 μL) of each cycle sequencing reaction from the 96-well plate into the sequencing reaction plate with the premixed XTerminator solution

(g) Seal the sequencing reaction plate with clear adhesive film and vortex the plate for 30 min
(h) When the plate has finished shaking, centrifuge at 1000 × *g* for 2 min
(i) Carefully remove adhesive film and replace with a septum on the plate. Immediately transfer the sequencing reaction plate to the SeqStudio and begin the capillary electrophoresis run

21. Analyze sequencing data using SnapGene as follows:
    (a) Open .ab1 files using SnapGene
    (b) Analyze the chromatogram and look for insertion or deletion of genetic materials (indels), and insertion of the desired mutations

4.8 Off-target genotyping

4.8.1 Timing: 1 week

When generating mice via microinjection of SpCas9 mRNA transcript, off-target cleavage rarely occurs unless guides with less than two mismatches are used. The following procedure describes the steps to characterize potential off-target cleavage sites. We recommend analyzing five potential off-target sites to start; if no off-target cleavage is observed, it is not necessary to continue the analysis.

22. For each potential off-target, setup PCR reactions using specific off-target primer pairs, as described in Table 17. Analyze genomic DNA from each potential founder. Also include:
    (a) Positive control (wild-type mouse genomic DNA)
    (b) Negative control (No DNA)
23. Use the cycling conditions described in Table 18
24. Following amplification, put aside:
    (a) 10 μL to run on a 2% TAE-agarose gel supplemented with 0.5 × GelRed®
    (b) 25 μL for sample preparation using ExoSAP-IT PCR Product Cleanup Reagent (Affymetrix, 78,200) and DNA sequencing
25. Run 10 μL on a 2% TAE-agarose gel supplemented with 0.5 × GelRed® at 120 V for 40–60 min in 1 × TAE buffer
26. Image gel using a UV transilluminator (ChemiDoc™ MP Imaging System, Bio-Rad) and keep for your records
27. Set up ExoSAP-IT PCR Product Cleanup (Affymetrix, 78,200) and DNA sequencing reactions, as detailed in the "On-target genotyping" section of this chapter

4.9 Expected outcomes

If the genetic modification is not deleterious to mouse development, 10–50% of animals will harbor the desired missense mutation or large in-frame deletion, and 5–20% of mice will harbor the in-frame insertion of genetic material. For the generation of the *Scyl1*^{K454A} mutation, we produced 20 mice, 2 of which had the desired mutation (10%). For the generation of the *Scyl1*$^{\Delta HEATs}$ mutation, we produced 7 mice and 1 had the desired mutation (14%). For the generation of the *Scyl1*$^{N\text{-}FLAG}$ mutation, we produced 30 mice from zygote microinjection and 3 had the desired mutation (10%).

4.10 Limitations

There are two important limitations to the use of this protocol:

1. The possibility of introducing off-target mutations within the mouse genome
2. The possibility of not finding a suitable guide sequence near the site of the desired genetic alteration

The first problem can be resolved by outbreeding the undesirable mutation(s) or by selecting a more specific guide. The second problem can be solved by using an RNA guided endonuclease with a different PAM requirement as an alternative to SpCas9.

4.11 Safety considerations and standards

Mice were housed in an Association for Assessment and Accreditation of Laboratory Animal Care accredited facility and maintained in accordance with the National Institutes of Health *Guide for the Care and Use of Laboratory Animals*. Animal experiments were reviewed and approved by the Indiana University Institutional Animal Care and Use Committee.

References

Behringer, R. (2014). *Manipulating the mouse embryo: A laboratory manual* (4th ed.). Cold Spring Harbor, New York: Cold Spring Harbor Laboratory Press.

Blot, S., Poirier, C., & Dreyfus, P. A. (1995). The mouse mutation muscle deficient (mdf) is characterized by a progressive motoneuron disease. *Journal of Neuropathology and Experimental Neurology*, *54*(6), 812–825.

Doench, J. G., Fusi, N., Sullender, M., Hegde, M., Vaimberg, E. W., Donovan, K. F., et al. (2016). Optimized sgrna design to maximize activity and minimize off-target effects of crispr-cas9. *Nature Biotechnology*, *34*(2), 184–191. https://doi.org/10.1038/nbt.3437.

Gingras, S., Earls, L. R., Howell, S., Smeyne, R. J., Zakharenko, S. S., & Pelletier, S. (2015). Scyl2 protects ca3 pyramidal neurons from excitotoxicity during functional maturation of the mouse hippocampus. *The Journal of Neuroscience*, *35*(29), 10510–10522. https://doi.org/10.1523/JNEUROSCI.2056-14.2015.

Gingras, S., Kuliyev, E., & Pelletier, S. (2017). Scyl1 does not regulate rest expression and turnover. *PLoS One*, *12*(6), e0178680. https://doi.org/10.1371/journal.pone.0178680.
Kuliyev, E., Gingras, S., Guy, C. S., Howell, S., Vogel, P., & Pelletier, S. (2018). Overlapping role of scyl1 and scyl3 in maintaining motor neuron viability. *The Journal of Neuroscience*, *38*(10), 2615–2630. https://doi.org/10.1523/JNEUROSCI.2282-17.2018.
Lenz, D., McClean, P., Kansu, A., Bonnen, P. E., Ranucci, G., Thiel, C., et al. (2018). Scyl1 variants cause a syndrome with low gamma-glutamyl-transferase cholestasis, acute liver failure, and neurodegeneration (calfan). *Genetics in Medicine*, *20*(10), 1255–1265. https://doi.org/10.1038/gim.2017.260.
McNiven, V., Gattini, D., Siddiqui, I., Pelletier, S., Brill, H., Avitzur, Y., & Mercimek-Andrews, S. (2021). SCYL1 disease and liver transplantation diagnosed by reanalysis of exome sequencing and deletion/duplication analysis of SCYL1. *American Journal of Medical Genetics. Part A*, *185*(4), 1091–1097. https://doi.org/10.1002/ajmg.a.62079.
Milione, M., Ardini, E., Christiansen, J., Valtorta, E., Veronese, S., Bosotti, R., et al. (2017). Identification and characterization of a novel scyl3-ntrk1 rearrangement in a colorectal cancer patient. *Oncotarget*, *8*(33), 55353–55360. https://doi.org/10.18632/oncotarget.19512.
Murphy, J. M., Zhang, Q., Young, S. N., Reese, M. L., Bailey, F. P., Eyers, P. A., et al. (2014). A robust methodology to subclassify pseudokinases based on their nucleotide-binding properties. *The Biochemical Journal*, *457*(2), 323–334. https://doi.org/10.1042/BJ20131174.
Pelletier, S. (2016). Scyl pseudokinases in neuronal function and survival. *Neural Regeneration Research*, *11*(1), 42–44. https://doi.org/10.4103/1673-5374.175040.
Pelletier, S., Gingras, S., & Green, D. R. (2015). Mouse genome engineering via crispr-cas9 for study of immune function. *Immunity*, *42*(1), 18–27. https://doi.org/10.1016/j.immuni.2015.01.004.
Pelletier, S., Gingras, S., Howell, S., Vogel, P., & Ihle, J. N. (2012). An early onset progressive motor neuron disorder in scyl1-deficient mice is associated with mislocalization of tdp-43. *The Journal of Neuroscience*, *32*(47), 16560–16573. https://doi.org/10.1523/JNEUROSCI.1787-12.2012.
Popp, M. W., & Maquat, L. E. (2016). Leveraging rules of nonsense-mediated mrna decay for genome engineering and personalized medicine. *Cell*, *165*(6), 1319–1322. https://doi.org/10.1016/j.cell.2016.05.053.
Richardson, C. D., Ray, G. J., DeWitt, M. A., Curie, G. L., & Corn, J. E. (2016). Enhancing homology-directed genome editing by catalytically active and inactive crispr-cas9 using asymmetric donor DNA. *Nature Biotechnology*, *34*(3), 339–344. https://doi.org/10.1038/nbt.3481.
Schmidt, W. M., Kraus, C., Hoger, H., Hochmeister, S., Oberndorfer, F., Branka, M., et al. (2007). Mutation in the scyl1 gene encoding amino-terminal kinase-like protein causes a recessive form of spinocerebellar neurodegeneration. *EMBO Reports*, *8*(7), 691–697. https://doi.org/10.1038/sj.embor.7401001.
Seidahmed, M. Z., Al-Kindi, A., Alsaif, H. S., Miqdad, A., Alabbad, N., Alfifi, A., et al. (2020). Recessive mutations in scyl 2 cause a novel syndromic form of arthrogryposis in humans. *Human Genetics*, *139*(4), 513–519. https://doi.org/10.1007/s00439-020-02117-7.
Shohet, A., Cohen, L., Haguel, D., Mozer, Y., Shomron, N., Tzur, S., et al. (2019). Variant in scyl1 gene causes aberrant splicing in a family with cerebellar ataxia, recurrent episodes of liver failure, and growth retardation. *European Journal of Human Genetics*, *27*(2), 263–268. https://doi.org/10.1038/s41431-018-0268-2.
Spagnoli, C., Frattini, D., Salerno, G. G., & Fusco, C. (2019). On calfan syndrome: Report of a patient with a novel variant in scyl1 gene and recurrent respiratory failure. *Genetics in Medicine*, *21*(7), 1663–1664. https://doi.org/10.1038/s41436-018-0389-6.
Xu, H., Xiao, T., Chen, C. H., Li, W., Meyer, C. A., Wu, Q., et al. (2015). Sequence determinants of improved crispr sgrna design. *Genome Research*, *25*(8), 1147–1157. https://doi.org/10.1101/gr.191452.115.

Kolibri 360 GmbH
Ferdinand-Jühlke-Straße 7
99095 Erfurt, Deutschland
produktsicherheit@kolibri360.de